Dairy Fat Products and Functionality

Tuyen Truong • Christelle Lopez
Bhesh Bhandari • Sangeeta Prakash
Editors

Dairy Fat Products and Functionality

Fundamental Science and Technology

Springer

Editors
Tuyen Truong
School of Science
College of Science Engineering and Health
RMIT University
Melbourne, VIC, Australia

Christelle Lopez
INRAE
French National Research Institute for Agriculture, Food & Environment
Research Unit BIA
Nantes, France

Bhesh Bhandari
ARC Dairy Innovation Hub
School of Agriculture and Food Sciences
The University of Queensland
St Lucia, QLD, Australia

Sangeeta Prakash
School of Agriculture and Food Sciences
The University of Queensland
St Lucia, QLD, Australia

ISBN 978-3-030-41660-7 ISBN 978-3-030-41661-4 (eBook)
https://doi.org/10.1007/978-3-030-41661-4

This Springer imprint is published by the registered company Springer Nature Switzerland AG
The registered company address is: Gewerbestrasse 11, 6330 Cham, Switzerland

Preface

Bovine milk fat is an important dairy component which is consumed by human infants and adults via various dairy products such as milk, milk powder, cream, cheeses, and butter. Apart from the major physiological role of milk fat in delivering human nutrition and providing health benefits, milk fat also plays as a pivotal functionality controlling role in many dairy products. The biological membrane surrounding milk fat globules has been found to have interesting applications in food formulation innovations, human nutrition, and paediatrics. Therefore, understanding the fundamental science and technology of milk fat is of great importance not only from a scientific point of view but also with regard to the economic impact of milk fat, especially in fat-rich products. The intention of this book is to provide readers with recent and outstanding literature in important areas of dairy fat containing products and functionality. This book represents a valuable source of information for students, academics, and industry personnel who work in the fields of dairy lipids and dairy fat products.

In this book, 24 chapters have been organised into three main themes: (1) dairy fat chemistry and nutrition, (2) dairy fat physics and materials science, and (3) dairy fat processes and products. In the first theme, fascinating insights into milk fat globule membrane such as structure, lipidomic characterisation, and applications are presented. Physiochemical characterisation, lipase action, adulteration, and digestibility of milk lipids are also covered. The second theme provides excellent fundamentals in physical properties of milk fat such as crystallisation, rheology, and microstructure aspects. It also highlights relatively new research areas driven by material science approach in dairy fats, where innovative dairy products and the new generation of dairy ingredients can be tailor-made. Cutting-edge topics on tribology of dairy fat products, oil structuring in dairy fat products, manipulation of differentiated-sized milk fat globules, and production of human milk fat substitutes from dairy fats are featured in dedicated chapters. The third theme presents comprehensive reviews on the relationship among processes—structure—functionality in dairy fat products towards new developments in manufacturing processes and health-promoting dairy products. The last theme also focuses on the importance of milk fat as a key quality controlling factor in the manufacturing of butter, ghee, dairy cream, dairy powder, cheeses, and aerated products.

The expert contributions of 43 authors from 10 countries are greatly acknowledged with gratitude. Special acknowledgements are also extended to Springer for their support during the course of this book project.

Melbourne, VIC, Australia — Tuyen Truong
Nantes, France — Christelle Lopez
St Lucia, QLD, Australia — Sangeeta Prakash
St Lucia, QLD, Australia — Bhesh Bhandari

Contents

Part II Dairy Fat Physics and Materials Science

Contributors

Nidhi Bansal The University of Queensland, School of Agriculture and Food Sciences, St Lucia, QLD, Australia

Kate M. Barry Teagasc Food Research Centre, Moorepark, Fermoy, Co, Cork, Ireland

Department of Psychiatry, University College Cork, Cork, Ireland

Surendra Nath Battula Southern Regional Station, ICAR-National Dairy Research Institute, Bengaluru, India

Bhesh Bhandari ARC Dairy Innovation Hub, School of Agriculture and Food Sciences, The University of Queensland, St Lucia, QLD, Australia

M. Pillar Castro-Gómez Institute of Food Science Research (CIAL, CSIC-UAM), Madrid, Spain

Hilton C. Deeth School of Agriculture and Food Sciences, University of Queensland, St Lucia, QLD, Australia

Pramesh Dhungana ARC Dairy Innovation Hub, School of Agriculture and Food Sciences, The University of Queensland, St Lucia, QLD, Australia

Timothy G. Dinan Department of Psychiatry, University College Cork, Cork, Ireland

Alimentary Pharmabiotic Centre, University College Cork, Cork, Ireland

Qiying Feng Department of Food Science, University of Guelph, Guelph, ON, Canada

Javier Fontecha Institute of Food Science Research (CIAL, CSIC-UAM), Madrid, Spain

Sophie Gallier Fonterra Research and Development Centre, Fonterra Co-operative Group Ltd, Palmerston North, New Zealand

Kamal Gandhi Dairy Chemistry Division, ICAR-National Dairy Research Institute, Karnal, Haryana, India

Thao M. Ho ARC Dairy Innovation Hub, School of Agriculture and Food Sciences, The University of Queensland, St Lucia, QLD, Australia

Philip M. Kelly Teagasc Food Research Centre, Moorepark, Fermoy, Co, Cork, Ireland

Bal Kumari Sharma Khanal The University of Queensland, School of Agriculture and Food Sciences, St Lucia, QLD, Australia

Mette K. Larsen Department of Food Science, Aarhus University, Aarhus, Denmark

Arla Foods Amba, Videbaek, Denmark

Thomas S. H. Leong ARC Dairy Innovation Hub, University of Melbourne, Parkville, VIC, Australia

Dasong Liu State Key Laboratory of Food Science and Technology, Jiangnan University, Wuxi, Jiangsu Province, China

Christelle Lopez INRAE, UMR STLO, Rennes, France

INRAE, UR BIA, Nantes, France

Braulio A. Macias-Rodriguez Department of Food Science, University of Guelph, Guelph, ON, Canada

Bimlesh Mann ICAR-National Dairy Research Institute, Karnal, Haryana, India

Alejandro G. Marangoni Department of Food Science, University of Guelph, Guelph, ON, Canada

Marie-Caroline Michalski INRAE, CarMeN laboratory, UMR1397, INSERM, U1060, Université Claude Bernard Lyon 1, INSA-Lyon, Université de Lyon, Pierre-Bénite, France

Centre de Recherche en Nutrition Humaine Rhône-Alpes, Univ-Lyon, CarMeN Laboratory, Université Claude Bernard Lyon1, Hospices Civils de Lyon, CENS, FCRIN/FORCE Network, Pierre-Bénite, France

Olimpio Montero Spanish National Research Council, CSIC, Valladolid, Spain

N. Laxmana Naik Southern Regional Station, ICAR-National Dairy Research Institute, Bengaluru, India

Phuong Nguyen School of Agriculture and Food Sciences, The University of Queensland, St Lucia, QLD, Australia

Dian Widya Ningtyas School of Agriculture and Food Sciences, The University of Queensland, St Lucia, QLD, Australia

Bhavesh Panchal ARC Dairy Innovation Hub, School of Agriculture and Food Sciences, The University of Queensland, St Lucia, QLD, Australia

Ashok R. Patel Guangdong Technion Israel Institute of Technology, Shantou, China

Sangeeta Prakash School of Agriculture and Food Sciences, The University of Queensland, St Lucia, QLD, Australia

Pere R. Ramel Department of Food Science, University of Guelph, Guelph, ON, Canada

Laurie-Eve Rioux STELA Dairy Research Centre, Institute of Nutrition and Functional Foods, Université Laval, Québec City, QC, Canada

Luis Miguel Rodríguez-Alcalá Universidade Católica Portuguesa, CBQF – Centro de Biotecnologia e Química Fina – Laboratório Associado, Escola Superior de Biotecnologia, Porto, Portugal

Pierre Schuck INRA Agrocampus Ouest, UMR 1253, STLO, Rennes, France

Rajan Sharma ICAR-National Dairy Research Institute, Karnal, India

Harjinder Singh Riddet Institute and Massey Institute of Food Science and Technology, Massey University, Palmerston North, New Zealand

Tuyen Truong ARC Dairy Innovation Hub, The University of Queensland, St Lucia, QLD, Australia

School of Science, RMIT University, Melbourne, VIC, Australia

Sylvie L. Turgeon STELA Dairy Research Centre, Institute of Nutrition and Functional Foods, Université Laval, Québec City, QC, Canada

Martin R. Weisbjerg Department of Animal Science, Aarhus University, Tjele, Denmark

Lars Wiking Department of Food Science, Aarhus University, Aarhus, Denmark

Yun Yu State Key Laboratory of Food Science and Technology, Jiangnan University, Wuxi, Jiangsu Province, China

Peng Zhou State Key Laboratory of Food Science and Technology, Jiangnan University, Wuxi, Jiangsu Province, China

Yang Zhu School of Agriculture and Food Sciences, The University of Queensland, St Lucia, QLD, Australia

Part I
Dairy Fat Chemistry and Nutrition

Chapter 1
Manipulation of Dietary and Physiological Factors on Composition and Physicochemical Characteristics of Milk Fat

Lars Wiking, Mette K. Larsen, and Martin R. Weisbjerg

1 Introduction

Fatty acids (FA) in milk origin partly from FA in the feed and partly from de novo synthesis in the mammary gland, commonly with half from feed and half from de novo synthesis. De novo synthesized FA are generally short or medium chain length, maximum C16, and are saturated. Milk FA originating from feed will mirror the feed ration FA according to chain length; however, due to rumen microbial biohydrogenation, the feed FA supplied to the mammary gland will be much more saturated and contain *trans* and conjugated linoleic acid (CLA) intermediates from the rumen biohydrogenation or later desaturation in the mammary glands. The proportion of milk FA from the two sources can deviate considerably from the half-half, depending mainly on FA supply via the feed but also on nutrient supply for the de novo synthesis, and on FA supply/demand from mobilisation/deposition.

Knowledge on how the FA composition of milk is affected by feeding and by physiological factors creates possibilities for manipulation of milk fat composition. The purpose of such manipulations could be nutritional or technological. Nutritionally some FA are regarded as beneficial to human health, whereas other FA are regarded as detrimental, and increasing the former and/or decreasing the latter would create a healthier milk fat. Technologically, more saturated and longer chain FA give a firmer texture of dairy products such as butter and cheese, whereas more

L. Wiking (✉)
Department of Food Science, Aarhus University, Aarhus, Denmark
e-mail: Lars.Wiking@food.au.dk

M. K. Larsen
Department of Food Science, Aarhus University, Aarhus, Denmark

Arla Foods Amba, Videbaek, Denmark

M. R. Weisbjerg
Department of Animal Science, Aarhus University, Tjele, Denmark

T. Truong et al. (eds.), *Dairy Fat Products and Functionality*,
https://doi.org/10.1007/978-3-030-41661-4_1

unsaturated FA and shorter chain FA give a softer texture. Higher content of unsaturated FA, in particular, polyunsaturated FA, makes the milk fat more prone to oxidation.

Intensive milk production relies on specialized dairy cow breeds. Holstein is globally by far the most dominating breed, and Jersey is the second most popular breed (http://www.survey-icar.org/cow_survey/). Jersey cows are lower yielding than the Holsteins. However, the Jersey milk has a higher content of fat and protein than the Holsteins. The milk production varies throughout lactation. A normal calving interval is one year, resulting in lactation profiles for milk yield and composition as shown in Fig. 1.1 for the first 30 weeks of lactation. Dairy management and production systems vary depending on climatic conditions, local traditions and development in technology and trade. Traditionally, milk production has been based on grazing of natural grasslands, where calving and milk production were seasonal

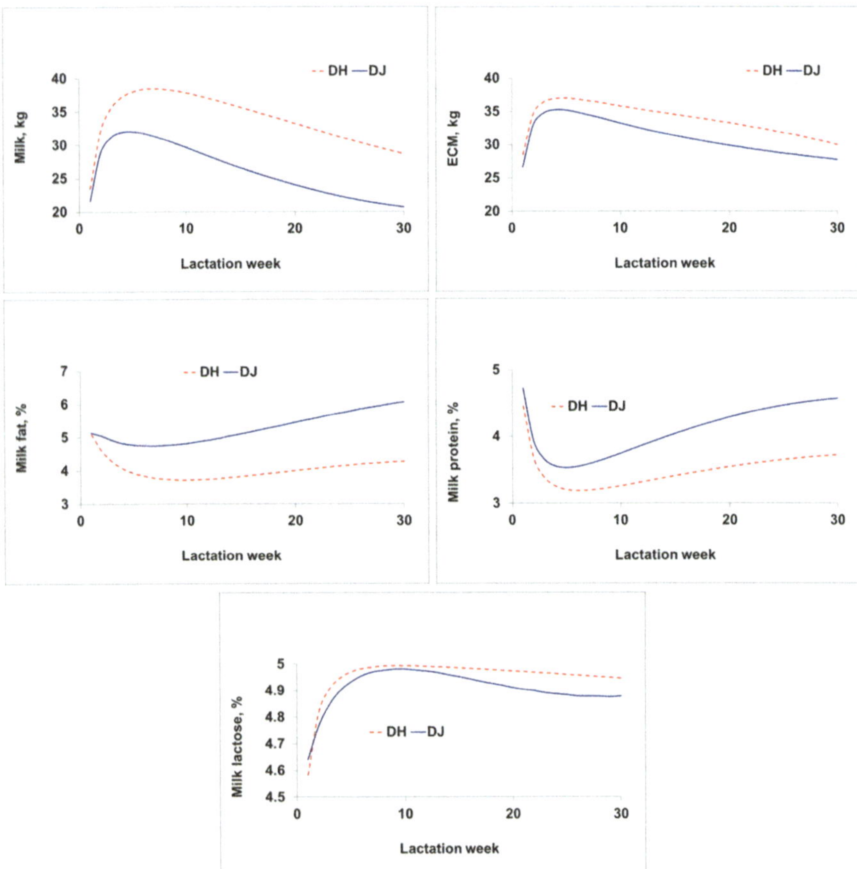

Fig. 1.1 Profiles for daily milk yield, 3.14 MJ/kg energy corrected milk (ECM) and milk concentration of fat and protein during the first 30 weeks of lactation for similar fed Danish Holstein and Danish Jersey (modified from Weisbjerg et al., 2013)

following the abundance of grass. Grazing is still an important part of dairy cow feeding, although with increased production intensity, the extent of grazing declines and is partly or fully replaced with all-year barn feeding with conserved forage and concentrates, typically fed as a total mixed ration (TMR).

2 Manipulation of Dietary and Physiological Factors on Milk Fat Composition

2.1 *Forages*

Green forages like grass and legumes are a major part of dairy cows' diets, and in some countries, e.g. New Zealand, milk production highly relies on grazing grass clover pastures. In North America, lucerne is the major green supplement to yellow forages like whole crop maize silage forage (Thoma et al., 2013). In Western Europe, the situation is more variable and range from grass and grass-clover as the sole forage and major part of total feed to high maize silage rations, and finally to rations with a high proportion of concentrate (De Vries, Debruyne, & Aarts, 2013; Kristensen et al., 2015).

Green forages are significant sources of FA, mainly linolenic acid (Larsen, Kidmose, Kristensen, Beaumont, & Mortensen, 2013). The yellow whole crop forages like maize silage supply linoleic acid which is the main fatty acid in maize silage as well as in whole cereal crops with developed kernels (Larsen, Hymøller, Brask-Pedersen, & Weisbjerg, 2012).

2.2 *Energy and Protein Concentrates*

Concentrates used for dairy cows are mainly cereal grains, legume seeds and by-products from the cereal, bioethanol, sugar and oilseed industry. Oilseed meals have been solvent extracted and are characterised by a low residual fat content, whereas oilseed cakes have been pressed and are higher in residual fat content. The fatty acid composition varies a lot among oilseeds, i.e. palm kernel and coconut being rich in C12 and C14, rapeseed being rich in C18:1, sunflower, soybean and cottonseed rich in C18:2 and linseed in C18:3 (NorFor, 2017). Furthermore, fat supplements such as palm fatty acid distillate coming from the oilseed industry are used as cattle feed (Weisbjerg, Wiking, Kristensen, & Lund, 2008). FA composition of some feedstuffs are presented in Table 1.1.

Table 1.1 Fat concentration and FA composition in some feedstuffs. FA composition in by-products like cakes, meals, distillers grains etc. will generally reflect composition in original seed/grain (NorFor, 2017)

	Crude fat	FA	<C12	C12:0	C14:0	C16:0	C18:0	C18:1	C18:2	C18:3	Other FA
	g/kg DM	g FA/ kg crude fat	g/100 g FA								
Palmkernel cake	79	800	9.7	46.9	14.1	8.8	1.3	18.5	0.7	0.0	0.0
Rape seed	461	900	0	0	0	4.4	1.4	54.3	19.5	11.4	9
Cotton seed	210	900	0	0	1.4	23.4	1.1	22.9	47.8	0	3.4
Soya beans	222	900	0	0.2	0.1	9.8	2.4	28.9	50.7	6.5	1.4
Sun flower seed	480	900	0	0	0	5.6	2.2	25.1	66.2	0	0.9
Linseed	390	900	0	0	0	5.5	4.3	21.1	13.3	55.7	0.1
Wheat grain	25	700	0	0	0.1	17.8	0.8	15.2	56.4	5.9	3.8
Maize grain	46	900	0	0	0	11.4	1.9	25.3	60.7	0.7	0
Grass	39	540	0	0	0	20.2	2.7	2.7	14	56.9	3.5
White clover	36	480	0	0.2	0.4	17.2	2.9	5	16	55.4	2.9
Red clover	32	480	0	0.3	0.5	18.8	3.7	8	23.2	42.9	2.6

Table 1.2 Nutrients eaten and nutrients absorbed by ruminants (Volden, 2011; Weisbjerg et al., 1992)

Nutrients in feed	Nutrients absorbed
Carbohydrates	VFA (acetic, propionic and butyric acid), microbial matter, glucose
Protein	Microbial protein rumen escape feed protein
Fat	Hydrogenated fatty acids from feed, microbial fatty acids

2.3 *Response to Energy and Nutrient Supply*

Due to the rumen metabolism, nutrients absorbed by ruminants such as dairy cows deviate considerably from the nutrients eaten (Table 1.2), in contrast to nutrition of monogastric animals, where the nutrients from the feed are transferred more directly to the animal. The microbial community in the rumen (bacteria, protozoa and fungi) will ferment both fibrous and non-fibrous carbohydrates to be used for their growth, and they will degrade protein to supply nitrogen and to a lesser degree energy for their growth. Short-chain volatile FA mainly acetic, propionic and

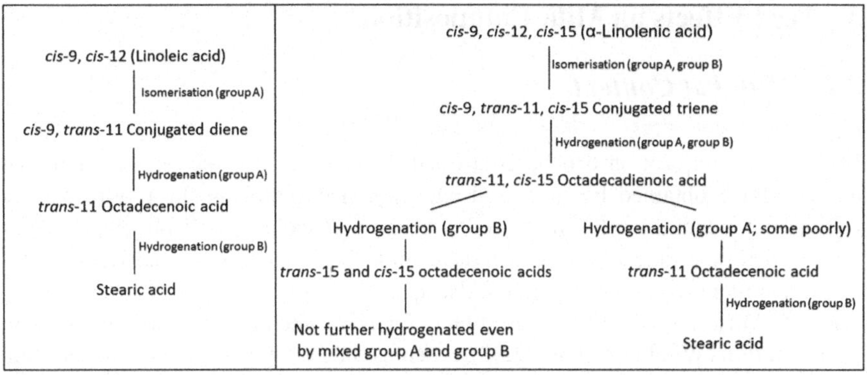

Fig. 1.2 Rumen metabolism of linoleic and linolenic acid. Group A and B refer to two classes of biohydrogenating bacteria (adapted from Harfoot & Hazlewood, 1997)

butyric are produced from the carbohydrates during their fermentation and the short-chain FA are absorbed over the rumen wall as the major energy source for the cow. FA are not fermented in the rumen, but glycerides are hydrolysed, the glycerol part being fermented, and unsaturated FA are extensively (70–90% efficiency) hydrogenated (Weisbjerg, Børsting, & Hvelplund, 1992). Hydrogenation results in a large number of intermediates, which can be seen in Fig. 1.2 (Harfoot & Hazlewood, 1997).

2.4 De Novo FA Synthesis

The substrate for the de novo FA synthesis in the mammary gland is mainly acetate and β-hydroxybutyric acid (Harvatine, Boisclair, & Bauman, 2009). β-hydroxybutyric acid originate from butyrate (Table 1.2) which is metabolised to β-hydroxybutyric acid during absorption over the rumen wall. The FA synthesis is catalyzed by acetyl-CoA carboxylase and fatty acid synthase (Wakil, Stoops, & Joshi, 1983) and result in FA with chain length from 4 to 16 (Harvatine et al., 2009).

De novo synthesized FA are saturated, and FA of feed origin transported to the mammary glands are also mainly saturated due to rumen biohydrogenation. Saturated FA can be desaturased in the mammary gland catalyzed by stearoyl-CoA desaturase. Stearoyl-CoA desaturase is located in the endoplasmic reticulum with C18 and C16 FA as primary substrates (Palmquist, 2006). This enzyme adds a *cis9* double bond in FA and C18:0 is the main substrate. This enzymatic activity is higher in Holstein cows compared to Jersey, and the differences between animals are mainly genetic related (Poulsen et al., 2012).

3 Feed Effects on Milk Composition

3.1 Milk Fat Content

Milk gross composition can be controlled by feeding to some extent. Increased energy intake obtained by increased energy concentration in the ration, e.g. by increased concentrate offer, increased concentrate/forage ratio in mixed rations or increased forage digestibility can reduce milk fat concentration and consistently increase milk protein concentration (Alstrup et al., 2014; Hymøller, Alstrup, Larsen, Lund, & Weisbjerg, 2014). Increased fat in feed consistently decreases milk protein concentration (Weisbjerg et al., 2008), whereas the effect on milk fat concentration is more variable. Often small increases in fat supplementations to rations with low fat concentrations will increase milk fat production, and thereby severely decrease protein/fat ratio in the milk, whereas higher inclusions to >4–5% FA in ration DM might reduce milk fat concentration (Østergaard, Danfær, Daugaard, Hindhede, & Thysen, 1981). However, the use of saturated fat with chain length ≥ C16 seems to increase milk fat also at high concentrations in the ration (Weisbjerg et al., 2008). The effect of specific carbohydrates can mainly be ascribed to their effect on energy intake and rumen fermentation pattern, where non-dietary fibre (NDF) reduce the energy intake and increase the acetate proportion of rumen VFA, thereby increasing milk fat and reducing milk protein (Alstrup et al., 2014). Starch will increase the energy intake and the propionate ratio in the rumen, subsequently increasing milk protein and reducing milk fat (Stensig, Weisbjerg, & Hvelplund, 1998). Sugar may increase butyrate levels in the rumen, resulting in higher milk fat concentration and possibly increased milk protein (Stensig et al., 1998).

3.2 Fatty Acids

Roughly, FA with a chain length up to C16 originate from de novo synthesis, which to some extent is genetically controlled, whereas the longer chained FA to a large extent is determined by the content and composition in the feed. A very large number of studies have reported how different feeding regimes affect milk fatty acid composition, and several attempts have been made to perform meta-analyses to predict FA composition in milk fat based on feed composition. Such models seem valid as long as they are used in feed conditions similar to those that are included in the meta-analysis. In Table 1.3, typical values for FA distribution of the main FA in milk are shown.

The content of C4–C14 in milk fat is mainly dependent on the de novo synthesis of these FA; although some feed items contain minor amounts. Feeding fat supplements to dairy cows decreases de novo synthesis and use of fat sources containing more than 80% C16 has decreased the content of C4–C14 in milk fat by 22–43 g/kg FA (Piantoni, Lock, & Allen, 2013; Weisbjerg et al., 2013), corresponding to about

Table 1.3 Variation in composition of major fatty acids (FA) in milk as affected by different treatments

Treatment	Proportion of fresh grass	Hay or grazing	Oilseed supplements	Fat supplements
Reference	Couvreur et al. (2006)	Coppa et al. (2011)	Larsen et al. (2012)	Weisbjerg et al. (2013)
Fatty acids (g/kg FA)				
C4–C14	262–284	211–264	200–244	212–260
C16:0	241–310	230–326	217–337	242–397
C18:0	103–112	75–116	101–180	74–152
C18:1 *cis*9	194–211	174–223	162–266	147–256
C18:1 *trans*	24–59	18–75	11–31	12–30
C18:2 n6	13–16	8–14	14–18	15–20
C18:3 n3	2–7	6–10	5–8	4–8
CLA *cis*9 *trans*11	5–17	6–24	4–10	3–9
VLCFA[a]	1.5–2.8	3.2–4.4		

[a]Very long chained (>C18) polyunsaturated FA

half of the observed increase of C16 in milk fat. Feeding oil or oilseed supplements affected milk content of C4–C14 differently in different studies. Thus, 200–300 g oils or 1–1.5 kg oilseeds resulted in no reductions in the content of C4–C14 in milk (McNamee, Fearon, & Pearce, 2002), reductions corresponding to half of the reductions in C16 content (Larsen et al., 2012; Weisbjerg et al., 2013), or reductions similar to the reductions in C16 content (Fearon, Mayne, Beattie, & Bruce, 2004). The distribution within C4–C14 is feed dependent and when the de novo synthesis is reduced due to increasing fat supplements, C12–C14 FA are decreasing more rapidly than C4–C10 FA in the produced milk (Weisbjerg et al., 2008).

Palmitic acid, C16:0, is the dominant FA of milk and originates from feed as well as from the mammary de novo synthesis. The distribution between these two sources of C16 varies and in milk from high yielding dairy cows, the output of C16 in milk has been reported as 2.6–5.6 times the input (Larsen et al., 2012), where the lower values were obtained at high levels of fat supplementation from feed. Thus, to reduce the C16 content in milk fat it is important to reduce the amount in feed but controlling the de novo synthesis may be even more important. An obvious source of C16 in feed is fractionated palm fat, which has been used extensively to increase milk production, and supplementation of a fat source containing more than 80% C16 by 2–3% of dry matter intake (DMI) has increased C16 in milk fat by 50–90 g/kg FA compared to no fat supplement (Piantoni et al., 2013; Weisbjerg et al., 2013). The de novo synthesis is reduced when serum concentrations of C18 FA are increased e.g. by feeding cows oils or oilseeds. A decrease in C16 content of 50–100 g/kg FA is reported after feeding 1–1.5 kg per cow per day oilseeds of various types (Larsen et al., 2012; McNamee et al., 2002; Weisbjerg et al., 2013). Such feeding rich in C18 FA also increase the content of C18 FA in milk fat.

C18 FA in milk is derived from feed or from body mobilization, which means that higher contents are obtained when feed is rich in C18 FA or when the cow is in negative energy balance. C18 FA in feed are to a wide extent unsaturated and are extensively hydrogenated by the rumen microorganisms. The content and distribution of a vast range of C18 FA in milk are frequently reported and such results are used to document and explain how rumen processes are affected by feeding. Oleic acid (C18:1 *cis*9) is the second most abundant fatty acid in milk. Almost all C18:1 *cis*9 in milk is formed from C18:0 by mammary stearoyl-CoA desaturase. Some C18:1 *cis*9 can be transferred to milk from feed as rumen hydrogenation is incomplete, and a major feed source of C18:1 *cis*9 is rapeseed products (Larsen et al., 2012).

Linoleic acid, C18:2 n6, is the most abundant polyunsaturated FA in milk. The only source of C18:2 n6 in milk is C18:2 n6 from feed. In diets without fat supplements, grains and maize silage are significant sources, legumes and herbs are typically richer sources than grass species, and the richest sources of C18:2 n6 are oilseed products of soy and sunflower. The recovery from feed to milk depends very much on the degree of biohydrogenation and recoveries of 7–12% have been reported after feeding different levels of oilseeds (Larsen et al., 2012), whereas feeding freshly cut grass or herbs instead of TMR increased the transfer efficiency of n6 FA from 16% to 50–64% (Petersen, Søegaard, & Jensen, 2011).

Linolenic acid, C18:3 n3 is another essential FA present in milk. The primary source of C18:3 n3 in milk is grass-based feed, but linseed and rapeseed may be used as an alternative source. The level of biohydrogenation of C18:3 n3 from feed is usually high and recoveries in milk in the range of less than 10% are normal (Larsen et al., 2012; Petersen et al., 2011). The relative differences in C18:3 n3 concentrations in milk fat can be large although the absolute differences are small. Differences may be caused by differences in C18:3 n3 content in feed; however, in some cases, differences are mainly due to differences in biohydrogenation. Havemose, Weisbjerg, Bredie, and Nielsen (2004) fed cows maize silage or grass silage and found a threefold increase in the C18:3n3 concentration in milk, from 2 to 6 g/kg FA in milk from cows fed grass silage, which is due to the higher content of C18:3 n3 in grass compared to maize. Petersen et al. (2011) increased the n3 FA content from 7 g/kg FA in milk from TMR feed, over 11 g/kg FA% in milk from cows fed pure fresh cut white clover/rye grass mixture, to 21 g/kg in milk from cows fed species-rich fresh cut herbs. At the same time, the transfer efficiency of n3 FA increased from 7% in TMR fed cows over 8% for clover/grass fed cows to 16% for the herb fed cows. Feeding increased amounts of linseed resulting in increased contents of C18:3 n3 in milk, and Jersey cows have lower initial levels compared to Holstein (Larsen et al., 2012). The increased C18:3 n3 level is accompanied by reduced transfer efficiencies from 7% at no linseed supplement to 4% at the highest level of linseed (Larsen et al., 2012).

The main CLA present in milk is CLA *cis*9 *trans*11 (normally 80–90% of CLA) accompanied by smaller amounts of CLA *trans*7 *cis*9 and CLA *trans*10 *cis*12

(Shingfield, Bonnet, & Scollan, 2013). The main source of CLA *cis*9 *trans*11 is C18:1 *trans*11, which is desaturated by mammary desaturase. Thus, the content of CLA *cis*9 *trans*11 is mainly dependent on the C18:1 *trans*11 supply to the udder and the desaturase activity. Content of CLA *cis*9 *trans*11 in milk is increased when high amounts of grazing are used and very high correlations between dry matter intake from pasture and CLA *cis*9 *trans* 11 content in milk have been reported (Couvreur, Hurtaud, Lopez, Delaby, & Peyraud, 2006). Also, fat-rich feed can increase the content of CLA *cis*9 *trans*11 as the fat-rich feed may reduce the final rumen hydrogenation and thus result in a higher amount of C18:1 *trans*11. In an oilseed feeding experiment (Larsen et al., 2012) there was a higher increase in CLA *cis*9 *trans*11 content for Holstein cows at highest level of oilseed feeding compared to other oilseed levels and compared to Jersey cows. However, if large amounts of linoleic acid are fed together with high starch rations, the acidic rumen fermentation can favour production of the CLA isomer C18:2 *trans*10 *cis*12. This CLA isomer is known to cause milk fat depression and will, therefore, be accompanied by a lowering of the milk fat percentage (Bauman, Mather, Wall, & Lock, 2006).

A range of *trans* FA are formed during rumen biohydrogenation and those which are not finally hydrogenated can be *trans*ferred to milk. The C18:1 *trans* FA include all positions from 4 to 16, of which C18:1 *trans* 11 (vaccenic acid) is the most abundant under most practical feeding conditions. However, certain feeding conditions involving high levels of polyunsaturated FA and high levels of concentrate may impair normal rumen function to such an extent that formation of C18:1 *trans* 10 is favoured (Bauman et al., 2006; Shingfield et al., 2013). Also, branched and odd-chain FA are synthesized microbially in the rumen and the content in milk can be used as a diagnostic tool of rumen function (Vlaeminck, Fievez, Cabrita, Fonseca, & Dewhurst, 2006).

The very long chained polyunsaturated FA (>C18, (*VLCFA*)) constitute less than 1% of total FA in normal milk fat, but in the milk fat globule membrane (*MFGM*) they constitute 3–3.5% of total FA. Arachidonic acid (ARA, C20:4n-6) is the dominating VLCFA and constitute around one-third of these FA. The n-3 VLCFA (C20:5n-3 and C22:5n-3) constitute 20–25% of VLCFA in both milk fat and MFGM. Docosahexaenoic acid (DHA, C22:6n-3) is absent in cow milk (Jensen & Nielsen, 1996).

Besides feeding, lactation stage also affects FA composition in milk. During mobilization the concentrations of C16 and C18 FA in milk fat increase, because body fat is used as a fat source (Stoop, Bovenhuis, Heck, & van Arendonk, 2009). The stage of lactation also affects the fatty acid composition, where C16 increases during early and mid-lactation and remains at the same level at mid to late lactation, and there is a corresponding decrease in C18 FA. The amount of unsaturated FA follows the content of C18 FA until late lactation, where there is an increase in unsaturated FA (Stoop et al., 2009).

4 Physicochemical Attributes of Milk Fat Affected by Feed and Management

4.1 Effect of Feed on Milk Fat Globules

It has been known for more than 300 years that fat exists as globules in milk (Leewenhoeck, 1674). The diameter of the milk fat globule (MFG) ranges from 0.1 to 12 μm with an average of around 4.5 μm. The average diameter of the MFG depend on breed of cow. Cows like Jersey and Gurneys produce milk with not only a higher fat content but also larger MFG than cows like Frisians and Holsteins. Figure 1.3 shows the difference in MFG diameter between the three most frequent breeds in Denmark.

The feeding also influences the diameter of MFG. Increasing the dietary fat supplements in cows increase the average diameter of the MFGs especially when the fat supplements is based on most of saturated FA (Weisbjerg et al., 2008; Wiking, Bjorck, & Nielsen, 2003). This seems to be regulated through an increased daily fat yield, which is positively correlated to increased average diameter of the MFGs (Wiking, Stagsted, Björck, & Nielsen, 2004; Wiking, Theil, Nielsen, & Sørensen, 2010). In contrast, fresh grass feeding resulting in lower fat yield is followed by a reduction in MFG size (Couvreur, Hurtaud, Marnet, Faverdin, & Peyraud, 2007). A theory about the mechanisms behind the increase in MFG diameter when the cow produces more fat is that it can be due to a limited amount of milk fat globule membrane (MFGM) material present during milk fat synthesis, so the fat is enveloped in large globules which require less surface material. This is indicated by a decreased activity of the MFGM enzyme, γ-glutamyl transpeptidase, and lower concentration of polar lipids in whole milk with increasing MFG size (Liu, Logan, Cocks, & Rochfort, 2017; Wiking et al., 2004). Oppositely, the level of another membrane component, cholesterol, correlated positively with daily fat yield (Larsen, 2012). Thus, another simple theory could be that a larger milk fat synthesis increases the

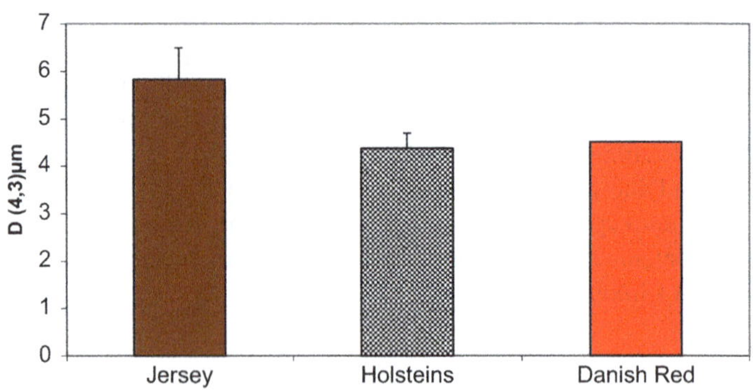

Fig. 1.3 Volume weighted diameter ($D_{4,3}$) of the three Danish cattle breeds

fusion of oil droplets during the transport from the basal to the apical boundaries in the mammary epithelial cells. It is especially the population of larger fat globules (average diameter >8 μm) which increase when the cow produces more fat (Wiking et al., 2004). During the lactation period of cows, the fat yield often decreases which result in smaller average MFG (Altenhofer et al., 2015; Walstra, 1969).

Besides affecting the physical size of MFGs, feeding also have some effect on the lipid composition of the milk fat globule membrane (MFGM). As previously described, feeding highly affect the fatty acid composition in the core milk fat. The effects on the FA composition of MFGM are lesser. MFGM phospholipids contain high levels of palmitic and oleic acid, while the short and medium-chain FA are present in very low levels (Mcpherson & Kitchen, 1983). Smith, Bianco, and Dunkley (1977) found that feeding a supplement rich in linoleic acid increased the unsaturation of the phospholipids in the MFGM. However, this unsaturation was less than that of the core lipids. Likewise, feeding a fresh pasture based spring diet results in more unsaturated FA in the MFGM versus a corn silage based diet (Lopez, Briard-BIon, & Ménard, 2014). Palmquist and Schanbacher (1991) observed that by feeding palmitic acid to the cow, it is possible to increase the saturation of the lipids in the membrane.

Recently, several papers have studied the difference in FA composition between small and large MFGs. These indicate that small MFGs contain less short-chained FA (Mesilati-Stahy, Mida, & Argov-Argaman, 2011; Rahmatyar & Wiking, 2012; Timm & Patton, 1988) and some studies also reported more content of polyunsaturated FA in the small MFG (Mesilati-Stahy et al., 2011; Rahmatyar & Wiking, 2012). However, literature is contradictory. For example, Briard, Leconte, Michel, and Michalski (2003) found the opposite that the methods used for size separating MFGs differ among studies, thus different size group are obtained which might explain some of the variation. The purpose of studying such compositional variations are driven by finding unique functionality of MFG subpopulations. Several studies have produced cheeses from milk enriched with either large or small MFG and especially the lipolysis increase in cheeses with increased MFG size (Michalski & Briard, 2004; O'Mahoney, Auty, & McSweeney, 2005). Current focus is on large differences between MFG and the small extracellular vesicles, that contain no triglycerides, e.g the extracellular vesicles are much richer in sphingomyelin (Blans et al., 2017).

4.2 *Effect of Cow Feeding on Crystallization and Melting of Milk Fat*

Since the FA composition of bovine milk fat is highly affected by feeding, it also changes the crystallization and melting behavior of milk fat (Bertram, Wiking, Nielsen, & Andersen, 2005). The high diversity of FA in milk fat gives a high diversity of triacylglycerides which promote a very broad melting range of milk fat, from −40 to 40 °C. Further, the melting profile is characterized by three melting fractions, a low melting fraction from −25 to 10 °C, a medium fraction from 10 to 19 °C, and the high melting fraction from above 20 °C (Timms, 1984). It is well known that butter produced during summer, when cows are grazing, is softer than butter produced from winter milk. This effect has been correlated to the changes in the FA composition over the season, where the unsaturated-to-saturated FA ratio increases during the summer, which is related to grazing. To balance a more uniform texture of butter throughout the year, manufacturers can give the cream a season-dependent temperature treatment to manipulate fat crystal size and composition, which defines the spreadability and hardness of the butter. Larger effects can be obtained by feeding high amounts of unsaturated oilseed, for example feeding extruded linseeds can lower the level of saturated FA from 71 to 61% in milk, which result in 14% less crystalized fat at 5 °C (Smet et al., 2010).

Recent studies indicate that it is mainly the lowest melting fraction, which is affected by the feeding of the cows (Buldo, Larsen, & Wiking, 2013; Larsen, Andersen, Kaufmann, & Wiking, 2014). This fraction is important for the functionality, e.g. spreadability and hardness, for dairy products stored in the fridge. Buldo et al. (2013) increased the C16:0 in milk through feeding and observed that the melting point of the lowest melting fraction is positively correlated to the concentration of C16:0 and negatively correlated to the concentration of C18:1 *cis*-9 in milk fat. By comparing organic and conventional milk, Larsen et al. (2014) also reported that only the lowest melting fraction varied between dairies and season and this was positively correlated to the proposition of C16 and negatively correlated to C18:1 *cis*-9.

5 Free Fatty Acids in Milk

A serious sensory quality defect related to milk fat is the formation of free fatty acids (FFA) due to hydrolysis of triglycerides. Especially the short-chained FFA (C4–C8) contribute with a distinct rancid flavour. FFA is caused by hydrolysis of the triglycerides by mainly lipoprotein lipase (LPL) in raw milk. It originates from the mammary gland, where it is involved in the uptake of blood lipids for milk synthesis. The enzyme is active in lipid-water interfaces. Its optimum temperature is 33 °C, and pH optimum is about 8.5. It is a relatively heat labile enzyme which is mostly inactivated by a high temperature-short time heat treatment. In milk, LPL is mainly associated with the casein micelles. LPL is first brought into contact with the triglycerides when the MFGM is disrupted and casein will recoat the formed lipid-water interface. In fact, just cooling the raw milk will change the configuration of the MFGM and bring LPL together with caseins in contact with the triglycerides (Dickow, Larsen, Hammershøj, & Wiking, 2011). The enzyme is activated by apolipoprotein CII, which is normally found in the blood and therefore can be transferred to the milk. This activator assists LPL to bind onto the fat globule. In spite of the high amount of LPL in milk, lipolysis is limited since milk fat is protected by the membrane and raw milk is normally stored at temperatures far below the optimum temperature of LPL. Furthermore, the products of the hydrolysis of the triglycerides, the FFA, inhibit the enzyme presumably due to FFA bindings to the LPL. The level of FFA in milk can be affected by milking systems and set-up at the farm, milking frequency and to some extent feeding.

5.1 *Effect of Cow Feeding on Milk FFA*

Changes in the composition of the cow's feed can cause milk to be more susceptible to lipolysis. Feeding concentrate with a large amount of saturated lipids results in an elevated fat percentage and large-sized fat globules, which are very prone to lipolysis (Wiking et al., 2003) and milk originating from feeding unsaturated fat or from diets stimulating de novo synthesis caused a reduced fat percentage and this milk fat is more stable during pumping. Underfeeding of the cow can also lead to increased FFA concentration in the milk (Thomson, Van der Poel, Woolford, & Auldist, 2005).

5.2 *Milking Frequency*

A higher milk yield can be obtained by increasing the milking frequency, i.e. from twice to three times per day. However, this also increases the content of FFA in milk (Klei et al., 1997; Wiking, Nielsen, Båvius, Edvardsson, & Svennersten-Sjaunja, 2006). The mechanism behind this accumulation of FFA is rather unknown. It has been observed that the increase in FFA is not found immediately after milking, but first after storage, indicating that a weaker membrane is formed (Wiking et al., 2006). Moreover, MFG size increases when the cows are milked more often (Wiking et al., 2006). Milk fat content increases during milking (Nielsen, Larsen, Bjerring, & Ingvardtsen, 2005), so at higher milking frequencies, the residual high-fat milk from the previous milking contributes to a higher proportion of the cisternal milk, this may explain the higher milk fat content at short intervals between milkings (Nielsen et al., 2005) which might also have some effect on FFA.

5.3 *Milking Systems*

The contribution from specific milking systems on FFA formation has always drawn attention. The introduction of pipeline systems resulted in increased FFA content in milk due to high air intake from the claw or the pumps. In addition, the milk is often lifted in the pipes above cow level, which increases the pressure on the MFG and thereby increases the degree of damages (Fleming, 1979). The contact between a MFG and an air bubble results in rupture of the MFG. Herring bone parlors and rotary parlors, normally provide a better quality regarding FFA since air intake and milk flows are better controlled.

The introduction of automatic milking system (AMS) in the late 1990s brought the attention on FFA again. The increased milking frequency and the continuous pumping of warm milk into the tank are factors contributing to elevated FFA levels in these systems. The average milking frequency in AMS is between 2.4 and 2.6 a day. This implies that a large share of the cows attends the milking robot three or four times per day.

6 Conclusion

This chapter has described the tools and mechanisms to manipulate composition and some physiochemical attributes of the milk fat. Focus is on the effect of the feed ration. Especially, the effect of the feed ration on the milk FA composition have been intensively studied. In general, the de novo FA, C4–16, in milk increase by starch-based concentrates and decrease by feeding fat supplements. The long-chain FA, >C16, can to some extent be manipulated by changed concentration and composition of FA in the diet. For instance, a large relative increase in C18:3 n3 in milk can be obtained by feeding linseeds rich in C18:3 n3, or by feeding fresh grass or herbs. The hydrogenation in the rumen, however, limits some of the effects from polyunsaturated FA in feedstuff. The fat composition and amount in feed also affects physical attributes like the size of the MFG and the fat crystallization/melting behaviors. Besides feeding, FA composition and MFG size are highly affected by cow breed and modestly affected by lactation state. Finally, milking management that include choice of milking systems and milking frequency, also affect the quality of the milk fat, i.e. formation of FFA.

References

Alstrup, L., Weisbjerg, M. R., Hymøller, L., Larsen, M. K., Lund, P., & Nielsen, M. O. (2014). Milk production response to varying protein supply is independent of forage digestibility in dairy cows. *Journal of Dairy Science, 97*, 4412–4422.

Altenhofer, C., Holzmüller, W., Wolfertstetter, F., Wolfschoon, R. D., Kulozik, U., Pfaffl, M. W., et al. (2015). Temporal variation of milk fat globule diameter, fat and cholesterol content and milk epithelial cell gene expression in dairy cows. *International Journal of Dairy Technology, 68*(4), 519–526.

Bauman, D. E., Mather, I. H., Wall, R. J., & Lock, A. L. (2006). Major advances associated with the biosynthesis of milk. *Journal of Dairy Science, 89*, 1235–1243.

Bertram, H. C., Wiking, L., Nielsen, J. H., & Andersen, H. J. (2005). Direct measurement of phase transitions in milk fat during cooling of cream - A low-field NMR approach. *International Dairy Journal, 15*, 1056–1063.

Blans, K., Hansen, M. S., Sørensen, L. V., Hvam, M. L., Howard, K. A., Möller, A., et al. (2017). Pellet-free isolation of human and bovine milk extracellular vesicles by size-exclusion chromatography. *Journal of Extracellular Vesicles, 6*, 1294340.

Briard, V., Leconte, N., Michel, F., & Michalski, M. C. (2003). The fatty acid composition of small and large naturally occurring milk fat globules. *European Journal of Lipid Science and Technology, 105*, 677–682.

Buldo, P., Larsen, M. K., & Wiking, L. (2013). Multivariate data analysis for finding the relevant fatty acids contributing to the melting fractions of cream. *Journal of Science and Food Agriculture, 93*, 1620–1625.

Coppa, M., Verdier-Metz, I., Ferlay, A., Pradel, P., Didienne, R., Farruggia, A., et al. (2011). Effect of different grazing systems on upland pastures compared with hay diet on cheese sensory properties evaluated at different ripening times. *International Dairy Journal, 21*, 815–822.

Couvreur, S., Hurtaud, C., Lopez, C., Delaby, L., & Peyraud, J. L. (2006). The linear relationship between the proportion of fresh grass in the cow diet, milk fatty acid composition, and butter properties. *Journal of Dairy Science, 89*, 1956–1969.

Couvreur, S., Hurtaud, C., Marnet, P. G., Faverdin, P., & Peyraud, J. L. (2007). Composition of milk fat from cows selected for milk fat globule size and offered either fresh pasture or a corn silage-based diet. *Journal of Dairy Science, 90*(1), 392–403.

De Vries, M., Debruyne, L., & Aarts, F. (2013). *Sustainability of dairy farming and the implementation of EU environmental directives in the northwest of Europa*. Wageningen, The Netherlands: Dairyman.

Dickow, J. A., Larsen, L. B., Hammershøj, M., & Wiking, L. (2011). Cooling causes changes in the distribution of lipoprotein lipase and milk fat globule membrane proteins between the skim milk and cream phase. *Journal of Dairy Science, 94*, 646–656.

Fearon, A. M., Mayne, C. S., Beattie, J. A. M., & Bruce, D. W. (2004). Effect of level of oil inclusion in the diet of dairy cows at pasture on animal performance and milk composition and properties. *Journal of Science and Food Agriculture, 84*, 497–504.

Fleming, M. G. (1979). Lipolysis in bovine milk as affected by mechanical and temperature activation – a review. *Irish Journal of Food Science and Technology, 3*, 111–129.

Harfoot, C. G., & Hazlewood, G. P. (1997). Lipid metabolism in the rumen. In P. N. Hobson & C. S. Stewart (Eds.), *The rumen microbial ecosystem* (pp. 382–426). London, UK: Chapman & Hall.

Harvatine, K. J., Boisclair, Y. R., & Bauman, D. E. (2009). Recent advances in the regulation of milk fat synthesis. *Animal, 3*, 40–54.

Havemose, M. S., Weisbjerg, M. R., Bredie, W. L. P., & Nielsen, J. H. (2004). Influence of feeding different types of roughage on the oxidative stability of milk. *International Dairy Journal, 14*, 563–570.

Hymøller, L., Alstrup, L., Larsen, M. K., Lund, P., & Weisbjerg, M. R. (2014). High quality forage can replace concentrate when cows enter deposition phase without negative consequences for milk production. *Journal of Dairy Science, 97*, 4433–4443.

Jensen, S. K., & Nielsen, K. N. (1996). Distribution of tocopherols, retinol, ß-carotene and fatty acids among fat globule membrane and fat globule core in cow milk. *The Journal of Dairy Research, 63*, 565–574.

Klei, L. R., Lynch, J. M., Barbano, D. M., Oltenacu, P. A., Lednor, A. J., & Bandler, D. K. (1997). Influence of milking three times a day on milk quality. *Journal of Dairy Science, 80*, 427–436.

Kristensen, T., Jensen, C., Østergaard, S., Weisbjerg, M. R., Aaes, O., & Nielsen, N. (2015). Feeding, production, and efficiency of Holstein-Friesian, Jersey, and mixed-breed lactating dairy cows in commercial Danish herds. *Journal of Dairy Science, 98*, 263–274.

Larsen, M. K., Andersen, K. K., Kaufmann, N., & Wiking, L. (2014). Seasonal variation in the composition and melting behaviour of milk fat. *Journal of Dairy Science, 97*, 4703–4712.

Larsen, M. K., Hymøller, L., Brask-Pedersen, D. B., & Weisbjerg, M. R. (2012). Milk fatty acid composition and production performance of Danish Holstein and Danish Jersey cows fed different amounts of linseed and rapeseed. *Journal of Dairy Science, 95*, 3569–3578.

Larsen, M. K., Kidmose, U., Kristensen, T., Beaumont, P., & Mortensen, G. (2013). Chemical composition and sensory quality of bovine milk as affected by type of forage and proportion of concentrate in the feed ration. *Journal of Science and Food Agriculture, 93*, 93–99.

Larsen, T. (2012). Enzymatic-fluorometric quantification of cholesterol in bovine milk. *Food Chemistry, 135*(3), 1261–1267.

Leewenhoeck, M. (1674). More microscopical observations. *Philosophical Transactions, 9*, 23–24.

Liu, Z. Q., Logan, A., Cocks, B. G., & Rochfort, S. (2017). Seasonal variation of polar lipid content in bovine milk. *Food Chemistry, 237*, 865–869.

Lopez, C., Briard-BIon, V., & Ménard, O. (2014). Polor lipids, sphingomyelin and long-chain unsaturated fatty acids from the milk fat globule membrane are increased in milks produced by cows fed pasture based diet during spring. *Food Research International, 58*, 59–68.

McNamee, B. F., Fearon, A. M., & Pearce, J. (2002). Effect of feeding oilseed supplements to dairy cows on ruminal and milk fatty acid composition. *Journal of Science and Food Agriculture, 82*, 677–684.

Mcpherson, A. V., & Kitchen, B. J. (1983). Reviews of the progress of dairy sciences: the bovine milk fat globule membrane – its formation, composition, structure and behaviour in milk and dairy products. *The Journal of Dairy Research, 50*, 107–133.

Mesilati-Stahy, R., Mida, K., & Argov-Argaman, N. (2011). Size-dependent lipid content of bovine milk fat globule and membrane phospholipids. *Journal of Agricultural and Food Chemistry, 59*, 7427–7435.

Michalski, M. C., & Briard, V. (2004). Fatty acid composition of total fat from cammembert cheeses with small and large native milk fat globules. *Milchwissenschaft, 59*, 273–277.

Nielsen, N. I., Larsen, T., Bjerring, M., & Ingvardtsen, K. L. (2005). Quarter health, milking interval, and sampling time during milking affect the concentration of milk constituents. *Journal of Dairy Science, 88*, 3186–3200.

NorFor. (2017). *NorFor feedtable*. Retrieved 21 August, 2017, from http://feedstuffs.norfor.info/

O'Mahoney, J. A., Auty, M., & McSweeney, P. L. H. (2005). The manufacture of miniature Cheddar-type cheeses from milks with different fat globule size distributions. *The Journal of Dairy Research, 72*, 338–348.

Østergaard, V., Danfær, A., Daugaard, J., Hindhede, J., Thysen, I. (1981). *Foderfedtets indflydelse på malkekøernes produktion*. Beretning 508, Statens Husdyrbrugsforsøg, Copenhagen, Denmark.

Palmquist, D. L. (2006). Milk fat: Origin of fatty acids and influence of nutritional factors thereon. In P. F. Fox & P. L. H. McSweeney (Eds.), *Advanced dairy chemistry, volume 2: Lipids* (3rd ed., pp. 43–92). New York, NY: Springer.

Palmquist, D. L., & Schanbacher, F. L. (1991). Dietary-fat composition influences fatty-acid compositon of milk-fat globule-membrane in lactating cows. *Lipids, 26*, 718–722.

Petersen, M. B., Søegaard, K., & Jensen, S. K. (2011). Herb feeding increases n-3 and n-6 fatty acids in cow milk. *Livestock Science, 141*, 90–94.

Piantoni, P., Lock, A. L., & Allen, M. S. (2013). Palmitic acid increased yields of milk and milk fat and nutrient digestibility across production level of lactating cows. *Journal of Dairy Science, 96*, 7143–7154.

Poulsen, N. A., Gustavsson, F., Glantz, M., Paulsson, M., Larsen, L. B., & Larsen, M. K. (2012). The influence of feed and herd on fatty acid composition in 3 dairy breeds (Danish Holstein, Danish Jersey, and Swedish Red). *Journal of Dairy Science, 95*, 6362–6371.

Rahmatyar, Z., & Wiking, L. (2012). Fatty acid composition and thermal behavior of small and large milk fat globules. *Milchwissenschaft, 67*, 34–38.

Shingfield, K. J., Bonnet, M., & Scollan, N. D. (2013). Recent developments in altering the fatty acid composition of ruminant-derived foods. *Animal, 7*, 132–162.

Smet, K., Coudijzer, K., Fredrick, E., De Campeneere, S., De Block, J., Wouters, J., et al. (2010). Crystallization behavior of milk fat obtained from linseed-fed cows. *Journal of Dairy Science, 93*, 495–505.

Smith, L. M., Bianco, D. H., & Dunkley, W. L. (1977). Composition of milk fat globules with increased linoleic-acid. *Journal of the American Oil Chemists Society, 54*, 132–137.
Stensig, T., Weisbjerg, M. R., & Hvelplund, T. (1998). Digestion and passage kinetics of fibre in dairy cows as affected by the proportion of wheat starch or sucrose in the diet. *Acta Agriculturae Scandinavica, Section A — Animal Science, 48*, 129–140.
Stoop, W. M., Bovenhuis, H., Heck, J. M. L., & van Arendonk, J. A. M. (2009). Effect of lactation stage and energy status on milk fat composition of Holstein-Friesian cow. *Journal of Dairy Science, 92*, 1469–1478.
Thoma, G., Popp, J., Nutter, D., Shonnard, D., Ulrich, R., Matlock, M., et al. (2013). Greenhouse gas emissions from milk production and consumption in the United States: A cradle-to-grave life cycle assessment circa 2008. *International Dairy Journal, 31*, S3–S14.
Thomson, N. A., Van der Poel, W. C., Woolford, M. W., & Auldist, M. J. (2005). Effect of cow diet on free fatty acid concentrations in milk. *New Zealand Journal of Agricultural Research, 48*(3), 301–310.
Timm, H., & Patton, S. (1988). Milk fat globules: Fatty acid composition, size and in vivo regulation of fat liquidity. *Lipids, 23*, 685–689.
Timms, R. E. (1984). Phase behavior and polymorphism of milk fat, milk fat fractions and fully hardened milk fat. *Australian Journal of Dairy Technology, 35*, 47–53.
Vlaeminck, B., Fievez, V., Cabrita, A. R. J., Fonseca, A. J. M., & Dewhurst, R. J. (2006). Factors affecting odd- and branched-chain fatty acids in milk: A review. *Animal Feed Science and Technology, 131*, 389–417.
Volden, H. (2011). Norfor – The nordic feed evaluation system. In *Paper presented at the 62nd Annual Meeting of European Federation of Animal Science (EAAP) Stavanger, Norway, 29 Aug–2 Sep. Publication No. 130* (pp. 41–54). Wageningen, The Netherlands: Wageningen Academic Publishers.
Wakil, J., Stoops, J. K., & Joshi, V. C. (1983). Fatty acid synthesis and its regulation. *Annual Review of Biochemistry, 52*, 537–579.
Walstra, P. (1969). Studies on milk fat dispersion II. The globule-size distribution of cow's milk. *Netherlands Milk and Dairy Journal, 23*, 99–110.
Weisbjerg, M. R., Børsting, C. F., & Hvelplund, T. (1992). Fatty acid metabolism in the digestive tract of cows when tallow is fed in increasing amounts at two feed levels. *Acta Agriculturae Scandinavica, 42*, 106–114.
Weisbjerg, M. R., Larsen, M. K., Hymøller, L., Thorhauge, M., Kidmose, U., Nielsen, J. H., et al. (2013). Milk production and composition in Danish Holstein, Danish Red, and Danish Jersey cows supplemented with saturated or unsaturated fat. *Livestock Science, 155*, 60–70.
Weisbjerg, M. R., Wiking, L., Kristensen, N. B., & Lund, P. (2008). Effects of supplemental dietary fatty acids on milk yield and fatty acid composition in high and medium yielding cows. *The Journal of Dairy Research, 75*, 142–152.
Wiking, L., Bjorck, L., & Nielsen, J. H. (2003). Influence of feed composition on stability of fat globules during pumping of raw milk. *International Dairy Journal, 13*, 797–803.
Wiking, L., Nielsen, J. H., Båvius, A.-K., Edvardsson, A., & Svennersten-Sjaunja, K. (2006). Impact of milking frequencies on the level of free fatty acids in milk, fat globule size and fatty acid composition. *Journal of Dairy Science, 89*, 1004–1009.
Wiking, L., Stagsted, J., Björck, L., & Nielsen, J. H. (2004). Milk fat globule size is affected by fat production in dairy cows. *International Dairy Journal, 14*, 909–913.
Wiking, L., Theil, P., Nielsen, J. H., & Sørensen, M. T. (2010). Effect of grazing fresh legumes or feeding silage on enzymes involved in the synthesis of milk fat in dairy. *The Journal of Dairy Research, 77*, 337–342.

Chapter 2
Lipase Action on Milk Fat

Hilton C. Deeth

1 Introduction

Lipases have important functions in almost all living organisms and play a major role in the metabolism of lipids. True lipases (EC 3.1.1.1) act on triglycerides (TG), diglycerides (DG) and monoglycerides (MG) (tri-, di- and monoacylglycerols), with triglycerides being the major substrate. Other lipases act on other types of lipids, for example, phospholipases act on phospholipids. This chapter largely concerns true lipases.

While lipases are best known for catalysing hydrolysis (lipolysis) of triglycerides, they can also catalyse acyl transfer reactions. Lipolysis reactions occur in aqueous conditions but acyl-transfer reactions occur in organic solvents containing just enough water to maintain the activity of the enzyme but minimising the hydrolysis reaction. Acyl transfer reactions include alcoholysis, acidolysis and interesterification (Fig. 2.1). Alcoholysis occurs where an alcohol is incorporated as the nucleophilic receptor and a new ester is formed; if methanol is the alcohol, methyl esters are formed whereas if glycerol is incorporated, di- and monoglycerides are formed. Acidolysis occurs where an introduced carboxylic acid replaces one of the fatty acids of the TG, and interesterification occurs where acyl transfer occurs between an ester and a TG, often between two TGs or TG mixtures. Bourlieu, Bouhallab, and Lopez (2009) reviewed 28 recent publications on the use of enzymes on milk fat and found 44% on interesterification, 36% on hydrolysis, 11% on acidolysis and 7% on alcoholysis (glycerolysis). In a previous survey of 38 publications, Balcão and Malcata (1998b) found 46% concerned interesterification and 43% hydrolysis.

H. C. Deeth (✉)
School of Agriculture and Food Sciences, University of Queensland,
St Lucia, QLD, Australia
e-mail: h.deeth@uq.edu.au

T. Truong et al. (eds.), *Dairy Fat Products and Functionality*,
https://doi.org/10.1007/978-3-030-41661-4_2

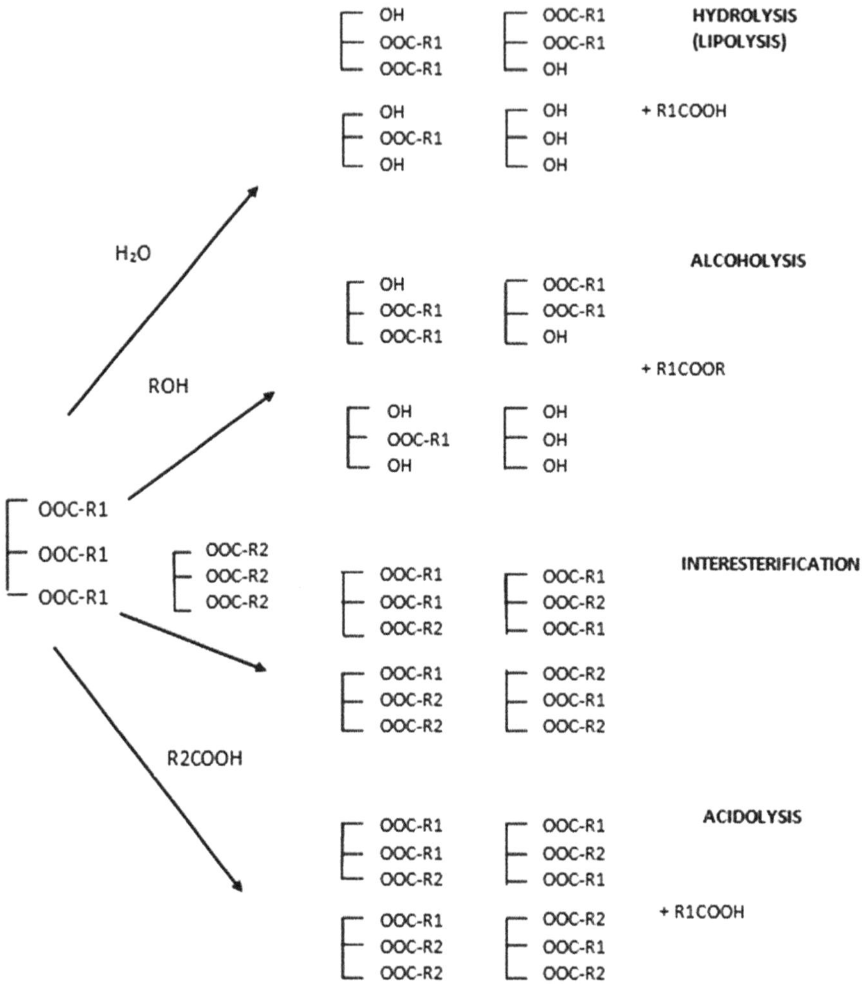

Fig. 2.1 Schematic representation of hydrolysis (lipolysis) and acyl transfer reactions catalysed by lipases

Like most enzymes, lipases are characterised by their substrate specificity. A major type of specificity is in relation to the position of the fatty acid on the glyceride. Since most natural triglycerides are asymmetric, the three positions on the glycerol backbone are different; they are designated *sn*-1, *sn*-2 and *sn*-3, where *sn* stands for stereospecific numbering. Most commonly, lipases have a strong preference for the primary esters, those in *sn*-1 and *sn*-3 positions, although a few lipases, e.g., from *Candida parapsilosis* (Akoh, Sellaopoan, Fomuso, & Yankah, 2003; Riaublanc, Ratomahenina, Galzy, & Nicolas, 1993), have a preference for esters in the *sn*-2 position. A summary of the specificities of some enzymes is given in Table 2.1.

Table 2.1 Specificities of some lipases and esterases

Specificity	Enzyme name/source
Non-specific	*Candida rugosa, Pseudomonas fluorescens, Pseudomonas cepacia, Candida antarctica, Chromobactrerium viscosum, Penicillium cyclopium, Candida cyclindracea*
sn-1,3	*Rhizopus niveus, Aspergillus niger, Rhizopus oryzae, Penicillium roqueforti, Rhizopus delemar, Mucor javanicus, Thermomyces lanuginosa, Rhizomucor miehei, Mucor miehei, Candida deformans*, pancreatic lipase
sn-1,3; *sn*-1 > *sn*-3	Lipoprotein lipase
sn-1,3; *sn*-3 > *sn*-1	Pregastric esterases
sn-2	*Candida parapsilosis*
MG/DG	*Penicillium camemberti*
Cis *n*-9 unsaturated FAs	*Geotricum candidum*
Linoleic acid	*Aspergillus lipolyticum*
Medium/long chain FAs	*Rhizopus niveus, Aspergillus niger, Rhizopus oryzae*
Small/medium chain FAs	*Penicillium roqueforti*
Short chain FAs	Pregastric esterases

From: Morley and Kuksis (1977), Balcão and Malcata (2002), Bourlieu et al. (2009), Jooyandeh et al. (2009)

As a consequence of the various functions of lipases, this chapter is divided into sections relating to the hydrolysis reaction (lipolysis) and its consequences for the dairy and other food industries, and the acyl transfer reactions which relate more to the biotechnology industry than to the dairy industry. Reviews related to the first have been published by Sørhaug and Stepaniak (1997), Olivecrona, Vilaró, and Olivecrona (2003) and Deeth and Fitz-Gerald (2006) while reviews related to the biotechnological topics have been published by Seitz (1974), Balcão and Malcata (1998b, 2002), Bourlieu et al. (2009), Jooyandeh, Kaur, and Minhas (2009), Kontkanen et al. (2011) and Javed et al. (2018).

2 Lipolysis

Lipolysis, the term used for lipase-catalysed hydrolysis of lipids, occurs at the ester bond between a fatty acid and an alcohol group. The products of lipolysis of TGs are therefore an unesterified or free fatty acid (FFA) and a DG. The DG can be further hydrolysed to a MG which, in some cases, can be further hydrolysed to glycerol, each time releasing a FFA, as shown below.

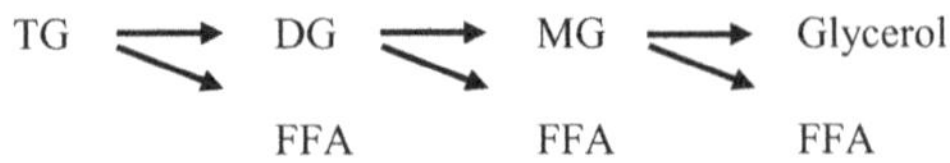

DGs and MGs are sometimes collectively categorised as partial glycerides, and, as discussed below, have significant surface-active properties.

The hydrolysis action catalysed by lipases differs from that of most other enzymes in that it occurs at a surface rather than in a solution of the lipid substrates. Commonly the lipid surface is a part of an oil-in-water emulsion. Enzymes which act on the ester bond of soluble esters are referred to as esterases. In general, lipids which are substrates for esterases tend to be glycerides of short-chain fatty acids, although esterases also act on a wide range of soluble esters. However, there is some overlap between lipases and esterases with some lipases acting on glycerides of short-chain FFAs such as tributyrin. Several types of esterase have been identified according to their preferred substrate, e.g., carboxylesterase, arylesterase.

The lipases of most relevance for their hydrolytic action on the fat in milk and milk products are the natural milk lipase, a lipoprotein lipase (LPL), and bacterial lipases produced by a wide range of bacteria. The bacterial lipases of significance in the dairy industry are those produced by bacteria which contaminate milk and dairy products and cause flavour and other defects. Another important category of lipases is those added in the manufacture of some products such as feta, romano and parmesan cheeses. The enzymes traditionally used for this purpose originate from the stomachs of young ruminant animals, calves, lambs or kids; they are sometimes referred to as pre-gastric esterases because of their preference for releasing short-chain FFAs.

2.1 The Lipases Responsible for Lipolysis in Milk and Milk Products

2.1.1 Milk Lipoprotein Lipase: The Natural Milk Lipase

Cows' milk and the milk of other species, including humans, naturally contain LPL (Olivecrona et al., 2003). The enzyme is so-called because it is activated by lipoproteins or their fragments (apo-lipoproteins). This can be demonstrated with milk LPL by adding blood serum to milk and observing the consequent lipolysis.

Milk LPL is the same enzyme as the one in the mammary gland where its function is the synthesis of milk fat triglycerides (Barber, Clegg, Travers, & Vernon, 1997). Its presence in milk appears to be due to spill-over from the mammary gland as its biological function in milk is unclear. It may play a role in fat digestion in the young animal, but this does not appear to be a significant function because it is inactivated at low pH, such as encountered in the stomach.

In cows' milk, LPL accounts for virtually all of the lipase activity. The LPL content is high and, under ideal conditions, is capable of causing the hydrolysis of a large proportion of the fat in milk. Cows' milk also contains some minor esterase activity which has been designated aryl esterase (Marquardt & Forster, 1965) and carboxylesterase (Fitz-Gerald, Deeth, & Kitchen, 1981). The esterase activity is elevated in colostrum and mastitic milk (milk with high somatic cell count). However, the total esterase activity in these abnormal milks is small compared with LPL activity and has little if any relevance to lipolysis of milk fat (Deeth, 1978; Marquardt & Forster, 1962). By contrast, human milk contains a bile-salt-stimulated lipase, in addition to LPL, which is believed to have a major role in digestion of milk fat by the newborn (Hernell & Bläckberg, 1991, 1994). The following discussion relates specifically to LPL in cows' milk.

Milk LPL is a dimeric glycoprotein with a molecular weight of ~100,000 Da (Kinnunen, Huttunen, & Ehnholm, 1976). Most of it is normally associated with the casein micelle in milk through both electrostatic and hydrophobic bonding. It is physically separated from its substrate, milk fat, which is present in the form of fat globules, encased within the milk fat globule membrane (MFGM).

LPLs bind strongly to heparin and this feature is utilised in the isolation of LPL from milk using heparin-sepharose affinity chromatography (Iverius, Olivecrona, Egelrud, & Lindahl, 1972). LPL has maximum activity at a pH of ~9.0 and at a temperature of ~37 °C. It is quite heat-labile, being almost completely inactivated by pasteurisation at 72 °C for 15 s (Andrews, Anderson, & Goodenough, 1987).

LPL releases fatty acids attached to the *sn*-1 and *sn*-3 positions of the triglyceride molecule with a preference for the *sn*-1 position. It does not have a fatty acid specificity (Morley & Kuksis, 1977); however, it does tend to release a higher proportion of short-chain fatty acids that are present in the parent milk fat. This is attributable to the higher proportion of these fatty acids in the *sn*-3 position of cows' milk triglycerides. LPL action normally only progresses as far as 2-monoglycerides; it can only release fatty acids from 2-monoglycerides if they first rearrange to 1(3)-monoglycerides (Nilsson-Ehle, Egelrud, Belfrage, Olivecrona, & Borgstrom, 1973).

2.1.2 Bacterial Lipases from Milk-Contaminating Bacteria

Bacterial exocellular lipases are the other major type of lipases which act on milk fat in milk and dairy products. While there is potentially a huge range of such enzymes, the main ones of relevance to the quality of milk and dairy products are those produced during growth of psychrotrophic bacterial contaminants, principally in raw milk. These enzymes differ from the natural milk lipase (LPL) in two major ways which make them highly significant in the dairy industry: firstly, they are much more heat-stable and secondly, the MFGM is much less of a barrier to accessing triglycerides inside native, intact fat globules.

The major sources of bacterial lipases in milk and milk products are pseudomonads, particularly *Pseudomonas fluorescens* and *P. fragi* (Shelley, Deeth, & MacRae, 1986). Others include *Serratia*, *Acinetobacter*, *Achromobacter*, *Aeromonas, Alcaligenes*, *Bacillus, Flavobacterium, Micrococcus* and *Moraxella* (Muir, Phillips, & Dalgleish, 1979; Stewart, Murray, & Neill, 1975). The occurrence and significance of lipases produced by psychrotrophic bacteria in milk and other foods have been reviewed by McKellar (1989) and Sørhaug and Stepaniak (1997).

Psychrotrophic bacteria produce lipase during the late log and early stationary phases of growth; the activity of the lipase produced sometimes reaches a peak and then decreases (Rowe, Johnston, Kilpatrick, Dunstall, & Murphy, 1990). Little lipase is produced before cell numbers reach 10^6–10^7/mL. However, different species and strains of bacteria produce lipase at different cell densities and the rate of bacterial growth is not a reliable guide to lipase production. McKellar (1989) reviewed the effects of several environmental and nutritional factors on the production of enzymes by psychrotrophic bacteria.

The molecular weights of lipases from psychrotrophic bacteria are mostly in the range 14,000–50,000 Da. For example, *P. fluorescens* lipases with molecular weights of ~16,000 Da (isolated under dissociating conditions) (Dring & Fox, 1983; Stepaniak, Birkeland, Sørhaug, & Vagias, 1987), 33,000 Da (Sugiura, Oikawa, Hirano, & Inukai, 1977), and ~50,000 Da (Chung, Lee, Jeohn, Yoo, & Rhee, 1991; Dieckelmann, Johnson, & Beacham, 1998) have been reported.

The pH optimum of the lipases is usually between 7 and 9 and the optimum temperature is 40–50 °C, although the optimum temperature may vary according to the assay conditions used (Fitz-Gerald & Deeth, 1983). Of interest in relation to chilled dairy products is that many of these lipases are active at low temperatures, e.g., ≤10 °C (Landaas & Solberg, 1978; Te Whaiti & Fryer, 1978).

One of the most important properties of these lipases is their heat stability. This varies with the species and strain of the bacteria but many are stable to pasteurization (Fitz-Gerald, Deeth, & Coghill, 1982; Kalogridou-Vassiliadou, 1984; Law, Sharpe, & Chapman, 1976) and even after UHT treatment (Christen, Wang, & Ren, 1986; Kishonti, 1975; Mottar, 1981). It is particularly significant for UHT milk in which only a small residual activity can cause lipolysis during room-temperature storage for several months. Some of the lipases are susceptible to 'low-temperature inactivation' (LTI), e.g., heating at 55 °C for 1 h, which is also effective for inactivating the corresponding proteases (Barach, Adams, & Speck, 1976). However, in fat-containing media, considerable lipolysis can occur during prolonged heating at 55 °C and hence treatment at this temperature may have limited value for eliminating these lipases from milk products (Fitz-Gerald et al., 1982; Griffith, Phillips, & Muir, 1981). Bucky, Hayes, and Robinson (1987) combined UHT treatment with a subsequent LTI treatment of 60 °C for 5 min in a patented process which is reported to considerably enhance the effectiveness of UHT treatment in reducing lipase activity.

Bacterial lipases are usually capable of acting on triglycerides in intact milk fat globules (Fitz-Gerald & Deeth, 1983), a property not exhibited by the indigenous milk lipase because of the protection afforded by the MFGM (Danthine, Blecker, Paquot, Innocente, & Deroanne, 2000). This may be because the lipases can penetrate the MFGM or their access to the substrate is facilitated by activating factors or by their enzymes such as glycosidases, proteases and phospholipases (Alkanhal, Frank, & Christen, 1985; Griffiths, 1983; Marin, Mawhinney, & Marshall, 1984). Based on the behaviour of three bacterial enzymes preparations, Bourlieu, Rousseau, Briard-Bion, Madec, and Bouhallab (2012) concluded that access by the lipases to globular fat was facilitated by the action of annex phospholipase activity.

2.2 Lipolysis in Milk

In milk, the fat is contained within fat globules which vary in diameter from <0.1 to 15 μm. Each globule is covered by a biological membrane, the MFGM. The fat contained inside the membrane is neutral lipid, of which about 98% is triglyceride and about 2% is other lipids including DGs, MGs, FFAs and cholesterol. The lipid component of the MFGM consists of phospholipids, sphingolipids, glycolipids and neutral lipids; the complex lipids—phospholipids, sphingolipids and glycolipids—which make up about 35% of the MFGM material (but only about 0.25% of the milk) are important for the stability of the MFGM (Deeth, 1997).

Lipolysis encountered in raw milk is almost always caused by the native milk LPL. Normally it is prevented from accessing its substrate, triglycerides, by the MFGM. However, there are situations where the triglycerides come into contact with the milk LPL and lipolysis results. The lipolysis can be divided into two categories: spontaneous and induced (Deeth, 2006; Deeth & Fitz-Gerald, 2006).

Spontaneous lipolysis occurs at the farm in raw milk from certain cows. In general, cows on poor quality feed and in late lactation are the most likely to produce milk which is susceptible to this type of lipolysis. The lipolysis is initiated when the milk is cooled, at which time some of the LPL moves from the casein micelle to the MFGM (Dickow, Larsen, Hammershoj, & Wiking, 2011; Hohe, Dimick, & Kilara, 1985). When attached to the MFGM, it is able to cause lipolysis of the triglycerides in the fat globule. The exact mechanism for this has not been fully elucidated but it is clear that at least three factors determine the extent of spontaneous lipolysis: the total LPL in the milk; the nature of the MFGM; and the balance between inhibiting and activating factors (Cartier & Chilliard, 1990; Deeth & Fitz-Gerald, 1975a; Sundheim, 1988). The last factor is probably the most important; addition of normal milk to milk susceptible to spontaneous lipolysis causes a reduction in the expected level of lipolysis, indicating the inhibitor-activator balance has been altered towards inhibition. This is beneficial as it means that herd bulk milks will have a lower overall lipolysis level than would be expected from the susceptibilities of individual milks. From a practical viewpoint, it has been shown that the bulk milk from certain producers is more likely to present with high levels of lipolysis, i.e., with high FFA, than other producers. The reason for this is not entirely clear.

Induced lipolysis can occur on farm or in factory. It occurs when the MFGM is disrupted in some way, thereby allowing the LPL access to the exposed triglyceride. It can be caused by agitation and pumping of raw milk, especially if foaming is involved as the fat globules at the interfacial surfaces of the foam become damaged when the bubbles are formed. A most effective way of inducing lipolysis is by homogenisation of raw whole milk which provides a triglyceride substrate with a high surface area. Similarly, mixing raw milk with (pasteurised) homogenised milk is an effective means of inducing lipolysis and hence should never occur in the dairy industry.

Above a certain level, FFAs produced by lipolysis in milk are detrimental to its quality. The major defects caused are the development of a rancid, unclean off-flavour and a reduction in the foaming capacity of milk. The latter is very important in milk for making cappuccino coffee. There is little effect on the properties of milk when the FFA level is <1 mmol/L; an off-flavour will be noticed by many people and the foaming capacity will be noticeably reduced when it reaches 1.5 mmol/L, and when it reaches 2 mmol/L almost all people will detect the rancid flavour and the foaming capacity will be negligible.

Reduced foaming capacity of milk is an ongoing issue for the dairy industry and makers of cappuccino coffee (Huppertz, 2010). Lipolysis of the milk is known to be a major cause of reduced foaming capacity through reducing its surface tension (Buchanan, 1965; Deeth & Smith, 1983; Kamath, Wulandewi, & Deeth, 2008) but other factors, including milk compositional factors, may also be involved (Corrandini & Innocente, 1994; Gambini, Castagnetti, & Losi, 1995). The effect of lipolysis is mainly through the surface-active partial glycerides produced during lipolysis in milk caused by milk LPL although some FFAs also have surface-active properties. Commercial milk-processing operations, pasteurization and homogenization, markedly enhance the steam-frothing capacity of milk (Deeth & Smith, 1983) and hence pasteurised milk foams much better than raw milk; however, heat treatment and homogenisation are unable to significantly enhance the foaming ability of milk with a high level of lipolysis, e.g., with FFA ≥2.0 mmol/L.

Since native milk LPL is almost completely inactivated by the minimum pasteurisation conditions, 72 °C for 15 s, any lipolysis which occurs after heat processing is due to bacterial lipases. Such lipolysis is rare in pasteurised milk but can occur in UHT milk which typically has a long storage time and is stored at room temperature. The lipases responsible for lipolysis in this type of milk are the heat-stable lipases produced by psychrotrophic bacteria in raw milk as discussed above.

2.3 *Lipolysis in Milk Products*

As for lipolysis in milk, lipolysis in milk products can be detrimental; however, it can also be beneficial. For example, lipolysis is important in the ripening of most cheese varieties. It results from the action of added lipases, lipases produced by microorganisms, and, in raw milk cheese, milk LPL. In some pasteurised cheese varieties, such as cheddar, lipolysis is usually not extensive but is an important contributor to flavour, especially of ripened cheese (Law, 1984). However, excessive lipolysis can render the cheese unacceptable. This occurs when the FFA level reaches 2.8–3.0 mmol/100 g fat (Deeth & Fitz-Gerald, 1975b). High FFA levels in cheddar cheese are usually caused by heat-stable enzymes produced by psychrotrophic bacteria in the cheese-milk before pasteurisation, as discussed above for UHT milk.

In some cheeses, such as blue vein and hard Italian varieties, high FFA levels are characteristic of the variety (Fox & Law, 1991; Gripon, 1987). For example, a FFA content of >66,000 mg/kg is acceptable in blue vein cheese (Horwood, Lloyd, & Stark, 1981) (compared to <4000 mg/kg for good cheddar (Bills & Day, 1964) and <~400 mg/kg for milk. The most flavoursome of the FFAs is butyric acid and high levels of this acid are characteristic of Italian hard cheeses, certain pickled cheeses (Fox & Guinee, 1987) and feta cheese. Butyric acid levels can be >3000 mg/kg in

Romano (Woo & Lindsay, 1984) and up to 520 mg/kg for feta cheese (Horwood et al., 1981). The high FFA levels in these cheese varieties are due to the action of added pregastric esterases which have a preference for hydrolysis of short-chain fatty acids such as butyric and caprylic acids.

In Swiss cheese varieties, FFAs can arise through the action of lipases of propionibacteria (Oterholm, Ordal, & Witter, 1970) while in internal mould-ripened cheeses such as gorgonzola, roquefort and stilton, the FFAs result from the action of *Penicillium roqueforti* lipase. These FFAs contribute to the flavour of these cheese varieties but are also precursors for production of methyl ketones to which the peppery taste of such cheeses is attributable (Kinsella & Hwang, 1976). Lipase produced by *P. camemberti* in surface mould cheeses such and brie and camembert performs a similar function.

The level of lipolysis in raw milk cheese is generally greater than that in pasteurised milk cheese. For example, Hickey, Kilcawley, Beresford, and Wilkinson (2007) reported that after 168 days of ripening, raw milk cheddar cheese had a total FFA level of 1223 mg/kg while corresponding pasteurised cheese had a level of 856 mg/kg. The difference is largely attributable to the action of milk LPL. Buffa, Guamis, Pavia, and Trujillo (2001) also obtained higher FFA levels in cheese made from raw milk than in cheese made from pasteurised milk. Interestingly they found a similar level of lipolysis in cheese made from high-pressure-processed milk and cheese made from raw milk, indicating that high pressure does not inactivate the native milk LPL.

2.4 Lipolysis for Production of Dairy Flavours

Besides the detrimental effects of off-flavour and reduced foaming capacity of milk, lipolysed milk fat can have several benefits. At levels below those which impart a definite rancid flavour, FFAs provide richness and creaminess, and can, therefore, contribute positively to the flavour of foods. Even at higher concentrations which may make milk undrinkable, FFAs contribute a characteristic flavour to some food products.

Controlled lipolysis of milk fat is used to produce creamy and buttery flavours for bakery and cereal products, confectionery (milk chocolate, fudge), coffee whiteners and other imitation dairy products (Arnold, Shahani, & Dwivedi, 1975; Fox, 1980; Kilara, 1985). Lipolysed products with the most flavour tend to be those produced with lipases having *sn*-3 or *sn*-1,3 regioselectivity since the most flavoursome, short-chain fatty acids are largely located in the *sn*-3 position of milk fat TGs. These products are particularly useful for enhancing the flavour of reduced-fat products such as cheese (Noronha, Cronin, O'Riordan, & O'Sullivan, 2008). A good example of a commercial lipolysed milk fat product is Butter Buds® which is a highly flavoured powder made from butter, cream and cheese which can be used to enhance the dairy flavour of food products (http://www.butterbuds.com/). Another example of the use of lipolysed fat in a food product is chocolate; it accounts for the characteristic flavour of some brands.

Balcão and Malcata (2002) reviewed the potential of lipolysis of milk fat to produce dairy-type flavours. They pointed out that the fatty acid profile produced is dependent on the source of lipase and hence judicious choice of the lipase used for the hydrolysis can enable tailoring of the flavour in the product. They also reported studies showing that lipase produced by *Aspergillus lipolyticum, Penicillium roqueforti,* and *Candida candidum* produced soapy and sometimes musty flavours when used in bread, kid and lamb pregastric esterases produce rancid flavours and *P. roqueforti* lipase produces a cheese-like product when used to lipolyse milk fat emulsified in lactic-fermented condensed skim milk. *P. roqueforti* lipases have been shown to be largely responsible for the flavour of blue-veined cheese (Seitz, 1974).

Lipolysis in cheese can also occur through the addition of lipases or esterases to accelerate flavour development in varieties such as cheddar (Jooyandeh et al., 2009; Law & Wigmore, 1985). Furthermore, preparations with flavours typical of particular varieties of cheese can be produced with the aid of lipases of appropriate specificities (Kilara, 1985). Such flavours, sometimes called enzyme-modified cheese flavours, are used in processed cheese, analogue cheese, biscuits, sauces, soups, dips and spreads (Jolly & Kosikowski, 1975a, 1975b; Jooyandeh et al., 2009).

3 Alcoholysis

Alcoholysis is closely related to hydrolysis in that an alcohol replaces water in the lipase reaction. As for hydrolysis, the reaction results in an alcohol group replacing a fatty acid on the glycerol backbone; however, instead of a fatty acid being also produced, an ester of the fatty acid is produced (Fig. 2.1).

One application of alcoholysis involves the use of glycerol as the alcohol. The products of the lipase-catalysed alcoholysis, in this case, are DGs and MGs. Because of their surface-active properties, DG/MG mixtures are widely used in the food industry for stabilising emulsions; they account for about 75% of all emulsifiers used (Balcão & Malcata, 2002). Yang, Harper, Parkin, and Chen (1994) screened nine lipases for their ability to catalyse glycerolysis of milk fat. They found a range of abilities among the lipases, with two lipases from *Pseudomonas* species presenting the highest yield of MG (55–60%) and DG (24–39%) and three lipases, from *Geotricum candidum, Mucor javanicus* and *Candida cylindracea* (*C. rugosa*), yielding no partial glycerides.

Lubary, Hofland, and ter Horst (2011) adopted a different approach to producing a DG-rich milk fat. DG-rich oils have some physiological advantages over TG-rich oils. They used ethanolysis catalysed by a *Pseudomonas fluorescens* lipase. The reaction product mix contained 1,2-DGs together with fatty acid ethyl esters. The latter were removed with supercritical carbon dioxide and the 1,2-DGs isomerised using the same *Pseudomonas* lipase to give mostly (63%) 1,3-DGs.

Schmid, Bornscheuer, Soumanou, McNeill, and Schmid (1998) used alcoholysis with ethanol as the first step in the synthesis of structured triglycerides with palmitic acid in the *sn*-2 position. Such a structure is very significant as it occurs predominantly in human milk fat. This is physiologically preferable to having such a long-chain saturated fatty acid in either the *sn*-1 or *sn*-3 positions because fatty acids removed from the primary positions by 1,3-regiospecific lipases such as pancreatic lipase form calcium salts in the body which are difficult to absorb. By contrast, fatty acids in the *sn*-2 position are readily absorbed as the 2-MG. The alcoholysis reaction was carried out on tripalmitin (PPP) in a solvent, such as acetone or methyl tert-butyl ether, with anhydrous ethanol at 3.5 mM with a 1,3-regiospecific lipase. The highest yield (88%) of 2-monopalmitin was obtained with a *Rhizopus delemar* lipase immobilised on celite in the solvent methyl tert-butyl ether. The targeted TG (OPO) was then produced by acidolysis with oleic acid using the same lipase.

An interesting application of the alcoholysis reaction was recently proposed by Andrewes (2018) for the analysis of lipase activity in UHT milk. The principle was demonstrated by the addition of methanol and a commercial enzyme preparation, produced by *Rhizopus oryzae*, to the milk and, after incubation, measuring the fatty acid methyl esters formed. The amount of methyl esters formed is taken as a measure of the lipase activity in the milk.

4 Acidolysis

Acidolysis involves the reaction of a TG with a carboxylic acid, usually a fatty acid, in the presence of a lipase. The added acid substitutes some of the fatty acids on the TG, thus producing TGs enriched in the acid added and releasing some of the original fatty acids as FFAs. It can also be used to esterify MGs or DGs, as used by Schmid et al. (1998) to esterify monopalmitin in the example given above.

A good example of acidolysis is the work of Balcão and Malcata (1997, 1998a, 1998b) and Balcão, Kemppinen, Malcata, and Kalo (1998) who reported acidolysis of milk fat with oleic acid using lipases produced by *Mucor circinelloides* and *M. javanicus* immobilized by on hydrophobic hollow fibers. After acidolysis, the milk fat contained 27–30% more oleic acid, 8% less lauric acid, and 2–6% less myristic acid. For the reaction catalysed by the *Mucor circinelloides* lipase, the modified milk fat product had 19% more low-melting fraction and 83% less high-melting fraction than the original milk fat. The authors concluded that even though there was some hydrolysis, the enzymatic acidolysis was able to produce milk fat with improved nutritional quality (Balcão et al., 1998).

One of the most important applications of the acidolysis reactions involving milk fat is modifying its structure to resemble that of human milk fat; this is to "humanise" the fat for use in infant formulae. The modification is aimed at changing not only the fatty acids but also their position on the TG. To simulate human milk fat, the *sn*-2 position of the humanised fat needs to contain mostly saturated fatty acids such as palmitic acid, and the *sn*-1 and *sn*-3 positions need to carry all of the unsaturated fatty acids. For example, Christensen and Holmer (1993) interesterified bovine milk fat with unsaturated fatty acids using the *sn*-1,3-regiospecific lipase of *Rhizomucor miehei* and produced a modified milk fat enriched in unsaturated fatty acids in the *sn*-1 and -3 positions, as occurs in human milk. Similarly, Sørensen et al. (2010) interesterified milk fat with a 7:3 mixture of rapeseed and soybean oil fatty acids using Lipozyme® RM IM lipase (from *Rhizomucor miehei*). These authors fractionated the interesterified fat and retained a solid fraction as a human milk fat substitute (HMFS). It contained 56% palmitic acid in the *sn*-2 position compared with 72% in human milk fat. Interestingly, these authors also produced a similar HMFS by interesterifying a solid milk fat fraction with rapeseed and soybean oil fatty acids. Production of HMFSs has been reviewed by Soumanou, Perignon, and Villeneuve (2013).

Acidolysis can be used for incorporating fatty acids with particular therapeutic properties. Conjugated linoleic acid (CLA) is one such fatty acid as it is considered to be an anticarcinogenic agent (Parodi, 1999). Garcia, Keough, Arcos, and Hill (2000) used six commercial immobilised lipases, of which they found *Candida antarctica* lipase to exhibit the best activity, in a batch reactor to effect acidolysis of milk fat with CLA in a solvent-free system containing water contents of 0.15 to 2%. The optimum temperature for the reaction was 50 °C. Similar results were reported by Sehanputri and Hill (2003) using the same enzyme in a packed bed reactor.

The spreadability of butter has been a serious issue for the dairy industry. Several approaches have been used to improve this property, including mixing milk fat or cream with an unsaturated oil such as canola oil before churning into butter. Another approach is to interesterify milk fat with unsaturated fatty acids in order to substitute some long-chain saturated fatty acids with these acids. An example of this was reported by Kim et al. (2002) who used *Rhizopus arrhizus* lipase to incorporate α-linolenic acid from perilla oil into milk fat. They achieved an incorporation of up to ~24% (w/w) in the milk fat. The modified milk fat had better spreadability than the original milk fat.

5 Interesterification

As shown in Fig. 2.1, enzymic interesterification of milk fat involves reacting it with a lipase, usually in the presence of a different TG or TG mixture. If carried out in the absence of water or with a very low water content, the hydrolysis reaction is minimised and the fatty acyl interchange is maximised. The major product is a modified milk fat. This product will still contain the fatty acids from the original milk fat TGs as well as incorporated fatty acids from any added TG. Separation of the newly formed TGs from TGs containing the original fatty acids is very difficult as they have similar properties. This contrasts with the situation with alcoholysis and acidolysis where the non-TG products differ in properties from the modified TG and can be separated from it. However, where the mixed TG is a desirable end-product, this is not an issue.

As for alcoholysis and acidolysis, the intended changes can be tailored by the regiospecificity of the lipase used. Therefore, if a *sn*-1,3 specific lipase is used, the fatty acids on the *sn*-1 and -3 positions will be exchanged. In some circumstances, however, small changes in the fatty acids on the *sn*-2 may occur. Rønne, Yang, Mu, Jacobsen, and Xu (2005) reported that this occurred when a *Thermomyces lanugi-*

nosus lipase was used in interesterifying milk fat and rapeseed oil but it did not happen when *Rhizimucor miehei* lipase was used. If a non-specific lipase is used, random exchange of fatty acids will result. Such random exchange occurs during chemical interesterification. However, if random exchange is required, enzymic interesterification is preferable to chemical as it can be carried out under relatively mild conditions.

A major reason why interesterification is carried out is to change the physical properties of the fat. If milk fat alone is reacted with a lipase, the product has different melting characteristics. For example, Hayati et al. (2000) reported that after interesterification with a *sn*-1,3-specific lipase, the solid fat content at various temperatures was altered; it was lower at 5 and 10 °C but higher at 15–35 °C. This is consistent with the finding of Chmura, Staniewski, Panfil-Kuncewicz, Szpendowski, and Zawadzka (2008) that interesterification of milk fat with a *sn*-1,3-specific lipase of *Rhizomucor miehei* resulted in a decrease in C34 to C42 TGs with a simultaneous increase of C44 to C52 TGs, and an increase in penetration value at 5 °C. By contrast, chemical interesterification of milk fat with sodium methoxide decreases the solid fat content at all temperatures from 5 to 40 °C (Rousseau, Forestiere, Hill, & Marangoni, 1996).

Blending of fats is used commercially for modifying their melting behaviour and other properties. This method can be used for overcoming some of the problems associated with milk fat's lack of solid content at high temperature, which limits its use in certain food products such as bakery and confectionery. In their study, Hayati et al. (2000) interesterified binary mixtures of milk fat and two palm oil fractions, soft and hard stearin fractions. They reported that the interesterified blend generally had a higher maximum peak temperature (T_p) than the untreated mixture, and that the DSC curves were smoother than those of milk fat (milk fat typically exhibits three peaks) indicating a more homogeneous mixture of TGs. Interestingly, the interesterified blends of milk fat-soft palm stearin had similar melting behaviour to that of milk fat high melting fractions obtained by dry fractionation, leading the authors to suggest that they have commercial potential in several food applications such as in pastry and confectionery. While blends with between 25 and 75% soft palm stearin were considered to have this potential, the authors recommended that blends containing no more than 25% of the hard palm stearin have similar potential as they retained the desired milk fat flavour in the final product.

Interesterification of milk fat with unsaturated fats/oils has been reported by several authors (with canola, Rousseau, Hill, & Marangoni, 1996, Morais Nunes, de Paula, de Castro, & dos Santos, 2011; rapeseed, Kalo, Kemppinen, & Antila, 1987, Rønne et al., 2005, Giet et al., 2009, linseed (flaxseed), Aguedo et al., 2008; Shin, Akoh, & Lee, 2010; Giet et al., 2009). These studies were mainly aimed at modifying the milk fat to have a purportedly nutritionally better fatty acid profile (Kontkanen et al., 2011). In the study by Shin et al. (2010), this was assessed by the so-called "atherogenic index". Nutritionally, another benefit of employing enzymic rather than chemical methods for the interesterification is to produce a low-*trans* fat suitable for spreadable spreads (Shin et al., 2010). Enzymic interesterification does not cause *cis*/*trans* isomerisation of double bonds (Aguedo et al., 2008). This is because of the milder conditions used for enzymic interesterification, as exemplified in the study by De et al. (2007) in which chemical interesterification was carried out at 180 °C for 4 h while enzymic interesterification (using a *Mucor miehei* lipase) was carried out at 60 °C for 4 h. Milder reaction conditions have been reported by other authors for chemical interesterification, e.g., ~80 °C for 2 h (Rousseau, Forestiere, et al., 1996) and shorter reaction times have been reported for enzymic, reactions e.g., 30 and 60 min by Rønne et al. (2005) reactions. Even milder conditions may be possible for enzymic interesterifications carried out under ultrasonication (Lerin et al., 2014).

HMFSs can also be produced by interesterification of milk fat. For example, Kalo et al. (1987) prepared a HMFS by interesterifying a milk fat solid fraction with rapeseed oil. However, starting with bovine milk fat is not a common pathway for producing HMFS. More commonly, unsaturated oils such as marine and vegetable oils are interesterified with TGs containing a high percentage of palmitic acid in the *sn*-2 position, even tripalmitin. For example, Karabulut, Turan, Vural, and Kayahan (2007) interesterified palm oil, palm kernel oil, olive oil, sunflower oil, and marine oil blend, in the ratio of 4.0:3.5:1.0:1.5:0.2, with Lipozyme® TL IM, a lipase from *Thermomyces lanuginosa*, and produced a product with mostly unsaturated fatty acids in the *sn*-1 and *sn*-3 positions and saturated fatty acids in the *sn*-2 position of the TGs, similar to that in human milk fat TGs.

6 Conclusion

The effects of lipases on milk fat are of ongoing interest to the dairy and wider food industry. In the dairy industry, they can cause flavour problems but also contribute substantially to the desirable flavour of some products. Several commercial lipases are now available and their different specificities enable them to be used for a range of modifications of milk fat. These include changes to the fatty acid composition and the structure of the TGs; these can be tailored to meet nutritional and physical functionality criteria. Lipase reactions occur under relatively mild conditions and this makes them attractive for use in a range of applications. The potential for their future commercial use with milk fat is substantial provided the costs and availability of the enzymes and the required processing equipment are favourable.

References

Aguedo, M., Hanon, E., Danthine, S., Paquot, M., Lognay, G., Thomas, A., et al. (2008). Enrichment of anhydrous milk fat in polyunsaturated fatty acid residues from linseed and rapeseed oils through enzymatic interesterification. *Journal of Agricultural and Food Chemistry, 56*, 1757–1765.

Akoh, C. C., Sellaopoan, S., Fomuso, L. B., & Yankah, V. V. (2003). Enzymatic synthesis of structured lipids. In T. M. Kuo & H. W. Gardner (Eds.), *Lipid biotechnology* (pp. 433–460). New York, NY: Marcel Dekker, Inc.

Alkanhal, H. A., Frank, J. F., & Christen, G. L. (1985). Microbial protease and phospholipase C stimulate lipolysis of washed cream. *Journal of Dairy Science, 68*, 3162–3170.
Andrewes, P. (2018). Indirect detection of lipase in UHT milk by measuring methyl ester formation. *International Dairy Journal, 79*, 1–4.
Andrews, A. T., Anderson, M., & Goodenough, P. W. (1987). A study of the heat stabilities of a number of indigenous milk enzymes. *The Journal of Dairy Research, 54*, 237–246.
Arnold, R. G., Shahani, K. M., & Dwivedi, B. K. (1975). Application of lipolytic enzymes to flavor development in dairy products. *Journal of Dairy Science, 58*, 1127–1143.
Balcão, V. M., Kemppinen, A., Malcata, F. X., & Kalo, P. J. (1998). Lipase-catalyzed acidolysis of butterfat with oleic acid: Characterization of process and product. *Enzyme and Microbial Technology, 23*, 118–128.
Balcão, V. M., & Malcata, F. X. (1997). Lipase-catalyzed modification of butterfat via acidolysis with oleic acid. *Journal of Molecular Catalysis B: Enzymatic, 3*, 161–169.
Balcão, V. M., & Malcata, F. X. (1998a). Interesterification and acidolysis of butterfat with oleic acid by *Mucor javanicus* lipase: Changes in the pool of fatty acid residues. *Enzyme and Microbial Technology, 22*, 511–519.
Balcão, V. M., & Malcata, F. X. (1998b). Lipase catalyzed modification of milkfat. *Biotechnology Advances, 16*, 309–341.
Balcão, V. M., & Malcata, F. X. (2002). Enzyme-mediated modification of milkfat. In T. M. Kuo & H. W. Gardner (Eds.), *Lipid biotechnology* (pp. 479–492). New York, NY: Marcel Dekker.
Barach, J. T., Adams, D. M., & Speck, M. L. (1976). Low temperature inactivation in milk of heat-resistant proteases from psychrotrophic bacteria. *Journal of Dairy Science, 59*, 391–395.
Barber, M. C., Clegg, R. A., Travers, M. T., & Vernon, R. G. (1997). Lipid metabolism in the lactating mammary gland. *Biochimica et Biophysica Acta, Lipids and Lipid Metabolism, 1347*, 101–126.
Bills, D. D., & Day, E. A. (1964). Determination of the major free fatty acids in Cheddar cheese. *Journal of Dairy Science, 47*, 733–738.
Bourlieu, C., Bouhallab, S., & Lopez, C. (2009). Biocatalyzed modifications of milk lipids: Applications and potentialities. *Trends in Food Science and Technology, 20*, 458–469.
Bourlieu, C., Rousseau, F., Briard-Bion, V., Madec, M. N., & Bouhallab, S. (2012). Hydrolysis of native milk fat globules by microbial lipases: Mechanisms and modulation of interfacial quality. *Food Research International, 49*, 533–544.
Buchanan, R. A. (1965). Lipolysis and the frothing of milk. *Australian Journal of Dairy Technology, 20*, 62–66.
Bucky, A. R., Hayes, P. R., & Robinson, D. S. (1987). A modified ultra-high temperature treatment for reducing microbial lipolysis in stored milk. *The Journal of Dairy Research, 54*, 275–282.
Buffa, M., Guamis, B., Pavia, M., & Trujillo, A. J. (2001). Lipolysis in cheese made from raw, pasteurized or high-pressure-treated goats' milk. *International Dairy Journal, 11*, 175–179.
Cartier, P., & Chilliard, Y. (1990). Spontaneous lipolysis in bovine milk: Combined effects of nine characteristics in native milk. *Journal of Dairy Science, 73*, 1178–1186.
Chmura, M., Staniewski, B., Panfil-Kuncewicz, H., Szpendowski, J., & Zawadzka, J. (2008). The effects of enzymatic interesterification on selected physicochemical characteristics of milk fat. *Milchwissenschaft, 63*, 37–40.
Christen, G. L., Wang, W.-C., & Ren, T.-J. (1986). Comparison of the heat resistance of bacterial lipases and proteases and the effect on ultra-high temperature milk quality. *Journal of Dairy Science, 69*, 2769–2778.
Christensen, T. C., & Holmer, G. (1993). Lipase-catalyzed acyl-exchange reactions of butter oil - synthesis of a human-milk fat substitute for infant formulas. *Milchwissenschaft, 48*, 543–548.
Chung, G. H., Lee, Y. P., Jeohn, G. H., Yoo, O. J., & Rhee, J. S. (1991). Cloning and nucleotide sequence of thermostable lipase gene from *Pseudomonas fluorescens* SIK W1. *Agricultural and Biological Chemistry, 55*, 2359–2365.
Corrandini, C., & Innocente, N. (1994). Influence of proteose peptone fraction on milk foaming capacity. *Scienza Technica Lattiero-Caseria, 45*, 107–113.
Danthine, S., Blecker, C., Paquot, M., Innocente, N., & Deroanne, C. (2000). Progress in milk fat globule membrane research: A review. *Le Lait, 80*, 209–222.

De, B. K., Hakimji, M., Patel, A., Sharma, D., Desai, H., & Kumar, T. (2007). Plastic fats and margarines through fractionation, blending and interesterification of milk fat. *European Journal of Lipid Science and Technology, 109*, 32–37.

Deeth, H. C. (1978). Fluorimetric detection of carboxyl-esterase activity in milk. In *Proceedings of 20th International Dairy Congress E* (pp. 364–365). Brussels, Belgium: International Dairy Federation.

Deeth, H. C. (1997). The role of phospholipids in the stability of the milk fat globule membrane. *Australian Journal of Dairy Technology, 52*, 44–46.

Deeth, H. C. (2006). Lipoprotein lipase and lipolysis in milk. *International Dairy Journal, 16*, 555–562.

Deeth, H. C., & Fitz-Gerald, C. H. (1975a). Factors governing the susceptibility of milk to spontaneous lipolysis. In *IDF document 86* (pp. 24–34). Brussels, Belgium: International Dairy Federation.

Deeth, H. C., & Fitz-Gerald, C. H. (1975b). The relevance of milk lipase activation to rancidity in Cheddar cheese. *Australian Journal of Dairy Technology, 30*, 74–76.

Deeth, H. C., & Fitz-Gerald, C. H. (2006). Lipolytic enzymes and hydrolytic rancidity. In P. F. Fox & P. McSweeney (Eds.), *Advanced dairy chemistry, volume: Lipids* (pp. 481–556). New York, NY: Springer.

Deeth, H. C., & Smith, R. A. D. (1983). Lipolysis and other factors affecting the steam frothing capacity of milk. *Australian Journal of Dairy Technology, 38*, 14–19.

Dickow, J. A., Larsen, L. B., Hammershoj, M., & Wiking, L. (2011). Cooling causes changes in the distribution of lipoprotein lipase and milk fat globule membrane proteins between the skim milk and cream phase. *Journal of Dairy Science, 94*, 646–656.

Dieckelmann, M., Johnson, L. A., & Beacham, I. R. (1998). The diversity of lipases from psychrotrophic strains of *Pseudomonas*: A novel lipase from a highly lipolytic strain of *Pseudomonas fluorescens*. *Journal of Applied Microbiology, 85*, 527–536.

Dring, R., & Fox, P. F. (1983). Purification and characterization of a heat-stable lipase from *Pseudomonas fluorescens* AFT 29. *Irish Journal of Food Science and Technology, 7*, 157–171.

Fitz-Gerald, C. H., & Deeth, H. C. (1983). Factors influencing lipolysis by skim milk cultures of some psychrotrophic microorganisms. *Australian Journal of Dairy Technology, 38*, 97–103.

Fitz-Gerald, C. H., Deeth, H. C., & Coghill, D. M. (1982). Low-temperature inactivation of lipases from psychrotrophic bacteria. *Australian Journal of Dairy Technology, 37*, 51–54.

Fitz-Gerald, C. H., Deeth, H. C., & Kitchen, B. J. (1981). The relationship between the levels of free fatty acids, lipoprotein lipase, carboxylesterase, *N*-acetyl-β-D-glucosaminidase, somatic cell count and other mastitis indices in bovine milk. *The Journal of Dairy Research, 48*, 253–265.

Fox, P. F. (1980). Enzymes other than rennets in dairy technology. *Journal of the Society of Dairy Technology, 33*, 118–128.

Fox, P. F., & Guinee, T. P. (1987). Italian cheeses. In P. F. Fox (Ed.), *Cheese: Chemistry, physics and microbiology* (Vol. 2, pp. 221–255). London, UK: Elsevier Applied Science.

Fox, P. F., & Law, B. A. (1991). Enzymology of cheese ripening. *Food Biotechnology, 5*, 239–262.

Gambini, G., Castagnetti, G. B., & Losi, G. (1995). Influence of somatic cell count and heat treatments on milk foam formation and stability. *Industrie Alimentari, 34*, 247–252.

Garcia, H. S., Keough, K. J., Arcos, J. A., & Hill, C. G. (2000). Interesterification (acidolysis) of butterfat with conjugated linoleic acid in a batch reactor. *Journal of Dairy Science, 83*(3), 371–377.

Giet, J.-M., Aguedo, M., Danthine, S., Paquot, M., Thomas, A., Vandenbol, M., et al. (2009). Enzymatic interesterification of anhydrous milk fat with rapeseed and/or linseed oil: Oxidative stability. *Journal of Agricultural and Food Chemistry, 57*, 6787–6794.

Griffith, M. W., Phillips, J. D., & Muir, D. D. (1981). Thermostability of proteases and lipases number of species of psychrotrophic bacteria of dairy origin. *The Journal of Applied Bacteriology, 50*, 289–303.

Griffiths, M. W. (1983). Synergistic effects of various lipases and phospholipase C on milk fat. *Journal of Food Technology, 18*, 495–505.

Gripon, J. C. (1987). Mould-ripened cheeses. In P. F. Fox (Ed.), *Cheese: Chemistry, physics and microbiology* (Vol. 2, pp. 121–149). London, UK: Elsevier Applied Science.

Hayati, I. N., Aminah, A., Mamot, S., Aini, I. N., Lida, H. M. N., & Sabariah, S. (2000). Melting characteristic and solid fat content of milk fat and palm stearin blends before and after enzymatic interesterification. *Journal of Food Lipids, 7*, 175–193.

Hernell, O., & Bläckberg, L. (1991). Digestion and absorption of human milk lipids. In R. Dulbecco (Ed.), *Encyclopedia of human biology* (Vol. 3, pp. 47–56). New York, NY: Academic Press.

Hernell, O., & Bläckberg, L. (1994). Human milk bile salt-stimulated lipase: Functional and molecular aspects. *The Journal of Pediatrics, 125*(Suppl), S56–S61.

Hickey, D. K., Kilcawley, K. N., Beresford, T. P., & Wilkinson, M. G. (2007). Lipolysis in cheddar cheese made from raw, thermized, and pasteurized milks. *Journal of Dairy Science, 90*, 47–56.

Hohe, K. A., Dimick, P. S., & Kilara, A. (1985). Milk lipoprotein-lipase distribution in the major fractions of bovine-milk. *Journal of Dairy Science, 68*, 1067–1073.

Horwood, J. F., Lloyd, G. T., & Stark, W. (1981). Some flavour components of Feta cheese. *Australian Journal of Dairy Technology, 36*, 34–37.

Huppertz, T. (2010). Foaming properties of milk: A review of the influence of composition and processing. *International Journal of Dairy Technology, 63*, 477–488.

Iverius, P. H., Olivecrona, T., Egelrud, T., & Lindahl, U. (1972). Effects of heparin on lipoprotein-lipase from bovine milk. *The Journal of Biological Chemistry, 247*, 6610.

Javed, S., Azeem, F., Hussain, S., Rasul, I., Siddique, M. H., Riaz, M., et al. (2018). Bacterial lipases: A review on purification and characterization. *Progress in Biophysics and Molecular Biology, 132*, 23–34.

Jolly, R. C., & Kosikowski, F. V. (1975a). Flavor development in pasteurized milk blue cheese by animal and microbial lipase preparations. *Journal of Dairy Science, 58*, 846–852.

Jolly, R. C., & Kosikowski, F. V. (1975b). A new blue cheese food material from ultrafiltrated skim milk and microbial enzyme - Mold spore reacted fat. *Journal of Dairy Science, 58*, 272–1275.

Jooyandeh, H., Kaur, A., & Minhas, K. S. (2009). Lipases in dairy industry: A review. *Journal of Food Science and Technology, 46*, 181–189.

Kalo, P., Kemppinen, A., & Antila, M. (1987). *Candida cylindracae* lipase-catalyzed interesterification of butterfat and butterfat solid fraction rapeseed oil mixtures. *Journal of the American Oil Chemists' Society, 64*, 1263–1263.

Kalogridou-Vassiliadou, D. (1984). Lipolytic activity and heat resistance of extracellular lipases of some Gram-negative bacteria. *Milchwissenschaft, 39*, 601–603.

Kamath, S., Wulandewi, A., & Deeth, H. C. (2008). Relationship between surface tension, free fatty acid concentration and foaming properties of milk. *Food Research International, 41*, 623–929.

Karabulut, I., Turan, S., Vural, H., & Kayahan, M. (2007). Human milk fat substitute produced by enzymatic interesterification of vegetable oil blend. *Food Technology and Biotechnology, 45*, 434–438.

Kilara, A. (1985). Enzyme-modified lipid food ingredients. *Process Biochemistry, 20*, 35–45.

Kim, J. S., Maeng, I. K., Lee, B. O., Kim, C. K., Kwon, Y. H., & Kim, Y. J. (2002). Lipase-catalyzed acidolysis of butterfat with α-linolenic acid from perilla oil. *Food Science and Biotechnology, 11*, 66–70.

Kinnunen, P. K. J., Huttunen, J. K., & Ehnholm, C. (1976). Properties of purified bovine milk lipoprotein lipase. *Biochimica et Biophysica Acta, 450*, 342–351.

Kinsella, J. E., & Hwang, D. (1976). Biosynthesis of flavours by *Penicillium roqueforti*. *Biotechnology and Bioengineering, 18*, 927–938.

Kishonti, E. (1975). Influence of heat resistant lipases and proteases in psychrotrophic bacteria on product quality. In *IDF document 86* (pp. 121–124). Brussels, Belgium: International Dairy Federation.

Kontkanen, H., Rokka, S., Kemppinen, A., Miettinen, H., Hellstrom, J., Kruus, K., et al. (2011). Enzymatic and physical modification of milk fat: A review. *International Dairy Journal, 21*, 3–13.

Landaas, A., & Solberg, P. (1978). Production and characterization of lipase from a fluorescent pseudomonad. In *Proceedings of 20th International Dairy Congress (Paris) E* (pp. 304–305). Brussels, Belgium: International Dairy Federation.

Law, B. A. (1984). Flavour development in cheeses. In F. L. Davies & B. A. Law (Eds.), *Advances in the microbiology and biochemistry of cheese and fermented milk* (pp. 187–208). London, UK: Elsevier Applied Science.

Law, B. A., Sharpe, M. E., & Chapman, H. R. (1976). The effect of lipolytic Gram-negative psychrotrophs in stored milk on the development of rancidity in Cheddar cheese. *The Journal of Dairy Research, 43*, 459–468.
Law, B. A., & Wigmore, A. S. (1985). Effect of commercial lipolytic enzymes on flavor development in cheddar cheese. *Journal of the Society of Dairy Technology, 38*, 86–88.
Lerin, L. A., Loss, R. A., Remonatto, D., Zenevicz, M. C., Balen, M., Oenning Netto, V., et al. (2014). A review on lipase-catalyzed reactions in ultrasound-assisted systems. *Bioprocess and Biosystems Engineering, 37*, 2381–2394.
Lubary, M., Hofland, G. W., & ter Horst, J. H. (2011). A process for the production of a diacylglycerol-based milk fat analogue. *European Journal of Lipid Science and Technology, 113*, 459–468.
Marin, A., Mawhinney, T. P., & Marshall, R. T. (1984). Glycosidic activities of *Pseudomonas fluorescens* on fat-extracted skim milk, buttermilk and milk fat globule membranes. *Journal of Dairy Science, 67*, 52–59.
Marquardt, R. R., & Forster, T. L. (1962). Arylesterase activity of bovine milk as related to incidence of mastitis. *Journal of Dairy Science, 45*, 653.
Marquardt, R. R., & Forster, T. L. (1965). Milk A-esterase levels as influenced by stage of lactation. *Journal of Dairy Science, 48*, 1526–1528.
McKellar, R. C. (Ed.). (1989). *Enzymes of psychrotrophs in raw food*. Boca Raton, FL: CRC Press.
Morais Nunes, G. F., de Paula, A. V., de Castro, H. F., & dos Santos, J. C. (2011). Optimization of the enzymatic interesterification of milk fat and canola oil blends using immobilized *Rhizopus oryzae* lipase by response surface methodology. *Food Technology and Biotechnology, 49*, 385–390.
Morley, N., & Kuksis, A. (1977). Lack of fatty-acid specificity in lipolysis of oligo-saturated and polyunsaturated triacylglycerols by milk lipoprotein-lipase. *Biochimica et Biophysica Acta, 487*, 332–342.
Mottar, J. (1981). Heat resistant enzymes in UHT milk and their influence on sensoric changes during uncooled storage. *Milchwissenschaft, 36*, 87–91.
Muir, D. D., Phillips, J. D., & Dalgleish, D. G. (1979). The lipolytic and proteolytic activity of bacteria isolated from blended raw milk. *Journal of the Society of Dairy Technology, 32*, 19–23.
Nilsson-Ehle, P., Egelrud, T., Belfrage, P., Olivecrona, T., & Borgstrom, B. (1973). Positional specificity of purified milk lipoprotein-lipase. *The Journal of Biological Chemistry, 248*, 6734–6737.
Noronha, N., Cronin, D., O'Riordan, D., & O'Sullivan, M. (2008). Flavouring reduced fat high fibre cheese products with enzyme modified cheeses (EMCs). *Food Chemistry, 110*, 973–978.
Olivecrona, T., Vilaró, S., & Olivecrona, G. (2003). Lipases in milk. In P. F. Fox & P. L. H. McSweeney (Eds.), *Advanced dairy chemistry: Proteins* (Vol. 1, 3rd ed., pp. 473–494). New York, NY: Kluwer Academic.
Oterholm, A., Ordal, Z. J., & Witter, L. D. (1970). Glycerol ester hydrolase activity of *Propionibacterium shermanii*. *Journal of Dairy Science, 53*, 592–593.
Parodi, P. W. (1999). Conjugated linoleic acid and other anticarcinogenic agents of bovine milk fat. *Journal of Dairy Science, 82*, 1339–1349.
Riaublanc, A., Ratomahenina, R., Galzy, P., & Nicolas, M. (1993). Peculiar properties of lipase from *Candida-parapsilosis* (Ashford) Langeron and Talice. *Journal of the American Oil Chemists' Society, 70*, 497–500.
Rønne, T. H., Yang, T. K., Mu, H. L., Jacobsen, C., & Xu, X. B. (2005). Enzymatic interesterification of butterfat with rapeseed oil in a continuous packed bed reactor. *Journal of Agricultural and Food Chemistry, 53*, 5617–5624.
Rousseau, D., Forestiere, K., Hill, A. R., & Marangoni, A. G. (1996). Restructuring butterfat through blending and chemical interesterification. 1. Melting behavior and triacylglycerol modifications. *Journal of the American Oil Chemists' Society, 73*, 963–972.
Rousseau, D., Hill, A. R., & Marangoni, A. G. (1996). Restructuring butterfat through blending and chemical interesterification. 3. Rheology. *Journal of the American Oil Chemists' Society, 73*, 983–989.

Rowe, M. T., Johnston, D. E., Kilpatrick, D. J., Dunstall, G., & Murphy, R. J. (1990). Growth and extracellular enzyme production by psychrotrophic bacteria in raw milk stored at a low temperature. *Milchwissenschaft, 45*, 495–499.

Schmid, U., Bornscheuer, U. T., Soumanou, M. M., McNeill, G. P., & Schmid, R. D. (1998). Optimization of the reaction conditions in the lipase-catalyzed synthesis of structured triglycerides. *Journal of the American Oil Chemists' Society, 75*, 1527–1531.

Sehanputri, P. S., & Hill, C. G. (2003). Lipase-mediated acidolysis of butteroil with free conjugated linoleic acid in a packed bed reactor. *Biotechnology and Bioengineering, 83*, 608–617.

Seitz, E. W. (1974). Industrial application of microbial lipases - Review. *Journal of the American Oil Chemists' Society, 51*, 12–16.

Shelley, A. W., Deeth, H. C., & MacRae, I. C. (1986). Growth of lipolytic pschrotrophic pseudomonads in raw and ultra-heat-treated milk. *The Journal of Applied Bacteriology, 61*, 395–400.

Shin, J.-A., Akoh, C. C., & Lee, K.-T. (2010). Enzymatic interesterification of anhydrous butterfat with flaxseed oil and palm stearin to produce low-trans spreadable fat. *Food Chemistry, 120*, 1–9.

Sørensen, A. D. M., Xu, X., Zhang, L., Kristensen, J. B., Jacobsen C. (2010) Human Milk Fat Substitute from Butterfat: Production by Enzymatic Interesterification and Evaluation of Oxidative Stability. *Journal of the American Oil Chemists' Society, 87*, (2):185–194.

Sørhaug, T., & Stepaniak, L. (1997). Psychrotrophs and their enzymes in milk and dairy products: Quality aspects. *Trends in Food Science and Technology, 8*, 35–41.

Soumanou, M. M., Perignon, M., & Villeneuve, P. (2013). Lipase-catalyzed interesterification reactions for human milk fat substitutes production: A review. *European Journal of Lipid Science and Technology, 115*, 270–285.

Stepaniak, L., Birkeland, S.-E., Sørhaug, T., & Vagias, G. (1987). Isolation and partial characterization of heat stable proteinase, lipase and phospholipase C from *Pseudomonas fluorescens* P1. *Milchwissenschaft, 42*, 75–79.

Stewart, D. B., Murray, J. G., & Neill, S. D. (1975). Lipolytic activity of organisms isolated from refrigerated bulk milk. In *IDF document 86* (pp. 38–50). Brussels, Belgium: International Dairy Federation.

Sugiura, M., Oikawa, T., Hirano, K., & Inukai, T. (1977). Purification, crystallization and properties of triacylglycerol lipase from *Pseudomonas fluorescens*. *Biochimica et Biophysica Acta, 488*, 353–358.

Sundheim, G. (1988). Spontaneous lipolysis in bovine milk: Combined effects of cream, skim milk, and lipoprotein lipase activity. *Journal of Dairy Science, 71*, 620–626.

Te Whaiti, I. E., & Fryer, T. F. (1978). Production and heat stability in milk of proteinases and lipases of psychrotrophic pseudomonads. In *Proceedings of 20th International Dairy Congress (Paris) E* (pp. 303–304). Brussels, Belgium: International Dairy Federation.

Woo, A. H., & Lindsay, R. C. (1984). Concentration of major free fatty acids and flavor development in Italian cheese varieties. *Journal of Dairy Science, 67*, 960–968.

Yang, B., Harper, W. J., Parkin, K. L., & Chen, J. (1994). Screening of commercial lipases for production of mono- and diacylglycerols from butteroil by enzymic glycerolysis. *International Dairy Journal, 4*, 1–13.

Chapter 3
Structure of the Milk Fat Globule Membrane: New Scientific Advances Revealing the Role of Sphingomyelin in Topographical and Mechanical Heterogeneities

Christelle Lopez

1 Introduction

Milk fat globules are enveloped by a biological membrane called the milk fat globule membrane (MFGM). The MFGM results from the mechanisms of milk fat globule secretion by the mammary secretory cells (Heid & Keenan, 2005), that tailor its composition and structure. The MFGM is derived mainly from the apical plasma membrane of the epithelial cells but also from the Golgi vesicles (Heid & Keenan, 2005; Wooding & Sargeant, 2015). This biological membrane is a thin layer (10–50 nm thick in cross section) that accounts for 2–6% of the milk fat globule mass (Heid & Keenan, 2005). The MFGM represents about 80 m^2/L of milk and is, therefore, an important interface between the core of triacylglycerols and the surrounding aqueous phase (Lopez, 2011). This biological interface is naturally involved in the physical stability of milk fat globules, and MFGM-rich ingredients can be used for technological applications such as the preparation of emulsions. The MFGM is also involved in the mechanisms of milk fat globule digestion. Recent research studies provided pieces of evidence that the MFGM, or components thereof, play roles in brain development and cognitive functions, immunity and gut physiology, and cardiovascular health. The MFGM has raised high attention in the past 10 years since understanding of the relationship between its chemical composition, its structure, and its functions will open perspectives and innovation in food formulation, human nutrition, and pediatrics (Lopez, Cauty, & Guyomarc'h, 2019). For example, the new generation of infant milk formula will undoubtedly contain MFGM in order to improve the composition and structure of the emulsion by mimicking human fat globules in breast milk.

C. Lopez (✉)
INRAE, UMR STLO, Rennes, France

INRAE, UR BIA, Nantes, France
e-mail: christelle.lopez@inrae.fr

T. Truong et al. (eds.), *Dairy Fat Products and Functionality*,
https://doi.org/10.1007/978-3-030-41661-4_3

In this chapter, I have highlighted new scientific advances about the structure and mechanical properties of the MFGM, in relation with the specific composition of this biological membrane and the physical properties of milk polar lipids that are mainly due to the high amount of milk-sphingomyelin.

2 Composition of the Milk Fat Globule Membrane

The MFGM that surrounds milk fat globules is mainly derived from the apical plasma membrane of the secretory cells in the lactating mammary glands (Heid & Keenan, 2005). The MFGM is therefore chemically very similar to the plasma membrane. The MFGM is composed of membrane-specific proteins, enzymes, a variety of polar lipids including the sphingolipids (mainly milk-sphingomyelin), cholesterol and minor components.

2.1 Protein Composition of the MFGM

The proteins represent 25–70% of MFGM mass. Based on the size of milk fat globules, the surface of MFGM where the proteins are in contact with the surrounding aqueous environment is important. Up-to-date proteomic techniques revealed over 100 MFGM proteins (Reinhardt & Lippolis, 2006). Among well-described proteins present in high concentrations in the MFGM are Butyrophilin, Mucin 1, Mucin 15, CD36 (also called PAS IV), Lactadherin (also called PAS-6/7 or MFG-E8), Adipophilin, and the enzyme Xanthine dehydrogenase/oxidase (Mather, 2000). The structures, amino acid sequences and properties of major MFGM proteins are reviewed and discussed in detail elsewhere (Mather, 2000; Singh, 2006). The proteome of the human MFGM was characterised over 12 months of lactation (Liao, Alvarado, Phinney, & Lönnerdal, 2011). Most of the MFGM proteins that protrude in the aqueous phase are heavily glycosylated (e.g. Mucin 1, Mucin 15, Butyrophilin) and form a glycocalyx around fat globules.

2.2 Lipid Composition of the MFGM

Cholesterol stands for about 90% of the sterols identified in the MFGM. Cholesterol content is about 300 mg/100 g fat and represents ~30 wt% of the MFGM lipid fraction (Et-Thakafy, Guyomarc'h, & Lopez, 2017; Mesilati-Stahy & Argov-Argaman, 2014; Yao et al., 2016).

The MFGM polar lipids account for about 0.5–1% of milk lipids. The amount of polar lipids found in milk depends (1) on the size of fat globules and then on the amount of surface covered by the biological membrane (Fig. 3.1), since polar lipids

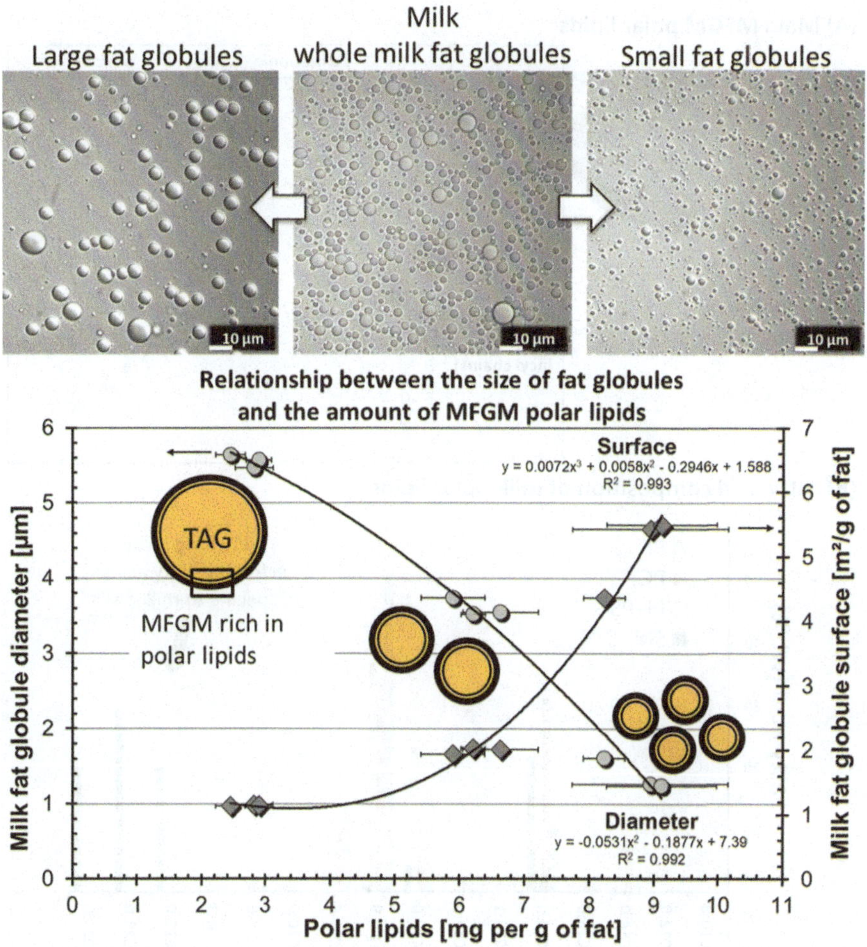

Fig. 3.1 Relationship between the amount of polar lipids from the milk fat globule membrane (MFGM) and the structural parameters of milk fat globules of various sizes selected from milk using a multi-stage microfiltration process; the mean diameter (left y-scale) and the surface (right y-scale) of milk fat globules. Microscopy images of the milks containing fat globules of various sizes are presented. Adapted with permission from Lopez et al. (2011)

are mainly located in the MFGM (i.e. for a similar amount of total fat, small fat globules are more numerous and contain a higher amount of MFGM and polar lipids compared to large fat globules (Lopez et al., 2011)) and (2) on the total amount of fat globules. Studies reported a positive relationship between the amount of fat in the milk and the size of fat globules; milks with a small amount of fat contain small fat globules and then a high amount of MFGM and polar lipids (Lopez, Briard-Bion, & Ménard, 2014). The availability of membrane material able to stabilize fat globules upon their secretion governs their size in milk. In mammal species such as buffalo that produce a high amount of milk fat (7% vs. 4% for bovine milk), the size

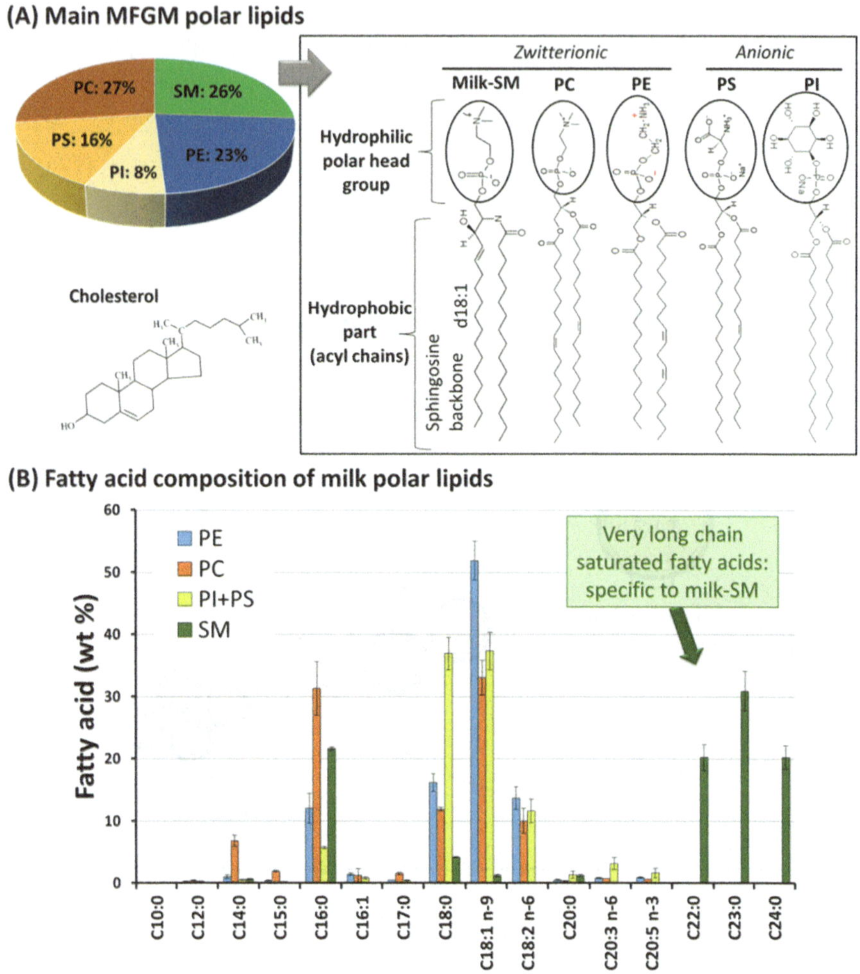

Fig. 3.2 Main polar lipid species found in the milk fat globule membrane (MFGM). (**A**) Relative proportion of polar lipid species found in bovine milk: phosphatidylethanolamine (PE), phosphatidylinositol (PI), phosphatidylserine (PS), phosphatidylcholine (PC) and sphingomyelin (milk-SM). Chemical structure of polar lipids with their polar head group and their hydrophobic part, molecule of cholesterol. (**B**) fatty acid composition of the polar lipid species. Adapted with permission from Lopez et al. (2019)

of fat globules is large compared to bovine milk, i.e. 5 μm vs. 3.5 μm (Ménard et al., 2010).

The MFGM contains two main classes of polar lipids, i.e. the glycerophospholipids and the sphingolipids (Fig. 3.2). The glycerophospholipids have fatty acids at positions *sn*-1 and *sn*-2, and a phosphate linked to a polar head-group on the *sn*-3 position. The major glycerophospholipids are the phosphatidylethanolamine (PE) and the phosphatidylcholine (PC) while phosphatidylserine (PS) and phosphatidylinositol (PI) are minor quantitatively (Fig. 3.2A). About one-third to one-half of the polar lipids in the MFGM are sphingolipids. Sphingolipids are based on a sphingosine backbone and according to the head group can be differentiated into sphingomyelin (phosphocholine), ceramide (H), glycosylceramides (glucose; galactose for cerebrosides), lactosylceramides (lactose) and with more complex glycosyl residues gangliosides (combination of glucose, monosaccharides, N-acetylgalactosamine, sialic acid and others) (Nilsson, 2016). The main sphingolipids are the milk sphingomyelin (milk-SM; Fig. 3.2B) and the glycosphingolipids. The relative proportion of polar lipids depends on the size of milk fat globules (Lopez et al., 2011), and on mammal species (Garcia et al., 2012). For example, milk-SM weights for 20–45 wt% of milk polar lipids depending on the mammal species, and milk-SM is the most abundant polar lipid species in human milk (Giuffrida et al., 2013; Lopez & Ménard, 2011; Yao et al., 2016; Zou et al., 2012). The MFGM surrounding fat globules in milk, as well as by-products of the butter industry (i.e. buttermilk, butter serum) are major and convenient sources of exogenous PS and sphingolipids in the human diet, particularly milk-SM and gangliosides.

While polar lipids constitute a very small proportion of the total milk lipids, they play an important structural role because of their mixed hydrophilic and hydrophobic nature. This is the reason why they are the key structural elements of the MFGM. Milk polar lipids are also efficient to form liposomes and to stabilize emulsions. Milk-SM, PC and PE are zwitterionic polar lipids while PI and PS are anionic polar lipids (Fig. 3.2A). They are involved in electrostatic interactions and are affected by changes in pH and ionic strength. The isoelectric point of the MFGM is pI ~ 4.2 (Lopez et al., 2017). In their hydrophobic part, milk polar lipids are characterised by a wide diversity of saturated and unsaturated acyl residues (Sanchez-Juanes, Alonso, Zancada, & Hueso, 2009; Yao et al., 2016) (Fig. 3.2B). Each class of glycerophospholipid (PC, PE, PI, PS) corresponds to several molecular species combining fatty acids onto the *sn*-1 and *sn*-2 positions of the glycerol backbone. The fatty acid composition of milk-SM is completely different from the fatty acid composition of the other polar lipids found in the MFGM. Milk-SM contains several molecular species with long-chain saturated fatty acids (C16:0, C18:0, C22:0, C23:0, C24:0) (Fig. 3.2B). This unique composition of milk-SM leads to specific biophysical properties (e.g. high phase transition temperature Tm; formation of ordered domains, interaction with cholesterol), that are discussed in the next section of this chapter. Also, the complex chemical composition of milk-SM molecular species in comparison to other dietary sources of SM such as brain-SM (~50% C18:0-SM) or egg-SM (~85% C16:0-SM) could be involved in specific biological functions and health benefits (e.g. hypocholesterolemic effect; Noh & Koo, 2004).

The fatty acid composition of milk polar lipids is not constant. Changes in the fatty acid composition of MFGM polar lipids have been reported in relation to cow diet (Lopez et al., 2008, 2014). A diet rich in unsaturated fatty acids (e.g. grass, addition of linseeds) leads to an increased amount of unsaturated fatty acids in MFGM polar lipids. It has also been reported that the relative proportion of MFGM polar lipid classes (i.e. milk-SM, PC, PE, PI, PS) is not affected by the diet.

The MFGM is a source of many molecules with structural properties as well as bioactive compounds of nutritional and health interest, for example, the unsaturated fatty acids provided by the MFGM polar lipids and the high amount of sphingolipids, mainly milk-SM.

3 Biophysical Properties of MFGM Polar Lipids: Specific Role Played by Sphingomyelin and Cholesterol

Lipid species with different chemical compositions have individual intrinsic properties such as the melting transition temperature Tm and may, therefore, exhibit different phases depending on the temperature T (i.e. T < Tm or T > Tm). As a result, lipids experience different types of interactions with each other, which in turn dictate the membrane fluidity, lateral phase segregation, and the membrane mechanical properties and functions.

Gathering information about milk polar lipid properties and interactions as a function of temperature is essential in the understanding of their functions in the MFGM and in functional ingredients containing milk polar lipids and MFGM fragments.

3.1 Thermotropic and Structural Behavior of Saturated MFGM Polar Lipids

The polar lipids located in the MFGM correspond to a wide range of lipid species, including various polar heads and acyl chains (Fig. 3.2). In fully hydrated conditions, polar lipids organize as lamellar structures that can exhibit different packing of the acyl chains as a function of temperature (Fig. 3.3). The unsaturated milk polar lipids are fluid above 0 °C (e.g. Tm-DOPC = −20 °C; DOPC: dioleoyl-PC) while some polar lipids containing at least one saturated acyl chain exhibit high gel to fluid Lα phase transition temperatures (e.g. Tm-POPE = 25.8 °C with P = palmitic acid and O = oleic acid; Tm-DPPC = 41.1 °C; dipalmitoyl-PC) (Fig. 3.3A). It is important to note that major PC species composed of a saturated acyl chain (e.g. P: palmitic acid C16:0) and an unsaturated acyl chain (e.g. O: oleic acid C18:1 n-9) are in the fluid state above 0 °C, such as POPC that exhibits a phase transition temperature at Tm = −4.7 °C.

Among the saturated milk polar lipids, the thermotropic phase behavior of milk-SM has recently been the focus of interest of several studies since it is present in high amount (20–45% of milk polar lipids) and plays major structural and metabolic roles. Bilayers of milk-SM exhibit a broad asymmetric multicomponent endothermic peak composed by a sharp peak that determines the melting transition temperature Tm, with broad components on both sides of the main peak (thermograms recorded by differential scanning calorimetry, DSC) (Fig. 3.3A). The phase transition temperature of milk-SM depends on the hydration level. For fully hydrated

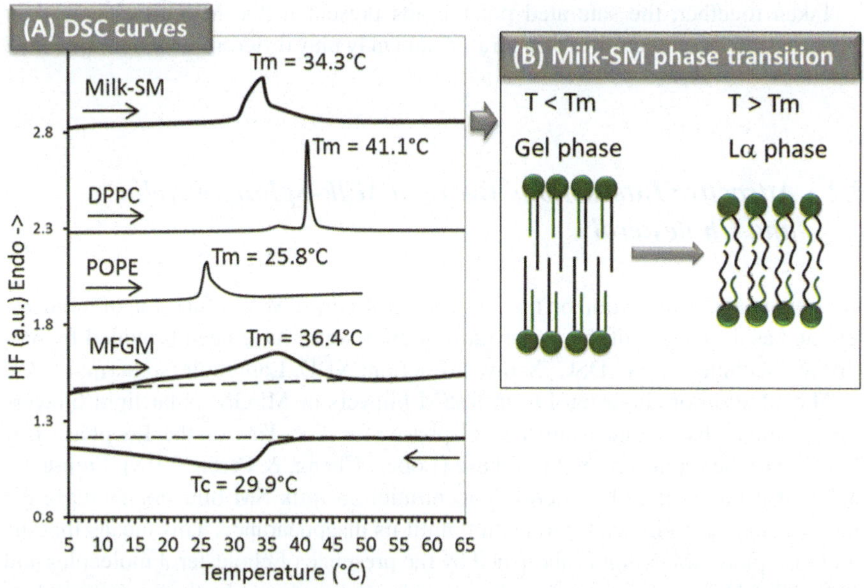

Fig. 3.3 Thermotropic properties of polar lipids. (**A**) thermograms recorded by differential scanning calorimetry (DSC) on heating of polar lipids, showing the temperature Tm of gel to fluid Lα phase transition; (**B**) Schematic representation of the organization of milk-sphingomyelin molecules in lamellar structures in the gel phase for T below Tm and in the Lα phase for T above Tm. Adapted with permission from Murthy et al. (2015)

milk-SM bilayers (i.e. below 50 mol% milk-SM), the phase transition temperature is Tm = 34.3 ± 0.1 °C (Fig. 3.3A) (Cheng, Ropers, & Lopez, 2017; Malmsten, Bergenståhl, Nyberg, & Odham, 1994; Murthy, Guyomarc'h, Paboeuf, Vié, & Lopez, 2015). Interestingly, Tm of milk-SM is close to the physiological temperature and then the physical state of milk-SM could be involved in biological functions in the gastro-intestinal tract. The width of the endothermic event recorded between the onset and completion temperatures spanned from about 30–40 °C. This complex thermal behaviour of milk-SM results from the large heterogeneity in the hydrocarbon chains (from C16:0 to C24:0; Fig. 3.2B), the successive melting of the individual SM species, and to structural reorganizations occurring on heating of the bilayers. Structural analysis of fully hydrated milk-SM bilayers showed the formation of a gel phase below Tm and a fluid liquid-crystalline Lα phase above Tm (Cheng et al., 2017; Malmsten et al., 1994; Murthy et al., 2015) (Fig. 3.3B).

The biophysical properties of the MFGM polar lipids are affected by the temperature. High Tm saturated polar lipids (milk-SM, DPPC) are involved in the complex thermotropic phase behavior of the complex mixture of MFGM polar lipids, with a main phase transition at Tm = 36.4 ± 0.2 °C (Murthy et al., 2015) (Fig. 3.3A). Structural analysis of fully hydrated mixture of polar lipids extracted from the MFGM showed a gel phase below Tm and a fluid liquid-crystalline Lα phase over Tm (Murthy et al., 2015).

Taken together, the saturated polar lipids present in the MFGM change their physical state as a function of temperature and may play different structural roles and functions.

3.2 *Attractive Interactions Between Milk-Sphingomyelin and Cholesterol*

In recent years, information on the phase state of milk-SM as a function of temperature and as a function of the molar ratio of cholesterol have been provided by biophysical techniques (i.e. DSC, X-ray diffraction; XRD, Langmuir isotherms).

The addition of cholesterol to milk-SM bilayers or MFGM polar lipid bilayers change their phase state from the gel phase (for T < Tm) or the Lα phase (for T > Tm) to the liquid-ordered Lo phase (Lopez, Cheng, & Perez, 2018). Figure 3.4 shows that increasing cholesterol concentration in milk-SM bilayers decrease the endothermic gel to Lα phase transition until its disappearance. This means that the gel to Lα phase transition is abolished by the presence of cholesterol molecules and that milk-SM bilayers are in the liquid-ordered Lo phase. In the binary mixture milk-SM/cholesterol, the liquid-ordered Lo phase is completed for 40% mol cholesterol in the bilayers (Fig. 3.4). In the complex MFGM polar lipid extract, the

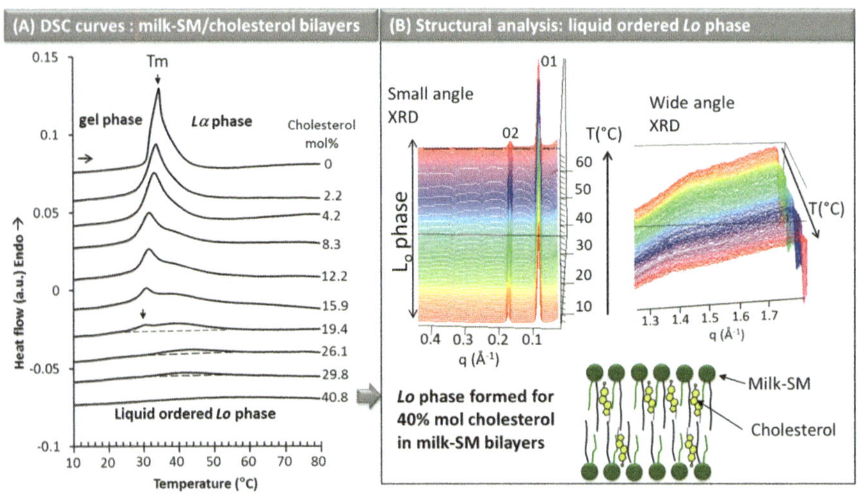

Fig. 3.4 Role of cholesterol in milk-SM membranes leading to the formation of the liquid-ordered Lo phase. (**A**) Thermograms recorded by differential scanning calorimetry showing the impact of various % mol of cholesterol in the abolishment of the gel to fluid Lα phase of milk-SM membranes until the formation of the Lo phase for 40 % mol cholesterol; (**B**) X-ray diffraction patterns of milk-SM/cholesterol bilayers showing the absence of structural evolution as a function of temperature, schematic representation of milk-SM bilayers in the Lo phase in presence of cholesterol. Adapted with permission from Lopez et al. (2018)

liquid-ordered Lo phase is completed for 27% mol cholesterol, which corresponds to milk-SM/cholesterol molar ratio of 50/50 (Murthy, Guyomarc'h, & Lopez, 2016a). The XRD signature of the liquid-ordered Lo phase is a lamellar structure with a lateral disorder in the acyl chains organization (Fig. 3.4B). The structural parameters of the lamellar liquid ordered Lo phase do not evolve as a function of temperature, in accordance with the absence of phase transition (Fig. 3.4). Furthermore, the presence of cholesterol induces attractive interactions with milk-SM molecules with a decrease in the area occupied per phospholipid leading to a condensing effect of cholesterol, as demonstrated using Langmuir isotherms (Cheng et al., 2017; Murthy et al., 2015).

This means that ordered lipid domains composed of milk-SM and cholesterol in the Lo phase can be present in the MFGM whatever the temperature. Furthermore, the attractive interactions between milk-SM and cholesterol that lead to the formation of complexes may be involved in the decrease of cholesterol absorption in the intestine as demonstrated in in vivo studies (Noh & Koo, 2004; Nyberg, Duan, & Nilsson, 2000).

4 New Research Advances in the Structure of the MFGM and Polar Lipid Assemblies Revealing Topographical and Mechanical Heterogeneities

From more than 50 years, the structure of the MFGM has been the focus of attention of research groups. The first morphological studies performed using electron microscopy with negatively stained preparations provided details about the structure of the MFGM and remain very informative (Wooding, 1971; Wooding & Mather, 2017). Authors agree on the fact that the MFGM is organized as a trilayer of polar lipids, where the internal monolayer originates from the endoplasmic reticulum of the epithelial cells while the external bilayer of the MFGM derives from the apical plasma membrane during secretion of the fat globule by the mammary cell. A thin layer of cytoplasmic material is entrapped between the monolayer and the bilayer membrane and can lead to the formation of cytoplasmic remnants attached to milk fat globules (Lopez & Ménard, 2011).

For many years, authors were mainly interested in the proteins associated with the MFGM, in terms of localization and bioactive properties (Dewettinck et al., 2008; Rasmussen, 2009). Some of the MFGM proteins are integral (e.g. butyrophilin), or transmembrane proteins integrated in the outer bilayer of the MFGM (e.g. Mucin 1, Mucin 15, CD36), others are reported to be loosely attached in outer part of the membrane (e.g. lactadherin and PP3). The intermembrane space between the monolayer and the bilayer of polar lipids is occupied with the cytosolic parts of the transmembrane proteins and "coat"-proteins (e.g. Xanthine oxidoreductase and Adipophilin) (Fig. 3.5). The MFGM glycoproteins (mainly the mucins and the butyrophilin), together with the MFGM glycolipids, form a glycocalyx that protrudes in the aqueous phase surrounding fat globules in milk. The glycocalyx is

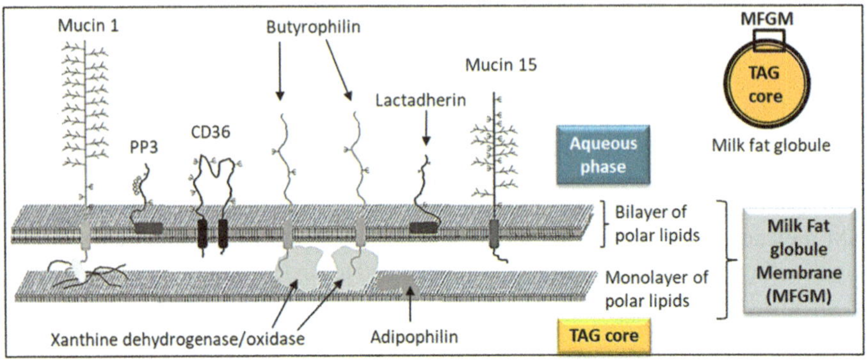

Fig. 3.5 Schematic drawing of the milk fat globule membrane (MFGM) showing the insertion of proteins within the trilayers of polar lipids. The transmembrane proteins (Mucin 1, Mucin 15, Butyrophilin and CD 36) and peripheral bound proteins (Lacadherin and PP3) are embeded in the outer bilayer of polar lipids. The glycoproteins protrude in the aqueous phase and form the glycocalix. The cytosolic parts of the transmembrane proteins and "coat"-proteins (Xanthine dehydrogenase/oxidase and Adipophilin) are located in the space between the bilayer and the monolayer of polar lipids. In this drawing, the polar lipids are organized as a homogeneous fluid matrix. Adapted with permission from Rasmussen (2009)

involved in the physical stability of milk fat globules in milk and in interactions with bacteria, toxins and viruses. These MFGM glycosylated components may participate in the defense against infections in human body by preventing pathogen adhesion to the epithelium in the intestine (Douellou, Montel, & Sergentet, 2017; Sprong, Hulstein, Lambers, & van der Meer, 2012).

The localization of polar lipids within the three layers of the MFGM is still a debate. As for other biological membranes, authors reported an asymmetry in the localization of polar lipids in the MFGM. Milk-SM, PC and cholesterol have been reported to be preferably located in the outer bilayer of the MFGM, while PE, PI, PS are mainly concentrated on the inner monolayer of the MFGM (Deeth, 1997). Other studies reported the localization of PS in the outer bilayer of the MFGM (Zheng, Jimenez-Flores, & Everett, 2014). The negative zeta potential values of bovine and breast milk fat globules (−12 mV and −7 mV, respectively) and the high sensitivity of the MFGM to calcium may correspond to the localization of the anionic polar lipids PS and PI at the surface of the MFGM in contact with the aqueous environment.

The outer bilayer of the MFGM has long been described as a homogeneous membrane according to the concept of fluid mosaic membrane, as reviewed in Dewettinck et al. (2008). As the concept of lipid phase separation and presence of domains called "rafts" started to emerge in cellular biology (Simons & Ikonen, 1997), questions have raised on whether the MFGM obeyed similar rules as other biological membranes. Moreover, the different physical properties of milk polar lipids (saturated high Tm polar lipids such as milk-SM vs. unsaturated low Tm polar lipids) raised questions about their homogeneous accommodation in the MFGM.

In recent years, the development of methods, innovative protocols as well as the combination of microscopy techniques permitted to increase the knowledge about the structure of the MFGM and evidenced lateral phase separation of polar lipids with formation of SM-rich microdomains, as well as topographical and mechanical heterogeneities.

This part of the chapter summarizes the main recent research advances in the structure of the MFGM and polar lipid assemblies and highlights the specific role of milk-SM and milk-SM/cholesterol complexes.

4.1 Ordered Lipid Domains Rich in Sphingomyelin and the Heterogeneous Distribution of Proteins Observed in situ in the MFGM Around Fat Globules at the Microscopic Level

A new aspect of MFGM organization has emerged in the last decade thanks to the potentialities of confocal laser scanning microscopy (CLSM) combined with the specific labeling of membrane components (e.g. lipids, proteins, sugars) that permit structural observations of the MFGM in situ in milk.

Using lipophilic and lectin fluorescent dyes for CLSM observations, Evers et al. (2008) showed the heterogeneous distribution of lipids and proteins in the MFGM. The presence of non-fluorescent areas around fat globules in milk raised the question of the possible absence of a bilayer membrane at these locations. Using the exogenous phospholipid DOPE head-labelled with the fluorescent dye rhodamine (i.e. Rhodamine-DOPE) that selectively partitions in fluid phospholipids, Lopez's group revealed the presence of non-fluorescent ordered lipid domains on the surface of milk fat globules (Lopez, Briard-Bion, Beaucher, & Ollivon, 2008; Lopez, Madec, & Jimenez-Flores, 2010) (Fig. 3.6). These ordered lipid domains observed at room temperature have circular shapes and their size is in the range of the µm. It is also possible that such domains with a smaller size exist in the MFGM but they are not detected by CLSM. These micro-domains were interpreted as the lateral segregation of milk-SM in the gel phase, or to the formation of complexes between milk-SM and cholesterol in the liquid-ordered Lo phase. This interpretation was formulated on the following basis: (1) milk-SM and cholesterol are major lipid components present in the outer bilayer of the MFGM, (2) milk-SM and cholesterol exhibit attractive interactions involved in the formation of lipid domains (Cheng et al., 2017). The double labelling of fluid phospholipids (with Rh-DOPE) and membrane proteins (with fast green FCF, and various lectins such as WGA-488) confirmed that the non-fluorescent domains do not correspond to the localization of proteins in the MFGM (Lopez et al., 2010). The lectin WGA-488 showed that the glycosylated molecules (i.e. glycoproteins and glycolipids), are heterogeneously distributed in the MFGM, protrude in the surrounding aqueous phase to form the glycocalyx and are not located in the ordered lipid domains (Fig. 3.7). The presence of ordered lipid domains enriched in milk-SM and cholesterol in the MFGM has

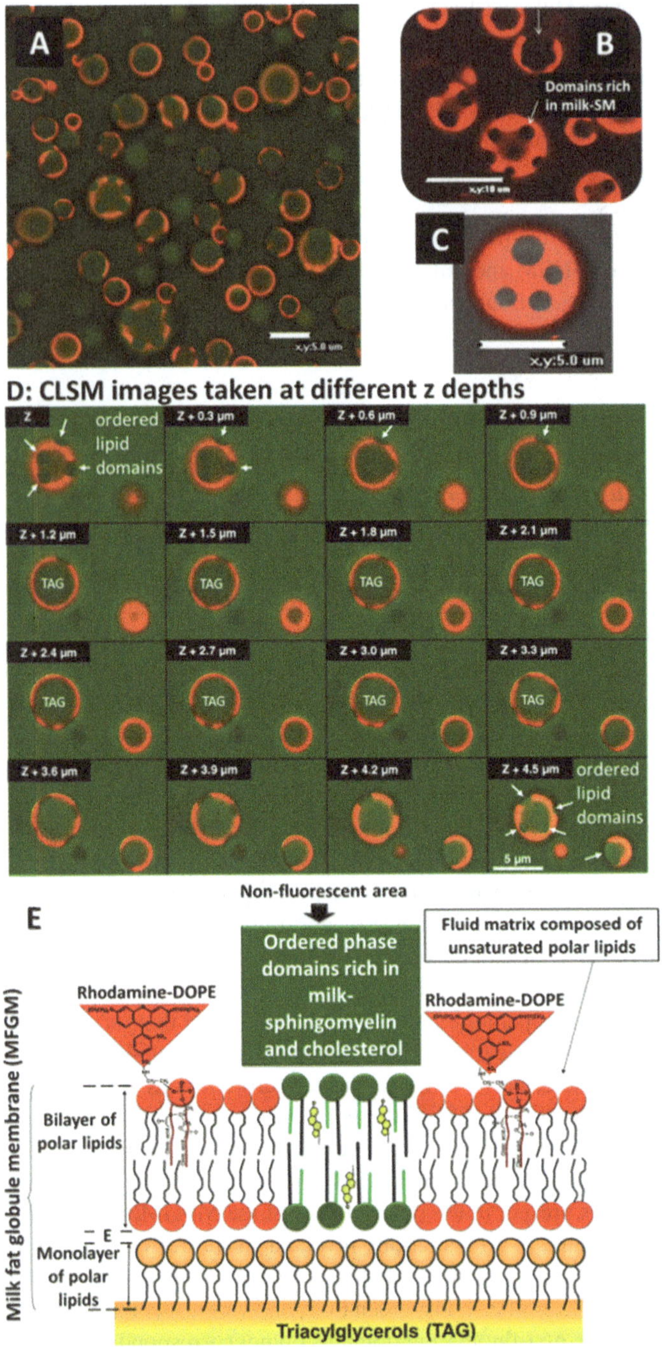

Fig. 3.6 Confocal laser scanning microscopy (CLSM) images showing the heterogeneous organization of polar lipids in the milk fat globule membrane (MFGM) surrounding fat globules in milk. (**A–C**) CLSM images; (**D**) The CLSM images were taken at different z depths as indicated in the figure. The polar lipids were labelled with the fluorescent dye rhodamine-DOPE (red colour) that integrates the fluid matrix of polar lipids and reveal the non-fluorescent areas that correspond to ordered lipid domains rich in milk sphingomyelin and cholesterol. (**E**) Schematic representation of the lateral phase separation of milk-sphingomyelin and cholesterol in the outer bilayer of the MFGM. Adapted with permission from Lopez et al. (2010)

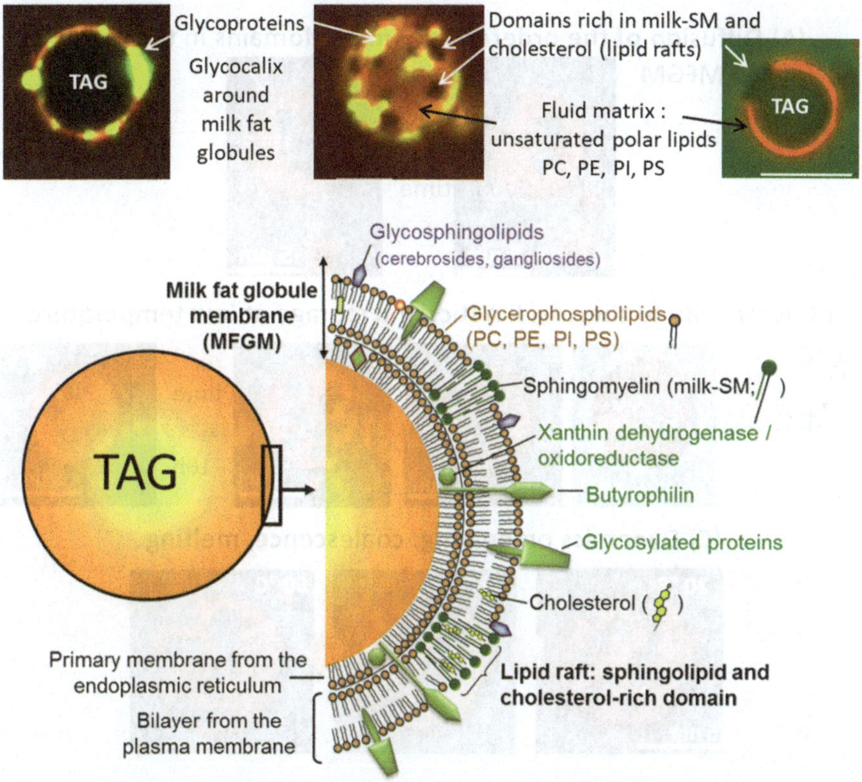

Fig. 3.7 Confocal laser scanning microscopy (CLSM) images of the milk fat globule membrane observed in situ around fat globules in milk. The CLSM images show the heterogeneous distribution of (glyco)proteins embedded in the fluid matrix of polar lipids (red colour), in coexistence with ordered lipid domains rich in milk-sphingomyelin and cholesterol. *Bottom*: schematic representation of the MFGM showing the lateral phase separation of milk sphingomyelin and cholesterol in lipid domains. Adapted with permission from Lopez, Briard-Bion, Menard, et al. (2008); Lopez et al. (2010)

been reported in the MFGM whatever the size of milk fat globules (i.e. 1.6 vs. 6.3 μm) even if the amount of milk-SM is lower in small fat globules (Lopez et al., 2011). Moreover, CLSM observations revealed the diffusion of the ordered lipid domains in the plane of the MFGM, without any coalescence (Et-Thakafy et al., 2017; Lopez et al., 2010; Nguyen et al., 2016) (Fig. 3.8A).

On the basis of these new scientific findings obtained from CLSM observations, an updated model of the structure of the MFGM was proposed in 2008 by Lopez's group (INRAE, France), with additional information about the lateral phase separation of milk-SM and cholesterol to form ordered domains in the outer bilayer of the MFGM and the heterogeneous distribution of proteins in the fluid matrix of the MFGM (Figs. 3.6E and 3.7) (Lopez, 2011; Lopez et al., 2010; Lopez, Briard-Bion, Menard, et al., 2008). These ordered lipid domains found in the outer bilayer of the MFGM were called lipid rafts (Lopez et al., 2010), in analogy with rafts in mammalian plasma membranes (Simons & Ikonen, 1997).

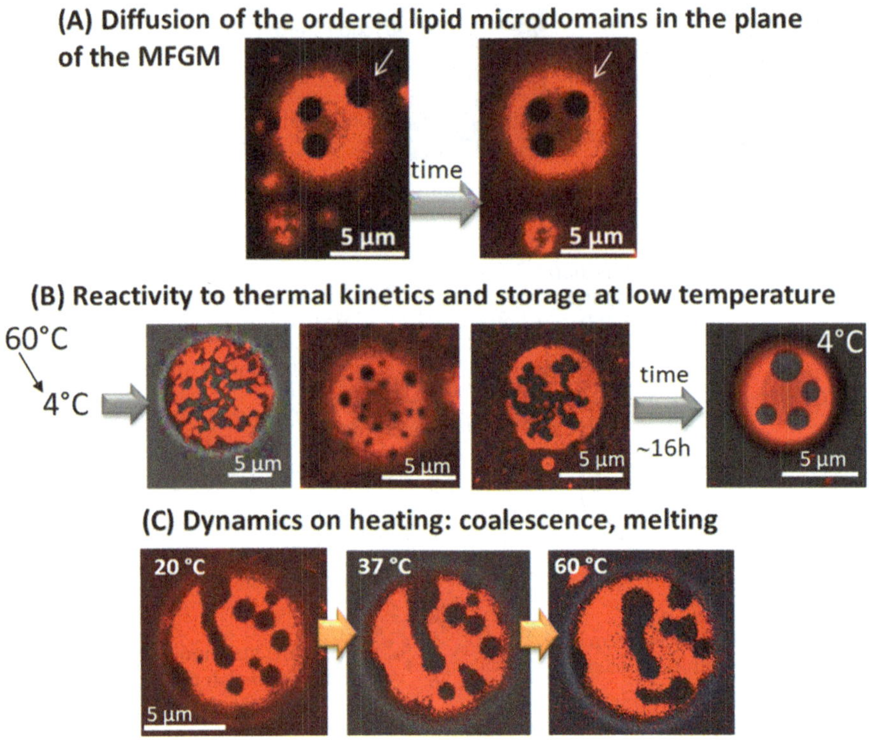

Fig. 3.8 Confocal laser scanning microscopy (CLSM) images showing the dynamics of the polar lipids in the milk fat globule membrane (MFGM). The polar lipids were labelled with the fluorescent dye rhodamine-DOPE (red colour) that integrates the fluid matrix of polar lipids and reveal the non-fluorescent areas that correspond to ordered lipid domains rich in milk sphingomyelin and cholesterol. Adapted with permission from Et-Thakafy et al. (2017)

The presence of ordered lipid domains in the MFGM was confirmed by other groups using CLSM and the fluorescent dye Rh-DOPE to label the fluid matrix of the MFGM. These ordered lipid domains have been observed in situ in milks from various species: bovine milk (Et-Thakafy et al., 2017; Gallier, Gragson, Jiménez-Flores, & Everett, 2010a; Lopez et al., 2010; Zou et al., 2015), human breast milk (Gallier et al., 2015; Lopez & Ménard, 2011; Zou et al., 2012), buffalo milk (Nguyen et al., 2015), goat and sheep milks (Et-Thakafy et al., 2017) and yack milk (Luo, Huang, Liu, Zhang, & Ren, 2018). Differences in the number, size and shape of the ordered lipid domains have been characterized at the surface of milk fat globules and could be related to specific lipid composition of the MFGM from various mammal species, e.g. the ratio between milk-SM and cholesterol.

Most of the CLSM observations of the MFGM have been performed at room temperature after cooling the milk from the physiological temperature, or after storage at 4 °C. However, investigations of the MFGM structure as a function of temperature are of primary importance to better understand its function at the physiological temperature of 37°C upon digestion of milk fat globules and in the range spanning from 4 to 60 °C for technological applications. Authors showed that the ordered lipid domains formed in the MFGM are present in a wide range of temperatures, i.e. from 4 to 60 °C (Fig. 3.8), including the physiological temperature 37 °C (Et-Thakafy et al., 2017; Lopez & Ménard, 2011; Nguyen et al., 2016; Zou et al., 2015). This means that these ordered lipid domains are present at the surface of milk fat globules at the physiological temperature and then upon digestion in the gastro-intestinal tract. The ordered lipid domains rich in milk-SM and cholesterol and/or the difference in composition, height and mechanical properties between the domains and the surrounding fluid phase could, therefore, be involved in the mechanisms of milk fat globule digestion by the lipolytic enzymes. The presence of the ordered domains above the phase transition temperature of the saturated MFGM polar lipids also shows that these domains should be in the liquid-ordered Lo phase, i.e. correspond to interactions between milk-SM and cholesterol (Fig. 3.4). However, the determination of the exact composition and physical phase of these lipid domains has not yet be determined in situ in the MFGM.

The dynamics of the lipid domains were investigated during kinetics in temperature (Fig. 3.8). Cooling of milk fat globules from 60 to 4 °C enhances the mechanisms of nucleation in the MFGM, which leads to the formation of small numerous domains or elongated domains (Fig. 3.8B). Extended storage at low temperature after rapid cooling induces lipid reorganization within the MFGM with growth, leading to the formation of microdomains with a circular shape. Heating of milk leads to the melting of some ordered lipid domains, with a phase transition from the gel phase to the fluid Lα phase (Fig. 3.8C).

All these structural investigations performed using CLSM showed that the organization of polar lipids and proteins is heterogeneous and that the MFGM is a highly dynamic system.

4.2 Topography and Mechanical Properties of MFGM Model Membranes Examined with a Nanoscale Resolution

The observation of ordered lipid domains in situ in the MFGM surrounding fat globules in milk raised questions about their physical properties and the consequences on the topography and the mechanical properties of the MFGM. The high spatial resolution of atomic force microscopy (AFM) permitted investigations of model lipid membranes at the nanometer scale.

4.2.1 Milk-Sphingomyelin Molecules are Responsible for Topographical and Mechanical Heterogeneities in Membranes

The main role played by milk-SM on the topography and mechanics of the MFGM was first evidenced using AFM with the observation of hydrated planar supported lipid bilayers (SLB) composed of equimolar mixture of milk-SM and dioleoylphosphatidylcholine (DOPC; unsaturated polar lipid mimicking the fluid phase of the MFGM) (Guyomarc'h et al., 2014).

AFM images recorded at 20 °C (i.e. for T < Tm of milk-SM) show the lateral segregation of milk-SM rich domains in the gel phase surrounded by a DOPC-rich fluid phase (Fig. 3.9). In the bilayer, the domains are thicker than the surrounded fluid phase, due to elongation of the acyl chains of milk-SM in the ordered gel phase. The difference in height between the milk-SM domains and the fluid phase is H ~ 1 nm. This height mismatch between polar lipids in the gel or in the fluid states creates disorder that is minimized through phase separation and the formation of domains. At complete equilibrium, the domains tend to circularity to minimize the phase boundary.

The gel phase domains are also more rigid than the fluid phase (Fig. 3.9). The milk-SM-rich domains exhibited breakthrough force, $F_B = 31.6 \pm 1.9$ *versus* 5.7 ± 2.3 nN for the continuous DOPC-rich phase (Murthy, Guyomarc'h, & Lopez, 2018). The higher F_B values exhibited in the milk-SM domains reflect the tight acyl chain packing of high Tm polar lipids in the gel phase, as compared to the continuous phase where the acyl chains are in disordered state.

Interestingly, the presence of sub-domains in the gel phase was occasionally reported in bilayers containing milk-SM (Guyomarc'h et al., 2014; Guyomarc'h, Chen, Et-Thakafy, Zou, & Lopez, 2017) (Fig. 3.9; bottom). The two levels characterized by AFM demonstrated the existence of gel–gel phase separation between the SM molecular species present in milk-SM. These two levels of gel phase domains were interpreted as the segregation of individual species composed by different N-acyl chain lengths and/or as the co-existence of two interdigitated organizations of the long SM molecules (Carbon atoms > 20). By using the ternary system, milk-SM/DOPC/chol (3:7:1), (Bhojoo, Chen, & Zou, 2018) also observed sub-domains in the bilayers, while they did not observe these subdomains with egg-SM/DOPC/chol (2:2:1). These different levels in the ordered phase domains could be specific to natural mixtures of milk-SM composed by individual molecular species with chain lengths varying from C16:0 to C24:0. This natural complexity of milk-SM has also been shown to alter the elasticity and the mechanical stability of the ordered lamellar structures (Et-Thakafy, Delorme, Guyomarc'h, & Lopez, 2018). Thus, membranes with as complex composition as that of the MFGM are expected to be more compliant to mechanical stress than membranes with saturated polar lipids with homogeneous acyl chain length such as DPPC.

4.2.2 Milk Polar Lipid Membranes Exhibit Topographical and Mechanical Heterogeneities Due to the Formation of Milk-Sphingomyelin Rich Domains

Using the Langmuir-Blodgett technique, Prof. R. Jimeénez-Flores' group (CalPoly, San Luis Obispo, CA) first pioneered the deposition of MFGM components isolated from buttermilk powder to form a monolayer on a mica surface and performed observations using AFM imaging in air (Jiménez-Flores & Brisson, 2008). The same group used this approach to show that polar lipids isolated from milk fractions or processed milk phase separate and form domains at various temperatures and

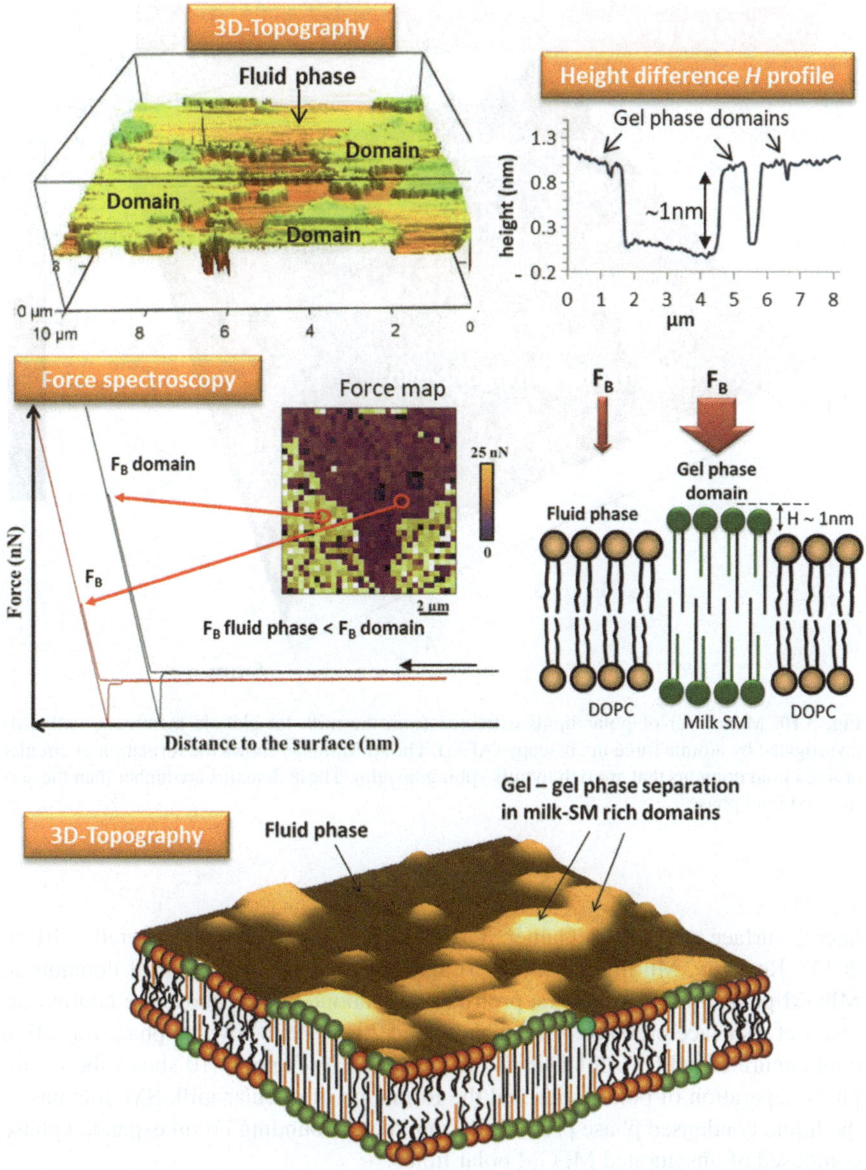

Fig. 3.9 Membranes of polar lipids investigated by atomic force microscopy (AFM) imaging and force spectroscopy. The bilayers were composed of an equimolar proportion of milk sphingomyelin (milk-SM) and dioleoylphosphatidylcholine (DOPC) to mimic the fluid phase of polar lipids in the milk fat globule membrane (MFGM). AFM images show the lateral phase separation of milk-sphingomyelin in ordered gel phase domains that are higher and more rigid than the surrounding fluid phase (F_B: breakthrough forces; the size of the arrows is proportional to the rigidity). Bottom: 3D topography AFM image combined with a schematic representation of polar lipids in the MFGM, showing the lateral phase segregation of milk-sphingomyelin in domains and the existence of a gel–gel phase separation. Adapted with permission from Guyomarc'h et al. (2017); Lopez, Cauty, and Guyomarc'h (2015); Lopez et al. (2019)

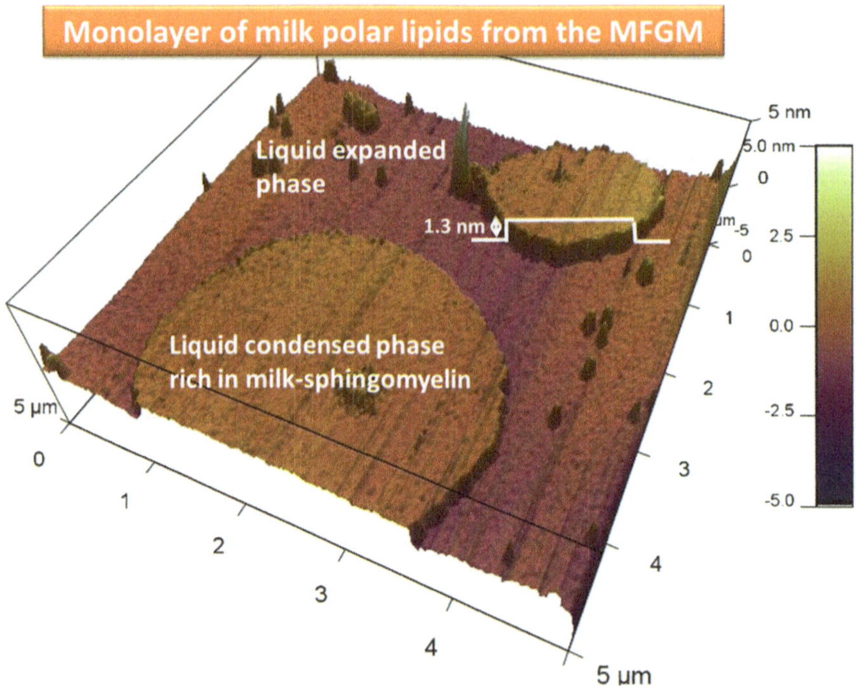

Fig. 3.10 Monolayer of polar lipids extracted from the milk fat globule membrane (MFGM) investigated by atomic force microscopy (AFM). The AFM image shows the formation of circular ordered lipid domains that are rich in milk sphingomyelin. These domains are higher than the surrounded fluid phase

lateral surface pressures (Gallier, Gragson, Jiménez-Flores, & Everett, 2010b, 2012). Recently, Murthy et al. (2015) also reported the formation of domains in MFGM polar lipid monolayers, protruding by about 1.5 nm above the continuous phase at a surface pressure of 30 $mN.m^{-1}$ and 20°C (i.e. below the phase transition temperature of MFGM polar lipids, Tm = 36.4 °C). Figure 3.10 shows the lateral phase separation of polar lipids and the formation of circular milk-SM domains in the liquid condensed phase protruding from the surrounding liquid expanded phase composed of unsaturated MFGM polar lipids.

To take account of the natural complexity of the MFGM, the topography and mechanical properties of hydrated SLB composed of polar lipids extracted from the MFGM were examined (Murthy, Guyomarc'h, & Lopez, 2016b). As in milk-SM/DOPC bilayers, SLB made of MFGM polar lipids exhibit at 20 °C a lateral phase separation with the formation of milk-SM rich domains protruding by 1 nm from the surrounding fluid phase composed of the unsaturated MFGM polar lipids PE, PC, PI and PS. The lipid domains exhibited a higher breakthrough force F_B than the surrounding fluid phase, i.e. 30 versus 15 nN (Fig. 3.11).

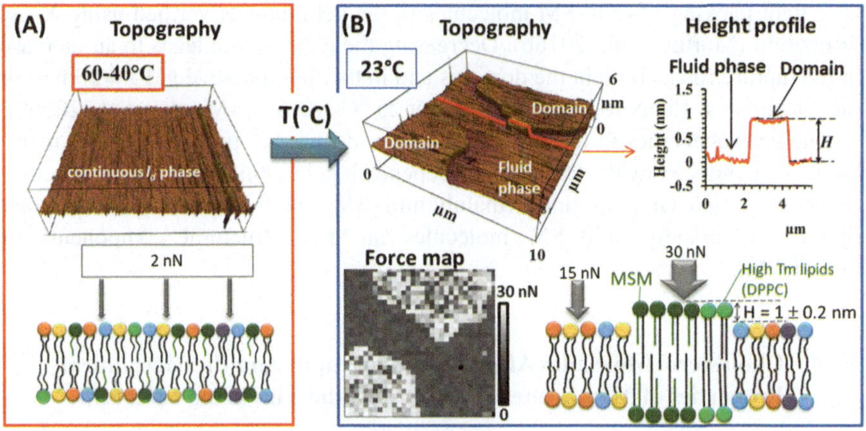

Fig. 3.11 Impact of temperature on the topography and mechanical properties of membranes composed of polar lipids from the milk fat globule membrane (MFGM): (**A**) temperature above the melting temperature of MFGM polar lipids, (**B**) temperature below the fluid to gel phase transition temperature. Atomic force microscopy (AFM) topographical images and force maps showing, below 35 °C, the formation of ordered lipid domains rich in milk-sphingomyelin (MSM) that are higher and more rigid than the surrounding fluid phase. Adapted with permission from Murthy et al. (2016b)

4.2.3 The Temperature Governs the Physical State of Polar Lipids with Consequences on the Topography and Mechanical Properties of the Membrane

In multicomponent lipid bilayers, such as the outer bilayer of the MFGM, the presence of a variety of saturated and unsaturated polar lipids with distinct Tm makes these systems more prone to temperature variations. The impact of temperature on the topography and nano-mechanics of membranes was studied using AFM temperature-controlled imaging and force mapping in the range from 60 °C to 6 °C upon cooling of SLB composed of milk-SM/DOPC (50/50 mol%) or the complex blend of milk polar lipids extracted from the MFGM (Guyomarc'h et al., 2014; Murthy et al., 2016b). These studies showed that milk-SM and MFGM polar lipid bilayers are dynamic systems exhibiting changes in their topography and mechanical properties as a function of temperature (Fig. 3.11). Above 35 °C (i.e. above Tm of milk-SM) and in the absence of cholesterol, the polar lipid molecules are homogeneously distributed in the membrane, and the rupture force of the fluid membrane was $F_B = 2$ nN (Fig. 3.11A). The segregation of the milk SM-rich domains into salient features protruding by about 1 nm above the fluid phase was detected upon phase transition at 35 °C (Murthy et al., 2015) and below that temperature (Guyomarc'h et al., 2014; Murthy et al., 2016b) (Fig. 3.11B). At 35 °C the formation of lipid domains was associated with heterogeneities in the mechanical properties since the rupture forces were $F_B = 23$ nN in the gel phase milk-SM rich domains and $F_B = 6$ nN in the fluid phase. The higher force rupture recorded for the domains is due to the very tight packing of milk-SM molecules in the gel phase as verified using X-ray diffraction (Murthy et al., 2016b). Decreasing the temperature leads to an increase in the rupture forces both in the domains and in the fluid phase, e.g. $F_B = 30$ nN for the domains vs. 15 nN for the fluid phase at 25 °C and 6 °C (Murthy et al., 2016b).

The heterogeneities in the structure and the mechanical properties of the membrane were induced by the temperature-dependent gel to fluid Lα phase immiscibility between high Tm polar lipids (mainly milk-SM) and the unsaturated fluid polar lipids. Undoubtedly, milk-SM molecules are key structural components of the MFGM.

4.2.4 Cholesterol Molecules Affect the Topography and the Mechanical Properties of Membranes: Condensing and Fluidizing Effects

The functional role of cholesterol in the MFGM was poorly understood, despite the fact that cholesterol is known to play a crucial role in the lateral organization of lipids and formation of lipid rafts in mammalian cell membranes. Research studies performed by Lopez's group at INRAE (France) revealed the main role played by cholesterol in milk-SM containing membranes, i.e. changes in the topography of the ordered lipid domains and in the mechanics of the membrane.

In the ternary system composed of milk-SM/DOPC/cholesterol (40/40/20% mol), the AFM height images showed the coexistence of the milk-SM/cholesterol-enriched liquid ordered (Lo) phase and DOPC-enriched Lα phase both on mono or bilayers (Guyomarc'h et al., 2014). The presence of cholesterol yielded narrow and more elongated milk-SM/cholesterol Lo domains compared to those formed by the gel phase milk-SM in the binary milk-SM/DOPC (50/50% mol) system, as revealed in both Langmuir-Blodgett monolayers and hydrated SLB (Guyomarc'h et al., 2014). Introduction of 20% mol cholesterol in the binary system milk-SM/DOPC induced a significant decrease in the height difference H between the ordered domains and the Lα phase. In hydrated bilayers, this decrease was from $H_{gel} = 0.8–1.1$ nm to $H_{Lo} = 0.4–0.6$ nm. In Langmuir-Blodgett monolayers, the decrease was from 1.4 ± 0.2 nm for the gel phase milk-SM domains down to 1.1 ± 0.1 nm for the Lo phase domains in the presence of 20% mol cholesterol. Preferred partitioning of cholesterol into milk-SM domains resulted in the gel to Lo phase transition of these domains, associated with a reduction of their thickness (thinning of the membrane). As cholesterol amount increases, the DOPC-rich Lα phase could also progressively increase in thickness (thickening of the membrane) and could further contribute to reduce H values, i.e. the height mismatch between the ordered domains and the surrounding fluid phase. AFM force spectroscopy experiments performed on hydrated SLB showed that cholesterol decreases the rupture force F_B of the membrane. In SLB composed of milk-SM/DOPC/cholesterol (50/50/0 and 40/40/20% mol), the addition of cholesterol significantly lowered the breakthrough force F_B corresponding to a decreased resistance of the bilayer to perforation, and revealed the fluidizing effect of cholesterol on milk-SM/DOPC bilayers (Guyomarc'h et al., 2014). The addition of cholesterol in a complex mixture of polar lipids extracted from the MFGM has also been demonstrated to affect the topography of the membranes, both in Langmuir-Blodgett monolayers (Murthy et al., 2015) and in hydrated bilayers (Murthy et al., 2016a).

In monolayers composed of the MFGM polar lipid extract, Murthy et al. (2015) revealed that cholesterol plays two major roles on the lateral packing of the polar lipids, first through reducing their molecular area (attractive interactions leading to a condensing effect) and second through dispersing the domains into scattered subunits in the continuous phase (fluidizing effect on the domains). The increased addition of cholesterol up to 30% mol of polar lipids (milk-SM/cholesterol; 50/50% mol) induces a decrease in the area of the gel phase (liquid-condensed) milk-SM-rich domains, as a result of the condensing effect of the cholesterol, and the formation of a number of small domains, as a result of the fluidizing and therefore dispersing effect of the cholesterol. Moreover, the height difference H between the domains and the continuous fluid (liquid-expanded) phase decreased. These results were interpreted in terms of nucleation effects of cholesterol and decrease of the line tension between the domains and the fluid phase, hence increasing the number of the domains and decreasing their size, as revealed both in monolayers and in bilayers (Murthy et al., 2015, 2016a). As a consequence of the fluidizing effect of cholesterol on the milk-SM rich domains, the addition of cholesterol in a complex mixture of polar lipids extracted from the MFGM affects the mechanical properties of the membranes, by decreasing the resistance of the membrane, and especially the domains, to rupture (Murthy et al., 2016b).

5 Updated Model of the MFGM

Figure 3.12 shows an updated model for the organization of polar lipids and proteins in the MFGM. Recent findings revealed that the MFGM is a dynamic system and a patchwork of polar lipids. The complex composition of MFGM polar lipids leads to lipid lateral phase separation in the outer bilayer of the membrane according to their physical state, i.e. ordered lipid domains rich in milk-SM and cholesterol surrounded by a fluid phase of the unsaturated polar lipids PC, PE, PI and PS. These ordered lipid domains are higher and more rigid than the surrounding fluid matrix. The phase separation of polar lipids also corresponds to heterogeneities in the charge provided by the different classes of polar lipids, milk-SM and PC are zwitterionic while PS and PI are anionic (Fig. 3.2). The proteins are embedded in the fluid matrix composed of the unsaturated polar lipids. These heterogeneities in the phase state and charge of polar lipids in the outer bilayer of the MFGM could modulate the interactions with the proteins such as milk the milk proteins (casein micelles, whey proteins) or enzymes.

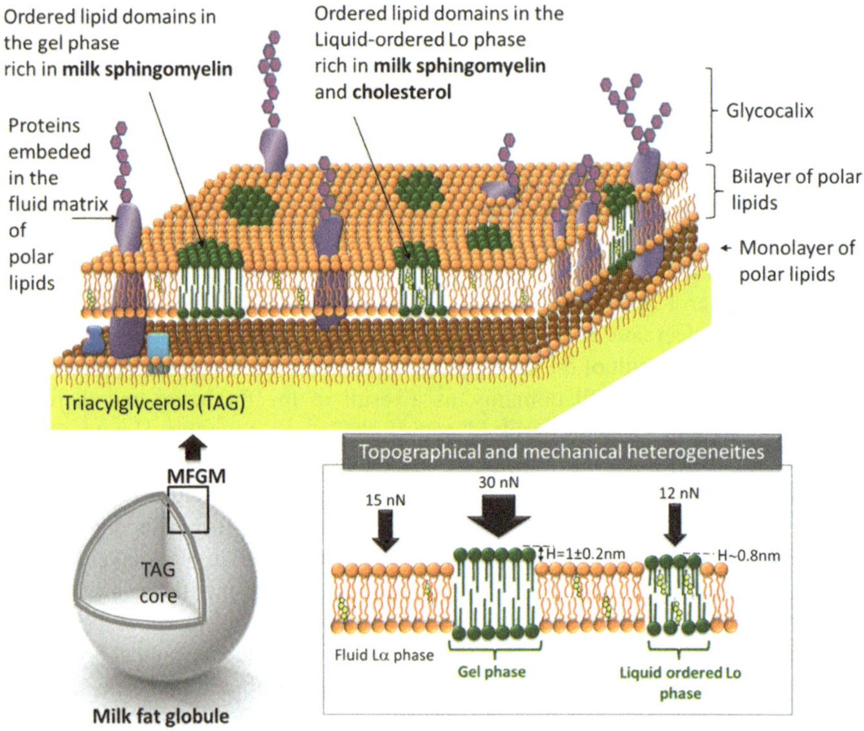

Fig. 3.12 Updated model of the milk fat globule membrane (MFGM). The MFGM is a patchwork of polar lipids with a lateral phase separation and formation of ordered lipid domains rich in milk-sphingomyelin and cholesterol. The outer bilayer of the MFGM exhibits topographical and mechanical heterogeneities. The proteins are embedded in the fluid matrix composed of the unsaturated polar lipids. Adapted with permission from Lopez et al. (2015)

6 Conclusions

The biological membrane surrounding fat globules in milk, the MFGM, is a complex and dynamic assembly of polar lipids and membrane-specific proteins. In the last 10 years, the development of microscopy techniques, such as confocal microscopy combined with pertinent fluorescent probes and atomic force microscopy, associated to an increased knowledge of the physical properties of MFGM polar lipids and their interactions with cholesterol has permitted huge advances in the understanding of the structure of the MFGM. It has been considered for a long time that milk polar lipids have similar properties and that they form a continuous fluid matrix embedding proteins in the MFGM. Scientific advances highlighted differences between low Tm polar lipids rich in unsaturated fatty acids that remain fluid and high Tm polar lipids rich in saturated fatty acids, mainly milk-sphingomyelin, that change their physical state as a function of temperature and can interact strongly with cholesterol. The major role played by milk-sphingomyelin in the formation of ordered lipid domains, leading to heterogeneities in the topography and in the mechanical properties of the MFGM, has been evidenced. Further scientific studies are required for a better understanding of the impact of the structure of the MFGM, particularly the presence of ordered domains rich in milk-sphingomyelin, on its biological functions.

References

Bhojoo, U., Chen, M., & Zou, S. (2018). Temperature induced lipid membrane restructuring and changes in nanomechanics. *Biochimica et Biophysica Acta (BBA) - Biomembranes, 1860*, 700–709.

Cheng, K., Ropers, M.-H., & Lopez, C. (2017). The miscibility of milk sphingomyelin and cholesterol is affected by temperature and surface pressure in mixed Langmuir monolayers. *Food Chemistry, 224*, 114–123.

Deeth, H. C. (1997). The role of phospholipids in the stability of milk fat gobules. *Australian Journal of Dairy Technology, 52*, 44–46.

Dewettinck, K., Rombaut, R., Thienpont, N., Le, T. T., Messens, K., & Van Camp, J. (2008). Nutritional and technological aspects of milk fat globule membrane material. *International Dairy Journal, 18*, 436–457.

Douellou, T., Montel, M. C., & Sergentet, D. T. (2017). Anti-adhesive properties of bovine oligosaccharides and bovine milk fat globule membrane-associated glycoconjugates against bacterial food enteropathogens. *Journal of Dairy Science, 100*, 3348–3359.

Et-Thakafy, O., Delorme, N., Guyomarc'h, F., & Lopez, C. (2018). Mechanical properties of milk sphingomyelin bilayer membranes in the gel phase: Effects of naturally complex heterogeneity, saturation and acyl chain length investigated on liposomes using AFM. *Chemistry and Physics of Lipids, 210*, 47–59.

Et-Thakafy, O., Guyomarc'h, F., & Lopez, C. (2017). Lipid domains in the milk fat globule membrane: Dynamics investigated in situ in milk in relation to temperature and time. *Food Chemistry, 220*, 352–361.

Evers, J. M., Haverkamp, R. G., Holroyd, S. E., Jameson, G. B., Mackenzie, D. D. S., & McCarthy, O. J. (2008). Heterogeneity of milk fat globule membrane structure and composition as observed using fluorescence microscopy techniques. *International Dairy Journal, 18*, 1081–1089.

Gallier, S., Gragson, D., Jiménez-Flores, R., & Everett, D. (2010a). Using confocal laser scanning microscopy to probe the milk fat globule membrane and associated proteins. *Journal of Agricultural and Food Chemistry, 58*, 4250–4257.

Gallier, S., Gragson, D., Jiménez-Flores, R., & Everett, D. W. (2010b). Surface characterization of bovine milk phospholipid monolayers by Langmuir isotherms and microscopic techniques. *Journal of Agricultural and Food Chemistry, 58*, 12275–12285.

Gallier, S., Gragson, D., Jiménez-Flores, R., & Everett, D. W. (2012). β-Casein–phospholipid monolayers as model systems to understand lipid–protein interactions in the milk fat globule membrane. *International Dairy Journal, 22*, 58–65.

Gallier, S., Vocking, K., Post, J. A., Van de Heijning, B., Acton, D., Van der Beek, E. M., et al. (2015). A novel infant milk formula concept: Mimicking the human milk fat globule structure. *Colloids and Surfaces. B, Biointerfaces, 136*, 329–339.

Garcia, C., Lutz, N. W., Confort-Gouny, S., Cozzone, P. J., Armand, M., & Bernard, M. (2012). Phospholipid fingerprints of milk from different mammalians determined by 31P NMR: Towards specific interest in human health. *Food Chemistry, 135*, 1777–1783.

Giuffrida, F., Cruz-Hernandez, C., Flück, B., Tavazzi, I., Thakkar, S. K., Destaillats, F., et al. (2013). Quantification of phospholipids classes in human milk. *Lipids, 48*, 1051–1058.

Guyomarc'h, F., Chen, M., Et-Thakafy, O., Zou, S., & Lopez, C. (2017). Gel-gel phase separation within milk sphingomyelin domains revealed at the nanoscale using atomic force microscopy. *Biochimica et Biophysica Acta, Biomembranes, 1859*, 949–958.
Guyomarc'h, F., Zou, S., Chen, M., Milhiet, P.-E., Godefroy, C., Vie, V., et al. (2014). Milk sphingomyelin domains in biomimetic membranes and the role of cholesterol: Morphology and nanomechanical properties investigated using AFM and force spectroscopy. *Langmuir, 30*, 6516–6524.
Heid, H. W., & Keenan, T. W. (2005). Intracellular origin and secretion of milk fat globules. *European Journal of Cell Biology, 84*, 245–258.
Jiménez-Flores, R., & Brisson, G. (2008). The milk fat globule membrane as an ingredient: why, how, when? *Dairy Science & Technology, 88*, 5–18.
Liao, Y., Alvarado, R., Phinney, B., & Lönnerdal, B. (2011). Proteomic characterization of human milk fat globule membrane proteins during a 12 month lactation period. *Journal of Proteome Research, 10*, 3530–3541.
Lopez, C. (2011). Milk fat globules enveloped by their biological membrane: Unique colloidal assemblies with a specific composition and structure. *Current Opinion in Colloid & Interface Science, 16*, 391–404.
Lopez, C., Briard-Bion, V., Beaucher, E., & Ollivon, M. (2008). Multiscale characterization of the organization of triglycerides and phospholipids in emmental cheese: From the microscopic to the molecular level. *Journal of Agricultural and Food Chemistry, 56*, 2406–2414.
Lopez, C., Briard-Bion, V., & Ménard, O. (2014). Polar lipids, sphingomyelin and long-chain unsaturated fatty acids from the milk fat globule membrane are increased in milks produced by cows fed fresh pasture based diet during spring. *Food Research International, 58*, 59–68.
Lopez, C., Briard-Bion, V., Ménard, O., Beaucher, E., Rousseau, F., Fauquant, J., et al. (2011). Fat globules selected from whole milk according to their size: Different compositions and structure of the biomembrane, revealing sphingomyelin-rich domains. *Food Chemistry, 125*, 355–368.
Lopez, C., Briard-Bion, V., Menard, O., Rousseau, F., Pradel, P., & Besle, J.-M. (2008). Phospholipid, sphingolipid, and fatty acid compositions of the milk fat globule membrane are modified by diet. *Journal of Agricultural and Food Chemistry, 56*, 5226–5236.
Lopez, C., Cauty, C., & Guyomarc'h, F. (2015). Organization of lipids in milks, infant milk formulas and various dairy products: Role of technological processes and potential impacts. *Dairy Science & Technology, 95*, 863–893.
Lopez, C., Cauty, C., & Guyomarc'h, F. (2019). Unraveling the complexity of milk fat globules to tailor bioinspired emulsions providing health benefits: The key role played by the biological membrane. *European Journal of Lipid Science and Technology, 121*, 1800201.
Lopez, C., Cauty, C., Rousseau, F., Blot, M., Margolis, A., & Famelart, M.-H. (2017). Lipid droplets coated with milk fat globule membrane fragments: Microstructure and functional properties as a function of pH. *Food Research International, 91*, 26–37.
Lopez, C., Cheng, K., & Perez, J. (2018). Thermotropic phase behavior of milk sphingomyelin and role of cholesterol in the formation of the liquid ordered phase examined using SR-XRD and DSC. *Chemistry and Physics of Lipids, 215*, 46–55.
Lopez, C., Madec, M.-N., & Jimenez-Flores, R. (2010). Lipid rafts in the bovine milk fat globule membrane revealed by the lateral segregation of phospholipids and heterogeneous distribution of glycoproteins. *Food Chemistry, 120*, 22–33.
Lopez, C., & Ménard, O. (2011). Human milk fat globules: Polar lipid composition and in situ structural investigations revealing the heterogeneous distribution of proteins and the lateral segregation of sphingomyelin in the biological membrane. *Colloids and Surfaces B: Biointerfaces, 83*, 29–41.
Luo, J., Huang, Z., Liu, H., Zhang, Y., & Ren, F. (2018). Yak milk fat globules from the Qinghai-Tibetan Plateau: Membrane lipid composition and morphological properties. *Food Chemistry, 245*, 731–737.
Malmsten, M., Bergenståhl, B., Nyberg, L., & Odham, G. (1994). Sphingomyelin from milk-characterization of liquid crystalline, liposome and emulsion properties. *Journal of the American Oil Chemists' Society, 71*, 1021–1026.

Mather, I. H. (2000). A review and proposed nomenclature for major proteins of the milk-fat globule membrane1,2. *Journal of Dairy Science, 83*, 203–247.

Ménard, O., Ahmad, S., Rousseau, F., Briard-Bion, V., Gaucheron, F., & Lopez, C. (2010). Buffalo vs. cow milk fat globules: Size distribution, zeta-potential compositions in total fatty acids and in polar lipids from the milk fat globule membrane. *Food Chemistry, 120*, 544–551.

Mesilati-Stahy, R., & Argov-Argaman, N. (2014). The relationship between size and lipid composition of the bovine milk fat globule is modulated by lactation stage. *Food Chemistry, 145*, 562–570.

Murthy, A. V. R., Guyomarc'h, F., & Lopez, C. (2016a). Cholesterol decreases the size and the mechanical resistance to rupture of sphingomyelin rich domains, in lipid bilayers studied as a model of the milk fat globule membrane. *Langmuir, 32*, 6757–6765.

Murthy, A. V. R., Guyomarc'h, F., & Lopez, C. (2016b). The temperature-dependent physical state of polar lipids and their miscibility impact the topography and mechanical properties of bilayer models of the milk fat globule membrane. *Biochimica et Biophysica Acta - Biomembranes, 1858*, 2181–2190.

Murthy, A. V. R., Guyomarc'h, F., & Lopez, C. (2018). Palmitoyl ceramide promotes milk sphingomyelin gel phase domains formation and affects the mechanical properties of the fluid phase in milk-SM/DOPC supported membranes. *Biochimica et Biophysica Acta (BBA) - Biomembranes, 1860*, 635–644.

Murthy, A. V. R., Guyomarc'h, F., Paboeuf, G., Vié, V., & Lopez, C. (2015). Cholesterol strongly affects the organization of lipid monolayers studied as models of the milk fat globule membrane: Condensing effect and change in the lipid domain morphology. *Biochimica et Biophysica Acta (BBA) - Biomembranes, 1848*, 2308–2316.

Nguyen, H. T. H., Madec, M.-N., Ong, L., Kentish, S. E., Gras, S. L., & Lopez, C. (2016). The dynamics of the biological membrane surrounding the buffalo milk fat globule investigated as a function of temperature. *Food Chemistry, 204*, 343–351.

Nguyen, H. T. H., Ong, L., Beaucher, E., Madec, M.-N., Kentish, S. E., Gras, S. L., et al. (2015). Buffalo milk fat globules and their biological membrane: In situ structural investigations. *Food Research International, 67*, 35–43.

Nilsson, A. (2016). Role of sphingolipids in infant gut health and immunity. *The Journal of Pediatrics, 173*, S53–S59.

Noh, S. K., & Koo, S. I. (2004). Milk sphingomyelin is more effective than egg sphingomyelin in inhibiting intestinal absorption of cholesterol and fat in rats. *The Journal of Nutrition, 134*, 2611–2616.

Nyberg, L., Duan, R.-D., & Nilsson, Å. (2000). A mutual inhibitory effect on absorption of sphingomyelin and cholesterol. *The Journal of Nutritional Biochemistry, 11*, 244–249.

Rasmussen, J. T. (2009). Bioactivity of milk fat globule membrane proteins. *Australian Journal of Dairy Technology, 64*, 63–67.

Reinhardt, T. A., & Lippolis, J. D. (2006). Bovine milk fat globule membrane proteome. *Journal of Dairy Research, 73*, 406–416.

Sanchez-Juanes, F., Alonso, J. M., Zancada, L., & Hueso, P. (2009). Distribution and fatty acid content of phospholipids from bovine milk and bovine milk fat globule membranes. *International Dairy Journal, 19*, 273–278.

Simons, K., & Ikonen, E. (1997). Functional rafts in cell membranes. *Nature, 387*, 569–572.

Singh, H. (2006). The milk fat globule membrane—A biophysical system for food applications. *Current Opinion in Colloid & Interface Science, 11*, 154–163.

Sprong, R. C., Hulstein, M. F. E., Lambers, T. T., & van der Meer, R. (2012). Sweet buttermilk intake reduces colonisation and translocation of Listeria monocytogenes in rats by inhibiting mucosal pathogen adherence. *British Journal of Nutrition, 108*, 2026–2033.

Wooding, F. B. P. (1971). The structure of the milk fat globule membrane. *Journal of Ultrastructure Research, 37*, 388–400.

Wooding, F. B. P., & Mather, I. H. (2017). Ultrastructural and immunocytochemical evidence for the reorganisation of the milk fat globule membrane after secretion. *Cell and Tissue Research, 367*, 283–295.

Wooding, F. B. P., & Sargeant, T. J. (2015). Immunocytochemical evidence for golgi vesicle involvement in milk fat globule secretion. *The Journal of Histochemistry and Cytochemistry, 63*, 943–951.

Yao, Y., Zhao, G., Xiang, J., Zou, X., Jin, Q., & Wang, X. (2016). Lipid composition and structural characteristics of bovine, caprine and human milk fat globules. *International Dairy Journal, 56*, 64–73.

Zheng, H., Jimenez-Flores, R., & Everett, D. W. (2014). Lateral lipid organization of the bovine milk fat globule membrane is revealed by washing processes. *Journal of Dairy Science, 97*, 5964–5974.

Zou, X., Guo, Z., Jin, Q., Huang, J., Cheong, L., Xu, X., et al. (2015). Composition and microstructure of colostrum and mature bovine milk fat globule membrane. *Food Chemistry, 185*, 362–370.

Zou, X.-Q., Guo, Z., Huang, J.-H., Jin, Q.-Z., Cheong, L.-Z., Wang, X.-G., et al. (2012). Human milk fat globules from different stages of lactation: A lipid composition analysis and microstructure characterization. *Journal of Agricultural and Food Chemistry, 60*, 7158–7167.

Chapter 4
Preparation and Applications of Milk Polar Lipids/MFGM

Kate M. Barry, Timothy G. Dinan, and Philip M. Kelly

1 Introduction

Milk fat is a mosaic encompassing a large variety of chemically different compounds including tri-glycerides (TG), di-glycerides (DG), and mono-glycerides (MG), cholesterol and cholesterol esters, free fatty acids (FA), glycolipids and phospholipids (PLs) (Jensen, 2002; Pimentel, Gomes, Pintado, et al., 2016). This diversity in lipid species owes to the complexity of the milk fat fraction within the whole milk matrix. A more simplistic categorisation of these different milk fat compounds is based on their individual polarity, i.e. polar or non-polar. Polar lipids are essential for the emulsification of fat in milk (Contarini & Povolo, 2013). PLs constitute the polar lipid fraction and while only accounting for between 0.2% and 1.0% of the total milk lipids (Lopez, 2011) they make a major functional contribution through their inherent ability to emulsify/stabilise the fat in milk.

PLs are primarily located in the biological membrane surrounding milk fat globules i.e. milk fat globule membrane (MFGM) (Lopez, 2011), and in other membranous material of the skim milk phase (Contarini & Povolo, 2013). PLs represent a class of complex bioactive and functional lipids that are subdivided into two primary constituent groups: glycerophospholipids and sphingolipids (Barry, Dinan, Murray, et al., 2016). Extensive investigation and revision of the bio- and techno-functionalities of these complex lipids in recent decades (Contarini & Povolo, 2013;

K. M. Barry
Teagasc Food Research Centre, Moorepark, Fermoy, Co, Cork, Ireland

Department of Psychiatry, University College Cork, Cork, Ireland

T. G. Dinan
Department of Psychiatry, University College Cork, Cork, Ireland

Alimentary Pharmabiotic Centre, University College Cork, Cork, Ireland

P. M. Kelly (✉)
Teagasc Food Research Centre, Moorepark, Fermoy, Co, Cork, Ireland

T. Truong et al. (eds.), *Dairy Fat Products and Functionality*,
https://doi.org/10.1007/978-3-030-41661-4_4

Dewettinck, Rombaut, Thienpont, et al., 2008; Pimentel et al., 2016) has generated considerable interest in these dairy PLs due to their higher individual concentrations of phosphatidylserine (PS) and presence of sphingomyelin (SM) compared to other sources. For this reason, recent advances in the preparation, analytical evaluation and technological and biological applications of these lipids is a key focus of this review.

2 Milk: A General Overview

Milk is a nutritionally complete whole food source naturally produced and secreted by the female mammal to provide all the nutritional requirements for the growing neonate. This complex biological fluid is primarily composed of water, lipids, protein and lactose, with numerous other minor constituents including vitamins, minerals, amino acids and enzymes (Fox & McSweeney, 1998). Milk production is shaped by mammalian genotype-environment interactions, which explains why the compositions of its constituents vary considerably between species and within species due to factors such as seasonality, stage of lactation and feeding; lipids 2–55%, protein 1–20% and lactose 0–10%. This compositional variability is a reflection of both the growth rate and energy requirement of the growing neonate (O'Mahoney & Fox, 2014; Singh & Gallier, 2017). Bovine milk, for example, contains approximately 3.3% fat, 3.2% protein and 5.3% lactose (Fox & McSweeney, 1998).

Dietary energy required by the neonate is provided primarily by the lipid portion of the whole milk and is, thus, a key contributor to the growth and development of the infant. The lipid portion is represented by a huge diversity of compounds. Ninety eight percentage of milk lipids are represented by triglycerides (Lopez, 2011) and are present in the fluid system as spherical entities referred to as milk fat globules, 4–5 μm in diameter, which are enclosed in a biological membrane known as the milk fat globule membrane (MFGM) (Lopez, 2011). The composition and structural integrity of these milk fat globules attributes to their importance for the growth and development of the neonate from both a nutritional and health perspective. Polar lipids and proteins associated with the MFGM contribute to the immunological defence of the neonate exposed to infection from various bacterial and viral origins (Lopez, 2011).

3 Milk Fat Secretion and Formation of Milk Fat Globules

The origin and secretion of milk fat globules has been the focus of many reviews. Milk lipids originate from cytoplasmic lipid droplets, although the mechanism by which this occurs is not yet fully understood. The accepted theory is that triglycerides, synthesised in or on the surface of the endoplasmic reticulum (ER), accumulate in the cytoplasm forming lipid droplets which migrate toward the apical pole of a cell where they are coated in proteins and polar lipids from the ER (Lopez, 2011). From this point onwards, there are two possible mechanisms of fat globule release from the mammary cell. In one case lipid droplets closely approach or come in contact with the apical plasma membrane from where they are gradually completely enveloped by the plasma membrane and are pinched off and dissociate from the cell (Bargmann & Knoop, 1959; Heid & Keenan, 2005). The second proposed mechanism of fat globule secretion involves exocytosis of both lipid droplets and casein micelles in vesicles from the apical surface of the cell whereby the entire surface of the milk fat globule is derived from the secretory vesicle membrane (Keenan & Mather, 2006; Lopez, 2011). The occurrence of sterol and sphingolipid-rich domains on the surface of MFGM, known as lipid rafts, among a liquid-disordered matrix of unsaturated glycerophospholipids owe their origin to the apical plasma membrane where they are known to accumulate (Lopez, 2011; Lopez & Ménard, 2011).

Proteins and phospholipids (PLs) associated with secretory cells are known to play pivotal roles in milk fat secretion. As already mentioned with milk fat globules originating from cytoplasmic lipid droplets, Wu, Howell, Neville, et al. (2000) determined that the protein content of cytoplasmic lipid droplets isolated from secreting cells of mice was very similar to that of the milk fat globule secreted by the mice. Adipophilin, a member of the PAT (perilipin/ADRP/TIP47) family of lipid droplet proteins, is a prominent member of cytoplasmic lipid droplets and has been hypothesised to play an important role in the formation and secretion of milk lipids and also in relation to cellular lipid storage (Chong, Reigan, Mayle-Combs, et al., 2011). Chong et al. (2011) reported that adipophilin acts as an adapter by initiating contact between the cytoplasmic lipid droplet and the inner leaflet of the plasma membrane, thus, allowing for the eventual secretion of the cytoplasmic lipid droplet as a milk lipid globule. Butyrophilin and xanthine oxidoreductase have also been implicated as necessary components for secretion of milk fat from the secretory cell. Protein complexes formed through interactions of butyrophilin and xanthin oxidoreductase concentrated at the apical plasma membrane are thought to be responsible for the secretion of the milk fat globules (Robenek, Hofnagel, Buers, et al., 2006). Utilising free-fracture immunocytochemistry techniques, Robenek et al. (2006) proposed that interactions between plasma membrane butyrophilin and butyrophilin in the PL monolayer were responsible for milk fat globule secretion rather that butyrophilin/xanthine oxidoreductase interactions.

Availability of PLs has been demonstrated as the limiting factor determining the size of milk fat globule secreted. Smoczyński (2017) reported on the correlation between PL availability and milk fat globule size—the greater the level of available PLs, the smaller the milk fat globule produced. This perspective is confirmed by Lopez (2011) who reported a higher MFGM/TAG ratio during the secretion of smaller milk fat globules.

4 Milk Fat Globule Membrane (MFGM)

4.1 MFGM Structure

MFGM, as mentioned earlier, is a biological membrane surrounding the milk fat globule and is 10–20 nm in thickness (Nguyen, Ong, Hoque, et al., 2017). The representation of MFGM as a tripartite structure is well established (Heid & Keenan, 2005) based on an inner surface-active monolayer composed primarily of proteins and polar lipids originating from the endoplasmic reticulum and cytoplasm of the lactating cell and an outer bilayer of polar lipids and proteins from the apical plasma membrane of the secretory mammary epithelial cell (Evers, Haverkamp, Holroyd, et al., 2008; Lopez, 2011). MFGM proteins represent 40% of the overall composition of the membrane, while lipids constitute the final 60% (Dewettinck et al., 2008). Numerous reviews have been published on the composition and structure of the MFGM (Anderson and Caeston 1975; Dewettinck et al., 2008; Evers et al., 2008; Keenan & Dylewski, 1995; Mather & Keenan, 1998; McPherson & Kitchen, 1983); however, values are highly variable due to the use of different isolation and purification methods as well as the analytical protocols conducted (Dewettinck et al., 2008).

Microscopy techniques have been utilised to gain insights into the microstructure of the MFGM since the 1950s, but it required the capability of the electron microscope to reveal MFGM as a true bilayer that originated during milk secretion (Bargmann & Knoop, 1959; Bauer, 1971). In more recent years, non-destructive microscopic techniques such as confocal laser scanning microscopy (CLSM) performed in situ in milk have revealed more extensive details of the MFGM microstructure. Utilising CLSM and with fluorescent lipophilic probes and lectin, Evers et al. (2008) revealed structural and chemical heterogeneity of the MFGM within individual globules and also between species (bovine and human). Lopez, Briard-Bion, Ménard, et al. (2011) employed Rh-DOPE, a fluorescent phospholipid probe to determine the distribution of polar lipids and MFGM proteins in the MFGM structure. This approach revealed the lateral segregation of SM, highly saturated PLs and cholesterol in liquid ordered domains surrounded by glycerophospholipids in a liquid disordered phase (Lopez, 2011; Lopez et al., 2011; Zou, Guo, Jin, et al., 2015). Full details of MFGM composition and structure are presented in Chap. 3 of this book.

4.2 Proteins of MFGM

Adoption of the intricate science of proteomics has enabled extensive characterisation of the protein fraction of the MFGM particularly with respect to bovine milk (Affolter, Grass, Vanrobaeys, et al., 2010; Fong & Norris, 2009; Reinhardt & Lippolis, 2006). As many as forty, mainly glycosylated proteins have been identified utilising a combination of gel electrophoresis and isoelectric focusing. Some of these proteins have been further characterised by means of sequence analytic capabilities (Singh, 2006). More recently, Le, Debyser, Gilbert, et al. (2013) identified a total of 225 proteins associated with the MFGM by utilising liquid chromatography-tandem mass spectrometry (LC-MS/MS) (Le et al., 2013). Mucin 1 (MUC 1), butyrophilin (BTN), mucin 15 (MUC 15), and cluster of differentiation 36 (CD36) are heavily glycosylated transmembrane proteins (Dewettinck et al., 2008; Mather, 2000; Singh & Gallier, 2017). Periodic acid/ Schiff 6/7 (PAS 6/7) while also heavily glycosylated are only loosely attached to the MFGM (Vanderghem, Francis, Danthine, et al., 2011). Adipophilin (ADPH), located on the inner face of the bilayer, xanthine dehydrogenase/oxidase (XDH/OH) located in dense proteinaceous layer of the MFGM and fatty acid binding protein (FABP) located close to the lipid core represent the other three major MFGM proteins (Dewettinck et al., 2008; Singh & Gallier, 2017). MFGM proteins, while small (1–2%) in overall representation of the total milk protein concentration (Riccio, 2004) possess important biofunctionalities including anti-cancer activity (Dewettinck et al., 2008; Spitsberg & Gorewit, 2002), anti-adhesion effects and inhibition of pathogenic bacteria growth (Clare, Zheng, Hassan, et al., 2008; Hirmo, Kelm, Iwersen, et al., 1998) as well as regulatory roles in both milk fat globule secretion and lipid metabolism (Harmon & Abumrad, 1993; Nguyen et al., 2017; Vorbach, Scriven, & Capecchi, 2002).

4.3 Lipids of MFGM

The lipid composition of the MFGM has been the focus of numerous reviews (Dewettinck et al., 2008; Keenan & Mather, 2006; Lopez, Briard-Bion, Ménard, et al., 2008; MacGibbon & Taylor, 2006; Mather & Keenan, 1998) (Table 4.1). MFGM is represented by both polar and neutral lipids, the latter being represented by mono-, di- and triglycerides, cholesterol and cholesterol esters. The amount of neutral lipid present is variable, and concentration very much depends on the isolation method used when separating MFGM (Dewettinck et al., 2008; Singh, 2006; Walstra, 1985).

The polar lipid fraction of the MFGM is constituted by PLs which are inherently amphiphilic in nature due to their unique structural design featuring a hydrophobic fatty acid tail and a hydrophilic head (Barry et al., 2016). This unique attribute facilitates the formation of bilayers, thus underpinning their structural and functional ability to emulsify fat in milk (Barry et al., 2016; Deeth, 1997; Dewettinck et al., 2008; Heid & Keenan, 2005). The major PLs of MFGM: phosphatidylethanolamine (PE), phosphatidylcholine (PC) and sphingomyelin (SM) represent the zwitterionic species, while the anionic phosphatidylserine (PS) and phosphatidylinositol (PI) species represent the minor PLs (Availli & Contarini, 2005; Barry et al., 2016; Contarini & Povolo, 2013; Singh, 2006). Reports of phosphatidic acid (PA) are rare as its presence is mostly attributed to phospholipase activity or poor sample handling or storage (Barry et al., 2016). SM represents the major

Table 4.1 Lipid composition of bovine MFGM (data adapted from Keenan and Mather (2006))

Lipid class	% of total lipid
Triglycerides	62
Diacylglycerides	9
Monoacylgylcerides	Traces
Sterols	0.2–2
Sterol esters	0.1–0.3
Free fatty acids	0.6–6
Phospholipids	26–31
Phosphatidylethanolamine	27
Phosphatidylcholine	36
Sphingomyelin	22
Phosphatidylserine	4
Phosphatidylinositol	11

sphingolipid while all other PLs are categorised as glycerophospholipids (Barry, Dinan, & Kelly, 2017a; Sánchez-Juanes, Alonso, Zancada, et al., 2009).

Glycerophospholipid Structure The structural conformation of glycerophospholipids consists of a glycerol backbone to which two fatty acids are esterified at the sn-1 and sn-2 positions. These fatty acids are generally unsaturated in nature with and O-acyl, O-alkyl or O-alk-1'enyl residues at sn-1 and an O-acyl at sn-2 (Verardo, Gómez-Caravaca, Arráez-Román, et al., 2017). At the sn-3 position, a phosphate head group is attached, and it is here that the distinguishing polar head group ethanolamine, choline, serine etc. may be linked (Barry et al., 2016). The fatty acids attached to the glycerol backbone are generally longer chain with C14:0 being the shortest chain occurring in significant amounts (MacGibbon & Taylor, 2006). Saturated fatty acids, C16:0, C18:0 and C18:1 are predominantly positioned at the sn-1 position while polyunsaturated, C18:2 and C18:3 are attached at the sn-2 position (Dewettinck et al., 2008; MacGibbon & Taylor, 2006). In order of degree of saturation, PE is highly unsaturated (26%), followed by PI and PS (37–50%) with PC having the greater degree (50%) of saturation (Dewettinck et al., 2008; MacGibbon & Taylor, 2006). The unsaturated nature of PLs yields a low melting point and affords fluidity in the membrane "liquid-disordered state" (Lopez et al., 2011). Utilising TLC and gas chromatography, Morrison, Jack, and Smith (1965) determined the distribution of the fatty acids in the diacylglycerophospholipids of bovine milk PLs by phospholipase A hydrolysis (Table 4.2) (Morrison et al., 1965).

Sphingolipid Structure The sphingoid base, e.g. sphingosine, forms the foundation block of sphingolipids (Fig. 4.1). Sphingosine is a long-chain aliphatic amine with two or three hydroxyls (Barry et al., 2017a; Verardo et al., 2017). A ceramide is then formed through attachment of two fatty acid molecules to the amino group of the base. An organophosphate group in the case of SM, phosphocholine attaches to form a sphingophospholipid (Availli & Contarini, 2005; Donato, Cacciola, Cichello, et al., 2011) (Fig. 4.1). SM is highly saturated in comparison to the glyc-

Table 4.2 Fatty acid distribution of bovine diacyglcerophospholipids (% mol)

	Phosphatidylethanolamine		Phosphatidylserine		Phosphatidylcholine	
Fatty acid	sn-1	sn-2	sn-1	sn-2	sn-1	sn-2
C14:0	1.9	1.3	4.0	2.2	5.6	10.8
C16:0	19.7	4.7	24.5	7.6	41.9	30.6
C16:1	1.2	2.2	1.7	2.3	0.6	1.2
C18:0	19.0	1.3	28.0	3.9	17.5	2.4
C18:1	45.8	47.8	25.8	46.2	20.3	27.8
C18:2	2.9	21.4	1.6	18.3	2.7	9.2
C18:3	1.1	4.5	1.2	3.0	0.8	1.8
C20:3	0.2	2.2	1.2	2.0	–	1.6
C20:4	0.2	3.0	1.1	2.7	0.2	1.2

Data adapted from Morrison et al. (1965)

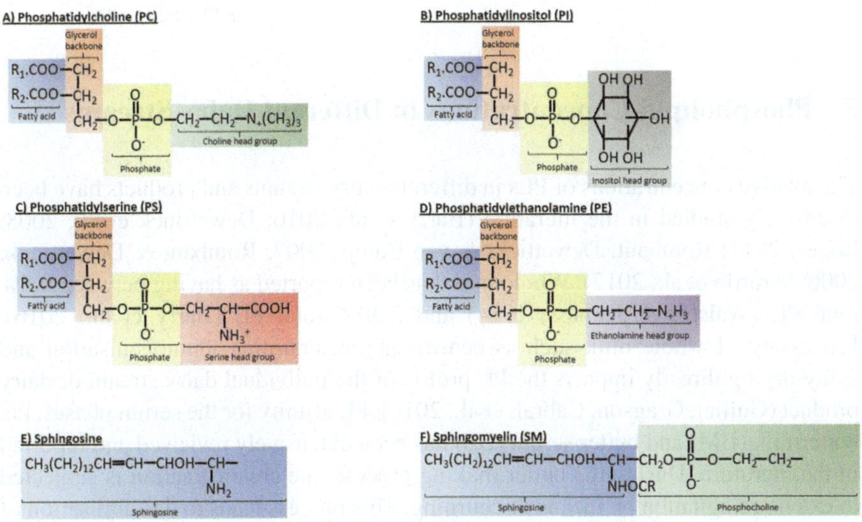

Fig. 4.1 Structure of glycerophospholipids (**a–d**) and sphingoid backbone (**e, f**) (Contarini & Povolo, 2013)

erophospholipids with approximately 97% of the fatty acids attached represented by saturated fatty acids (C16:0, C22:0, C23:0) and is even more unique due to the presence of the long-chain saturated fatty acid C23:0 (Table 4.3) which is not readily identifiable in the glycerophospholipid structure. (Bitman & Wood, 1990). This high degree of saturation enables SM to pack tightly together in a "liquid ordered state" to form rigid domains referred to as lipid rafts which are known to be involved in numerous cellular mechanisms such as signal transduction, cholesterol trafficking and endocytosis (Brown & London, 2000; Dewettinck et al., 2008; Van Meer & Lisman, 2002).

Table 4.3 Fatty acid composition of sphingomyelin

Fatty acid	% w w^{-1}
C14:0	4.1
C15:0	1.1
C16:0	36.1
C16:1	0.6
C17:0	1.5
C18:0	8.7
C18:1	1.5
C18:2	0.2
C20:0	1.2
C21:0	1.5
C22:0	14.4
C23:0	17.4
C24:0	11.3

Data adapted from Bitman and Wood (1990)

5 Phospholipid Concentrations in Different Dairy Streams

The relative concentrations of PLs in different dairy streams and products have been extensively studied in the literature (Barry et al., 2016; Dewettinck et al., 2008; Lopez, 2011; Rombaut, Dewettick, & van Camp, 2007; Rombaut & Dewettinck, 2006; Verardo et al., 2017). Whole milk has been reported as having between 0.8% total PL (Walstra & Jenness, 1984) and 2.30% total PL (Barry et al., 2016). Processing of whole milk such as centrifugation, churning, homogenisation and spray-drying directly impacts the PL profile of the individual dairy stream or dairy product (Gallier, Gragson, Cabral, et al., 2010). PL affinity for the serum phases, i.e. buttermilk (BM) and butter serum (BS) has been extensively reviewed and reported in the literature. During the butter making process, the cream fraction is subjected to extensive agitation by means of churning. This process leads to destabilisation of the cream emulsion and complete disruption of MFGM resulting in a phase separation of butter and buttermilk. The water-soluble components of the MFGM, including the PLs, are no longer associated with the membrane but instead separate into the aqueous buttermilk phase while the triglycerides and other non-polar components form the butter phase (Barry et al., 2016; Gallier et al., 2010). In a recent study (Barry et al., 2016) using simulated processing of unpasteurised whole milk, PL migration into the different dairy streams beginning with whole milk were followed during the various separation processes (Table 4.4). Preferential distribution of the PLs into the aqueous or serum phase streams such as the BM and BS occurred as a PL concentration effect was observed in these fractions, 35.32% total PL in the BM and 46.09% total PL in the BS (Barry et al., 2016).

While BS has the highest concentration of PLs, it is not a widely available dairy stream. For this reason, BM is a more attractive source from which to obtain or enrich dairy PLs. Worldwide BM production is estimated to be equivalent to that of

Table 4.4 Phospholipid composition of different dairy streams

Sample	% fat	% PL (as a % of fat)
Milk	4.08	2.30
Skim milk	0.74	11.07
Cream	31.53	0.37
Butter	78.17	0.09
Buttermilk	2.80	35.32
Butter oil	99.82	0.01
Butter serum	2.15	46.09

Data adapted from Barry et al. (2016)

butter manufacture which was reported as 5.17 million tonnes in 2014 (FAOSTAT, 2015). At present, BM in the form of spray dried buttermilk powder, though widely utilised in the food and beverage industry because of its emulsification properties and enhancement of dairy flavours, is generally classed as a low value by-product of butter making. Hence, the opportunity exists to harness BM as a unique source of PLs in order to generate PL enriched dairy ingredients or fractions for functional and technological applications as well as for nutritional and health benefitting purposes. For the most part, unless otherwise stated, the succeeding sections of this review will be related to BM as a source of PLs.

6 Analytical, Preparation, Determination and Quantification of Dairy Phospholipids

Due to the growing interest in dairy PLs and their applications, greater attention is being paid to the development and utilisation of improved analytical procedures in recent decades. Quantitative determination of PLs generally requires a two-pronged approach; firstly, extraction of the crude lipid by means of a solvent-based extraction method and secondly, quantitative determination of the PL species by means of chromatographic separation. Characterisation and quantification of dairy PLs from different dairy streams has been the focus of numerous publications except that there is a high degree of variability in the measured values most likely due to different analytical approaches utilised for both the total lipid extraction procedure and the quantification step.

6.1 Total Lipid Extraction Procedure

The importance of the initial extraction procedure is paramount to accurate quantification of PLs. Milk, due to its high water content and presence of protein, fat and lactose, is by its nature a very complex food matrix. The amphiphilic nature of PLs and the fact that they interact with both protein and lipids further exacerbates the importance of the extraction procedure (Contarini & Povolo, 2013). The Röse Gottlieb-based official ISO standard for total fat extraction from milk (IDF, 1987) is based on disruption of the lipo-protein bonds by ammonia allowing complete dissolution of the lipids in ether. However, this method of lipid extraction is unreliable for quantitative analysis of PLs as losses, particularly in the case of PI and PS, have been reported in the literature (Barry et al., 2016; Le, Miocinovic, Nguyen, et al., 2011). These losses have been attributed to the role of the ammonia in increasing the water solubility of these more acidic PLs (Availli & Contarini, 2005; Barry et al., 2016; Pimentel et al., 2016). The thermal load applied in this extraction procedure is also a contributory factor for PL losses due to induced hydrolysis and oxidation of the PLs (Barry et al., 2016; Gallier et al., 2010). The preferred extraction procedures for PL analysis are based on that of Folch, Lees, and Stanley (1957) utilising a cold chloroform/methanol-based solvent extraction system. This protocol utilises an initial binary system of chloroform and methanol in a 2:1 ratio (2:1 v v^{-1}) to extract the majority of the lipid. A second washing step can include chloroform alone or a salt solution to extract the remaining lipid material. While the salt induces a phase separation referred to as a "salting out" effect (Le et al., 2011), it can result in losses of PLs as they may be retained in the upper phases as has been previously reported (Akoh & Min, 2008; Barry et al., 2016; Christe, 2003). Barry et al. (2016) demonstrated that an adaptation of the Folch method published by Rodrigues-Alcalá and Fontech (2010) utilising the initial binary system of chloroform:methanol (2:1 v v^{-1}) followed by a secondary chloroform wash yielded the greatest PL recovery when compared to Rose Gottlieb (IDF, 1987) or Folch et al. (1957) with an incorporated salt solution. The BD method, developed originally by Bligh and Dryer (1959), is a similar chloroform/methanol-based extraction protocol except that it takes into account the amount of water in the system (both added water and that present in the sample). Other lipid extraction methods have been developed to overcome the issue of toxicity when utilising chloroform. Solvents such as dichloromethane (Cequier-Sánchez, Rodríguez, Ravelo, et al., 2008), hexane/isopropanol (Hara & Radin, 1978) and methyl tert-butyl ether (MTBE) (Matyash, Liebisch, Kurzchalia, et al., 2008) have all been utilised successfully for lipid extraction however to date the Folch method is still the most preferred method for downstream analysis of PLs.

6.2 *Fractionation of Lipid Species*

Prior to quantitative analysis, there may be a need to further isolate or purify the PL fraction from the crude lipid extract due to a low PL concentration or as a prerequisite to the analytical procedure. Isolation or "clean up" procedures most frequently utilised to separate complex lipid mixtures include thin layer chromatography (TLC) and open or solid phase chromatography (SPE) (Contarini & Povolo, 2013). The principle of TLC is a plate separation based upon polarity and the preferential affinity of the compound of interest for either the stationary (polar) or solvent mobile (non-polar) phases (Fuchs, Süß, Teuber, et al., 2011) and is employed to separate the PL fraction from the non-polar or neutral lipids (Gentner, Bauer, & Dieterich, 1981). A silica precoat is the most commonly used stationary phase. TLC is relatively inexpensive but has drawbacks in that compounds of interest can become oxidised due to exposure of a large surface area and contamination of the sample can occur during scraping whilst recovering the fraction from the TLC plate (Fuchs et al., 2011; Pimentel et al., 2016). Two-dimensional TLC may also be utilised to improve separation of different PL species as demonstrated by MacKenzie, Vyssotski, and Nekrasov (2009).

Solid phase extraction (SPE) as mentioned above is also commonly utilised to separate the PL fraction from complex lipid mixtures. This method of separation is based upon application of the lipid mixture to a column packed with a sorbent e.g. silica and NH_2 for normal phase and C8 or C18 for reversed phase, from which different lipid classes are eluted by changing the relative polarity of the elution system with different solvents. Generally, the non-polar compounds are eluted early in the procedure with the more polar compounds eluting in the later stages with increasing solvent polarity during normal phase SPE. For reverse phase, the opposite occurs in that the polar compounds elute at the beginning (Contarini & Povolo, 2013). Availli and Contarini (2005) achieved 96% PL yield compared to 47.5% PL for a normal phase silica column and reversed phase C8 column, respectively (Availli & Contarini, 2005). Furthermore, the authors demonstrated the importance of the eluting solvents polarity in recovering the PLs from the column. Unsatisfactory results (61% PL yield) were obtained when methanol only was applied to the column. However, subsequent application of a chloroform/ methanol/ water mixture (3:5:2, vvv^{-1}) to the column increased the polarity of the eluting solvent and, thus, improved PL recovery to 96% PL yield (Availli & Contarini, 2005). Contrary to this, Caboni, Menotta, and Lercker (1996) demonstrated that a C8 reverse phase column yielded an even higher recovery (97% PL) compared to 70% PL using a silica column (Caboni et al., 1996).

6.3 Identification and Quantification of Dairy PLs

The demand for quick and reliable analytical methods for quantification of dairy PLs and PLs from other food sources has propelled the development of different screening strategies in recent decades. Quantitative determination of PLs and individual PL species in a dairy sample is most commonly conducted by high performance liquid chromatography (HPLC). Numerous applications of HPLC for PL analysis have been developed employing different detection and solvent systems. HPLC coupled to an evaporative light scattering detection (ELSD) system is the widely used for chromatographic separation, identification and quantification of PLs (Barry et al., 2016; Restuccia, Spizzirri, Puoci, et al., 2011) and has been regularly featured in publications focusing on PL determination in milk and dairy products (Availli & Contarini, 2005; Caboni et al., 1996; Donato et al., 2011; Fagan & Wijesundera, 2004; Le et al., 2011; Lopez et al., 2008; Rodrigues-Alcalá & Fontech, 2010; Rombaut, Dewettick, & van Camp, 2007; Rombaut, van Camp, & Dewettick, 2005). The working principle of the ELSD is based upon generation of an aerosol of analyte particles utilising a nebulising or inert carrier gas, usually N_2, which is then passed through a light beam and the level of light scattering is related to the amount of analyte present. However, to overcome the disadvantages of the ELSD such as low sensitivity and selectivity (Verardo et al., 2017), a detector based on a similar principle has been developed in recent years i.e. charged aerosol detector (CAD). The CAD utilises corona discharge to charge the aerosol of analyte particles which gives better precision and sensitivity compared to the ELSD during analysis of dairy PLs (Barry et al., 2016; Contarini & Povolo, 2013; Kielbowicz, Micek, & Wawrzenczyk, 2013). However, one limitation with both the ELSD and CAD is the requirement for expensive pure PL standards in order to generate a calibration curve from which a concentration determination can be made.

Other analytical strategies employed for quantitative determination of dairy PLs include ^{31}P NMR and mass spectrometry techniques such as matrix-assisted laser-desorption/ ionisation (MALDI)-time of flight (TOF) mass spectrometry. ^{31}P NMR, due to its specificity for the measurement of phosphorus-containing compounds such as PLs, is becoming an ever more popular method for routine determination of PLs as it does not require prior sample extraction using detergents and, consequently, provides quick sample throughput and turn-around of results. (Gallier, Gordon, & Singh, 2012; Garcia, Lutz, Confort-Gount, et al., 2012; MacKenzie et al., 2009). Another advantage of employing NMR for PL analysis is that it does not require expensive standards for determination. However, ^{31}P NMR is limited when applied to samples with a low concentration of PLs (MacKenzie et al., 2009). Calvano, de Ceglie, Aresta, et al. (2013) employed MALDI-TOF for rapid analysis of PLs in different dairy streams as a means of monitoring milk fraud in the dairy industry (Calvano et al., 2013).

7 Enrichment of Dairy Phospholipids

As mentioned earlier, PLs are very attractive compounds due to both their techno-functional capabilities and nutritional potentialities. For this reason, in the past number of years, there has been a determined effort to investigate and develop technologies and processes to enrich dairy PLs using different dairy streams. BM, as already discussed, is naturally rich in PLs and has thus been the starting substrate in many enrichment studies. Numerous enrichment strategies have been explored at both the laboratory and pilot-scale with much of the focus primarily concentrated on membrane filtration-based separation strategies. Complementary measures such as enzymatic digestion, cream washing, and precipitation/coagulation have also been employed in combination or alone to enrich dairy PLs. More recently, alternative separation strategies such as supercritical fluid extraction have been utilised to enrich dairy PLs.

Microfiltration (MF), due to its porosity and selectivity to remove caseins, has been the most widely utilised membrane process for the purposes of PL enrichment and has been the kingpin of many membrane filtration studies and extensive reviews (Astaire, Ward, German, et al., 2003; Corredig, Roesch, & Dalgleish, 2003; Dewettinck et al., 2008; Holzmüller & Kulozik, 2016; Morin, Jiménez-Flores, & Pouliot, 2004; Morin, Pouliot, & Jiménez-Flores, 2006; Rombaut, Dejonckheere, & Dewettinck, 2007; Surel & Flamelart, 1995). However, even allowing for its greater porosity, MF was incapable of differentiation due to the near similarity in size between casein micelles and MFGM. It has been postulated that interactions between the casein micelles and MFGM may be responsible for the mediocre separation and enrichment of PLs by means of MF (Barry et al., 2017a; Morin, Britten, Jiménez- Flores, et al., 2007; Morin, Jiménez- Flores, & Pouliot, 2007; Sachdeva & Buchheim, 1997). Other factors such as the processing temperature (Astaire et al., 2003; Morin et al., 2004), the membrane pore size and membrane configuration (Morin et al., 2004), the pH at the time of processing (Rombaut, Dejonckheere, & Dewettinck, 2007) and the substrate type, i.e. different types of BM utilised (Astaire et al., 2003; Morin et al., 2006; Morin, Britten, et al., 2007) also impact on the transmission efficiency during MF of PL-based feedstocks.

To overcome the drawbacks experienced by MF for PL enrichment, complementary pre-treatments of the substrate prior to membrane filtration have been explored. Addition of citrate prior to filtration to disrupt the casein micelle has been demonstrated to improve the casein transmission during MF (Corredig et al., 2003; Roesch, Rincon, & Corredig, 2004; Rombaut, Dejonckheere, & Dewettinck, 2006); however losses of PL material due to membrane fouling has been encountered (Rombaut et al., 2006). Utilisation of rennet coagulation of the caseins as seen in cheese making has been adapted for the purposes of PL enrichment upstream of the membrane filtration process however significant PL losses (20%) in the rennet curd were observed (Sachdeva & Buchheim, 1997). Another approach that has been exploited is cream washing prior to churning into butter and buttermilk. Utilising skim milk ultrafiltrate to wash the cream has been demonstrated to successfully reduce the protein content of the resulting BM by 74–80% (Britten, Lamothe, & Robitaille, 2008; Holzmüller, Müller, Himbert, et al., 2016; Morin, Jiménez- Flores, & Pouliot, 2007), however, this approach also results in severe losses of PLs (Britten et al., 2008).

More recently, enzymatic hydrolysis of the native skim milk proteins present in the BM source prior to ultrafiltration (UF) has been investigated. Konrad, Kleinschmidt, and Lorenz (2013) combined peptic hydrolysis of a whey BM substrate with UF and generated a PL concentrate with a 2.5% increase in PL content compared to the starting material. In a study performed by Barry et al. (2017a) the authors exploited enzymatic hydrolysis to extensively digest the skim milk proteins present in a reconstituted buttermilk powder substrate to produce smaller peptides that could permeate a 50 kDa UF membrane and generate a PL enriched retentate with a 7.8-fold increase in PL material compared to the starting material.

Supercritical fluid extraction (SFE), a more advanced separation technology, has also been explored as a means of "clean extraction" to selectively enrich dairy PLs (Astaire et al., 2003; Barry, Dinan, & Kelly, 2017b; Costa, Elias-Argote,

Jiménez- Flores, et al., 2010; Spence, Jiménez-Flores, Qian, et al., 2009). SFE utilises a non-organic extraction solvent, typically carbon dioxide, above its critical temperature and pressure to selectively isolate a compound of interest. In the case of isolating lipid mixtures with different inherent polarities, the solvating power of the extraction solvent can be altered by addition of co-solvents (e.g. ethanol) to increase the polarity of the extraction solvent (Barry et al., 2017b). Reduction of the non-polar lipid content by SFE has yielded PL concentrates with a 3.5-fold and a 4.3-fold increase in PL content following MF of sweet and whey cream BMs, respectively (Spence et al., 2009). Costa et al. (2010) obtained a PL concentrate with a fivefold increase in PL by utilising SFE following UF of a whey buttermilk substrate. Barry et al. (2017b) pushed the boundaries of previous PL enrichment by combining SFE with other technologies to generate a highly-enriched extract, containing 56.24% PL in a total MFGM lipid isolate (Barry et al., 2017b).

As mentioned earlier, the BM source also impacts the level of PL enrichment obtained. Whey BM, a by-product of cheese production, while desirable in that it is devoid of micellar casein has increased susceptibility to lipid oxidation (Morin, Britten, et al., 2007). Differences in processing history of BMs, fresh, pasteurised and reconstituted, has also been shown to impact membrane permeation of MFGM components (Morin et al., 2004).

8 Applications of Dairy Phospholipids

Dairy PLs are gaining considerable interest due to the ever-expanding knowledge of their unique techno- and bio-functional properties. For this reason, dairy PLs have been the focus of many reviews and numerous investigations exploring their health/nutritional benefits and techno-functional potentials in different food applications (Contarini & Povolo, 2013; Dewettinck et al., 2008; Singh & Gallier, 2017; Verardo et al., 2017). The distinctive amphiphilic nature of PLs is responsible for many of these functionalities.

8.1 Techno-Functional Applications

Given their primary function in stabilising fat dispersal in milk, dairy PLs are natural emulsifiers (Corredig & Dalgleish, 1997; Dewettinck et al., 2008; Kanno, 1989). This emulsifying capability was confirmed in a study by Phan, Asaduzzaman, Le, et al. (2013) where comparing the stability of emulsions prepared with buttermilk powder (BMP), skim milk powder (BMP), microfiltered BM (MF-BM) or sodium caseinate (SC). The emulsions prepared with MF-BM demonstrated excellent emulsion stability, lower viscosity and a Newtonian-like flow behaviour which was attributed to the presence of more milk fat globule components, 9.30 ± 0.31% PL, compared to the other emulsions (Phan et al., 2013).

PLs have also been demonstrated to improve the heat stability of dairy emulsions during sterilisation (Kasinos, Le, & Van der Meeren, 2014). Kasinos et al. (2014) compared the effect of a cream residue powder (CRP) and a sweet buttermilk powder (SBP) on the heat stability of recombined evaporated milk emulsions during sterilisation at 121 °C for 15 min. Viscosity and particle size distribution data indicated that addition of CRP and SBP greatly improved the heat stability of the emulsions during heating. CRP, with its higher 6.4% PL content, had a more pronounced effect than SBP with its 2.9% PL (Kasinos et al., 2014). A similar finding was observed by Tran Le, El-Bakry, Neirynck, et al. (2007) after addition of hydroxylated and hydrolysed lecithin's minimised the heat-induced whey protein-casein micelle aggregation.

PLs may be readily used as delivery systems for nutraceuticals through the formation of liposomes. Liposomes are vesicles formed by the self-assembly of amphiphilic molecules and have many potential applications in both the food and pharmaceutical industry for enhancing the efficacy of certain food additives and protection of sensitive compounds, e.g. vitamins (Dewettinck et al., 2008). Utilising a microfluidisation technique, Thompson and Singh (2006) successfully generated liposomes from a PL-enriched MFGM fraction—Phospholac600 (source: Fonterra, NZ), demonstrating that liposomes generated from MFGM derived PLs were structurally more stable than those produced from soy or egg lecithins. The efficacy of PL liposomes as effective encapsulation and delivery modules has already been demonstrated with the encapsulation of tea polyphenols and ascorbic acid (Farhang, Kakuda, & Corredig, 2012; Gülseren, Guri, & Corredig, 2012) thus confirming that these natural vesicles may support numerous potential applications, including drug delivery systems (Singh, 2006).

8.2 Health and Nutritional Applications

The putative health and nutritional benefits of PLs has been articulated in recent years particularly in relation to brain health and development (Contarini & Povolo, 2013; Dewettinck et al., 2008; Singh & Gallier, 2017; Verardo et al., 2017). Dairy PLs, derived from MFGM, are particularly attractive due to their higher levels of phosphatidylserine (PS) compared with other PL sources and the presence of sphingomyelin (SM), which is completely absent from plant and vegetable sources (Burling & Gaverholt, 2008).

8.2.1 Anticancer Activity

SM has been associated with cell regulation and tumour suppression. This action is mediated via the bioactive metabolites, ceramide and sphingosine, which arrest cell growth and induce apoptosis (Parodi, 2003). The anticarcinogenic activity has been most notably demonstrated in the incidence of colon carcinogenesis (Berra,

Colombo, Scottocornola, et al., 2002; Hertervig, Nilsson, Cheng, et al., 2003; Kuchta-Noctor, Murray, Stanton, et al., 2016). However, it must be noted that investigation of the anticancer activity has only been conducted in vitro and that no human clinical trials have been conducted.

8.2.2 Neurological Development and Repair of the Central Nervous System

Numerous studies have demonstrated the importance of PLs in cognitive development and repair. Recent studies in the area of infant nutrition have demonstrated the importance of PLs in cognitive development of the infant (Timby, Domellöf, Hernell, et al., 2014). In a double-blind, randomised controlled trial, Timby et al. (2014) found that infants that were fed a formula supplemented with MFGM had improved cognitive development at year one compared to those fed a standard formula as tested using the Bayley Scales of Infant and Toddler Development. Enhanced neurodevelopment at 4 weeks was observed in piglets fed a diet that included MFGM, prebiotics and lactoferrin (Mudd, Alexander, Berding, et al., 2016). In vitro studies have also been conducted to demonstrate the response elicited when neuronal cells are treated with dairy PLs. Recently Barry, Dinan, Stanton, et al. (2018) conducted a study demonstrating a 43% increase in neuronal stimulation following treatment of neuronal cells with a highly-enriched PL extract compared to those exposed to the control (12%).

Similarly, PLs have been shown to be important components in cognitive repair. Guan, MacGibbon, Fong, et al. (2015) demonstrated improved spatial reference learning and memory in young adult rats fed a diet supplemented with beta serum concentrate (BSC) PLs. A reduction in the level of SM in particular has been associated with the incidence of neurodegenerative diseases such as Alzheimer's Disease and Parkinson's Disease while PS has been demonstrated as a treatment for age-related cognitive decline in Alzheimer's disease (Burling & Gaverholt, 2008; de Chaves & Sipione, 2009; Dewettinck et al., 2008; Parodi, 2003; Pepeu, Pepeu, & Amanducci, 1996).

SM is a prominent component of cell membranes (Cultler & Mattson, 2001) and is a key factor involved in the myelination of the central nervous system (Oshida, Shimizu, Takase, et al., 2003). In fact, SM and its metabolites, cerebroside, are markers for myelination of the sheath surrounding axons (Oshida et al., 2003).

8.2.3 Cardiovascular Disease and Cholesterol Absorption

A study carried out by Wat, Tandy, Kaper, et al. (2009) demonstrated the therapeutic advantages of a high fat diet enriched in dairy PLs against cardiovascular diseases. The authors determined that mice fed a high fat diet supplemented with a phospholipid-rich dairy milk extract (PLRDME) had reduced liver weight, and total liver lipid content compared to those fed a control diet suggesting the hepatoprotective or cardioprotective benefits of a PL rich diet in humans (Wat et al., 2009). Given the strong correlation between the incidence of cardiovascular disease and increased levels of LDL cholesterol, Watanabe, Takahashi, Tanaka, et al. (2011) demonstrated reduced levels of plasma and hepatic cholesterol in mice fed a lipid concentrated-butter serum and a ceramide fraction compared to those fed a control. Milk ceramides were regarded as responsible for down regulation of stearoyl-CoA desaturase—resulting in decreased levels of triacylglycerol and cholesterol in the liver and plasma (Watanabe et al., 2011).

Inhibition of cholesterol intestinal uptake and absorption by SM has been demonstrated (Nyberg, Duan, & Nilsson, 2000) in vivo in rats using a dual-isotope plasma ratio method to monitor the metabolism of a mixture of cholesterol and SM following ingestion. Noh and Koo (2004) demonstrated that milk SM was a significantly more potent inhibitor of intestinal cholesterol absorption than egg SM, most likely due to the higher degree of saturation and longer fatty acid chains of milk SM compared to egg SM.

8.2.4 Bactericidal Effects and Infections

Considering that many bacteria, including bacterial toxins and viruses, utilise glycosphingolipids to bind to cells, it is plausible that sphingolipids have a protective effect against bacterial infections (Dewettinck et al., 2008). Sphingosine was deemed, in an earlier study (Sprong, Hulstein, & van der Meer, 2001), to be a more potent bactericide with sphingolipid exhibiting a significant toxicity against *C. jejuni* and *L. monocytogenes* and to a lesser degree against *E. coli* and *S. enteritidis*.

Dietary supplementation of MFGM/PL-rich and complex lipid ingredients in infants (Timby, Hernell, Vaarala, et al., 2015) and toddlers (Poppitt, McGregor, Wiessing, et al., 2014) have been demonstrated to have protective effects against gastrointestinal infections. In a randomised controlled trial conducted by Timby et al. (2015), it was determined that infants fed a diet supplemented with MFGM had reduced incidence of acute otitis media compared to those fed a control formula. Poppitt et al., 2014 demonstrated a reduced persistence of rotavirus diarrhoea in infants fed a high-ganglioside milk lipid mixture during a preliminary controlled double-blind randomised controlled trial. Thus, while results to date are promising, there is a need for further studies to generate data into the effects of MFGM and dairy PLs against infections.

9 Conclusion

Dairy PLs represent a distinguishable and intriguing fraction of milk fat. Over the past number of years, our knowledge of these complex polar lipids has significantly increased through the comprehensive research undertaken on understanding the origin, analysis, isolation and application of these bioactive and functional lipids.

Advances in food imaging capabilities has unlocked the structural complexity of these amphiphilic molecules and provided insights into their arrangement within the overall MFGM physico-chemical architecture. Improvements in analytical techniques are now supporting efforts to track and quantify PLs in different dairy streams and inter-species variation.

While numerous techno-functional and nutritional applications have been highlighted, boundaries still exist to be explored further in order to exploit the full potential of these bioactive, functional lipids with the emergence of novel technologies and the scope for optimisation or adaptation of existing technologies. With an extensive array of putative health and nutritional benefits already flagged, there is a need for continuing investigation and exploration in both cellular and mammalian systems to uncover the full potential of dairy PLs.

References

Affolter, M., Grass, L., Vanrobaeys, F., Casado B., & Kussmann, M. (2010). Qualitative and quantitative profiling of the bovine milk fat globule membrane proteome. *Journal of Proteomics, 6*, 1079–1088.

Akoh, C. C., & Min, D. B. (2008). *Food lipids: Chemistry, nutrition and biotechnology*. New York, NY: CRC Press.

Astaire, J. C., Ward, R., German, J. B., & Jiménez-Flores, R. (2003). Concentration of polar MFGM lipids from buttermilk by microfiltration and supercritical fluid extraction. *Journal of Dairy Science, 86*, 2297–2307.

Availli, A., & Contarini, G. (2005). Determination of phospholipids in dairy products by SPE/HPLC/ELSD. *Journal of Chromatography. A, 1*, 185–190.

Bargmann, W., & Knoop, A. (1959). Über die morpjologie der milchsekretion lichte-und elekronemikroskopische studien an der milchdrüse der ratte. *Zeitschrift für Zellforschung und Mikroskopische Anatomie, 49*, 344–388.

Barry, K. M., Dinan, T. G., & Kelly, P. M. (2017a). Selective enrichment of dairy phospholipids in a buttermilk substrate through investigation of enzymatic hydrolysis of milk proteins in conjunction with ultrafiltration. *International Dairy Journal, 68*, 80–87.

Barry, K. M., Dinan, T. G., & Kelly, P. M. (2017b). Pilot scale production of a phospholipid-enriched dairy ingredient by means of an optimised integrated process employing enzymatic hydrolysis, ultrafiltration and super-critical fluid extraction. *Innovative Food Science and Emerging Technologies, 41*, 301–306.

Barry, K. M., Dinan, T. G., Murray, B. A., & Kelly, P. M. (2016). Comparison of dairy phospholipid preparative extraction protocols in combination with analysis by high performance liquid chromatography couple to a charged aerosol detector. *International Dairy Journal, 56*, 179–185.

Barry, K. M., Dinan, T. G., Stanton, C., & Kelly, P. M. (2018). Investigation of the neurotrophic effect of dairy phospholipids on cortical neuron outgrowth and stimulation. *Journal of Functional Foods, 40*, 60–67.

Bauer, H. (1971). Ultrastructural observations on the milk fat globule envelope of cow's milk. *Journal of Dairy Science, 55*, 1375–1387.

Berra, B., Colombo, I., Scottocornola, E., & Giacosa, A. (2002). Dietary sphingolipids in colorectal cancer prevention. *European Journal of Cancer Prevention, 1*, 193–197.

Bitman, J., & Wood, D. L. (1990). Changes in milk fat phospholipids during lactation. *Journal of Dairy Science, 73*, 1208–1216.

Bligh, E. G., & Dryer, W. (1959). A raid method of total lipid extraction and purification. *Canadian Journal of Biochemistry and Physiology, 37*, 911–917.

Britten, M., Lamothe, S., & Robitaille, G. (2008). Effect of cream treatment on phospholipids and protein recovery in butter-making process. *International Journal of Food Science and Technology, 43*, 651–657.

Brown, D. A., & London, E. (2000). Structure and function of sphingolipids and cholesterol rich membrane rafts. *The Journal of Biological Chemistry, 275*, 17221–17224.

Burling, H., & Gaverholt, G. (2008). Milk-a new source for bioactive phospholipids for use in food formulations. *Lipid Technology, 20*, 229–231.

Caboni, M. F., Menotta, S., & Lercker, G. (1996). Separation and analysis of phospholipids in different foods with a light-scattering detector. *Journal of the American Oil Chemists' Society, 73*, 1561–1566.

Calvano, C. D., de Ceglie, C., Aresta, A., Facchini L. A., & Zambonin C. G. (2013). MALDI-TOF mass spectrometric determination of intact phospholipids as markers of illegal bovine milk adulteration of high-quality milk. *Analytical and Bioanalytical Chemistry, 405*, 1641–1649.

Cequier-Sánchez, E., Rodríguez, C., Ravelo, Á., & Zárate, R. (2008). Dichloromethane as a solvent for lipid extraction and assessment of lipid classes and fatty acids from samples of different natures. *Journal of Agricultural and Food Chemistry, 56*, 4297–4303.

Chong, B. M., Reigan, P., Mayle-Combs, K. D., Orlicky, D. J., & McManaman, J. L. (2011). Determinants of adipophilin function in milk lipid formation and secretion. *Trends in Endocrinology and Metabolism, 22*, 211–217.

Christe, W. W. (2003). Lipids: Their structure and occurrence. In W. W. Christie (Ed.), *Lipid analysis. Isolation, separation, identification and structural analysis of lipids* (3rd ed., pp. 3–36). Bridgewater, UK: The Oily Press.

Clare, D. A., Zheng, Z., Hassan, H. M., Swaisgood, H. E., & Catignani, G. L.(2008). Antimicrobial properties of milkfat globule membrane fractions. *Journal of Food Protection, 71*, 126–133.

Contarini, G., & Povolo, M. (2013). Phospholipids in milkfat: Composition, biological and technological significance, and analytical strategies. *International Journal of Molecular Sciences, 14*, 2808–2831.

Corredig, M., & Dalgleish, D. G. (1997). Isolates from industrial buttermilk: Emulsifying properties of materials derived from the milk fat globule membrane. *Journal of Agricultural and Food Chemistry, 45*, 4595–4600.

Corredig, M., Roesch, R. R., & Dalgleish, D. G. (2003). Production of a novel ingredient from buttermilk. *Journal of Dairy Science, 86*, 2744–2750.

Costa, M. R., Elias-Argote, X. E., Jiménez- Flores, R., & Gigante, M. L. (2010). Use of ultrafiltration and supercritical fluid extraction to obtain a whey buttermilk powder enriched in milk fat globule membrane phospholipids. *International Dairy Journal, 20*, 598–602.

Cultler, R. G., & Mattson, M. P. (2001). Sphingomyelin and ceramide as regulators of development and lifespan. *Mechanisms of Ageing and Development, 122*, 895–908.

de Chaves, E. P., & Sipione, S. (2009). Sphingolipids and gangliosides of the nervous system in membrane function and dysfunction. *FEBES Letters, 584*, 1748–1759.

Deeth, H. C. (1997). The role of phospholipids in the stability of milk fat globules. *Australian Journal of Dairy Technology, 52*, 44–46.

Dewettinck, K., Rombaut, R., Thienpont, N., Le, T. T., Messens, K., & van Camp, J. (2008) Nutritional and technological aspects of milk fat globule membrane material. *International Dairy Journal, 18*, 436–457.

Donato, P., Cacciola, F., Cichello, F., Russo, M., Dugo, P., & Mondello, L. (2011). Determination of phospholipids in milk samples by means of hydrophilic interaction liquid chromatography coupled to evaporative light scattering and mass spectrometry detection. *Journal of Chromatography. A, 1218*, 6476–6482.

Evers, J. M., Haverkamp, R. G., Holroyd, S. E., Jameson, G. B., Mackenzie, D. D. S., & McCarthy, O. J. (2008). Heterogeneity of milk fat globule membrane structure and composition as observed using fluorescence microscopy techniques. *International Dairy Journal, 18*, 1081–1089.

Fagan, P., & Wijesundera, C. (2004). Liquid chromatographic analysis of milk phospholipids with online pre-concentration. *Journal of Chromatography. A, 1054*, 241–249.

FAOSTAT. (2015). *World cow butter production*. Retreived from http://faostat3.fao.org/download/O/OP/E

Farhang, B., Kakuda, Y., & Corredig, M. (2012). Encapsulation of ascorbic acid in liposomes prepared with milk fat globule membrane-derived phospholipids. *Dairy Science & Technology, 92*, 353–366.

Folch, J., Lees, M., & Stanley, G. H. S. (1957). A simple method for the isolation and purification of total lipids from animal tissue. *The Journal of Biological Chemistry, 226*, 497–509.

Fong, B. Y., & Norris, C. S. (2009). Quantification of milk fat globule membrane proteins using selected reaction monitoring mass spectrometry. *Journal of Agricultural and Food Chemistry, 57*, 6021–6028.

Fox, P. F., & McSweeney, P. L. H. (1998). *Dairy chemistry and biochemistry*. New York, NY: Springer.

Fuchs, B., Süß, R., Teuber, K., Eibisch, M., & Schiller, J. (2011). Lipid analysis by thin-layer chromatography-A review of the current state. *Journal of Chromatography. A, 1218*, 2754–2774.

Gallier, S., Gordon, K. C., & Singh, H. (2012). Chemical and structural characterisation of almond oil bodies and bovine milk fat globules. *Food Chemistry, 132*, 1996–2006.

Gallier, S., Gragson, D., Cabral, C., Jiménez-Flores, R., & Everett, D. W. (2010). Composition and fatty acid distribution of bovine milk phospholipids from processed milk products. *Journal of Agricultural and Food Chemistry, 58*, 10503–10511.

Garcia, C., Lutz, N. W., Confort-Gouny, S., Cozzone, P. J., Armand, M, & Bernard, M. (2012). Phospholipid fingerprints of milk from different mammalians determined by ^{31}P NMR: Towards specific interest in human health. *Food Chemistry, 165*, 1777–1783.

Gentner, P. R., Bauer, M., & Dieterich, I. (1981). Thin-layer chromatography of phospholipids: Separation of major phospholipid classes of milk without previous isolation from total lipid extracts. *Journal of Chromatography. A, 206*, 200–204.

Guan, J., MacGibbon, A., Fong, B., Zhang, R., Liu, K., Rowan, A., et al. (2015). Long-term supplementation with beta serum concentrate (BSC), a complex of milk lipids, during post-natal brain development improves memory in rats. *Nutrients, 7*, 4526–4541.

Gülseren, I., Guri, A., & Corredig, M. (2012). Encapsulation of tea polyphenols in nanoliposomes prepared with milk phospholipids and their effect on the viability of HT-29 human carcinoma cells. *Food Digestion, 3*, 36–45.

Hara, A., & Radin, N. S. (1978). Lipid extraction of tissues with a low-toxicity solvent. *Analytical Biochemistry, 90*, 420–246.

Harmon, C. M., & Abumrad, N. A. (1993). Binding of sulfosuccinimidyl fatty acids to adipocyte membrane proteins: Isolation and ammo-terminal sequence of an 88-kD protein implicated in transport of long-chain fatty acids. *The Journal of Membrane Biology, 133*, 43–49.

Heid, H. W., & Keenan, T. W. (2005). Intracellular origin and secretion of milk fat globules. *European Journal of Cell Biology, 84*, 245–258.

Hertervig, E., Nilsson, A., Cheng, Y., & Duan, R. D. (2003). Purified intestinal alkaline sphingomyelinase inhibits proliferation without inducing apoptosis in HT-29 colon carcinoma cells. *Journal of Cancer Research and Clinical Oncology, 129*, 577–582.

Hirmo, S., Kelm, S., Iwersen, M., Hotta, K., Goso, Y., Ishihara, K., et al (1998). Inhibition of helicobacter pylorisialic acid-specific hemagglutination by human gastrointestinal mucins and milk glycoproteins. *Pathogens and Disease, 20*, 275–281.

Holzmüller, W., & Kulozik, U. (2016). Technical difficulties and future challenges in isolating membrane material from milk fat globules in industrial settings- A critical review. *International Dairy Journal, 61*, 51–66.

Holzmüller, W., Müller, M., Himbert, D., & Kulozik, U. (2016). Impact of cream washing on fat globules and milk fat globule membrane proteins. *International Dairy Journal, 59*, 52–61.

IDF. (1987). *Dried milk, dried whey, dried buttermilk and dried butter serum. Determination of fat content (Röse Gottlieb reference method)*. Brussels, Belgium: International IDF Standard 9C.

Jensen, R. G. (2002). The composition of bovine milk lipids: January 1995 to december 2000. *Journal of Dairy Science, 85*, 295–350.

Kanno, C. (1989). Emulsifying properties of bovine milk fat globule membrane in milk fat emulsion: Conditions for the reconstitution of milk fat globules. *Journal of Food Science, 57*, 1534–1539.

Kasinos, M., Le, T. T., & Van der Meeren, P. (2014). Improved heat stability of recombined evaporated milk emulsions upon addition of enriched dairy by-products. *Food Hydrocolloids, 34*, 112–118.

Keenan, T. W., & Dylewski, D. P. (1995). Intracellular origin of the milk lipid globules and the nature and structure of the milk lipid globule membrane. In P. F. Fox (Ed.), *Advanced dairy chemistry 2: Lipids* (2nd ed., pp. 89–130). London, UK: Chapman and Hall.

Keenan, T. W., & Mather, I. H. (2006). Intracellular origin of milk fat globules and the nature of the milk fat globule membrane. In P. F. Fox & P. L. H. McSweeney (Eds.), *Advanced dairy chemistry 2: Lipids* (3rd ed., pp. 137–171). New York, NY: Springer.

Kielbowicz, G., Micek, P., & Wawrzenczyk, C. (2013). A new liquid chromatography method with charge aerosol detector (CAD) for the determination of phospholipid classes. Application to milk phospholipids. *Talanta, 105*, 28–363.

Konrad, G., Kleinschmidt, T., & Lorenz, C. (2013). Ultrafiltration of whey buttermilk to obtain a phospholipid concentrate. *International Dairy Journal, 30*, 39–44.

Kuchta-Noctor, A. M., Murray, B. A., Stanton, C., Devery, R., & Kelly, P. M. (2016). Anticancer activity of buttermilk against SW480 colon cancer cells is associated with caspase-independent cell death and attenuation of Wnt, Akt, and ERK signalling. *Nutrition and Cancer, 68*, 1234–1246.

Le, T. T., Debyser, G., Gilbert, W., Struijs, K., van Camp, J., de Wiele, T. V., et al. (2013). Distribution and isolation of milk fat globule membrane proteins during dairy processing as revealed by proteomic analysis. *International Dairy Journal, 32*, 110–120.

Le, T. T., Miocinovic, J., Nguyen, T. M., Rombaut, R., van Camp, J., & Dewettinck, K. (2011). Improved solvent extraction procedure and high-performance liquid chromatography-evaporative light-scattering detector method for analysis of polar lipids from dairy materials. *Journal of Agricultural and Food Chemistry, 59*, 10407–10413.

Lopez, C. (2011). Milk fat globules enveloped by their own biological membrane: Unique colloidal assemblies with a specific composition and structure. *Current Opinion in Colloid & Interface Science, 16*, 391–404.

Lopez, C., Briard-Bion, V., Ménard, O., Rousseau, F., Pradel, P., & Besle, J. M. (2008). Phospholipid, Sphingolipid, and fatty acid composition of the milk fat globule membrane are modified by diet. *Journal of Agricultural and Food Chemistry, 56*, 5226–5236.

Lopez, C., Briard-Bion, V., Ménard, O., Beaucher, E., Rousseau, F., Fauquant, J., et al. (2011). Fat globules selected from whole milk according to their size: Different compositions and structure of the biomembrane, revealing sphingomyelin-rich domains. *Food Chemistry, 125*, 355–368.

Lopez, C., & Ménard, O. (2011). Human milk fat globules: Polar lipid composition and *in situ* structural investigations revealing the heterogeneous distribution of proteins and the lateral segregation of sphingomyelin in the biological membrane. *Colloids and Surfaces. B, Biointerfaces, 83*, 29–41.

MacGibbon, A. K. H., & Taylor, M. W. (2006). Composition and structure of bovine milk lipids. In P. F. Fox & P. L. H. McSweeney (Eds.), *Advanced dairy chemistry 2: Lipids* (3rd ed., pp. 1–42). New York, NY: Springer.

MacKenzie, A., Vyssotski, M., & Nekrasov, E. (2009). Quantitative analysis of dairy phospholipids by ^{31}P NMR. *Journal of the American Oil Chemists' Society, 86*, 757–763.

Mather, I. H. (2000). A review and proposed nomenclature for major proteins of the milk-fat globule membrane. *Journal of Dairy Science, 83*, 203–247.

Mather, I. H., & Keenan, T. W. (1998). Origin and secretion of milk lipids. *Journal of Mammary Gland Biology and Neoplasia, 3*, 259–273.

Matyash, V., Liebisch, G., Kurzchalia, T. V., Shevchenko, A., & Schwudke, D. (2008). Lipid extraction by methyl-tert-butyl ether for high-throughput lipidomics. *Journal of Lipid Research, 49*, 1137–1146.

McPherson, A. V., & Kitchen, B. J. (1983). Reviews of the progress of dairy science: The bovine milk fat globule membrane–its formation, composition, structure and behaviour in milk and dairy products. *The Journal of Dairy Research, 50*, 107–133.

Morin, P., Britten, M., Jiménez- Flores, R., & Pouliot, Y. (2007). Microfiltration of buttermilk and washed cream buttermilk for concentration of milk fat globule membrane components. *Journal of Dairy Science, 90*, 2132–2140.

Morin, P., Jiménez- Flores, R., & Pouliot, Y. (2007). Effect of processing on the composition and microstructure of buttermilk and its milk fat globule membranes. *International Dairy Journal, 17*, 1179–1187.

Morin, P., Jiménez-Flores, R., & Pouliot, Y. (2004). Effect of temperature and pore size on the fractionation of fresh and reconstituted buttermilk by microfiltration. *Journal of Dairy Science, 87*, 267–273.

Morin, P., Pouliot, Y., & Jiménez-Flores, R. (2006). A comparative study of the fractionation of regular buttermilk and whey buttermilk by microfiltration. *Journal of Food Engineering, 77*, 521–528.

Morrison, W. R., Jack, E. L., & Smith, L. M. (1965). Fatty acids of bovine milk glycolipids and phospholipids and their specific distribution in the diacylglycerophospholipids. *Journal of the American Chemical Society, 45*, 1142–1147.

Mudd, A. T., Alexander, L. S., Berding, K., Waworuntu, R. V., Berg, B. M., Donovan, S. M., et al. (2016). Dietary prebiotics, milk fat globule membrane, and lactoferrin affects structural neurodevelopment in the young piglet. *Frontiers in Pediatrics, 4*, 1–10.

Nguyen, H. T. H., Ong, L., Hoque, A., Kentish, S. E., Williamson, N., Ang, C.-S., et al. (2017). A proteomic characterization shows differences in the milk fat globule membrane of buffalo and bovine milk. *Food Bioscience, 19*, 7–16.

Noh, S. K., & Koo, S. I. (2004). Milk sphingomyelin is more effective than egg sphingomyelin in inhibiting intestinal absorption of cholesterol and fat in rats. *The Journal of Nutrition, 134*, 2611–2616.

Nyberg, L., Duan, R. D., & Nilsson, A. (2000). A mutual inhibitory effect on absorption of sphingomyelin and cholesterol. *The Journal of Nutritional Biochemistry, 11*, 244–249.

O'Mahoney, J. A., & Fox, P. F. (2014). Milk an overview. In M. Boland, H. Singh, & A. Thompson (Eds.), *Milk proteins: From expression to food* (2nd ed., pp. 19–73). San Diego, CA: Elsevier.

Oshida, K., Shimizu, T., Takase, M., Tamura, Y., Shimizu, T., & Yamashiro, Y. (2003). Effects of dietary sphingomyelin on central nervous system myelination in developing rats. *Pediatric Research, 53*, 589–593.

Parodi, P. W. (2003). Anti-cancer agents in milkfat. *Australian Journal of Dairy Technology, 58*, 114–118.

Pepeu, G., Pepeu, I. M., & Amanducci, I. (1996). A review of phosphatidylserine pharmacological and clinical effects. I. Phosphatidylserine a drug for the ageing brain? *Pharmaceutical Research, 33*, 73–80.

Phan, T. T. Q., Asaduzzaman, M., Le, T. T., Fredrick, E., Van der Meeren, P., & Dewettinck, K. (2013). Composition and emulsifying properties of a milk fat globule membrane enriched material. *International Dairy Journal, 29*, 99–106.

Pimentel, L., Gomes, A., Pintado, M., & Rodríguez-Alcalá, L. M. (2016). Isolation and analysis of phospholipids in dairy foods. *Journal of Analytical Methods in Chemistry, 2016*, 9827369.

Poppitt, S. D., McGregor, R. A., Wiessing, K. R., Goyal, V. K., Chitkara, A. J., Gupta, S., et al. (2014). Bovine complex milk lipid containing gangliosides for prevention of rotavirus infection and diarrhoea in northern Indian infants. *Journal of Pediatric Gastroenterology and Nutrition, 56*, 167–171.

Reinhardt, T. A., & Lippolis, J. D. (2006). Bovine milk fat globule membrane proteome. *The Journal of Dairy Research, 73*, 406–416.

Restuccia, D., Spizzirri, G., Puoci, F., Cirillo, G., Vinci, G., & Picci, N. (2011). Determination of phospholipids in food samples. *Food Review International, 28*, 1–46.

Riccio, P. (2004). The proteins of the milk fat globule membrane in the balance. *Trends in Food Science and Technology, 15*, 458–461.

Robenek, H., Hofnagel, O., Buers, I., Lorkowski, S., Schnoor, M., Robenek, M. J., et al (2006). Butyrophilin controls milk fat globule secretion. *Proceedings of the National Academy of Sciences of the United States of America, 103*, 10385–10390.

Rodrigues-Alcalá, L. M., & Fontech, J. (2010). Major lipid classes separation of buttermilk, and cows, goats and ewes milk by high performance liquid chromatography with an evaporative light scattering detector focused on the phospholipid fraction. *Journal of Chromatography. A, 1217*, 3063–3066.

Roesch, R. R., Rincon, A., & Corredig, M. (2004). Emulsifying properties of fractions prepared form commercial buttermilk by microfiltration. *Journal of Dairy Science, 87*, 4080–4087.

Rombaut, R., Dejonckheere, V., & Dewettinck, K. (2006). Microfiltration of butter serum upon casein micelle destabilisation. *Journal of Dairy Science, 89*, 1915–1925.

Rombaut, R., Dejonckheere, V., & Dewettinck, K. (2007). Filtration of milk fat globule membrane fragments from acid buttermilk cheese whey. *Journal of Dairy Science, 90*, 1662–1673.

Rombaut, R., Dewettick, K., & van Camp, J. (2007). Phospho- and sphingolipid content of selected dairy products as determined by HPLC coupled to an evaporative light scattering detector (HPLC-ELSD). *Journal of Food Composition and Analysis, 20*, 308–312.

Rombaut, R., & Dewettinck, K. (2006). Properties, analysis and purification of milk polar lipids. *International Dairy Journal, 16*, 1362–1373.

Rombaut, R., van Camp, J., & Dewettick, K. (2005). Analysis of phosphor- and sphingolipids on dairy products by a new HPLC method. *Journal of Dairy Science, 88*, 482–488.

Sachdeva, S., & Buchheim, W. (1997). Recovery of phospholipids from buttermilk using membrane processing. *Kieler Milchwirtschaftliche Forschungsberichte, 49*, 47–68.

Sánchez-Juanes, F., Alonso, J. M., Zancada, L., & Hueso, P. (2009). Distribution and fatty acid content of phospholipids from bovine milk and bovine milk fat globule membranes. *International Dairy Journal, 19*, 237–278.

Singh, H. (2006). The milk fat globule membrane-A biophysical system for food applications. *Current Opinion in Colloid & Interface Science, 11*, 154–163.

Singh, H., & Gallier, S. (2017). Nature's complex emulsion: The fat globules of milk. *Food Hydrocolloids, 68*, 81–89.

Smoczyński, M. (2017). Role of phospholipid flux during milk secretion in the mammary gland. *Journal of Mammary Gland Biology and Neoplasia, 22*, 117–129.

Spence, A. J., Jiménez-Flores, R., Qian, M., & Goddik, L. (2009). Phospholipid enrichment in sweet and whey cream buttermilk powders using supercritical fluid extraction. *Journal of Dairy Science, 92*, 2373–2381.

Spitsberg, V. L., & Gorewit, R. C. (2002). Isolation, purification and characterization of fatty-acid-binding protein form milk fat globule membrane. Effect of bovine growth hormone treatment. *Pakistan Journal of Nutrition, 1*, 43–48.

Sprong, C., Hulstein, M. F. E., & van der Meer, R. (2001). Bactericidal activities of milk lipids. *Antimicrobial Agents and Chemotherapy, 45*, 1298–1301.

Surel, O., & Flamelart, M. H. (1995). Ability of ceramic membranes to reject lipids of dairy products. *Australian Journal of Dairy Technology, 88*, 2289–2294.

Thompson, A. K., & Singh, H. (2006). Preparation of liposomes from milk fat globule membrane phospholipids using a microfluidizer. *Journal of Dairy Science, 89*, 410–419.

Timby, N., Domellöf, E., Hernell, O., Lönnerdal, B., & Domellöf, M. (2014). Neurodevelopment, nutrition, and growth until 12 mo of age in infants fed a low-energy, low-protein formula supplemented with bovine milk fat globule membranes: A randomised controlled trial. *The American Journal of Clinical Nutrition, 99*, 860–868.

Timby, N., Hernell, O., Vaarala, O., Melin, M., Lönnerdal, B., & Domellöf, M. (2015). Infections in infants fed formula supplemented with bovine milk globule membranes. *Journal of Pediatric Gastroenterology and Nutrition, 60*, 348–389.

Tran Le, T., El-Bakry, M., Neirynck, N., Bogus, M., Hoa, H. D., & Van der Meeren. P. (2007). Hydrophilic lecithins protect milk proteins against heat induced aggregation. *Colloids and Surfaces. B, Biointerfaces, 60*, 167–173.

Van Meer, G., & Lisman, Q. (2002). Sphingolipid transport: Rafts and translocators. *The Journal of Biological Chemistry, 277*, 25855–25858.

Vanderghem, C., Francis, F., Danthine, S., Deroanne, C., Paquot, M., De Pauw, E., & Blecker, C. (2011). Study of the susceptibility of the bovine milk fat globule membrane proteins to enzymatic hydrolysis and organization of some of the proteins. *International Dairy Journal, 21*, 312–318.

Verardo, V., Gómez-Caravaca, A. M., Arráez-Román, D., & Hettinga, K. (2017). Recent advances in phospholipids from colostrum, milk and dairy by-products. *International Journal of Molecular Sciences, 18*, 173–196.

Vorbach, C., Scriven, A., & Capecchi, M. R. (2002). The housekeeping gene xanthine oxidoreductase is necessary for milk fat droplet enveloping and secretion: Gene sharing in the lactating mammary gland. *Genes & Development, 16*, 3223–3235.

Walstra, P. (1985). Some comments on the isolation of fat globule membrane material. *The Journal of Dairy Research, 52*, 309–312.

Walstra, P., & Jenness, R. (1984). *Dairy chemistry and physics*. New York, NY: Wiley.

Wat, E., Tandy, S., Kaper, E., Kamili, A., Chung R. W., Brown A., et al. (2009). Dietary phospholipid-rich dairy milk extract reduces hepatomegaly, hepatic steatosis and hyperlipidemia in mice fed a high-fat diet. *Atherosclerosis, 205*, 144–150.

Watanabe, S., Takahashi, T., Tanaka, L., Haruta-Ono, Y., Shiota, M., Hosokawa, M., et al. (2011). The effect of milk polar lipids separated from butter serum on the lipid levels in the liver and the plasma of obese-model mouse (KK-A^y). *Journal of Functional Foods, 3*, 313–320.

Wu, C. C., Howell, K. E., Neville, M. C., Yates, J. R. 3rd, & McManaman. J. L. (2000). Proteomics reveal a link between the endoplasmic reticulum and lipid secretory mechanisms in mammary epithelial cells. *Electrophoresis, 21*, 3470–3482.

Zou, X., Guo, Z., Jin, Q., Huang, J., Cheong, L., Xu X., et al. (2015). Composition and microstructure of colostrum and mature bovine milk fat globule membrane. *Food Chemistry, 185*, 362–370.

Chapter 5
Lipidomic Characterization of the Milk Fat Globule Membrane Polar Lipids

Olimpio Montero, Javier Fontecha, M. Pillar Castro-Gómez, and Luis Miguel Rodríguez-Alcalá

1 Introduction

The milk fat globule structure is of colloidal nature with a triacylglycerol (TAG) rich core surrounded by the milk fat globule membrane (MFGM), emulsified and dispersed into the aqueous phase. MFGM is a unique trilayer structure, with the inner monolayer stemming from the endoplasmic reticulum and the outer bilayer from the apical cells of the mammary gland. Therefore, as biological membrane, MFGM is a highly complex ensemble of self-associating lipids, mainly phospholipids (PL), into which a diverse array of proteins and other biomolecules are embedded. Although PL constitute a small percentage of the milk total lipids (0.5–2% in cow's milk), they may represent a substantial part of the total lipids in buttermilk, the aqueous phase released during the elaboration of butter, owing to the presence of significant quantities of MFGM components in this by-product.

Milk PL and sphingolipids have aroused great interest during the last years, not only because of the acknowledged technological qualities of these polar lipids, but also because of their biological activities. Recent review articles have reported that PL seem to develop important metabolic functions of interest as bioactive agents because of their biological activities which are potentially beneficial for human health on reducing the risk of numerous diseases (Crespo, Tomé-Carneiro, Gómez-Coronado, et al., 2018; Rodríguez-Alcalá, Castro-Gómez, Pimentel, & Fontecha, 2017). It has

O. Montero
Spanish National Research Council, CSIC, Valladolid, Spain

J. Fontecha (✉) · M. P. Castro-Gómez
Institute of Food Science Research (CIAL, CSIC-UAM), Madrid, Spain
e-mail: j.fontecha@csic.es

L. M. Rodríguez-Alcalá
Universidade Católica Portuguesa, CBQF – Centro de Biotecnologia e Química Fina – Laboratório Associado, Escola Superior de Biotecnologia, Porto, Portugal

T. Truong et al. (eds.), *Dairy Fat Products and Functionality*,
https://doi.org/10.1007/978-3-030-41661-4_5

also been reported that dietary supplementation with MFGM obtained from buttermilk produced an improvement in lowering cholesterol and TAG levels in plasma of moderately hypercholesterolemic humans with normal blood pressure (Conway, Couture, Gauthier, et al., 2014) and suppression of postprandial inflammatory markers in obese and overweight adults (Rogers, Demmer, Rivera, et al., 2017). Moreover, it had already been reported that some polar lipid fractions isolated from MFGM have antiproliferative activity in cell cultures of various human tumors (Castro-Gómez, Rodríguez-Alcalá, Monteiro, et al., 2016). All this has made it possible to consider the MFGM as a potential nutraceutical.

As commented above, the main polar lipid constituents of MFGM are PL and sphingolipids and these include, according to the polar head, phosphatidylethanolamine (PE), phosphatidylinositol (PI), phosphatidylserine (PS), and phosphatidylcholine (PC) and sphingomyelin (SM).

Milk cream, buttermilk, butter serum or even cheese whey are the dairy products that may be used as source material for the isolation of MFGM. However, different methods and protocols have been described (Astaire, Ward, German, & Jiménez-Flores, 2003; Castro-Gómez et al., 2016; Holzmüller & Kulozik, 2016; Konrad, Kleinschmidt, & Lorenz, 2013; Le, Debyser, Gilbert, et al., 2013; Morin, Jiménez-Flores, & Pouliot, 2007; Rombaut, Dejonckheere, & Dewettinck, 2006).

2 Recent Advances of Lipidomic Analyses of Polar Lipid Composition of MFGM

The MFGM analyses involve lipid extraction in a non-polar solvent. Currently the chloroform:methanol method developed by Folch, Lees, and Stanley (1957) as well as different variations of this method are the most suitable for lipid extraction in dairy samples containing high amount of phospho- and sphingolipids (Avalli & Contarini, 2005; Barry, Dinan, Murray, & Kelly, 2016; Bourlieu, Cheillan, Blot, et al., 2018; Luo, Huang, Liu, et al., 2018; Rodríguez-Alcalá & Fontecha, 2010).

Due to the recent attention paid to the health properties of MFGM associated to the biological effects of the polar lipids, great interest has been focused on the selective isolation of such compounds (Pimentel, Fontes, Salsinha, et al., 2018). Thin layer chromatography (TLC) is one of the most common techniques used for purification purposes although this separation technique is time-consuming and has disadvantages like poor separation yield and oxidation of unsaturated acyl chains. Improvement of PL species resolution was achieved with the development of high-performance thin layer chromatography (HPTLC), with valuable results (Lopez, Blot, Briard-Bion, et al., 2017; Lu, Pickova, Vázquez-Gutiérrez, & Langton, 2018; Sánchez-Juanes, Alonso, Zancada, & Hueso, 2009). However, determination of the acyl chains esterified to the glycerol backbone is needed for further analysis as fatty acid methyl ester (FAME) derivatives by GC after the spot of each PL class is scraped off. Separation of PL using HPLC with detection by evaporative light scattering (ELSD), charged aerosol (CAD), or flame ionization (FID) largely improves

PL species resolution within each class, but full identification of the acyl chains remains incomplete. Nonetheless, since these methods have recently been reviewed (Gallo & Ferranti, 2016; Pimentel, Gomes, Pintado, & Rodríguez-Alcalá, 2016; Verardo, Gómez-Caravaca, Arráez-Román, & Hettinga, 2017), the most recent advances in milk PL analysis will be the focus of this section.

Identification of the full set of individual PL species along with quantification at a sensitivity level reaching picomols (~70 pmol for less abundant species, i.e. PI) has become suitable after the development of ultra-high resolution liquid chromatography coupled to time-of-flight mass spectrometry (UHPLC-ToF) or Orbitrap mass spectrometers. In addition to highly improved separation of the diverse phospholipid species within each class with the new available solid phases of very narrow bore columns (2.1 mm or less), mass resolutions ([m/z]/[Δm/z]) of about 20,000 for the ToF instruments or even >70,000 for the Orbitrap mass spectrometers allow to distinguish extremely close signals in the mass spectrum (Fong, Norris, Lowe, & McJarrow, 2009; Li, Zhao, Zhu, et al., 2017). The newly developed narrow bore columns can be operated with a shorter total run duration, thus gaining time in sample number analysis and reducing the solvent volume. Both reversed phase (RP) and normal-like phase chromatographic modes still are available for UHPLC analysis; however, hydrophilic interaction liquid chromatography (HILIC) is presently used instead of normal (silica) phases as it can be employed with the polar solvents currently used in reversed phase (Buszewski & Noga, 2012). RP chromatography separates the PL according to the non-polar fatty acyl substituents, whilst HILIC separation is led by the polarity of the head groups as in normal phase chromatography (Buszewski & Noga, 2012). Regarding RP chromatography, Alexandre-Gouabau, Moyon, Cariou, et al. (2018) recently reported 21 species of PC (including 3 plasmalogens), 24 species of PE (including 5 plasmalogens), 5 PG species, 1 PI species and 2 PS species, using a CSH® C18 column (Waters) and Orbitrap mass spectrometer operated under positive ionization, in a study of the human breast milk lipidome with Folch's extraction and no further treatment. A modified RP column, concretely an Acquity™ HSS T3 column (WATERS) was used by Castro-Gómez, Montero, and Fontecha (2017) to characterize the lipidome, including PL, of the buttermilk; these authors carried out a pressurized liquid extraction of fat before PL fractionation and isolation by flash chromatography for later being separately analysed. Using an Acquity™ UPLC chromatographic system (WATERS, Manchester, UK) and a QToF mass spectrometer (SYNAPT HDMS G2, WATERS, Manchester, UK), which was operated in both positive and negative ionization modes, these authors determined 24 PC and lysoPC species, 14 PE and lysoPE species, 2 PS species and 3 PI species. CORTECS C18 and XSelect CSH C18 columns, both driving RP chromatography, were used by Li et al. (2017) to report 25 PC and lysoPC species, 18 PE and lysoPE species, 20 PI species, 14 PS species, 9 PG species, and 6 PA species, with Orbitrap MS detection, in goat, bovine and soy milks. Methods of milk fat analysis using HILIC separation have also been reported. A first method was proposed by Donato, Cacciola, Cichello, et al. (2011); these authors used a fused-core Ascentis Express HILIC column (150 × 2.1 mm I.D.), and eluted the PL with a gradient of acetonitrile and acetonitrile:water (2:1, v/v), with PL being previously purified by SPE after chloroform:methanol extraction though a substantial amount of TAG still remained with PL; detection was conducted with both ELS and MS, and with the exception of PI the other PL, namely PS, PE, PC and SM, were detected under positive ionization. Elution order was according to decreasing polarity with increasing retention time for PI < PS < PE < PC < SM < lysoPC, and the limit of quantification (LOQ) ranged between 1.70 µg/mL for PS and 4.05 µg/mL for PI. With this methodology, Donato et al. (2011) identified 3 PE and PS species, 4 PI species, 2 SM and 10 PC and lysoPC species; given that more than 20 PC species and 10 PE species have previously been reported by other authors, it is likely that some PL were lost in the SPE purification step. An ethylene bridge hybrid (BEH) HILIC column has recently been used by Ali, Zou, Huang, et al. (2017) to assess the PL profile in diverse mammalian milk powders; the Folch's method with further SPE purification was the extraction method used by these authors. Acetonitrile and 50 mM ammonium acetate with 0.1% formic acid were the elution solvents. The elution order, with increasing retention time, was PS < PI < PE < PC < SM < lysoPC. With MS detection in both positive and negative ionization, Ali et al. (2017) reported 22 PC species and 9 lysoPC species, 15 PE species, 8 PI species, 8 PS species, and 9 SM species, with fatty acyls ranging from C12:0 to C22:5. HILIC (Kinetex HILIC 100A, Phenomenex, Sydney, Australia) separation of PL from cow milk was also accomplished by Craige Trenerry, Akbaridoust, Plozza, et al. (2013), who detected 7 PI, 12 PE, 18 PC, and 13 SM species, but no PS species were found by these authors using an LTQ Ion-Trap mass spectrometer (Thermo Fisher, San Jose, CA). Other authors have used HILIC for PL separation in other food matrices with comparable results (Zhao, Xiong, & Curtis, 2011). A possible advantage of HILIC against RP chromatography using mass spectrometric detection is the fact that PL of the same class elute together over a narrow time range, which limits the potential ion suppression effect of the PL species more prone to ionization, even though this effect is likely to take place to some extent for the more abundant species. Within each PL class, diverse species are eluted with short elution time difference, therefore possibly overlapping due to the combined effect of FA length and FA unsaturation, two factors producing contrary effects to each other. Taking into consideration the relative abundance of two co-eluting PL species of the same class, ascription of the fatty acyls may be easy. Conversely, RP chromatography allows well-defined separation of species belonging to the same PL class, with PC species eluting before the fatty acyl equivalent PE species (i.e. PC 36:2 elutes earlier than PE 36:2). Accurate PL profiling is likely bound to a combined analysis using both HILIC and RP chromatography.

Even though ELS detection is useful for most studies on PL profiling, the chance of detection with mass spectrometry (MS) is gaining the favor of the majority of researchers. MS with electrospray ionization (ESI) affords excellent results when coupled to UHPLC or even HPLC, and both HILIC and RP chromatography can be run. ESI permits phospholipids to be ionized positively and negatively though the more anionic PL are preferably ionized in negative mode. Positive ionization currently yields abundant and highly intense signals whereas negative ionization offers the opportunity of clear, low noisy, spectra and easy fatty acyl determination in MS/MS operating mode. Accordingly, most studies have been accomplished using only negative ionization but then there is the risk of losing detection of the less abundant

PC and lysoPC and SM species. Up to date, three types of mass spectrometers have mainly been utilized for PL profiling, namely hybrid ion-trap-time of flight (IT-ToF), quadruple-ToF (QToF), and Orbitrap mass spectrometers. IT-ToF and QToF instruments can provide a mass resolution of 15,000–25,000, whereas Orbitrap instruments easily reach a resolution of 70,000 in full scan (and c.a. 15,000 in MS^2). The high cost of the Orbitrap instruments may sometimes impose a constraint to selection. Both IT-ToF and Orbitrap instruments can be operated in tandem mass spectrometry (MS^2 and MS^3) with parent ion selection for the assessment of the fatty acyls, whereas a MS^E method can be run in some commercial QToF instruments (SYNAPT or XEVO-ToF from WATERS) for fragmentation of the base peak at each elution time in one run. IT-ToF and Orbitrap instruments likely pose a higher specificity as tandem mass spectrometry (e.g. MS^2 and MS^3) can be applied to defined parent ions, but the disadvantage exists regarding that more than one run is necessary with concurrent time and sample consuming. Multiple reaction monitoring (MRM) methods may partially overcome this drawback by defining a list of precursor ions as complete as possible (Ikeda, Shimizu, & Taguchi, 2008), the extent of the list influencing however the resolution capability of the instrument and computer guided data acquisition.

Acetonitrile (ACN) and acetonitrile: water or methanol:water are the preferred elution solvents for compatibility with MS detection. Ammonium acetate or ammonium formate (10–50 mM) plus 0.1% formic acid are currently added as modifiers for better chromatographic peak shape and facilitate electrospray ionization. Under these operating conditions, the PL belonging to the different classes (and sphingomyelins, SM) are detected as the diverse adducts shown in Table 5.1 (Alexandre-Gouabau et al., 2018; Ali et al., 2017; Castro-Gómez et al., 2017; Craige Trenerry

Table 5.1 Most common adduct ions observed in mass spectrometry analysis of phospholipids as reported in diverse studies

Phospholipid	MS positive ionization (ESI+)	MS^2 (MS^E)/MS^3 fragments (ESI+)	MS negative ionization (ESI−)	MS^2 (MS^E)/MS^3 fragments (ESI−)
PC	$[M+H]^+$	$[M-R_2+OH+H]^+$, m/z 184.07/ $[M-R_2-H_2O+H]^+$	$[M+FA-H]^-$	$[M-FA-CH_3-H]^-$, $[R_1-H]^-$, $[R_2-H]^-$/ $[M-R_2-H]^-$,$[R_1-H]^-$,$[R_2-H]^-$
PE	$[M+H]^+$	$[M-R_2+OH+H]^+$/ $[M-R_2-H_2O+H]^+$	$[M-H]^-$	m/z 196.0, $[M-R_2-H]^-$, $[M-R_1-H]^-$,$[R_1-H]^-$,$[R_2-H]^-$
PS	$[M+H]^+$, $[M+Na]^+$	$[M-R_2+OH+H]^+$, $[M-185.01+H]^+$/ $[M-R_2-H_2O+H]^+$	$[M-H]^-$	NL = 88.04, $[M-R_2-H]^-$, $[M-R_1-H]^-$,$[R_1-H]^-$,$[R_2-H]^-$
PI	$[M+NH_4]^+$, $[M+Na]^+$		$[M-H]^-$	m/z 241.01, $[M-R_2-H]^-$, $[M-R_1-H]^-$,$[R_1-H]^-$,$[R_2-H]^-$
PG	$[M+NH_4]^+$		$[M-H]^-$	m/z 227.02, $[M-R_2-H]^-$, $[M-R_1-H]^-$,$[R_1-H]^-$,$[R_2-H]^-$
SM	$[M+H]^+$	m/z 184.07, $[M-264.2+H]^+$/ $[M-264.2+H]^+$	$[M+FA-H]^-$	$[M-FA-CH_3-H]^-$

MS^2 and MS^3 apply to IT-ToF and Orbitrap instruments, and MS^2 and MS^E apply to diverse commercial QToF instruments. *FA* formic acid, *NL* neutral loss. R_1 and R_2, fatty acyls esterifying the *sn-1* and *sn-2* positions of the glycerol backbone et al., 2013; Donato et al., 2011; Li et al., 2017). PL analysis using mass spectrometry has extensively been studied (Murphy & Harrison, 1994; Pulfer & Murphy, 2003). These studies have established characteristic fragments after MS^2 for every PL classes (Table 5.1), and their results have been validated by a number of further studies. How these characteristic fragments are obtained is dependent upon the type of instrument. Ion-trap instruments need to run specific MS^3 transitions to obtain the R_1 and R_2 fatty acyl negative ions of PC since MS^2 only yields the negative ion $[M\text{-}HCOOH\text{-}CH_3]^-$ with an m/z value of M-60.02 Da, whereas both transitions happen simultaneously in the QToF instruments operated in automatic MS/MS or MS^E (WATERS QToFs), even though the resulting fragment intensity and the spectrum are generally lower and less clear than in the IT or Orbitrap instruments. The typical loss of 60.02 Da in MS/MS can also be applied to SM in negative ionization. Nonetheless, both PC and SM can accurately be identified by the characteristic fragment of m/z 184.07, corresponding to the PC polar head, which is obtained in MS/MS experiments under positive ionization (Fig. 5.1). PS exhibit a behavior similar to PC in positive ionization but to a lower intensity, rendering an ion that corresponds to the loss of the polar head PS (m/z 185.01); however, PS account for a more accurate identification through the neutral loss (NL) of 88.04 Da (serine) in negative MS/MS (see for example figure 4C in Ali et al. (2017)). Characteristic fragments of PI, PE and PG in negative ionization are m/z 241.01, m/z 196.04, and m/z 227.00, respectively. These PL also render the negative ions of the fatty acyls along with the resulting lysoPL ($-R_1$ or $-R_2$); however PI fragment intensities are currently low under the collision induced disociation (CID) energies used in PL analyses.

Mass spectrometry analysis of PL has also been accomplished by direct infusion, without previous separation by liquid chromatography (LC), in the so-called 'shotgun' approach (Ejsing, Sampaio, Surendranath, et al., 2009; Han & Gross, 2005). Regarding milk analysis, a method taking as reference the 'shotgun' methodology was proposed (Sokol, Ulven, Faergeman, & Ejsing, 2015). These and other authors have claimed 'shotgun' MS offers a high throughput and wide dynamic range in addition to the required analytical sensitivity and specificity (Ejsing et al., 2009; Ståhlman, Ejsing, Tarasov, et al., 2009). Nonetheless, two known disadvantages of the 'shotgun' methodology as compared to UHPLC-MS are the enhanced ion suppression effect and deconvolution strategies are necessary for resolution of isomeric (isobaric) species. In the study by Sokol et al. (2015) the lipidome was analyzed by a QToF analyzer and with a Triversa Nanomate ion source, after chloroform:methanol extraction; 77 molecular glycerolipid species, they including PC, PE, PS and PI, were identified by the authors. Triacylglycerides (TAG) were also measured in the same run, but SM were not reported.

Fig. 5.1 (continued) chain as well as the lysoPS resulting from the loss of the C22:6 acyl chain; (**e**) MS^E spectrum of PE(16:0/20:4) showing the carboxy anion of each acyl chain as well as the lysoPE at m/z 452.2755 resulting from the loss of the C20:4 acyl chain (insert), and the typical fragment of PEs at m/z 196.0377 corresponding to glycerophosphoethanolamine; (**f**) MS^E spectrum of PI(18:1/20:4) at m/z 883.5383 showing the carboxy anion of each acyl chain as well as the lysoPI at m/z 579.2923 resulting from the loss of the C20:4 acyl chain, and the typical fragment of PIs at m/z 241.0116 corresponding to phosphoinositol

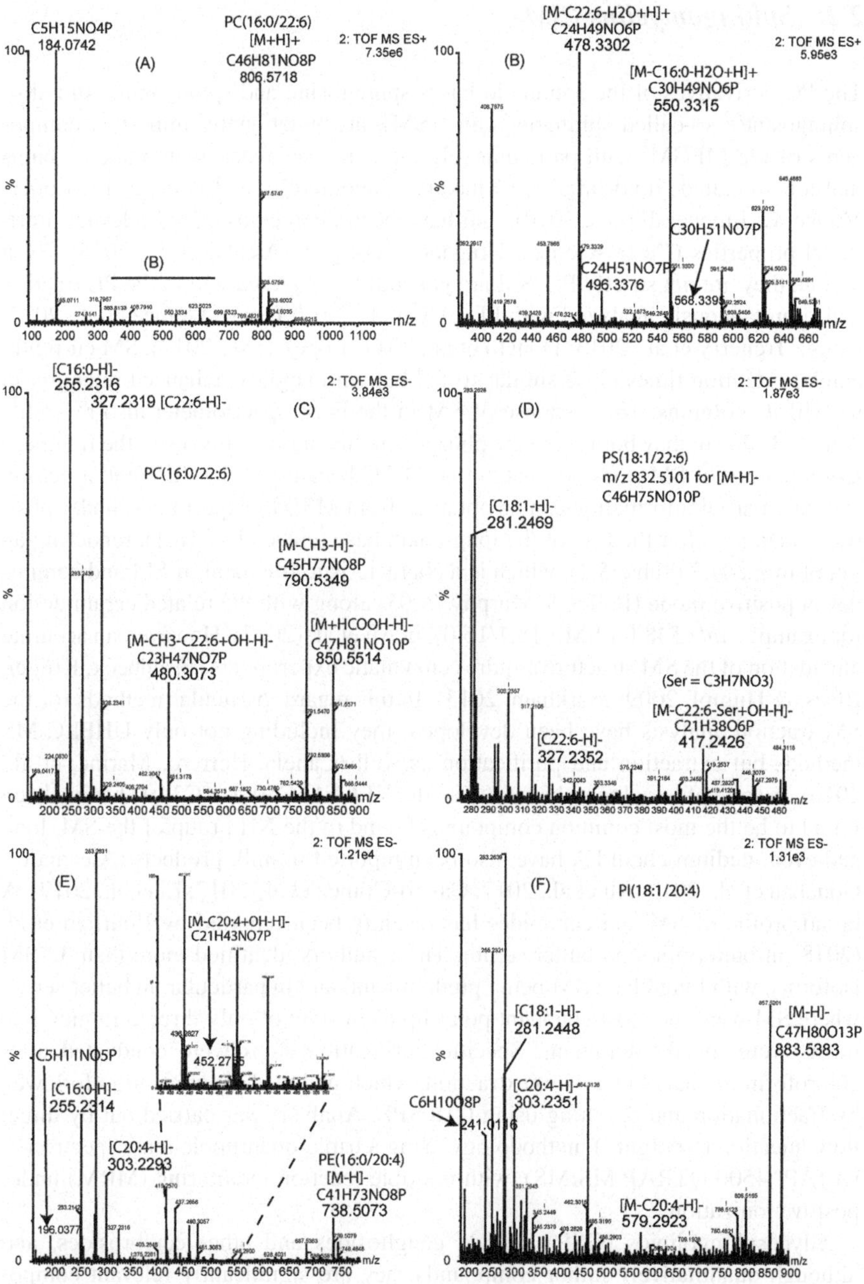

Fig. 5.1 Examples of fragmentation patterns (high energy MS^E spectra) of glycerophospholipid clases as measured at authors' laboratory with a UPLC-QToF-MS (Acquity UPLC and SYNAPT HDMS G2, from WATERS). (**a**) $[M+H]^+$ ion of PC(16:0/22:6) at m/z 806.5718; (**b**) amplification of the region where the lysoPCs resulting from the loss of each acyl chain can be observed; (**c**) $[M+H_2COO\text{-}H]^-$ ion of PC(16:0/22:6) at 850.5514, along with the carboxy anions ($[R\text{-}COO]^-$) of the acyl chains and the ion resulting from the loss of a methyl group plus the formate (−60.02 Da) with m/z 790.5349; (**d**) MS^E spectrum of PS(18:1/22:6) showing the carboxy anion of each acyl

2.1 Sphingomyelins (SM)

The PC derivatives of the sphingoid bases sphingosine and sphinganine (dihydrosphingosine), so-called sphingomyelins (SM), are functionally important components of the MFGM, with particular relevance in specialized membrane domains that are so-called 'lipid-rafts' (Et-Thakafy, Guyomarc'h, & Lopez, 2017; Lopez, Madec, & Jimenez-Flores, 2010), and have been claimed to afford relevant nutritional properties (Castro-Gómez, Holgado, Rodríguez-Alcalá, et al., 2015). Even though they are not strictly PL, SM are generally considered within the PL fraction and simultaneously analyzed with PL (Ali et al., 2017; Castro-Gómez et al., 2017; Craige Trenerry et al., 2013; Donato et al., 2011; Lopez et al., 2017). SM currently exhibit retention times close similar to PC but with slightly enhanced retention in the HILIC columns. The response of SM in the mass spectrometer also resemble that of PC due to they both share the phosphocholine moiety; however, the fragmentation pattern of SM does not match that of PC because of the different structure, and only partial information can be obtained from MS/MS experiments under positive ionization after the loss of the long-chain base (currently C18:1), rendering an ion of m/z 264.2 (Table 5.1), which is a characteristic fragment of SM and ceramides in positive mode (Pulfer & Murphy, 2003) along with the related ceramide ion for example, m/z 548 for SM(d18:1/16:0), in Ali et al. (2017). However, an accurate elucidation of the SM structure requires enzymatic experiments (Fischbeck, Krüger, Blaas, & Humpf, 2009; Markham, 2013). In this regard, particular methods for the SM fraction analysis have been developed, they including not only UHPLC-MS methods but extraction and purification as well (Canela, Herrero, Mariné, et al., 2016). In spite of very long-chain fatty acids (VLCFA; C20:0 to C24:1) having been found to be the most common compounds bound to the NH group of the SM, long and even medium-chain FA have also been reported in milk products (Alexandre-Gouabau et al., 2018; Ali et al., 2017; Castro-Gómez et al., 2017; Li et al., 2017). A broad profile of SM and ceramides has recently been reported by Bourlieu et al. (2018) in buttermilk and butter serum. These authors identified more than 32 SM isoforms, with long-chain SM being predominant, and in particular, in butter serum where SM were the most abundant polar lipids in spite of only three ceramide isoforms were mostly dominant. Specific purification steps were conducted after chloroform:methanol (1:2, v/v) extraction, which included saponification followed by fractionation and desalting using C18 SPE. Analysis was carried out by direct flow injection ('shotgun') methodology using a triple-quadrupole mass spectrometer (API 4500 QTRAP MS/MS), with multiple reaction monitoring (MRM) under positive ionization.

Glycosphingolipids (GSL) include gangliosides and other cerebrosides, and although quantitatively minor compounds they are significantly relevant components of the MFGM polar lipids (Koletzko, 2017). In cell membrane and cell organelle membranes GSL participate in micro-domain structures like caveolae and lipid-rafts (Hakomori, 1990; Malisan & Testi, 2002; Sisu, Flangea, Serb, et al., 2011). The analysis of GSL is by far the most challenging within MFGM compounds. The GSL structure is composed of two moieties, one constituted by a hydrophilic and polar oligosaccharide chain and other by a hydrophobic non-polar ceramide. The oligosaccharide chain currently contains N-acetylneuraminic acid (Neu5Ac), also called sialic acid, in humans, and the position of the sialic acid together with the sugar composition serve to classify the GSL. Major GSL in human neuronal tissue are the gangliosides (GL) GD1a and GT1b, whereas other human tissues mostly account for the GL GM3 and GD3. Full elucidation of the GL implies to decipher the sequence of the different sugars and their relative position plus ascertaining the sphingoid base and the acyl chains of the ceramide. As in other lipids, TLC followed by electron impact (EI) MS was the exclusive analytical approach available for GL composition. Furthermore, the extremely different polarity between the oligosaccharide moiety and the ceramide moiety becomes a drawback in developing analytical methods, even at the extraction step as GL keep at the water-chloroform interface, thus having to recover GL molecules from both extraction phases (Christiansen, 1975; Kolarovic & Fournier, 1986; Reis, Rudnitskaya, Blackburn, et al., 2013). Intact GL molecules started to be analyzed in complex samples after coupling of GC and HPLC to MS with soft ionization sources like ESI. The presence of the sialic acid prompts GL to be analyzed in negative mode, rendering the $[M\text{-}H]^-$ ion of the intact molecule. Nonetheless, GC analysis implies, as for other lipids, derivatization, which was a constraint in GL analysis until the permethylation method was developed (Ciucanu & Kerek, 1984; Hakomori, 1990). Conversely, GC has the advantage over HPLC of providing evidence on the sphingoid moiety, *cis/trans* isomerism of the acyl chains and even regio-isomers. A comparison of GL from milk and MGFM from Holstein-Friesian cows was accomplished using TLC and GC-EI-MS (Sanchez-Juanes, Alonso, Zancada, & Hueso, 2009); in this study the GL composition was found to be similar between both raw products but GM3 content was higher and GD3 content was lower in milk than in MGFM, as well, LCFAs exhibited a lower content in MGFM with concurrent increase in unsaturated medium-chain FAs (C18:1 and C18:2). LC-MS/MS analysis of GL was firstly attempted by Sørensen (2006) in bovine milk and infant formulae. These authors used the extraction method of Svennerholm and Fredman (1980), an HPLC coupled to a triple quadrupole with TurboIonSpray ion source for ganglioside separation in a Luna C_5 100A. Quantification at pmol level was carried out at three concentration levels (blank, 10 and 20 μg/mL for GM3, and 100 and 200 μg/mL for GD3) using commercial standards, and taking as reference the transition from parent ion to m/z 290 (MRM). In full scan, clusters of each ganglioside for the $[M\text{-}H]^-$ and $[M\text{-}2H]^{2-}$ ions were obtained. Ganglioside structure was investigated by collision induced dissociation (CID) MS/MS. The dominant long chain of the sphingosine base was dihydroxy-C18:1, with C22:0, C23:0, and C24:0 being the dominant FA in the ceramide moiety. The ganglioside concentrations in bovine milk measured in this study were 9.3 and 17.5 μg/mL for GD3, and 0.10 and 0.18 μg/mL for GM3, for Holstein and Jersey breed, respectively. Using a LC-ESI-MS/MS method with an ODS column the gangliosides standards GM1, GM2, GM3, GD1, and GT1, and gangliosides GD2, GD3, GT2, GT3, and GQ1 from porcine and mouse brains could be separated an quantified at picomolar to femtomolar levels

(Ikeda et al., 2008). However, these authors pointed out that with RP chromatography isomers like GD1a and GD1b could not be resolved. In further research works, Ikeda and Taguchi (2010) successfully accomplished this isomer separation using a HILIC-NH_2 column taking chance of the interaction of the glyco moiety with the amino group; using acetonitrile with ammonium formate instead of methanol the double charged ions of both isomers were separated, the GD1a species eluting earlier than the GD1b species, a fact the authors attributed to the increase in the hydrophobic interaction driven by the ammonium formate. Preparative methods for on-line separation before scanning in a QToF mass spectrometer (Zarei, Müthing, Peter-Katalinić, & Bindila, 2010), or improved resolution combining LTQ-Orbitrap after glyco-based separation in hydrophilic hypersil columns of gangliosides purified in a RPTrap C18 column (Fong et al., 2009), have also been proposed. Afterwards, assying this method, up to 27 GD3 compounds were mesured by Fong, Norris, and McJarrow (2011) in buttermilk powder and infant formulae. On the other hand, further authors assayed a QTRAP mass spectrometer instead of the LTQ-Orbitrap to quantify the gangliosides in breast milk and serum of Malaysian mothers during breast feeding period (Ma, MacGibbon, Jan Mohamed, et al., 2015). According to the reported results, an increase in ganglosides content was observed from second to the third assayed trimester; ganglioside species of GM1, GM2, GM3, GD3, $GD_{1a/b}$, GT_{1b}, and GQ_{1b} were detected in this study.

Elsewhere, gangliosides of 12 classes (GM1, GM2, GM3, GM4, GD1, GD2, GD3, GD4, GT1, GT2, GT3, and GT4) from bovine buttermilk powder were analyzed and up to 600 different species were identified and quantified (Rivas-Serna, Polakowski, Shoemaker, et al., 2015). In this study, a low resolution triple quadrupole (TQ) operated in MRM mode for the transition parent ion to m/z 290 was used instead of high resolution mass spectrometers (QToF, Orbitrap), with a library of 600 potential ganglioside species being previously generated for the MRM mode; linearity was obtained within the range 0.6–5 μM with RP HPLC separation in a C18 column (3.0 × 50 mm, 2.7 μm particle size). Enhancement of MS signal of gangliosides from milk pasteurized as compared to raw milk was assayed and MS response was substantially higher for GD3 from buttermilk than from colostrum or total milk (Rivas-Serna et al., 2015).

MALDI MS has also been applied to the analysis of gangliosides (Lee, An, Lerno Jr., German, & Lebrilla, 2011). Fourier transform ion cyclotron resonance (FTICR) mass spectrometry, an extremely high resolution instrument was used for detection and accurate assignment of chain length and degree of unsaturation of the ceramide moiety. MS/MS experiments were conducted using CID and infrared multiphoton dissociation (IRMPD). The results of this study confirm that only in human milk the ceramide moiety contains even-numbered fatty acyls. For detailed analysis of species within a cluster, these authors considered the Kendrick mass defect due to consecutive CH_2. Up to 10 GM3 and 10 GD3 species were identified by these authors in human milk, whereas 7 GM3 and 13 GD3 species along with 3 glycosphingolipid species were identified in bovine milk.

Using UPLC-MS methods for the analysis of phospholipids, Castro-Gómez et al. (2017) reported 17 species of gangliosides and GSL in buttermilk, but the exact species could not be determined; most species were detected in positive mode as the double of triple charged species. As well, using an untargeted approach, Montero, Velasco, Rodríguez-Lázaro, and Hernández (2013) detected a number of gangliosides and glycosphingomyelins in cheese as multi charged ions in both positive and negative ionization; the results of this study showed that polarity of the solvent mixture used for the extraction may render distinct extraction yields for these compounds. A combination of extraction methods and analysis with different platforms, perhaps in a multicenter study, may serve to unequivocally and accurately describe the full ganglioside composition of a given milk product.

3 Future Lipidomic Approaches for Determining MFGM Components

Within food technology and, in particular, lipidomics, MS based platforms have gained an unquestionable predominant role because of their potential to profile complex mixtures of biomolecules with minimal manipulation (Gallo & Ferranti, 2016). Nonetheless, new approaches are being accomplished for lipid analysis using methodologies developed with techniques that have accounted for improvement regarding sensitivity and capability as it is ion mobility coupled to mass spectrometry (Castro-Perez, Roddy, Nibbering, et al., 2011; Lintonen, Baker, Suoniemi, et al., 2014; Maccarone, Duldig, Mitchell, et al., 2014; Šala, Lísa, Campbell, & Holčapek, 2016; Shvartsburg, Isaac, Leveque, et al., 2011).

3.1 Structural Analysis and Quantification of Polar Lipid Species

Further challenges in structural analysis of the PL, SM and particular gangliosides is mainly concerned with the accurate assessment of regio-isomers, double bond position in the acyl chains as well as the *cis/trans* isomers. For this task, the combination of different approaches and platforms is likely to be necessary. Development or improvement of NMR coupled to UHPLC analytical methods may aid to this task. Meanwhile, GC-MS analysis is likely to be the gold technique for gaining information on such issues.

3.1.1 Phospholipids (PL)

The structural elucidation of PL is well stablished regarding both PL class and esterified acyl chains, with the *sn*-1 and *sn*-2 positions being assigned in base to the acyl chain rendering the more intense lysophospholipid that is esterified at the *sn*-1 position (for further details see (Pulfer & Murphy, 2003)). However, this rule is devoted of specific MS/MS experiments as in (U)HPLC-MS/MS analysis of complex mixtures there may be overlay of species that hinders each species fragments or accumulation results, with ion suppression leading also to confounding results. To a certain extent, MRM methods or MS^3 experiments with ion trap (IT) or Orbitrap instruments may overcome this drawback. Direct determination of double position and *cis/trans* isomers in the acyl chains in (U)HPLC-MS analysis keeps still challenging for these compounds.

3.1.2 Glycolipids

As indicated above, the complexity of gangliosides and related GLS needs of combined analysis using different methodologies. From our knowledge, there is not yet a single method to elucidate the structure of these compounds.

Quantification by MS has the disadvantage that no linear response can be expected for different molecules of the same class of PL, SM, or gangliosides; accordingly, quantification has to be performed with regard to a generic molecule standard since no standard of each species can be used for quantification. This drawback is likely less important in molecules with the same ionizable group, i.e. phosphocholine in PC, but it may become relevant in ganglioside species. Accordingly, relative quantifications are currently provided, or correction factors have to be calculated. A correlation of the fatty acyls determined by GC-MS (FAME) with the fatty acyls expected in the intact molecule identified by (U)HPLC-MS analysis for PL could be drawn, this is a consequence of the fact that a molecule of PL or TAG has to yield a molecule of FA, then raising correction factors that can be used in further analysis of the intact molecule. The shotgun methodology has perhaps an advantage in this issue over HPLC-MS analysis in that all the molecules are analyzed under the same ionic conditions (the same solvent), and, consequently, all the molecules of the same type may be expected to yield the same response. However, ion suppression may be considered using the shotgun methodology, thus the less abundant species being likely underestimated. A recent work reports a comparison of quantification of PL with three methodologies, they including HPLC-ELS, HPLC-MS/MS and ^{31}P-NMR (Tavazzi, Fontannaz, Lee, & Giuffrida, 2018). The authors claim improved robustness in quantification as compared to previous methods although PS species quantification became limited by deficient chromatographic resolution of these species.

3.2 Comparative Lipidomic Approaches for Analysis of Molecular Lipid Species

In order to determine the advantages and drawbacks of each methodology, a multicenter with multi-technique analysis should be conducted over the same sample. Taking into consideration the diverse polarity and hydrophobicity/hydrophilicity of the main constituents of the milk lipid profile, from the less polar TAGs to the relatively polar glycolipids (gangliosides), analysis of milk lipids is to require a combination of a set of techniques and methodologies while a single analysis approach for all the compounds is likely unfeasible.

Application of multivariate statistical methods to the raw data from two or more sample groups is becoming a current methodology to find out differential features in regard to nutritional value (infant formulae for example), adulteration, or milk and dairy product processing (Alexandre-Gouabau et al., 2018; Li et al., 2017; Lu et al., 2018; Montero et al., 2013). Two statistical methods are currently applied in the comparison of sample groups, principal component analysis (PCA) and partial-least square (or projection of latent structures) discriminant analysis (PLS-DA); furthermore, sequential application of PCA first to define the model in an unsupervised mode, and subsequent application of this model through PLS-DA, which is a supervised multivariate method, is a current workflow as well. Thus, it does not need of initial full compound characterization, but only the compounds associated to the features that have been shown as differential in the multivariate statistical analysis (loadings-plots). This type of analysis can be performed in a targeted or untargeted mode. A targeted mode can use the TAG or the PL profile, for example, whereas in an untargeted mode there is no previous selection of the compound class over which the analysis is conducted.

The seasonal variation of cow milk composition as well as the effect of ultra-high temperature processing (UHT) was studied (Lu et al., 2018). In spite of raw and UHT milk samples could not be grouped separately and some overlay existed, the loadings plot pointed out that there were a higher proportion of polar lipids, FFA and 1-3 DAG and lower proportion of TAG in UHT milk.

Other studies concluded that PLS-DA analysis spotted breast milk lipids that could be used as biomarkers determining a clear-cut between the neonates growing faster or slower (Alexandre-Gouabau et al., 2018). A set of 14 lipid compounds were proposed by Li et al. (Li et al., 2017) using HPLC-Orbitrap-MS analysis as biomarkers to distinguish soy, goat and bovine milks, the compound set being used to screen potential adulterations.

4 Conclusions

Although during the last decade big progress has been made in the field of lipidomics toward establishing the determinants of specificity for lipid molecules of MFGM, standardization of analytical measurements together with development of more sensitive and specific methods is a demanding future issue. Accordingly, suitability of a milk standard reference material as for human plasma would be valuable for inter-laboratory and inter-instrument comparison of results (Simón-Manso, Lowenthal, Kilpatrick, et al., 2013). Additionally, accurate characterization of regio- and *cis/trans* isomers may be relevant for assessing bioavailability and health-related functionality of these compounds. To this end, as discussed above, it is expected a great advanced on the understanding of the diverse biological roles of these complex lipids after development of integrated tools. Application of multivariate statistical methods, i.e. PCA or PLS-DA, to comparative studies between treated and control groups may help to the correct determination of which are the actual functional compounds without extensive identification of the full compound set, or otherwise whether MFGM supplementation allows to distinguish between groups for a given physiological response (Alexandre-Gouabau et al., 2018; Bhinder, Allaire, Garcia, et al., 2017). Therefore, such biostatistical methods can also be useful for biomarker-based characterization of any given product, adulteration detection or nutritional deficiencies in infant formulae (Li et al., 2017).

References

Alexandre-Gouabau, M.-C., Moyon, T., Cariou, V., Antignac, J.-P., Qannari, E., Croyal, M., et al. (2018). Breast milk lipidome is associated with early growth trajectory in preterm infants. *Nutrients, 10*(2), 164. https://doi.org/10.3390/nu10020164

Ali, A. H., Zou, X., Huang, J., Abed, S. M., Tao, G., Jin, Q., et al. (2017). Profiling of phospholipids molecular species from different mammalian milk powders by using ultra-performance liquid chromatography-electrospray ionization-quadrupole-time of flight-mass spectrometry. *Journal of Food Composition and Analysis, 62*, 143–154. https://doi.org/10.1016/j.jfca.2017.05.007

Astaire, J. C., Ward, R., German, J. B., & Jiménez-Flores, R. (2003). Concentration of polar MFGM lipids from buttermilk by microfiltration and supercritical fluid extraction. *Journal of Dairy Science, 86*(7), 2297–2307.

Avalli, A., & Contarini, G. (2005). Determination of phospholipids in dairy products by SPE/HPLC/ELSD. *Journal of Chromatography A, 1071*(1–2), 185–190. https://doi.org/10.1016/j.chroma.2005.01.072

Barry, K. M., Dinan, T. G., Murray, B. A., & Kelly, P. M. (2016). Comparison of dairy phospholipid preparative extraction protocols in combination with analysis by high performance liquid chromatography coupled to a charged aerosol detector. *International Dairy Journal, 56*, 179–185. https://doi.org/10.1016/j.idairyj.2016.01.022

Bhinder, G., Allaire, J. M., Garcia, C., Lau, J. T., Chan, J. M., Ryz, N. R., et al. (2017). Milk fat globule membrane supplementation in formula modulates the neonatal gut microbiome and normalizes intestinal development. *Scientific Reports, 7*, 45274. https://doi.org/10.1038/srep45274

Bourlieu, C., Cheillan, D., Blot, M., Daira, P., Trauchessec, M., Ruet, S., et al. (2018). Polar lipid composition of bioactive dairy co-products buttermilk and butterserum: Emphasis on sphingolipid and ceramide isoforms. *Food Chemistry, 240*, 67–74. https://doi.org/10.1016/J.FOODCHEM.2017.07.091

Buszewski, B., & Noga, S. (2012). Hydrophilic interaction liquid chromatography (HILIC)–a powerful separation technique. *Analytical and Bioanalytical Chemistry, 402*(1), 231–247. https://doi.org/10.1007/s00216-011-5308-5

Canela, N., Herrero, P., Mariné, S., Nadal, P., Ras, M. R., Rodríguez, M. Á., et al. (2016). Analytical methods in sphingolipidomics: Quantitative and profiling approaches in food analysis. *Journal of Chromatography A, 1428*, 16–38. https://doi.org/10.1016/j.chroma.2015.07.110

Castro-Gómez, M. P., Holgado, F., Rodríguez-Alcalá, L. M., Montero, O., & Fontecha, J. (2015). Comprehensive study of the lipid classes of krill oil by fractionation and identification of Triacylglycerols, Diacylglycerols, and phospholipid molecular species by using UPLC/QToF-MS. *Food Analytical Methods, 8*(10), 2568–2580. https://doi.org/10.1007/s12161-015-0150-6

Castro-Gómez, P., Montero, O., & Fontecha, J. (2017). In-depth lipidomic analysis of molecular species of triacylglycerides, diacylglycerides, glycerophospholipids, and sphingolipids of buttermilk by GC-MS/FID, HPLC-ELSD, and UPLC-QTOF-MS. *International Journal of Molecular Sciences, 18*(3), 605. https://doi.org/10.3390/ijms18030605

Castro-Gómez, P., Rodríguez-Alcalá, L. M., Monteiro, K. M., Ruiz, A. L. T. G., Carvalho, J. E., & Fontecha, J. (2016). Antiproliferative activity of buttermilk lipid fractions isolated using food grade and non-food grade solvents on human cancer cell lines. *Food Chemistry, 212*, 695–702. https://doi.org/10.1016/j.foodchem.2016.06.030

Castro-Perez, J., Roddy, T. P., Nibbering, N. M. M., Shah, V., McLaren, D. G., Previs, S., et al. (2011). Localization of fatty acyl and double bond positions in phosphatidylcholines using a dual stage CID fragmentation coupled with ion mobility mass spectrometry. *Journal of the American Society for Mass Spectrometry, 22*(9), 1552–1567. https://doi.org/10.1007/s13361-011-0172-2

Christiansen, K. (1975). Lipid extraction procedure for in vitro studies of glyceride synthesis with labeled fatty acids. *Analytical Biochemistry, 66*(1), 93–99. https://doi.org/10.1016/0003-2697(75)90728-9

Ciucanu, I., & Kerek, F. (1984). A simple and rapid method for the permethylation of carbohydrates. *Carbohydrate Research, 131*(2), 209–217. https://doi.org/10.1016/0008-6215(84)85242-8

Conway, V., Couture, P., Gauthier, S., Pouliot, Y., & Lamarche, B. (2014). Effect of buttermilk consumption on blood pressure in moderately hypercholesterolemic men and women. *Nutrition (Burbank, Los Angeles County, Calif.), 30*(1), 116–119. https://doi.org/10.1016/j.nut.2013.07.021

Craige Trenerry, V., Akbaridoust, G., Plozza, T., Rochfort, S., Wales, W. J., Auldist, M., et al. (2013). Ultra-high-performance liquid chromatography-ion trap mass spectrometry characterisation of milk polar lipids from dairy cows fed different diets. *Food Chemistry, 141*(2), 1451–1460. https://doi.org/10.1016/j.foodchem.2013.04.023

Crespo, M. C., Tomé-Carneiro, J., Gómez-Coronado, D., Burgos-Ramos, E., Garciá-Serrano, A., Martín-Hernández, R., et al. (2018). Modulation of miRNA expression in aged rat hippocampus by buttermilk and krill oil. *Scientific Reports, 8*(1). https://doi.org/10.1038/s41598-018-22148-5

Donato, P., Cacciola, F., Cichello, F., Russo, M., Dugo, P., & Mondello, L. (2011). Determination of phospholipids in milk samples by means of hydrophilic interaction liquid chromatography coupled to evaporative light scattering and mass spectrometry detection. *Journal of Chromatography A, 1218*(37), 6476–6482. https://doi.org/10.1016/j.chroma.2011.07.036

Ejsing, C. S., Sampaio, J. L., Surendranath, V., Duchoslav, E., Ekroos, K., Klemm, R. W., et al. (2009). Global analysis of the yeast lipidome by quantitative shotgun mass spectrometry. *Proceedings of the National Academy of Sciences, 106*(7), 2136–2141. https://doi.org/10.1073/pnas.0811700106

Et-Thakafy, O., Guyomarc'h, F., & Lopez, C. (2017). Lipid domains in the milk fat globule membrane: Dynamics investigated in situ in milk in relation to temperature and time. *Food Chemistry, 220*, 352–361. https://doi.org/10.1016/j.foodchem.2016.10.017

Fischbeck, A., Krüger, M., Blaas, N., & Humpf, H.-U. (2009). Analysis of sphingomyelin in meat based on hydrophilic interaction liquid chromatography coupled to electrospray ionization–tandem mass spectrometry (HILIC-HPLC-ESI-MS/MS). *Journal of Agricultural and Food Chemistry, 57*(20), 9469–9474. https://doi.org/10.1021/jf9025376

Folch, J., Lees, M., & Stanley, G. H. S. (1957). A simple method for the isolation and purification of total lipids from animal tissues. *The Journal of Biological Chemistry, 226*(1), 497–509. http://www.jbc.org

Fong, B., Norris, C., Lowe, E., & McJarrow, P. (2009). Liquid chromatography-high-resolution mass spectrometry for quantitative analysis of gangliosides. *Lipids, 44*(9), 867–874. https://doi.org/10.1007/s11745-009-3327-1

Fong, B., Norris, C., & McJarrow, P. (2011). Liquid chromatography-high-resolution electrostatic ion-trap mass spectrometric analysis of GD3 ganglioside in dairy products. *International Dairy Journal, 21*(1), 42–47. https://doi.org/10.1016/j.idairyj.2010.07.001

Gallo, M., & Ferranti, P. (2016). The evolution of analytical chemistry methods in foodomics. *Journal of Chromatography A, 1428*, 3–15. https://doi.org/10.1016/j.chroma.2015.09.007

Hakomori, S. (1990). Bifunctional role of glycosphingolipids. Modulators for transmembrane signaling and mediators for cellular interactions. *In Journal of Biological Chemistry, 265*(31), 18713–18716.

Han, X., & Gross, R. W. (2005). Shotgun lipidomics: Electrospray ionization mass spectrometric analysis and quantitation of cellular lipidomes directly from crude extracts of biological samples. *Mass Spectrometry Reviews, 24*(3), 367–412. https://doi.org/10.1002/mas.20023

Holzmüller, W., & Kulozik, U. (2016). Isolation of milk fat globule membrane (MFGM) material by coagulation and diafiltration of buttermilk. *International Dairy Journal, 63*, 88–91. https://doi.org/10.1016/j.idairyj.2016.08.002

Ikeda, K., Shimizu, T., & Taguchi, R. (2008). Targeted analysis of ganglioside and sulfatide molecular species by LC/ESI-MS/MS with theoretically expanded multiple reaction monitoring. *Journal of Lipid Research, 49*(12), 2678–2689. https://doi.org/10.1194/jlr.D800038-JLR200

Ikeda, K., & Taguchi, R. (2010). Highly sensitive localization analysis of gangliosides and sulfatides including structural isomers in mouse cerebellum sections by combination of laser microdissection and hydrophilic interaction liquid chromatography/electrospray ionization mass spectrom. *Rapid Communications in Mass Spectrometry, 24*(20), 2957–2965. https://doi.org/10.1002/rcm.4716

Kolarovic, L., & Fournier, N. C. (1986). A comparison of extraction methods for the isolation of phospholipids from biological sources. *Analytical Biochemistry, 156*(1), 244–250. https://doi.org/10.1016/0003-2697(86)90179-X

Koletzko, B. (2017). Human milk lipids. *Annals of Nutrition and Metabolism, 69*(2), 28–40. https://doi.org/10.1159/000452819

Konrad, G., Kleinschmidt, T., & Lorenz, C. (2013). Ultrafiltration of whey buttermilk to obtain a phospholipid concentrate. *International Dairy Journal, 30*(1), 39–44. https://doi.org/10.1016/j.idairyj.2012.11.007

Le, T. T., Debyser, G., Gilbert, W., Struijs, K., Van Camp, J., Van de Wiele, T., et al. (2013). Distribution and isolation of milk fat globule membrane proteins during dairy processing as revealed by proteomic analysis. *International Dairy Journal, 32*(2), 110–120. https://doi.org/10.1016/j.idairyj.2013.05.002

Lee, H., An, H. J., Lerno Jr., L. A., German, J. B., & Lebrilla, C. B. (2011). Rapid profiling of bovine and human milk gangliosides by matrix-assisted laser desorption/ionization Fourier transform ion cyclotron resonance mass spectrometry. *International Journal of Mass Spectrometry, 305*(2–3), 138–150. https://doi.org/10.1016/j.ijms.2010.10.020

Li, Q., Zhao, Y., Zhu, D., Pang, X., Liu, Y., Frew, R., et al. (2017). Lipidomics profiling of goat milk, soymilk and bovine milk by UPLC-Q-Exactive orbitrap mass spectrometry. *Food Chemistry, 224*, 302–309. https://doi.org/10.1016/j.foodchem.2016.12.083

Lintonen, T. P. I., Baker, P. R. S., Suoniemi, M., Ubhi, B. K., Koistinen, K. M., Duchoslav, E., et al. (2014). Differential mobility spectrometry-driven shotgun lipidomics. *Analytical Chemistry, 86*(19), 9662–9669. https://doi.org/10.1021/ac5021744

Lopez, C., Blot, M., Briard-Bion, V., Cirié, C., & Graulet, B. (2017). Butter serums and buttermilks as sources of bioactive lipids from the milk fat globule membrane: Differences in their lipid composition and potentialities of cow diet to increase n-3 PUFA. *Food Research International, 100*, 864–872. https://doi.org/10.1016/J.FOODRES.2017.08.016

Lopez, C., Madec, M. N., & Jimenez-Flores, R. (2010). Lipid rafts in the bovine milk fat globule membrane revealed by the lateral segregation of phospholipids and heterogeneous distribution of glycoproteins. *Food Chemistry, 120*(1), 22–33. https://doi.org/10.1016/j.foodchem.2009.09.065

Lu, J., Pickova, J., Vázquez-Gutiérrez, J. L., & Langton, M. (2018). Influence of seasonal variation and ultra high temperature processing on lipid profile and fat globule structure of Swedish cow milk. *Food Chemistry, 239*, 848–857. https://doi.org/10.1016/J.FOODCHEM.2017.07.018

Luo, J., Huang, Z., Liu, H., Zhang, Y., & Ren, F. (2018). Yak milk fat globules from the Qinghai-Tibetan plateau: Membrane lipid composition and morphological properties. *Food Chemistry, 245*, 731–737. https://doi.org/10.1016/J.FOODCHEM.2017.12.001

Ma, L., MacGibbon, A. K. H., Jan Mohamed, H. J. B., Loy, S., Rowan, A., McJarrow, P., et al. (2015). Determination of ganglioside concentrations in breast milk and serum from Malaysian mothers using a high performance liquid chromatography-mass spectrometry-multiple reaction monitoring method. *International Dairy Journal, 49*, 62–71. https://doi.org/10.1016/j.idairyj.2015.05.006

Maccarone, A. T., Duldig, J., Mitchell, T. W., Blanksby, S. J., Duchoslav, E., & Campbell, J. L. (2014). Characterization of acyl chain position in unsaturated phosphatidylcholines using differential mobility-mass spectrometry. *Journal of Lipid Research, 55*(8), 1668–1677. https://doi.org/10.1194/jlr.M046995

Malisan, F., & Testi, R. (2002). GD3 ganglioside and apoptosis. *Biochimica et Biophysica Acta - Molecular and Cell Biology of Lipids, 1585*(2–3), 179–187. https://doi.org/10.1016/S1388-1981(02)00339-6

Markham, J. E. (2013). Detection and quantification of plant sphingolipids by LC-MS. *In Methods in molecular biology (Clifton, N.J.), 1009*, 93–101. https://doi.org/10.1007/978-1-62703-401-2_10

Montero, O., Velasco, M., Rodríguez-Lázaro, D., & Hernández, M. (2013). Analysis of cheese small molecules by UPLC-QToF-MS and multivariate statistical methods using several extraction procedures. *Food Analytical Methods, 6*(6), 1497–1507. https://doi.org/10.1007/s12161-013-9661-1

Morin, P., Jiménez-Flores, R., & Pouliot, Y. (2007). Effect of processing on the composition and microstructure of buttermilk and its milk fat globule membranes. *International Dairy Journal 17*(10), 1179–1187. https://doi.org/10.1016/j.idairyj.2007.03.010

Murphy, R. C., & Harrison, K. A. (1994). Fast atom bombardment mass spectrometry of phospholipids. *In Mass Spectrometry Reviews, 13*(1), 57–75. https://doi.org/10.1002/mas.1280130105

Pimentel, L., Fontes, A. L., Salsinha, S., Machado, M., Correia, I., Gomes, A. M., et al. (2018). Suitable simple and fast methods for selective isolation of phospholipids as a tool for their analysis. *Electrophoresis, 39*(15), 1835–1845. https://doi.org/10.1002/elps.201700425

Pimentel, L., Gomes, A., Pintado, M., & Rodríguez-Alcalá, L. M. (2016). Isolation and analysis of phospholipids in dairy foods. *Journal of Analytical Methods in Chemistry, 2016*, 12, (Article ID 9827369), https://doi.org/10.1155/2016/9827369

Pulfer, M., & Murphy, R. C. (2003). Electrospray mass spectrometry of phospholipids. *Mass Spectrometry Reviews, 22*(5), 332–364. https://doi.org/10.1002/mas.10061

Reis, A., Rudnitskaya, A., Blackburn, G. J., Mohd Fauzi, N., Pitt, A. R., & Spickett, C. M. (2013). A comparison of five lipid extraction solvent systems for lipidomic studies of human LDL. *Journal of Lipid Research, 54*(7), 1812–1824. https://doi.org/10.1194/jlr.M034330

Rivas-Serna, I. M., Polakowski, R., Shoemaker, G. K., Mazurak, V. C., & Clandinin, M. T. (2015). Profiling gangliosides from milk products and other biological membranes using LC/MS. *Journal of Food Composition and Analysis, 44*, 45–55. https://doi.org/10.1016/j.jfca.2015.06.006

Rodríguez-Alcalá, L. M., & Fontecha, J. (2010). Major lipid classes separation of buttermilk, and cows, goats and ewes milk by high performance liquid chromatography with an evaporative light scattering detector focused on the phospholipid fraction. *Journal of Chromatography A, 1217*(18), 3063–3066. https://doi.org/10.1016/j.chroma.2010.02.073

Rodríguez-Alcalá, L. M., Castro-Gómez, M. P., Pimentel, L. L., & Fontecha, J. (2017). Milk fat components with potential anticancer activity - A review. *Bioscience Reports, 37*(6), BSR20170705. https://doi.org/10.1042/BSR20170705

Rogers, T. S., Demmer, E., Rivera, N., Gertz, E. R., German, J. B., Smilowitz, J. T., et al. (2017). The role of a dairy fraction rich in milk fat globule membrane in the suppression of postprandial inflammatory markers and bone turnover in obese and overweight adults: An exploratory study. *Nutrition & Metabolism, 14*(1), 36. https://doi.org/10.1186/s12986-017-0189-z

Rombaut, R., Dejonckheere, V., & Dewettinck, K. (2006). Microfiltration of butter serum upon casein micelle destabilization. *Journal of Dairy Science, 89*(6), 1915–1925. https://doi.org/10.3168/jds.S0022-0302(06)72259-7

Šala, M., Lísa, M., Campbell, J. L., & Holčapek, M. (2016). Determination of triacylglycerol regioisomers using differential mobility spectrometry. *Rapid Communications in Mass Spectrometry, 30*(2), 256–264. https://doi.org/10.1002/rcm.7430

Sánchez-Juanes, F., Alonso, J. M. M., Zancada, L., & Hueso, P. (2009). Distribution and fatty acid content of phospholipids from bovine milk and bovine milk fat globule membranes. *International Dairy Journal, 19*(5), 273–278. https://doi.org/10.1016/j.idairyj.2008.11.006

Sanchez-Juanes, F., Alonso, J. M., Zancada, L., & Hueso, P. (2009). Glycosphingolipids from bovine milk and milk fat globule membranes: A comparative study. Adhesion to enterotoxigenic Escherichia coli strains. *Biological Chemistry, 390*(1), 31–40.

Shvartsburg, A. A., Isaac, G., Leveque, N., Smith, R. D., & Metz, T. O. (2011). Separation and classification of lipids using differential ion mobility spectrometry. *Journal of the American Society for Mass Spectrometry, 22*(7), 1146–1155. https://doi.org/10.1007/s13361-011-0114-z

Simón-Manso, Y., Lowenthal, M. S., Kilpatrick, L. E., Sampson, M. L., Telu, K. H., Rudnick, P. A., et al. (2013). Metabolite profiling of a NIST standard reference material for human plasma (SRM 1950): GC-MS, LC-MS, NMR, and clinical laboratory analyses, libraries, and web-based resources. *Analytical Chemistry, 85*(24), 11725–11731. https://doi.org/10.1021/ac402503m

Sisu, E., Flangea, C., Serb, A., Rizzi, A., & Zamfir, A. D. (2011). High-performance separation techniques hyphenated to mass spectrometry for ganglioside analysis. *In Electrophoresis, 32*(13), 1591–1609. https://doi.org/10.1002/elps.201100067

Sokol, E., Ulven, T., Faergeman, N. J., & Ejsing, C. S. (2015). Comprehensive and quantitative profiling of lipid species in human milk, cow milk and a phospholipid-enriched milk formula by GC and MS/MS supALL/sup. *European Journal of Lipid Science and Technology, 117*(6), 751–759. https://doi.org/10.1002/ejlt.201400575

Sørensen, L. K. (2006). A liquid chromatography/tandem mass spectrometric approach for the determination of gangliosides GD3 and GM3 in bovine milk and infant formulae. *Rapid Communications in Mass Spectrometry, 20*(24), 3625–3633. https://doi.org/10.1002/rcm.2775

Ståhlman, M., Ejsing, C. S., Tarasov, K., Perman, J., Borén, J., & Ekroos, K. (2009). High-throughput shotgun lipidomics by quadrupole time-of-flight mass spectrometry. *In Journal of Chromatography B: Analytical Technologies in the Biomedical and Life Sciences, 877*(26), 2664–2672. https://doi.org/10.1016/j.jchromb.2009.02.037

Svennerholm, L., & Fredman, P. (1980). A procedure for the quantitative isolation of brain gangliosides. *Biochimica et Biophysica Acta (BBA)/Lipids and Lipid Metabolism, 617*(1), 97–109. https://doi.org/10.1016/0005-2760(80)90227-1

Tavazzi, I., Fontannaz, P., Lee, L. Y., & Giuffrida, F. (2018). Quantification of glycerophospholipids and sphingomyelin in human milk and infant formula by high performance liquid chromatography coupled with mass spectrometer detector. *Journal of Chromatography B: Analytical Technologies in the Biomedical & Life Sciences, 1072*, 235–243. http://10.0.3.248/j.jchromb.2017.10.067

Verardo, V., Gómez-Caravaca, A. M., Arráez-Román, D., & Hettinga, K. (2017). Recent advances in phospholipids from colostrum, milk and dairy by-products. *International Journal of Molecular Sciences, 18*(1), 173. https://doi.org/10.3390/ijms18010173

Zarei, M., Müthing, J., Peter-Katalinić, J., & Bindila, L. (2010). Separation and identification of GM1b pathway Neu5Ac- and Neu5Gc gangliosides by on-line nanoHPLC-QToF MS and tandem MS: Toward glycolipidomics screening of animal cell lines. *Glycobiology, 20*(1), 118–126. https://doi.org/10.1093/glycob/cwp154

Zhao, Y.-Y., Xiong, Y., & Curtis, J. M. (2011). Measurement of phospholipids by hydrophilic interaction liquid chromatography coupled to tandem mass spectrometry: The determination of choline containing compounds in foods. *Journal of Chromatography A, 1218*(32), 5470–5479. https://doi.org/10.1016/j.chroma.2011.06.025

Chapter 6
Detection of Milk Fat Adulteration

Rajan Sharma, Kamal Gandhi, Surendra Nath Battula, and Bimlesh Mann

1 Introduction

Milk fat has been acclaimed as an indispensable superfood as its nutritional and sensory attributes offer plenty of health benefits (Achaya, 1997). It possesses good flavour, pleasant aroma, high calorific value, besides being a source of valuable nutrients such as fat-soluble vitamins and essential fatty acids. The prices of milk fat have shown upward trend due to the growing demand for it in developed countries which has been attributed to the shift in the opinion of the health concern related to its consumption (OECD/FAO, 2018). International Dairy Federation (IDF) has also noted that over the years there has been a shift in demand from vegetable oil based substitutes to butter and dairy fat due to positive health assessment of milk fat and its sensory properties (IDF, 2018). Increased price and fluctuation in its seasonal availability offers an advantage to the milk fat manufacturers to fraudulently adulterate it with cheaper oils/fats to reduce production costs and increase profit margins. Economic advantage of replacing high-priced fats and oils with low-priced oils without labeling the product accordingly escalates the adulteration of expensive oils and fats such as milk fat. This also poses a risk to human health and decreases its functional value (Ntakatsane, Liu, & Zhou, 2013). Characterization of milk fat for its purity is an absolute necessity in order to ensure a constant well-defined quality. Detection of adulterants in milk fat has always been a challenge because of the variable composition of the triglycerides present. The challenge to detect foreign fats in milk become bigger because of the seasonal, species or breeds related

R. Sharma (✉) · K. Gandhi · B. Mann
Dairy Chemistry Division, ICAR-National Dairy Research Institute, Karnal, Haryana, India
e-mail: rajan.sharma@icar.gov.in

S. N. Battula
Southern Regional Station, ICAR-National Dairy Research Institute, Bengaluru, India

T. Truong et al. (eds.), *Dairy Fat Products and Functionality*,
https://doi.org/10.1007/978-3-030-41661-4_6

variation in the properties of milk fat. Further, the advent of hydrogenated vegetable oil (HVO) industry in the middle of the twentieth century led to large scale adulteration of milk fat with HVO due to the matching physical properties of both fats. Studies related to the detection and quantification of foreign fats in dairy products have been conducted for many decades and constitute priority areas in many research centers (Fontecha, Mayo, Toledano, & Juárez, 2006; Lipp, 1996; Parodi, 1971; Rebechi, Vélez, Vaira, & Perotti, 2016; Timms, 1980).

Methods that have been exploited for the detection of foreign fats in milk fat includes determination of physico-chemical properties, constituents of unsaponifiable matter and evaluation of water-soluble and/or insoluble volatile fatty acids. Further, thin layer chromatography (TLC) of whole and unsaponifiable matter of milk fat, gas chromatography (GC) analysis of triacylglycerol (TAG) or fatty acid profile and HPLC analysis for TAG and marker sterols of milk fat in combination with multivariate statistical data processing have also been utilized to detect adulteration in milk and milk products with foreign fats. However, the majority of the above-mentioned parameters are successful only when a higher quantity of the adulterants are used and are not capable to detect type and the level of added adulterants. This is because of the wide variations in the physico-chemical makeup of milk fat owing to diverse factors like animal species, feeding practices and nutritional management etc. One particular factor which requires attention is the feed of the animal and in this regard, it is pertinent to mention that feeding of cotton seed oil/cake has a profound effect on physico-chemical properties of milk fat. In India, there are separate regulatory requirements (FSSAI Rules, 2011) for milk fat originating in those areas (cotton tract area), where it is a practice to feed animal with cotton seed based products.

This chapter summarizes the existing knowledge in the area of establishing milk fat authenticity employing different techniques for the detection of adulterants like cheaper fats and oils in bovine milk fat.

2 Methods Based on Physical Properties

2.1 Melting Point

Anhydrous milk fat consists of a varied composition of triglycerides. Therefore, it does not have a sharp melting point. The melting point of milk fat ranges from −40 °C to +40 °C (Viriato, de Queirós, da Gama, et al., 2018). Therefore, for practical purposes, it is possible to obtain a temperature range at which a column of fat in a capillary tube starts melting and when it melts completely. The temperature at which milk fat just starts melting is known as slip point. Melting point of various oils and fats varies over a wide range because of variable composition of triglycerides present in them. The melting point of animal body fat (36–51.3 °C) and hydrogenated vegetable oil (HVO) (37.8–38 °C) is slightly higher as compared to that of the vegetable oils (Kumar, Lal, Seth, & Sharma, 2005). Buffalo milk fat

(32.4–34.2 °C) possess higher melting point as compared to that of cow milk fat (30.6–31.2 °C). Melting point of the milk fat from cotton tract area (43.0–44.0 °C) resembles with that of animal body fats. The addition of different adulterants in milk fat causes deviation in the normal range of the melting point, enabling their detection. The melting point method can detect the adulteration of milk fat with body fats (buffalo, goat, pig and sheep) when added at a level higher than 20% (Sharma & Singhal, 1995).

2.2 *Opacity Test*

Opacity test was developed by Singhal (1980) to detect the adulteration of milk fat with animal body fats. To perform this test, 5 g of clear melted fat sample is taken in a test tube (8 × 1.5 cm) and maintained at 50 ± 1 °C for 30 min. Test tube is then transferred to 23 °C water-bath and the opacity time (time taken by the clear melted fat sample to become opaque i.e. O.D. ≥ 0.5) is recorded at 590 nm in a colorimeter. Opacity time of the normal milk fat is around 14–15 min, while that of the animal body fats (buffalo, goat and sheep) is only 10–20 s. The adulteration with buffalo, goat and sheep body fats at 5% level and above could be safely detected by using this test. The detection of body fats, particularly lard in ghee derived from cotton tract area is difficult. This test also suffers from the limitation that adulteration of both body fats and vegetable oils in milk fat could not be detected simultaneously (Singhal, 1980). Dilip, Bindal, and Panda (1998) modified the procedure and recorded the opacity time as the time required by a fat sample at 23 °C to acquire the optical density in the range of 0.14–0.16 and consequent transmittance of 68–72%. The test though simple for field application, sometimes has ambiguity with the onset of opacity and also there are issues with respect to reproducibility owing to polymorphism phenomenon in milk fat.

2.3 *Apparent Solidification Time*

The time taken by the melted fat to get apparently solidified at a given temperature is referred as apparent solidification time (AST). A method has been developed by Kumar, Kahlon, and Chaudhary (2011) to analyse AST of milk fat sample with a view to overcome the limitations of opacity test. To perform this test, 3 g completely melted fat sample is taken into a test tube (10 × 1.0 cm I.D.) and maintained at 60 °C for 5 min followed by maintaining the test tube at 18 ± 0.2 °C in a water-bath. The test tubes are observed constantly till the apparent solidification of fat samples takes place, which is confirmed by non-movement of fat samples on tilting the test tube. At this stage, when the fat sample has apparently solidified, the time taken for the same is recorded as AST using stopwatch. Based on the apparent solidification time of different fats/oils at various temperatures, 18 °C has been recommended by the authors for recording AST. The method can detect adulteration of milk fat with animal body fats at levels ≥10%. Recently, Gandhi, Kumar, and Lal (2018) applied this test to solid fraction obtained at 15 °C (S_{15}) of milk fat obtained after solvent fractionation and indicated that detection of the combination of palm olein and sheep body fat added in milk fat at a level of 6 + 14 (20)% can be detected, which otherwise is difficult to detect.

2.4 *Complete Liquefaction Time*

The time taken by solid milk fat to become completely liquified at a given temperature is defined as complete liquefaction time (CLT) and has been used for the detection of adulteration of milk fat. CLT is expressed in minutes and seconds (min:s) and temperature range of 44–48 °C has been recommended for liquefaction of sample (Kumar, 2008). Recently, this approach has been combined with solvent fraction of milk fat, wherein a solid fraction of fat sample obtained at 15 °C has been subjected to CLT test at 46 °C and 48 °C (Upadhyay, Kumar, Goyal, & Lal, 2017). The results indicated that adulteration in milk fat with goat body fat and groundnut oil could be detected at around 10% level, either singly or in combination.

2.5 *Crystallization Test*

The difference in crystallizing behavior (time) of pure milk fat and adulterants dissolved in a solvent at a particular temperature has been exploited for the development of a field type of test to check milk fat adulteration. Panda and Bindal (1998) studied the crystallization behaviour of fat dissolved in a solvent mixture of acetone and benzene (3.5:1) at 17 °C. They reported that pure milk fat, milk fat adulterated with body fats (10% level) and milk fat adulterated with vegetable oils and fats (10% level) took 19, 3–15 and 22–23 min, respectively to crystallize. They concluded from the study that even a low level of adulteration with animal body fats and vegetable oils and fats in milk fat could be detected. Vegetable oils and body fats could individually be detected at a level of 10% in milk fat based on this test, but their mixture was not detectable.

2.6 *Differential Scanning Calorimetry*

Differential Scanning Calorimetry (DSC) technique is a thermoanalytical technique in which change in heat capacity of a sample is tracked as a function of temperature during heating or cooling. Each fat or oil has a characteristic melting or crystallization curve due to its unique fatty acid and TAG composition. DSC has been exploited by different workers for the detection of adulterants in milk fat, particularly for the detection of animal body fats. Lambelet and Ganguli (1983) observed that the DSC technique is suitable for detection of adulteration of milk fat with body fats as such adulteration results in the appearance of an additional peak in the melting and crystallization curves. It was also highlighted that DSC technique is not suitable for detection of adulteration of milk fat with vegetable oils. Similar observations were also observed by Coni, Di Pasquale, Coppolelli, and Bocca (1994), who developed a DSC based protocol for detection of chicken fat added in milk fat up to 20% level. In their protocol, authors have used the instrument software to generate a thermograph by subtracting cooling data curve of pure milk fat from the mixture pure milk and chicken fat and observed an additional exothermic peak at −12 °C, the area of which increases in proportion to level of adulteration. Later, Aktaş and Kaya (2001) applied this technique for detection of adulteration of butterfat with beef body fat and margarine and indicated an increase in the area of certain peaks in crystallization curves. Recently, Upadhyay, Goyal, Kumar, and Lal (2017) exploited this technique for the detection of caprine body fat and groundnut oil in milk fat (cow as well as buffalo) either individually or mixture thereof as there is difference in crystallizing and melting curves of pure milk fat and adulterant (Fig. 6.1). In adulterated milk fat and pure milk fat, although, the total number of crystallizing and melting peaks were same, difference in cooling and melting curves of pure milk fat and adulterated fat were significantly different in mid-range temperature enabling detection of adulterants in milk fat. Further, shapes of endothermic peaks of pure and adulterated milk fat were also different and can be used for detection of adulteration at more than 5% level. DSC in combination with multivariate statistical analysis technique has also been employed for the detection of adulteration of butter with palm oil (Tomaszewska-Gras, 2016) and lard (Nurrulhidayah, Arieff, Rohman, et al., 2015).

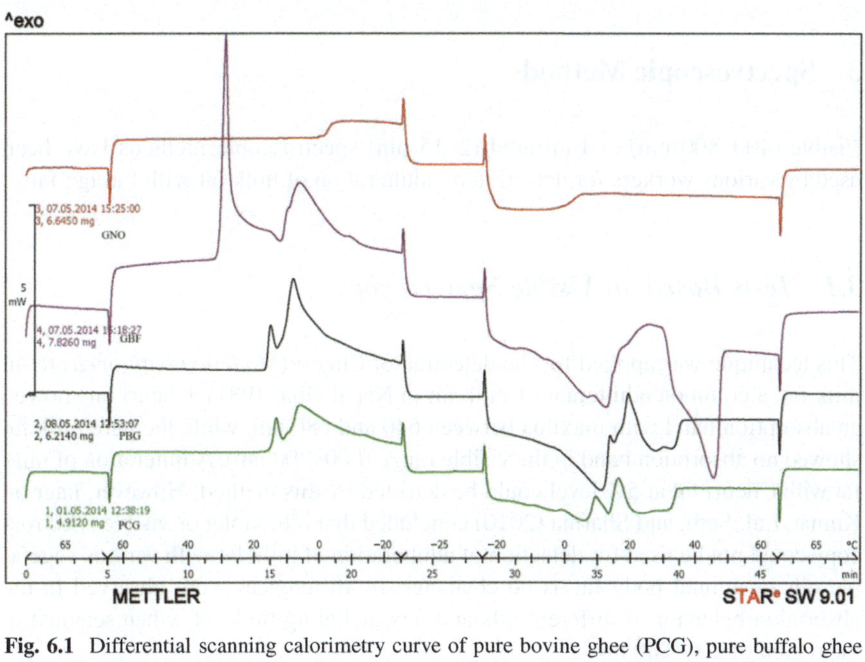

Fig. 6.1 Differential scanning calorimetry curve of pure bovine ghee (PCG), pure buffalo ghee (PBG), caprine body fat (CBF) and groundnut oil (GNO). Adapted from Upadhyay, Goyal, et al. (2017), with permission

2.7 Butyro-Refractometer Reading

The refractive index of milk fat is either analysed in Abbe refractometer, which gives the true refractive index or in a butyro-refractometer (BR reading), which is more convenient to interpret. There exists a mathematical relationship, wherein refractive index can be converted to BR reading and vice versa (BIS, 1966). As refractive index varies with temperature (decreases with a rise and increases with a fall in temperature), refractive index/BR reading is noted at constant temperature. Since milk fat is often solid or semi-solid at room temperature, the BR reading is usually taken at 40 °C, at which the milk fat sample is clear and transparent. The general values for BR reading of milk fat (40–45) and vegetable oils and fats (above 50) are so wide apart that this property can be employed as an indicator for milk fat adulteration with vegetable oils and fats, except coconut oil (38–39) and palm oil (39–40). The BR reading of animal body fats is in the range of 44–51 (Kumar et al., 2005). An increase in BR reading in ghee is also caused by a decrease in the content of lower chain fatty acids, or by an increase either in higher saturated or unsaturated fatty acids, particularly the later (Rangappa & Achaya, 1974). Depending upon the nature of adulterant, 5–20% presence of vegetable fat can be detected in milk using BR reading (Sharma & Singhal, 1995; Singhal, 1980). Practically, BR reading based method is simple and rapid and is widely used as a preliminary screening method for knowing the quality of milk fat. The method has been included as regulatory standard in India (FSSAI Rules, 2011). One of the drawbacks of the method is that the BR reading changes with the animal feed particularly cotton seed oil/cake which limits its application.

3 Spectroscopic Methods

Visible (400–800 nm) and infrared (2–15 μm) spectroscopic methods have been used by various workers for detection of adulteration of milk fat with foreign fats.

3.1 Tests Based on Visible Spectroscopy

This technique was applied for the detection of Cheuri (*Madhuca butyracea*) fat in milk fat, a common adulterant of milk fat in Nepal (Jha, 1981). Cheuri fat showed an absorption band with maxima between 640 and 680 nm, while the pure milk fat showed no absorption band in the visible range (600–700 nm). Adulteration of milk fat with Cheuri fat at 5% level could be detected by this method. However, later on Kumar, Lal, Seth, and Sharma (2010) concluded that ultraviolet or visible spectroscopy could not be used for detection of adulteration of milk fat with various vegetable oils or animal body fat as no characteristic differences were observed in the absorption behavior of different oils and fats including milk fat, when scanned in the region of 200–800 nm.

3.2 Tests Based on Infrared Spectroscopy

Infrared (IR) light is electromagnetic radiation with longer wavelengths than those of visible light, extending from the nominal red edge of the visible spectrum at 14,000–10 cm^{-1} (~0.71–1000 μm). IR light is emitted or absorbed by molecules when they change their rotational-vibrational movements. IR range is further divided into three regions; (1) near IR (NIR) ranges from 14,000 to 4000 cm^{-1}(~0.71–2.5 μm), (2) mid IR (MIR) ranges from 4000 to 400 cm^{-1} (~2.5–25 μm) and far IR (FIR) ranges from 400 to 10 cm^{-1} (~25–1000 μm). Most of the useful diagnostic information from the IR spectrum is provided in the MIR range; however, NIR range has also been used in some of the studies of milk components. IR absorption has been widely used to characterize lipids especially for their cis and trans isomers. Natural vegetable oils and fats contain unsaturated fatty acids, which are in cis configuration and are non-conjugated. Trans-isomers are produced as a result of partial hydrogenation or oxidation. Small amounts of natural trans-isomers have also been reported in animal and marine fats (Akoh, 2017; Kirk & Sawyer, 1991). HVO may contain as high as 50% trans fatty acids due to non-stereospecific hydrogenation in comparison to bovine milk fat, which contains a low level (less than 5%) of trans fat (Fox, Uniacke-Lowe, Mcsweeney, et al., 2015). Absorption maxima at 10.36 μm gets increased by the addition of hydrogenated fats containing iso-oleic acids (trans-octadecenoic acids) in milk fat (Bartlet & Chapman, 1961). Konevets, Roganoua, and Smolyanski (1987) also reported that addition of up to 10% of animal, vegetable and hydrogenated fats in milk fat could be detected using IR spectroscopy. NIR spectroscopic method after second derivatization has been used for the detection of as little as 3% margarine in milk fat (Sato, Kawano, & Iwamoto, 1990).

The full potential of IR spectroscopy has been realized only after the advent of Fourier transforms Infrared (FTIR) spectrometers. This had made the scanning process faster due to the development of an optical device called an interferometer. The interferometer produces a unique type of signal, which has all of the infrared frequencies "encoded" into it. The signal can be measured very quickly, usually in the order of one second or so. Thus, the time element per sample is reduced to a matter of a few seconds rather than several minutes. A further expansion in IR spectroscopy is development of a simple sample holding device called attenuated total reflectance (ATR) which has made the application of sample for FTIR measurement quite simple as just one drop or a few micrograms of sample is required without any pre-treatment. In the last few years, a lot of publications appeared, wherein ATR-FTIR in combination with chemometrics tools such as principal component analysis (PCA) have been used for ascertaining the authenticity of a food sample. In a recent study, Upadhyay, Jaiswal, and Jha (2018) have used ATR-FTIR along with PCA for the detection of pig body fat in milk fat. The authors observed that the absorption values in the spectral region of wavenumbers 3030–2785 cm^{-1}, 1786–1680 cm^{-1} and 1490–919 cm^{-1} were different for pure milk fat, pig body fat and the samples spiked with pig body fat in pure milk fat (Fig. 6.2). The data was further processed using PCA and limit of detection of 3% could be achieved. Earlier also ATR-FTIR coupled with multivariate analysis has been used for the detection of adulteration of milk fat with various body fats. Fadzlillah, Rohman, Ismail, et al. (2013) used the

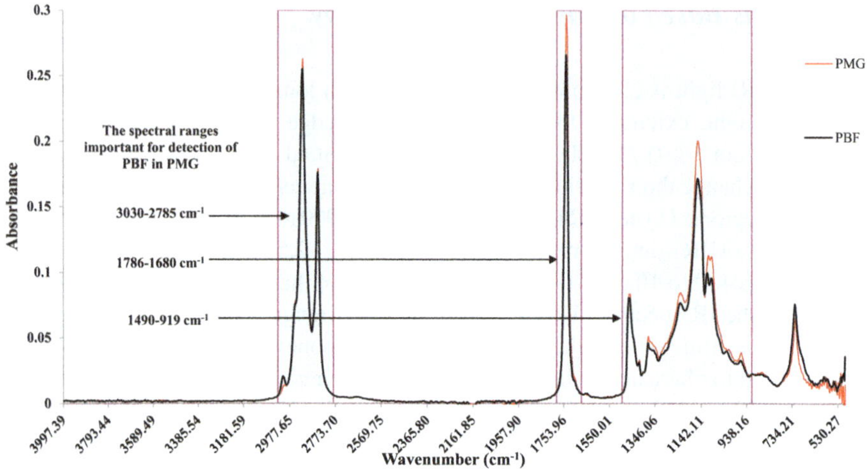

Fig. 6.2 Spectra of pure mixed ghee along with the addition of pig body fat in the wavenumber range 4000–500 cm^{-1}. PMG- Pure Mixed Ghee (1:1 pure cow ghee and pure buffalo ghee) PBF pig body fat. Pure mixed ghee used in the study is combination (1:1) of cow and and buffalo milk fat. Adapted from Upadhyay et al. (2018), with permission

frequency region of 3910–710 cm^{-1} for the detection of mutton fat in milk fat. Upadhyay, Jaiswal, and Jha (2016) employed the difference in absorbance values of pure milk fat and goat body fat in the spectral region of 1786–1680, 1490–919 and 1260–1040 cm^{-1} and indicated that 5% level of adulteration could be detected.

4 Methods Based on Chemical Properties

Several chemical methods based on the nature and content of different constituents of fats such as fatty acids, TAG, unsaponifiable matter, tracers and specific tests using chromatography based instruments have been used to characterize the various fats and oils with a view to check the purity of milk fat.

4.1 Tests Based on Chemical Constants

Certain well-known chemical constants that have been used for the purpose of characterization are described briefly as follows:

(a) **Reichert-Meissl (RM) value**: It is defined as the volume (mL) of 0.1N NaOH required to neutralise steam volatile and water-soluble fatty acid distilled from 5 g of fat under specified conditions (BIS, 1966). This constant is a measure of butyric acid and caproic acid in milk fat. The value for milk fat ranges between 17 and 35, which is well above the values for all other fats and oils (generally less than 1) except coconut oil and palm kernel oil for which the value ranges between 4 and 8 (Singhal, 1980). RM value of milk fat obtained from milk of milch animals fed on cottonseed cake is reduced by 5–6 units (Rangappa & Achaya, 1974). In India, RM value has been a regulatory parameter for milk fat quality as prescribed by a FSSAI Rules (2011) and generally, it should be more than 28. Although, the lower RM value indicates milk fat adulteration, but due to natural variation in milk fat (sometimes RM value is as high as 36), the method cannot be used for detection of milk fat adulteration and generally detection limit of this method is high (Gandhi, Upadhyay, Aghav, et al., 2014; Kumar et al., 2005).

(b) **Polenske value**: Polenske value is the millilitres of 0.1N aqueous alkali solution required to neutralize the steam volatile and water-insoluble fatty acids distilled from 5 g of fat under specified conditions. For milk fat, this value varies from 1.5 to 3.0. Capric acid (C10:0) contributes 3/4th and caprylic acid (C8:0) contributes 1/4th to Polenske value. Polenske value of milk fat obtained from the milk of milch animals fed on cottonseed cake is reduced by 0.3–0.7 units (Rangappa & Achaya, 1974). This method can be used for the detection of adulteration of milk fat with coconut oil as literature indicates high (15–20) Polenske value of coconut oil (Kumar et al., 2005).

(c) **Iodine value**: Iodine value is expressed as the grams of iodine absorbed per 100 g of lipid. It gives a measure of the degree of unsaturation of a lipid. Iodine value is normally used to determine the degree of unsaturation of oils, and to follow processes such as hydrogenation and oxidation that involve changes in the degree of unsaturation. Iodine value of milk fat lies in the range of 30–37. Animal fats like tallow and lard have iodine value of similar order. Many common vegetable oils have high iodine value; coconut oil is unique in having an iodine value in the range of 6–10. HVO usually exhibits higher iodine value (more than 50). Thus, iodine value can be used for the detection of adulteration of milk fat with HVO and some vegetable oils.

(d) **Saponification value**: Saponification value is expressed as the number of milligrams of KOH required to saponify one gram of fat. Saponification value is related to the molecular weight of milk fat and from it can be calculated the saponification equivalent, which is the amount of milk fat saponified by one gram equivalent of potassium hydroxide, and is equal to 56,100 divided by the saponification value. Since milk fat contains high proportions of fatty acids of low molecular weight, its saponification value is exceedingly high (>225). Most of the other fats, which contain palmitic (C16:0), stearic (C18:0), oleic (C18:1) and linoleic (C18:2) fatty acids have a saponification value of around 190. Coconut oil is exceptional in having a saponification value of 255 due to its high content of lauric (C12:0) and myristic (C14:0) fatty acids. Saponification value is inversely proportional to molecular weight. Therefore, high saponification value will result from either an increase in the proportion of lower chain fatty acids or decrease in higher chain fatty acids. Saponification value is of a little use in detection of common adulterants in milk fat. However, the method can be used for detecting the presence of mineral oils, such as liquid paraffin in milk fat as they are not acted upon by alkali and such a sample does not form a homogeneous solution on saponification. This forms the basis of Holde test used for the detection of mineral oil in milk fat (Kumar et al., 2005).

The above-mentioned chemical constants are easy to do and can be used as a screening method to know the quality of milk fat. However, one has to careful in arriving at conclusion with respect to negative sample as these chemical constants can be manipulated by adjusting the mixing of milk fat with foreign fat. In an earlier work done by Sharma and Singhal (1995), it has been concluded that these chemical constants were not suitable to detect the animal body fats (buffalo, goat, sheep and pig) added to buffalo and cow milk fat when adulterated even at a 20% level. Zachariah, Parmar, Bhavadasan, and Nath (2010) also reported that the test was not useful for the detection of the palm and coconut oil.

Attempts have been made to enrich fraction(s) of milk fat rich in adulterants by employing dry or solvent fractionation followed by measuring chemical constants. Keeping fat under different time and temperature combinations, animal body fat and vegetable fat can be partitioned into solid and liquid fractions. Dry and solvent fractionation techniques have been coupled with tests like iodine value (Gandhi, Kumar, & Lal, 2015), RM value (Gandhi et al., 2014), AST (Kumar, Kumar, Lal, et al., 2010), fatty acid profile using GC (Kumar, Upadhyay, Prabhakar, et al., 2015), BR reading (Gandhi & Lal, 2017; Kumar, Upadhyay, Gandhi, et al., 2017) etc. for the authentication of milk fat and it was reported that the technique had improved the sensitivity of the these tests. The time-temperature combination for solvent fractionation using acetone had been optimized (Gandhi et al., 2018) in order to enrich sheep body fat in the first solid fraction and palm olein in the last liquid fraction. In various studies using the fractionation technique, it has been concluded that the technique had increased the sensitivity of iodine value (Gandhi, 2014), RM value (Gandhi et al., 2014), Polenske value (Gandhi, 2014), apparent solidification time (Gandhi et al., 2018) and BR reading (Gandhi & Lal, 2017) to detect adulteration of milk fat with various fats/oils.

4.2 Tests Based on Fatty Acid Profile Using Gas Chromatography

Milk fat is a complex fat composed of approximately 400 different fatty acids (Lindmark Månsson, 2008), many of them are in traces. Table 6.1 indicates the major fatty acids in milk fat and common vegetable oils and animal body fats which may be used as adulterants. Fatty acids are usually analyzed by GC equipped with flame ionization detector (FID). Fatty acid values in literature have been reported as a % of total fatty acids and thus, individual fatty acid values may vary in different studies. By using GC-MS, as more number of fatty acids are detected, the individual fatty acid value may go down. A cursory look at Table 6.1 indicates that milk fat is the only fat which contains short-chain fatty acids like butyric acid (C4:0), caproic acid (C6:0) etc, while vegetable oils/fats are rich in linoleic acid (C18:2), palmitic acid (C:16:0) and stearic acid (C18:0) (Kim, Kim, Lee, et al., 2010; Kumar et al., 2015). Coconut oil is a well-known exception, containing lauric (C12:0) and myristic (C14:0) acids in very large amount (Rangappa & Achaya, 1974). Animal body

Table 6.1 Fatty acids profile of different vegetable oils and animal body fats

	Milk fat		Body fats				Vegetable oils							
Fatty acids	Cow	Buffalo	Goat	Sheep	Pig	Buffalo	Soybean	Palm	Palm olein	RBO	GNO	Sunflower oil	Coconut oil	Mustard oil
4:0	3.2	4.4	–	–	–	–	–	–	–	–	–	–	–	–
6:0	2.1	1.5	–	–	–	–	–	–	–	–	–	–	–	–
8:0	1.2	0.8	–	–	–	–	–	–	–	–	–	–	6.21	–
10:0	2.6	1.3	–	–	–	–	–		–	–	–	–	6.15	–
12:0	2.8	1.8	–	–	–	–	–	–	0.23	–	–	–	51.02	–
14:0	12	10.8	2.6	1.7	1.6	3.1	–	1	1.02	0.5	1.0	0.1	18.94	–
16:0	30	33.1	30.3	23.4	23.9	24.5	10	42	39.66	18	11.0	9.1	8.62	6.52
16:1	2.2	2.0	1.6	2.1	2.4	0.6	–	–	0.18	–	–	0.1	–	–
18:0	10.1	12	18.5	39.8	12.8	44.6	4.0	4	3.81	2	3.0	4.6	1.94	1.98
18:1	27.4	27.2	38.5	25.7	35.8	25.5	25	43	44.01	42	49.0	26.4	5.84	45.39
18:2	1.5	1.5	9.2	2.6	14.3	1.7	52	8	10.73	37	29.0	57.8	1.28	46.02
18:3	0.6	0.5	–	1.1	1.4	–	7	–	–	–	1.0	0.1	–	0.12
Others	4.5	3.6	–	3.8	0.2	–	2	2	–	0.5	6.0	0.3	–	–

RBO-rice bran oil, pig body fat (Wood et al., 2008); mustard and coconut oil (Chowdhury et al., 2007); other fats and oils (Gandhi, 2014); GNO-groundnut oil fats like tallow and lard mostly contain oleic acid. Therefore, considering the varied fatty acid composition of different oils/fats, efforts were made by different researchers to detect adulteration in milk fat with cheaper oils/fats using GC technique. Apart from the content, the ratio of fatty acids has also been exploited for the detection of adulteration of milk fat. Table 6.2 depicts the basis of detection of adulteration with different adulterants in milk fat using GC.

Table 6.2 Gas chromatography (GC-FID) of milk fats for their fatty acids and triglycerides composition to detect adulteration

Sample type	Adulterant	Marker	Main findings	References
Milk fat	Vegetable oils such as sunflower, canola, maize, and rice bran	Long chain unsaturated fatty acids and saturated fatty acids along with principal component analysis	The PCA revealed that the pure and adulterated samples were discriminated according to the total concentration of SFA and UFA respectively	Ntakatsane et al. (2013)
Cow milk fat, buffalo milk fat	Soybean oil, buffalo depot fat	Ratios of sum of C4:0 to C14:1/sum of C15:0 to C20:0 fatty acids and vice-versa	5% and above levels are detected	Kumar et al. (2015)
Milk fat	Non-milk fat Constituents	Two fatty acids (C18:1 and C18:2) along with TAG (C52 and C54 and cholesterol content	These bio-markers enabled the detection of as low as 10% adulteration of various types of non-milk (both vegetable oils into pure milk fat	Kim, Kim, and Park (2015)
Milk fat	Tallow, lard	A multiple linear regression (MLR) model with fatty acids such as C4:0, C8:0, C10:0, C10:1, C12:0, C14:0, C14:1, C16:0, C16:1, C18:3 and $18:2_{conj}$ has been developed	The models were able to detect adulterations of milk fat at levels greater than 10% for tallow and 5% for lard	Rebechi et al. (2016)
Milk Fat	Vegetable oils, pork lard and beef tallow	Using C14:0, C16:0, C18:0, C18:1, and C18:2 fatty acids	The result showed that qualitative determination of the milk fat samples adulterated with different vegetable oils and animal fats was possible by a comparison of these fatty acids	Park, Kim, Yang, et al. (2014)
Butter	Lard	Using C12:0, C14:0, C16:0, C15:0, C18:1 and C18:2	Due to the addition of lard into butter, fatty acids such as namely lauric (12:0), myristic (14:0) and palmitic (C16:0) decreased linearly. The increased concentration of lard in adulterated butter sample caused the fatty acids levels of C18:0, C18:1 and C18:2 to increase linearly	Fadzlillah, Rohman, Rosman, et al. (2016)

Fatty acid ratio ($C_{10:0}/C_{8:0}$, $C_{12:0}/C_{10:0}$, $C_{14:0}/C_{12:0}$ and $C_{14:0}/C_{18:1}$) of milk fat is being used as an official method in Argentina for ascertaining the quality of milk fat (ANMAT, 2011). Normal range values of these fatty acid ratios have been proposed and any sample outside this normal range is considered adulterated. These values are 1.85–2.30 for $C_{10:0}/C_{8:0}$, 0.95–1.30 for $C_{12:0}/C_{10:0}$, 3.00–4.10 for $C_{14:0}/C_{12:0}$, and >0.30 for $C_{14:0}/C_{18:1}$. However, in a recent study (Rebechi et al., 2016), it was observed that these ratios failed to identify non-milk fat such as tallow and lard in milk fat at a level less than 10%. These authors have suggested the use of Multiple Linear Regression (MLR), and prediction model was chosen and validated. These developed models were able to detect adulterations of milk fat at a level greater than 5% and 10% for tallow and lard, respectively.

4.3 Tests Based on Triacylglycerol Profile Using Gas Chromatography

Milk fat is composed predominantly of TAG with 26–52 carbon number, while animal body fats and common vegetable oils other than coconut and palm kernel oils have triglycerides with 50–54 carbon number. Coconut and palm kernel oil contain short and medium chain length triglycerides with 30–52 carbon number, a range almost similar to butterfat (Parodi, 1969; Rangappa & Achaya, 1974). TAG profile of milk fat has been used as a tool for detection of adulteration of milk fat with foreign fats. TAG profiling can be achieved by HPLC, hyphenated with evaporative light-scattering detection (ELSD) or by GC using either packed or capillary column. As there is a linearity problem with ELSD (Destaillats, de Wispelaere, Joffre, et al., 2006), most of the literature on TAG profiling of milk fat samples is available on capillary column based GC. Timms (1980) analysed 76 Australian milk fat by GC and derived an equation ($14.197C_{40} - 36.396C_{42} + 32.364C_{44} = R$) using MLR in which the contents of the TAG C_{40}, C_{42} and C_{44} were inserted. It was shown that Australian milk fat and milk fat originated from 10 different countries fitted into this equation with a range of $R = 98.12$–101.88. It was concluded that as little as 5% of any non-milk fat can be detected with >99% confidence. In 1992, Precht (1992) analysed 755 different milk fats and 13 foreign fats (vegetable as well as animal body fats) including their combination for TAG composition. TAG formulae were derived using statistical procedures and weightage was also given to TAG with carbon number 26, 28, 30, 32, 34, 46 apart from the TAG proposed by Timms (2005). This allowed the detection of foreign fats in milk fats with enhanced sensitivity. Soybean oil, sunflower oil, olive oil, coconut oil, palm oil, palm kernel oil, lard and beef tallow with limits of detection of 1.3, 1.2, 1.7, 2.5, 2.7, 2.9, 1.9 and 4.1%, respectively were established with 95% confidence and only slightly higher values with 99% confidence. The regression equation developed by Precht (1992) was based on the TAG content analysed from European milk fat. This method was included as an official method in Germany in 1994 (DIN 10336: 1994-09, 1994) and as EU reference method in 1995, currently included in Commission Regulation (EC) No 213/2001 (2001).

In subsequent studies, Molkentin (2007) confirmed that the regression equation developed by Precht (1992) can be used for authentication of milk fat originating from New Zealand and South Africa. Later in 2010, the method was adopted by IDF (2010) wherein S-values (calculated from TAG percentage) were slightly different from those proposed by (Precht, 1992). IDF (2010) recommends the use of both packed as well as capillary column with FID to determine the TAG content. In this method, a range of S-values of pure milk fat is prescribed by five different equations and if S-values deviate from the given range, the sample is considered adulterated. In this method, mass fraction of TAG with carbon number 24, 26, 28, 30, 32, 34, 36, 38, 40, 42, 44, 46, 48, 52 and 54 is taken. The applicability of regression equations prescribed by various official bodies to ascertain the genuineness of non-bovine milk fat has been questioned and there are reports in this regard (Amrutha Kala, 2013; Hazra, Sharma, Sharma, & Arora, 2017; Povolo, Pelizzola, & Contarini, 2008). IDF (2010) in its scope also indicated that the method is not applicable to milk fat (a) obtained from bovine milk other than cow's milk; (b) obtained from single cows; (c) obtained from cows, which received an exceptionally high feeding of pure vegetable oils such as rapeseed oil; (d) obtained from colostrum; (e) subjected to technological treatment such as removal of cholesterol or fractionation; (f) obtained from skim milk or buttermilk and (g) extracted by using the Gerber, Weibull–Berntrop or Schmid–Bondzynski–Ratzlaff methods, or that has been isolated using detergents. Further, the suitability of this method for the detection of tallow in milk fat adulterated with palm olein has also been questioned due to similarity in TAG profile (Sharma, Hazra, Kandhol, et al., 2018).

4.4 Tests Based on the Type of Sterols

The characteristic sterol in milk fat (or in animal fat) is cholesterol, while type of sterol present in vegetable oils is a mixture of various other sterols (β-sitosterol, stigmasterol, campesterol, brassicasterol, etc.) collectively referred as phytosterols (González-Larena, García-Llatas, Vidal, et al., 2011). Phytosterols contain an extra methyl group, ethyl group, or double bond. The presence of phytosterols in a milk fat sample thus indicates its adulteration with vegetable oil. Among the various phytosterols, β-sitosterol has been reported as major phytosterol in most of the vegetable oils followed by campesterol and stigmasterol (Chen, Wesley, Shamburek, et al., 2005). Various analytical methods have been developed to detect the presence of β-sitosterol in suspected milk samples. IDF (2006a) has recommended a routine method, wherein individual sterols in unsaponifiable matter are detected by capillary based GC system. The individual sterols are detected by retention time and adulteration of milk fat can be detected by the appearance of phytosterols in the chromatogram. In another IDF recommended reference method (IDF, 2006b), sterols in unsaponifiable matter are separated by thin layer chromatography (TLC) and then they are converted to their silyl derivatives followed by analysis in GC using capillary column. The introduction of TLC step removes most of the impurities from the unsaponifiable matter. Later on, researchers have recommended use of reversed phase thin layer chromatography (RP-TLC) alone also for the detection of vegetable oil in milk fat (Rani, Sharma, Arora, et al., 2015; Upadhyay, Kumar, Rathod, et al., 2015). In this approach extracted unsaponifiable matter is applied on alumina backed silica gel RP-18 plate followed by the separation of sterols using a solvent system consisting of petroleum ether, acetonitrile and methanol in a specific ratio. The separated sterols on the plate are visualized using phosphomolybdic acid solution. The adulterated sample is distinguished by the presence of two (cholesterol and β-sitosterol) bands on the developed TLC plate, while in case of pure milk fat, only one distinct band of cholesterol appears (Fig. 6.2). The method has been claimed to detect the presence of most of the oils at 1% level in milk fat. In the case of coconut oil, sensitivity is around 7.5%. These methods suffer from the fact the separation of cholesterol and β-sitosterol band is not very clear and retention factor is very close, often overlapping with each other. Apart from the GC and TLC based method, liquid chromatography has also been suggested for the quantification of sterols in dairy products (Kamm, Dionisi, Fay, et al., 2001; Rani, Sharma, Arora, et al., 2016).

5 Miscellaneous Tests

5.1 Tests Based on the Identification of DNA in Milk Fat

DNA based analytical methods are in use for the detection of species adulteration of milk products (Poonia, Jha, Sharma, et al., 2017) or milk products (Di Pinto, Terio, Marchetti, et al., 2017; Liao, Liu, Ku, et al., 2017). In all these methods, DNA is extracted from milk/milk product sample having somatic cells followed by the amplification of specific region of DNA using primers by polymerase chain reaction (PCR). The amplified product has a specific size, which is visualized on gel electrophoresis and results can be interpreted. Usually, such methods are quite sensitive and the main task is to get good quality DNA, which sometimes may be problematic due to prior processing of the dairy sample. Isolation of intact DNA in sufficient quantity form high fat product is difficult and is one of the critical steps (Costa, Mafra, & Oliveira, 2012). A method has been standardized by (Kumar et al., 2011) using ASL (a stool lysis) buffer QIAmp DNA Stool Kit for the isolation of DNA from vegetable oils. The same method has been applied by Sharma et al. (2018) for the isolation of DNA from cow milk fat and using buffalo species specific DNA primers, the authors were able to demonstrate the detection of buffalo tallow in milk fat. The final amplified PCR product of 226-bp size absent in control sample and was present in milk fat sample containing 10% buffalo tallow. The method was validated using IDF reference method (IDF, 2010) for detection of purity of milk using TAG profiling and in principle can be used for detection of adulteration of milk fat with any vegetable oil. However, more work is required to be done with special reference to repeatability of isolation of DNA to establish the utility of DNA based methods for detection of milk fat adulteration.

5.2 Bomer Value

Bomer value is defined as the sum of the melting point of saturated triglycerides (isolated by diethyl ether method) and twice the difference between this melting point and that of the fatty acids obtained. The Bomer value of animal body fats, e.g., goat, sheep and buffalo (68–69) and pig body (75–76) is higher as compared to that of cow and buffalo milk fat (63–64). Singhal (1980, 1987) reported that the Bomer value of milk fat increased on adulteration with body fats even in the presence of vegetable oils but not when vegetable oils were added alone. The method was suggested as the confirmatory test for the detection of pig body fat (lard) in milk fat (Sharma & Singhal, 1996).

5.3 Detection of Vegetable Fat in Milk by Phytosterol Acetate Test

Phytosteryl acetate test is based on the principle that the melting point of cholesteryl acetate (114–115 °C) is lower than that of phytosteryl acetate (126–137 °C). Further, the crystals of acetates can be differentiated under microscope as the shapes of crystals of cholesterol (parallelogram with an obtuse angle of 100°) and phytosterol (elongated hexagonal form with an apical angle of 108°) differ. In this test, milk fat sample is saponified and the unsaponifiable matter is extracted and sterols present are isolated with the help of digitonin and then converted into their acetates followed by determination of their melting point. If the melting point of steryl acetate obtained is more than 117 °C, then the adulteration with vegetable oil is confirmed. Alternatively, sterol crystals can be viewed under a microscope, presence of sterol crystals having a re-entry angle (swallow's tail) confirmed the presence of vegetable fat in milk fat. This was an official method of International Organization of Standardization (ISO, 1976) and BIS (BIS, 1966). ISO in 2007 has withdrawn this method on the recommendation of the technical committee as the method was cumbersome and lacked repeatability.

5.4 Methods Based on Tracer Components of Fats and Oils

Tracer components can be defined as those compounds, which are present in adulterant oils and fats, either naturally or by addition as per law, but absent in pure milk fat. Addition of some tracer component in the likely adulterant of milk fat has been suggested as a rapid and reliable tool to identify them in adulterated milk fat. A tracer can be a latent colour which is not detectable visually but can be identified by its colour reaction with certain chemicals or a colouring matter (natural or synthetic), which may impart direct colouration distinct from that of the natural colour of butterfat.

5.4.1 Baudouin Test

In India, it is mandatory to add sesame oil (5% by weight) to oil used for the preparation of HVO according to FSSAI Rules (2011) to enable its detection in milk fat by Baudouin test. Sesamolin (present in sesame oil) on hydrolysis in the presence of concentrated HCl gives sesamol, which reacts with furfural to develop a permanent crimson colour. Hydrofuramide (Kapur, Srinivasan, & Subrahmanyan, 1960) or p-hydroxybenzaldehyde (Singhal, 1973) has also been suggested in place of furfural for this test. To perform this test 5 mL of milk fat sample is melted in a test tube. To which 5 mL of concentrated hydrochloric acid followed by 0.4 mL of 2% furfural solution in alcohol is added. Mixture is then shaken vigorously for 2 min. The mixture is then allowed to separate by keeping it undisturbed. Positive test is indicated by the development of permanent pink/red color in the acid layer. Confirmation is done by addition of 5.0 mL of distilled water and shaking again. If the color in acid layer persists, then the presence of HVO in milk fat is confirmed. If the color disappears, then it is absent in milk fat. In India, this test has been quite effective in controlling the adulteration of milk fat with HVO. However, the availability of partially HVO (which does not require the addition of sesame oil) in recent times may further aggravate the problem of mixing of such oils to milk fat.

5.4.2 Halphen Test

Cottonseed oils contain cyclopropene ring containing fatty acid, viz., malvalic (C18:0) and sterculic (C19:0) acids, which are altogether absent in milk fat. The presence of cyclopropene ring containing fatty acids is used as a tool for the detecting cottonseed oil in milk fat and also to distinguish cotton tract milk fat from normal milk fat. This test is based on the reaction between cyclopropenoic acids (constituent fatty acids of cottonseed oil) and Halphen reagent (1% sulphur solution in carbon disulphide + equal volume of iso-amyl alcohol) after incubation for an hour in a boiling bath of saturated sodium chloride solution, which results in the development of a crimson colour. This test is used to distinguish the cotton tract milk fat from normal milk fat (Singhal, 1980) as well as for the detection of cottonseed oil in milk fat.

5.4.3 Test for the Detection of Palm Oil

This test is based on the assumption that tannins present as traces in palm oil give Prussian blue colour with potassium ferricyanide and ferric chloride reagent. To perform this test, 2 mL of melted fat sample is mixed with 5 mL acetonitrile. The contents are shaken and centrifuged. Acetonitrile layer (lower layer) is mixed with 1 mL of 0.008M ferric chloride and 0.3 mL of 0.003M potassium ferricyanide solution. The appearance of Prussian blue colour indicates the presence of palm oil in milk fat. Limitation of this test is that antioxidants present in oils may also be responsible for the positive test. Hence, the test is not specific for the presence of palm oil. The test is valid for the detection of vegetable oils in milk fat samples which are free from any added antioxidants including Butylated Hydroxy Anisole (BHA).

5.4.4 Modified Bieber Test

In this method (Sharma, Lal, & Sharma, 2007), 1 ml of milk fat is dissolved in 1.5 mL hexane and is mixed with 1 mL of freshly prepared colour reagent. The colour reagent is prepared by mixing distilled water, sulphuric acid (sp. gr. 1.835) and nitric acid (sp. gr. 1.42) in the ratio of 20:6:14. The appearance of distinct orange brownish tint in the upper layer indicates the presence of vegetable oil/fat in milk. Through experimentation, it has been hypothesized that fat exclusively from vegetable origin is responsible for this orange brownish tint. Further, it was inferred that some allied component of β-sitosterol present in vegetable oil might be responsible for the color. The method has been validated using TLC based method as indicated in IDF (2006a).

5.5 *Use of Pixel Intensity, Fractal Dimension and Skeleton Parameters*

Image analysis of milk fat crystals has been also been attempted to detect its adulteration with HVO. Wasnik, Menon, Nath, et al. (2017) determined the pixel intensity, fractal dimension and skeleton parameters of the images of the ghee samples adulterated with HVO at various levels (5, 10, 15 and 20%) using a flatbed scanner and concluded that these parameters increased with the level of adulteration enabling its detection at 10% level. Wasnik, Menon, Surendra Nath, et al. (2017) also measured the equivalent particle diameter, yellowness index and whiteness index by using image analysis technique and concluded that the detection of the HVO at 5% and above level was possible. The methods used in the above studies require flatbed scanner and images have been processed using Image open J software.

6 Conclusion

The detection of foreign fats in milk fat is a very intricate phenomenon. Wide variation in physico-chemical constants limits their use as such for the detection of adulteration of milk fat. The application of physico-chemical constants to fractionated milk fat appears to have some merit. For the detection of vegetable oil, test based on phytosterols needs attention and sophisticated techniques based on mass spectrophotometry may be attempted. The proposed IDF method based on the profiling of TAG using GC needs more validation, particularly for its application in milk fat of non-bovine species. Some of the recent literature indicates the use of direct analysis in real-time ionization–high resolution mass spectrometry (Hrbek, Vaclavik, Elich, et al., 2014) for TAG profiling appears to be promising. Among the rapid techniques, which have been explored for detection of milk fat adulteration include Raman spectroscopy (Karacaglar, Bulat, Boyaci, et al., 2018) wherein the timeline of 30 min for sample analysis has been claimed. Detection of adulteration of milk fat with animal body fats and their combination with vegetable oils continues to remain a challenge. In this regard, it is pertinent to emphasize that no single strategy may work for detection of adulteration of milk fat and a combination of analytical methods needs to be used to ascertain the purity of milk fat.

References

Achaya, K. T. (1997). *Ghee, vanaspati and special fats in India*. New York, NY: Marcel Dekker.

Akoh, C. (2017). *Food lipids: Chemistry, nutrition, and biotechnology*. Boca Raton, FL: CRC Press Taylor and Francis Group.

Aktaş, N., & Kaya, M. (2001). Detection of beef body fat and margarine in butterfat by differential scanning calorimetry. *Journal of Thermal Analysis and Calorimetry, 66*, 795–801. https://doi.org/10.1023/A:1013196106365

Amrutha Kala, A. L. (2013). Detection of possible adulteration in commercial ghee samples using low-resolution gas chromatography triglyceride profiles. *International Journal of Dairy Technology, 66*, 346–351. https://doi.org/10.1111/1471-0307.12049

ANMAT (2011). *Código Alimentario Argentino. (Chap. 8) Alimentos Lácteos*. Retrieved 28 October, 2018, from http://www.anmat.gov.ar/codigoa/caa1.htm

Bartlet, J. C., & Chapman, D. G. (1961). Butter adulteration, detection of hydrogenated fats in butter fat by measurement of cis-trans conjugated unsaturation. *Journal of Agricultural and Food Chemistry, 9*, 50–53.

BIS. (1966). *Indian standards. Methods for sampling and test for ghee (butterfat). IS 3508: 1960, Amended 2013*. New Delhi, India: Bureau of Indian Standards.

Chen, J. T., Wesley, R., Shamburek, R. D., Pucino, F., & Csako, G. (2005). Meta-analysis of natural therapies for hyperlipidemia: Plant sterols and stanols versus policosanol. *Pharmacotherapy, 25*, 171–183. https://doi.org/10.1016/j.chroma.2006.07.040

Chowdhury, K., Banu, L. A., Khan, S., & Latif, A. (2007). Studies on the fatty acid composition of edible oil. *Bangladesh Journal of Scientific and Industrial Research, 42*(3), 311–316.

Commission Regulation (EC) No 213/2001. (2001). Methods for the analysis and quality evaluation of milk and milk products. Commission Regulation (EC) No. 213/01 of 9 January. *Official Journal of the European Community, L37*, 42–99.

Coni, E., Di Pasquale, M., Coppolelli, P., & Bocca, A. (1994). Detection of animal fats in butter by differential scanning calorimetry: A pilot study. *Journal of the American Oil Chemists' Society, 71*, 807–810. https://doi.org/10.1007/BF02540453

Costa, J., Mafra, I., & Oliveira, M. B. P. P. (2012). Advances in vegetable oil authentication by DNA-based markers. *Trends in Food Science and Technology, 26*, 43–55.

Destaillats, F., de Wispelaere, M., Joffre, F., Golay, P. A., Hug, B., Giuffrida, F., Fauconnot, L., & Dionisi, F. (200 6). Authenticity of milk fat by fast analysis of triacylglycerols: Application to the detection of partially hydrogenated vegetable oils. *Journal of Chromatography. A, 1131*, 227–234. https://doi.org/10.1016/j.chroma.2006.07.040

Di Pinto, A., Terio, V., Marchetti, P., Bottaro, M., Mottola, A., Bozzo, G., Bonerba, E., Ceci, E., & Tantillo, G. (2017). DNA-based approach for species identification of goat-milk products. Food Chemistry, 229, 93–97. https://doi.org/10.1016/j.foodchem.2017.02.067

Dilip, P., Bindal, M. P., & Panda, D. (1998). Detection of adulteration in ghee with animal body fats and vegetable oils using opacity test. *Journal of Dairying, Foods and Home Sciences, 17*, 31–36.

DIN 10336: 1994-09. (1994). *Nachweis und Bestimmung von Fremdfetten in Milchfett anhand einer gaschromatographischen Triglyceridanalyse*. Berlin, Germany: BeuthVerlag GmbH.

Fadzlillah, N. A., Rohman, A., Rosman, A. S., Yusof, F. M., Ismail, A., Mustaffa, S., Minhat, A. E., & Khatib, A. (2016). Differentiation of fatty acid composition of butter adulterated with lard using gas chromatography mass spectrometry combined with principal component analysis. *Jurnal Teknologi, 78*(2), 71–177.

Fadzlillah, N. A., Rohman, A., Ismail, A., Mustafa, S., & Khatib, A. (2013). Application of FTIR-ATR spectroscopy coupled with multivariate analysis for rapid estimation of butter adulteration. *Journal of Oleo Science, 62*, 555–562. https://doi.org/10.5650/jos.62.555

Fontecha, J., Mayo, I., Toledano, G., & Juárez, M. (2006). Triacylglycerol composition of protected designation of origin cheeses during ripening. Authenticity of milk fat. *Journal of Dairy Science, 89*, 882–887.

Fox, P. F., Uniacke-Lowe, T., Mcsweeney, P. L., H. & O'Mahony, J. A. (2015). *Dairy chemistry and biochemistry* (2nd ed.). Cham, Switzerland: Springer.

FSSAI Rules. (2011). *Akalank's food safety and standards act, rules and regulation*. New Delhi, India: Akalank Publication.

Gandhi, K. (2014). *Detection of palm olein and sheep body fat adulteration in ghee using solvent fractionation technique*. Karnal, India: NDRI.

Gandhi, K., Kumar, A., & Lal, D. (2015). Iodine value integrated with solvent fractionation technique as a tool for detecting palm olein and sheep body fat adulteration in ghee (clarified milk fat). *Indian Journal of Dairy Science, 68*, 347–351.

Gandhi, K., Kumar, A., & Lal, D. (2018). Solvent fractionation technique paired with apparent solidification time (AST) test as a method to detect palm olein and sheep body fat in ghee (clarified milk fat). *Indian Journal of Dairy Science, 71*, 246–251.

Gandhi, K., & Lal, D. (2017). Butyro-refractometer (BR) reading linked with solvent fractionation technique as an aid to detect adulteration of palm olein and sheep body fat in ghee. *Indian Journal of Natural Products and Resources, 8*, 276–281.

Gandhi, K., Upadhyay, N., Aghav, D., Sharma, V., & Lal, D. (2014). Detection of adulteration of ghee (clarified milk fat) with palmolein and sheep body fat using Reichert-Meissl (RM) value coupled with solvent fractionation technique. *Indian Journal of Dairy Science, 67*, 387–393.

González-Larena, M., García-Llatas, G., Vidal, M. C., Sánchez-Siles, L. M., Barberá, R., & Lagarda, M. J. (2011). Stability of plant sterols in ingredients used in functional foods. *Journal of Agricultural and Food Chemistry, 59*(8), 3624–3631.

Hazra, T., Sharma, V., Sharma, R., & Arora, S. (2017). A species specific simplex polymerase chain reaction-based approach for detection of goat tallow in heat clarified milk fat (ghee). *International Journal of Food Properties, 20*, S69–S75. https://doi.org/10.1080/10942912.2017.1289542

Hrbek, V., Vaclavik, L., Elich, O., & Hajslova, J. (2014). Authentication of milk and milk-based foods by direct analysis in real time ionization–high resolution mass spectrometry (DART–HRMS) technique: A critical assessment. *Food Control, 36*(1), 138–145.

IDF. (2006a). *Anhydrous milk fat – Determination of sterol composition by gas liquid chromatography (Routine method). IDF 200: 2006*. Brussels, Belgium: International Dairy Federation.

IDF. (2006b). *Anhydrous milk fat – Determination of sterol composition by gas liquid chromatography (Reference method). IDF 159: 2006*. Brussels, Belgium: International Dairy Federation.

IDF. (2010). *Milk and milk products – Determination of milk fat purity by gas chromatographic analysis of triglycerides (Reference method). IDF 202:2010*. Brussels, Belgium: International Dairy Federation.

IDF. (2018). *IDF Press Release. OECD expects dairy fat consumption to increase*. July 3, 2018. Brussels, Belgium.

ISO. (1976). *Milk fat - Detection of vegetable fat by the phytosteryl acetate test. ISO 3595: 1976*. Geneva, Switzerland: International Organization for Standardization.

Jha, J. S. (1981). Spectrophotometric studies of cheuri (*Madhucabutyracea*) fat and ghee mixtures: I. *Journal of the American Oil Chemists' Society, 58*, 843–845. https://doi.org/10.1007/BF02672954

Kamm, W., Dionisi, F., Fay, L. B., Hischenhuber, C., Schmarr, H. G., & Engel, K. H. (2001). Analysis of steryl esters in cocoa butter by on-line liquid chromatography–gas chromatography. *Journal of Chromatography. A, 918*(2), 341–349.

Kapur, O. P., Srinivasan, M., & Subrahmanyan, V. (1960). Colouring of vanaspati with curcumin from turmeric. *Current Science, 29*, 350–351.

Karacaglar, N. N. Y., Bulat, T., Boyaci, I. H., & Topcu, A. (2018). Raman spectroscopy coupled with chemometric methods for the discrimination of foreign fats and oils in cream and yogurt. *Journal of Food and Drug Analysis, 27*(1), 101–110. https://doi.org/10.1016/j.jfda.2018.06.008

Kim, J., Kim, D. N., Lee, S. H., Yoo, S. H., & Lee, S. (2010). Correlation of fatty acid composition of vegetable oils with rheological behaviour and oil uptake. *Food Chemistry, 118*, 398–402. https://doi.org/10.1016/j.foodchem.2009.05.011

Kim, J. M., Kim, H. J., & Park, J. M. (2015). Determination of milk fat adulteration with vegetable oils and animal fats by gas chromatographic analysis. *Journal of Food Science, 80*, 1945–1951. https://doi.org/10.1111/1750-3841.12979

Kirk, R. S., & Sawyer, R. (1991). *Pearson's composition and analysis of foods*. Boston, MA: Addison-Wesley Longman Ltd.

Konevets, V. I., Roganoua, Z. A., & Smolyanski, A. L. (1987). Use of infra-red spectroscopy for analysing fats of different origin. *Izv Vyrshikh O Chebnykh Zavad Pischevaya Tekhnologiya, 1*, 64–67.

Kumar, A. (2008). *Detection of adulterants in ghee*. PhD Thesis. Submitted to National Dairy Research Institute (Deemed University), Karnal, India.

Kumar, A., Kumar, A., Lal, D., Seth, R., & Sharma, V. (2010). Validation of ultra-violet and visible spectroscopic methods for detection of milk fat adulteration. *Journal of Dairy and Food, 29*(1), 8–14.

Kumar, A., Lal, D., Seth, R., & Sharma, R. (2005). Approaches for detection of adulteration in milk fat-an overview. *Indian Dairyman, 57*, 31.

Kumar, A., Lal, D., Seth, R., & Sharma, V. (2010). Detection of milk fat adulteration with admixture of foreign oils and fats using a fractionation technique and the apparent solidification time test. *International Journal of Dairy Technology, 63*, 457–462. https://doi.org/10.1111/j.1471-0307.2010.00562.x

Kumar, A., Upadhyay, N., Gandhi, K., Naik, S. N., & Sharma, V. (2017). Detection of adulteration in anhydrous milk fat using season variation in Butyro-refractometer reading studied by employing dry fractionation technique. *Indian Journal of Dairy Science, 70*, 563–570.

Kumar, A., Upadhyay, N., Padghan, P. V., Gandhi, K., Lal, D., & Sharma, V. (2015). Detection of vegetable oil and animal depot fat adulteration in anhydrous milk fat (ghee) using fatty acid composition. *MOJ Food Processing & Technology, 1*, 13–19. https://doi.org/10.15406/mojfpt.2015.01.00013

Kumar, S., Kahlon, T., & Chaudhary, S. (2011). A rapid screening for adulterants in olive oil using DNA barcodes. *Food Chemistry, 127*, 1335–1341. https://doi.org/10.1016/j.foodchem.2011.01.094

Lambelet, P., & Ganguli, N. C. (1983). Detection of pig and buffalo body fat in cow and buffalo ghees by differential scanning calorimetry. *Journal of the American Oil Chemists' Society, 60*, 1005–1008. https://doi.org/10.1007/BF02660216

Liao, J., Liu, Y. F., Ku, T., Liu, M. H., & Huang, Y. (2017). Qualitative and quantitative identification of adulteration of milk powder using DNA extracted with a novel method. *Journal of Dairy Science, 100*, 1657–1663. https://doi.org/10.3168/jds.2016-11900

Lipp, M. (1996). Determination of the adulteration of butter fat by its triglyceride composition obtained by GC. A comparison of the suitability of PLS and neural networks. *Food Chemistry, 55*(4), 389–395.

Månsson, L. (2008). Fatty acids in bovine milk fat. *Food & Nutrition Research, 52*, 1821.

Molkentin, J. (2007). Detection of foreign fat in milk fat from different continents by triacylglycerol analysis. *European Journal of Lipid Science and Technology, 109*, 505–510.

Ntakatsane, M. P., Liu, X. M., & Zhou, P. (2013). Short communication: Rapid detection of milk fat adulteration with vegetable oil by fluorescence spectroscopy. *Journal of Dairy Science, 96*, 2130–2136. https://doi.org/10.3168/jds.2012-6417

Nurrulhidayah, A. F., Arieff, S. R., Rohman, A., Amin, I., Shuhaimi, M., & Khatib, A. (2015). Detection of butter adulteration with lard using differential scanning calorimetry. *International Food Research Journal, 22*, 832–839. https://doi.org/10.1007/s10973-011-1913-y

OECD/FAO. (2018). *OECD-FAO agricultural outlook 2018-2027*. Rome, Italy: OECD Publishing, Paris/Food and Agriculture Organization of the United Nations.

Panda, D., & Bindal, M. P. (1998). Detection of adulteration in ghee with animal body fat and vegetable oils using crystallization test. *Indian Dairyman, 50*, 13–20.

Park, J. M., Kim, N. K., Yang, C. Y., Moon, K. W., & Kim, J. M. (2014). Determination of the authenticity of dairy products on the basis of fatty acids and triacylglycerols content using GC analysis. *Korean Journal for Food Science of Animal Resources, 34*, 316–324. https://doi.org/10.5851/kosfa.2014.34.3.316

Parodi, P. W. (1969). Detection of a synthetic milkfat. *Australian Journal of Dairy Technology, 24*, 56.

Parodi, P. W. (1971). Detection of synthetic and adulterated butterfat 3. Triglyceride fatty acid analysis. *Australian Journal of Dairy Technology, 26*, 155.

Poonia, A., Jha, A., Sharma, R., Singh, H. B., Rai, A.K., Sharma, N. (2017). Detection of adulteration in milk: A review. *International Journal of Dairy Technology, 70*, 23–42.

Povolo, M., Pelizzola, V., & Contarini, G. (2008). Directly resistively heated-column gas chromatography for the evaluation of cow milk fat purity. *European Journal of Lipid Science and Technology, 110*, 1050–1057. https://doi.org/10.1002/ejlt.200800010

Precht, D. (1992). Detection of foreign fat in milk fat II. Quantitative evaluation of foreign fat mixtures. *Zeitschrift für Lebensmittel-Untersuchung und -Forschung, 194*, 107–114. https://doi.org/10.1007/BF01190178

Rangappa, K. S., & Achaya, K. T. (1974). *Indian dairy products*. Mumbai, India: Asia Publishing House.

Rani, A., Sharma, V., Arora, S., Lal, D., & Kumar, A. (2015). A rapid reversed-phase thin layer chromatographic protocol for detection of adulteration in ghee (clarified milk fat) with vegetable oils. *Journal of Food Science and Technology, 52*, 2434–2439.

Rani, A., Sharma, V., Arora, S., Ghai, D. L. (2016). Comparison of rapid reversed phase high-performance liquid chromatography (RP-HPLC) method with rapid reversed phase thin layer chromatography method for detecting vegetable oils in ghee (clarified milk fat). *International Journal of Food Properties, 19*(5), 1154–1162.

Rebechi, S. R., Vélez, M. A., Vaira, S., & Perotti, M. C. (2016). Adulteration of Argentinean milk fats with animal fats: Detection by fatty acids analysis and multivariate regression techniques. *Food Chemistry, 192*, 1025–1032. https://doi.org/10.1016/j.foodchem.2015.07.107

Sato, T., Kawano, S., & Iwamoto, M. (1990). Detection of foreign fat adulteration of milk fat by near infrared spectroscopic method. *Journal of Dairy Science, 73*, 3408–3413. https://doi.org/10.3168/jds.S0022-0302(90)79037-6

Sharma, R., & Singhal, O. P. (1995). Physico-chemical constants of ghee prepared from milk adulterated with foreign fat. *Indian Journal of Dairy and Biosciences, 6*, 51–53.

Sharma, R., & Singhal, O. P. (1996). Fatty acid composition, Bomer value and opacity profile of ghee prepared from milk adulterated with foreign fats. *Indian Journal of Dairy Science, 49*, 62–67.

Sharma, V., Hazra, T., Kandhol, R., Sharma, R., & Arora, S. (2018). Confirmation of buffalo tallow in anhydrous cow milk fat using gas liquid chromatography in tandem with species-specific polymerase chain reaction. *International Journal of Dairy Technology, 71*, 158–163. https://doi.org/10.1111/1471-0307.12390

Sharma, V., Lal, D., & Sharma, R. (2007). Color based platform test for the detection of vegetable oils/fats in ghee. *Indian Journal of Dairy Science, 60*, 16–18.

Singhal, O. P. (1973). *Studies on ghee clarified butterfat and animal body fats with a view to detect adulteration*. Karnal, India: NDRI.

Singhal, O. P. (1980). Adulterants and methods for detection. *Indian Dairyman, 32*, 771–774.

Singhal, O. P. (1987). *Detection of adulteration in butterfat. NDRI annual report* (pp. 64–65). Karnal, India: National Dairy Research Institute.

Timms, R. E. (1980). The phase behaviour and polymorphism of milk fat, milk fat fractions and fully hardened milk fat. *Australian Journal of Dairy Technology, 35*, 47–53.

Timms, R. E. (2005). Fractional crystallisation - the fat modification process for the 21st century. *European Journal of Lipid Science and Technology, 107*, 48–57. https://doi.org/10.1002/ejlt.200401075

Tomaszewska-Gras, J. (2016). Rapid quantitative determination of butter adulteration with palm oil using the DSC technique. *Food Control, 60*, 629–635. https://doi.org/10.1016/j.foodcont.2015.09.001

Upadhyay, N., Goyal, A., Kumar, A., & Lal, D. (2017). Detection of adulteration by caprine body fat and mixtures of caprine body fat and groundnut oil in bovine and buffalo ghee using differential scanning calorimetry. *International Journal of Dairy Technology, 70*, 297–303. https://doi.org/10.1111/1471-0307.12336

Upadhyay, N., Jaiswal, P., & Jha, S. N. (2016). Detection of goat body fat adulteration in pure ghee using ATR-FTIR spectroscopy coupled with chemometric strategy. *Journal of Food Science and Technology, 53*, 3752–3760. https://doi.org/10.1007/s13197-016-2353-2

Upadhyay, N., Jaiswal, P., & Jha, S. N. (2018). Application of attenuated total reflectance Fourier Transform Infrared spectroscopy (ATR–FTIR) in MIR range coupled with chemometrics for detection of pig body fat in pure ghee (heat clarified milk fat). *Journal of Molecular Structure, 1153*, 275–281. https://doi.org/10.1016/j.molstruc.2017.09.116

Upadhyay, N., Kumar, A., Goyal, A., & Lal, D. (2017). Complete liquification time test coupled with solvent fractionation technique to detect adulteration of foreign fats in ghee. *International Journal of Dairy Technology, 70*, 110–118. https://doi.org/10.1111/1471-0307.12323

Upadhyay, N., Kumar, A., Rathod, G., Goyal, A., & Lal, D. (2015). Development of a method employing reversed-phase thin-layer chromatography for establishing milk fat purity with respect to adulteration with vegetable oils. *International Journal of Dairy Technology, 68*, 207–217. https://doi.org/10.1111/1471-0307.12178

Viriato, R. L. S., de Souza Queirós, M., da Gama, M. A. S., Ribeiro, A. P. B., & Gigante, M. L. (2018). Milk fat as a structuring agent of plastic lipid bases. *Food Research International, 111*, 120–129. https://doi.org/10.1111/1750-3841.14728

Wasnik, P. G., Menon, R. R., Surendra Nath, B., Balasubramanyam, B. V., Manjunatha, M., & Sivaram, M. (2017). Application of pixel intensity, fractal dimension and skeleton parameters for detection of adulteration of cow ghee with vanaspati derived from image analysis. *Indian Journal of Dairy Science, 70*, 331–337.

Wasnik, P. G., Menon, R. R., Surendra Nath, B., Balasubramanyam, B. V., Manjunatha, M., & Sivaram, M. (2017). Application of particle analysis and colour parameters for detection of adulteration of cow ghee with vanaspati derived from image analysis. *Indian Journal of Dairy Science, 70*, 1–9.

Wood, J. D., Enser, M., Fisher, A. V., Nute, G. R., Sheard, P. R., Richardson, R. I., Hughes, S. I., Whittington, F. M. (2008). Fat deposition, fatty acid composition and meat quality: A review. *Meat science, 78*(4), 343–358.

Zachariah, S. P., Parmar, S. C., Bhavadasan, M. K., & Nath, B. S. (2010). Detection of adulteration of ghee with coconut oil or palm oil. *Indian Journal of Dairy Science, 63*, 278–282.

Chapter 7
Modification of Milk Fat Globules During Processing and Gastrointestinal Digestion

Sophie Gallier and Harjinder Singh

1 Introduction

Milk is an important source of nutrition in the human diet across all life stages, providing proteins (mainly caseins and whey proteins), carbohydrates (mainly lactose but also human milk oligosaccharides for infants) and lipids. Milk contains between 3 and 5% lipids, which are complex not only in composition but also in structure (Jensen & Newsburg, 1995); 95–98% of milk lipids are triglycerides, the remaining being sterols and polar lipids, such as phospholipids and glycosphingolipids. Milk triglycerides are packaged into globules of 0.1–15 μm in diameter and are surrounded by a unique and complex trilayer of phospholipids that embeds membrane-specific proteins and cholesterol, called the milk fat globule membrane (MFGM) (Gallier, Gragson, Jimenez-Flores, & Everett, 2010a; Lopez, Madec, & Jimenez-Flores, 2010).

Milk has evolved to support the healthy growth and development of the neonate and to be easily digested for optimal delivery of its nutrients. Although the composition of the milk fat globule varies among mammalian species, reflecting the nutritional requirements of the neonate, fascinatingly, its complex structure is similar across species studied to date, despite some differences in fatty acid composition and ratios of the MFGM components (Contarini & Povolo, 2013; Oftedal, 2012). In the last decade, some progress has been made in our understanding of the structure of the milk fat globule and how common dairy processing treatments

S. Gallier
Fonterra Research and Development Centre, Fonterra Co-operative Group Ltd, Palmerston North, New Zealand

H. Singh (✉)
Riddet Institute and Massey Institute of Food Science and Technology, Massey University, Palmerston North, New Zealand
e-mail: H.Singh@massey.ac.nz

T. Truong et al. (eds.), *Dairy Fat Products and Functionality*,
https://doi.org/10.1007/978-3-030-41661-4_7

affect this structure to deliver dairy products with improved functional properties (Gallier, Acton, Manohar, & Singh, 2017; Lopez, Cauty, & Guyomarc'h, 2015). However, the repackaging or structural alteration of milk fat may affect the way in which it is digested, absorbed and metabolized (Bourlieu & Michalski, 2015; Michalski et al., 2006; Singh & Gallier, 2014; Vors et al., 2013). In addition, the surrounding matrix has been shown to play a role in the bioaccessibility of raw or processed milk fat globules, which may affect access of the gastrointestinal (GI) enzymes to the globules and, therefore, may influence the bioavailability of milk fat and its stabilizing membrane (Gallier et al., 2017; Ye, Cui, Dalgleish, & Singh, 2016). In this review, we discuss recent advances in the structure of milk fat globules, the effect of dairy processing on the bioaccessibility and bioavailability of milk fat globules in adults and infants and the development of milk-fat-globule-like biomimetics.

2 Secretion and Structure of Milk Fat Globules

The secretion of milk fat globules in the secretory cells of the mammary gland is a very intricate process (Masedunskas, Chen, Stussman, Weigert, & Mather, 2017; Masedunskas, Weigert, & Mather, 2014). The rough endoplasmic reticulum releases small cytoplasmic droplets, which are covered by a monolayer of phospholipids and proteins, into the cytosol of the secretory cell (Keenan & Dylewski, 1995; Masedunskas et al., 2014, 2017). The cytoplasmic droplets migrate towards the apical part of the cell and grow in size by fusion, resulting in droplets of various sizes (Masedunskas et al., 2014, 2017). When the droplets reach the apical part, their monolayer and the bilayer of the plasma membrane interact together, and the droplets are released by pinocytosis into the alveolar lumen. This results in milk fat globules that are covered by a monolayer originating from the endoplasmic reticulum and a bilayer from the plasma membrane (Fig. 7.1). Cytoplasmic crescents (Fig. 7.2a), containing proteinaceous material and secretory vesicles

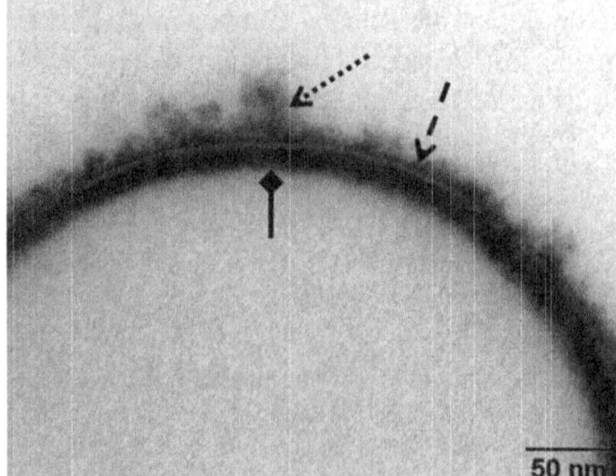

Fig. 7.1 Transmission electron microscopy (TEM) image of human MFGM showing the trilayer structure and the glycocalyx. From Gallier et al. (2015), with permission from Elsevier

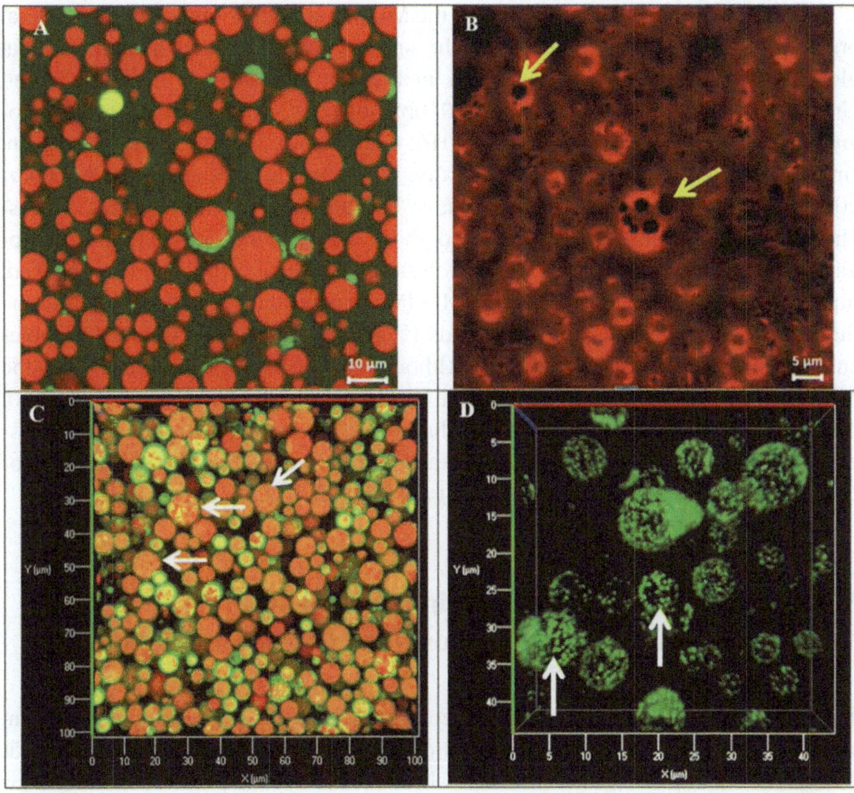

Fig. 7.2 Confocal laser scanning microscopy (CLSM) images of human milk fat globules showing the fat core stained with Nile Red and cytoplasmic crescents stained with Fast Green FCF (**a**), the lateral organization of the phospholipids using a fluorescent phospholipid analogue (**b**), the fat core stained with Nile Red and glycosylated proteins and lipids stained with Alexa Fluor-488 wheat germ agglutinin (**c**) and cholesterol stained with filipin (**d**). From Gallier et al. (2015), with permission from Elsevier

(Gallier et al., 2015, 2017; Keenan & Dylewski, 1995), may be present. These are sandwiched between the monolayer and the bilayer, and their formation may be due to a defect in the "zipping" mechanism between the monolayer and the bilayer (Huston & Patton, 1990; Patton & Huston, 1988).

The diameter of bovine and human milk fat globules ranges from 0.1 to 15 µm, with a mean diameter of between 3 and 5 µm (Singh & Gallier, 2016). In bovine milk, the diameter of the milk fat globules is affected by the breed, the stage of lactation and feeding (Couvreur, Hurtaud, Lopez, Delaby, & Peyraud, 2006; Martini, Salari, & Altomonte, 2016; Walstra, 1995). Regardless of whether the secretion of milk fat globules of different diameters is a random process or is directed by a cellular process, globules of different diameters may play a specific role in controlling the digestion rate of milk fat in addition to delivering energy and lipo-soluble nutrients for the growth and development of the newborn (Bourlieu & Michalski, 2015).

The composition and the structure of the MFGM are rather intriguing. The lateral organization of the phospholipids at the surface of bovine and human milk fat globules has been extensively studied in the last decade both in situ (Gallier, Gragson, Jimenez-Flores, & Everett, 2010b; Lopez et al., 2010) and using monolayer model systems (Bourlieu et al., 2016; Gallier et al., 2010b; Gallier, Gragson, Jimenez-Flores, & Everett, 2012; Murthy, Guyomarc'h, Paboeuf, Vié, & Lopez, 2015; Zheng, Jiménez-Flores, Gragson, & Everett, 2014). The unsaturated phospholipids form a liquid-disordered, fluid phase, whereas the saturated phospholipids, such as sphingomyelin, and cholesterol segregate in a tightly packed, liquid-ordered phase (Fig. 7.2b). The MFGM glycoproteins are located within the fluid phase of unsaturated phospholipids (Fig. 7.2c). MFGM glycosylated molecules form the MFGM glycocalyx. Although cholesterol is located within the MFGM (Fig. 7.2d), the formation of a complex between filipin and cholesterol induces a local deformation of the membrane, such that the location of the fluorescence emitted by filipin may not be exactly representative of the distribution of cholesterol within the MFGM (Gallier et al., 2015, 2017).

3 Impact of Common Processing Treatments on the Structure and Composition of Milk Fat Globules

The MFGM stabilizes the fat in the aqueous phase of milk, preventing flocculation, coalescence and enzymatic degradation of the fat globules (Keenan & Dylewski, 1995). The beating of air into milk, high shear caused by high pressure homogenization and high velocity gradients induce disruption of the fat globules. Milk and other dairy products, such as cream, are also often homogenized to prevent creaming and to improve stability during storage. Milk is commonly heat treated (i.e. pasteurization or ultra-high temperature treatment) to extend its shelf life and to ensure its safety and quality (Tunick et al., 2016). However, common milk processing treatments tend to damage the structure and the composition of the MFGM and to affect the interfacial properties and stability of the fat globules (Singh, 2006). Whereas the safety and quality of dairy products are improved with processing, the microstructure of the milk fat globules may be dramatically altered.

MFGM proteins denature and associate with whey proteins via disulphide bonding at temperatures above 60 °C, i.e. during pasteurization and ultra-high temperature treatment (Corredig & Dalgleish, 1998; Kim & Jimenez-Flores, 1995; Ye, Singh, Taylor, & Anema, 2004). Heating appears to lead to the release of PAS 6/7 (Ye, Singh, Taylor, & Anema, 2002) and phospholipids (Houlihan, Goddard, Kitchen, & Masters, 1992) into the serum phase of milk. Cooling of milk also results in the release of MFGM material, including about 20% of phospholipids, into its serum phase (Walstra, Wouters, & Geurts, 2006). For human milk, Holder pasteurization, i.e. heating the milk for 30 min at 62.5 °C, is a common practice that is used to kill bacteria and viruses (de Oliveira et al., 2016; Sousa, Delgadillo, & Saraiva, 2016). Buchheim, Welsch, Huston, and Patton (1988) studied the microstructure of fresh and heat-treated human milk samples and observed that heat treatment induced the release of highly glycosylated MFGM proteins. This release started at 50–60 °C and became substantial at 70–80 °C. Glycoproteins provide steric repulsion between globules because of the negative charge of some glycan residues, specifically sialic acid. The release of glycoproteins into the serum phase will have an impact on the surface properties and the digestion behaviour of the globules. Buchheim et al. (1988) hypothesized that the mucus-like properties of these high molecular weight glycoproteins may help the globules to adhere on to the intestinal mucosal wall and thus slow their movement down the GI tract for a more efficient digestion of the fat. These glycoproteins may also interact with receptors on the mucosal cells or act like receptors for anchoring the lipases to the globule surface. Therefore, heat treatment of human milk and subsequent loss of the glycocalyx could contribute to a less efficient digestion of human milk fat in infants. In this context, a non-thermal pasteurization appears to be more suitable for retaining the structural and nutritional properties of human milk. Although high pressure processing has been shown to preserve the nutritional quality of human milk, it has not yet been shown if it has an impact on the structure of the milk fat globules (Sousa et al., 2016).

Freezing of milk or cream induces crystallization of the triglycerides and therefore rupture of the MFGM and destabilization of the fat globules upon thawing. Freezing and thawing of milk are common practices in human milk banks (Sousa et al., 2016). In addition, Lepri, Del Bubba, Maggini, Donzelli, and Galvan (1997) observed lipolysis of human milk fat globules not only after freezing for a period of time and thawing of the milk but also after Holder pasteurization. Pasteurization and the freeze/thaw cycle disrupt the human MFGM, affecting the digestion and the delivery of bioactive compounds. Destabilization of bovine milk fat globules by damaging the MFGM also occurs during the manufacture of whipped cream and butter, which involves the rapid incorporation of air.

Homogenization ruptures the MFGM and markedly reduces the size of the fat globules. The increased surface area of the homogenized globules is then covered by additional skim milk proteins, preferentially caseins, because of the lack of sufficient MFGM material (Walstra, 1995). A two-stage homogenization disrupts any fat globule clusters formed during homogenization. The replacement of MFGM components with other milk and exogenous emulsifiers may alter the ζ-potential of the globules (Tunick et al., 2016; Ye, Cui, Dalgleish, & Singh, 2017) and their stability. Therefore, the processing of milk fat globules is likely to affect their digestion behaviour.

4 Gastrointestinal Digestion of Milk Fat Globules in Adults

The GI tract is a succession of complex compartments, all of which help to digest and deliver dietary nutrients efficiently at the site of absorption. The reader is referred to other reviews for details of the GI processing of foods (N'Goma, Amara, Dridi, Jannin, & Carrière, 2012; Singh & Gallier, 2014). The physico-chemical properties of lipids play a role in the digestive process in all compartments (Singh & Gallier, 2016). In the mouth, the salivary mucins interact with the surface of the fat globules, leading to their flocculation under certain conditions (Sarkar, Goh, & Singh, 2009). However, the mechanism and the extent of the interaction of salivary mucins with the MFGM have not been well explored. Milk, as a fluid, will be in contact with saliva only briefly before being swallowed; however, thicker dairy products such as cream and ice cream and solid dairy products such as cheese will be in contact with saliva for a longer time. The interfacial behaviour of fat globules in thicker dairy products during oral processing has not been investigated.

In the stomach, the gastric acidity and enzymatic degradation affect the structure of raw and processed fat globules. The acidic gastric juice mixes with milk but the buffering capacity of milk increases the intragastric pH. The continuous secretion of gastric juice decreases the pH slowly, causing aggregation of the fat globules near the isoelectric point of the interfacial components. For example, homogenized milk fat globules, coated with caseins, aggregate at pH 4.6 whereas native milk fat globules are more stable under gastric conditions and do not aggregate as rapidly (Armand et al., 1996; Gallier et al., 2013b). In addition, pepsin is able to hydrolyse most proteins; however, some MFGM glycoproteins, such as MUC1 and PAS 6/7, are resistant to digestion because of their degree of glycosylation (Le et al., 2012). Therefore, the surface of native milk fat globules appears to be less affected by the hydrolytic action of pepsin than homogenized milk fat globules, which are covered by intact milk proteins and their peptides formed by pepsin digestion (Gallier, Cui, et al., 2013a). The repulsive force between fat globules is reduced by the interfacial accumulation of free fatty acids from gastric lipolysis in addition to the hydrolysis of interfacial proteins (Gallier, Cui, et al., 2013a; Ye, Cui, & Singh, 2011).

Using a human gastric simulator, Ye et al. (2016, 2017) showed that gastric acidity and pepsin hydrolytic activity induced coagulation of the caseins in whole milk and the formation of a clot that trapped fat globules (Fig. 7.3). The curds formed by homogenized and heat-treated milk had a more crumbled and porous structure than the soft curd formed by raw milk under dynamic gastric conditions (Ye et al., 2017). A few fat globules remained in the aqueous phase of the gastric chyme whereas some of the trapped fat globules had coalesced within the close-knit structure. A longer residence time in the stomach compartment resulted in an increased density of the clot and coalescence of the fat globules. The breakdown of the proteins allowed the release of the fat globules over time (Ye et al., 2016).

During the digestion of raw milk or raw cream, the MFGM helps the fat globule to maintain its structure because of the stability of phospholipids and glycoproteins under gastric conditions (Armand et al., 1996; Gallier et al., 2012; Gallier, Cui, et al., 2013a). This allows slowing down of the formation of an aggregated protein matrix and the trapping of fat globules, compared with homogenized fat globules, as seen in vitro (Bourlieu et al., 2015) and in vivo (Gallier, Cui, et al., 2013a). This delayed separation of the solid phase (i.e. the clot) and the aqueous phase is likely to have an impact on gastric emptying, as demonstrated in vivo in humans (Marciani et al., 2007) and in an animal model (Gallier, Rutherfurd, Moughan, & Singh, 2014a). Therefore, raw milk is hypothesized to empty from the stomach more slowly than homogenized milk. The accumulation of lipolytic products was evidenced as spherical protrusions at the surface of raw, pasteurized and pasteurized–homogenized fat

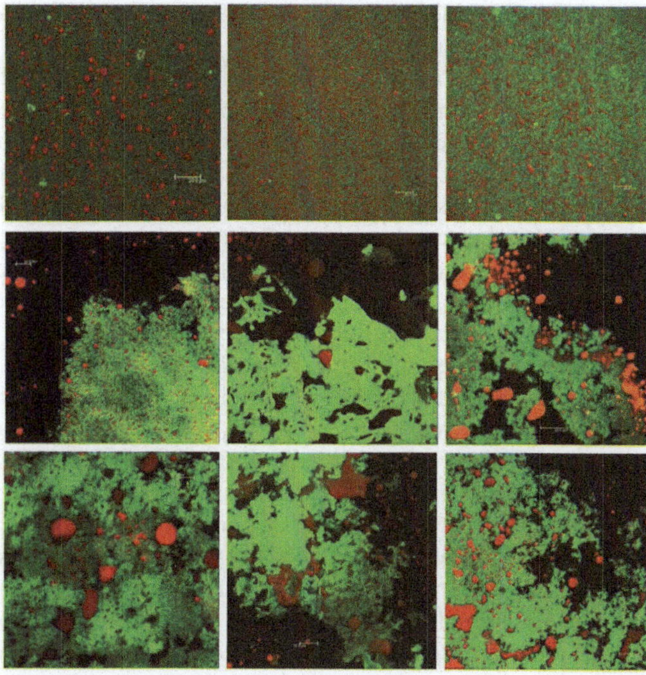

Fig. 7.3 CLSM images of untreated whole milk (left), homogenized milk (middle) and heated and homogenized milk (right) at different times during gastric digestion from 0 to 80 min. Nile Red (red) was used to stain the fat and Fast Green FCF (green) was used to stain the protein. From Ye et al. (2017), with permission from Elsevier

globules at different times of in vivo gastric digestion in the rat (Gallier, Cui, et al., 2013a). Although all fat globules increased in size, homogenized globules remained smaller than raw globules, as seen in vitro (Bourlieu et al., 2015) and in vivo (Armand et al., 1996; Gallier et al., 2013b; Gallier, Cui, et al., 2013a).

The emulsification of fat globules with milk phospholipids, as in native milk fat globules, appears to have a positive impact on lipid metabolism. Rats fed a dairy cream with small protein-coated fat droplets had a slower lipid metabolism than rats fed a dairy cream with large phospholipid-stabilized fat droplets, which was probably induced by a slower rate of gastric emptying of the former, caused by the formation of a curd in the rat stomach (Michalski et al., 2006; Michalski, Briard, Desage, & Geloen, 2005). In other rat (Lecomte et al., 2015) and in vitro (Mathiassen et al., 2015) studies, the emulsification of fat droplets with milk phospholipids, in comparison with soy phospholipids, led to greater digestion and absorption of lipids and a sharper postprandial lipaemia. The presence of milk phospholipids at the fat droplet surface may prevent or slow down aggregation of the fat droplets under gastric conditions and modulate gastric emptying. In addition, under intestinal conditions, the phospholipid species and the degree of saturation in milk phospholipids, both of which govern the physico-chemical properties of the membrane (Gallier et al., 2014b), may favour the interfacial adsorption of digestive lipases (Bourlieu et al., 2016; Chu et al., 2010). For example, the addition of phosphatidylserine, a negatively charged phospholipid, enhanced the absorption of gastric lipase (Bourlieu et al., 2016).

The small intestine is the main site for the digestion of fat globules and the absorption of nutrients. Pancreatic juices increase the pH of the incoming chyme to near neutral. The juices also contain several proteases and specific and non-specific lipases. The change in pH, the enzymatic hydrolytic activity, the ionic environment and the presence of surface-active components, such as bile salts, influence greatly the internal and external structure of the fat globules in the chyme (Gallier, Cui, et al., 2013a; Gallier, Zhu, et al., 2013b). The pancreatic lipase forms a complex with colipase, which adsorbs on to bile-salt platforms at the surface of the fat globules and initiates the digestion of triglycerides at the *sn-1* and *sn-3* positions (Borgstrom & Erlanson-Albertsson, 1982). This results in the formation of monoglycerides and free fatty acids, which migrate to the surface of the globules. Bile salts can efficiently displace the MFGM, interfacial proteins and peptides formed during gastric proteolysis (Gallier, Cui, et al., 2013a; Gallier, Shaw, et al., 2014b; Maldonado-Valderrama, Wilde, Macierzanka, & Mackie, 2011; Patton, Borgstrom, Stemberger, & Welsch, 1986). Bile salts are also able to solubilize lipolytic products and transport them within micelles for absorption through the small intestinal wall. Most MFGM proteins and peptides are digested by pancreatic proteases, but some glycoproteins are able to resist pancreatic digestion to some extent (Le et al., 2012). MFGM glycerophospholipids are digested by phospholipase A_2 in the small intestine whereas sphingomyelin is digested by sphingomyelinase in the large intestine. This may be a mechanism for allowing the formation of a complex between cholesterol and sphingomyelin in the small intestine to limit the absorption of cholesterol (Kuchta, Kelly, Stanton, & Devery, 2012). Regardless of whether milk fat globules were untreated, pasteurized or homogenized, a liquid–crystalline lamella (Fig. 7.4) formed during the small intestinal digestion of cream in rats and was associated with the interfacial accumulation of lipolytic products, calcium and possibly bile salts (Gallier, Cui, et al., 2013a).

As lipolysis is an interfacial process, the interfacial composition and structure and the size of the fat globules influence the rate and extent of lipid GI digestion (Bourlieu & Michalski, 2015; Singh & Gallier, 2016). The proteins and lipids from the MFGM or the newly formed globule interface after processing interact or compete with gastric and pancreatic lipases and bile salts. Homogenized globules, with a larger available surface area for lipase anchoring, are digested at a faster rate than non-homogenized globules, provided that their interfacial compositions are similar (Berton, Sebban-Kreuzer, Rouvellac, Lopez, & Crenon, 2009; Bourlieu et al., 2015; Ye, Cui, & Singh, 2010). Tunick et al. (2016) showed that homogenization had a greater effect than heat treatment on the in vitro intestinal lipolysis of fat globules. Using an ex vivo model of GI digestion, Islam et al. (2017) did not observe any significant gastric lipolysis of raw and processed milk samples and reported no difference in duodenal lipolysis after pasteurization of the raw milk. However, the higher was the homogenization pressure, the greater was the degree of duodenal lipolysis. Native milk fat globules vary in size and the curvature of the surface of differently sized globules is associated with a different composition of the MFGM phospholipids

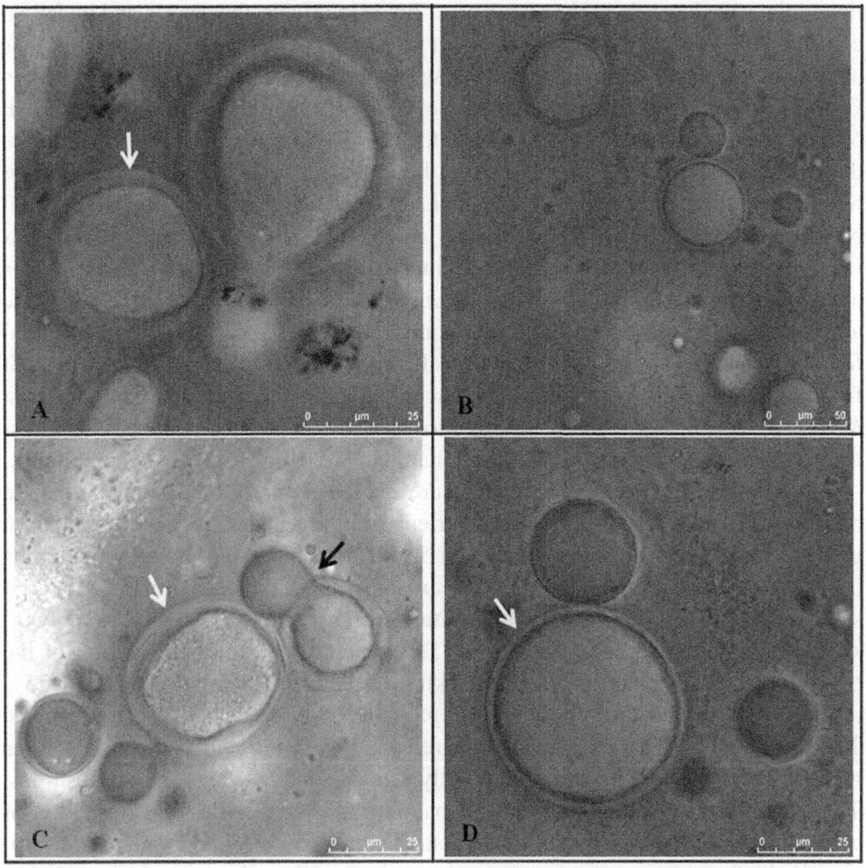

Fig. 7.4 Differential interference contrast images of upper (**a**, **c**) and lower (**b**, **d**) small intestinal digesta from rats fed raw milk. From Gallier, Cui, et al. (2013a), with permission from Elsevier

(Michalski et al., 2006). This feature may allow controlled and sustained digestion and absorption of lipids, with smaller globule populations being digested at a faster rate than larger globule populations. This controlled digestion process is unlikely to be replicated in homogenized or uniformly sized emulsion products.

5 Gastrointestinal Digestion of Milk Fat Globules in Infants

As the GI tract of infants is immature at birth, the digestion process in infants differs from that in adults. The activity of some digestive enzymes, the secretion of gastric and pancreatic juices, the concentration of bile salts and the gastric motility are very low at birth (Bourlieu et al., 2015; Ménard et al., 2018; Shani-Levi et al., 2017). Unlike in adults, where pancreatic lipase is mainly responsible for lipolysis, gastric lipase plays the key role in the digestion of lipids in infants, the gastric and intestinal environments of whom are favourable for its activity. The human gastric lipase has an optimal activity at pH 5.6 and hydrolyses lipids at pHs from 2 to 7 (Sams, Paume, Giallo, & Carriere, 2016). The intragastric pH in infants remains high (i.e. pH 5.3 at 60 min of gastric digestion) and close to the pH of optimal activity of gastric lipase during the residence time of the milk in the infant's stomach (Billeaud, Guillet, & Sandler, 1990; Cavell, 1981; Ménard et al., 2018). Gastric lipase is also able to penetrate the MFGM interface to access the triglyceride core and can hydrolyse lipids in the presence of bile salts in the small intestine (Sams et al., 2016). Other lipases aid lipid digestion in infants; the endogenous pancreatic carboxyl ester hydrolase and pancreatic lipase-related protein 2 contribute to intestinal lipolysis in formula-fed and breast-fed infants (Bourlieu et al., 2015). Breast-fed infants also benefit from the endogenous lipases that are present in unprocessed human milk (Lindquist & Hernell, 2010). The reader is referred to other reviews for the details of GI digestion in infants (Bourlieu et al., 2015; Ménard et al., 2018; Shani-Levi et al., 2017).

Nevertheless, it appears that the structure and the composition of human milk are optimized for efficient digestion or nutrient handling under the GI conditions in infants. Human milk lipids provide about 50% of the energy required for an infant's healthy growth and development. Human milk lipids typically contain 35–40% saturated fatty acids, 45–50% monounsaturated fatty acids and about 15% polyunsaturated fatty acids (Koletzko, 2016). The structure and the composition of the triglycerides in human milk are quite striking, as palmitic acid represents about 25% of the total fatty acids and about 70% of palmitic acid is in the *sn-2* position (Innis, 2011; Koletzko, 2016). This leads to the release of palmitic acid as monoglycerides because the digestive lipases preferentially cleave at the *sn-1* and/or *sn-3* positions. This allows for efficient absorption of palmitic acid by limiting the formation of calcium fatty acid soaps, which harden stools and reduce fat and calcium absorption. In the formulation of infant formula products, palm oil is commonly added as a source of palmitic acid; however, the palmitic acid in vegetable oils is commonly at the *sn-1* and *sn-3* positions of the triglycerides (Innis, 2011). Therefore, it is released as free fatty acids during intestinal digestion and can complex with calcium to form calcium fatty acid soaps. Some infants may experience digestive discomfort, which may be alleviated by feeding with infant formula containing 1,3-dioleo yl-2-palmitoylglycerol (OPO)-rich oil, obtained by the enzymatic transesterification of palm oil. This OPO-rich oil contains a higher proportion of palmitic acid in the *sn-2* position to mimic the structure of human milk lipids (Carnielli et al., 1996).

Another cause of digestive discomfort in infants is regurgitation or spit up. In the infant's stomach, human milk tends to remain as a stable emulsion, i.e. the human milk fat globules maintain their size and shape because the MFGM glycosylated proteins and lipids, which are resistant to digestion, can continue to provide steric repulsion (Armand et al., 1996). In contrast, whey-protein-dominant and casein-dominant infant formulae tend to form a curd rapidly in the stomach, which leads to a slower gastric emptying and may cause regurgitation, especially in infants with a high intragastric pH and a low gastric motility (Cavell, 1981). The early micro-phase separation of infant formula under an infant's gastric conditions, the consequent delay in gastric emptying and the absence of bile-salt-stimulated lipase in infant formula may contribute to the difference in lipid and protein handling in breast-fed and formula-fed infants after a feed (Armand et al., 1996). Similarly, the impact of heat treatment on proteins, the inactivation of the endogenous enzymes in human milk by Holder pasteurization and the destabilization of human milk fat globules during frozen storage contribute to a difference in protein and lipid digestion and absorption in term and preterm infants fed either unprocessed human milk or processed (i.e. heat-treated and stored) human milk (de Oliveira et al., 2015, 2016, 2017).

Indeed, at home and in human milk banks, Holder pasteurization is commonly used for the preservation of human milk for later feeding (Sousa et al., 2016; Vieira, Mendes Soares, Porto Pimenta, Abranches, & Lopes Moreira, 2011). This heat treatment inactivates endogenous milk enzymes, reduces the levels of some nutrients (e.g. immune factors) and may affect the structure of the MFGM. The storage of fresh or pasteurized milk at cold or frozen temperatures induces the crystallization of milk fat and the rupture of the MFGM, which, upon thawing of the human milk, destabilize the fat globule structure (Sousa et al., 2016; Vieira et al., 2011). Holder pasteurization of preterm human milk resulted in aggregation of the fat globules under gastric in vitro preterm conditions, the formation of large aggregates and slower lipolysis under intestinal in vitro preterm conditions (de Oliveira et al., 2016). Similarly, Holder pasteurization of term human milk led to the protein-induced aggregation of fat globules in a simulated gastric phase and lower GI lipolysis compared with that of unprocessed human milk under in vitro term conditions (de Oliveira et al., 2015).

When the in vitro digestion of a minimally processed milk emulsion (i.e. native milk fat globules and non-denatured proteins) was compared with those of a homogenized model infant formula (i.e. small fat droplets covered with proteins) and a homogenized and pasteurized model infant formula (i.e. small fat droplets covered with denatured proteins) under infant GI conditions, Bourlieu et al. (2015) showed that the proteolysis and the lipolysis of the minimally processed milk emulsion were slower and that the differences in globule size and microstructure were maintained in the gastric phase. The GI digestion of fat and proteins was controlled by globule size, interfacial structure, protein composition and degree of protein denaturation. Using an updated static in vitro model of GI digestion in full-term infants, Ménard et al. (2018) showed that caseins were digested mainly in the gastric phase but at a slower rate than under adult gastric conditions. Only up to 20% of whey proteins were digested in the gastric phase under infant conditions whereas they were extensively hydrolysed under adult conditions. The slow proteolysis in the gastric phase may contribute to the limited lipolysis of the fat droplets in infant formula, because of the trapping of the fat droplets within the protein aggregates, limiting access of the gastric lipase to the fat droplets (Ménard et al., 2018).

Manufacturing infant formulae with hydrolysed proteins or fermenting infant formulae with lactic-acid-producing bacteria appears to limit the formation of protein-induced aggregates in the gastric phase (Cavell, 1981; van de Heijning, Berton, Bouritius, & Goulet, 2014). Fermented and hydrolysed infant formulae tend to form a softer curd in the stomach, resulting in a gastric emptying rate that is faster than that of a standard infant formula with intact milk proteins and that is closer to that of human milk.

6 Mimicking the Structure of Raw Milk Fat Globules

Because of the common use of vegetable oils in infant formula and the homogenization applied to improve their incorporation, the microstructure of infant formula fat droplets, in particular, the interface, is significantly different from that of human milk fat globules (Gallier et al., 2015). Infant formula fat droplets are sub-micronic and are covered with milk proteins and other emulsifiers such as soy lecithin. As the size and the interface of the droplets have an impact on the digestive process, the digestion of infant formula fat droplets is different from that of human milk fat globules. Recently, MFGM has been added back to infant formulae to match the composition and the structure of human milk more closely. Gallier et al. (2015) showed that the structure and the composition of the human milk fat globules can be mimicked by adding an MFGM-rich ingredient during the manufacture of infant formula and using more gentle processing conditions to obtain MFGM fragments in the aqueous phase and large fat droplets stabilized by MFGM components (Fig. 7.5). This MFGM-containing infant formula was shown to reduce adiposity and to improve the metabolic response despite a similar food intake in an animal model with rats fed this infant formula or a control infant formula and subsequently challenged to a high fat Western-style diet (Baars et al., 2016; Oosting et al., 2012, 2014).

By adding MFGM-rich ingredients, such as buttermilk, beta-serum, butter serum and whey buttermilk, it is possible to increase the levels of MFGM complex milk lipids (i.e. phospholipids and gangliosides), cholesterol and MFGM proteins to close to those found in human milk (Cilla, Quintaes, Barberá, & Alegría, 2016; Claumarchirant et al., 2016; Claumarchirant, Matencio, Sanchez-Siles, Alegría, & Lagarda, 2015; Fong, Ma, & Norris, 2013; Fong & Norris, 2009). Human milk provides about 90–150 mg cholesterol/L, mostly in the free form, whereas infant formula contains very little to none (Koletzko, 2016). The dietary cholesterol supply regulates the synthesis rate of endogenous cholesterol. In an infant nutrition intervention, 4-month-old infants were fed their mother's milk (high cholesterol and low phytoestrogen), a cow's-milk-based formula (low cholesterol and low phytoestrogen) or a soy-based infant formula (no cholesterol and high phytoestrogen) (Cruz et al., 1994). Infants fed breast milk had the lowest synthesis rate of endogenous cholesterol, infants fed cow's-milk-based formula had an intermediate rate and infants fed soy-based formula had the highest rate. Addition of cholesterol to a soy-based infant formula resulted in a synthesis rate that was similar to that observed in infants fed a cow's-milk-based formula. The dietary supply of cholesterol via human milk in infancy may upregulate cholesterol homeostasis in adulthood, by lowering total and low density lipoprotein serum cholesterol (Koletzko, 2016).

Rosqvist et al. (2015) investigated the effect of the stabilization of milk fat with MFGM on plasma lipoproteins, gene expression and cholesterol metabolism in an 8-week randomized, controlled trial in overweight healthy adults fed MFGM-coated milk fat or a control non-coated milk fat. MFGM-stabilized milk fat did not induce an increase in plasma lipids, including plasma cholesterol, as seen in subjects fed a control non-coated milk fat. Reduced cholesterol absorption, probably caused by

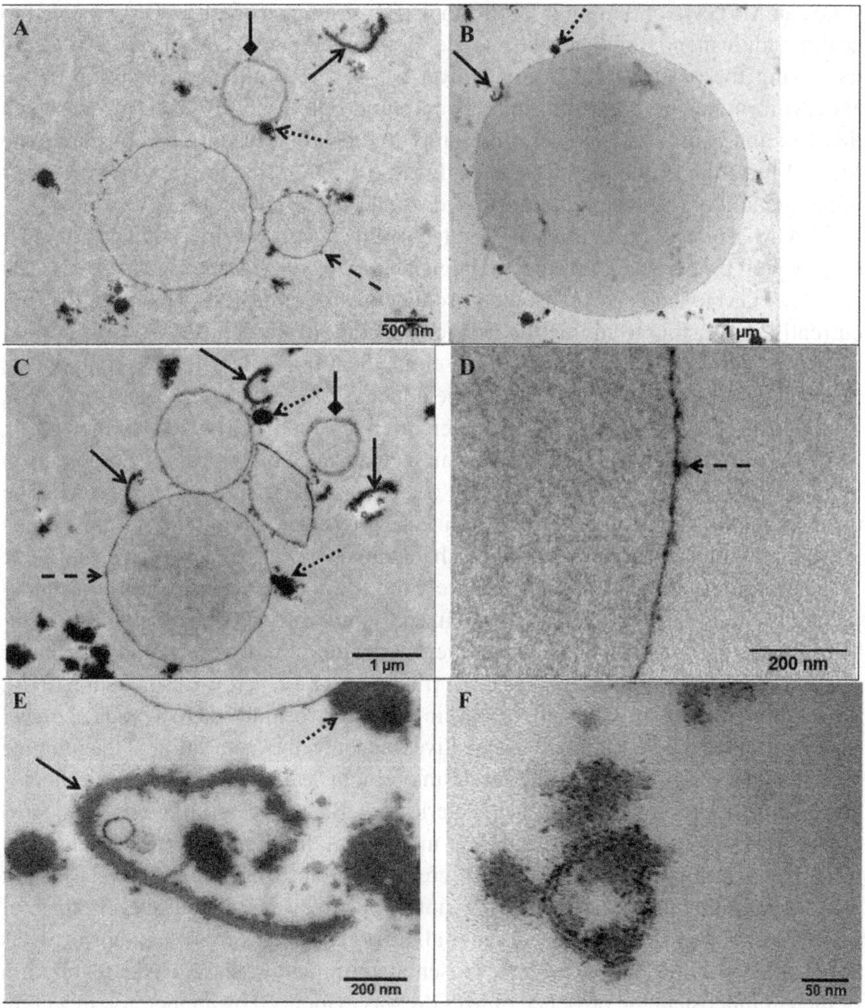

Fig. 7.5 TEM images of an infant formula manufactured with an MFGM-rich ingredient and adapted processing conditions. From Gallier et al. (2015), with permission from Elsevier

the presence of MFGM sphingomyelin, and phospholipid-induced alterations in hepatic gene expression may have contributed to the improved lipid handling despite the presence of saturated fats in the meal (Rosqvist et al., 2015). Similarly, Demmer et al. (2016) studied the impact of the addition of MFGM (a cream-derived beta-serum concentrate) to a meal that was high in saturated fats on postprandial inflammation and insulin resistance in overweight and obese adults. Palm oil was coarsely emulsified with whey proteins and MFGM-enriched beta-serum concentrate. The addition of MFGM resulted in a reduction in postprandial cholesterol, inflammatory markers and insulin response in both overweight and obese subjects. In an acute study, an MFGM-containing infant formula (Oosting et al., 2012, 2014) was fed to healthy adult men (Baumgartner, van de Heijning, Acton, & Mensink, 2017). By measuring the postprandial response over a 5-h period after consumption of the MFGM formula or a control formula (small fat droplets, no MFGM), it was shown that fats and carbohydrates were more rapidly absorbed when men consumed the MFGM formula, which may have been due to a difference in the rapid gastric emptying rate and/or the access of digestive enzymes to the substrates.

MFGM isolates, buttermilk and butter serum have been investigated for their emulsifying properties (Corredig & Dalgleish, 1997, 1998; Lopez et al., 2017). The use of MFGM to emulsify fat droplets produces stable emulsions. However, it is still currently impossible to mimic the trilayer structure of the MFGM. Even if present as an intact MFGM fragment before emulsification, only MFGM components released from the intact fragments in the aqueous phase during processing will be available for adsorption at the oil–water interface of the droplets (Gallier et al., 2015). In addition, considering the common occurrence of cytoplasmic crescents in human milk (Huston & Patton, 1990), it can be assumed that they play a role in the development of the neonate but the engineering of a true biomimetic of the human milk fat globule appears to be an unreachable goal.

In an animal study, neonatal piglets were fed sow's milk, an infant formula manufactured with either vegetable oils stabilized by milk proteins or vegetable oils stabilized with proteins and MFGM fragments or a mixture of milk fat and vegetable oils stabilized with proteins and MFGM fragments (Le Huërou-Luron et al., 2018). The combination of milk fat and MFGM in the infant formula led to a reduced digestion of casein and β-lactoglobulin, a positive impact on the maturation of the immune system and a modification of the faecal microbiota, with a reduction in Firmicutes and a higher prevalence of Proteobacteria and Bacteroides. Of note, Wang and Donovan (2015) reported similar faecal microbiota differences between breast-fed infants and formula-fed infants. The difference in protein digestion in the absence of MFGM may be due to the greater amount of adsorbed proteins at the interface of vegetable-oil-only droplets, as Sarkar et al. (2009) showed a more rapid proteolysis of adsorbed proteins than of proteins present in the aqueous phase. Faster maturation of the gut mucosa in the presence of MFGM was evidenced by an increase in weight and density of the jejunum and the ileum (Le Huërou-Luron et al., 2018). Sphingolipids, such as sphingomyelin and gangliosides, are likely contributors as they are known to promote gut maturation (Kuchta et al., 2012; Le Huërou-Luron et al., 2018; Miklavcic, Schnabl, Mazurak, Thomson, & Clandinin, 2012).

There is a growing body of research that shows the biological importance of the MFGM in early life and childhood nutrition (Hernell, Timby, Domellof, & Lonnerdal, 2016; Koletzko, 2016; Mendez-Otero, Pimentel-Coelho, Ukraintsev, & McJarrow, 2013; Timby, Domellof, Lonnerdal, & Hernell, 2017). The MFGM plays a role not only in the healthy development and growth of infants but also in brain development, gut maturation and protection against infection (Gurnida, Rowan, Idjradinata, Muchtadi, & Sekarwana, 2012; Hernell et al., 2016; Koletzko, 2016; Tanaka et al., 2013; Timby et al., 2015, 2017; Timby, Domellof, Hernell, Lonnerdal, & Domellof, 2014; Veereman-Wauters et al., 2012; Zavaleta et al., 2011). Two clinical trials in infants (Gurnida et al., 2012; Timby et al., 2014) recently showed that supplementation with MFGM-enriched ingredients, as a source of complex milk lipids, in early life improved an infant's cognitive scores in the first year of life. A clinical trial is currently underway to investigate the impact of MFGM supplementation during pregnancy to support brain development of the foetus and cognitive outcome up to 1 year of age (Huang et al., 2017). It is highly likely that more and more research that unravels the true potential of the MFGM not only in early life but also across all life stages will be published in the next decade.

7 Conclusions

There has been an increase in research on the effect of the processing of dairy products on the digestion of lipids and proteins. The difficulty in comparing all in vitro studies is the use of different models: (1) static, semi-dynamic or dynamic, (2) with different concentrations of digestive enzymes and compositions of gastric and pancreatic juices, (3) under fasted or fed conditions and (4) with or without a gastric lipase or a fungal lipase. However, one consensus from all in vitro and in vivo studies is the impact of processing on the digestion of milk fat globules because of alteration of the size of the globules and their interfacial structure and composition. The growing evidence around the food structure–function relationship is focusing research and product development towards minimization of the impact of processing on the natural structure of the milk fat globules and/or reassembling ingredients to mimic their native but intricate structure. The infant formula category is the most active space because of the recognition of the difference in structure between human milk fat globules and infant formula fat droplets and the increasing understanding of the role of the MFGM in the healthy development of infants.

References

Armand, M., Hamosh, M., Mehta, N. R., Angelus, P. A., Philpott, J. R., Henderson, T. R., et al. (1996). Effect of human milk or formula on gastric function and fat digestion in the premature infant. *Pediatric Research, 40*, 429–437.

Baars, A., Oosting, A., Engels, E., Kegler, D., Kodde, A., Schipper, L., et al. (2016). Milk fat globule membrane coating of large lipid droplets in the diet of young mice prevents body fat accumulation in adulthood. *British Journal of Nutrition, 115*, 1930–1937.

Baumgartner, S., van de Heijning, B. J. M., Acton, D., & Mensink, R. P. (2017). Infant milk fat droplet size and coating affect postprandial responses in healthy adult men: A proof-of-concept study. *European Journal of Clinical Nutrition, 71*, 1108–1113.

Berton, A., Sebban-Kreuzer, C., Rouvellac, S., Lopez, C., & Crenon, I. (2009). Individual and combined action of pancreatic lipase and pancreatic lipase-related proteins 1 and 2 on native versus homogenized milk fat globules. *Molecular Nutrition & Food Research, 53*, 1592–1602.

Billeaud, C., Guillet, J., & Sandler, B. (1990). Gastric emptying in infants with or without gastro-oesophageal reflux according to the type of milk. *European Journal of Clinical Nutrition, 44*, 577–583.

Borgstrom, B., & Erlanson-Albertsson, C. (1982). Hydrolysis of milk fat globules by pancreatic lipase – Role of colipase, phospholipase A2, and bile salts. *Journal of Clinical Investigation, 70*, 30–32.

Bourlieu, C., & Michalski, M.-C. (2015). Structure–function relationship of the milk fat globule. *Current Opinion in Clinical Nutrition & Metabolic Care, 18*, 118–127.

Bourlieu, C., Menard, O., De La Chevasnerie, A., Sams, L., Rousseau, F., Madec, M. N., et al. (2015). The structure of infant formulas impacts their lipolysis, proteolysis and disintegration during in vitro gastric digestion. *Food Chemistry, 182*, 224–235.

Bourlieu, C., Paboeuf, G., Chever, S., Pezennec, S., Cavalier, J. F., Guyomarc'h, F., et al. (2016). Adsorption of gastric lipase onto multicomponent model lipid monolayers with phase separation. *Colloids and Surfaces B: Biointerfaces, 143*, 97–106.

Buchheim, W., Welsch, U., Huston, G. E., & Patton, S. (1988). Glycoprotein filament removal from human milk fat globules by heat treatment. *Pediatrics, 81*, 141–146.

Carnielli, V. P., Luijendijk, I. H., Van Goudoever, J. B., Sulkers, E. J., Boerlage, A. A., Degenhart, H. J., et al. (1996). Structural position and amount of palmitic acid in infant formulas: Effects on fat, fatty acid, and mineral balance. *Journal of Pediatric Gastroenterology and Nutrition, 23*, 553–560.

Cavell, B. (1981). Gastric emptying in infants fed human milk or infant formula. *Acta Paediatrica Scandinavica, 70*, 639–641.

Chu, B. S., Gunning, A. P., Rich, G. T., Ridout, M. J., Faulks, R. M., Wickham, M. S. J., et al. (2010). Adsorption of bile salts and pancreatic colipase and lipase onto digalactosyldiacylglycerol and dipalmitoylphosphatidylcholine monolayers. *Langmuir, 26*, 9782–9793.

Cilla, A., Quintaes, K. D., Barberá, R., & Alegría, A. (2016). Phospholipids in human milk and infant formulas: Benefits and needs for correct infant nutrition. *Critical Reviews in Food Science and Nutrition, 56*, 1880–1892.

Claumarchirant, L., Matencio, E., Sanchez-Siles, L. M., Alegría, A., & Lagarda, M. J. (2015). Sterol composition in infant formulas and estimated intake. *Journal of Agricultural and Food Chemistry, 63*, 7245–7251.

Claumarchirant, L., Cilla, A., Matencio, E., Sanchez-Siles, L. M., Castro-Gomez, P., Fontecha, J., et al. (2016). Addition of milk fat globule membrane as an ingredient of infant formulas for resembling the polar lipids of human milk. *International Dairy Journal, 61*, 228–238.

Contarini, G., & Povolo, M. (2013). Phospholipids in milk fat: Composition, biological and technological significance, and analytical strategies. *International Journal of Molecular Science, 14*, 2808–2831.

Corredig, M., & Dalgleish, D. G. (1997). Isolates from industrial buttermilk: Emulsifying properties of materials derived from the milk fat globule membrane. *Journal of Agricultural and Food Chemistry, 45*, 4595–4600.

Corredig, M., & Dalgleish, D. G. (1998). Effect of heating of cream on the properties of milk fat globule membrane isolates. *Journal of Agricultural and Food Chemistry, 46*, 2533–2540.

Couvreur, S., Hurtaud, C., Lopez, C., Delaby, L., & Peyraud, J. L. (2006). The linear relationship between the proportion of fresh grass in the cow diet, milk fatty acid composition, and butter properties. *Journal of Dairy Science, 89*, 1956–1969.

Cruz, M. L., Wong, W. W., Mimouni, F., Hachey, D. L., Setchell, K. D., Klein, P. D., et al. (1994). Effects of infant nutrition on cholesterol synthesis rates. *Pediatric Research, 35*, 135–140.

de Oliveira, S. C., Deglaire, A., Menard, O., Bellanger, A., Rousseau, F., Henry, G., et al. (2015). Holder pasteurization impacts the proteolysis, lipolysis and disintegration of human milk under in vitro dynamic term newborn digestion. *Food Research International, 88*(Part B), 263–275.

de Oliveira, S. C., Menard, O., Bellanger, A., Henry, G., Rousseau, F., Dirson, E., et al. (2016). Impact of pasteurization of human milk on preterm newborn in vitro digestion: Gastrointestinal disintegration, lipolysis and proteolysis. *Food Chemistry, 211*, 171–179.

de Oliveira, S. C., Bellanger, A., Menard, O., Pladys, P., Le Gouar, Y., Dirson, E., et al. (2017). Impact of human milk pasteurization on gastric digestion in preterm infants: A randomized controlled trial. *American Journal of Clinical Nutrition, 105*, 379–390.

Demmer, E., Van Loan, M. D., Rivera, N., Rogers, T. A., Gertz, E. R., German, B., et al. (2016). Addition of a dairy fraction rich in milk fat globule membrane to a high-saturated fat meal reduces the postprandial insulinaemic and inflammatory response in overweight and obese adults. *Journal of Nutritional Science*, 5. https://doi.org/10.1017/jns.2015.42

Fong, B. Y., & Norris, C. S. (2009). Quantification of milk fat globule membrane proteins using selected reaction monitoring mass spectrometry. *Journal of Agricultural and Food Chemistry, 57*, 6021–6028.

Fong, B., Ma, L., & Norris, C. (2013). Analysis of phospholipids in infant formulas using high performance liquid chromatography–tandem mass spectrometry. *Journal of Agricultural and Food Chemistry, 61*, 858–865.

Gallier, S., Gragson, D., Jimenez-Flores, R., & Everett, D. W. (2010a). Using confocal laser scanning microscopy to probe the milk fat globule membrane and associated proteins. *Journal of Agricultural and Food Chemistry, 58*, 4250–4257.

Gallier, S., Gragson, D., Jimenez-Flores, R., & Everett, D. W. (2010b). Surface characterization of bovine milk phospholipid mono layers by Langmuir isotherms and microscopic techniques. *Journal of Agricultural and Food Chemistry, 58*, 12275–12285.

Gallier, S., Gragson, D., Jimenez-Flores, R., & Everett, D. W. (2012). β-Casein–phospholipid monolayers as model systems to understand lipid–protein interactions in the milk fat globule membrane. *International Dairy Journal, 22*, 58–65.

Gallier, S., Cui, J., Olson, T. D., Rutherfurd, S. M., Ye, A., Moughan, P. J., et al. (2013a). In vivo digestion of bovine milk fat globules: Effect of processing and interfacial structural changes. I. Gastric digestion. *Food Chemistry, 141*, 3273–3281.

Gallier, S., Zhu, X. Q., Rutherfurd, S. M., Ye, A., Moughan, P. J., & Singh, H. (2013b). In vivo digestion of bovine milk fat globules: Effect of processing and interfacial structural changes. II. Upper digestive tract digestion. *Food Chemistry, 141*, 3215–3223.

Gallier, S., Rutherfurd, S. M., Moughan, P. J., & Singh, H. (2014a). Effect of food matrix microstructure on stomach emptying rate and apparent ileal fatty acid digestibility of almond lipids. *Food and Function, 5*, 2410–2419.

Gallier, S., Shaw, E., Laubscher, A., Gragson, D., Singh, H., & Jimenez-Flores, J. (2014b). Adsorption of bile salts to milk phospholipid and phospholipid-protein monolayers. *Journal of Agricultural and Food Chemistry, 62*, 1363–1372.

Gallier, S., Vocking, K., Post, J. A., Acton, D., Van Der Beek, E. M., & Van Baalen, T. (2015). A novel infant milk formula concept: Mimicking the human milk fat globule structure. *Colloids and Surfaces B: Biointerfaces, 136*, 329–339.

Gallier, S., Acton, D., Manohar, G., & Singh, H. (2017). Natural and processed milk and oil body emulsions: Bioavailability, bioaccessibility and functionality. *Food Structure, 13*, 13–23.

Gurnida, D. A., Rowan, A. M., Idjradinata, P., Muchtadi, D., & Sekarwana, N. (2012). Association of complex lipids containing gangliosides with cognitive development of 6-month-old infants. *Early Human Development, 88*, 595–601.

Hernell, O., Timby, N., Domellof, M., & Lonnerdal, B. (2016). Clinical benefits of milk fat globule membranes for infants and children. *Journal of Pediatrics, 173*, S60–S65.

Houlihan, A. V., Goddard, P. A., Kitchen, B. J., & Masters, C. J. (1992). Changes in structure of the bovine milk fat globule membrane on heating whole milk. *Journal of Dairy Research, 59*, 321–329.

Huang, S., Mo, T. T., Norris, T., Sun, S., Zhang, T., Han, T. L., et al. (2017). The CLIMB (Complex Lipids In Mothers and Babies) study: Protocol for a multicentre, three-group, parallel randomised controlled trial to investigate the effect of supplementation of complex lipids in pregnancy, on maternal ganglioside status and subsequent cognitive outcomes in the offspring. *BMJ Open, 7*, e016637.

Huston, G. E., & Patton, S. (1990). Factors related to the formation of cytoplasmic crescents on milk fat globules. *Journal of Dairy Science, 73*, 2061–2066.

Innis, S. M. (2011). Dietary triacylglycerol structure and its role in infant nutrition. *Advances in Nutrition, 2*, 275–283.

Islam, M. A., Devle, H., Comi, I., Ulleberg, E. K., Rukke, E.-O., Vegarud, G. E., et al. (2017). Ex vivo digestion of raw, pasteurised and homogenised milk – effects on lipolysis and proteolysis. *International Dairy Journal, 65*, 14–19.

Jensen, R. J., & Newsburg, D. S. (1995). Bovine milk lipids. In G. J. Robert (Ed.), *Handbook of milk composition* (pp. 543–575). San Diego, CA: Academic.

Keenan, T. W., & Dylewski, D. P. (1995). Intracellular origin of milk lipid globules and the nature and structure of the milk lipid globule membrane. In P. F. Fox (Ed.), *Advanced dairy chemistry. Lipids* (Vol. 2, pp. 137–171). London: Chapman & Hall.

Kim, H. H. Y., & Jimenez-Flores, R. (1995). Heat-induced interactions between the proteins of milk fat globule membrane and skim milk. *Journal of Dairy Science, 78*, 24–35.

Koletzko, B. (2016). Human milk lipids. *Annals of Nutrition and Metabolism, 69*(Suppl 2), 28–40.

Kuchta, A. M., Kelly, P. M., Stanton, C., & Devery, R. A. (2012). Milk fat globule membrane – A source of polar lipids for colon health? A review. *International Journal of Dairy Technology, 65*, 315–333.

Le, T. T., Van de Wiele, T., Do, T. N. H., Debyser, G., Struijs, K., Devreese, B., et al. (2012). Stability of milk fat globule membrane proteins toward human enzymatic gastrointestinal digestion. *Journal of Dairy Science, 95*, 2307–2318.

Le Huërou-Luron, I., Bouzerzour, K., Ferret-Bernard, S., Menard, O., Le Normand, L., Perrier, C., et al. (2018). A mixture of milk and vegetable lipids in infant formula changes gut digestion, mucosal immunity and microbiota composition in neonatal piglets. *European Journal of Nutrition, 57*, 463–476.

Lecomte, M., Bourlieu, C., Meugnier, E., Penhoat, A., Cheillan, D., Pineau, G., et al. (2015). Milk polar lipids affect in vitro digestive lipolysis and postprandial lipid metabolism in mice. *Journal of Nutrition, 145*, 1770–1777.

Lepri, L., Del Bubba, M., Maggini, R., Donzelli, G. P., & Galvan, P. (1997). Effect of pasteurization and storage on some components of pooled human milk. *Journal of Chromatography B Biomedical Sciences and Application, 704*, 1–10.

Lindquist, S., & Hernell, O. (2010). Lipid digestion and absorption in early life: An update. *Current Opinion in Clinical Nutrition & Metabolic Care, 13*, 314–320.

Lopez, C., Madec, M. N., & Jimenez-Flores, R. (2010). Lipid rafts in the bovine milk fat globule membrane revealed by the lateral segregation of phospholipids and heterogeneous distribution of glycoproteins. *Food Chemistry, 120*, 22–33.

Lopez, C., Cauty, C., & Guyomarc'h, F. (2015). Organization of lipids in milks, infant milk formulas and various dairy products: Role of technological processes and potential impacts. *Dairy Science & Technology, 95*, 863–893.

Lopez, C., Cauty, C., Rousseau, F., Blot, M., Margolis, A., & Famelart, M.-H. (2017). Lipid droplets coated with milk fat globule membrane fragments: Microstructure and functional properties as a function of pH. *Food Research International, 91*, 26–37.

Maldonado-Valderrama, J., Wilde, P., Macierzanka, A., & Mackie, A. (2011). The role of bile salts in digestion. *Advances in Colloid and Interface Science, 165*, 36–46.

Marciani, L., Wickham, M., Sing, G., Bush, D., Pick, B., Cox, E., et al. (2007). Enhancement of intragastric acid stability of a fat emulsion meal delays gastric emptying and increases cholecystokinin release and gallbladder contraction. *American Journal of Physiology – Gastrointestinal and Liver Physiology, 292*, G1607–G1613.

Martini, M., Salari, F., & Altomonte, I. (2016). The macrostructure of milk lipids: The fat globules. *Critical Review in Food Science and Nutrition, 56*, 209–221.

Masedunskas, A., Weigert, R., & Mather, I. H. (2014). Chapter 9 – Intravital imaging of the lactating mammary gland in transgenic mice expressing fluorescent proteins. In R. Weigert (Ed.), *Advances in intravital microscopy: From basic to clinical research* (pp. 187–204). Dordrecht: Springer Science+Business Media.

Masedunskas, A., Chen, Y., Stussman, R., Weigert, R., & Mather, I. H. (2017). Kinetics of milk lipid droplet transport, growth, and secretion revealed by intravital imaging: Lipid droplet release is intermittently stimulated by oxytocin. *Molecular Biology of the Cell, 28*, 935–947.

Mathiassen, J. H., Nejrup, R. G., Frøkiær, H., Nilsson, Å., Ohlsson, L., & Hellgren, L. I. (2015). Emulsifying triglycerides with dairy phospholipids instead of soy lecithin modulates gut lipase activity. *European Journal of Lipid Science & Technology, 117*, 1522–1539.

Ménard, O., Bourlieu, C., De Oliveira, S. C., Dellarosa, N., Laghi, L., Carrière, F., et al. (2018). A first step towards a consensus static *in vitro* model for simulating full-term infant digestion. *Food Chemistry, 240*, 338–345.

Mendez-Otero, R., Pimentel-Coelho, P. M., Ukraintsev, S., & McJarrow, P. (2013). Chapter 9 – Role of gangliosides in neurological development and the influence of dietary sources. In R. R. Watson et al. (Eds.), *Nutrition in infancy* (Vol. 2, pp. 105–118). New York, NY: Nutrition and Health, Springer Science+Business Media.

Michalski, M.-C., Briard, V., Desage, M., & Geloen, A. (2005). The dispersion state of milk fat influences triglyceride metabolism in the rat – A (CO_2)-C^{13} breath test study. *European Journal of Nutrition, 44*, 436–444.

Michalski, M.-C., Soares, A. F., Lopez, C., Leconte, N., Briard, V., & Geloen, A. (2006). The supramolecular structure of milk fat influences plasma triacylglycerols and fatty acid profile in the rat. *European Journal of Nutrition, 45*, 215–224.

Miklavcic, J. J., Schnabl, K. L., Mazurak, V. C., Thomson, A. B. R., & Clandinin, M. T. (2012). Dietary ganglioside reduces proinflammatory signaling in the intestine. *Journal of Nutrition and Metabolism, 2012*, 280–286.

Murthy, A. V. R., Guyomarc'h, F., Paboeuf, G., Vié, V., & Lopez, C. (2015). Cholesterol strongly affects the organization of lipid monolayers studied as models of the milk fat globule membrane: Condensing effect and change in the lipid domain morphology. *Biochimica et Biophysica Acta, 1848*, 2308–2316.

N'Goma, J.-C. B., Amara, S., Dridi, K., Jannin, V., & Carrière, F. (2012). Understanding the lipid-digestion processes in the GI tract before designing lipid-based drug-delivery systems. *Therapeutic Delivery, 3*, 105–124.

Oftedal, O. T. (2012). The evolution of milk secretion and its ancient origins. *Animal, 6*, 355–368.

Oosting, A., Kegler, D., Woopereis, H. J., Teller, I. C., Van De Heijning, B. J. M., Verkade, H. J., et al. (2012). Size and phospholipid coating of lipid droplets in the diet of young mice modify body fat accumulation in adulthood. *Pediatric Research, 72*, 362–369.

Oosting, A., Van Vlies, N., Kegler, D., Schipper, L., Abrahamse-Berkeveld, M., Ringler, S., et al. (2014). Effect of dietary lipid structure in early postnatal life on mouse adipose tissue development and function in adulthood. *British Journal of Nutrition, 111*, 215–226.

Patton, S., Borgstrom, B., Stemberger, B. H., & Welsch, U. (1986). Release of membrane from milk-fat globules by conjugated bile-salts. *Journal of Pediatric Gastroenterology and Nutrition, 5*, 262–267.

Patton, S., & Huston, G. E. (1988). Incidence and characteristics of cell pieces on human milk fat globules. *Biochimica et Biophysica Acta – General Subjects, 965*, 146–153.

Rosqvist, F., Smedman, A., Lindmark-Mansson, H., Paulsson, M., Petrus, P., Straniero, S., et al. (2015). Potential role of milk fat globule membrane in modulating plasma lipoproteins, gene expression, and cholesterol metabolism in humans: A randomized study. *American Journal of Clinical Nutrition, 102*, 20–30.

Sams, L., Paume, J., Giallo, J., & Carriere, F. (2016). Relevant pH and lipase for in vitro models of gastric digestion. *Food & Function, 7*, 30–45.

Sarkar, A., Goh, K. K. T., & Singh, H. (2009). Colloidal stability and interactions of milk-protein-stabilized emulsions in an artificial saliva. *Food Hydrocolloids, 23*, 1270–1278.

Shani-Levi, C., Alvito, P., Andres, A., Assuncao, R., Barbera, R., Blanquet-Diot, S., et al. (2017). Extending in vitro digestion models to specific human populations: Perspectives, practical tools and bio-relevant information. *Trends in Food Science & Technology, 60*, 52–63.

Singh, H. (2006). The milk fat globule membrane – A biophysical system for food applications. *Current Opinion in Colloid & Interface Science, 11*, 154–163.

Singh, H., & Gallier, S. (2014). Chapter 2 – Processing of food structures in the gastrointestinal tract and physiological responses. In *Food structures, digestion and health* (pp. 51–81). San Diego, CA: Academic.

Singh, H., & Gallier, S. (2016). Nature's complex emulsion: The fat globules of milk. *Food Hydrocolloids, 68*, 81–89.

Sousa, S. G., Delgadillo, I., & Saraiva, J. A. (2016). Human milk composition and preservation: Evaluation of high-pressure processing as a nonthermal pasteurization technology. *Critical Reviews in Food Science and Nutrition, 56*, 1043–1060.

Tanaka, K., Hosozawa, M., Kudo, N., Yoshikawa, N., Hisata, K., Shoji, H., et al. (2013). The pilot study: Sphingomyelin-fortified milk has a positive association with the neurobehavioural development of very low birth weight infants during infancy, randomized control trial. *Brain & Development, 35*, 45–52.

Timby, N., Domellof, E., Hernell, O., Lonnerdal, B., & Domellof, M. (2014). Neurodevelopment, nutrition, and growth until 12 mo of age in infants fed a low-energy, low-protein formula supplemented with bovine milk fat globule membranes: A randomized controlled trial. *American Journal of Clinical Nutrition, 99*, 860–868.

Timby, N., Hernell, O., Vaarala, O., Melin, M., Lonnerdal, B., & Domellof, M. (2015). Infections in infants fed formula supplemented with bovine milk fat globule membranes. *Journal of Pediatric Gastroenterology and Nutrition, 60*, 384–389.

Timby, N., Domellof, M., Lonnerdal, B., & Hernell, O. (2017). Supplementation of infant formula with bovine milk fat globule membranes. *Advances in Nutrition, 8*, 351–355.

Tunick, M. H., Ren, D. X., Van Hekken, D. L., Bonnaillie, L., Paul, M., Kwoczak, R., et al. (2016). Effect of heat and homogenization on in vitro digestion of milk. *Journal of Dairy Science, 99*, 4124–4139.

van de Heijning, B. J. M., Berton, A., Bouritius, H., & Goulet, O. (2014). GI symptoms in infants are a potential target for fermented infant milk formulae: A review. *Nutrients, 6*, 3942–3967.

Veereman-Wauters, G., Staelens, S., Rombaut, R., Dewettinck, K., Deboutte, D., Brummer, R. J., et al. (2012). Milk fat globule membrane (INPULSE) enriched formula milk decreases febrile episodes and may improve behavioral regulation in young children. *Nutrition, 28*, 749–752.

Vieira, A. A., Mendes Soares, F. V., Porto Pimenta, H., Abranches, A. D., & Lopes Moreira, M. E. (2011). Analysis of the influence of pasteurization, freezing/thawing, and offer processes on human milk's macronutrient concentrations. *Early Human Development, 87*, 577–580.

Vors, C., Pineau, G., Gabert, L., Drai, J., Louche-Pelissier, C., Defoort, C., et al. (2013). Modulating absorption and postprandial handling of dietary fatty acids by structuring fat in the meal: A randomized crossover clinical trial. *American Journal of Clinical Nutrition, 97*, 23–36.

Walstra, P. (1995). Chapter 4 – Physical chemistry of milk fat globules. In P. F. Fox (Ed.), *Advanced dairy chemistry. Lipids* (Vol. 2, pp. 131–178). London: Chapman & Hall.

Walstra, P., Wouters, J. T. M., & Geurts, T. J. (2006). *Dairy science and technology* (p. 756). Boca Raton, FL: Taylor & Francis Group.

Wang, M., & Donovan, S. M. (2015). Human microbiota-associated swine: Current progress and future opportunities. *ILAR Journal, 56*, 63–73.

Ye, A., Singh, H., Taylor, M. W., & Anema, S. (2002). Characterization of protein components of natural and heat-treated milk fat globule membranes. *International Dairy Journal, 12*, 393–402.

Ye, A. Q., Singh, H., Taylor, M. W., & Anema, S. (2004). Interactions of whey proteins with milk fat globule membrane proteins during heat treatment of whole milk. *Le Lait, 84*, 269–283.

Ye, A., Cui, J., & Singh, H. (2010). Effect of the fat globule membrane on in vitro digestion of milk fat globules with pancreatic lipase. *International Dairy Journal, 20*, 822–829.

Ye, A., Cui, J., & Singh, H. (2011). Proteolysis of milk fat globule membrane proteins during in vitro gastric digestion of milk. *Journal of Dairy Science, 94*, 2762–2770.

Ye, A., Cui, J., Dalgleish, D., & Singh, H. (2016). Formation of a structured clot during the gastric digestion of milk: Impact on the rate of protein hydrolysis. *Food Hydrocolloids, 52*, 478–486.

Ye, A., Cui, J., Dalgleish, D., & Singh, H. (2017). Effect of homogenization and heat treatment on the behaviour of protein and fat globules during gastric digestion of milk. *Journal of Dairy Science, 100*, 36–47.

Zavaleta, N., Kvistgaard, A. S., Graverholt, G., Respicio, G., Guija, H., Valencia, N., et al. (2011). Efficacy of an MFGM-enriched complementary food in diarrhea, anemia, and micronutrient status in infants. *Journal of Pediatric Gastroenterology and Nutrition, 53*, 561–568.

Zheng, H., Jiménez-Flores, R., Gragson, D., & Everett, D. W. (2014). Phospholipid architecture of the bovine milk fat globule membrane using giant unilamellar vesicles as a model. *Journal of Agricultural and Food Chemistry, 62*, 3236–3243.

Chapter 8
Role of the Matrix on the Digestibility of Dairy Fat and Health Consequences

Marie-Caroline Michalski, Laurie-Eve Rioux, and Sylvie L. Turgeon

Abbreviations

AUC	Area under curve
Ca	Calcium
CCK	Cholecystokinin
CLA	Conjugated linoleic acid
CM	Casein micelles
CN	Casein
CRP	C-reactive protein
CVD	Cardiovascular disease
FA	Fatty acids
FG	Fat globule
GGT	Gamma glutamyl transferase
HDL	High density lipoprotein
IL	Interleukin
LAB	Lactic acid bacteria
LCFA	Long chain fatty acids
LDL	Low density lipoprotein
LPS	Lipopolysaccharides
MCFA	Medium chain fatty acids
MCP	Monocyte chemoattractant protein
MF	Milk fat

M.-C. Michalski
INRAE, CarMeN laboratory, UMR1397, INSERM, U1060, Université Claude Bernard Lyon 1, INSA-Lyon, Université de Lyon, Pierre-Bénite, France

Centre de Recherche en Nutrition Humaine Rhône-Alpes, Univ-Lyon, CarMeN Laboratory, Université Claude Bernard Lyon1, Hospices Civils de Lyon, CENS, FCRIN/FORCE Network, Pierre-Bénite, France
e-mail: marie-caroline.michalski@inrae.fr

L.-E. Rioux · S. L. Turgeon (✉)
STELA Dairy Research Centre, Institute of Nutrition and Functional Foods, Université Laval, Québec City, QC, Canada
e-mail: Sylvie.Turgeon@fsaa.ulaval.ca

T. Truong et al. (eds.), *Dairy Fat Products and Functionality*,
https://doi.org/10.1007/978-3-030-41661-4_8

MFGM	Milk fat globule membrane
ML	Milk lipids
NA	Not available
NEFA	Non-esterified fatty acids
PC	Phosphatidylcholine
PE	Phosphatidylethanolamine
PI	Phosphatidylinositol
PL	Phospholipid
PS	Phosphatidylserine
RTC	Randomized control trial
SCFA	Short chain fatty acids
FA	Saturated fatty acids
SM	Sphingomyelin
T2D	Type 2 diabetes
TCA	Trichloroacetic acid
TAG	Triacylglycerol
TNF-α	Tumor necrosis factor-α
VLDL	Very low-density lipoprotein
WAT	White adipose tissue
WP	Whey proteins

1 Introduction

Dairy products are basic products largely consumed in the population, from human milk which is the perfect meal for the newborn to a large variety of dairy products from cow and other mammalians. Dairy products consumption has been recommended for its richness in valuable nutrients, but some research some 30 years ago raised concern on dairy lipid possible health impacts. Since then, the scientific community has tried to decipher the intricate parameters of lipid metabolism in response to lipids varying in composition, structure, food source, in a meal, in a diet, etc. In this chapter, the knowledge coming from epidemiologic studies will be first reviewed to reveal the possible factors that should be studied to understand the lipid travel in the food and in the human body after consumption in order to try to understand their physiological role and health impact.

1.1 An Epidemiologic Perspective

After a few decades incriminating dairy products by considering them as a source of saturated fatty acids (SFA) there are new insights on the benefits of dairy products (de Oliveira Otto et al., 2012). Recent detailed reviews and perspectives state numerous meta-analyses and epidemiological studies about the health impact of different types of dairy products (Drouin-Chartier, Brassard, et al., 2016; Drouin-Chartier, Cote, et al., 2016; Lovegrove & Givens, 2016; Thorning et al., 2016). We will, therefore, summarize endpoints of interest regarding milk fat impact, namely type 2 diabetes (T2D), cardiovascular disease (CVD) risk and hypertension.

A meta-analysis of 22 cohort studies including about 580,000 people describes an inverse correlation between all kinds of dairy products and T2D (Gijsbers et al., 2016). A sub-analysis according to the dairy matrix reveals a similar effect for yogurt only, although the relationship was not linear anymore for the highest yogurt intakes. Notably, no association was found between skimmed fermented products and T2D risk, while a 12% decrease of the risk was observed for a consumption of 40 g of fermented products per day including both skimmed and full-fat products. In this study, skimmed-, half- skimmed- and full-fat milk, cheese, cream and full-fat dairy were not associated with T2D risk.

Regarding CVD risk, several recent meta-analyses conclude that there is either no association or an inverse correlation, between dairy products consumption and CVD risk (Alexander et al., 2016; Chen, Wang, et al., 2016; de Goede, Geleijnse, Ding, & Soedamah-Muthu, 2015; Qin et al., 2015). Looking closer at the type of dairy matrix, cheese consumption was often found inversely correlated with the risk of CVD and stroke. This apparent matrix effect is all the more interesting than cheeses are described as high contributors to salt intake, otherwise known to increase CVD risk. Two cohort studies using a different method show that the plasma concentrations of circulating biomarkers of dairy fat consumption, notably 15 and 17 carbon-chain fatty acids (FA), are not associated with the risk of stroke (Yakoob et al., 2014). Moreover, a recent meta-analysis suggests that the consumption of butter would not be associated with CVD and stroke risks (Pimpin, Wu, Haskelberg, Del Gobbo, & Mozaffarian, 2016). An analysis of three large cohort studies shows that milk fat is not associated with CVD and stroke risks compared to an equivalent carbohydrate intake (Chen, Li, et al., 2016). The authors highlight the need to elucidate whether different dairy products would exert different effects. Finally, nutritional intervention studies test the change in lipid markers of CVD risk. A decrease in low density lipoprotein (LDL)-cholesterol and an increase of high density lipoprotein (HDL)-cholesterol was observed in T2D volunteers after consumption of probiotic yogurts for 8 weeks (Mohamadshahi et al., 2014). Another study shows decreased total cholesterol after 6 weeks of consumption of a probiotic yogurt enriched in *Lactobacillus acidophilus* and *Bifidobacterium lactis* compared to a regular yogurt (Ataie-Jafari, Larijani, Alavi Majd, & Tahbaz, 2009). A meta-analysis confirmed the favorable effects of probiotics (Sun & Buys, 2015) and a review highlights the possible beneficial effects of yogurt consumption on metabolic inflammation and lipid markers of CVD risk in obesity (Pei, Martin, DiMarco, & Bolling, 2017).

The most recent meta-analysis about hypertension shows an inverse linear correlation between hypertension and total dairy products intake, skimmed dairy products intake, and milk (Soedamah-Muthu, Verberne, Ding, Engberink, & Geleijnse, 2012). Regarding lipids, full-fat dairy products were not associated with hypertension while skimmed products decreased hypertension risk by 4%. A recent review of randomized controlled trials suggests that there is no apparent risk of deleterious effects of dairy products consumption on cardiometabolic risk, regardless of fat content in the dairy products, on a wide panel of risk markers (blood lipids, blood pressure, inflammation, insulin resistance, vascular function) (Drouin-Chartier, Cote, et al., 2016). Authors suggest that the supposed effects of SFA on cardiometabolic risk would be "cancelled" when the latter are consumed as part of a complex matrix such as in cheese and dairy products in general. These authors, as well as a recent expert panel (Thorning et al., 2017), provide incentive to further explore the impact of the "dairy matrix effect" on metabolism.

1.2 Lipid Composition, Structure and Matrices

Lipids consumed in food and meals vary in nature and composition. They are found in foods naturally or through their addition during food formulation or meal preparation. As pointed out in the previous section, the complexity of each food matrix should be considered to better understand metabolic responses, and this also implies considering the transformation of food during digestion and the role of each digestion step on nutrient utilisation.

Milk is the building block of a large variety of dairy products. The complete lipid composition of milk will not be presented in this chapter as it can be found in other chapters of this book. However, some compositional characteristics are recalled as they may impact nutritional and health properties of dairy products. Milk fat is the most concentrated natural source of short-chain fatty acids (SCFA) which are mostly esterified on sn-3 of the triacylglycerol (TAG) (Fig. 8.1). Milk is a rich source of SFA and it contains a rich variety of polar lipids concentrated in the milk fat globule membrane (Jensen & Newburg, 1995). Milk being the first food of newborn mammals, lipids are naturally organized to facilitate digestion and utilization. Fat droplets are excreted by the mammalian cells with a complex membrane, the milk fat globule membrane (MFGM), made of phospholipids, and sphingolipids and containing several proteins and enzymes. Homogenization and pasteurization of milk ensure products stability and safety overtime but induce changes in the fat droplet interface. If milk is first homogenized and then pasteurized, 99% of the absorbed proteins are β- and κ-caseins and the remaining 1% corresponds to MFGM fragments (see reviews, e.g.: Michalski, 2009; Michalski & Januel, 2006). When milk is pasteurized and then homogenized, whey proteins are denatured and interact with casein micelles (mainly κ-caseins) and native proteins of the MFGM. These complexes are then absorbed on the surface of the fat globules upon homogenization.

As shown in Fig. 8.1, several levels of structuration can be considered to describe each dairy product. Each has a specific lipid composition organized in fat droplets with a specific size distribution and interface composition. Any process contributing to change the matrix composition, for example, acid gel formation, cheese making, or butter churning induces a different dairy food structure and lipid organization.

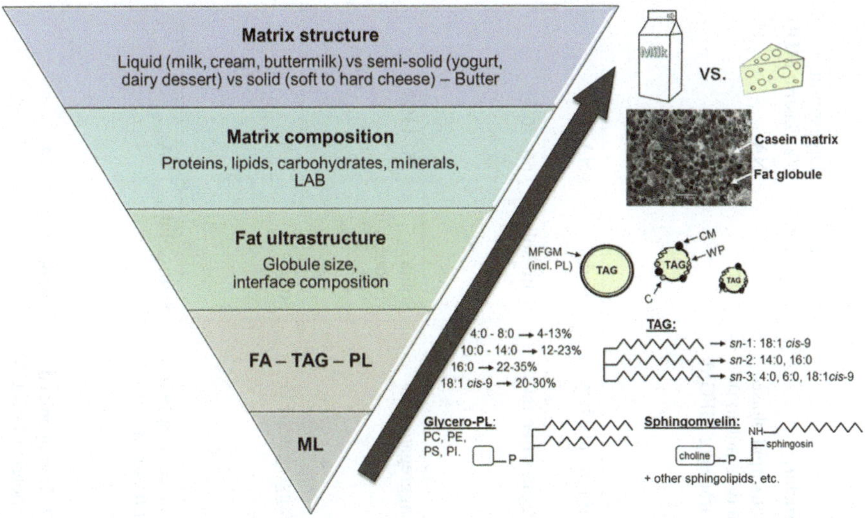

Fig. 8.1 Organization levels of lipids in dairy products. *CM* Casein micelles, *FA* Fatty acids, *LAB* Lactic acid bacteria, *MFGM* Milk fat globule membrane, *ML* Milk lipids, *PC* Phosphatidylcholine, *PE* Phosphatidylethanolamine, *PI* Phosphatidylinositol, *PL* Phospholipid, *PS* Phosphatidylserine, *TAG* Triacylglycerol, *WP* Whey proteins. The NH of sphingosin is shown to visualize the amide bond within sphingomyelin.

Most dairy products are emulsions, milk and buttermilk are liquid emulsion while cheese is a dispersion of lipid droplets or fat pools in a solid continuous phase. Butter is also an emulsion, but the aqueous droplets are found in a continuous lipid phase containing partially crystallized TAG. Each level of structuration is sometimes referred as microstructure and macrostructure (Barbé et al., 2014) and changes in structures due to different interactions may impact their digestive fate and metabolic outcome as demonstrated later in this chapter.

Natural processes as fermentation have emerged a long time ago to prolong storage time and are at the origin of a wide range of dairy products. From the initial milk composition, an intelligent use of processing tools as pH, enzymes (rennet), concentration, drying, etc. to induce interactions between dairy components are the basis of new dairy food structures (Table 8.1). These products are often a concentrated version of milk nutrients having different matrix organisation.

Table 8.1 classifies dairy products based on their structure and describes their composition in lipid, proteins and calcium, which influences their textural properties. Lipid contents varies from 0% to 90% depending on the dairy products. The emulsified state of milk is kept after its processing into yogurt and cheese in which lipid droplets are embedded in a gelled matrix. The properties of these casein matrices vary from a semi-solid acid gel in yogurt to a solid matrix in cheese. Depending on the cheese making process, the level of mineralization is different resulting in a semi-solid texture for Camembert with a lower calcium content compared to a solid/viscoelastic matrix as Cheddar cheese. Parmesan cheese has a stronger solid

Table 8.1 Dairy matrices composition, structure and digestibility

Dairy products	Proteins[a] (g/100 g)	Lipids[a] (g/100 g)	Ca[a] (mg/100 g)	LAB	Microstructure[b]	Lipid structure[c]	Textural properties	Digestibility[d] (% Gastric MDI)
Butter, salted	0.9	81.1	24	No/ Yes[e]	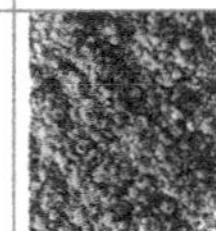 Reproduced from[f]	Continuous lipid phase (water in oil emulsion)/ residual traces of MFGM	–	NA
Buttermilk (1% MF)	3.3	0.9	116	No	NA	MFGM fragments, tiny MFG	Liquid	NA
Cheese (Camembert)	19.8	24.3	388	Yes	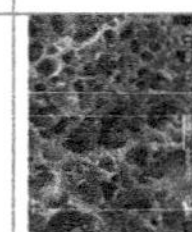	FG/aggregates/free fat	Semi-solid	66–76%[g]
Cheese (Cheddar)	24.0	33.8	675	Yes	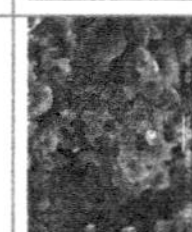	Free fat/aggregates/FG	Solid/ viscoelastic	25–43%[h]
Cheese (Parmesan)	35.8	25.8	1184	Yes		Free fat/aggregates/FG	Solid	NA
Cheese (Cream cheese)	5.9	34.2	98	Yes	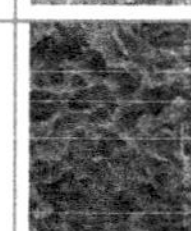	Homogenized milk FG/ potential fragments of MFGM	Semi-solid	NA
Cream (35% MF)	2.1	35.0	66	No	NA	Native FG/homogenized milk FG[i]/potential MFGM fragments	Liquid	NA

Milk (whole, 3.25% MF)	3.2	3.3	113	No	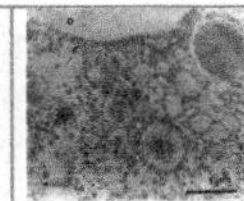 Reproduced from[f]	Native FG/homogenized milk FG/potential MFGM fragments	Liquid	100%[j, k]
Yogurt (greek, plain, 2% MF)	9.7	2.0	283	Yes		Native FG/homogenized milk FG/potential MFGM fragments	Gel/ viscoelastic	100%[j]
Yogurt (plain, 2–3.9% MF)	4.6	2.0	147	Yes	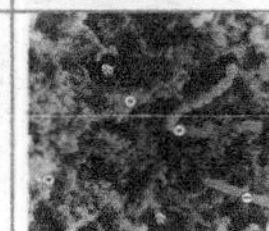 Reproduced from[l]	Native FG/homogenized milk FG/potential MFGM fragments	Gel/ viscoelastic	100%[i]

Adapted from Turgeon and Brisson (2019)
MFGM Milk fat globule membrane, *FG* Fat globule, *LAB* Lactic acid bacteria, *MF* Milk fat, *NA* Not available
[a]Retrieved from Health Canada (2015)
[b]Authors personal SEM images unless specified
[c]Based on Michalski (2009); Michalski et al. (2013); Lopez et al. (2015)
[d]Matrix degradation index determined at the end of the gastric digestion using an *in vitro* model adapted from Versantvoort et al. (2005)
[e]Based on the processing method used
[f]Heertje (2014)
[g]Fang et al. (2016a); Vallières (2016)
[h]Fang et al. (2016b); Lamothe et al. (2012)
[i]NB: for cream, milk and yogurts, homogenization and its intensity may vary among products/brands
[j]Lamothe et al. (2017)
[k]Milk with 2% MF
[l]Sandoval-Castilla, Lobato-Calleros, Aguirre-Mandujano, and Vernon-Carter (2004). Adapted from Turgeon and Brisson (2019)

character corresponding to a higher calcium content. Electron microscopy allows to see differences in lipid organisation, for example, the continuous lipid phase in butter or the fat globules size, distribution, and aggregation in cheese may be visualized.

In addition to lipid organisation in dairy products, it should also be pointed out that when these dairy products are incorporated in a meal their structure evolves due to cooking, mixing, etc. and this may as well modify their behaviour during digestion. These aspects have not been overlooked until now and most of the nutritional studies were standardized on a nutrient basis without considering a possible effect of the lipid structure and dairy products state on their nutritional properties. For example, how is lipemia changing if lipids come from cheese, butter, or melted cheese? What happens if they are part of a complex meal? How other foods does impact dairy products nutritional properties? Interactions between food and drug absorption has been studied, but not much is known on the relative importance of food interactions on nutritional properties.

1.3 Lipid Digestion

After its ingestion, food undertakes a travel in the digestive tract with several compartments involving mechanical, enzymatic and biochemical reactions aiming to break food into simple components that can pass through the gastrointestinal tract mucosa. The release of a specific component from a food matrix into the digestive juice and being ready to be absorbed is defined as the bioaccessibility. The absorption and passage into the systemic circulation to finally reach its metabolic target is named the bioavailability. Figure 8.2 presents the different steps lipids follow for

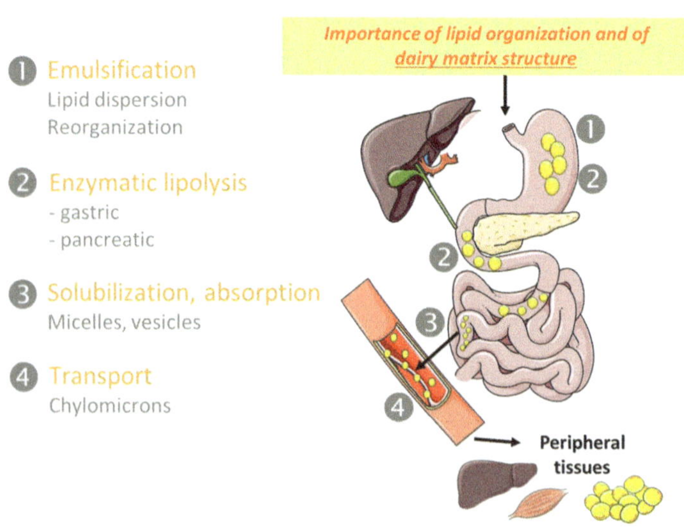

Fig. 8.2 Lipid digestion. Adapted from Michalski et al. (2013)

their digestion and absorption. After swallowing, the bolus formed in the mouth reaches the stomach, a reservoir where food disintegration continues. The physiological temperature increases the proportion of liquid fat, the acidic pH of stomach and the enzymatic attack of the protein network (by pepsin) promotes the release of lipids in the gastric lumen where they can be emulsified by the gastric mechanical movements. Lipids are dispersed and reorganized as oil-in-water emulsified droplets facilitating accessibility for the gastric lipase. Human gastric lipase acts preferentially on position *sn*-3 and generally leads to the hydrolysis of 10–30% of esterified FA, or 5–40% according to different reports (Armand, 2007; Carriere, Barrowman, Verger, & Laugier, 1993; Favé, Coste, & Armand, 2004; Favé, Peyrot, Hamosh, & Armand, 2007). SCFA are preferentially hydrolyzed from TAG than longer ones (German, 2008).

The emulsified chyme exits the stomach into the duodenum. Several factors regulate gastric emptying as the volume contained in the stomach, the degree of the food disintegration and chemical breakdown of the chyme (Stenson, 2006). Fat emptying is delayed compared to the aqueous phase because it floats over it. Furthermore, solid fat may stick to solid particles requiring grinding for their disintegration. Size reduction averaging 1–2 mm is necessary to pass into the small intestine (Kong & Singh, 2008; Stenson, 2006). If lipids are embedded in a solid food matrix, its disintegration is required for lipid release. The physical properties of the matrix may also influence lipid digestion (ex: liquid vs solid foods, emulsified fat or not).

Fat entering the duodenum consists of 70% TAG and a mixture of partially digested hydrolyzed products (Jones & Kubow, 2006). Fat hydrolysis continues with pancreatic lipase acting on lipids emulsified by bile (Favé et al., 2004). The efficiency of lipolysis is dependent on the emulsion droplet size and interface composition (Borel et al., 1994). Nonetheless, the absorption rate of TAG is higher than 95% (Sethi, Gibney, & Williams, 1993). Absorption mechanism is different for SCFA (<12 carbons) and long-chain FA (LCFA >12 carbons). SCFA are more hydrophilic and diffuse directly to the portal vein to be transported to the liver while LCFA requires assembly into mixed micelles to be absorbed (Bernard & Carlier, 1991; Duchateau & Klaffke, 2009; Jones & Kubow, 2006). Regarding LCFA, once they have been taken up by small intestinal enterocytes to be absorbed, FA are re-esterified into TAG and polar lipids and their accretion results in the formation of TAG-rich lipoproteins named chylomicrons. The latter are then secreted into the lymph to reach the bloodstream where they contribute to postprandial lipemia. In peripheral tissues, chylomicrons are hydrolysed by the lipoprotein lipase and released FA are taken up by tissues to be used as an energy source by beta-oxidation. In turn, the white adipose tissue (WAT) stores FA in the form of TAG, and chylomicron remnants remaining in the bloodstream get cleared by the liver (Lambert & Parks, 2012; Mu & Hoy, 2004; Wang, Liu, Portincasa, & Wang, 2013).

1.4 Lipid Metabolism

Lipemia is the amount of lipids present in the bloodstream. During the postprandial phase (i.e., the hours following the consumption of a meal), TAG-rich lipoproteins (firstly chylomicrons produced by the intestine, secondly liver-produced very low density lipoprotein (VLDL)) acutely increase and later decrease due to their hydrolysis by circulating lipases and their further clearance from the bloodstream. If postprandial lipemia is too high for a too long time, arteria are too much exposed to these TAG-rich lipoproteins (Vors, Nazare, Michalski, & Laville, 2014). This is why the kinetics of postprandial lipemia is important, i.e., not only the area under the curve (AUC) of plasma TAG during the postprandial phase, but also the time of appearance of the TAG peak, the maximum concentration reached and also the time for return to baseline lipemia. In fact, humans in Western countries spend most of the day in a postprandial state and it has been clearly established that a too high and too long postprandial lipemia is an independent risk factor to develop cardiovascular diseases (CVD) (Nordestgaard, Benn, Schnohr, & Tybjærg-Hansen, 2007), as observed in subject suffering from metabolic syndrome, obesity or type 2 diabetes (Lopez-Miranda, Williams, & Lairon, 2007). Therefore, studying how to modulate the kinetics of postprandial lipemia, notably by the food matrix, is a subject of growing research interest in the frame of prevention of metabolic diseases (Drouin-Chartier et al., 2017; Grundy, Lapsley, & Ellis, 2016; Keogh et al., 2011; Thorning et al., 2016; Vors et al., 2013).

2 Factors Impacting Lipid Metabolism: In Vivo Studies

2.1 Dairy State (Liquid vs Semi-Solid vs Solid)

Several reviews summarize results on the impact of differently structured dairy products (i.e. different matrixes) on postprandial lipid metabolism and associated metabolic disorders (German et al., 2009; Labonte, Couture, Richard, Desroches, & Lamarche, 2013; Lamarche, 2008; Nestel, 2008).

Acute postprandial studies were performed in vivo. In rats, lymphatic absorption of lipids is slower and lower after gavage with butter < cream cheese < regular or sour cream (Fruekilde & Hoy, 2004). Postprandial studies in humans with metabolic disorders or disease confirm this trend. In T2D subjects, consuming a meal containing 30 g lipid as butter induces a later increase of plasma TAG than using Mozzarella cheese or milk, despite a similar AUC over 6 h (Clemente et al., 2003). This appears consistent with a clinical study where obese men presented a relative delay in the digestion/absorption of spread (unemulsified) vs emulsified milkfat (lower area under the curve of chylomicron-TAG over 5 h), unlike normal-weight men (Vors et al., 2013). Regarding longer-term differences, in a clinical intervention study, 14 healthy men received for 3 weeks a diet containing 20% of daily energy as milk fat as butter, milk or hard cheese. After 4 days, no differences were observed on postprandial lipid metabolism among groups. However, at the end of the trial, plasma LDL-cholesterol was higher after the butter diet than after the cheese diet (Tholstrup, Hoy, Andersen, Christensen, & Sandstrom, 2004).

Regarding viscosity, structure and gastric emptying modulation human studies showed that ingestion of solid foods slows down gastric emptying compared with liquid foods (Hunt & Knox, 1968), which results in a later appearance of TAG peak in plasma (Dubois et al., 1994). Such phenomenon was also demonstrated using dairy products with different viscosities (Fruekilde & Hoy, 2004; Sanggaard et al., 2004). A study using (1) fresh curd vs cream and (2) fermented milk vs regular milk showed an increase gastric emptying time for viscous matrices. Moreover, fresh curd results in delayed peak of postprandial lipemia vs cream. Of note, a semi-solid or solid texture decreases hunger and induces satiety as observed using semi-solid yogurt and drinkable yogurt compared with a fully liquid dairy drink (Tsuchiya, Almiron-Roig, Lluch, Guyonnet, & Drewnowski, 2006). Moreover, recent studies report that dairy product viscosity or gelation and gel structure (acid, rennet) have an impact on protein digestion and bioavailability, however to date, skimmed milk was used so that the impact on lipid fate was not explored (Barbé et al., 2014; Dupont, Ménard, Le Feunteun, & Rémond, 2014). In humans, full-fat yogurt slows down gastric emptying and induces a slower and prolonged release of proteins in the jejunum compared with whole milk (Gaudichon et al., 1994; Gaudichon et al., 1995). Such approach should thus now be explored regarding the fate of milk lipids; notably homogenization of fat globules can indirectly impact matrix structure (because of decreased globule size and coverage with proteins that interact with the casein network) and possibly the lipid fate (Sect. 2.2) (Michalski & Januel, 2006).

During the digestion of fermented milk, a slower gastric emptying was observed, together with a higher peak of plasma TAG returning faster to baseline than with standard liquid milk (Sanggaard et al., 2004). This can be explained by the higher viscosity of fermented milk, which slows down gastric emptying and ultimately the appearance of plasma TAG. In healthy individuals, consuming a solid meal results in a peak of lipemia appearing 3–4 h after the meal, whereas the TAG peak occurs 2 h after a liquid meal (Dubois et al., 1994). Moreover, in humans, the aqueous phase of a meal empties fast from the stomach, whereas solid and lipid phases empty together after a lag time (Meyer et al., 1986). This is why a more or less solid dairy matrix can impact milk lipid digestion (e.g. hard cheese vs fresh curd). Notably, other components of the dairy matrix such as proteins and calcium can impact postprandial lipid metabolism (see below) (Lorenzen & Astrup, 2011; Mortensen et al., 2009), as well as matrix structure. Regarding hard pressed cheeses (cooked or uncooked), pressing the curd allows to drain the serum. This process disintegrates milk fat globules, notably by inducing coalescence and the formation of "free fat" inclusions, all the more than the pressure is high (Lopez, Cauty, & Guyomarc'h, 2015; Michalski et al., 2004; Michalski et al., 2007). Such structural changes in milk fat inside the product could contribute to cheese properties during digestion but this remains to be elucidated in detail as highlighted in different reviews (Michalski, 2009; Thorning et al., 2017).

2.2 Food Manufacturing Processes

As summarized in Table 8.1 and in Fig. 8.1, dairy processes such as homogenization and cheese press can greatly impact milk fat globule structure and functional properties. Recent research shed the light on a further impact on lipid digestion, absorption and postprandial metabolism. Notably, postprandial lipemia can be greatly modulated by the emulsified structure of fat (Armand et al., 1999; Couedelo et al., 2015; Keogh et al., 2011; Michalski et al., 2013; Vors et al., 2012; Vors et al., 2013). Regarding milk fat, the fat globules homogenized or not, and a milkfat emulsion covered either with proteins or with polar lipids, result in different kinetics of postprandial lipemia in rats (Michalski, Briard, Desage, & Geloen, 2005; Michalski, Soares, et al., 2006). A response of lower amplitude is observed with homogenized vs native globules and for droplets covered with caseins vs soy lecithin. Moreover, still in rats, appearance of absorbed FA in the lymph occurs faster after gavage with homogenized cream than butter, and cumulative absorption over 8 h is greater (Fruekilde & Hoy, 2004). A recent clinical trial shows in healthy men that emulsified milk fat in skimmed milk as part of a breakfast results in a faster and higher plasma TAG peak compared with unemulsified fat, revealing an easier intestinal absorption of lipids (Vors et al., 2013). This demonstrates that the enhancing effect of emulsification on lipolysis and absorption can occur in the frame of a mixed meal in humans. A novel aspect of this work was to elucidate the final metabolic fate of the ingested FA. By a $^{13}CO_2$ breath test, this study demonstrated a higher beta-oxidation of ingested FA when fed in emulsified fat structure (Vors et al., 2013). This was due to a faster influx of dietary non-esterified FA (NEFA) in plasma, thereby used as a priority energy source. In a recent postprandial study, the consumption of soft cream cheese (emulsified dairy fat in a semi-solid matrix), Cheddar cheese and butter (control) as part of a test meal was provided at breakfast to healthy participants in a randomized, crossover and controlled trial (Drouin-Chartier et al., 2017). Of note, the structure of the dairy food was not altered in this study in order to keep their matrix structure. After 2 h, plasma TAGs significantly increase for the cream cheese meal compared to the Cheddar cheese meal. At 6 h, the cream cheese meal induces a lower TAG response compared to the Cheddar cheese meal suggesting different kinetics of postprandial TAG response. In addition, the lower ApoB48 incremental AUC found when consuming cream cheese suggests lower and smaller chylomicron secretion than for cheddar cheese. These results suggest that cream cheese is digested faster than Cheddar cheese confirming that cheese matrix modulates postprandial lipemia.

Combination of homogenization and heat treatment was also studied. In minipigs, gastric lipolysis is lower when milk or fermented milk are pasteurized and homogenized (Timmen & Precht, 1984). In rats, homogenized pasteurized cream modifies fat structure and digestion in the stomach and small intestine compared with untreated cream (Gallier et al., 2013). In the small intestine, homogenization results in an apparently higher lipolysis and increased appearance of fat crystals in the second half of the small intestine. However, a lower postprandial lipemia was observed in another rat study with homogenized vs native milk fat globules (Michalski, Soares, et al., 2006). To our knowledge, only two clinical trial were performed to date on the impact of milk homogenization on postprandial lipid metabolism. One crossover trial in overweight men consisted in three postprandial tests after the consumption of 900 mL of whole milk unhomogenized, or homogenized, or of 44 g butter with skimmed milk (Masson, 2013). This PhD thesis study reports a lower rise of plasma TAG after homogenized milk than butter, but no difference with unhomogenized milk. The second trial was a pilot study in a small group of 11 healthy volunteers designed to investigate the impact of milk homogenization on gastrointestinal symptoms (Nuora et al., 2018). Milk homogenization did not impact postprandial lipaemia, but after 4 h, authors observed more of the major long-chain SFA (myristic, palmitic and stearic acids) in plasma after homogenized vs unhomogenized milk. Authors report no significant difference in the amount of gastrointestinal symptoms or in the intestinal pressure but point out that further studies in this area are needed with larger group size and longer exposure times to differently processed milk types (Nuora et al., 2018). Of note, another acute trial reports the postprandial response of healthy men to the consumption of a bolus of infant formula differing in fat globule structure: small droplets covered with milk proteins (i.e. similar to homogenized milk fat droplets) vs large droplets coated with phospholipids and proteins (designed to model native milk fat globule structure) (Baumgartner et al., 2013). Here the small droplets covered with proteins resulted in a slower rise in plasma TAG than the large droplets coated with PL and proteins (which is consistent with previous rodent studies; Michalski et al., 2005; Michalski, Soares, et al., 2006). More studies are needed to elucidate in humans the impact of milk homogenization on lipid fate in the body and metabolic consequences. Of note, Tholstrup (2006) used to highlight that the consumption of cheeses made from unhomogenized milk was high in France where coronary mortality was quite low, whereas in Scandinavian countries where milk is rather consumed as homogenized liquid milk, coronary mortality is high (Tholstrup, 2006). However, correlation is not causation and as discussed above, we now know that the specific impact of cheese would rather be due to an overall "matrix effect". Previous reviews on the impact of milk homogenization on health report that to date, no link was established between homogenization and CVD or T2D, but this should be further explored in humans due to the controversial nature of this topic (Michalski, 2007; Michalski & Januel, 2006). This is all the more interesting to explore than homogenization modifies dramatically the size of milk fat globules (see Sect. 1.2) and destroys the native structure of the MFGM (see section MFGM and milk PL content).

2.3 Dairy Matrix Composition

2.3.1 Lipid Content

Lipid Amount One major aspect of dairy matrix composition regarding lipids is the broad range of fat content, from low-fat and so-called/claimed "diet" products to full-fat ones including cheeses. Importantly, dietary guidelines in many countries such as Canada (Health Canada, 2016) or US (US Department of Health and Human Services and US Department of Agriculture, 2015) recommend consuming low fat dairy products. However, recent researches challenge such recommendations.

A study in pigs shows that a regular full-fat cheese results in increased HDL-cholesterol compared with butter after 14 days. However, this beneficial effect was not observed with low-fat cheese (Thorning et al., 2016). In this study though, low-fat cheese was consumed with butter in order to equalize lipid intake. Therefore, observed effects on cholesterol could be due to the presence of the MFGM in full-fat cheese and/or to a "matrix effect" as discussed recently by an expert panel (Thorning et al., 2017). In humans, another team also showed that for a similar consumption of full-fat or low-fat cheese without adjusting lipid and energy intakes, full-fat cheese induced a higher HDL-cholesterol concentration and no difference in LDL-cholesterol (Raziani et al., 2016). Two recent studies show an inverse association or a lack of correlation between the consumption of different dairy products and the appearance of clinical cardiovascular events, and the lack of association between the consumption of low-fat or full-fat dairy products and different metabolic risk factors, notably regarding lipid metabolism and inflammation (Drouin-Chartier, Brassard, et al., 2016; Drouin-Chartier, Cote, et al., 2016). They highlight the importance of studying in more details the metabolic impact of dairy products with different fat contents with an aim to argue on the current dietary guidelines advising to consume low-fat dairy. Long-term nutritional interventions are necessary to elucidate the link between full-fat dairy consumption and metabolic and cardiovascular risks.

FA Composition and Melting Points Milk fat has a relatively solid structure, whose melting point can vary according to its fatty acid profile, notably SFA content. The more or less solid state of TAG at body temperature (37 °C) can also impact their digestibility and absorption. In healthy adults, stearic acid (18:0) in a high melting point fat results in a lower lipid absorption, the latter being in fact inversely correlated with the proportion of solid fat at 37 °C (Berry, Miller, & Sanders, 2007; Berry & Sanders, 2005). Altogether, TAG containing long-chain saturated FA such as palmitate (16:0) and stearate (18:0) melt above 37 °C (e.g. in the range 68–73 °C for homogeneous TAG composed of palmitate and/or stearate only). Solid TAG in the digestive tract results in a limited action of lipases and therefore a lower postprandial lipemia compared with liquid oils (Berry & Sanders, 2005; Sanders, Filippou, Berry, Baumgartner, & Mensink, 2011). The hydrolysis rate by pancreatic lipases of a tripalmitin emulsion that is completely solid is lower than that of a similar emulsion with liquid tripalmitin (Bonnaire et al., 2008). In humans, vegetable fats of high melting points with a large proportion of solid TAG at 37 °C result in lower postprandial lipemia compared with fats liquid at 37 °C (Berry & Sanders, 2005). Regarding milk fat, Mekki et al. (2002) observed a lower postprandial accumulation of chylomicron triglycerides in plasma after a meal containing spread butter rather that vegetable oil in a sauce. In dairy products, the proportion of solid fat at 37 °C is lower than in high melting point vegetable fats, but still exists. Around 5% of milk TAG can be crystallized at 37 °C (Lopez, 2011) and this proportion can be of 3% in Emmental cheese (Lopez, Briard-Bion, Camier, & Gassi, 2006). In guinea pigs (Asselin et al., 2004) or rats (Lai & Ney, 1998), the plasma TAG concentration or AUC of plasma TAG was lower in animals fed with high melting point milkfat fraction (42–44 °C) compared with a low melting point fraction (13–14 °C). Similarly, rats fed Cheddar-type cheeses manufactured with anhydrous milk fat (AMF) of different melting points showed different postprandial TAG responses (Ayala-Bribiesca, Turgeon, Pilon, Marette, & Britten, 2018). Stearin AMF exhibited the lowest plasma TAG responses compared to the olein AMF. This could be explained by the high proportion of unsaturated saturated LCFA found in the stearin AMF resulting in a melting point higher than body temperature (42.3 vs 37 °C). Also, the mass fraction of the total FA recovered in the animal feces were significantly higher with the stearin AMF diet compared to the olein AMF diet. This was mainly attributed to the higher proportion of palmitic acid (16:0) which is prone to form calcium soap (detailed in the next section). Therefore, FA bioavailability was reduced in the presence of a stearin AMF diet.

MFGM and Milk PL Content Different dairy products can contain different amounts of MFGM and associated components (Fig. 8.1), which is an important structural component of the native MFG and source of bioactive molecules. When incorporated in animal diets, extracts of MFGM or of milk PL were shown to be able to decrease intestinal cholesterol absorption and hepatic lipids. In mice, adding a buttermilk extract rich in milk PL (1.2% PL in the diet) in a diet containing 21% lipid induces a decrease of hepatic total lipids, TAG and cholesterol, and a decrease in plasma lipids (TAG, total cholesterol, phospholipids (PL)). These effects were not observed when the milk PL-rich buttermilk extract was added in a diet containing 4.6% lipids. This suggests a beneficial effect of MFGM or milk PL in a deleterious/obesogenic dietary context (Wat et al., 2009). A hypocholesterolemic effect of the MFGM was also observed in rats with induction of lower hepatic cholesterol and TAG, but higher plasma TAG (Zhou, Hintze, Jimenez-Flores, & Ward, 2012). In mice fed a high-palm oil diet, incorporation of 1.2% of soybean PL induced higher hepatic lipids and higher WAT with larger adipocytes, which is a deleterious feature, compared with the diet devoid of PL. Conversely, when incorporating 1.2% of milk PL in the semi-synthetic high-fat diet, no increase of hepatic nor WAT lipids was observed (Lecomte et al., 2016). Incorporation of 1.6% of milk PL in a chow-based high-fat diet resulted in a lower body weight gain of mice after 8 weeks (Milard, Laugerette, et al., 2019).

A few studies also described a hypolipemic effect of buttermilk consumption in humans (Baumgartner et al., 2013; Conway et al., 2013). Consuming 45 g/day of a chocolate-flavored buttermilk drink during 4 weeks resulted in healthy men and women in decreased fasting plasma concentrations of total cholesterol (−3.1% vs placebo), LDL-cholesterol (−3.1% vs placebo) and TAG (−10.7% vs placebo) in a randomized controlled trial (Conway et al., 2013). Authors discussed the decreased cholesterolemia could be due to sphingomyelin (SM) present in the buttermilk, which could lower cholesterol absorption. Moreover, the decreased plasma TAG could be due to a lower hepatic TAG synthesis when consuming milk PL as observed in mice (Eckhardt, Wang, Donovan, & Carey, 2002; Reis et al., 2013). Baumgartner et al. (2013) studied the impact of the matrix on cholesterol metabolism. To this aim, 97 healthy volunteers were assigned either to the control group, which did not modify its dietary intakes, or to the group consuming one egg per day, or to the group consuming one egg yolk per day incorporated in a buttermilk drink (+215 mg cholesterol/day in women, +97 mg/day in men, in egg groups vs control group). In women consuming 1 egg/day, increased plasma concentrations of total- and LDL-cholesterol were observed vs control group. No increase was observed when the egg was associated with buttermilk (Baumgartner et al., 2013). Here again, authors suggest that the anti-hypercholesterolemic effect could be due to milk PL and notably SM.

A study with 11 healthy volunteers showed that the consumption of a "placebo buttermilk" (fermented skimmed milk) devoid of MFGM for 3 weeks did not modify plasma lipids, which supports the hypothesis that the potential hypolipemic impact of buttermilk could be due to MFGM or some of its components such as milk PL (Thompson et al., 1982). More recently, the impact of two dairy ingredients differing by their MFGM content and emulsified structure was explored on lipid metabolism. Volunteers consumed 40 g of milkfat per day incorporated in a muffin as whipping cream (containing 198 mg of milk PL) or as AMF (containing only 1.3 mg of milk PL) for 8 weeks. AMF-in-muffin diet increased total and LDL- cholesterol plasma concentrations, while whipping-cream-in-muffin diet did not exert such effect (Rosqvist et al., 2015). Authors discussed this could be due to the presence of MFGM in whipping cream. In this study, data indicate no difference in cholesterol absorption or synthesis among groups, suggesting another mechanism not yet elucidated.

Until 2019, clinical nutritional intervention studies led with milk PL or SM did not demonstrate significant effects on fecal excretion and intestinal absorption of cholesterol nor on plasma lipids in healthy volunteers (Ohlsson, Burling, & Nilsson, 2009; Ramprasath, Jones, Buckley, Woollett, & Heubi, 2013). In a parallel group randomized trial, healthy volunteers consumed daily during 4 weeks a dairy drink containing 2.8 g of egg PL or milk PL (by a buttermilk-derived concentrate). Volunteers of the egg PL group increased plasma lipids, while volunteers of the milk PL groups did not (Ohlsson et al., 2009). However, no real hypolipemic effect was observed in the milk PL group (i.e., no decrease of plasma lipids after intervention vs before intervention). Finally, a dietary supplementation with 1 g of milk SM per day for 14 days induced increased HDL-cholesterol but no change in absorption and synthesis of cholesterol in ten healthy volunteers (Ramprasath et al., 2013). The latter studies could have limited effects due to the low number of volunteers (<30) and/or because volunteers were healthy, without lipid metabolism disorders. In this respect, two recent studies in overweight or obese volunteers show an effect of milk PL consumption on a marker of hepatic steatosis, namely GGT (γ-glutamyl transferase) (Weiland, Bub, Barth, Schrezenmeir, & Pfeuffer, 2016). In the first study, increased GGT observed when consuming a control dairy drink was not observed when the drink was enriched with milk PL. However, no effect of milk PL was observed on plasma lipids and insulin resistance markers. In their second study, authors report a lower GGT concentration with milk PL than with soybean PL, devoid of SM.

Most recently, the first human trial reporting a significant impact of milk PL supplementation on reducing an array of cardiometabolic risk factors in a population at metabolic risk was published by Vors et al. (2020). Overweight postmenopausal women were subjected to a 4 week-dietary intervention with the daily intake of cream cheese either enriched with 3–5 g/day of milk PL via a butter serum concentrate rich in MFGM fragments, or control cream cheese devoid of milk PL (here fat ingredient was butteroil only) (parallel groups). Milk PL incorporated in cream cheese resulted in reduced total and LDL-C (−8.7% in 5 g group after vs before intervention), as well as decreased total/HDL-cholesterol and decreased ApoB/ApoA1 ratio (−6.8% in 5 g group), compared to the control group (no effect). This intervention with milk PL (1) decreased the plasma concentration of cholesterol carried by postprandial chylomicrons and (2) increased the coprostanol/cholesterol ratio in feces, suggesting an increased conversion of cholesterol to coprostanol, a non-absorbable metabolite of cholesterol, by the gut microbiota (Vors et al., 2020). In a complementary crossover clinical trial, four ileostomized subjects consumed these different cream cheeses with varying enrichment of milk PL. Here milk PL decreased intestinal absorption of cholesterol (lower cholesterol tracer concentration AUC in plasma and chylomicrons in the postprandial phase). Additionally, an increased ileal output of both total cholesterol (of dietary + endogenous origin) and of milk SM was observed in ileal efflux after both milk PL-rich meals, confirming previous observations of fecal excretion of milk SM in mice (Milard, Laugerette, et al., 2019). Moreover, these results in humans could have been enhanced by the fact that SM was consumed here as part of a complex milk PL mixture within the cream cheese matrix (from MFGM fragments of the butter serum ingredient), rather than as pure milk SM.

Altogether in animals, MFGM extracts and milk PL were shown to decrease intestinal cholesterol absorption and hepatic lipids in the long term. In humans, few clinical trials report a decrease in fasting lipids (notably cholesterol) or a prevention of increased lipids by consuming buttermilk. Authors suggest this is due to the presence of MFGM in buttermilk. However, to date, most human studies with MFGM extracts of milk PL at around 3 g/day do not show effects on intestinal cholesterol absorption or plasma lipids. Overall, the hypolipidemic effects of milk PL or milk SM observed in rodent studies have been reported in some human studies, although with smaller magnitude of the effects that were often non-significant, except in the recent trial using a real dairy food matrix and 3–5 g/day of milk PL that resulted in significant favorable effects (Vors et al., 2020). Of note, the small or neutral effects of lower milk PL intakes observed in the 4–12-week trials may contribute to maintaining a relative blood cholesterol homeostasis in the longer term, which would deserve further elucidation. Performing further clinical studies in patients with metabolic syndrome, moderate hypercholesterolemia (as in Vors et al., 2020) or high blood pressure could provide more insight on the potential beneficial effects of milk PL and MFGM and such perspectives are proposed in different articles and reviews (Castro-Gomez, Garcia-Serrano, Visioli, & Fontecha, 2015; Conway, Gauthier, & Pouliot, 2014; Norris, Milard, Michalski, & Blesso, 2019; Ohlsson et al., 2009).

2.3.2 Protein Content

The impact of proteins at the interface of milkfat emulsion droplets has been explained above (Chap. 7). Here we will focus on the impact of the presence of different amounts or quality of proteins in the dairy matrix or in the meal on the metabolic fate of lipids.

In a pilot study, healthy subjects consumed a bolus of dairy cream (30% fat) mixed with water and containing or not 50 g of sodium caseinate (Westphal et al., 2004). Caseinate addition induced a small delay of appearance of plasma TAG and decreased significantly plasma NEFA concentrations. Authors hypothesized this would be due to the increased insulin secretion observed with caseinate. However, another study in adults shows that milk proteins added in a meal do not acutely modify markers of postprandial lipid metabolism (Bortolotti, Schneiter, & Tappy, 2010). This study used complex meals with different types of dairy matrixes, with the control meal containing 19 g of proteins brought by cottage cheese only, while enriched meal brought 56 g of proteins by an additional intake of skimmed milk, buttermilk and cottage cheese. Both meals were isolipidic, 30 g of lipids among which a part was brought by butter in the control meal. Subjects then had a 4 days diet enriched in dairy proteins (by skimmed milk, cottage cheese and yogurts for 1.5 g proteins/kg body weight/day). This short period of protein enrichment of the diet induced increased postprandial chylomicronemia following the same hyperproteic test meal as above, and a lower beta-oxidation (i.e. use as an energy source) of ingested lipids. This suggests that the hyperproteic diet induced a worsening of chylomicron clearance from blood (Bortolotti, Dubuis, Schneiter, & Tappy, 2012). Discrepancies among these two studies could be partly explained by (1) the type of proteins: sodium caseinate vs total dairy proteins, and (2) the type of test meal: simple liquid meal vs realistic mixed meal containing a variety of real dairy products. This provides incentive to further explore the dairy matrix effect on lipemia modulation by different amounts and types of proteins.

The type of proteins presents in a product or a meal appears to be an important modulator of postprandial lipemia. Whey proteins, so-called "fast proteins", bring amino acids that are readily available in plasma compared to "slow" caseins (Boirie et al., 1997). Whey proteins also present benefits on some features of the metabolic syndrome, including an improvement of fasting plasma lipid profile (Pal & Radavelli-Bagatini, 2013). Digestion of casein micelles produces caseinomacropeptide, which in turn provokes the secretion of cholecystokinin (CCK), an inhibitor of gastric emptying. This can impact the entire digestion kinetics, as well as the satietogenic ileal brake mechanism (van Avesaat, Troost, Ripken, Hendriks, & Masclee, 2015). Several studies in rodents also describe a beneficial effect of whey proteins on lipid metabolism (Kawase, Hashimoto, Hosoda, Morita, & Hosono, 2000; Sautier et al., 1983; Zhang & Beynen, 1993) such as decreased serum total cholesterol after a supplementation with whey protein concentrate vs caseins (Sautier et al., 1983). Such favourable effects were also explored in humans. Recent clinical trials demonstrated that whey proteins in a meal can decrease postprandial lipemia and chylomicronemia compared to other types of proteins, especially in subjects with abdominal obesity, which can be considered beneficial (Bohl et al., 2015; Holmer-Jensen et al., 2013; Mortensen et al., 2009; Pal, Ellis, & Dhaliwal, 2010). A clinical trial in T2D volunteers compared the impact of a meal containing 100 g of butter (structure: melted in a non-energetic soup), 45 g of carbohydrates (white bread) and containing 45 g of proteins either in the form of calcium caseinate, whey proteins, fish proteins or gluten (altogether proteins in soup). Whey proteins result in a lower postprandial area under curve (AUC) of plasma TAG, NEFA and glucose vs other proteins, and the lipemia peak was delayed. Authors hypothesized a lower production and/or a better clearance of chylomicrons (Mortensen et al., 2009). Another study in non-diabetic obese subjects confirmed an impact of whey proteins on lowering postprandial lipemia compared with fish proteins and gluten (Holmer-Jensen et al., 2013). A 12-week intervention study compared the physiological response to a rich mixed meal including notably different dairy matrixes (butter, cheese, milk) after 4 different diets containing 63 g of milkfat per day (rich or low in short- and medium-chain SFA) and 60 g per day of proteins as whey proteins or caseinate. Regardless of milkfat composition, whey proteins induced a lower number of chylomicrons after a test meal compared to caseinate (Bohl et al., 2015). Here again, a lower production and/or better clearance of chylomicrons could occur, which might be favourable in the long term by limiting CVD risk. This remains to be elucidated with different types of proteins and dairy matrixes. Pal et al. observed in overweight and obese subjects a decrease in fasting plasma total- and LDL-cholesterol after 12 weeks of supplementation with whey proteins vs caseins (Pal et al., 2010). A lower exposition to TAG-rich lipoproteins was also observed in the postprandial phase (AUC of the TAG/ApoB48 ratio, ApoB48 being a marker of chylomicron number). Such decreased exposition of lower arteria to LDL-cholesterol and TAG-rich lipoproteins would be favourable. These studies suggest a hypolipemic effect of whey proteins in the postprandial phase compared with caseinate. Suggested mechanisms included impacts on hepatic de novo cholesterol synthesis, intestinal cholesterol absorption and intestinal FA transport.

However, we must highlight one study leading to different results that can be due to a matrix effect. Mariotti et al. observed an increase of postprandial TAG with total whey proteins or alpha-lactalbumin in the meal compared with caseins in the meal (Mariotti et al., 2015). Authors proposed a mechanism due to cream droplets being unstable in the casein meal, and to changes in meal viscoelasticity (coalescence and fat phase separation in gastric conditions with caseins, stability of fat droplets in the whey protein meal). We suggest that differences of source and structure of the milk fat in the test meal (cream in the latter study, melted butter in the former studies) could modify the relative impact of whey proteins vs caseins. There again this supports the impact of the food matrix and meal structure on metabolic outcomes.

Altogether, several clinical trials show that the type of dairy proteins can modify postprandial lipemia. Whey proteins would be hypolipemic but this remains to be confirmed using test meals of different protein composition and of different matrix composition and structure. This remains to be explored in detail in humans because an inter-connexion exists between the relative impact of protein amount, type, location at the interface of fat droplets or in the aqueous phase or proteinaceous network, denaturation and consequences relative to viscosity.

2.3.3 Calcium Content

An important aspect of the fate of SFA in the gut consists in their ability to form calcium soaps when they are located at the external positions of TAG and the meal contains calcium, which is naturally the case of most dairy matrixes. Milk fat contains a high proportion of palmitic acid on the sn-2 (internal) position of milk TAG, which ensures a high bioaccessibility of this FA. More generally, milk contains a huge diversity of molecular TAG species that can modify digestion kinetics compared to homogeneous TAG (Mu & Porsgaard, 2005). Notably in milk fat, a proportion of milk SFA are located at the external positions of TAG and can thus, when released by lipases, be associated with dairy calcium, resulting in fecal excretion of calcium soaps (Lorenzen et al., 2007; Lorenzen, Jensen, & Astrup, 2014; Soerensen, Thorning, Astrup, Kristensen, & Lorenzen, 2014). In a crossover clinical trial, 15 healthy men consumed for 2 weeks a control diet low in calcium (devoid of dairy products except butter, 500 mg Ca/day) then two diets enriched in calcium (Ca, 1700 mg/day) by either half-skimmed milk or by semi-hard cheese (Soerensen et al., 2014). The lipid fecal loss increased with the milk diet (5.2 g/day) and the cheese diet (5.7 g/day) compared with the control diet (3.9 g/day). Moreover, consuming both diets enriched in calcium via these dairy products induced a lower increase in fasting plasma concentrations of total- and LDL-cholesterol compared with the control diet. A correlation between lipid fecal loss and plasma cholesterol concentrations was even observed in this study. Such mechanism could contribute do decrease postprandial lipemia as shown in a study where 18 healthy volunteers consumed three isocaloric meals differing in calcium content by dairy products (yogurt and milk): low Ca (68 mg/meal), medium Ca (350 mg/meal) and high Ca (793 mg/meal) (Lorenzen et al., 2007; Lorenzen et al., 2014; Soerensen et al., 2014). Meals with medium and high Ca contents induced a lower AUC of chylomicron TAG compared with the low Ca meal (−17% and −19%, respectively). Most recently, rats were fed Cheddar-type cheese with regular (50 mg) and higher (66 mg) level of calcium (Ayala-Bribiesca et al., 2018). Higher plasma TAG were found with the high calcium fed group suggesting higher digestive lipolysis. Rat feces were also collected, and significantly higher fat excretion were found for rats fed the high calcium cheese diet. The analysis of the FA profile revealed that the differences were mainly attributed to LCFA that are more prone to precipitate in presence of calcium (detailed in in vitro section).

2.3.4 Lactic Acid Bacteria and Their Metabolites

Endogenous bacteria present in the dairy matrix can be able to metabolize milk lipids and produce more or less bioactive derivatives/molecules. For example, conjugated linoleic acid (CLA) were reported to be formed from linoleic acid (18:2 n−6) in organic yogurts and fermented milks (Florence et al., 2009). Moreover, some bacteria of the host gut microbiota can produce, from dietary lipids present in the intestinal lumen, CLA species that can present beneficial trophic bioactivity in the intestine (Druart et al., 2014; McIntosh, Shingfield, Devillard, Russell, & Wallace, 2009; Russell, Ross, Fitzgerald, & Stanton, 2011). This can be one mechanism explaining why epidemiological studies report overall beneficial health effects of fermented products.

Different reviews of epidemiological studies and meta-analyses of randomized control trials (RCT) highlight that dairy products intake, including fermented and full fat products, is associated with a lower or unaltered inflammatory profile (Labonte et al., 2013; Labonte et al., 2014; Vors, Gayet-Boyer, & Michalski, 2015). A decreased risk of T2D was recently shown with the consumption of total fermented food intake: a meta-analysis of 5 studies showed no association with T2D but a 12% risk reduction was observed when "high fat" products were included for an intake of 40 g/day; cheese intake was not associated with T2D risk (12 studies) and yogurt intake was inversely correlated with T2D risk (−14% for an intake of 80 g/day; 11 studies) (Gijsbers et al., 2016). Other studies report that cheese intake decreases stroke risk and both CHD and CVD risks (Alexander et al., 2016; Chen, Wang, et al., 2016; Qin et al., 2015).

Regarding intervention trials, a recent study demonstrated that milkfat consumed as cheese for 4 weeks reduced more LDL-cholesterol than butter, both products having similar effects on HDL- cholesterol (Brassard et al., 2017). The impact of two diets containing full-fat dairy products including either yogurt and cheese (fermented) or butter, cream and ice cream (not fermented) were compared to a diet containing low fat dairy products (milk and yogurt) in 12 overweight and obese volunteers (Nestel et al., 2013). No significant difference was observed on markers of inflammation and atherosclerosis, despite a tendency of the low fat diet to induce an increase of theses risk markers compared with the diet rich in fermented full-fat dairy products. These results suggest that (1) dairy lipids do not increase inflammation and atherosclerosis, (2) dairy matrix plays a major role in their metabolic effects and (3) fermentation could play a key role in these phenomena. As explained above, lactic bacteria can generate more or less bioactive lipids in fermented dairy products, but also several other bioactive mediators (Pessione & Cirrincione, 2016). The possible mechanisms for specific effects of fermented dairy products can also be due to fermentation effects notably on the gut microbiota (see below), and also to indirect effects due to matrix structural changes during fermentation and/or ripening.

The impact of fermentation on the digestion, absorption and metabolic impact of milk fat can indeed be due to an array of factors part of the matrix effect. For example, cheese making processes: modify milk fat globule structure by providing more or less aggregation, coalescence, free fat, released MFGM fragments etc.; modify the structure of the proteinaceous phase from gelation to solidification; concentrate more or less calcium and other minerals; and provide bacteria and various fermentation products. Notably, soft cheeses such as Camembert contain native milk fat globules as well as more or less aggregated or coalesced globules (Lopez et al., 2015; Michalski et al., 2003); while Emmental cheese rather contains "free fat" inclusions around which bacteria are located (Lopez et al., 2015; Michalski et al., 2004; Michalski et al., 2007) (see figures in these references). If milk fractions with small or large native milk fat globules collected by microfiltration are used (Michalski, Leconte, et al., 2006), such differences in milk fat globule size are still observed in lipid structures of fermented dairy products made therefrom, which modifies the overall structure of the gel or cheese matrix (Michalski et al., 2003; Michalski et al., 2004; Michalski et al., 2007; Michalski, Cariou, Michel, & Garnier, 2002) and the FA profile of the cheese (Briard & Michalski, 2004).

Importantly, evidence is growing about the impact of bacterial strains on lipid metabolism, notably cholesterolemia-lowering ability of lactic acid bacteria (Ito et al., 2015; Ivanovic et al., 2015; Jo, Choi, Lee, & Chang, 2015; Pan, Zeng, & Yan, 2011; Xie et al., 2011), as reviewed (Pereira & Gibson, 2002). Several studies have been performed in vitro and in rodents, and beneficial prebiotic effects are also documented in humans.

3 Towards a Modulation of Gut Microbiota and Metabolic Inflammation After Dairy Consumption

Beyond alterations of lipid metabolism, obesity and metabolic disorders such as T2D and metabolic syndrome are also characterized by a low-grade chronic inflammatory status named metabolic inflammation (Hotamisligil, 2006; Libby, 2002). Several clinical trials in obese subjects report moderate but chronic increase of proteins involved in inflammation such as C-reactive protein (CRP) or inflammatory cytokines such as interleukin-6 (IL-6) and tumor necrosis factor-α (TNF-α). The role of dietary lipids in metabolic inflammation is largely studied (Calder, 2002; Laugerette, Vors, Peretti, & Michalski, 2011; Schwab et al., 2014) as inflammation can be provoked by lipids derived from omega-6 FA for example. Moreover, recent works revealed the role of altered composition of the gut microbiota in metabolic inflammation. Altogether, the gut microbiota is a key player in health or disease development, notably regarding metabolic diseases including obesity and T2D.

3.1 Gut Microbiota and Associated Endotoxemia

The gut microbiota is now recognized as a major player in the development of obesity and T2D. The bacterial composition of the microbiota impacts energy balance (Turnbaugh et al., 2006), glucose metabolism (Cani et al., 2007) and metabolic inflammation (Cani et al., 2008). The amount and composition of dietary lipids can modulate the composition of the gut microbiota (Murphy, Velazquez, & Herbert, 2015). Lipids from milk, lard, or polyunsaturated FA-rich oil can modulate microbiota composition and inflammation of the adipose tissue (Huang et al., 2013). A recent study in pigs reports that a mixture of vegetable oils and milk fat including MFGM extracts, compared with vegetable oils alone, modifies gut microbiota composition (phylae: increased *Proteobacteria* and *Bacteroidetes*, decreased *Firmicutes*; genus: increased *Parabacteroides*, *Escherichia/Shigella* and *Klebsiella*), intestinal physiology and lymph node secretions (Le Huerou-Luron et al., 2016).

Among possible biomarkers of inflammation associated with the gut microbiota, endotoxemia is a player of particular interest (Laugerette et al., 2011). Bacterial lipopolysaccharides (LPS), so-called endotoxins, are inflammatory molecules naturally present in the digestive tract as endogenous components of the cell wall of Gram-negative bacteria of the gut microbiota. Microbiota is present all along the digestive tract, with an important gradient from 10 bacteria per g of stomach content to 10^{12} bacteria per g of colon content. Recent studies showed that endotoxins can cause a so-called metabolic endotoxemia in plasma after the consumption of unbalanced high-fat meals (Laugerette et al., 2011). Translocation mechanisms from the gut lumen to the blood would be due to (1) intestinal absorption of LPS with dietary lipids in the small intestine on the one hand, (2) paracellular translocation due to altered gut barrier in the colon on the other hand.

Interestingly regarding specific milk fat specificity, mice fed high-fat diets containing 0.25% milk sphingomyelin (SM) presented lower metabolic endotoxemia than after a high-fat diet devoid of milk SM (Norris, Jiang, Ryan, Porter, & Blesso, 2016). This was associated with a lower proportion of Gram-negative bacteria and a higher proportion of *Bifidobacterium* with milk SM. In line with these results, Milard, Laugerette, et al. (2019) recently reported that in mice fed a mixed HF diet, milk PL can limit HF-induced body weight gain and modulate gut physiology and the abundance in microbiota of bacteria of metabolic interest. Namely, mice fed high-fat diet with 1.1% of milk PL had increased *Bifidobacterium animalis* in cecal microbiota compared to control high-fat fed mice devoid of milk PL. Mice fed high-fat diet with 1.1% of milk PL had lower *Lactobacillus reuteri*, which correlated negatively with the fecal loss of milk SM-specific fatty acids. Medium chain fatty acids (MCFA) typical of milk, namely 10:0 and 12:0, and sphingolipids, have also shown direct bactericidal effects in vitro on *Listeria monocytogenes* and *Campylobacter jejuni*, and lower on *E. coli* O157:H7 and *Salmonella enteritidis* (Sprong, Hulstein, & Van der Meer, 2001; Sprong, Hulstein, & van der Meer, 2002). Therefore, the impact of differential release of milk lipids in the gut do to different structures of dairy matrixes on gut microbiota and endotoxemia regulation would deserve to be investigated in humans.

3.2 Metabolic Inflammation

Dairy product intake would be associated with a lower inflammatory status or would induce no inflammatory effect. As reviewed, dairy products can impact metabolic inflammation with a potential protective effect in normal-weight to obese subjects (Labonte et al., 2013; Labonte et al., 2014; Vors et al., 2015). Plasma markers of inflammation (CRP, IL-6 and TNF-α) were found lower amounts in high- dairy consumers (>14 servings/week) vs low-dairy consumers (<8 servings/week) in an observational study adjusting for confounding factors (Panagiotakos, Pitsavos, Zampelas, Chrysohoou, & Stefanadis, 2010). Moreover, fermented dairy products consumed during 3 weeks by overweight and obese men tended to lower inflammatory status in an intervention trial (Nestel et al., 2013). Yogurt consumption was also reported to reduce metabolic endotoxemia in elderly subjects (Schiffrin et al., 2009).

The presence of more or less MFGM in the dairy matrix might also contribute to impacts on metabolic inflammation. Mice fed diets base on AMF + MFGM and challenged with LPS injection had a lower inflammation (plasma concentrations of IL-6, IL-17, monocyte chemoattractant protein-1 (MCP-1) and TNF-α) than mice fed corn oil and receiving the same proinflammatory challenge; the mechanism would be due to a lower intestinal permeability with AMF+MFGM (Snow, Ward, Olsen, Jimenez-Flores, & Hintze, 2011). However, Zhou and Ward (2019) recently reported, using an ob/ob mice of preexisting obesity, that milk PL may have limited beneficial effects on gut barrier integrity, systemic inflammation, and lipid metabolism in the context of severe obesity. In a study mentioned above where mice were fed a semi-synthetic palm-oil based high fat diet for 8 weeks, or the same diet enriched with 1.2% or either soybean PL or milk PL (Lecomte et al., 2016), the WAT of mice fed with milk PL had a lower expression of markers of macrophage infiltration and of proinflammatory adipokines (MCP-1, IL-6 or TNF-α) compared with mice fed soy PL. This was associated to an increase of the number of goblet cells, that produce mucus in the colon, which could be explained by the presence of fatty acids specific of milk SM in feces (namely 22:0, 24:0) and suggest an improved gut barrier with milk PL. In this respect, Milard, Penhoat, et al., (2019) recently reported that unlike SM within the whole milk PL, pure milk SM can increase the expression of tight junction proteins in vitro in Caco-2 intestinal cells and revealed that IL-8 was a potential mediator of such effect. A modulation of gut barrier function could in the long term exert protective effects against metabolic inflammation induced notably by the translocation of endotoxins from the gut microbiota (Lecomte et al., 2016), thereby deserving further investigation.

3.3 *Newly Revealed Signalling Players: miRNA and Exosomes*

Moreover, it would be important to elucidate the fate in the digestive tract and the metabolic impact of complex milk lipid nanostructures after yogurt making or cheesemaking, e.g. lactosomes and exosomes and the mRNA and miRNA they contain (Argov, Lemay, & German, 2008; Argov-Argaman et al., 2010; Bourlieu & Michalski, 2015; Izumi et al., 2015). The metabolic impacts of milk miRNA are a promising emerging research topic (Auerbach, Vyas, Li, Halushka, & Witwer, 2016; Li, Dudemaine, Zhao, Lei, & Ibeagha-Awemu, 2016; Melnik & Schmitz, 2017; Title, Denzler, & Stoffel, 2015).

Altogether, these proofs of concept in rodents support the need to investigate the relative role of the MFGM, milk fat globule structure, and dairy matrix structure, according to different dairy processes, on the gut microbiota, endotoxemia and metabolic inflammation, because the latter can be impacted by dietary lipid fate in the gut through digestion kinetics and rate. New metabolic signalling players should also be taken into account.

4 How In Vitro Models Can Improve Our Understanding of Lipid Metabolism

Several characteristics of the dairy matrices combined with the activity of lactic acid bacteria (LAB) may modulate the cardiometabolic impact of dairy lipid content, but only very little is currently known on this topic. The dairy matrix represents a complex organization of nutrients, each having to first be released during digestion and then transported through the intestine epithelial membrane to be processed and metabolized (Versantvoort, Oomen, Van de Kamp, Rompelberg, & Sips, 2005). The dairy matrix microstructure and macrostructure have been shown to significantly influence kinetics of milk protein digestion in vivo (Barbé et al., 2013). Lipids are differentially incorporated depending on food structure (milk emulsion, cheese, etc.) and little information is available on how lipids behave in the gastric environment and how these behaviors affect digestive processes (Turgeon & Rioux, 2011). In addition, the extent to which dairy product structure and food matrix alter the appearance of FA in the blood circulation, and hence their metabolic fate and cardiometabolic effects, is poorly understood and needs further investigation. In vivo studies are expensive and sometimes difficult to achieve in order to improve our knowledge on dairy lipids mechanism. Notably, exploring lipid digestion in humans requires the use of naso-gastric and naso-duodenal cannulation (Armand et al.,

1999; Carriere et al., 2000), which is a heavy process and does not provide access to the most distal parts of the small intestine. Within the past 10 years, a multiplication of papers using in vitro studies is observed in the literature and standardized in vitro methods were published (Egger et al., 2016; Kopf-Bolanz et al., 2012; Minekus et al., 2014). In vitro models allow to compare food matrices with one another under controlled conditions and to improve our understanding of food behavior during digestion.

4.1 Impact of the Dairy Matrix Structure

Several reviews now suggest that food form and texture of dairy food modulate their nutritional properties (Thorning et al., 2017; Turgeon & Rioux, 2011). Limited number of in vitro studies focused on the impact of different dairy food on lipid bioaccessibility. (Lamothe, Rémillard, Tremblay, & Britten, 2017) evaluated the matrix degradation and FA release for milk, yogurt and cheese. One hour of gastric transit time was selected to compare matrices accurately. Protein hydrolysis was significantly higher in milk and yogurt than for cheese (~35 vs 20%). Milk and yogurt matrix degradation was fast while for cheese, the solid matrix was slowly disintegrated. When nutrients reach the duodenum, protein and lipid hydrolysis is quickly increased for each matrices, and cheese degradation reached values higher than 90% (Lamothe et al., 2017). During gastric digestion, solid matrices require strong shear forces to reduce their particle size to values smaller than 1–2 mm in order to be emptied in the duodenum (Kong & Singh, 2008). In human, the gastric transit time is adjusted based on the composition and the consistency (viscosity, texture) of the chyme (Hunt & Knox, 1968). This is of strong importance in regards to the protective effect of the cheese matrix against saturated FA intake. Indeed, the wide variety of existing cheeses (composition, texture, etc.) suggests potential heterogeneity in their digestive and absorptive processes. Lipid droplets size and their interface organization, TAG structure (position of FA on the TAG glycerol backbone), type and amount of phospholipids, are among the reported factors affecting FA bioavailability (Favé et al., 2004). In addition, fat globules are trapped in the cheese matrix which could delay or gradually release TAG during digestion and subsequently affect postprandial lipemia and also the fate of different lipid species along the gut. Studies investigating the impact of commercial cheeses with different composition and textural properties were performed (Fang, Rioux, Labrie, & Turgeon, 2016a; Fang, Rioux, Labrie, & Turgeon, 2016b; Guinot, Rioux, Labrie, Britten, & Turgeon, 2019; Lamothe, Corbeil, Turgeon, & Britten, 2012). Mild, low fat and aged cheddar along with mozzarella were submitted to an in vitro digestion system (Lamothe et al., 2012). The matrix degradation was investigated and at the end of the gastric digestion, aged cheddar was more disintegrated than low fat cheddar due to its low cohesiveness and elasticity. During the duodenal phase, different kinetics of matrix degradation was observed attributed to the protein matrix density and fat distribution. Mozzarella showed the greatest degradation at the end of the duodenal digestion due to its porous matrix (large fat pools) and its lower cohesiveness. The disruption of the solid matrix also regulates FA release rate and mozzarella showed the greatest increase in lipolysis within the first 90 min of duodenal digestion. Due to its high degradation, mozzarella lipid content was easily accessible to lipase increasing FA release. The degradation of regular and light cheeses like cheddar and mozzarella was studied using an in vitro model (Fang et al., 2016b). Although lipolysis during duodenal digestion was not the main target of the study, they analyzed the degradation of the matrix over time and cheese disintegration was ranked as followed: cheddar > light cheddar > mozzarella = light mozzarella independently of the digestion step (oral, gastric or duodenal). Using the composition and textural properties data, a prediction model was established to estimate cheese disintegration. Cheese fat content, proteolysis, hardness and chewiness was shown to positively affect the matrix disintegration. In a study including nine commercial cheeses, Guinot et al. (2019) have shown a correlation between texture parameters, cheese disintegration and lipid release from the cheese matrix during in vitro gastric digestion. Elastic and cohesive cheeses as Mozzarella and young Cheddar were disintegrated slowly as compared with cheeses which fractured more easily as aged cheddar. Therefore, the protein network and physicochemical properties of cheese may modulate the kinetics of digestion and lipid absorption.

4.2 *Impact of the Dairy Matrix Manufacturing Steps*

Dairy processing steps such as heat treatments, homogenization, etc. are required to ensure food safety and stability over time. These operations are known to modify the structure of proteins and lipids which subsequently affect the product composition and texture. For example, the denaturation of whey proteins after heat treatment of milk improve the quality of yogurts. Water retention (Harwalkar & Kalab, 1986) and textural (Cobos, Horne, & Muir, 1995; Rohm & Schmid, 1993) properties of yogurt are improved. The next section explores the possible impact on nutrients digestibility.

4.2.1 The Importance of the Homogenization/Pasteurization Sequence

Early work on the impact processing treatments was mainly focused on dairy proteins. For example, heat treatment promotes whey proteins (β-lactoglobulin) denaturation by increasing their proteolysis by gastric and duodenal enzymes (Mullally, Mehra, & FitzGerald, 1998). Later, the in vitro digestion of protein and lipid from milk products was shown to be affected by the processing treatments (Devle et al., 2014; Tunick et al., 2016; Van Hekken, Tunick, Ren, & Tomasula, 2017; Ye, Cui, Dalgleish, & Singh, 2016a). During digestion, milk proteins namely caseins clots in the gastric environment. This coagulum structure is altered when milk is heated and whey protein is denatured (Ye, Cui, Dalgleish, & Singh, 2016b). The clot was dense and tightly bound into a homogeneous mass for raw milk while for heated milk, it was more porous and fragmented. In addition, few whey proteins were seen in the clots of raw milk as opposed to the heat treated milk. Proteolysis was also improved with the heat treatment where at the end of the gastric digestion (2 h), no intact casein and whey proteins were visible on an SDS-PAGE as opposed to raw milk. In the presence of whole milk, the fat globules were trapped within these clots but how the structure of these clots affects FA release? In raw whole milk, the clots contained caseins, fat and few whey proteins while in heated milk, caseins, denatured whey protein and fat were fully included in the coagulum (Ye et al., 2016a). Although the coagulum structure was different in both milks, the release of the fat out of the coagulum was similar in both milks. Confocal images taken during gastric digestion revealed that fat globules in both milks were trapped within the coagulum but were not involved in the structure of the clots. This is an agreement with a previous study showing that native fat globules (covered with MFGM) were considered as inactive fillers in rennet gel formed with raw whole milk (Michalski et al., 2002). Therefore, as the clotted proteins are hydrolyzed by pepsin fat is slowly released in the chyme at similar rates in both milks. However, commercial fluid milk is usually homogenized and then pasteurized before its consumption and these treatments affect fat globule structure which could modify the clot structure. Ye, Cui, Dalgleish, and Singh (2017) investigated how these combined processing treatments affected the clot structure, degradation of dairy protein and fat release in the chyme. Raw whole milk was homogenized or homogenized-pasteurized. The degradation of the clot and the release of the fat in the chyme were higher for homogenized-pasteurized milk > homogenized > raw whole milk at the end of the gastric digestion (2 h) and this was mainly attributed to the structure of the clot. In raw whole milk, the coagulum was dense with a smooth cohesive mass while for homogenized-pasteurized milk, the structure was fragmented and brittle. Caseins and whey proteins were almost fully degraded in homogenized-pasteurized milk at the end of the gastric digestion in the clot and in the chyme as opposed to raw whole milk. In homogenized-pasteurized milk, the fat globule can interact with caseins and whey proteins altering the structure of the gel allowing a better diffusion of pepsin. Therefore, proteolysis in increased in homogenized-pasteurized milk promoting clots degradation and fat release. These studies investigated the behavior of milk clotting during gastric digestion and they showed that the processing treatment applied to the milk might affect the kinetics of lipid release.

Other studies have also been made with digestion system mimicking gastric and duodenal steps. (Tunick et al., 2016) proceeded to the digestion of raw whole and skim milks that were subsequently homogenized, pasteurized, homogenized-pasteurized or homogenized-sterilized using an in vitro digestion model. During gastric digestion (last 1 h), only modest differences in protein degradation were observed for milk despite the use of different manufacturing processes. Only homogenized- pasteurized and homogenized-sterilized whole milks showed darker bands in the peptides region (below 5 kDa). During duodenal digestion, the presence of fat in whole milk appears to delay protein digestion independently of the processing treatment. In whole milk, the homogenization treatment showed darker low molecular weight protein bands (<5 kDa) suggesting possible lipid-peptides associations (Michalski & Januel, 2006) compared to raw skim milk. When whole milks are pasteurized or homogenized-pasteurized, β-lactoglobulin persisted for a longer period of time as shown on SDS-PAGE gels delaying protein digestion compared to pasteurized skim milk. Similar observations were made in another study, where whole homogenized-pasteurized and skim pasteurized milks were digested in vitro using human gastrointestinal enzymes (Devle et al., 2014).

The authors suggested (1) possible FA-β-lactoglobulin associations preventing its hydrolysis or (2) the lack of bile salt secretion in the human duodenal juice. This last hypothesis is plausible since (Kopf-Bolanz et al., 2012) observed that when bile was omitted in their digestion model, β-lactoglobulin was still present and proteolysis was significantly reduced at the end of the duodenal digestion of pasteurized-homogenized whole milk.

Altogether, milk processing treatments clearly have an impact on protein digestion. But what are the consequences on the digestion of lipid? Early work on emulsion have previously shown that the fat droplet size affects their lipolysis by pancreatic lipase (Armand et al., 1992; Benzonana & Desnuelle, 1965). Milk fat homogenization is well known to decrease the fat globule size and an increase in lipolysis is therefore expected (Claeys et al., 2013). Tunick et al. (2016) showed that milks that had undergone a homogenization treatment had higher lipolysis during duodenal digestion due to the increase in free FA content. This was also confirmed by several other studies (Devle et al., 2014; Lamothe et al., 2017; Van Hekken et al., 2017). In addition, the whole homogenized-pasteurized milk exhibited the highest lipolysis at the end of the duodenal digestion (lasted 2 h) (Tunick et al., 2016). Following the homogenization of the milk, the structure of the fat globule is modified, and a new interface is created. The order in which the manufacturing processes (i.e. homogenization before or after the heat treatment) are applied will influence the nature of this new interface (Michalski, 2009). Conversely, whole raw milk that was either pasteurized or pasteurized-homogenized were digested in vitro using human gastrointestinal enzymes showed different protein digestion (Islam et al., 2017). The milk submitted to a homogenization treatment showed increased lipolysis at the end of the duodenal digestion compared to whole raw milk. These results are also in agreement with a previous study which only focused on the duodenal digestion (Ye, Cui, & Singh, 2010). The sequence of the processing step impacts protein hydrolysis kinetics but it is not clear if this has an influence on FA release and on lipemia in human.

Of note, contradictory results were found regarding FA bioaccessibility for homogenized, homogenized-pasteurized and homogenized-UHT milks compared to raw milk (Liang, Qi, Wang, Jin, & McClements, 2017). Raw and homogenized milks have similar FA release at the end of the duodenal digestion as opposed to the other studies (Devle et al., 2014; Lamothe et al., 2017; Tunick et al., 2016; Van Hekken et al., 2017). The authors suggested that the milk sample and the processing steps could be responsible for this discrepancy. However, the milk particle size after processing (raw ~3.6 μm and homogenized- pasteurized ~0.5 μm) were in similar range than the study of (Lamothe et al., 2017) which ruled out these two factors. The authors also suggest that the different in vitro model could account for these differences but without any further explanation. The selection of the in vitro model, therefore, appears to be of utmost importance and great care should be undertaken to carefully select the enzymes. The utilization of harmonized in vitro model allows better comparison between studies (Mat, Le Feunteun, Michon, & Souchon, 2016; Minekus et al., 2014).

4.2.2 Impact of Gelation/Cheesemaking

During dairy gel formation (acid or rennet), the structure of the dairy protein is altered and several factors such as milk pH, dairy formulation composition, LAB, temperature, etc. are known to influence the matrix structure (Aguilera, 2006). A limited number of studies have looked at the impact of dairy gels processing steps on lipid digestion. Stirred- and Greek-style yogurts were prepared from homogenized and heat treated milk (95 °C × 5 min) (Lamothe et al., 2017). Similar protein and lipid hydrolysis was observed during in vitro digestion. However, Greek-style yogurt tended to have a lower lipolysis attributed to its lower calcium content. Another study investigated how cream homogenization and cheesemaking pH (5.5 and 6.5) during cheddar manufacture will impact cheese nutrients release during digestion (Lamothe et al., 2017). The homogenization had no impact on the matrix degradation during in vitro digestion while the pH showed a significant impact. Cheese made with milk adjusted at pH 6.5 retained a higher proportion of calcium and moisture within the matrix while lower amounts of protein and fat were found when compared to cheese made with milk at pH 5.5. This favored the degradation of cheddar cheese made with milk adjusted to pH 6.5. The pores of the matrix are looser in high moisture cheese promoting enzyme diffusion and increasing protein hydrolysis. Indeed, the percentage of trichloroacetic acid (TCA) soluble proteins at the end of the gastric digestion was higher for cheeses made at pH 6.5. During duodenal digestion, a different behavior was observed. Both cheeses had a constant increase of degradation over time, but it was higher for cheese made at pH 5.5. The higher amount of calcium in cheddar made at pH 6.5 has promoted the contraction of the casein network due to the reduction of electrostatic interactions between casein molecules lowering the matrix degradation during duodenal digestion. This has resulted in a lower protein hydrolysis. Lipolysis was affected by both the homogenization treatment and the cheesemaking pH. FA release is faster for homogenized milk than for non-homogenized milk as expected. In non-homogenized milk, the pH had an important impact on lipolysis. In cheese made at pH 5.5 (low calcium content), the lipolysis rate was significantly lower than for cheese made at 6.5. Calcium ions are known to promote lipolysis due to their capacity to precipitate long-chain FA reducing the steric hindrance around the fat droplet. In homogenized milk, lipolysis was not affected by the milk pH. The smaller fat globules promoted the accessibility of lipases to their action site increasing dairy fat lipolysis. A recent publication investigating fat release and lipolysis of commercial cheeses during in vitro gastric and duodenal digestions (Guinot et al., 2019) revealed differences in free fatty acids released. The fat globule size (homogenized or not), fat and calcium content were the most influential factors on the lipolysis rate.

Additional research was conducted dairy protein emulsion gels and how their structure impact in vitro lipid digestion, as reviewed in (Guo, Ye, Bellissimo, Singh, & Rousseau, 2017). These studies are important because they allow to gain additional fundamental information on how isolated component of the matrix can impact lipid digestion. For example, gel firmness and protein density were shown to affect the kinetics of lipid release (Guo, Bellissimo, & Rousseau, 2017). Readers are referred to the previous chapter for more information on this topic (Chap. 7).

4.3 *Impact of the Dairy Matrix Composition*

4.3.1 Protein Content

The amount of dairy protein was shown to have an impact on lipid metabolism (Section 2.3.2). To our knowledge, no study investigated the impact of dairy protein concentration on lipid release as the main outcome. One example was provided in Sect. 4.2.2 where stirred- and Greek-style yogurts containing respectively 5.2 and 9.2% of proteins had no impact on lipolysis (Lamothe et al., 2017) and readers are referred to this section for more details. In addition, (Devle et al., 2014) studied FA bioaccessibility in homogenized-pasteurized whole milk and in milk permeate enriched with dairy fat. At the end of the duodenal step, the degree of lipolysis was low (16%) for ultra-filtered permeate containing milk fat (no protein) compared to full fat milk (37%). This may be due to the ability of milk proteins/peptides to act as emulsifiers in cooperation with bile acids, resulting in smaller fat globules with a larger surface area making them more available for attack by pancreatic lipase (Devle et al., 2014). Additional studies are necessary to confirm this hypothesis.

4.3.2 Fat Globule Size and Interface Composition

Smaller fat droplets improve lipase efficiency which results in vivo, in faster plasma TAG appearance (Vors et al., 2013). The fat globule size as influenced by different homogenization treatment or microfiltration procedure and its impact on FA release was investigated (Berton et al., 2012; Garcia, Antona, Robert, Lopez, & Armand, 2014; Islam et al., 2017). Whole pasteurized milk submitted to two homogenization treatments (5 vs 15 MPa) showed different rate of lipolysis (Islam et al., 2017). For the high pressure treated milk (average size: 0.58 μm), lipolysis reached a plateau after 30 min of duodenal digestion and 78% of FA were released after 120 min. The low pressure homogenized milk (average size: 1.45 μm) did not reach a steady state and only 56% of FA were released at the end of the duodenal digestion. These results suggest different kinetics in FA release.

Another study aimed to investigate the hydrolysis kinetics of pancreatic lipase of milk fat globules with different interface composition (Berton et al., 2012). Raw whole milk (average size: 3.9 μm) was fractionated by microfiltration to obtain milk with small (average size: 1.8 μm) and large (average size: 6.7 μm) native fat globules (Berton et al., 2012). Lipase catalytic efficiency (CE: mL/sec × mg of lipid) was determined based on the fat concentration. Data showed a significantly higher CE for milk with small native fat globules compared to the large native fat globules and raw whole milk. This is in accordance with another study (Garcia et al., 2014). In addition, milk homogenized (altered fat globule interface) at different pressures (HM: 50, 300 and 400 MPa) were studied. Particle size ranging from 1.5 to 0.14 μm for homogenized milks and 3.9 μm for raw whole milk was obtained. Homogenized milk displayed similar CE independently of the pressure used and their values were higher compared to the raw whole milk. Milk homogenized at 50 MPa and milk with the small native fat globule showed similar particle size but showed different behavior. Their data suggests that homogenized milk was hydrolyzed more quickly than milk with native fat globule of similar particle size. Therefore, the size of the fat globule is not the sole factor modulating lipase activity and composition of the interface play an important role. Of note though, in neonatal digestion conditions, it is the interfacial surface area that plays the major role as also pointed out by Bourlieu et al. (2015). Previous works on emulsions in adults' physiological conditions also confirmed the significance of the interface composition on fat digestion and additional information was reviewed in (Golding & Wooster, 2010).

4.3.3 Calcium Content

High calcium concentration in cheese may reduce the bioavailability of specific FA by precipitation as calcium salts (Lamothe et al., 2012). Calcium is known to increase lipase activity during digestion (Hu, Li, Decker, & McClements, 2010), but the formation of insoluble calcium soaps reduces fatty acids absorption and increases faecal fat excretion (Lorenzen & Astrup, 2011). In the duodenal environment, the solubility of calcium soaps decreases with increasing FA chain length and saturation (Graham & Sackman, 1983).

Depending on the calcium concentration in model cheese (cheddar type), lipolysis is increased (Ayala-Bribiesca, Lussier, Chabot, Turgeon, & Britten, 2016) and 4–23% of the fat were insolubilized as calcium soaps in duodenal environment (Ayala-Bribiesca, Turgeon, & Britten, 2017). However, the addition of calcium did not significantly reduce the bioaccessibility of individual long-chain FA in this study as opposed to the in vivo study (Ayala-Bribiesca et al., 2018). In addition, cheeses prepared with AMF having different melting points showed different matrix degradation rates. Indeed, at the end of the gastric digestion the stearin AMF cheese degradation was significantly lower than the control and olein AMF cheeses. As opposed to other studies, this was not associated with the initial texture since cheese hardness was not significantly different at 37 °C. However, the melting points of the AMF could account for these differences. Stearin AMF has a melting point higher than body temperature (42.3 °C) increasing the resistance to matrix degradation during gastric digestion. Lipolysis and subsequently the overall FA bioaccessibility were also reduced in the presence of stearin AMF due to a lower degradation of the matrix.

The calcium content was shown to promote lipolysis, but the concentration found within the matrix is also important. Casein emulsion gels (mimic cheese analogues) with two calcium concentrations (774 vs 357 mg Ca per 100 g) prepared with AMF were investigated (McIntyre, Osullivan, & Oriordan, 2017). Different kinetics of matrix degradation were found during gastric digestion where the high calcium gel exhibited a faster degradation. At the end of the duodenal digestion, the gels reached the same extent of degradation. However, no differences in lipolysis rate was observed as opposed to the study of (Ayala-Bribiesca et al., 2016) who studied cheeses with different calcium content. The discrepancy between both studies could be attributed to the dose of calcium found in each gel where the control and the high calcium cheeses respectively contained 595 and 1202 mg Ca per 100 g. The comparison of the matrices with the highest calcium levels in both studies shows a difference of 1.6 times lower for emulsified gels. Therefore, the amount of calcium is matrix dependent and further research is needed.

4.3.4 LAB and Cheese Microflora

For most varieties, the cheese microflora has an impact on the cheese FA availability through the action of microbial and/or fungal lipase during the ripening process. This activity is of utmost importance to produce flavour precursors (Collins, McSweeney, & Wilkinson, 2003; Das, Holland, Crow, Bennett, & Manderson, 2005). As an example, *Geotrichum candidum*, a yeast mainly found in washed-rind and mold-ripened cheeses, synthesize two lipase isoforms (GCLI and GCLII) having significant biases toward specific TAG (Baillargeon, Bistline Jr, & Sonnet, 1989; Bertolini et al., 1994; Bertolini et al., 1995). GCLII is more specific for the short-chain FA residues (C8–C14), while GCLI is specific for the C18:1, C18:2 and C18:3 FA residues (Bertolini et al., 1995; Charton & Macrae, 1992; Das, et al., 2005; Jacobsen & Poulsen, 1992, 1995). Moreover, the release of FA from TAG has been reported to be highly species/strain dependent (Collins et al., 2003). All together, this information suggests that the availability of the FA could differ greatly depending on the dairy matrices but to our knowledge, no study has investigated that topic.

5 Examples of Relationship Between In Vitro and In Vivo Models of Dairy Matrix Digestion

In vitro models are important to isolate one or more factors that influence the digestibility of dairy matrices. They are performed to understand the mechanism behind postprandial results or to plan in vivo studies. The literature covered in this chapter mainly focused on in vitro studies that aimed to understand the impact of the matrix structure, the dairy processing steps, the fat globule size and the calcium content. Correlation between in vitro and in vivo studies allowed to confirm the impact of different kinetics of digestion on postprandial lipemia. As explained previously, the kinetics of plasma TAG release was faster for a cream cheese meal than for a Cheddar meal in healthy volunteers (Drouin-Chartier et al., 2017). Cream cheese is a semi-solid matrix composed of emulsified dairy fat (average size: 0.5 μm) that was shown to easily melt at body temperature increasing fat bioaccessibility (Guinot et al., 2019). For Cheddar cheese, native fat droplets are found (average size: 3.0 μm) and the matrix must be disintegrated to allow dairy fat digestibility. Guinot et al. (2019) studied the matrix disintegration of the cheeses used in breakfast meals of the human clinical trial. The cream cheese was easily disintegrated while Cheddar disintegration was slow and this may have induced different gastric emptying rate and postprandial responses between cheeses. Other in vitro studies, showed that the Cheddar cheese matrix is slowly disintegrated and reached only 25–43% of degradation after 2 h (Fang et al., 2016b; Lamothe et al., 2012). Also, the fat droplet size is significantly lower for cream cheese. Emulsified dairy fat (alteration of the native interface) showed increase lipolysis in in vitro (Islam et al., 2017) and in vivo (Vors et al., 2013) studies. All these studies support the fact that the cheese matrix modulates postprandial lipemia. Another in vitro/in vivo correlation regards the calcium content and FA melting point towards the bioavailability of milk lipids.

Another example of correlation between in vitro and in vivo studies was performed recently with experimental cheese. Cheddar-type cheeses manufactured with different AMF (olein vs stearin) and calcium (regular: 50 mg and high: 66 mg) contents were studied in vitro (Ayala-Bribiesca et al., 2017) and in vivo (Ayala-Bribiesca et al., 2018). In cheddar-type cheese with regular and high calcium content, the stearin AMF exhibited lower plasma TAG responses compared to the olein AMF in the rat model. *In vitro* study revealed that lipolysis and the overall FA bioaccessibility were also reduced in the presence of stearin AMF due to a lower degradation of the matrix. Indeed, stearin AMF has higher melting point than body temperature (42.3 °C) increasing the resistance to gastric digestion. Also, the mass fraction of the total FA recovered in the animal feces were significantly higher with the stearin AMF meal compared to the olein AMF meal independently of the calcium concentration. The *in vitro* study revealed that 23% of the fat was insolubilized as calcium soaps in duodenal environment with the stearin AMF cheese (Ayala-Bribiesca et al., 2017). For the rats fed high calcium cheese, fat excretion was higher and long-chain FA represented 66% of the fat excreted. However, the reduction of the bioaccessibility of individual long-chain FA in the *in vitro* study was not corroborated (Ayala-Bribiesca et al., 2018). Therefore, both studies revealed that dairy fat bioaccessibility and bioavailability were reduced in the presence of a stearin AMF and high calcium concentration diet. This provides incentive to further study *in vivo* and notably in humans, the impact of different combinations of milk FA profiles, calcium content and matrix structures on digestion, absorption and metabolism.

Perspective on the Importance of In Vitro Models In vitro models have traditionally been used to study the survival of probiotic bacteria in food matrices, drug delivery, bioaccessibility of food contaminants and many other application (Adouard et al., 2016; Blanquet et al., 2004; Hernández-Galán et al., 2017; Versantvoort et al., 2005). Nowadays, these in vitro models are now well recognized in the literature since they improve our knowledge about the behavior of food matrices during digestion. Static and dynamic models have been used and each of them possess great advantages and weaknesses (Shani-Levi et al., 2017). Among their strengths, the controlled use of the amount of enzymes, pH, temperature, and stirring speed is reported. On the other hand, it is sometimes difficult to mimic the physiological conditions (children vs. adults vs elderly, fasted vs fed state, etc.), especially gastric emptying being the rate-limiting step of nutrients bioavailability. However, it is always important to keep in mind that for each of in vitro model it is important to perform adequate validation in vivo studies performed in animal model or in humans (Bohn et al., 2017).

In vitro cellular models are also used to understand absorption kinetics of several nutrients (Ex: Caco-2 cell line) and for example, to further understand post-prandial lipid absorption (Vors et al., 2012). The microbiota is also another hot topic since it might impact the host lipid metabolism (Bondia-Pons, Hyötyläinen, & Orešič, 2015). Several in vitro models were created to gain more information on how the host flora is modulated by food, probiotics, drugs, etc. (Aguirre et al., 2016; Fernandez, Savard, & Fliss, 2016; Ramadan et al., 2013). Readers are referred to review papers for more information (Guerra et al., 2012; Payne, Zihler, Chassard, & Lacroix, 2012).

6 Conclusion

Recently gained knowledge of the postprandial phase raises several mechanisms by which the dairy matrix structure can modulate the digestion, metabolic fate and physiological impact of milk lipids (Fig. 8.3), notably by its viscosity, nutrient composition and supramolecular organizations. According to the different types of products and processes, this "dairy matrix effect" has now to be elucidated in more details. Regarding milk processing, more studies are needed to elucidate in humans the impact of milk homogenization on lipid fate in the body and metabolic consequences. To date, reviews report no clear link between milk homogenization and health, but this subject of debate deserves further clinical studies as this process changes dramatically milk fat globule structure. Regarding the fat composition of the dairy matrix, two aspects deserve presently more attention: the amount of fat on the one hand, and the quality of fat regarding PL/MFGM presence and amount.

Different epidemiological studies and meta-analyses of clinical trials highlight the importance of studying in more details the metabolic impact of dairy products with different fat contents with an aim to challenge the current dietary guidelines

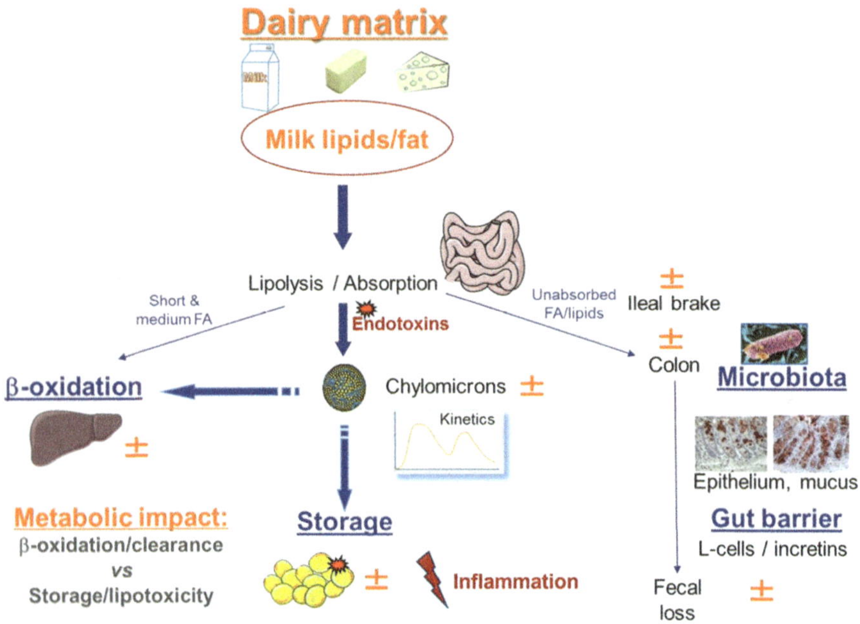

Fig. 8.3 Postprandial lipid metabolism. Intestinal lipid absorption modulates postprandial lipemia and the metabolic fate of absorbed lipids. The coabsorption of endotoxins from the gut microbiota can contribute to metabolic inflammation. Unabsorbed lipids in the gut can trigger satietogenic signals via the ileal brake and via incretin secretion by L-cells, impact gut microbiota and the intestinal barrier notably through a modulation of mucus-producing goblet cells

advising to consume mainly low-fat dairy. Long-term nutritional interventions are now necessary to elucidate the link between full-fat dairy consumption and metabolic and cardiovascular risks, as recent articles suggest a beneficial effect. Regarding the impact of the polar lipid fraction of milk fat, available supplementation studies with MFGM extracts in healthy humans with milk PL at around 3 g/day do not show significant effects on intestinal cholesterol absorption or plasma lipids. Performing studies in patients with metabolic syndrome, moderate hypercholesterolemia or high blood pressure could provide more insight on the potential beneficial effects of milk PL and MFGM, as recently demonstrated in overweight postmenopausal women using 3–5 g/day of milk PL in a cream cheese matrix. Adapted from Michalski (2009), Michalski et al. (2013), Bourlieu & Michalski (2015).

Dairy matrixes can also greatly vary in their protein composition. Several clinical trials show that the type of dairy proteins can modify postprandial lipemia: whey proteins would be hypolipemic but this remains to be confirmed using test meals of different protein composition and of different matrix composition and structure. Moreover, the relative impact of protein amount, type, location at the interface of fat droplets or in the aqueous phase or proteinaceous network, denaturation and consequences relative to viscosity, are related and their consequences should now be deciphered in humans. This provides incentive to further explore the dairy matrix effect on lipemia modulation by different amounts and types of proteins. Finally, the metabolic importance of the gut barrier and the gut microbiota has recently been revealed. Moreover, metabolic diseases such as obesity and type 2 diabetes are characterized not only by disorders of lipid metabolism but also by metabolic inflammation in which lipid metabolism, gut microbiota and gut barrier play a role. In this respect, the dairy matrix composition and structure can modulate these outcomes in many ways, notably via bioactive metabolites derived from the lactic acid bacteria, by bioactive peptides released during digestion or by components of the MFGM.

Altogether, proofs of concept performed in rodents support the need to investigate the relative role of the MFGM, milk fat globule structure, and dairy matrix structure, according to different dairy processes, on the gut microbiota, endotoxemia and metabolic inflammation. Metabolic inflammation can be impacted by dietary lipid fate in the gut through digestion kinetics. New metabolic signaling players should also be taken into account such as miRNAs. Moreover, the impact of the dairy matrix on other health outcomes such as, e.g., IBD, cancer, cognition, neurological diseases is still an open field of research.

To better screen and understand mechanisms related to matrix disintegration and digestion in the gut, in vitro models will continue to be the first approach as they are performed under controlled conditions, allow comparison between laboratories (when a standardized method is used) and the number of samples studied can be larger than with in vivo trials. Recently, addition of cellular models allows to apprehend absorption kinetics of specific nutrients. However, a validation with in vivo models remains an essential step because some physiological events such as gastric mechanical movements and gastric emptying cannot be mimicked efficiently by in vitro models. Furthermore, the real life feeding pattern is composed of a variety of foods consumed within and between meals. The study of individual foods is only the first step of a more inclusive approach of the digestive fate of meals. It is the variety of experimental approaches used that will permit to gain valuable knowledge on lipid metabolism and continue to explore the complex food-health axis.

References

Adouard, N., Magne, L., Cattenoz, T., Guillemin, H., Foligné, B., Picque, D., et al. (2016). Survival of cheese-ripening microorganisms in a dynamic simulator of the gastrointestinal tract. *Food Microbiology, 53*, 30–40.

Aguilera, J. M. (2006). Food product engineering: Building the right structures. *Journal of the Science of Food and Agriculture, 86*(8), 1147–1155.

Aguirre, M., Eck, A., Koenen, M. E., Savelkoul, P. H. M., Budding, A. E., & Venema, K. (2016). Diet drives quick changes in the metabolic activity and composition of human gut microbiota in a validated in vitro gut model. *Research in Microbiology, 167*(2), 114–125.

Alexander, D. D., Bylsma, L. C., Vargas, A. J., Cohen, S. S., Doucette, A., Mohamed, M., et al. (2016). Dairy consumption and CVD: A systematic review and meta- analysis. *The British Journal of Nutrition, 115*(04), 737–750.

Argov, N., Lemay, D. G., & German, J. B. (2008). Milk fat globule structure and function: Nanoscience comes to milk production. *Trends in Food Science and Technology, 19*(12), 617–623.

Argov-Argaman, N., Smilowitz, J. T., Bricarello, D. A., Barboza, M., Lerno, L., Froehlich, J. W., et al. (2010). Lactosomes: Structural and compositional classification of unique nanometer-sized protein lipid particles of human milk. *Journal of Agricultural and Food Chemistry, 58*(21), 11234–11242.

Armand, M. (2007). Lipases and lipolysis in the human digestive tract: Where do we stand? *Current Opinion in Clinical Nutrition & Metabolic Care, 10*(2), 156–164.

Armand, M., Borel, P., Ythier, P., Dutot, G., Melin, C., Senft, M., et al. (1992). Effects of droplet size, triacylglycerol composition, and calcium on the hydrolysis of complex emulsions by pancreatic lipase: An in vitro study. *The Journal of Nutritional Biochemistry, 3*(7), 333–341.

Armand, M., Pasquier, B., André, M., Borel, P., Senft, M., Peyrot, J., et al. (1999). Digestion and absorption of 2 fat emulsions with different droplet sizes in the human digestive tract. *The American Journal of Clinical Nutrition, 70*(6), 1096–1106.

Asselin, G., Lavigne, C., Bergeron, N., Angers, P., Belkacemi, K., Arul, J., et al. (2004). Fasting and postprandial lipid response to the consumption of modified milk fats by Guinea pigs. *Lipids, 39*(10), 985–992.

Ataie-Jafari, A., Larijani, B., Alavi Majd, H., & Tahbaz, F. (2009). Cholesterol-lowering effect of probiotic yogurt in comparison with ordinary yogurt in mildly to moderately hypercholesterolemic subjects. *Annals of Nutrition & Metabolism, 54*(1), 22–27.

Auerbach, A., Vyas, G., Li, A., Halushka, M., & Witwer, K. (2016). Uptake of dietary milk miRNAs by adult humans: A validation study. *F1000Res, 5*, 721.

Ayala-Bribiesca, E., Lussier, M., Chabot, D., Turgeon, S. L., & Britten, M. (2016). Effect of calcium enrichment of Cheddar cheese on its structure, in vitro digestion and lipid bioaccessibility. *International Dairy Journal, 53*, 1–9.

Ayala-Bribiesca, E., Turgeon, S. L., & Britten, M. (2017). Effect of calcium on fatty acid bioaccessibility during in vitro digestion of Cheddar-type cheeses prepared with different milk fat fractions. *Journal of Dairy Science, 100*(4), 2454–2470.

Ayala-Bribiesca, E., Turgeon, S. L., Pilon, G., Marette, A., & Britten, M. (2018). Postprandial lipemia and fecal fat excretion in rats is affected by the calcium content and type of milk fat present in Cheddar-type cheeses. *Food Research International, 107*, 589–595.

Baillargeon, M. W., Bistline Jr., R. G., & Sonnet, P. E. (1989). Evaluation of strains of Geotrichum candidum for lipase production and fatty acid specificity. *Applied Microbiology and Biotechnology, 30*(1), 92–96.

Barbé, F., Le Feunteun, S., Rémond, D., Ménard, O., Jardin, J., Henry, G., et al. (2014). Tracking the in vivo release of bioactive peptides in the gut during digestion: Mass spectrometry peptidomic characterization of effluents collected in the gut of dairy matrices fed mini-pigs. *Food Research International, 63*, 147–156.

Barbé, F., Ménard, O., Gouar, Y. L., Buffière, C., Famelart, M.-H., Laroche, B., et al. (2013). The heat treatment and the gelation are strong determinants of the kinetics of milk proteins digestion and of the peripheral availability of amino acids. *Food Chemistry, 136*(3- 4), 1203–1212.

Baumgartner, S., Kelly, E. R., van der Made, S., Berendschot, T. T., Husche, C., Lutjohann, D., et al. (2013). The influence of consuming an egg or an egg-yolk buttermilk drink for 12 wk on serum lipids, inflammation, and liver function markers in human volunteers. *Nutrition, 29*(10), 1237–1244.

Benzonana, G., & Desnuelle, P. (1965). Etude cinetique de l'action de la lipase pancreatique sur des triglycerides en emulsion. Essai d'une enzymologie en milieu heterogene. *Biochimica et Biophysica Acta (BBA) - Enzymology and Biological Oxidation, 105*(1), 121–136.

Bernard, A., & Carlier, H. (1991). Absorption and intestinal catabolism of fatty acids in the rat: Effect of chain length and unsaturation. *Experimental Physiology, 76*(3), 445–455.

Berry, S. E., Miller, G. J., & Sanders, T. A. (2007). The solid fat content of stearic acid-rich fats determines their postprandial effects. *The American Journal of Clinical Nutrition, 85*(6), 1486–1494.

Berry, S. E., & Sanders, T. A. (2005). Influence of triacylglycerol structure of stearic acid-rich fats on postprandial lipaemia. *The Proceedings of the Nutrition Society, 64*(2), 205–212.

Bertolini, M. C., Laramee, L., Thomas, D. Y., Cygler, M., Schrag, J. D., & Vernet, T. (1994). Polymorphism in the lipase genes of Geotrichum candidum strains. *European Journal of Biochemistry, 219*(1–2), 119–125.

Bertolini, M. C., Schrag, J. D., Cygler, M., Ziomek, E., Thomas, D. Y., & Vernet, T. (1995). Expression and characterization of Geotrichum candidum lipase I gene. Comparison of specificity profile with lipase II. *European Journal of Biochemistry, 228*(3), 863–869.

Berton, A., Rouvellac, S., Robert, B., Rousseau, F., Lopez, C., & Crenon, I. (2012). Effect of the size and interface composition of milk fat globules on their in vitro digestion by the human pancreatic lipase: Native versus homogenized milk fat globules. *Food Hydrocolloids, 29*(1), 123–134.

Blanquet, S., Zeijdner, E., Beyssac, E., Meunier, J.-P., Denis, S., Havenaar, R., et al. (2004). A dynamic artificial gastrointestinal system for studying the behavior of orally administered drug dosage forms under various physiological conditions. *Pharmaceutical Research, 21*(4), 585–591.

Bohl, M., Bjornshave, A., Rasmussen, K. V., Schioldan, A. G., Amer, B., Larsen, M. K., et al. (2015). Dairy proteins, dairy lipids, and postprandial lipemia in persons with abdominal obesity (DairyHealth): A 12-wk, randomized, parallel-controlled, double-blinded, diet intervention study. *The American Journal of Clinical Nutrition, 101*(4), 870–878.

Bohn, T., Carriere, F., Day, L., Deglaire, A., Egger, L., Freitas, D., et al. (2017). Correlation between in vitro and in vivo data on food digestion. What can we predict with static in vitro digestion models? *Critical Reviews in Food Science and Nutrition, 58*(13), 1–23.

Boirie, Y., Dangin, M., Gachon, P., Vasson, M.-P., Maubois, J.-L., & Beaufrère, B. (1997). Slow and fast dietary proteins differently modulate postprandial protein accretion. *Proceedings of the National Academy of Sciences of the United States of America, 94*(26), 14930–14935.

Bondia-Pons, I., Hyötyläinen, T., & Orešič, M. (2015). Role of microbiota in regulating host lipid metabolism and disease risk. In S. Kochhar & F.-P. Martin (Eds.), *Metabonomics and gut microbiota in nutrition and disease* (pp. 235–260). London: Springer.

Bonnaire, L., Sandra, S., Helgason, T., Decker, E. A., Weiss, J., & McClements, D. J. (2008). Influence of lipid physical state on the in vitro digestibility of emulsified lipids. *Journal of Agricultural and Food Chemistry, 56*(10), 3791–3797.

Borel, P., Armand, M., Pasquier, B., Senft, M., Dutot, G., Melin, C., et al. (1994). Digestion and absorption of tube-feeding emulsions with different droplet sizes and compositions in the rat. *Journal of Parenteral and Enteral Nutrition, 18*(6), 534–543.

Bortolotti, M., Dubuis, J., Schneiter, P., & Tappy, L. (2012). Effects of dietary protein on lipid metabolism in high fructose fed humans. *Clinical Nutrition, 31*(2), 238–245.

Bortolotti, M., Schneiter, P., & Tappy, L. (2010). Effects of dietary protein on post-prandial lipid metabolism in healthy humans. *Clinical Nutrition ESPEN, 5*(5), e191–e197.

Bourlieu, C., Menard, O., De La Chevasnerie, A., Sams, L., Rousseau, F., Madec, M. N., et al. (2015). The structure of infant formulas impacts their lipolysis, proteolysis and disintegration during in vitro gastric digestion. *Food Chemistry, 182*, 224–235.

Bourlieu, C., & Michalski, M. C. (2015). Structure-function relationship of the milk fat globule. *Current Opinion in Clinical Nutrition and Metabolic Care, 18*(2), 118–127.

Brassard, D., Tessier-Grenier, M., Allaire, J., Rajendiran, E., She, Y., Ramprasath, V., et al. (2017). Comparison of the impact of SFAs from cheese and butter on cardiometabolic risk factors: A randomized controlled trial. *The American Journal of Clinical Nutrition, 105*(4), 800–809.

Briard, V., & Michalski, M. C. (2004). Fatty acid composition of total fat from camembert cheeses with small and large native milk fat globules. *Milchwissenschaft-Milk Science International, 59*(5- 6), 273–277.

Calder, P. C. (2002). Dietary modification of inflammation with lipids. *Proceedings of the Nutrition Society, 61*(3), 345–358.

Cani, P. D., Amar, J., Iglesias, M. A., Poggi, M., Knauf, C., Bastelica, D., et al. (2007). Metabolic endotoxemia initiates obesity and insulin resistance. *Diabetes, 56*(7), 1761–1772.

Cani, P. D., Bibiloni, R., Knauf, C., Waget, A., Neyrinck, A. M., Delzenne, N. M., et al. (2008). Changes in gut microbiota control metabolic endotoxemia-induced inflammation in high-fat diet-induced obesity and diabetes in mice. *Diabetes, 57*(6), 1470–1481.

Carriere, F., Barrowman, J. A., Verger, R., & Laugier, R. (1993). Secretion and contribution to lipolysis of gastric and pancreatic lipases during a test meal in humans. *Gastroenterology, 105*(3), 876–888.

Carriere, F., Renou, C., Lopez, V., De Caro, J., Ferrato, F., Lengsfeld, H., et al. (2000). The specific activities of human digestive lipases measured from the in vivo and in vitro lipolysis of test meals. *Gastroenterology, 119*(4), 949–960.

Castro-Gomez, P., Garcia-Serrano, A., Visioli, F., & Fontecha, J. (2015). Relevance of dietary glycerophospholipids and sphingolipids to human health. *Prostaglandins, Leukotrienes, and Essential Fatty Acids, 101*, 41–51.

Charton, E., & Macrae, A. R. (1992). Substrate specificities of lipases A and B from Geotrichum candidum CMICC 335426. *Biochimica et Biophysica Acta (BBA)/Lipids and Lipid Metabolism, 1123*(1), 59–64.

Chen, G.-C., Wang, Y., Tong, X., Szeto, I. M. Y., Smit, G., Li, Z.-N., et al. (2016). Cheese consumption and risk of cardiovascular disease: A meta-analysis of prospective studies. *European Journal of Nutrition, 56*(8), 1–11.

Chen, M., Li, Y., Sun, Q., Pan, A., Manson, J. E., Rexrode, K. M., et al. (2016). Dairy fat and risk of cardiovascular disease in 3 cohorts of US adults. *The American Journal of Clinical Nutrition, 104*(5), 1209–1217.

Claeys, W. L., Cardoen, S., Daube, G., De Block, J., Dewettinck, K., Dierick, K., et al. (2013). Raw or heated cow milk consumption: Review of risks and benefits. *Food Control, 31*(1), 251–262.

Clemente, G., Mancini, M., Nazzaro, F., Lasorella, G., Rivieccio, A., Palumbo, A. M., et al. (2003). Effects of different dairy products on postprandial lipemia. *Nutrition, Metabolism, and Cardiovascular Diseases, 13*(6), 377–383.

Cobos, A., Horne, D. S., & Muir, D. D. (1995). Rheological properties of acid milk gels.1. Effect of composition, process and acidification conditions on products from recombined milks. *Milchwissenschaft, 50*(8), 444–448.

Collins, Y. F., McSweeney, P. L. H., & Wilkinson, M. G. (2003). Lipolysis and free fatty acid catabolism in cheese: A review of current knowledge. *International Dairy Journal, 13*, 841–866.

Conway, V., Couture, P., Richard, C., Gauthier, S. F., Pouliot, Y., & Lamarche, B. (2013). Impact of buttermilk consumption on plasma lipids and surrogate markers of cholesterol homeostasis in men and women. *Nutrition, Metabolism, and Cardiovascular Diseases, 23*(12), 1255–1262.

Conway, V., Gauthier, S. F., & Pouliot, Y. (2014). Buttermilk: Much more than a source of milk phospholipids. *Animal Frontiers, 4*(2), 44–51.

Couedelo, L., Amara, S., Lecomte, M., Meugnier, E., Monteil, J., Fonseca, L., et al. (2015). Impact of various emulsifiers on ALA bioavailability and chylomicron synthesis through changes in gastrointestinal lipolysis. *Food & Function, 6*(5), 1726–1735.

Das, S., Holland, R., Crow, V. L., Bennett, R. J., & Manderson, G. J. (2005). Effect of yeast and bacterial adjuncts on the CLA content and flavour of a washed-curd, dry-salted cheese. *International Dairy Journal, 15*, 807–815.

de Goede, J., Geleijnse, J. M., Ding, E. L., & Soedamah-Muthu, S. S. (2015). Effect of cheese consumption on blood lipids: A systematic review and meta-analysis of randomized controlled trials. *Nutrition Reviews, 73*(5), 259–275.

de Oliveira Otto, M. C., Mozaffarian, D., Kromhout, D., Bertoni, A. G., Sibley, C. T., Jacobs, D. R., et al. (2012). Dietary intake of saturated fat by food source and incident cardiovascular disease: The multi-ethnic study of atherosclerosis. *The American Journal of Clinical Nutrition, 96*(2), 397–404.

Devle, H., Ulleberg, E. K., Naess-Andresen, C. F., Rukke, E.-O., Vegarud, G., & Ekeberg, D. (2014). Reciprocal interacting effects of proteins and lipids during ex vivo digestion of bovine milk. *International Dairy Journal, 36*, 6–13.

Drouin-Chartier, J.-P., Brassard, D., Tessier-Grenier, M., Côté, J. A., Labonté, M.-È., Desroches, S., et al. (2016). Systematic review of the association between dairy product consumption and risk of cardiovascular-related clinical outcomes. *Advances in Nutrition, 7*(6), 1026–1040.

Drouin-Chartier, J. P., Cote, J. A., Labonte, M. E., Brassard, D., Tessier-Grenier, M., Desroches, S., et al. (2016). Comprehensive review of the impact of dairy foods and dairy fat on cardiometabolic risk. *Advances in Nutrition, 7*(6), 1041–1051.

Drouin-Chartier, J.-P., Tremblay, A. J., Maltais-Giguère, J., Charest, A., Guinot, L., Rioux, L.-E., et al. (2017). Differential impact of the cheese matrix on the postprandial lipid response: A randomized, crossover, controlled trial. *The American Journal of Clinical Nutrition, 106*(6), 1358–1365.

Druart, C., Neyrinck, A. M., Vlaeminck, B., Fievez, V., Cani, P. D., & Delzenne, N. M. (2014). Role of the lower and upper intestine in the production and absorption of gut microbiota-derived PUFA metabolites. *PLoS One, 9*(1), e87560.

Dubois, C., Armand, M., Azais-Braesco, V., Portugal, H., Pauli, A. M., Bernard, P. M., et al. (1994). Effects of moderate amounts of emulsified dietary fat on postprandial lipemia and lipoproteins in normolipidemic adults. *The American Journal of Clinical Nutrition, 60*(3), 374–382.

Duchateau, G. S. M. J. E., & Klaffke, W. (2009). Health food product composition, structure and bioavailability. In D. J. McClements & E. Decker (Eds.), *Designing functional foods: Measuring and controlling food structure breakdown and nutrient absorption. Vol 613.2 D457 2009* (pp. 647–675). Boca Raton, FL: CRC Press; Woodhead Publishing.

Dupont, D., Ménard, O., Le Feunteun, S., & Rémond, D. (2014). Comment la structure des gels laitiers régule-t-elle la biodisponibilité des acides aminés ? *Innovations Agronomiques, 36*, 57–68.

Eckhardt, E. R., Wang, D. Q., Donovan, J. M., & Carey, M. C. (2002). Dietary sphingomyelin suppresses intestinal cholesterol absorption by decreasing thermodynamic activity of cholesterol monomers. *Gastroenterology, 122*(4), 948–956.

Egger, L., Ménard, O., Delgado-Andrade, C., Alvito, P., Assunção, R., Balance, S., et al. (2016). The harmonized INFOGEST in vitro digestion method: From knowledge to action. *Food Research International, 88*, 217–225.

Fang, X., Rioux, L.-E., Labrie, S., & Turgeon, S. L. (2016a). Commercial cheeses with different texture have different disintegration and protein/peptide release rates during simulated in vitro digestion. *International Dairy Journal, 56*, 169–178.

Fang, X., Rioux, L.-E., Labrie, S., & Turgeon, S. L. (2016b). Disintegration and nutrients release from cheese with different textural properties during in vitro digestion. *Food Research International, 88*(Part B), 276–283.

Favé, G., Coste, T. C., & Armand, M. (2004). Physicochemical properties of lipids: New strategies to manage fatty acid bioavailability. *Cellular and Molecular Biology, 50*(7), 815–831.

Favé, G., Peyrot, J., Hamosh, M., & Armand, M. (2007). Digestion des lipides alimentaires: intérêt de la lipase gastrique humaine ? *Cahiers de Nutrition et de Diététique, 42*(4), 183–190.

Fernandez, B., Savard, P., & Fliss, I. (2016). Survival and metabolic activity of pediocin producer pediococcus acidilactici UL5: Its impact on intestinal microbiota and *Listeria monocytogenes* in a model of the human terminal ileum. *Microbial Ecology, 72*(4), 931–942.

Florence, A. C. R., da Silva, R. C., De Santo, A. P., Gioielli, L. A., Tamime, A. Y., & de Oliveira, M. N. (2009). Increased CLA content in organic milk fermented by bifidobacteria or yoghurt cultures. *Dairy Science & Technology, 89*(6), 541–553.

Fruekilde, M. B., & Hoy, C. E. (2004). Lymphatic fat absorption varies among rats administered dairy products differing in physiochemical properties. *The Journal of Nutrition, 134*(5), 1110–1113.

Gallier, S., Cui, J., Olson, T. D., Rutherfurd, S. M., Ye, A., Moughan, P. J., et al. (2013). In vivo digestion of bovine milk fat globules: Effect of processing and interfacial structural changes. I. Gastric digestion. *Food Chemistry, 141*(3), 3273–3281.

Garcia, C., Antona, C., Robert, B., Lopez, C., & Armand, M. (2014). The size and interfacial composition of milk fat globules are key factors controlling triglycerides bioavailability in simulated human gastro-duodenal digestion. *Food Hydrocolloids, 35*(0), 494–504.

Gaudichon, C., Laurent, C., Mahe, S., Marks, L., Tome, D., & Krempf, M. (1994). Rate of [15N]leucine incorporation and determination of nitrogenous fractions from gastro-jejunal secretion in fasting humans. *Reproduction, Nutrition, Development, 34*(4), 349–359.

Gaudichon, C., Mahe, S., Roos, N., Benamouzig, R., Luengo, C., Huneau, J. F., et al. (1995). Exogenous and endogenous nitrogen flow rates and level of protein hydrolysis in the human jejunum after [15N]milk and [15N]yoghurt ingestion. *The British Journal of Nutrition, 74*(2), 251–260.

German, J. B. (2008). Milk fats: A different perspective. *Sciences des Aliments, 28*, 176–186.

German, J. B., Gibson, R. A., Krauss, R. M., Nestel, P., Lamarche, B., van Staveren, W. A., et al. (2009). A reappraisal of the impact of dairy foods and milk fat on cardiovascular disease risk. *European Journal of Nutrition, 48*(4), 191–203.

Gijsbers, L., Ding, E. L., Malik, V. S., de Goede, J., Geleijnse, J. M., & Soedamah-Muthu, S. S. (2016). Consumption of dairy foods and diabetes incidence: A dose-response meta-analysis of observational studies. *The American Journal of Clinical Nutrition, 103*(4), 1111–1124.

Golding, M., & Wooster, T. J. (2010). The influence of emulsion structure and stability on lipid digestion. *Current Opinion in Colloid & Interface Science, 15*(1), 90–101.

Graham, D. Y., & Sackman, J. W. (1983). Solubility of calcium soaps of long-chain fatty-acids in simulated intestinal environment. *Digestive Diseases and Sciences, 28*(8), 733–736.

Grundy, M. M.-L., Lapsley, K., & Ellis, P. R. (2016). A review of the impact of processing on nutrient bioaccessibility and digestion of almonds. *International Journal of Food Science and Technology, 51*(9), 1937–1946.

Guerra, A., Etienne-Mesmin, L., Livrelli, V., Denis, S., Blanquet-Diot, S., & Alric, M. (2012). Relevance and challenges in modeling human gastric and small intestinal digestion. *Trends in Biotechnology, 30*(11), 591–600.

Guinot, L., Rioux, L. E., Labrie, S., Britten, M., & Turgeon, S. L. (2019). Identification of texture parameters influencing commercial cheese matrix disintegration and lipid digestion using an *in vitro* static digestion model. *Food Research International, 121*, 269–277.

Guo, Q., Bellissimo, N., & Rousseau, D. (2017). Role of gel structure in controlling in vitro intestinal lipid digestion in whey protein emulsion gels. *Food Hydrocolloids, 69*(Supplement C), 264–272.

Guo, Q., Ye, A., Bellissimo, N., Singh, H., & Rousseau, D. (2017). Modulating fat digestion through food structure design. *Progress in Lipid Research, 68*, 109–118.

Harwalkar, V. R., & Kalab, M. (1986). Relationship between microstructure and susceptibility to syneresis in yogurt made from reconstituted nonfat dry milk. *Food Microstructure, 5*(2), 287–294.

Health Canada. (2015). Canadian Nutrient File. Retrieved November 29, 2017, from http://www.hc-sc.gc.ca/fn-an/nutrition/fiche-nutri-data/cnf_downloads-telechargement_fcen-eng.php

Health Canada. (2016). Canada's food guide. Retrieved November 23, 2017, from http://www.hc-sc.gc.ca/fn-an/food-guide-aliment/index-eng.php

Heertje, I. (2014). Structure and function of food products: A review. *Food Structure, 1*(1), 3–23.

Hernández-Galán, L., Cattenoz, T., Le Feunteun, S., Canette, A., Briandet, R., Le-Guin, S., et al. (2017). Effect of dairy matrices on the survival of Streptococcus thermophilus, Brevibacterium aurantiacum and Hafnia alvei during digestion. *Food Research International, 100*, 477–488.

Holmer-Jensen, J., Mortensen, L. S., Astrup, A., de Vrese, M., Holst, J. J., Thomsen, C., et al. (2013). Acute differential effects of dietary protein quality on postprandial lipemia in obese non- diabetic subjects. *Nutrition Research, 33*(1), 34–40.

Hotamisligil, G. S. (2006). Inflammation and metabolic disorders. *Nature, 444*(7121), 860–867.

Hu, M., Li, Y., Decker, E. A., & McClements, D. J. (2010). Role of calcium and calcium-binding agents on the lipase digestibility of emulsified lipids using an in vitro digestion model. *Food Hydrocolloids, 24*(8), 719–725.

Huang, E. Y., Leone, V. A., Devkota, S., Wang, Y., Brady, M. J., & Chang, E. B. (2013). Composition of dietary fat source shapes gut microbiota architecture and alters host inflammatory mediators in mouse adipose tissue. *JPEN Journal of Parenteral and Enteral Nutrition, 37*(6), 746–754.

Hunt, J. N., & Knox, M. T. (1968). Control of gastric emptying. *The American Journal of Digestive Diseases, 13*(4), 372–375.

Islam, M. A., Devle, H., Comi, I., Ulleberg, E. K., Rukke, E.-O., Vegarud, G. E., et al. (2017). Ex vivo digestion of raw, pasteurised and homogenised milk – Effects on lipolysis and proteolysis. *International Dairy Journal, 65*, 14–19.

Ito, M., Oishi, K., Yoshida, Y., Okumura, T., Sato, T., Naito, E., et al. (2015). Effects of lactic acid bacteria on low-density lipoprotein susceptibility to oxidation and aortic fatty lesion formation in hyperlipidemic hamsters. *Beneficial Microbes, 6*(3), 287–293.

Ivanovic, N., Minic, R., Dimitrijevic, L., Radojevic Skodric, S., Zivkovic, I., & Djordjevic, B. (2015). Lactobacillus rhamnosus LA68 and Lactobacillus plantarum WCFS1 differently influence metabolic and immunological parameters in high fat diet-induced hypercholesterolemia and hepatic steatosis. *Food & Function, 6*(2), 558–565.

Izumi, H., Tsuda, M., Sato, Y., Kosaka, N., Ochiya, T., Iwamoto, H., et al. (2015). Bovine milk exosomes contain microRNA and mRNA and are taken up by human macrophages. *Journal of Dairy Science, 98*(5), 2920–2933.

Jacobsen, T., & Poulsen, O. M. (1992). Separation and characterization of 61- and 57-kDa lipases from Geotrichum candidum ATCC 66592. *Canadian Journal of Microbiology, 38*(1), 75–80.

Jacobsen, T., & Poulsen, O. M. (1995). Comparison of lipases from different strains of the fungus Geotrichum candidum. *Biochimica et Biophysica Acta - Lipids and Lipid Metabolism, 1257*(2), 96–102.

Jensen, R. G., & Newburg, D. S. (1995). Bovine milk lipids. In R. G. Jensen (Ed.), *Handbook of milk composition* (pp. 543–576). San Diego, CA: Academic.

Jo, S. Y., Choi, E. A., Lee, J. J., & Chang, H. C. (2015). Characterization of starter kimchi fermented with Leuconostoc kimchii GJ2 and its cholesterol-lowering effects in rats fed a high-fat and high- cholesterol diet. *Journal of the Science of Food and Agriculture, 95*(13), 2750–2756.

Jones, P. J. H., & Kubow, S. (2006). Lipids, sterols, and their metabolites. In M. E. Shils, M. Shike, A. C. Ross, B. Caballero, & R. Cousins (Eds.), *Modern nutrition in health and disease* (10th ed., pp. 92–122). Philadelphia: Lippincott Williams and Wilkins.

Kawase, M., Hashimoto, H., Hosoda, M., Morita, H., & Hosono, A. (2000). Effect of administration of fermented milk containing whey protein concentrate to rats and healthy men on serum lipids and blood pressure. *Journal of Dairy Science, 83*(2), 255–263.

Keogh, J. B., Wooster, T. J., Golding, M., Day, L., Otto, B., & Clifton, P. M. (2011). Slowly and rapidly digested fat emulsions are equally satiating but their triglycerides are differentially absorbed and metabolized in humans. *The Journal of Nutrition, 141*(5), 809–815.

Kong, F., & Singh, R. P. (2008). Disintegration of solid foods in human stomach. *Journal of Food Science, 73*(5), R67–R80.

Kopf-Bolanz, K. A., Schwander, F., Gijs, M., Vergeres, G., Portmann, R., & Egger, L. (2012). Validation of an *in vitro* digestive system for studying macronutrient decomposition in humans. *The Journal of Nutrition, 142*(2), 245–250.

Labonte, M. E., Couture, P., Richard, C., Desroches, S., & Lamarche, B. (2013). Impact of dairy products on biomarkers of inflammation: A systematic review of randomized controlled nutritional intervention studies in overweight and obese adults. *The American Journal of Clinical Nutrition, 97*(4), 706–717.

Labonte, M. E., Cyr, A., Abdullah, M. M., Lepine, M. C., Vohl, M. C., Jones, P., et al. (2014). Dairy product consumption has no impact on biomarkers of inflammation among men and women with low-grade systemic inflammation. *The Journal of Nutrition, 144*(11), 1760–1767.

Lai, H. C., & Ney, D. M. (1998). Gastric digestion modifies absorption of butterfat into lymph chylomicrons in rats. *The Journal of Nutrition, 128*(12), 2403–2410.

Lamarche, B. (2008). Review of the effect of dairy products on non-lipid risk factors for cardiovascular disease. *Journal of the American College of Nutrition, 27*(6), 741s–746s.

Lambert, J. E., & Parks, E. J. (2012). Postprandial metabolism of meal triglyceride in humans. *Biochimica et Biophysica Acta, 1821*(5), 721–726.

Lamothe, S., Corbeil, M.-M., Turgeon, S. L., & Britten, M. (2012). Influence of cheese matrix on lipid digestion in a simulated gastro-intestinal environment. *Food & Function, 3*(7), 724–731.

Lamothe, S., Rémillard, N., Tremblay, J., & Britten, M. (2017). Influence of dairy matrices on nutrient release in a simulated gastrointestinal environment. *Food Research International, 92*, 138–146.

Laugerette, F., Vors, C., Peretti, N., & Michalski, M. C. (2011). Complex links between dietary lipids, endogenous endotoxins and metabolic inflammation. *Biochimie, 93*(1), 39–45.

Le Huerou-Luron, I., Bouzerzour, K., Ferret-Bernard, S., Menard, O., Le Normand, L., Perrier, C., et al. (2016). A mixture of milk and vegetable lipids in infant formula changes gut digestion, mucosal immunity and microbiota composition in neonatal piglets. *European Journal of Nutrition, 57*(2), 463–476.

Lecomte, M., Couedelo, L., Meugnier, E., Plaisancie, P., Letisse, M., Benoit, B., et al. (2016). Dietary emulsifiers from milk and soybean differently impact adiposity and inflammation in association with modulation of colonic goblet cells in high-fat fed mice. *Molecular Nutrition & Food Research, 60*(3), 609–620.

Li, R., Dudemaine, P. L., Zhao, X., Lei, C., & Ibeagha-Awemu, E. M. (2016). Comparative analysis of the miRNome of bovine milk fat, whey and cells. *PLoS One, 11*(4), e0154129.

Liang, L., Qi, C., Wang, X.-G., Jin, Q., & McClements, D. J. (2017). Influence of homogenization and thermal processing on the gastrointestinal fate of bovine milk fat: In vitro digestion study. *Journal of Agricultural and Food Chemistry, 65*(50), 11109–11117.

Libby, P. (2002). Inflammation in atherosclerosis. *Nature, 420*(6917), 868–874.

Lopez, C. (2011). Milk fat globules enveloped by their biological membrane: Unique colloidal assemblies with a specific composition and structure. *Current Opinion in Colloid & Interface Science, 16*(5), 391–404.

Lopez, C., Briard-Bion, V., Camier, B., & Gassi, J. Y. (2006). Milk fat thermal properties and solid fat content in emmental cheese: A differential scanning calorimetry study. *Journal of Dairy Science, 89*(8), 2894–2910.

Lopez, C., Cauty, C., & Guyomarc'h, F. (2015). Organization of lipids in milks, infant milk formulas and various dairy products: Role of technological processes and potential impacts. *Dairy Science and Technology, 95*(6), 863–893.

Lopez-Miranda, J., Williams, C., & Lairon, D. (2007). Dietary, physiological, genetic and pathological influences on postprandial lipid metabolism. *The British Journal of Nutrition, 98*(03), 458–473.

Lorenzen, J. K., & Astrup, A. (2011). Dairy calcium intake modifies responsiveness of fat metabolism and blood lipids to a high-fat diet. *The British Journal of Nutrition, 105*(12), 1823–1831.

Lorenzen, J. K., Jensen, S. K., & Astrup, A. (2014). Milk minerals modify the effect of fat intake on serum lipid profile: Results from an animal and a human short-term study. *The British Journal of Nutrition, 111*(08), 1412–1420.

Lorenzen, J. K., Nielsen, S., Holst, J. J., Tetens, I., Rehfeld, J. F., & Astrup, A. (2007). Effect of dairy calcium or supplementary calcium intake on postprandial fat metabolism, appetite, and subsequent energy intake. *The American Journal of Clinical Nutrition, 85*(3), 678–687.

Lovegrove, J. A., & Givens, D. I. (2016). Dairy food products: Good or bad for cardiometabolic disease? *Nutrition Research Reviews, 29*(2), 249–267.

Mariotti, F., Valette, M., Lopez, C., Fouillet, H., Famelart, M.-H., Mathé, V., et al. (2015). Casein compared with whey proteins affects the Organization of dietary fat during digestion and attenuates the postprandial triglyceride response to a mixed high-fat meal in healthy, overweight men. *The Journal of Nutrition, 145*(12), 2657–2664.

Masson, C. (2013). *The effects of constituents and the food matrix of dairy products on postprandial metabolism in overweight subjects*. Maastricht: Maastricht University.

Mat, D. J. L., Le Feunteun, S., Michon, C., & Souchon, I. (2016). In vitro digestion of foods using pH-stat and the INFOGEST protocol: Impact of matrix structure on digestion kinetics of macronutrients, proteins and lipids. *Food Research International, 88*, 226–233.

McIntosh, F. M., Shingfield, K. J., Devillard, E., Russell, W. R., & Wallace, R. J. (2009). Mechanism of conjugated linoleic acid and vaccenic acid formation in human faecal suspensions and pure cultures of intestinal bacteria. *Microbiology, 155*(Pt 1), 285–294.

McIntyre, I., Osullivan, M., & Oriordan, D. (2017). Altering the level of calcium changes the physical properties and digestibility of casein-based emulsion gels. *Food & Function, 8*(4), 1641–1651.

Mekki, N., Charbonnier, M., Borel, P., Leonardi, J., Juhel, C., Portugal, H., et al. (2002). Butter differs from olive oil and sunflower oil in its effects on postprandial lipemia and triacylglycerol-rich lipoproteins after single mixed meals in healthy young men. *The Journal of Nutrition, 132*(12), 3642–3649.

Melnik, B. C., & Schmitz, G. (2017). Milk's role as an epigenetic regulator in health and disease. *Diseases, 5*(1), 12.

Meyer, J. H., Mayer, E. A., Jehn, D., Gu, Y., Fink, A. S., & Fried, M. (1986). Gastric processing and emptying of fat. *Gastroenterology, 90*(5 Pt 1), 1176–1187.

Michalski, M. C. (2007). On the supposed influence of milk homogenization on the risk of CVD, diabetes and allergy. *The British Journal of Nutrition, 97*(4), 598–610.

Michalski, M.-C. (2009). Specific molecular and colloidal structures of milk fat affecting lipolysis, absorption and postprandial lipemia. *European Journal of Lipid Science and Technology, 111*(5), 413–431.

Michalski, M.-C., Briard, V., Desage, M., & Geloen, A. (2005). The dispersion state of milk fat influences triglyceride metabolism in the rat. *European Journal of Nutrition, 44*(7), 436–444.

Michalski, M. C., Camier, B., Briard, V., Leconte, N., Gassi, J. Y., Goudedranche, H., et al. (2004). The size of native milk fat globules affects physico-chemical and functional properties of Emmental cheese. *Le Lait, 84*(4), 343–358.

Michalski, M. C., Camier, B., Gassi, J. Y., Briard-Bion, V., Leconte, N., Famelart, M. H., et al. (2007). Functionality of smaller vs control native milk fat globules in Emmental cheeses manufactured with adapted technologies. *Food Research International, 40*(1), 191–202.

Michalski, M. C., Cariou, R., Michel, F., & Garnier, C. (2002). Native vs. damaged milk fat globules: Membrane properties affect the viscoelasticity of milk gels. *Journal of Dairy Science, 85*(10), 2451–2461.

Michalski, M. C., Gassi, J. Y., Famelart, M. H., Leconte, N., Camier, B., Michel, F., et al. (2003). The size of native milk fat globules affects physico-chemical and sensory properties of camembert cheese. *Le Lait, 83*(2), 131–143.

Michalski, M. C., Genot, C., Gayet, C., Lopez, C., Fine, F., Joffre, F., et al. (2013). Multiscale structures of lipids in foods as parameters affecting fatty acid bioavailability and lipid metabolism. *Progress in Lipid Research, 52*(4), 354–373.

Michalski, M.-C., & Januel, C. (2006). Does homogenization affect the human health properties of cow's milk? *Trends in Food Science and Technology, 17*(8), 423–437.

Michalski, M. C., Leconte, N., Briard-Bion, V., Fauquant, J., Maubois, J. L., & Goudedranche, H. (2006). Microfiltration of raw whole milk to select fractions with different fat globule size distributions: Process optimization and analysis. *Journal of Dairy Science, 89*(10), 3778–3790.

Michalski, M. C., Soares, A. F., Lopez, C., Leconte, N., Briard, V., & Geloen, A. (2006). The supramolecular structure of milk fat influences plasma triacylglycerols and fatty acid profile in the rat. *European Journal of Nutrition, 45*(4), 215–224.

Milard, M., Laugerette, F., Durand, A., Buisson, C., Meugnier, E., Loizon, E., et al. (2019). Milk polar lipids in a high-fat diet can prevent body weight gain: Modulated abundance of gut bacteria in relation with fecal loss of specific fatty acids. *Molecular Nutrition & Food Research, 63*(4), e1801078. https://doi.org/10.1002/mnfr.201801078

Milard, M., Penhoat, A., Durand, A., Buisson, C., Loizon, E., Meugnier, E., et al. (2019). Acute effects of milk polar lipids on intestinal tight junction expression: Towards animpact of sphingomyelin through the regulation of IL-8 secretion? *The Journal of Nutritional Biochemistry, 65*, 128–138.
Minekus, M., Alminger, M., Alvito, P., Ballance, S., Bohn, T., Bourlieu, C., et al. (2014). A standardised static in vitro digestion method suitable for food - an international consensus. *Food & Function, 5*(6), 1113–1124.
Mohamadshahi, M., Veissi, M., Haidari, F., Javid, A. Z., Mohammadi, F., & Shirbeigi, E. (2014). Effects of probiotic yogurt consumption on lipid profile in type 2 diabetic patients: A randomized controlled clinical trial. *Journal of Research in Medical Sciences: The Official Journal of Isfahan University of Medical Sciences, 19*(6), 531–536.
Mortensen, L. S., Hartvigsen, M. L., Brader, L. J., Astrup, A., Schrezenmeir, J., Holst, J. J., et al. (2009). Differential effects of protein quality on postprandial lipemia in response to a fat-rich meal in type 2 diabetes: Comparison of whey, casein, gluten, and cod protein. *The American Journal of Clinical Nutrition, 90*(1), 41–48.
Mu, H., & Hoy, C. E. (2004). The digestion of dietary triacylglycerols. *Progress in Lipid Research, 43*(2), 105–133.
Mu, H., & Porsgaard, T. (2005). The metabolism of structured triacylglycerols. *Progress in Lipid Research, 44*(6), 430–448.
Mullally, M. M., Mehra, R., & FitzGerald, R. J. (1998). Thermal effects on the conformation and susceptibility of beta-lactoglobulin to hydrolysis by gastric and pancreatic endoproteinases. *Irish Journal of Agricultural and Food Research, 37*(1), 51–60.
Murphy, E. A., Velazquez, K. T., & Herbert, K. M. (2015). Influence of high-fat diet on gut microbiota: A driving force for chronic disease risk. *Current Opinion in Clinical Nutrition and Metabolic Care, 18*(5), 515–520.
Nestel, P. J. (2008). Effects of dairy fats within different foods on plasma lipids. *Journal of the American College of Nutrition, 27*(6), 735S–740S.
Nestel, P. J., Mellett, N., Pally, S., Wong, G., Barlow, C. K., Croft, K., et al. (2013). Effects of low-fat or full-fat fermented and non-fermented dairy foods on selected cardiovascular biomarkers in overweight adults. *The British Journal of Nutrition, 110*(12), 2242–2249.
Nordestgaard, B. G., Benn, M., Schnohr, P., & Tybjærg-Hansen, A. (2007). Nonfasting triglycerides and risk of myocardial infarction, ischemic heart disease, and death in men and women. *Journal of the American Medical Association, 298*(3), 299–308.
Norris, G. H., Jiang, C., Ryan, J., Porter, C. M., & Blesso, C. N. (2016). Milk sphingomyelin improves lipid metabolism and alters gut microbiota in high fat diet-fed mice. *The Journal of Nutritional Biochemistry, 30*, 93–101.
Norris, G. H., Milard, M., Michalski, M. C., & Blesso, C. N. (2019). Protective properties of milk sphingomyelin against dysfunctional lipid metabolism, gut dysbiosis, and inflammation. *The Journal of Nutritional Biochemistry, 73*, 108244. https://doi.org/10.1016/j.jnutbio.2019.108224
Nuora, A., Tupasela, T., Tahvonen, R., Rokka, S., Marnila, P., Viitanen, M., et al. (2018). Effect of homogenised and pasteurised versus native cows' milk on gastrointestinal symptoms, intestinal pressure and postprandial lipid metabolism. *International Dairy Journal, 79*, 15–23.
Ohlsson, L., Burling, H., & Nilsson, A. (2009). Long term effects on human plasma lipoproteins of a formulation enriched in butter milk polar lipid. *Lipids in Health and Disease, 8*, 44.
Pal, S., Ellis, V., & Dhaliwal, S. (2010). Effects of whey protein isolate on body composition, lipids, insulin and glucose in overweight and obese individuals. *The British Journal of Nutrition, 104*(5), 716–723.
Pal, S., & Radavelli-Bagatini, S. (2013). The effects of whey protein on cardiometabolic risk factors. *Obesity Reviews, 14*(4), 324–343.
Pan, D. D., Zeng, X. Q., & Yan, Y. T. (2011). Characterisation of lactobacillus fermentum SM-7 isolated from koumiss, a potential probiotic bacterium with cholesterol-lowering effects. *Journal of the Science of Food and Agriculture, 91*(3), 512–518.

Panagiotakos, D. B., Pitsavos, C. H., Zampelas, A. D., Chrysohoou, C. A., & Stefanadis, C. I. (2010). Dairy products consumption is associated with decreased levels of inflammatory markers related to cardiovascular disease in apparently healthy adults: The ATTICA study. *Journal of the American College of Nutrition, 29*(4), 357–364.

Payne, A. N., Zihler, A., Chassard, C., & Lacroix, C. (2012). Advances and perspectives in in vitro human gut fermentation modeling. *Trends in Biotechnology, 30*(1), 17–25.

Pei, R., Martin, D. A., DiMarco, D. M., & Bolling, B. W. (2017). Evidence for the effects of yogurt on gut health and obesity. *Critical Reviews in Food Science and Nutrition, 57*(8), 1569–1583.

Pereira, D. I., & Gibson, G. R. (2002). Effects of consumption of probiotics and prebiotics on serum lipid levels in humans. *Critical Reviews in Biochemistry and Molecular Biology, 37*(4), 259–281.

Pessione, E., & Cirrincione, S. (2016). Bioactive molecules released in food by lactic acid bacteria: Encrypted peptides and biogenic amines. *Frontiers in Microbiology, 7*, 876.

Pimpin, L., Wu, J. H. Y., Haskelberg, H., Del Gobbo, L., & Mozaffarian, D. (2016). Is butter back? A systematic review and meta-analysis of butter consumption and risk of cardiovascular disease, diabetes, and total mortality. *PLoS One, 11*(6), e0158118.

Qin, L. Q., Xu, J. Y., Han, S. F., Zhang, Z. L., Zhao, Y. Y., & Szeto, I. M. (2015). Dairy consumption and risk of cardiovascular disease: An updated meta-analysis of prospective cohort studies. *Asia Pacific Journal of Clinical Nutrition, 24*(1), 90–100.

Ramadan, Q., Jafarpoorchekab, H., Huang, C., Silacci, P., Carrara, S., Koklu, G., et al. (2013). NutriChip: Nutrition analysis meets microfluidics. *Lab on a Chip, 13*(2), 196–203.

Ramprasath, V. R., Jones, P. J., Buckley, D. D., Woollett, L. A., & Heubi, J. E. (2013). Effect of dietary sphingomyelin on absorption and fractional synthetic rate of cholesterol and serum lipid profile in humans. *Lipids in Health and Disease, 12*, 125.

Raziani, F., Tholstrup, T., Kristensen, M. D., Svanegaard, M. L., Ritz, C., Astrup, A., et al. (2016). High intake of regular-fat cheese compared with reduced-fat cheese does not affect LDL cholesterol or risk markers of the metabolic syndrome: A randomized controlled trial. *The American Journal of Clinical Nutrition, 104*(4), 973–981.

Reis, M. G., Roy, N. C., Bermingham, E. N., Ryan, L., Bibiloni, R., Young, W., et al. (2013). Impact of dietary dairy polar lipids on lipid metabolism of mice fed a high-fat diet. *Journal of Agricultural and Food Chemistry, 61*(11), 2729–2738.

Rohm, H., & Schmid, W. (1993). Influence of dry-matter fortification on flow properties of yogurt 1. Evaluation of flow curves. *Milchwissenschaft, 48*(10), 556–560.

Rosqvist, F., Smedman, A., Lindmark-Mansson, H., Paulsson, M., Petrus, P., Straniero, S., et al. (2015). Potential role of milk fat globule membrane in modulating plasma lipoproteins, gene expression, and cholesterol metabolism in humans: A randomized study. *The American Journal of Clinical Nutrition, 102*(1), 20–30.

Russell, D. A., Ross, R. P., Fitzgerald, G. F., & Stanton, C. (2011). Metabolic activities and probiotic potential of bifidobacteria. *International Journal of Food Microbiology, 149*(1), 88–105.

Sanders, T. A., Filippou, A., Berry, S. E., Baumgartner, S., & Mensink, R. P. (2011). Palmitic acid in the sn-2 position of triacylglycerols acutely influences postprandial lipid metabolism. *The American Journal of Clinical Nutrition, 94*(6), 1433–1441.

Sandoval-Castilla, O., Lobato-Calleros, C., Aguirre-Mandujano, E., & Vernon-Carter, E. J. (2004). Microstructure and texture of yogurt as influenced by fat replacers. *International Dairy Journal, 14*(2), 151–159.

Sanggaard, K. M., Holst, J. J., Rehfeld, J. F., Sandstrom, B., Raben, A., & Tholstrup, T. (2004). Different effects of whole milk and a fermented milk with the same fat and lactose content on gastric emptying and postprandial lipaemia, but not on glycaemic response and appetite. *The British Journal of Nutrition, 92*(3), 447–459.

Sautier, C., Dieng, K., Flament, C., Doucet, C., Suquet, J. P., & Lemonnier, D. (1983). Effects of whey protein, casein, soya-bean and sunflower proteins on the serum, tissue and faecal steroids in rats. *The British Journal of Nutrition, 49*(3), 313–319.

Schiffrin, E. J., Parlesak, A., Bode, C., Bode, J. C., van't Hof, M. A., Grathwohl, D., et al. (2009). Probiotic yogurt in the elderly with intestinal bacterial overgrowth: Endotoxaemia and innate immune functions. *The British Journal of Nutrition, 101*(7), 961–966.

Schwab, U., Lauritzen, L., Tholstrup, T., Haldorssoni, T., Riserus, U., Uusitupa, M., et al. (2014). Effect of the amount and type of dietary fat on cardiometabolic risk factors and risk of developing type 2 diabetes, cardiovascular diseases, and cancer: A systematic review. *Food & Nutrition Research, 58*, 25145.

Sethi, S., Gibney, M. J., & Williams, C. M. (1993). Postprandial lipoprotein metabolism. *Nutrition Research Reviews, 6*(01), 161–183.

Shani-Levi, C., Alvito, P., Andrés, A., Assunção, R., Barberá, R., Blanquet-Diot, S., et al. (2017). Extending in vitro digestion models to specific human populations: Perspectives, practical tools and bio- relevant information. *Trends in Food Science & Technology, 60*(Supplement C), 52–63.

Snow, D. R., Ward, R., Olsen, A., Jimenez-Flores, R., & Hintze, K. J. (2011). Membrane-rich milk fat diet provides protection against gastrointestinal leakiness in mice treated with lipopolysaccharide. *Journal of Dairy Science, 94*(5), 2201–2212.

Soedamah-Muthu, S. S., Verberne, L. D., Ding, E. L., Engberink, M. F., & Geleijnse, J. M. (2012). Dairy consumption and incidence of hypertension: A dose-response meta-analysis of prospective cohort studies. *Hypertension, 60*(5), 1131–1137.

Soerensen, K. V., Thorning, T. K., Astrup, A., Kristensen, M., & Lorenzen, J. K. (2014). Effect of dairy calcium from cheese and milk on fecal fat excretion, blood lipids, and appetite in young men. *The American Journal of Clinical Nutrition, 99*, 984–991.

Sprong, R. C., Hulstein, M. F., & Van der Meer, R. (2001). Bactericidal activities of milk lipids. *Antimicrobial Agents and Chemotherapy, 45*(4), 1298–1301.

Sprong, R. C., Hulstein, M. F. E., & van der Meer, R. (2002). Bovine milk fat components inhibit food-borne pathogens. *International Dairy Journal, 12*(2–3), 209–215.

Stenson, W. F. (2006). The esophagus and stomach. In M. E. Shils, M. Shike, A. C. Ross, B. Caballero, & R. Cousins (Eds.), *Modern nutrition in health and disease* (10th ed., pp. 1179–1188). Philadelphia: Lippincott Williams and Wilkins.

Sun, J., & Buys, N. (2015). Effects of probiotics consumption on lowering lipids and CVD risk factors: A systematic review and meta-analysis of randomized controlled trials. *Annals of Medicine, 47*(6), 430–440.

Tholstrup, T. (2006). Dairy products and cardiovascular disease. *Current Opinion in Lipidology, 17*(1), 1–10.

Tholstrup, T., Hoy, C. E., Andersen, L. N., Christensen, R. D., & Sandstrom, B. (2004). Does fat in milk, butter and cheese affect blood lipids and cholesterol differently? *Journal of the American College of Nutrition, 23*(2), 169–176.

Thompson, L. U., Jenkins, D. J., Amer, M. A., Reichert, R., Jenkins, A., & Kamulsky, J. (1982). The effect of fermented and unfermented milks on serum cholesterol. *The American Journal of Clinical Nutrition, 36*(6), 1106–1111.

Thorning, T. K., Bertram, H. C., Bonjour, J.-P., de Groot, L., Dupont, D., Feeney, E., et al. (2017). Whole dairy matrix or single nutrients in assessment of health effects: Current evidence and knowledge gaps. *The American Journal of Clinical Nutrition, 105*(5), 1033–1045.

Thorning, T. K., Raben, A., Bendsen, N. T., Jørgensen, H. H., Kiilerich, P., Ardö, Y., et al. (2016). Importance of the fat content within the cheese-matrix for blood lipid profile, faecal fat excretion, and gut microbiome in growing pigs. *International Dairy Journal, 61*, 67–75.

Timmen, H., & Precht, D. (1984). Zum Einfluss unterschiedlicher technologischer Behandlung von Milch auf die Verdauungsvorgange im Magen. V. Lipolyse im Magen. [influence of different technological treatments of milk on digestion in the stomach. V. Lipolysis in the stomach]. *Milchwissenschaft, 39*(5), 276–280.

Title, A. C., Denzler, R., & Stoffel, M. (2015). Uptake and function studies of maternal Milk-derived MicroRNAs. *The Journal of Biological Chemistry, 290*(39), 23680–23691.

Tsuchiya, A., Almiron-Roig, E., Lluch, A., Guyonnet, D., & Drewnowski, A. (2006). Higher satiety ratings following yogurt consumption relative to fruit drink or dairy fruit drink. *Journal of the American Dietetic Association, 106*(4), 550–557.

Tunick, M. H., Ren, D. X., Van Hekken, D. L., Bonnaillie, L., Paul, M., Kwoczak, R., et al. (2016). Effect of heat and homogenization on in vitro digestion of milk. *Journal of Dairy Science, 99*(6), 4124–4139.

Turgeon, S. L., & Brisson, G. (2019). The dairy matrix: Bioaccessibility and bioavailability of nutrients and physiological effects. *Journal of Dairy Science (accepted)*. https://doi.org/10.3168/jds.2019-17308

Turgeon, S. L., & Rioux, L.-E. (2011). Food matrix impact on macronutrients nutritional properties. *Food Hydrocolloids, 25*(8), 1915–1924.

Turnbaugh, P. J., Ley, R. E., Mahowald, M. A., Magrini, V., Mardis, E. R., & Gordon, J. I. (2006). An obesity- associated gut microbiome with increased capacity for energy harvest. *Nature, 444*(7122), 1027–1031.

US Department of Health and Human Services, US Department of Agriculture. (2015). 2015–2020 Dietary Guidelines for Americans. Retrieved December, 2015, from http://health.gov/dietaryguidelines/2015/guidelines/

Vallières, C. (2016). Effet de réduction en sodium sur la texture et la bioaccessibilité des protéines d'un fromage à pâte molle à croûte fleurie. Mémoire (M. Sc.), Université Laval, Québec, QC, Canada

van Avesaat, M., Troost, F. J., Ripken, D., Hendriks, H. F., & Masclee, A. A. (2015). Ileal brake activation: Macronutrient-specific effects on eating behavior? *International Journal of Obesity, 39*(2), 235–243.

Van Hekken, D. L., Tunick, M. H., Ren, D. X., & Tomasula, P. M. (2017). Comparing the effect of homogenization and heat processing on the properties and in vitro digestion of milk from organic and conventional dairy herds. *Journal of Dairy Science, 100*(8), 6042–6052.

Versantvoort, C. H. M., Oomen, A. G., Van de Kamp, E., Rompelberg, C. J. M., & Sips, A. J. A. M. (2005). Applicability of an in vitro digestion model in assessing the bioaccessibility of mycotoxins from food. *Food and Chemical Toxicology, 43*(1), 31–40.

Vors, C., Capolino, P., Guerin, C., Meugnier, E., Pesenti, S., Chauvin, M.-A., et al. (2012). Coupling in vitro gastrointestinal lipolysis and Caco-2 cell cultures for testing the absorption of different food emulsions. *Food & Function, 3*(5), 537–546.

Vors, C., Gayet-Boyer, C., & Michalski, M.-C. (2015). Produits laitiers et inflammation métabolique: Quels liens en phase postprandiale et à long terme ? *Cahiers de Nutrition et de Diététique, 50*(1), 25–38.

Vors, C., Joumard-Cubizolles, L., Lecomte, M., Combe, E., Ouchchane, L., Drai, J., et al. (2020). Milk polar lipids reduce lipid cardiovascular risk factors in overweight postmenopausal women: Towards a gut sphingomyelin-cholesterol interplay. *Gut, 69*, 487–501. https://doi.org/10.1136/gutjnl-2018-318155

Vors, C., Nazare, J. A., Michalski, M. C., & Laville, M. (2014). Intérêt de la phase postprandiale pour la santé de l'Homme. *Obésité, 9*(1), 31–41.

Vors, C., Pineau, G., Gabert, L., Drai, J., Louche-Pelissier, C., Defoort, C., et al. (2013). Modulating absorption and postprandial handling of dietary fatty acids by structuring fat in the meal: A randomized crossover clinical trial. *The American Journal of Clinical Nutrition, 97*(1), 23–36.

Wang, T. Y., Liu, M., Portincasa, P., & Wang, D. Q. (2013). New insights into the molecular mechanism of intestinal fatty acid absorption. *European Journal of Clinical Investigation, 43*(11), 1203–1223.

Wat, E., Tandy, S., Kapera, E., Kamili, A., Chung, R. W. S., Brown, A., et al. (2009). Dietary phospholipid-rich dairy milk extract reduces hepatomegaly, hepatic steatosis and hyperlipidemia in mice fed a high-fat diet. *Atherosclerosis, 205*(1), 144–150.

Weiland, A., Bub, A., Barth, S. W., Schrezenmeir, J., & Pfeuffer, M. (2016). Effects of dietary milk- and soya- phospholipids on lipid-parameters and other risk indicators for cardiovascular diseases in overweight or obese men - two double-blind, randomised, controlled, clinical trials. *Journal of Nutritional Science, 5*, e21.

Westphal, S., Kastner, S., Taneva, E., Leodolter, A., Dierkes, J., & Luley, C. (2004). Postprandial lipid and carbohydrate responses after the ingestion of a casein-enriched mixed meal. *The American Journal of Clinical Nutrition, 80*, 284–290.

Xie, N., Cui, Y., Yin, Y. N., Zhao, X., Yang, J. W., Wang, Z. G., et al. (2011). Effects of two lactobacillus strains on lipid metabolism and intestinal microflora in rats fed a high-cholesterol diet. *BMC Complementary and Alternative Medicine, 11*, 53.

Yakoob, M. Y., Shi, P., Hu, F. B., Campos, H., Rexrode, K. M., Orav, E. J., et al. (2014). Circulating biomarkers of dairy fat and risk of incident stroke in U.S. men and women in 2 large prospective cohorts. *The American Journal of Clinical Nutrition, 100*(6), 1437–1447.

Ye, A., Cui, J., Dalgleish, D., & Singh, H. (2016a). The formation and breakdown of structured clots from whole milk during gastric digestion. *Food & Function, 7*(10), 4259–4266.

Ye, A., Cui, J., Dalgleish, D., & Singh, H. (2016b). Formation of a structured clot during the gastric digestion of milk: Impact on the rate of protein hydrolysis. *Food Hydrocolloids, 52*, 478–486.

Ye, A., Cui, J., Dalgleish, D., & Singh, H. (2017). Effect of homogenization and heat treatment on the behavior of protein and fat globules during gastric digestion of milk. *Journal of Dairy Science, 100*, 36–47.

Ye, A., Cui, J., & Singh, H. (2010). Effect of the fat globule membrane on in vitro digestion of milk fat globules with pancreatic lipase. *International Dairy Journal, 20*(12), 822–829.

Zhang, X., & Beynen, A. C. (1993). Lowering effect of dietary milk-whey protein v. casein on plasma and liver cholesterol concentrations in rats. *The British Journal of Nutrition, 70*(1), 139–146.

Zhou, A. L., Hintze, K. J., Jimenez-Flores, R., & Ward, R. E. (2012). Dietary fat composition influences tissue lipid profile and gene expression in Fischer-344 rats. *Lipids, 47*(12), 1119–1130.

Zhou, A. L., & Ward, R. E. (2019). Milk polar lipids modulate lipid metabolism, gut permeability, and systemic inflammation in high-fat-fed C57BL/6J ob/ob mice, a model of severe obesity. *Journal of Dairy Science, 102*(6), 4816–4831.

Part II
Dairy Fat Physics and Materials Science

Chapter 9
Crystallization and Melting Properties of Milk Fat

Christelle Lopez

1 Introduction

Fat is secreted in milk in the form of oil in water (O/W) emulsion droplets called the milk fat globules (size from 0.1 to 10 μm, mean diameter around 4 μm in bovine milk), that correspond to a core of triacylglycerols (TAGs) enveloped by a biological membrane rich in phospholipids, cholesterol and proteins (Lopez, 2011). Milk TAGs provide more than 50% of the dietary energy intake that is essential for the growth of newborns and also provide bioactive molecules involved in infant health. Milk fat is also consumed by infants and adults in various dairy products, i.e. milk, cream, whipped cream, cheeses, butter and as an ingredient in many bakery and confectionary applications in which it can be found in anhydrous state and as O/W or W/O emulsions.

Milk fat can be in a semi-solid state (i.e. a mixture of crystal network and liquid oil) over a wide range of temperatures, including the temperature of storage (4–7 °C), consumption and digestion (37 °C) of food products. This crystallization behaviour of milk fat results from its complex composition with a high amount of saturated fatty acids (FAs) and polymorphism of TAGs. The crystallization properties of milk fat can be affected by many parameters such as the cooling rates and thermal history, the shear, the presence of minor lipid components, the dispersion state (anhydrous *versus* emulsified), the changes in the FA and TAG composition. Milk TAG crystals are involved in the rheological and sensorial properties of products (e.g. butter, cream, cheese). Besides solid fat content, the functionality of a fat is also affected by the crystal structure and networks formed by TAG molecules. Understanding the crystallization and melting behaviours of milk TAGs, as well as

C. Lopez (✉)
INRAE, UMR STLO, Rennes, France

INRAE, UR BIA, Nantes, France
e-mail: christelle.lopez@inrae.fr

T. Truong et al. (eds.), *Dairy Fat Products and Functionality*,
https://doi.org/10.1007/978-3-030-41661-4_9

the polymorphism of milk TAGs, is of great importance from a scientific point of view and with regard to the economic impact of milk fat, especially in fat-rich products.

2 A Wide Diversity of Fatty Acids and Triacylglycerols

Milk fat consists of 97–98% TAGs, i.e. molecules composed of a glycerol backbone with three fatty acid moieties esterified onto it. Bovine milk fat is the most complex fat found in the nature, with more than 400 different fatty acids (FAs) identified of which 12 are present in proportions greater than 1% (Fig. 9.1a). Milk fat contains short-chain (C4–C8), medium-chain (C10–C12) and long-chain (C14–C18) length FAs. About 70% of milk fat corresponds to saturated FAs and 25% to monounsaturated FAs mainly oleic acid (C18:1c9), with variations as a function of season and diet (Fig. 9.1A). Bovine milk fat contains a low amount of natural *trans* FAs produced by biohydrogenation in the cows. About 200 different TAGs have been identified (Fig. 9.1B). Milk fat contains many asymmetrical TAGs, i.e. TAGs of the SSU or UUS type, in which the single unsaturated (U) or saturated (S) FA resides in either the *sn*-1 or *sn*-3 position, or TAGs with differences between the FA chain length larger than two atoms of carbon, i.e. BPP (B: butyric acid C4:0, P: palmitic acid C16:0). This diversity of FAs and TAGs confers to milk fat specific physical properties, in particular, the presence of a solid TAG phase over a wide range of temperatures. The TAG composition induces a complex crystallization behaviour and a wide melting range with multiple melting points. Furthermore, individual TAG molecules are characterised by a complex thermal behaviour in relation to their polymorphism of monotropic type.

3 Crystallization and Melting Properties of Bovine Anhydrous Milk Fat

3.1 Thermal Properties

Bovine anhydrous milk fat (AMF), which is the fat isolated from butter, has a broad melting range from −40 to +40 °C (Fig. 9.2). AMF must be warmed at least to +40 °C to ensure complete melting of TAGs and is not completely solid until it reaches a temperature below −40 °C. Many authors have studied crystallization and melting behaviour of AMF by differential scanning calorimetry (DSC) and showed that it crystallises and melts in several steps (Lopez, Bourgaux, Lesieur, & Ollivon, 2007; Ten Grotenhuis, van Aken, van Malssen, & Schenk, 1999; Timms, 1980). A typical crystallization curve of AMF shows two main exotherms in the cooling DSC measurement, corresponding to groups of TAG molecules with high crystallization

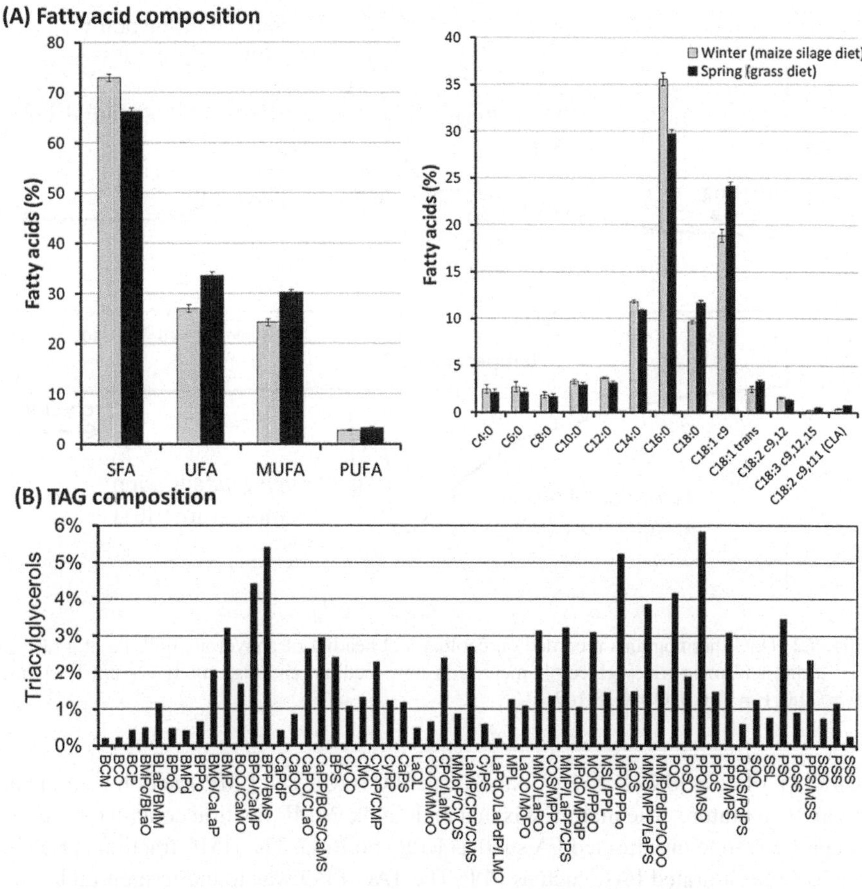

Fig. 9.1 Milk fat composition. (**A**) fatty acid (FA) composition with variations as a function of cow diet and season (**B**) triacylglycerol composition. Abreviations: *SFA* = saturated FA, *UFA* = unsaturated FA, *MUFA* = monounsaturated FA, *PUFA* = polyunsaturated FA, *B* = butyric acid (C4:0); *Ca* = caproic acid (C6:0), *Cy* = caprylic acid (C8:0), *C* = capric acid (C10:0), *La* = lauric acid (C12:0), *M* = myristic acid (C14:0), *Pd* = pentadecanoic acid (C15:0), *P* = palmitic acid (C16:0), *Po* = palmitoleic acid (C16:1), *S* = stearic acid (C18:0), *O* = oleic acid (C18:1c9). Adapted from Lopez, Bourgaux, et al. (2006) and Lopez et al. (2014)

temperature (HCT) and low crystallization temperature (LCT) segregated on cooling. A typical melting curve of AMF shows three overlapped endothermic peaks that have been called on the basis of their melting point, i.e. low melting point (LMP), medium melting point (MMP) and high melting point (HMP) (Fig. 9.2). The melting of TAGs in the MMP fraction is highly important for the sensory properties of milk fat in the mouth. The three melting peaks correspond to large groups of TAGs that melt separately and behave as solid solutions. The LMP endotherm corresponds to the melting of TAGs with a high content of long-chain unsaturated FAs and short-chain saturated FAs such as BOO, BPO and PPO (B: butyric acid

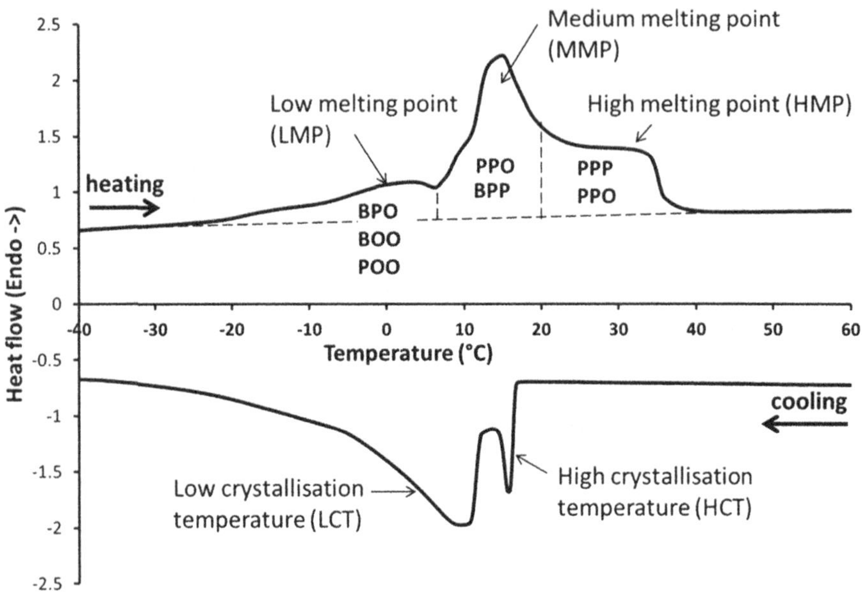

Fig. 9.2 DSC thermograms recorded on cooling and heating of anhydrous milk fat at a rate of 2 °C/min, and main triacylglycerols melting in low, medium and high melting point fractions. B = C4:0, P = C16:0, O = C18:1c9

C4:0; O: oleic acid C18:1c9; P: palmitic acid C16:0). The LMP fraction is liquid at room temperature. The main TAGs that melt in the MMP fraction contain one short-chain FA or one unsaturated FA such as BPP and PPO. The HMP fraction is rich in long-chain saturated FAs, such as PPP. The TAG PPO was found to melt under the three endotherms.

Since 1950, DSC was used to demonstrate the presence of polymorphism in AMF and X-ray diffraction (XRD) was used to identify the polymorphic forms obtained after rapid or slow cooling. The number of thermal transitions in DSC thermograms, the partial overlapping of the melting peaks, their respective enthalpies and transition temperatures depend strongly on the thermal treatments (e.g. heating and cooling rates, tempering) and on the entire thermal history of the sample. The wide melting range of AMF results in a wide range of plasticity where both solid and liquid TAGs are present in various proportions, i.e. increase in the solid fat content as a function of decrease in temperature.

3.2 *Effect of Cooling Rate*

Characterization of milk fat crystals at different length scales is important in the description and understanding of their functions in food products. The structure of TAG crystals networks is observed at the microscale level using polarized light

Fig. 9.3 Polarized light microscopy images taken at −8 °C after (**A**) slow cooling at 0.2 °C/min and (**B**) quenching anhydrous milk fat from 60 °C

microscopy because they are birefringent, i.e. TAG crystals appear bright whereas the liquid TAGs appear black. Milk fat crystals and crystal networks have been described at the microstructural level using polarised light microscopy (Campos, Narine, & Marangoni, 2002; Lopez, Lesieur, Bourgaux, & Ollivon, 2005). Milk fat crystals organize as needles or spherulites, depending on the cooling rate (Fig. 9.3). Cooling AMF below 5 °C at low cooling rate leads to crystals with a spherulitic microstructure, due to extensive crystal growth (Campos et al., 2002; Lopez, Lesieur, et al., 2005). When the same milk fat is cooled rapidly from the melt, a more granular microstructure is observed (Campos et al., 2002; Lopez, Lesieur, et al., 2005). Crystallization proceeds more quickly, and nucleation events predominate over crystal growth processes. The result is a large number of small microstructural features distributed in a less orderly fashion than in the case of slow cooling. Ramel, Peyronel, and Marangoni (2016) explored the structural features of high melting point milk TAG at the nanoscale, using the combination of ultra-small angle X-ray scattering and cryo-TEM, and described smooth crystalline nano-platelets.

Most of the studies dedicated to AMF crystallization properties have focused on the description of TAG crystals at a molecular level and investigated the impact of different factors such as cooling rate, TAG and FA composition, and shear (Bugeat et al., 2015; Lopez, Lesieur, et al., 2005; Lopez & Ollivon, 2009; Mazzanti, Marangoni, & Idziak, 2009; Ten Grotenhuis et al., 1999). The identification of the crystalline structures formed by TAG molecules is possible by using XRD recorded at both small and wide angles to have information on the longitudinal organization of TAG molecules, e.g., double chain length (2*L*) or triple chain length (3*L*) and on the packing of the FA chains (polymorphic forms: α, β′, β), respectively. The high energy flux of synchrotron radiation XRD (SR-XRD) allows characterization of the solid TAG phase as a function of temperature or time. Detailed information about the crystallization properties of milk fat has been provided by Michel Ollivon's group (CNRS, France) using the coupling of DSC and XRD as a function of temperature (Ollivon et al., 2006). The AMF samples were heated to 60 °C for at least

5 min to ensure that all crystals and nuclei were melted and to erase the thermal history of TAGs. Then, the crystallization properties of AMF were investigated with cooling rates in the range of 0.1 to ~1000 °C/min. The melting behaviour of AMF was characterized on heating at 2 °C/min. The main results are presented below.

Slow Cooling (0.1 °C/min) of AMF On slow cooling (dT/dt = 0.1 °C/min) of AMF from the melt, XRD patterns revealed the successive formation and the coexistence of four different TAG crystals (Lopez, Lavigne, Lesieur, Keller, & Ollivon, 2001a; Fig. 9.4). The first TAG crystals formed at 24 °C correspond to lamellar structures with a two-chain length organization 2*L* (41.5 Å) and a packing of β′ type. Then, the formation of β′-2*L* (48 Å) crystals was characterised. From 13 °C, the formation of a three-chain length organization 3*L* (62 Å) of α type (α-3*L*) was observed. From 5.5 °C, β′-2*L* (39 Å) crystals were formed. WAXS patterns also showed the formation of traces of β crystals at low temperature (Fig. 9.4). The DSC thermogram recorded simultaneously to XRD experiments during the decrease in temperature showed three main exothermic events that correspond to the successive formation of

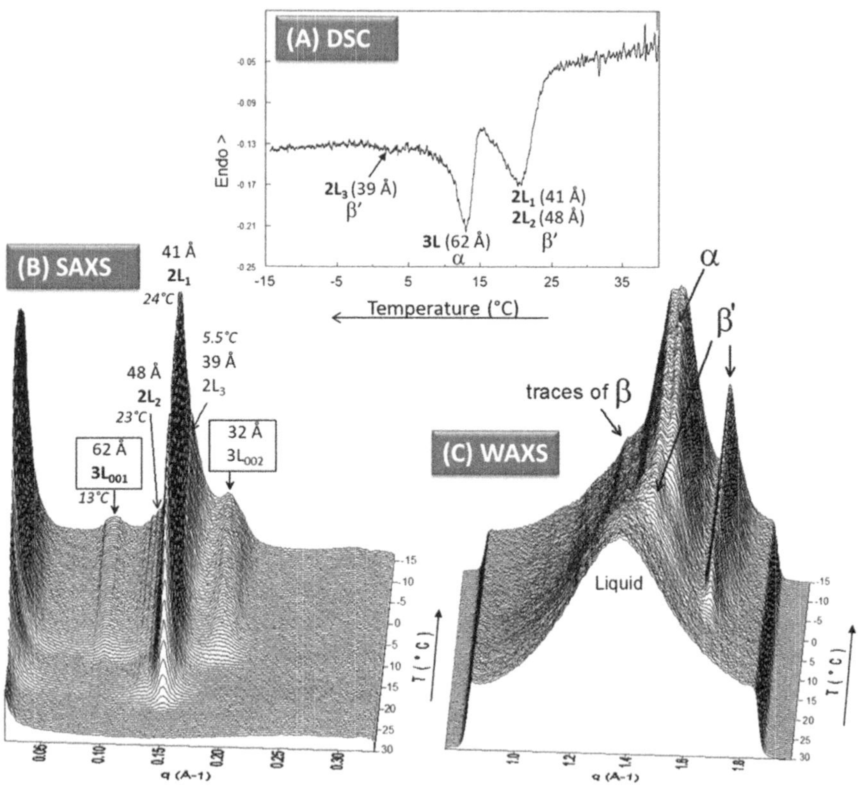

Fig. 9.4 Crystallization of anhydrous milk fat characterized on slow cooling at 0.1 °C/min. (**A**) DSC thermogram, and synchrotron radiation X-ray diffraction patterns recorded (**B**) at small angles (SAXS) and (**C**) at wide angles (WAXS). Adapted from Lopez, Lavigne, et al. (2001a)

TAG crystals present in AMF (Fig. 9.4A). On subsequent heating of the AMF sample, the α-3*L* crystals melt first around 13 °C, then the three different β′-2*L* crystals successively melt without any changes in the structural parameters. The high melting point TAG crystals correspond to a β′-2*L* (41.5 Å) organization that disappears around 40 °C. The absence of polymorphic transformation on heating of AMF is the signature of stable TAG crystals close to equilibrium. This study showed that a TAG molecular segregation occurs in milk fat on slow cooling from the melt. AMF crystallizes and melts in several independent steps corresponding to the phase separation of several groups of TAGs. Each TAG fraction acts as an independent solid solution. The ability of dry fractionation to separate such TAG crystal species into so-called olein and stearin fractions after slow cooling rates confirms that they correspond to different TAG compositions (Kaylegian & Lindsay, 1995).

Cooling of AMF at 1–3 °C/min On cooling of AMF from the melt at the rates of 1–3 °C/min (Fig. 9.5A), TAG molecules sequentially crystallize in α form under three different lamellar structures (Lopez, Lesieur, et al., 2005). From about 17 °C, the successive formation of the α-2*L* (46 Å) and α-2*L* (38 Å) crystals has been characterized and from 14 °C TAG molecules crystallize in a α-3*L* (72 Å) organization. A time-dependent sub-α ↔ α reversible transition was observed at −10 °C. DSC recordings show two successive exotherms that have been attributed to the successive crystallization of TAG in α-2*L* crystals and α-3*L* crystals (Fig. 9.5C). Subsequent heating at 2 °C/min has shown numerous structural rearrangements of the α-2*L* and α-3*L* crystals and the formation of β′-3*L* (67 Å) and β′-2*L* (40–41.5 Å) crystals that takes advantage of the melting of the other crystalline structures (Fig. 9.5B). The recording of an exotherm in the DSC thermogram confirms α to β′ polymorphic transition occurring on heating of AMF (Fig. 9.5D). For temperatures above 20 °C, the remaining β′-2*L* (40–41.5 Å) crystals progressively melt until their disappearance at about 39 °C (Fig. 9.5D). Such rearrangements of TAG molecules into a β′ lamellar organization are facilitated by the presence of the liquid TAG phase. The complex melting behaviour confirms that the TAG crystals formed on cooling at 1–3 °C/min are not thermodynamically stable forms.

Rapid Cooling of AMF The crystallization properties of milk TAGs were studied after quenching from the melt (cooling at ~1000 °C/min), to characterize the most unstable crystalline structures and their reorganization as a function of time (Lopez, Bourgaux, Lesieur, & Ollivon, 2002; Lopez, Lavigne, Lesieur, Keller, & Ollivon, 2001b). The samples were cooled rapidly from 60 to 4 °C or −8 °C to ensure crystallization of TAG molecules.

In the set of experiments performed after quenching of AMF to −8 °C, it was possible to identify the formation of α-2*L* (47 Å) and α-3*L* (70 Å) crystals thanks to the brightness of synchrotron X-ray beam which allows fast recordings (Fig. 9.6A and B). The α-2*L* (47 Å) crystals were very unstable since they disappeared during 20 min isothermal recording and progressively converted into 3*L* crystals (Fig. 9.6C; Lopez, Lavigne, et al., 2001b). On subsequent heating of the AMF sample, the α-3*L*

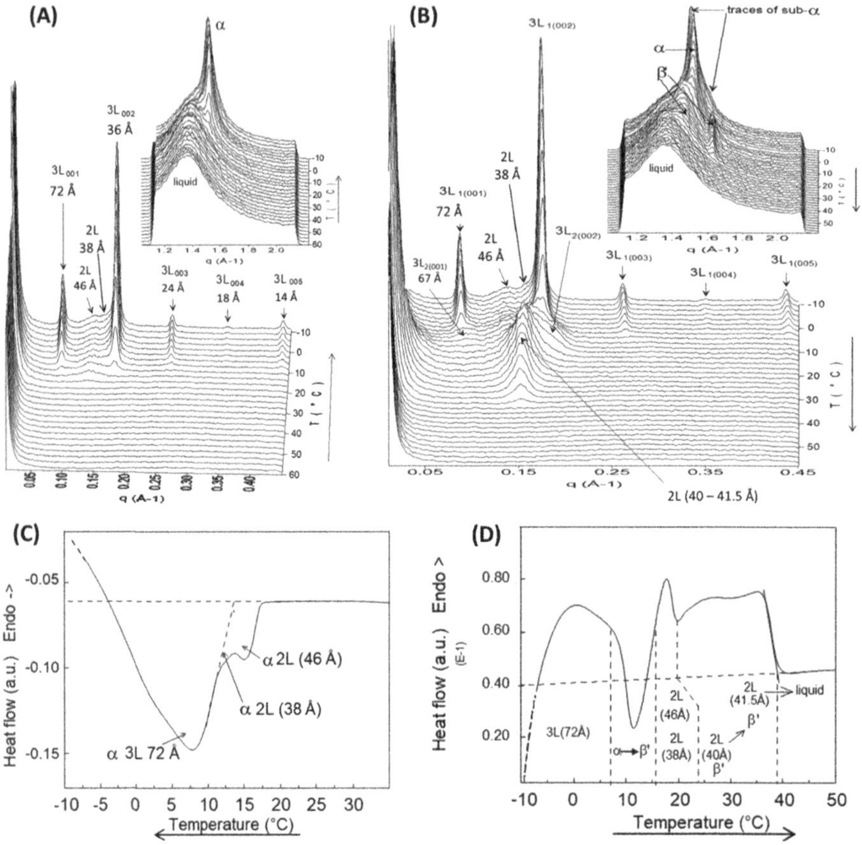

Fig. 9.5 Crystallization properties of anhydrous milk fat. SR-XRD patterns recorded at small and wide (insert) angles (**A**) on cooling at 3 °C/min from 60 to −10 °C, and (**B**) subsequent heating at 2 °C/min. DSC thermograms recorded (**C**) on cooling and (**D**) on heating. DSC thermograms and SR-XRD patterns were recorded simultaneously as a function of temperature. Adapted from Lopez, Lesieur, et al. (2005)

(70 Å) crystals melted. From about 11 °C, TAG molecules in the solid phase formed β′-2*L* (37 Å) crystals (Fig. 9.7). The exotherm recorded at around 11 °C in the DSC thermogram confirms the α to β′ polymorphic transition that occurs on heating of AMF (Fig. 9.7B). The thickness of this β′-2*L* lamellar structure increased up to 41 Å until its final melting, showing structural reorganizations as a function of the increase in temperature.

In a second set of experiments, AFM was quenched at 4 °C and the thermal and structural changes were characterized as a function of time in isothermal conditions, as shown Fig. 9.8A (Lopez, Bourgaux, Lesieur, & Ollivon, 2002). During the 30 min following quenching, structural changes occurred, the α-3*L* (70 Å) structure melted and the 3*L* (66 Å) and 2*L* (39 Å) structures corresponding to β′ and β polymorphs were formed. Isothermal DSC recorded simultaneously to XRD experiments

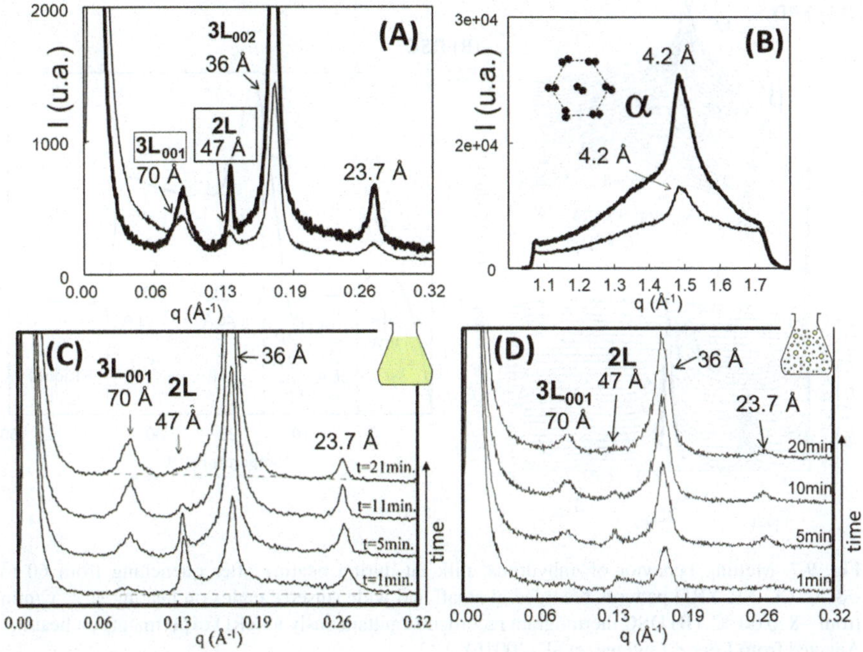

Fig. 9.6 Crystalline structures formed at −8°C by milk triacylglycerols after quenching from 60 °C and evolutions as a function of time in isothermal conditions. SR-XRD patterns recorded (**A**) at small angles and (**B**) at wide angles for anhydrous milk fat (thick lines) and cream (thin lines). Selected SR-XRD patterns recorded at small angles as a function of time at −8 °C just after quenching from 60 °C of (**C**) anhydrous milk fat and (**D**) cream samples. Adapted from Lopez, Lesieur, Keller, & Ollivon (2000) and Lopez, Lavigne, et al. (2001b)

showed exothermic signals with a release of the heat of crystallization as a function of time, corresponding to polymorphic transition from α to β′ and β of milk TAGs (Fig. 9.8C). The nucleation time (the time at which a peak starts forming), time of maximum crystallization rate (the time of peak maximum) and heat of crystallization (proportional to peak area) can all be determined from the DSC thermogram. After 4 days storage of AMF at 4 °C, the crystals were organized in 2*L* (40 Å) and 3*L* (54 Å) lamellar structures with the coexistence of α, β'_1, β'_2 and β polymorphic forms, as shown Fig. 9.8D (Lopez, Bourgaux, Lesieur, & Ollivon, 2002).

As a summary, milk TAGs segregate as a function of the decrease in temperature and display a complex polymorphism, as revealed by DSC and XRD investigations. Depending on the cooling rate, six different types of crystals were identified, several of them in coexistence, and their time- and temperature-dependent evolutions were quantitatively monitored. They correspond to lamellar structures with 2*L* (40–48 Å) and 3*L* (54–72 Å) organizations of TAGs (Fig. 9.9A). At least five crystalline sub-cell species were observed at wide angles: α and sub-α, two β′ and one β (Fig. 9.9B). All these crystalline structures coexist with a liquid phase even at low temperature, i.e. 4 °C. The comparison of the small number of crystal type formed to the large

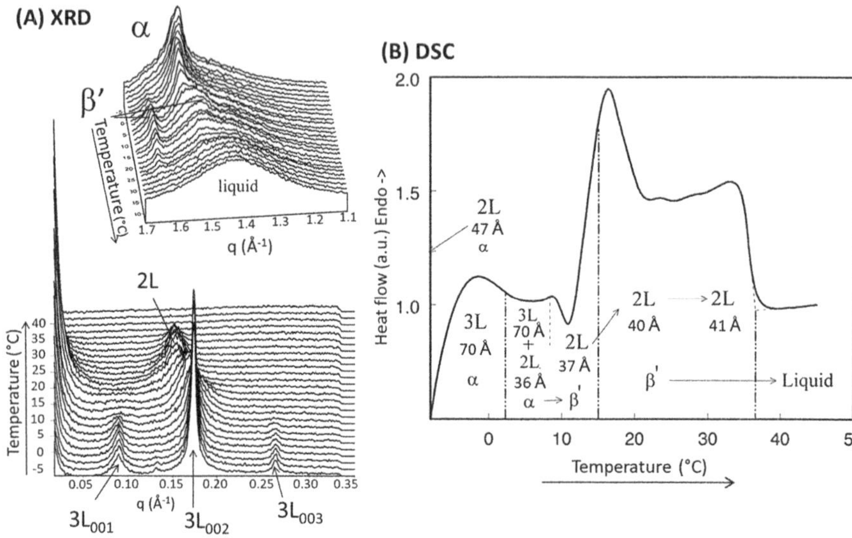

Fig. 9.7 Melting behavior of anhydrous milk fat during heating after quenching from 60 to −8 °C. (**A**) SR-XRD patterns recorded at small and wide (insert) angles on heating at 2 °C/min from −8 to 60 °C. (**B**) DSC thermogram recorded simultaneously to XRD experiments on heating. Adapted from Lopez, Lavigne, et al. (2001b)

number of TAGs present in milk fat (Fig. 9.1B) provides evidence that mixed crystals are formed in AMF (i.e. co-crystallization of different TAG molecules). On cooling, the first longitudinal organizations of TAG molecules correspond to *2L* structures with long spacings of about 38–48 Å. These crystalline varieties may correspond to crystallization of HMP fractions of TAGs. These *2L* crystals are generally formed by TAGs with saturated and similar chain length FAs, such as MPP and PPP (Fig. 9.9C). Then, crystallization of *3L* structures (about 62–72 Å) occur. Triple chain length 3L stackings may correspond to crystallization of unsaturated TAG molecules (e.g. PPO) or to that of TAGs with FA chains of different lengths, like BPP (Fig. 9.9C). On heating, the crystals formed on cooling (*2L*, *3L*) melt and recrystallization take place with the formation of a β′-*2L* (40–41.5 Å) structure. The increase in the thickness of the β′-*2L* crystals on heating is attributed to progressive and selective melting of the TAGs with the shortest FA chains so that the solid phase composition enriches with longer FA chains and higher melting points TAGs. Recrystallization also occurs during isothermal storage. The stable crystalline structures take advantages of the melting of the metastable crystals which melt first or Ostwald ripening occurring thanks to the liquid phase that coexists with the solid phases. The mixed TAG crystals formed in milk fat organize in the less polymorphic forms β′ and α because molecular packing is not very dense. Low amount of β crystals have been identified in milk fat, even after long time storage at low temperature. It has been reported on milk fats from individual cows with large differences in their

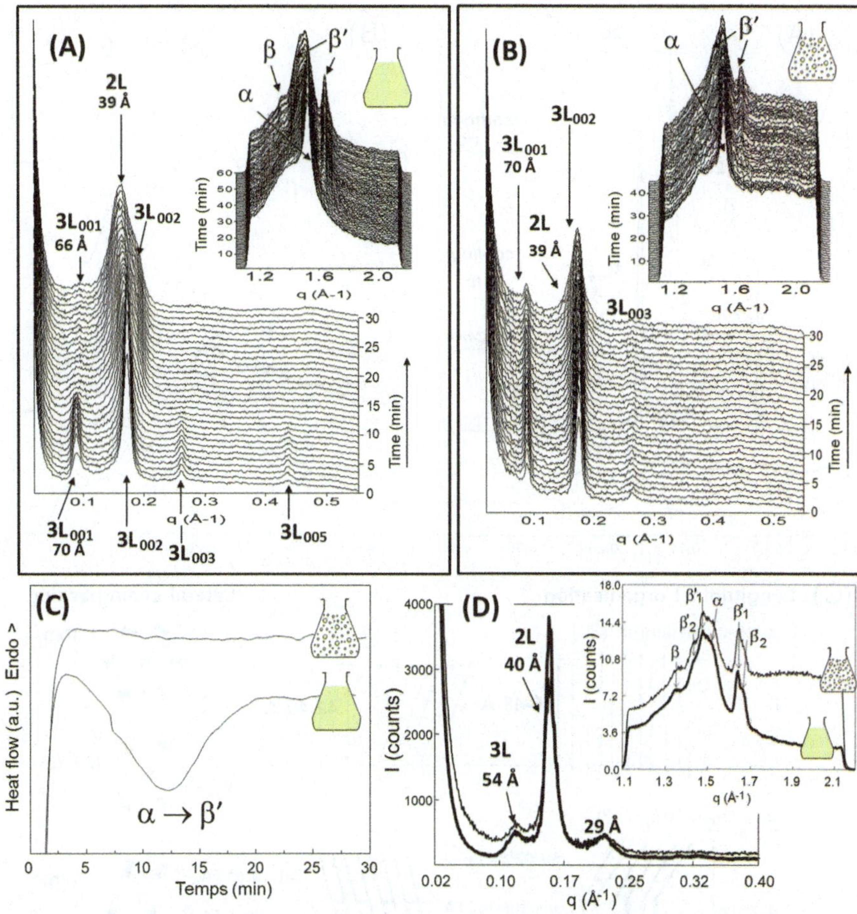

Fig. 9.8 Isothermal evolutions of SR-XRD patterns recorded at small and wide angles at 4 °C after rapid quenching from 60 °C of (**A**) anhydrous milk fat and (**B**) cream samples. (**C**) DSC recordings of AMF and cream samples obtained simultaneously with XRD experiments. The signal jumps observed on the left side of the DSC recordings correspond to the equilibration of the microcalorimeter after sample introduction. (**D**) Small and wide (insert) angle SR-XRD patterns recorded at 4 °C after storage of cream (thin line) and anhydrous milk fat (thick line) samples at this temperature for 135 and 105 h, respectively, following a rapid quenching from 60 °C. Adapted from Lopez, Bourgaux, Lesieur, and Ollivon (2002)

TAG profiles that the presence of β polymorphs is dependent on TAG composition and that cooling rate and tempering are not critical factors in the formation of β polymorphs (Tzompa-Sosa, Ramel, van Valenberg, and van Aken (2016). The high concentration of unsaturated TAGs with a carbon number 52–54 and the presence of a substantial amount of liquid TAGs may be equally important for the formation of β polymorphs.

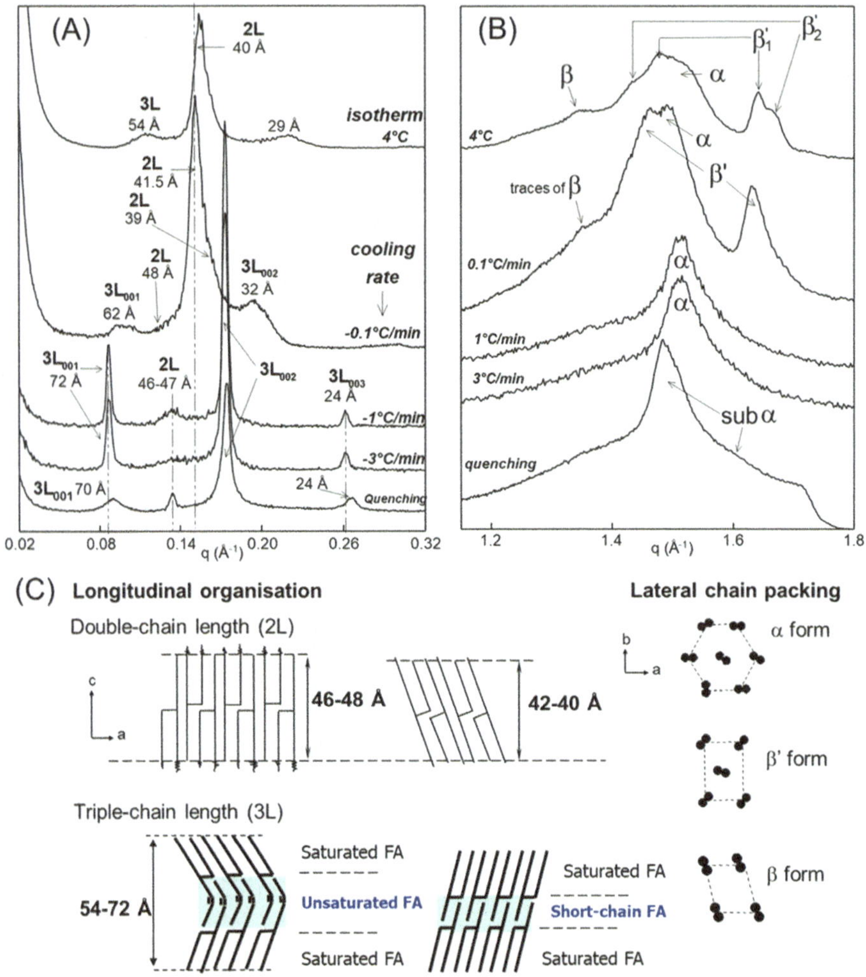

Fig. 9.9 SR-XRD patterns recorded at −8 °C (**A**) at small angles and (**B**) at wide angles after either cooling of anhydrous milk fat at different rates as indicated on the figure or isothermal conditioning at 4 °C during 5 days. (**C**) Lamellar structures formed by anhydrous milk fat molecules in the solid state (fatty acids are drawn as straight lines) and acyl chain lateral packings corresponding to α, β′ and β polymorphic forms. Adapted from Lopez, Lesieur, et al. (2005)

3.3 Effect of Minor Lipid Compounds

TAGs represent generally 97–98% of milk fat. The balance is composed mainly by minor lipids such as free fatty acids (FFAs), monoacylglycerols (MAGs), diacylglycerols (DAGs) and phospholipids that can influence the mechanisms of TAG crystallization in AMF i.e. the nucleation stage, the crystal growth and/or the polymorphic behaviour. Most of the experiments investing the effect of minor components on milk fat crystallization are performed under isothermal conditions and the crystallization behaviour is monitored by DSC and pulsed nuclear magnetic resonance (pNMR). The crystallization process is described by the Avrami and the Gometz models which are fitted by non-linear regression.

It has been shown that removal of minor lipid components from milk fat has no effect on melting point, equilibrium solid fat content, polymorphic forms, microstructural crystal network and mechanical properties (Wright, Hartel, Narine, & Marangoni, 2000; Wright & Marangoni, 2003). However, the minor components affect the crystallization kinetics of milk fat and delay the onset of crystallization at low degrees of supercooling (Wright et al., 2000). Milk fat DAGs have been reported to have an inhibitory effect on the crystallization of milk TAGs without modifying the microstructure of crystals. It was suggested that structural complementary between DAGs and crystallizing TAGs allowed the TAGs to co-crystallize within early seed crystals and subsequent further delay TAG crystallization (Wright et al., 2000). Other studies showed that the effect of DAGs and MAGs on the crystallization behaviour of milk fat depends on temperature and concentration. Moreover, the type of esterified FAs and the polar head of the amphiphilic molecules determine to what extent the partial glycerides MAGs and DAGs influence the nucleation and crystal growth of TAGs (Foubert, Vanhoutte, & Dewettinck, 2004). For example, stearic acid (C18:0) based MAGs and DAGs enhance heterogeneous nucleation at low temperatures, while at higher temperatures an interaction with the crystal growth predominates. Oleic acid (C18:1c9) based MAGs and DAGs have an effect on the nucleation process while no interaction with the crystal growth was observed (Foubert et al., 2004). The effects of MAGs on milk TAG crystallization in recombined cream revealed differences as a function of the FA esterified (Fredrick et al., 2013). MAGs with C18:0 formed upon cooling a two-dimensional crystal at the surface of the emulsion droplet which induced interfacial heterogeneous nucleation and an acceleration of TAG crystal growth and α to β′ polymorphic transition. On the contrary, MAGs with C18:1c9 did not affect the crystallization behaviour while MAGs with C12:0 showed an intermediate behaviour. None of the MAGs influenced the solid fat content after storage for 5 days at 5 °C. The observed differences in nucleation mechanisms and crystallization kinetics may influence the microstructural arrangement of the milk TAG inside the emulsion droplets and consequently the partial coalescence rate and hence the whipping properties of recombined creams. Phospholipids were shown to delay the onset time of AMF crystallization under isothermal conditions at 25 °C by their fast adsorption on the growth sites of crystals inducing steric hindrance (Vanhoutte, Dewettinck, Foubert, Vanlerberghe, & Huyghebaert, 2002). The effect of FFAs on milk fat crystallization has been demonstrated under isothermal conditions at 25 °C. The addition of short-chain FFAs increased the induction time of milk fat crystallization while the addition on saturated long-chain FFAs accelerated the crystallization kinetics with consequences on the microstructure of milk fat crystals (Bayard, Leal-Calderon, & Cansell, 2017). As a conclusion, the molecular interactions between the minor lipids and milk TAGs able to affect milk fat crystallization depend on the temperature, on their concentration, and on the FA composition (mainly the chain length similarity) of the phospholipids, FFAs, MAGs and DAGs.

3.4 *Effect of Shear on AMF Crystals*

AMF crystallization studies carried out under shear are of particular interest (Grall & Hartel, 1992; Mazzanti et al., 2009; Van Aken & Visser, 2000; Vanhoutte et al., 2002). Shear affects the crystallization process as a whole by enhancing molecular diffusion and favouring the rearrangement of TAG molecules in the melt, which helps to overcome the kinetic barriers for nucleation and growth. The studies on shear effects demonstrated the formation of smaller TAG crystals and narrower distribution sizes at higher shear rates and attributed this to higher nucleation rates and breakdown of milk fat crystals. Under very slow cooling rates and mixing speeds, little effect was found from the agitation speed on the kinetics of milk fat crystallization (Vanhoutte et al., 2002). A detailed synchrotron XRD study on the kinetics of crystallization of AMF performed in a Couette cell showed a shear-induced acceleration of the α to β′ form transition and the presence of crystalline orientation (Mazzanti et al., 2009). Shear, i.e. agitation of milk fat, affects the crystallization kinetics and the microstructure with consequences on the mechanical properties of the fat crystal network obtained. Agitation enhances nucleation, which leads to the formation of numerous small crystals with a softening of the material.

4 Crystallization of TAGs in Bovine Milk Fat Globules and Emulsion Droplets

Milk and many dairy products are O/W emulsions (e.g. cream, cheeses). Studying the crystallization properties of milk TAGs in milk fat globules and processed emulsion droplets is of prime importance because it can affect many properties, such as: (1) the resistance of fat globules to disruption and then to coalescence, (2) the susceptibility of fat globules to churning for the manufacture of butter, (3) the partial coalescence and stability of whipped cream, and (4) the texture, rheological properties and mouth feel of high-fat content food products, e.g. cream, butter, cheeses products. It is therefore important to understand better the physical properties of TAGs in milk fat globules, e.g. their thermal and crystallographic properties, for industrial applications and to improve the quality of food products. Moreover, it is interesting to compare crystallization of milk TAGs dispersed in emulsion, such as milk or cream (which is the high concentration of fat globules from milk to reach at least 30% fat) in which fat globules are surrounded by a biological membrane rich in phospholipids, with crystallization of bulk anhydrous milk fat (AMF). It is also important to know the crystallization properties of TAG dispersed in processed O/W emulsion droplets of various sizes to better control the quality of food products.

The crystallization properties of milk TAGs in O/W emulsions were studied at different scale levels by using polarized light microscopy and transmission electron microscopy (TEM), DSC and XRD. Microscopy techniques showed that the

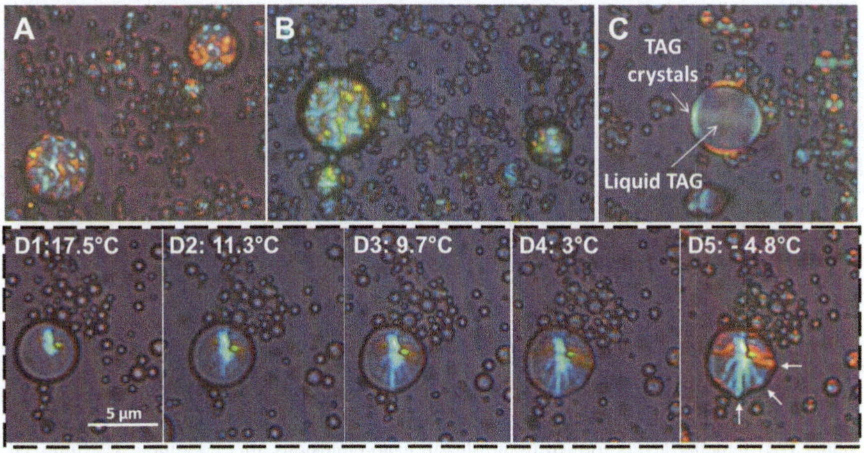

Fig. 9.10 Polarized light microscopy images of TAG crystals within milk fat globules formed after cooling at different rates from 60 to −5 °C, (**A**) cooling at 5 °C/min, (**B**) quenching, (**C**) cooling at 0.5 °C/min. (**D**) Images taken during cooling of milk at 1 °C/min, the intermediate temperatures are indicated in the figure. TAG crystals can deform the milk fat globule membrane as indicated by arrows. Adapted from Lopez (2011)

morphology and the location of TAG crystals within milk fat globules are affected by the cooling rates and tempering. After rapid cooling of milk, numerous small crystals of needle type are formed with no preferred orientation. At slow cooling rate, fat globules display needle-shaped crystals that can deform the biological membrane surrounding milk fat globules (Fig. 9.10). After long storage at low temperature (4–7 °C) or after tempering, TAG crystals are of needle type with a radially organized crystallization. Using cryo-TEM, Prof. Bhesh Bhandari's group showed the stacking of individual lamellar layers with 38–42 Å thickness formed by milk TAG molecules at the periphery of droplets in nano-emulsions upon cooling at 4 °C (Fig. 9.11; Truong et al., 2015).

The examination of the crystallization behaviour of TAGs and TAG polymorphism in emulsion by using XRD is much more challenging than for AMF and especially difficult since (1) both small and wide angle XRD should be considered at the same time and compared to determine the evolution of each of the species as a function of time, (2) the X-ray intensity diffracted by each of the crystalline structures is proportional to the fraction of particular crystal in the structure, (3) the whole XRD signal is largely absorbed by the surrounding water and its solutes (e.g. casein micelles, minerals, lactose) and (4) the small-angle XRD peak broadening results from the crystallization constraints in dispersed systems and to the lower size of the crystals. The use of DSC coupled to synchrotron radiation XRD is a suitable way allowing identification of the crystalline structures formed by TAG molecules as a function of temperature and time in dispersed systems such as milk fat globules (Lopez et al., 2000; Lopez et al., 2002; Lopez, Bourgaux, Lesieur, & Ollivon, 2002;

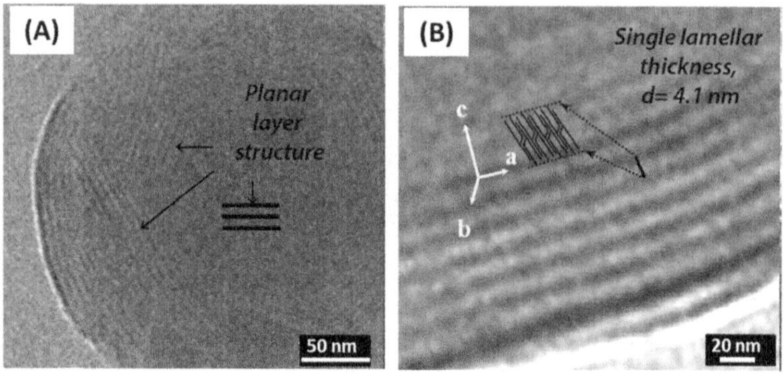

Fig. 9.11 Cryo-TEM micrographs of nano-emulsions containing crystalline particles of high melting point fraction of milk fat called stearin, upon crystallization at 4 °C. (**A**) stacking of individual TAG lamellar layers formed in the stearin nano-emulsions after cooling at 1 °C/min. (**B**) Thickness of single lamellae including light and dark layers in the stearin nano-emulsions. Adapted from Truong, Morgan, Bansal, Palmer, and Bhandari (2015)

Lopez, Lesieur, Bourgaux, Keller, & Ollivon, 2001) and emulsion droplets of various sizes (Bugeat et al., 2011; Lopez, Bourgaux, Lesieur, Bernadou, et al., 2002; Michalski, Ollivon, Briard, Leconte, & Lopez, 2004). The investigations of milk TAG crystallization within emulsion droplets is not performed below about −10 °C to avoid the formation of ice crystals that could alter the physical stability of the emulsion.

4.1 *Effect of Cooling Rate and Tempering*

The crystallization behaviour of milk fats in emulsion is influenced by changing the cooling rate, more remarkably than that in a bulk phase. It is also affected by tempering (i.e. successive cooling and heating). Therefore, we discuss here the effects of cooling rate on the crystallization properties of milk fat dispersed in milk fat globules or lipid droplets, which are crystallized at the rates of cooling in the range of 0.1–1000 °C/min from 60 °C and subjected to subsequent heating at the rate of 2 °C/min.

Slow Cooling of Milk Fat Globules Slow cooling of cream (0.1–0.15 °C/min) leads to the DSC recording of a single broad exotherm corresponding to crystallization of TAG in fat globules (Fig. 9.12B). However, precise XRD study allowed isolating four polymorphic forms that are successively formed during the cooling process (Fig. 9.12A, Lopez, Lesieur, et al., 2001). On cooling from the melt, the $2L$ (47 Å) crystals were first formed from 22 °C, then the $2L$ (40 Å) crystals were observed from 20 °C. From 16 °C, the formation of $3L$ (71 Å) and $3L$ (65 Å) organizations were reported. On slow cooling of cream (0.1–0.15 °C/min), nucleation

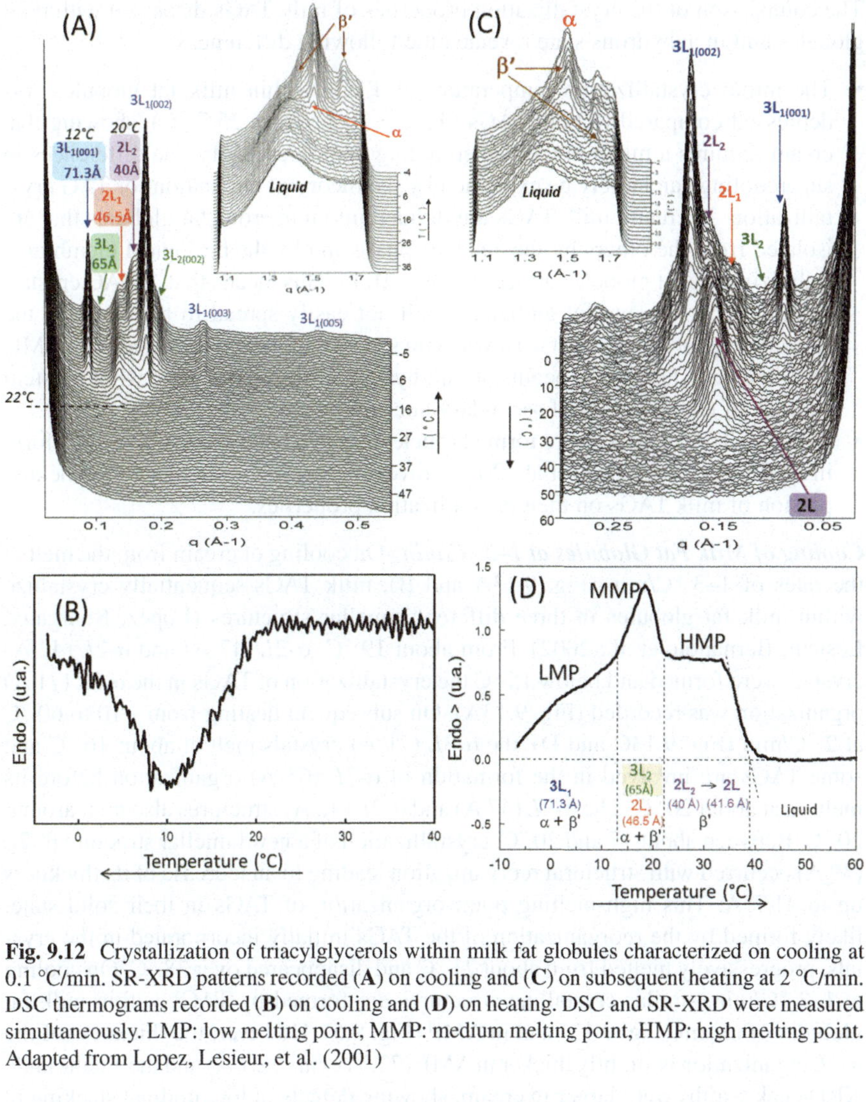

Fig. 9.12 Crystallization of triacylglycerols within milk fat globules characterized on cooling at 0.1 °C/min. SR-XRD patterns recorded (**A**) on cooling and (**C**) on subsequent heating at 2 °C/min. DSC thermograms recorded (**B**) on cooling and (**D**) on heating. DSC and SR-XRD were measured simultaneously. LMP: low melting point, MMP: medium melting point, HMP: high melting point. Adapted from Lopez, Lesieur, et al. (2001)

occurs in the α form, then both the α and β′ polymorphic forms coexist until the end of cooling. On subsequent heating of the cream, the successive melting of the 3*L* (71 Å), 3*L* (65 Å) and 2*L* (47 Å) organizations occurred (Fig. 9.12C). The β′-2*L* (40 Å) predominantly occurred during the cooling and heating processes, and its lamellar thickness increased to 416 Å on heating. The DSC curve recorded on heating of the cream showed three endotherms (Fig. 9.12D). The LMP endotherm corresponds to the melting of the 3*L* (71 Å) crystals, the MMP endotherm was related to the melting of both the 3*L* (65 Å) and the 2*L* (47 Å) crystals, the HMP endotherm was attributed to the melting of the β′-2*L* crystals (Fig. 9.12D).

The comparison of the crystallization properties of milk TAGs dispersed within fat globules and in anhydrous state revealed the following differences:

- The initial crystallization temperature of TAGs within milk fat globules was depressed compared to bulk TAGs (T_{onset} = 20.4 versus 25.7 °C), showing that cream requires a much higher supercooling than does AMF. The differences in supercooling can largely be explained by the theory of nucleation for TAG crystallization. In cream, milk TAGs are divided into numerous fat globules that are isolated from the others by the aqueous phase and by the biological membrane, stabilizing the fat globule surface (Lopez, 2011). This means that, if TAG crystallization occurs in one fat globule, it will not easily spread to the TAGs in the surrounding fat globules, at least when no shear is applied to the system. In AMF, once TAG crystallization begins, it rapidly spreads throughout the whole system because of the processes of secondary nucleation and crystal growth.
- Different TAG crystals were formed (nucleation in α form in cream vs. in β′ form in AMF; Lopez, Lesieur, et al., 2001), revealing the high role played by the dispersion of milk TAGs on their crystallization properties.

Cooling of Milk Fat Globules at 1–3 °C/min On cooling of cream from the melt at the rates of 1–3 °C/min (Fig. 9.13A and B), milk TAGs sequentially crystallize within milk fat globules in three different lamellar structures (Lopez, Bourgaux, Lesieur, Bernadou, et al., 2002). From about 19 °C, α-2*L* (47 Å) and α-2*L* (42 Å) crystals were formed and below 15 °C the crystallization of TAGs in the α-3*L* (71 Å) organization was recorded (Fig. 9.13A). On subsequent heating from −10 to 60 °C at 2 °C/min (Fig. 9.13C and D), the α-3*L* (71 Å) crystals melt at about 16 °C and some TAGs are involved in the formation of α-3*L* (65 Å) organization before its melting at about 20 °C. The α-2*L* (47 Å) and α-2*L* (42 Å) structures also melt around 20 °C. Between about 17 and 20 °C, crystallization of a new lamellar structure β′-2*L* (39 Å) occurred with structural reorganization leading to an increase of its thickness up to 41.7 Å. This high-melting point organization of TAGs in their solid state, likely formed by the reorganization of the TAGs initially incorporated in the crystals, progressively melted from about 21 °C and disappeared over 38 °C. On cooling at 1–3 °C/min, similar crystalline structures are formed by TAGs within milk fat globules of cream (Fig. 9.13) and in AMF (Fig. 9.5). However, the thickness of the α-3*L* organization is slightly thicker in AMF (72.5 *versus* 72.1 Å) and the small-angle XRD peak widths were larger in cream, showing defects of longitudinal stacking in α-3*L* crystals within milk fat globules. DSC curves recorded on cooling showed differences between cream and AMF (Fig. 9.13B, Fig. 9.5C). Crystallization of TAGs in AMF is induced at higher temperature with a sharp exotherm at about 18 °C, related to crystal growth of α-2*L* crystals.

Tempering of milk fat globules and emulsion droplets (i.e. successive cooling and heating) at controlled temperatures allows tailoring TAG crystals. In the previous paragraph, we discussed the formation of α-3*L* (71 and 65 Å) and α-2*L* (47 and 42 Å) crystals within milk fat globules after cooling at 1–3 °C/min from the melt. The subsequent heating of the cream to about 17–20 °C leads to the melting of these

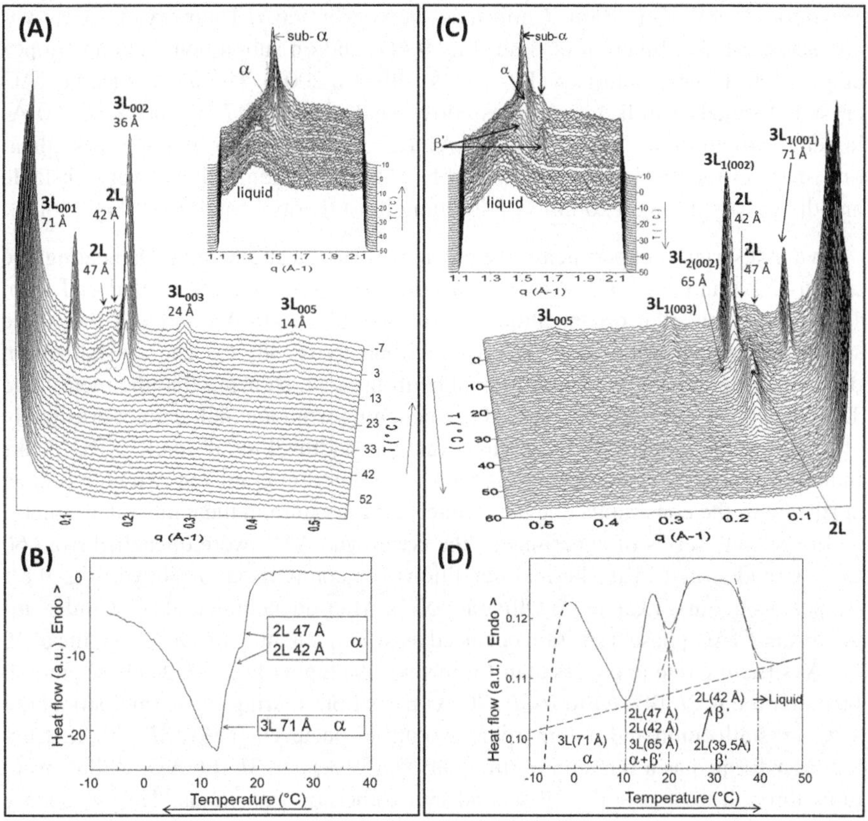

Fig. 9.13 Crystallization of triacylglycerols within milk fat globules characterized on cooling at 1 °C/min. SR-XRD patterns recorded at small and wide (insert) angles (**A**) on cooling at 1 °C/min from 60 to −10 °C, and (**C**) on subsequent heating at 2 °C/min. DSC thermograms recorded (**B**) on cooling and (**D**) on heating. DSC and SR-XRD were recorded simultaneously. Adapted from Lopez, Lesieur, et al. (2005)

TAG crystals and to crystallization of a new lamellar structure β′-2*L* (39 Å). The stabilization of temperature at about 20 °C and then cooling of the cream to 4 °C favours the growth of β′-2*L* (39 Å) crystals. Tempering of cream leads to reorganization of TAGs within emulsion droplets and can have implications on the physical stability and functional properties of milk fat globules (e.g. in the manufacture of butter, whipped cream).

Rapid Cooling of Milk Fat Globules In the dairy industry, milk and cream are often submitted to high-temperature thermal treatments (e.g. pasteurization) and rapidly cooled in a tank for storage before processing. Understanding the crystallization behaviour of milk TAGs within milk fat globules during this thermal history is therefore of industrial relevance. The most unstable crystalline structures of the TAGs formed within milk fat globules were studied after quenching from the melt down to −8 or 4 °C (~1000 °C/min), as shown Fig. 9.6A. Their organization was characterized as a function of time (Fig. 9.6D) and on subsequent heating (Lopez et al., 2000; Lopez, Bourgaux, Lesieur, & Ollivon, 2002). The most unstable TAG crystals formed in milk fat globules correspond to α-2*L* (47 Å) and α-3*L* (70 Å) lamellar structures, as in AMF (Fig. 9.6A). Due to the rapid cooling, the crystallization starts in the metastable α-polymorph. The α-2*L* organization is very unstable and disappears during a 20 min conditioning in isothermal conditions (Fig. 9.6D).

In a first series of experiments, the cream was heated at 2 °C/min. On heating, the α-2*L* (47 Å) and α-3*L* (70 Å) lamellar structures progressively melted and from 13 °C a new lamellar organization was formed, β′-2*L* (38 Å), and was the single solid TAG organization until its final melting above 38 °C. The crystallization occurring in emulsion is similar to in anhydrous state, showing that the metastable TAG molecular packings obtained after quenching from the melt are not drastically affected by the dispersion state. However, the width of small-angle XRD peaks indicated that TAG crystallization is more disordered in emulsion, which has been attributed to the constraints due to the interface curvature in the emulsion droplets.

In a second series of experiments, the cream and AMF were quenched from 60 to 4 °C and stored in isothermal conditions for comparison, as shown Fig. 9.8B (Lopez, Bourgaux, Lesieur, & Ollivon, 2002). After quenching at 4 °C, similar liquid to solid TAG phase transition occurred as after quenching to −8 °C, but the α-2*L* (47 Å) structure remained less than 1 min since the higher liquid TAG phase amount favoured the α-2*L* (47 Å) to α-3*L* (70 Å) transition. During isothermal storage at 4 °C, crystallization and polymorphic evolutions occurred. The α-3*L* (70 Å) structure transformed as a function of time into 2*L* (39 Å) and 3*L* (66 Å) lamellar structures through α → α + β′ + β secondary exothermic transitions. This secondary process is faster in AMF (within 30 min) than in cream (>1 h). The delayed polymorphic evolution observed in milk fat globules shows that the dispersion state of TAGs plays a role in the transition process and could be explained by a lack of stable nuclei in each fat globule at 4 °C as compared to AMF. Similar conclusions were drawn by comparing the isothermal crystallization behaviour at 5 °C of milk fat in bulk and emulsified state in natural and recombined creams (Fredrick et al., 2011). It is important to note that this exothermic transition leads to an increase in temperature during the storage of cream at the industrial level.

After 4 days storage at 4 °C, similar TAG crystals were characterized within milk fat globules and in AMF, i.e. coexistence of 2*L* (40 Å) and 3*L* (54 Å) lamellar structures corresponding to α, β'_1, β'_2 and β polymorphs (Fig. 9.8D). This is in line with Söderberg, Hernqvist, and Buchheim (1989) who reported that the main structure formed by the TAGs in milk fat globules after long-time storage at low temperature was a bilayer with a thickness of 3.9 nm. Also, these TAG crystals are similar to those characterized in butter (Buldo, Kirkensgaard, & Wiking, 2013; Ronholt, Kirkensgaard, Mortensen, & Knudsen, 2014). At 4 °C, the HMPF and the MMPF of milk fat will crystallize and constitute the solid TAG phase while the LMPF is present is the liquid TAG phase. Therefore, the solid TAG phase primarily consists of TAGs containing three long-chain saturated FAs and TAGs containing two long-chain saturated FAs and a long chain unsaturated FA or a short-chain saturated FA (Fig. 9.2).

The existence of an isothermal polymorphic evolution both in milk fat globules and in AMF was demonstrated (Lopez, Bourgaux, Lesieur, & Ollivon, 2002). Density measurements were sensitive to the α to β′ and β′ to β polymorphic transitions occurring within milk fat globules following the quenching of cream at 4 °C (Lopez, Bourgaux, Lesieur, & Ollivon, 2002). The increase in density, that corresponds to an increase in the compactness of milk TAGs, is explained by a reduction of the sub-cell volume from about 25.5 A^3 (α form) to 23.5 A^3 (β′ form) and a possible increase in the amount of TAGs crystallised.

After 6 days of storage at 4 °C, the cream and AMF were heated to 60 °C at 2 °C/min. The 3L crystals corresponding to α, β'_2 and β polymorphs melted first, and from about 23 °C the solid TAG phase corresponded to a β'_1-2*L* (40 Å) organization until its final melting. The DSC curves recorded simultaneous on heating showed to main endotherms corresponding to the successive melting of 3*L* and 2*L* crystals.

As a summary concerning the crystallization behaviour of milk TAGs in cream (i.e. concentrated milk fat globules), DSC curves recorded on cooling show two overlapped exothermal events, i.e. a small event due to the crystallization of 2*L* forms and a broad event related to crystallization of 3*L* structures identified thanks to the coupling with XRD. Whatever the cooling rate of cream, the DSC curve recorded on subsequent heating exhibits three main endotherms more or less overlapped corresponding to the LMP, MMP and HMP fractions of TAGs (Lopez, Bourgaux, Lesieur, Bernadou, et al., 2002). An exothermic event corresponding to α → β′ polymorphic evolution can be recorded between the first two endotherms after fast cooling of milk fat globules. As for AMF, the LMP fraction corresponds to melting of 3L lamellar structures, the MMP fraction to the melting of 2*L* lamellar structures formed on cooling. The HMP fraction corresponds to the progressive melting of the β′ 2*L* lamellar structures formed during heating until final melting of TAGs dispersed within fat globules of cream. These studies dedicated to the crystallization properties of milk TAGs in emulsion showed that milk TAGs are partially crystallized and that the solid TAG phase displays a complex polymorphism which is temperature and time dependent. The polymorphic transitions that occur in the solid TAG phase are favoured by the liquid TAG phase.

Figure 9.14 shows the crystalline structures formed within milk fat globules at −8 °C after cooling at different rates and the crystals formed upon storage at 4 °C. As for AMF, the comparison of the small number of crystal type formed to the large number of TAGs present in milk fat provides evidence that mixed crystals are formed within milk fat globules (i.e. co-crystallization of different TAG molecules). On cooling, the first longitudinal organizations of TAG molecules dispersed within fat globules correspond to 2*L* structures with long spacings of about 40–42 Å and 46–48 Å. These crystalline varieties may correspond to crystallization of HMP fractions of milk TAGs. These 2*L*-type crystals are generally formed by TAGs with saturated and similar chain length FAs, such as MPP and PPP (Fig. 9.9C). Then, crystallization of 3*L* structures (about 70–72 Å) occurs. The formation of a 3*L* (65 Å) structure was only observed on slow cooling of milk fat globules (0.15 °C/

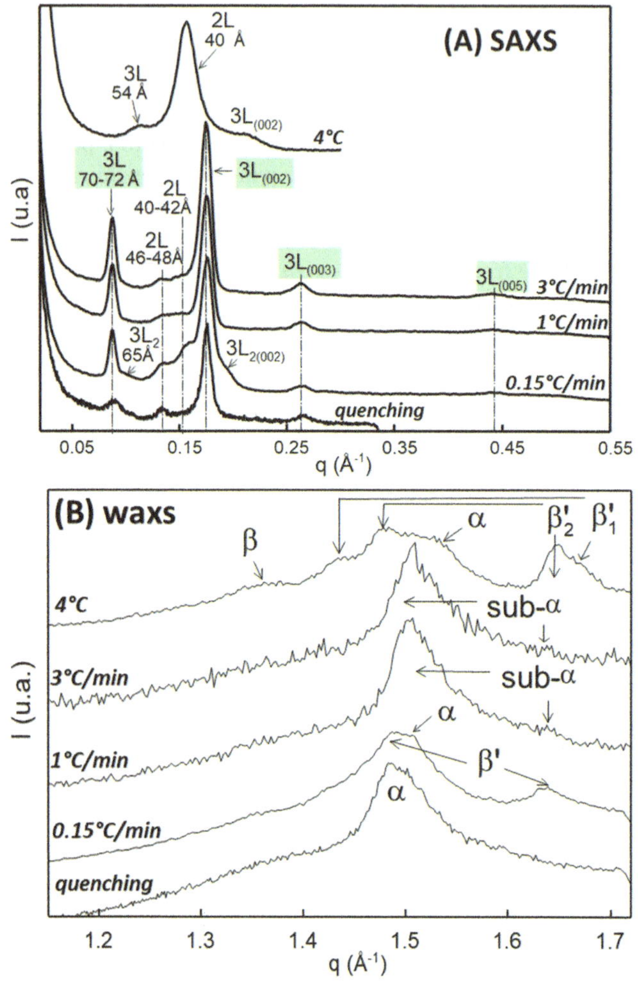

Fig. 9.14 SR-XRD patterns recorded at −8 °C: (**A**) at small angles (SAXS) and (**B**) at wide angles (WAXS) after either cooling of milk fat globules concentrated in cream at different rates as indicated on the figure or isothermal conditioning at 4 °C during 5 days. Adapted from Lopez, Bourgaux, Lesieur, Bernadou, et al. (2002)

min). Triple chain length (3*L*) stackings may correspond to crystallization of unsaturated TAG molecules or to that of TAGs with FA chains of different lengths, like BPP, OPP (Fig. 9.9C). Whatever the cooling rate, initial crystallization occurs in a hexagonal packing (α form). Then, as a function of the decrease in temperature, the formation of β′ form and the coexistence of α and β′ polymorphic forms was observed. Traces of β form crystals were only recorded on slow cooling of milk fat globules (0.15 °C/min) and after at least 3 days storage at 4 °C. According to Lopez, Bourgaux, Lesieur, and Ollivon (2002), stabilization of the TAG crystals is only attained after at least 4 days storage at 4 °C. On heating, the crystals formed on cooling (α 2*L*, α 3*L*) melt and reorganizations of crystals take place within milk fat globules with the formation of a β′-2*L* (40 Å) structure, accompanied or not by α-3*L* (54 Å). Recrystallization also occurs during isothermal storage. The stable crystalline structures take advantages of the melting of the metastable crystals which melt first or of some kind of Ostwald ripening occurring within the milk fat globules thanks to the liquid TAG phase that coexists with the solid TAG phases. The β′-2L mixed crystals selectively melt starting with the TAGs with shorter chains as shown by a progressive increase on their thickness. As for AMF, low amount of β crystals have been identified in milk fat globules, even after long time storage at low temperature. The mixed TAG crystals formed in milk fat organise in the less polymorphic forms β′ and α because molecular packing is not very dense.

4.2 *Effect of the Size of Milk Fat Globules and Lipid Droplets*

Studies on milk fat globules and protein-coated lipid droplets showed that the temperature of the initiation of TAG crystallization is lowered with decreasing size (Fig. 9.15), due to increased supercooling (Lopez, Bourgaux, Lesieur, Bernadou,

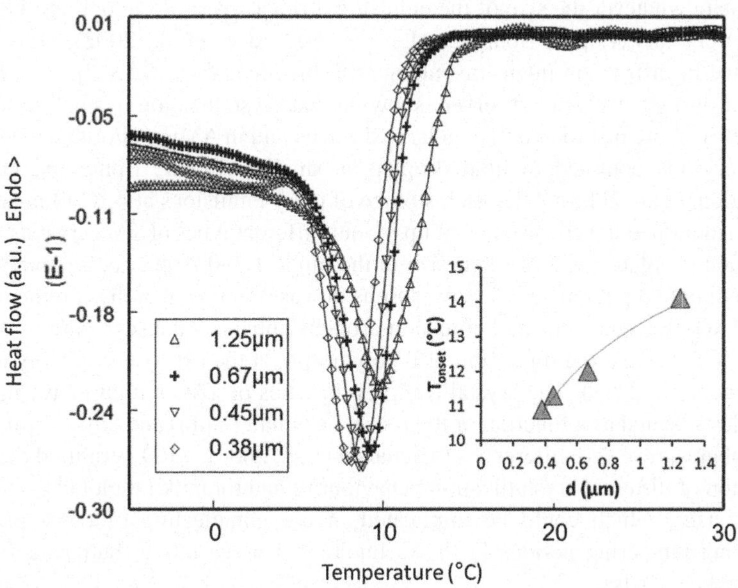

Fig. 9.15 Effect of the size of O/W emulsion droplets on the initial temperature of milk fat crystallization (T_{onset}). DSC thermograms recorded on cooling of protein-coated milk fat droplets of different sizes and changes in T_{onset} as a function of size (insert). Adapted from Lopez, Bourgaux, Lesieur, Bernadou, et al. (2002)

et al., 2002; Michalski et al., 2004; Truong, Bansal, Sharma, Palmer, & Bhandari, 2014). This can be explained by the theory of nucleation for crystallization. When the liquid TAGs are divided in droplets, as in an emulsion, not all the TAG droplets may contain the catalytic impurities required to start heterogeneous nucleation. The number of catalytic impurities per unit volume may be far too low to produce nuclei in every emulsion droplet, and considerable supercooling may occur. Then, it is accepted that, on cooling, TAG crystallization in the smaller globules is delayed compared to the larger ones. Assuming that there is no difference of composition from a milk fat globule to another, the major influence in TAG crystallization properties is that of fat droplet size.

It was reported that milk TAGs dispersed in small milk fat droplets have a lower final melting temperature compared to large droplets, in relation to the crystal structure (Bugeat et al., 2011). Furthermore, small droplets (0.18 μm) have a lower melting enthalpy than larger droplets (1.7 μm), which could be related to a lower solid TAG content in small droplets (Bugeat et al., 2011). Truong et al. (2014) also reported a strong tendency towards decreasing proportion of milk fat crystallinity with smaller droplet size stabilised by dairy proteins.

To elucidate the effects of O/W emulsion droplet size on the organization of TAG molecules, crystallization behaviour of protein-coated lipid droplets homogenized at different pressures were examined (Bugeat et al., 2011; Lopez, Bourgaux, Lesieur, Bernadou, et al., 2002; Truong et al., 2014). On cooling at 1 °C/min from 60 to −7 °C, similar lamellar structures were formed by milk TAG molecules in their solid state whatever the size of the emulsion droplets from 1.3 to 0.4 μm: i.e. α-3*L* (72 Å) crystals (Lopez, Bourgaux, Lesieur, Bernadou, et al., 2002). However, a decrease in diffraction intensities along with broaden of the SAXS peak width in smaller droplet size were observed, showing that crystallization in small milk TAG droplets is more disordered than in large droplets and in AMF and/or that the size of TAG crystals confined in lipid droplets is smaller (Lopez, Bourgaux, Lesieur, Bernadou, et al., 2002). After 48 h storage of dairy emulsions at 4 °C, Bugeat et al. (2011) identified the coexistence of up to four different types of TAG crystals within emulsion droplets whatever their size in the range 1.7–0.2 μm, i.e. 2*L* and 3*L* corresponding to β'_1, β'_2, β_1 and β_2 polymorphs. It was observed that the confinement of milk TAG molecules in small emulsion droplets enhanced the segregation of some types of TAGs and the formation of β polymorph, at the expense of β′ polymorphs (Bugeat et al., 2011). The crystallization properties of TAG within native milk fat globules selected as a function of their size, i.e. small (1 μm) and large (7 μm) milk fat globules, revealed differences (Michalski et al., 2004). XRD permitted the identification of different crystallization behaviour in natural milk fat globules with different sizes, which could be implicated in the manufacture of dairy products involving tempering periods in the technological process (e.g. butter, ice-cream, whipped products).

5 Crystallization Properties of Milk Fat in Dairy Products

The thermal properties and crystallization behaviour of milk fat in complex food products (e.g. butter, whipped cream, ice cream, cheeses) have been investigated since TAG crystals can impact the physical stability, texture, sensorial properties and acceptability by the consumer.

During manufacture of butter, milk fat globules concentrated in cream are first subjected to a specific time-temperature program to obtain partially crystallized fat globules and subsequently, the cream is exposed to a severe mechanical agitation in which fat globules destabilize through a mechanism known as partial coalescence for which TAG crystals are indispensable. Butter consists of a continuous fat phase in which water droplets, residual milk fat globules and a network of fat crystals are dispersed (Lopez, Cauty, & Guyomarc'h, 2015). The mechanical properties of butter (i.e. its consistency, spreadability, firmness), appearance and mouth-feel depend not only on the ratio of solid to liquid TAGs but also, to a large degree, on the size, shapes and spatial distribution of the TAG crystal network, that depend on the milk TAG composition and crystallization behaviour of TAG molecules. Both compositional and processing conditions influence crystallization of TAGs in butter.

The FA composition of butter, which can be affected by seasonal variations and cow diet (Fig. 9.1A), affects its crystallization and melting properties with consequences on the texture (Lopez et al., 2007). Butter produced during the period of the year when cows are fed a maize-silage based diet (i.e. during the winter) tends to have a higher level of palmitic acid C16:0 and less oleic acid C18:1c9 than butter produced when the cows are fed a grass-based diet (i.e. during spring; Fig. 9.1A). This results in a firmer consistency of butter in winter. Processing conditions (temperature, cooling rate, scale of operation, agitation, storage conditions) can impact crystallization and ultimately the rheology of butter (Ronholt et al., 2014). For example, rapid cooling of cream leads to the formation of many small TAG crystals, a higher solid fat content and a firmer texture of butter. It is well-known that thermal kinetics (i.e. tempering, cold-warm-cold processes also called physical ripening) are applied to cream in order to control the solid to liquid fat ratio and to govern the size and orientation of TAG crystals in milk fat globules before churning. Such treatment of cream prior to butter manufacturing largely determines the final textural characteristics of the butter. The rheological properties of milk fat and butter, in connection with the TAG crystal networks, are detailed elsewhere (Wright & Marangoni, 2006).

Among dairy products, cheeses have received special attention to better understand the role played by the physical properties of milk fat, especially the formation of TAG crystals. However, studies of the crystallization properties of milk fat in as such complex food products as cheeses remain scarce. It has been demonstrated that the liquid to solid milk TAG phase transition recorded by DSC on cooling of hard-type cheeses is sensitive to the microstructure of fat within the protein matrix, especially the destabilization of fat globules and the formation of non-emulsified fat during the manufacture of cheese (Lopez, Briard-Bion, Camier, & Gassi, 2006).

When cheese fat is dispersed in fat globules, the DSC profile is close to the recordings of cream, e.g. a single broad exotherm recorded on cooling from 60°C. When non-emulsified fat, also called free fat, is formed within the cheese matrix, the DSC profile evolves toward the behaviour of AMF, e.g. two successive exotherms recorded on cooling. Dairy products are stored at low temperature (e.g. 4–7 °C in the fridge), which raises questions about the crystallization of milk fat. Using synchrotron radiation XRD, Lopez, Briard-Bion, Beaucher, and Ollivon (2008) revealed the coexistence of several types of TAG crystals within hard-type cheese stored in the fridge and identified 2*L* (41 Å) and 3*L* (55 Å) longitudinal organizations of TAG molecules corresponding to the coexistence of α, β'_1, β'_2 and β polymorphic forms. These results obtained in cheese are in line with the TAG crystals formed upon long storage of AMF and cream at 4 °C. On heating of cheese previously stored at 4 °C, the DSC profile showed the three endotherms corresponding to the LMP, MMP and HMP fractions of milk fat (Fig. 9.2). Similar milk TAG crystals and melting behaviour have been characterized within the small size droplets of processed cheese (below 1 μm, Gliguem, Lopez, Michon, Lesieur, & Ollivon, 2011). The final melting point of milk fat within cheese, about 41 °C, is higher than the temperature of digestion in the gastro-intestinal tract of humans. The TAG that remain in their solid state above 37 °C, estimated to be about 3% of milk fat (Lopez, Briard-Bion, et al., 2006) and composed by long-chain saturated FAs such as palmitic acid (C16:0), could impact the digestibility of milk fat consumed in cheese.

The texture of processed cheese is a very important parameter affecting its acceptability by the consumer. Both microstructure and rheological properties of spreadable processed cheeses are strongly dependent on the properties of fat, mainly the amount and type of TAG crystals that can be formed at the temperature of storage and cheese consumption. A study combining DSC, XRD and rheology as a function of temperature demonstrated the influence of milk TAG crystallization, melting and polymorphism upon the viscoelastic properties of processed cheese (Gliguem et al., 2009). On cooling at 2 °C/min from 60 °C (Fig. 9.16), the crystals formed by milk TAG within processed cheese were observed from about 15 °C and corresponded to α-3*L* (72 Å) structures. These results were consistent with previous observations stating that crystallization in milk fat globules (4 μm in diameter) gives rise to a α-2*L* (42–47 Å) structure followed by α-3*L* (71 Å) structure, while in fat droplets ranging from 1.25 to 0.38 μm only the α-3*L* (72 Å) structure was observed (Lopez, Bourgaux, Lesieur, Bernadou, et al., 2002). The formation of α-3*L* (72 Å) crystals and the absence of α-2*L* crystals in processed cheese may be related to the small size of fat droplets (0.7 μm). The crystallization of milk TAG within the processed cheese matrix, characterized simultaneously by XRD and DSC (Fig. 9.16A and B), was related to an increase in the viscoelastic moduli G′ and G″ (Fig. 9.16C). These results showed that milk TAG crystals contribute to the firmness of processed cheese at low temperature, e.g. below 15 °C (Gliguem et al., 2009). On subsequent heating of the processed cheese, the α-3*L* (72 Å) crystals melted and from about 12 °C the formation of β′-2*L* (41 Å) crystals was characterized until its final melting over 38 °C. The successive melting of the 3*L* and 2*L* TAG crystals embedded in the

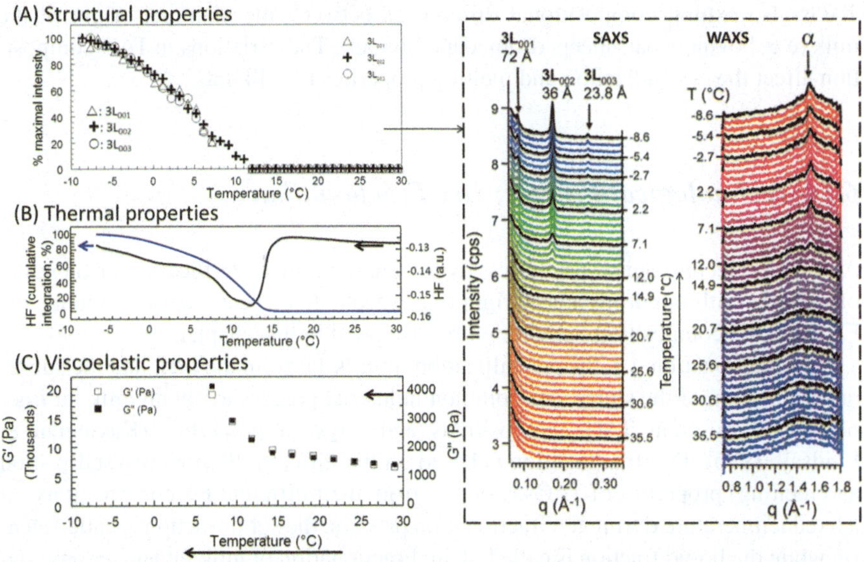

Fig. 9.16 Characterization of the crystallization of milk fat in processed cheese on cooling at 2 °C/min showing the consequences on the rheological properties. (**A**) Evolution as a function of temperature of the maximal intensities of the SR-XRD peaks recorded at small angles and corresponding to crystallization in α-3*L* (7.2 nm) structures as shown in the right part. (**B**) DSC curve recorded simultaneously on cooling and its cumulative integration as a function of temperature. (**C**) Changes in the viscoelastic moduli G′ and G″ of processed cheese on cooling. Adapted from Gliguem et al. (2009)

fat droplets was related to a successive decrease in the viscoelastic moduli recorded on heating of processed cheese.

These studies combining different biophysical techniques allowed the identification the milk TAG crystals formed within complex dairy products (i.e. butter, cheeses) and revealed polymorphic evolutions on heating. Moreover, the impact of milk fat crystals on the rheological properties fat-rich products was demonstrated.

6 The Crystallization Properties of Milk Fat Are Affected by the FA and TAG Compositions

The FA and TAG compositions of milk can be tailored for technological, nutritional and health reasons (e.g. increase in unsaturated FAs that are known to provide health benefits, and decrease in palmitic acid content that is involved in cardiovascular risk). In this respect, numerous techniques have been applied, including physical, chemical and dietary manipulation by means of feeding dairy animals. Technological treatments are often applied to have desired functionalities (e.g. improved cold spreadability of butter) and expend the use of milk fat in the food industry. Milk

TAGs also exhibit compositional differences between the mammal origin of the milk (e.g. bovine, goat, sheep, dromedary, human). The variations in TAG composition affect the crystallization and melting properties of milk fats.

6.1 Technological Process: Dry Fractionation

Among the technological processes, hydrogenation, interesterification or blending of milk fat with fats from other origin (e.g. vegetable oil) can be used to tailor the FA and TAG composition but will not be discussed in this chapter.

Dry fractionation, i.e. the crystallization of milk fat from the melt and the subsequent filtration of the slurry, is a common industrial process to obtain milk fat fractions with different TAG compositions and physical properties (Kaylegian & Lindsay, 1995). Dry fractionation is based on the different thermal (crystallization and melting) properties of TAGs resulting from their different FA compositions. At a fixed temperature during the fractionation process, the solid fraction is called stearin while the liquid fraction is called olein. Fractionation of milk fat and recombination of the fractions in various proportions allow to control and to improve the thermal and physical properties, e.g. its consistency and the development of cold spreadable butter (Kaylegian & Lindsay, 1995). Milk fat fractions are employed for pastry-making, as chocolate bloom inhibitor, butter flavour-rich concentrates and for improving the rheology of reduced fat cheese curds.

Many studies have focussed on the milk fat fractionation process, on the chemical and thermal characteristics of the separated fractions and on the phase behaviour of milk fat and its fractions (Marangoni & Lencki, 1998; Timms, 1980; Van Aken, ten Grotenhuis, van Langevelde, & Schenck, 1999). So far only a few studies have been reported about the structural characteristics of milk fat fractions investigated using time-resolved synchrotron radiation XRD as a function of temperature. The chemical composition, crystallization properties and melting behaviour of milk fat and its primary fractions, stearin and olein fractions, obtained by dry fractionation at 21 °C were characterized (Lopez, Bourgaux, Lesieur, Riaublanc, & Ollivon, 2006; Lopez & Ollivon, 2009). Compared to whole milk fat, the stearin fraction was enriched in TAGs with (1) three saturated long-chain FAs (SSS, PSS, PPS, PPP, MSS, MPS, MPP, SSL), (2) one or two saturated medium-chain FAs and a saturated long-chain FA (MMS, MMP, LaPS, LaPP, LaMP, CPS, CPP, CMS) and (3) one monounsaturated long-chain FA and two saturated long-chain FAs (PoSS, PPoS, LaOS, PSO). The olein fraction was enriched in TAGs with (1) two monounsaturated long-chain FA (SOO, PoSO, POO, MOO, PPoO), (2) one monounsaturated and two medium-chain FAs (CMO, CPO, LaMO, MMO, LaPO), (3) a short-chain FA and two saturated long-chain FAs (BMP, BPS, BPP, BMS, CaMP, CaPP, CaMS) or one monounsaturated FA (BPO, BMO) or two monounsaturated long-chain FAs (BOO). On cooling from the melt at 1 °C/min, milk fat showed the formation of two α-2*L* (47 and 42 Å) and one α-3*L* (72 Å) lamellar structures, as previously reported and discussed (Lopez, Lavigne, et al., 2001a; Fig. 9.5A). In similar experimental

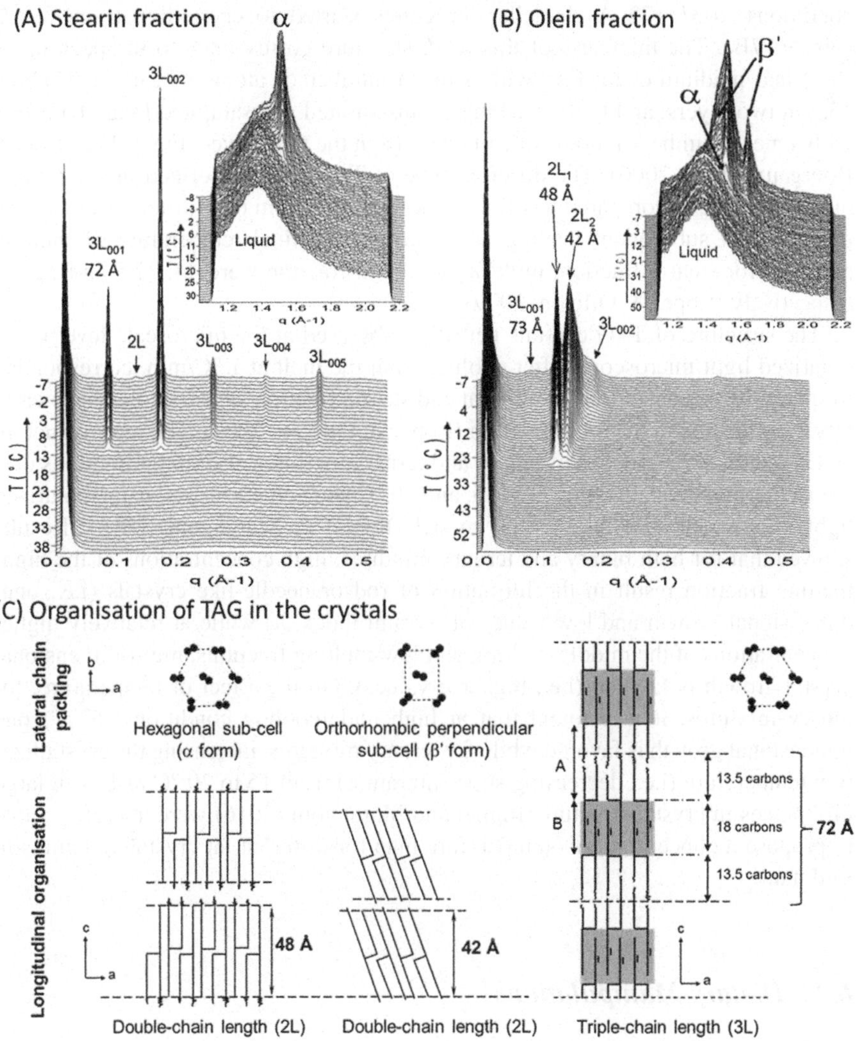

Fig. 9.17 TAG crystals formed in stearin fraction and olein fraction on cooling at 1 °C/min from 60 to −7 °C. SR-XRD patterns recorded at small and wide (insert) angles as a function of temperature (**A**) for stearin fraction and (**B**) for olein fraction. (**C**) Proposed TAG packing in the main crystals formed on cooling. The α-3L (7.2 nm) crystals in olein fraction correspond to the packing of short and medium-chain length fatty acids in layers A with unsaturated and saturated long-chain fatty acids in layers B. Adapted from Lopez, Bourgaux, et al. (2006)

conditions, the stearin fraction started to crystallize at 26 °C with the formation of two main lamellar structures, α-2*L* (48 Å; molecules arranged perpendicular to the methyl end group plane) and β′-2*L* (42 Å; tilt of the chains) (Fig. 9.17A). Then, from about 13 °C, a low amount of 3*L* (68 Å) crystals that may correspond either to α or β′ polymorphs were formed. In the olein fraction cooled in similar experimental conditions, α-3*L* (72 Å) lamellar structures started to crystallize from 13 °C (Fig. 9.17B). The thickness of this α-3*L* structure corresponds to the packing of short and medium-chain FAs with a mean number of atoms of carbon of about 13.5 in two layers, and to the packing of unsaturated and saturated long-chain FAs with a mean number of atoms of carbon of 18 in the third layer (Fig. 9.17C; Lopez, Bourgaux, et al., 2006). The different type of TAG crystals formed in stearin and olein fractions, as compared to whole milk fat, result from their different TAG composition. On subsequent heating at 2 °C/min, the final temperature of melting recorded for stearin fraction, milk fat and olein fraction were 44, 37.5 and 22 °C, respectively (Lopez & Ollivon, 2009).

The structure of TAG crystals networks observed at the microscale level using polarized light microscopy, after cooling from the melt at 1 °C/min, corresponded to spherulitic organizations in milk fat and stearin fraction while needle-shape crystals were formed in the olein fraction (Lopez & Ollivon, 2009). The microstructure and crystallization kinetics of binary and ternary mixtures of milk fat fractions during isothermal crystallization at 5, 15, and 20 °C were characterized using polarized light microscopy and the Avrami model (Ramel & Marangoni, 2016). Results showed that for both binary and ternary mixtures, high concentrations of the high-melting fraction result in the formation of rod or needle-like crystals (i.e., one-dimensional growth and low values of Avrami index, n) while at relatively higher concentrations of the middle-melting and low-melting fractions, multi-dimensional crystal growth is favored (i.e., higher n values). On the effect of temperature, for binary mixtures, it was found that at high undercooling conditions (5 °C) one dimensional growth is favored while for ternary mixtures, increasing the crystallization temperature (i.e., decreasing supersaturation) from 15 to 20 °C results in large differences in crystal structure. Ramel and Marangoni (2016) were, therefore, able to propose a concentration—temperature map for different fat crystal structures in milk fat.

6.2 Dietary Manipulations

Seasonal variations in the diet of cows naturally occurring in some countries, e.g. maize based diet in winter vs. fresh grass based diet in spring, affect the FA and TAG composition of milk (Fig. 9.1) and have consequences on the crystallization properties of milk fat and final texture of fat-rich products. Thus, the control of fat-rich product quality, e.g. butter, in different seasons is a real challenge for the industrials. The composition of milk fat can also be modified by specific feeding strategies and alter consequently the physical and functional properties of high-fat content dairy products. For example, feeding cows with highly unsaturated oils or whole oilseeds can reduce the level of saturated FAs while simultaneously increasing the unsaturated fatty acid (UFA) content. Several studies reported the improvement in the spreadability and softer texture of winter butter and milk fat in general through changes in the feed of the cow by adding unsaturated oils or fresh grass (Couvreur, Hurtaud, Lopez, Delaby, & Peyraud, 2006; Wright & Marangoni, 2006). Yoghurt, ice cream and cheeses made from milk enriched in UFAs have been reported to show a softer texture than the products made from control milk. The relationship between the FA composition of milk fat and the texture of dairy products has been demonstrated. However, few authors studied the effect of cow diet on the crystallization properties of milk fat enriched in UFAs (Bugeat et al., 2011; Bugeat et al., 2015; Smet et al., 2010). An increased amount of UFAs in milk TAGs was reported to decrease the solid fat content at 5 °C from 60% (control TAGs; 28% UFAs) down to 46% (UFA-enriched TAGs; 39% UFAs) (Smet et al., 2010). Upon isothermal crystallization monitored by pulsed NMR, higher content of the UFAs resulted in a slower nucleation, a longer induction time to crystallization and a lower solid fat content at the end of crystallization, although crystallization occurred according to similar α to β′ polymorphic transition (Smet et al., 2010).

Bugeat et al. (2015) compared the crystallization properties of control milk TAGs (29% UFAs) and UFA-enriched TAGs (51% UFAs obtained with a linseed oil rich diet) using the coupling of DSC with synchrotron radiation XRD (Fig. 9.18). On cooling from the melt at 3 °C/min, both milk TAG mixtures started to crystallize from about 16 °C in α-2*L* (45–49 Å) structures then formation of α-3*L* structures occurred with a higher thickness (75.5 versus 71.5 Å) and a delay (T_{onset} = 8.5 versus 12.1 °C) for UFA-enriched TAGs that result from a higher amount of C18:1c9. Groups of TAG molecules with high crystallization temperature (HCT; α-2*L* crystals) and low crystallization temperature (LCT; α-3*L* crystals) segregated on cooling. On subsequent heating, melting of TAG crystals and formation of a new 3*L* (65–80 Å) and β′-2*L* (40–44 Å) crystals associated with polymorphic reorganizations have been characterized. Increase in thickness of the lamellar structures was characterized for UFA-enriched TAGs as compared to control TAGs, demonstrating differences in the FA composition of the crystals. Interestingly, the melting profile of the UFA-enriched TAGs was mainly altered in the range 11–21 °C, corresponding to the MMP fraction, and not in the HMP fraction since the final melting temperatures of both the control and the UFA-enriched TAGs were similar.

It has also been demonstrated with the same milk fats that the enrichment of UFAs in TAGs decreases the solid fat content and affects the type of crystalline structures that are formed within O/W emulsion droplets upon storage at 4 °C (Bugeat et al., 2011). Control TAGs were crystallized in 2*L* (39.5 nm) and 3*L* (56.6 nm) lamellar structures with four polymorphic forms (β_1, β_2, β'_1, β'_2) while UFA-enriched TAGs were crystallized in 2*L* (41.8 nm) lamellar structures displaying three polymorphic forms (β_1, β'_1, β'_2). The absence of 3*L* crystals in the UFA-enriched TAG emulsions was due to decrease in the melting point of these TAG crystals rich in UFAs (Saturated-Unsaturated-Unsaturated TAGs vs. Saturated-Saturated-Unsaturated TAGs in control milk fat) that remain in the liquid TAG phase upon storage of the emulsion droplets at 4 °C.

As a conclusion, the enrichment of milk TAGs in UFAs affects both their crystallization and melting behaviours.

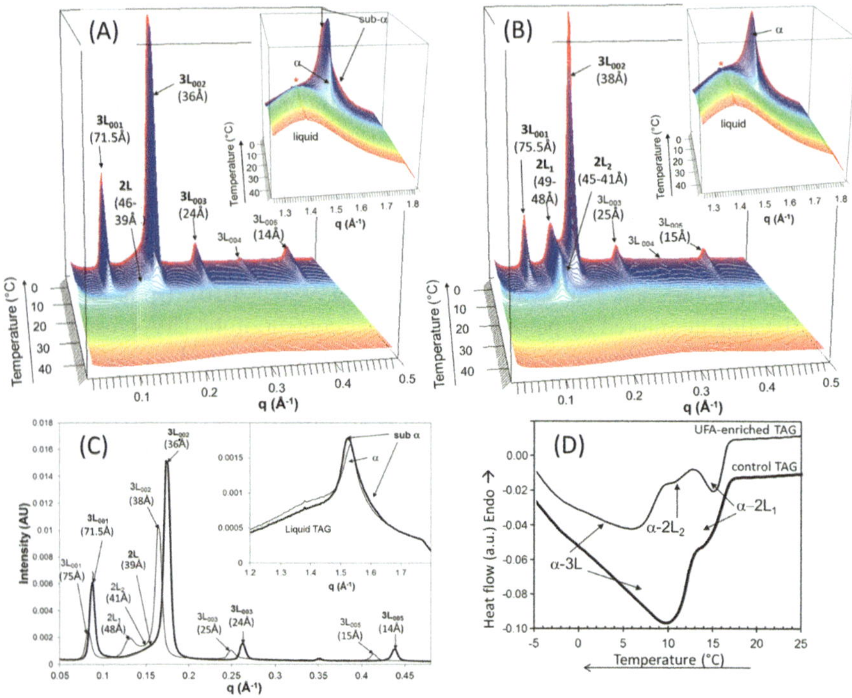

Fig. 9.18 Comparison of the crystallization behavior of unsaturated fatty acid (UFA)-enriched TAGs and control TAGs examined on cooling at 3 °C/min from 60 to −5 °C using the coupling of SR-XRD and DSC. SR-XRD patterns recorded at small and wide (insert) angles during cooling of (**A**) control TAGs and (**B**) UFA-enriched TAGs. (**C**) XRD patterns recorded at −5 °C for control (thin line) and UFA-enriched (thick line) TAGs after cooling. (**D**) DSC curves recorded simultaneously to XRD experiments. Adapted from Bugeat et al. (2015)

6.3 *Milk Fat from Various Mammal Species*

Most of the studies about milk fat crystallization have been performed with bovine milk fat since bovine milk represents about 84% of the global worldwide milk production (information from the International Dairy Federation). However, the FA and TAG compositions of milk depend on mammal species and changes in milk fat composition can affect the crystallization and melting properties between milk fats from various origins (i.e. goat, sheep, water buffalo, donkey, horse, camel; Smiddy, Huppertz, & van Ruth, 2012). The crystallization properties of milk TAGs have been characterized by the coupling of DSC and synchrotron radiation XRD in AMF and fat globules from goat milk (Ben Amara-Dali et al., 2007; Ben Amara-Dali, Lopez, Lesieur, & Ollivon, 2008), dromedary milk (Karray, Lopez, Lesieur, & Ollivon, 2005; Lopez, Karray, Lesieur, & Ollivon, 2005) and human milk (Lopez, Briard-Bion, Bourgaux, & Perez, 2013).

6.3.1 Crystallization Properties of Goat Milk Fat

Goat milk fat globules have a mean diameter of about 3–3.5 μm. They are rich in saturated FAs, about 70% of total FAs, and the five FAs C10:0, C14:0, C16:0, C18:0 and C18:1c9 account for more than 75% of total goat milk FAs (Ben Amara-Dali et al., 2008). The most important TAGs present in goat's milk fat contain medium-chain length saturated FAs (C8:0, C10:0, C12:0) and C18:1c9 as unsaturated FA. The molecular organization of the solid TAG phase formed within goat milk fat globules was investigated on cooling at the rates of 0.1 °C/min (slow cooling) and 1000 °C/min (quenching) and on subsequent heating at 1 °C/min. The lamellar structures 3*L* (69–70 Å) and 2*L* (37–45 Å) were characterised and the five polymorphic forms α, sub-α, β'_1, β'_2 and β were identified. The two main types of crystals correspond to a segregation of goat TAG molecules in the solid state as a result of different compositions, as observed for bovine TAGs. Polymorphic transitions were observed within goat's milk fat globules as a function of time after quenching from the melt and as a function of temperature on heating. Increasing the knowledge about the physical properties of goat's milk fat is essential to improve the quality of existing dairy products and to increase the technolocal application of goat's milk fat crystallization to contribute in the development of new food products.

6.3.2 Crystallization Properties of Dromedary Milk Fat

In camel milk, fat represents about 3.6% of the composition. Differences exist in the FA composition of camel milk compared to bovine milk. Short-chain FAs (C4:0–C12:0) are present in very small amounts and the amount of long-chain saturated FAs (C14:0–C22:0) is higher for camel milk fat than those for bovine milk fat. These differences lead to the formation of specific TAG crystals.

On cooling of anhydrous dromedary milk fat from the melt, the crystalline structures formed by TAG molecules correspond to 2*L* type crystals. The absence of short-chain FAs in dromedary milk fat prevents the formation of 3*L*-type crystals. On cooling at 1 °C/min, two successive 2*L* crystals are formed, α 2*L* (47 Å) from 24 °C and β′ 2*L* (42.2 Å) from 21 °C. These crystals successively melt on subsequent heating. On fast cooling at 5 °C/min a four-chain length longitudinal organization 4*L* (84.5 Å) was characterized (Karray et al., 2005). On slow cooling at 0.1 °C/min, the crystals formed from about 29.5 °C correspond to a lamellar structure with a double-chain length longitudinal organization of the TAG molecules (2*L* = 42.3 Å) associated with a β′ lateral packing of the chains. These crystals progressively melt on subsequent heating and disappear above 42 °C. Investigations of TAG crystallization within dromedary milk fat globules revealed the successive formation of two double-chain length (2*L*) lamellar structures: α $2L_1$ (46.7 Å) from 22 °C and β′ $2L_2$ (41.7 Å) from 9 °C, which coexist until the end of the cooling process. The same lamellar structures, α $2L_1$ and then β′ $2L_2$, are formed on cooling in the dispersed and bulk states. However, crystallization in the unstable α form is favoured in fat globules (Lopez, Karray, et al., 2005). Increasing the knowledge about dromedary milk fat crystallization contributes in developing the technological applications and textural properties of creams and dromedary milk fat-rich products, for example butter, that strongly depend on the thermal and structural properties of dromedary milk TAGs.

6.3.3 Crystallization Properties of Human Milk Fat

Human milk contains about 3–5% fat dispersed in fat globules having a mean diameter of about 5 μm. Human milk TAGs that contain 48–57% saturated FAs with about 28% of C16:0 contribute some 40–55% of the total energy intake for the breast-fed infants. The efficient digestion of TAGs is therefore of primary importance for the optimal growth of newborns. However, storage of breast milk in the fridge at 4 °C leads to the partial crystallization of TAGs within milk fat globules (Lopez et al., 2013). Microscopic observations of breast milk stored at 4 °C revealed the non-spherical distorted shape of fat globules due to the presence of TAG crystals (Fig. 9.19A). Synchrotron-radiation XRD experiments allowed the identification of

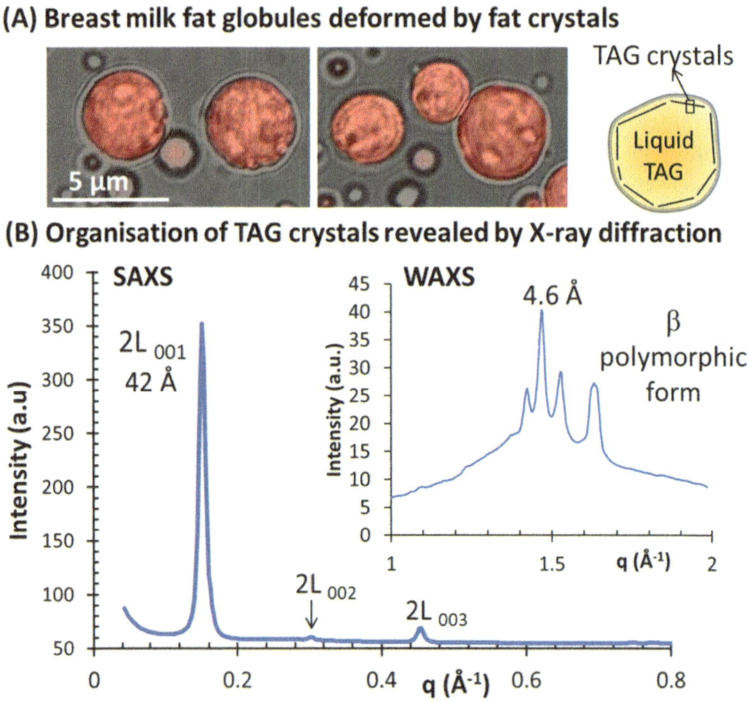

Fig. 9.19 TAG crystals formed within human milk fat globules upon storage at 4 °C. (**A**) Polarized light microscopy images showing the deformation of milk fat globules by TAG crystals, (**B**) Identification of the organization of TAG molecules performed using SR-XRD at both small and wide (insert) angles. Adapted from Lopez et al. (2013)

the TAG crystals that are formed within breast milk fat globules upon storage at 4 °C, i.e. β-2*L* (41.7 Å) lamellar structures (Fig. 9.19B). The β crystals correspond to the most thermodynamically stable polymorphic form of TAGs with a compact organization of the FA chains. The crystals formed in human milk fat globules upon storage at 4 °C are different from those characterized in bovine milk, which confirms that the chemical composition of milk TAGs govern their crystallization properties. The final melting point of the β-2*L* (41.7 Å) human TAG crystals was 41.1 ± 1.6 °C, which is above the in-body temperature of milk digestion by newborns.

The presence of solid TAGs in the core of breast milk fat globules after storage at 4 °C raises the question of the action of the digestive lipolytic enzymes on a solid substrate, on their solubilization and then on the absorption and metabolism of milk lipids. The influence of the physical state of TAGs and particularly the proportion of solid TAGs and type of crystals on lipid digestion and absorption remains poorly documented. Lopez et al. (2013) hypothesized that crystallization of milk TAGs could decrease the amount of utilizable fat for the recipient infant in the case breast milk is consumed at a temperature below the final melting temperature of TAG crystals. Warming breast milk at about 45–50 °C, i.e. above the final melting point of human β-2L TAG crystals, is then important to ensure optimal breast milk TAG digestibility.

7 Liquid TAG Phase

For temperatures above the melting point of milk TAGs, the liquid TAG phase exhibits an organization. Within milk fat globules, the synchrotron radiation X-ray patterns of milk TAGs in their liquid state correspond to scattering peaks at both small (SAXS) and wide (WAXS) angles, respectively centred at 22.4 nm and 4.5 Å (Lopez, Lavigne, et al., 2001a). In anhydrous milk fat and the primary fractions, stearin and olein, the synchrotron radiation X-ray scattering from the liquid-crystalline organization of milk TAGs in their liquid state recorded at 60 °C was centered at 23.3 Å (SAXS) and about 4.5 Å (WAXS) (Lopez & Ollivon, 2009). Differences were characterized as a function of the FA composition of TAGs, with a higher thickness *d* for UFA-enriched TAGs compared to control TAGs (d = 2.26 ± 0.01 vs. 2.21 ± 0.01 nm; Bugeat et al., 2015). The thickness value d of 22.1–23.3 Å supports the existence of liquid-crystalline like lamellae and corresponds to the stacking of TAGs in a single layer of the acyl chains along the long-chain axis integrating FAs with different chain length (from 4 to 18 atoms of carbon) and unsaturation (Fig. 9.20). The scattering peak recorded at about 4.5 Å corresponds to a disordered mesophase with short-range order of the FA chains. These synchrotron radiation XRD data support evidence that complex TAG blends such as milk TAGs display anisotropy with a lamellar ordering in the liquid state.

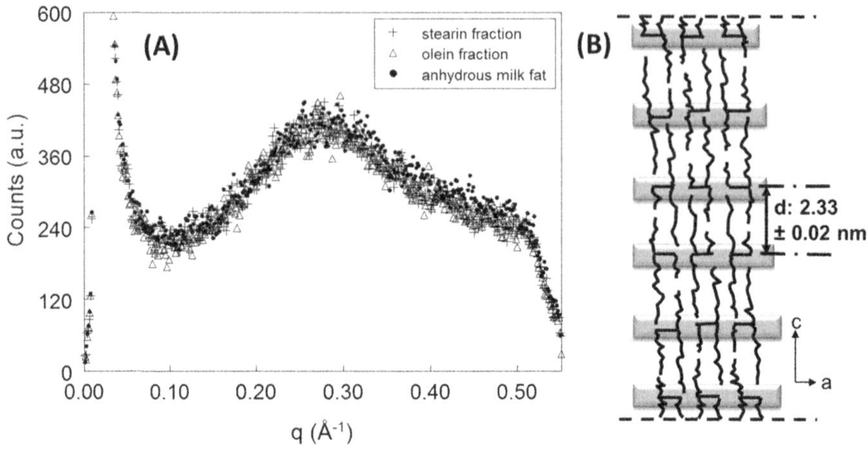

Fig. 9.20 Structural information on the liquid phase of milk TAG molecules. (**A**) SR-XRD patterns recorded at small angles with milk fat, olein fraction and stearin fractions at 60 °C. (**B**) Proposed structure for the molecular packing of TAG molecules in their liquid state, as seen in the ca projection. Short, medium and long-chain fatty acids, saturated and unsaturated fatty acids are stacked in monolayers, between glycerol groups. Adapted from Lopez and Ollivon (2009)

8 Conclusions

Crystallization of TAGs is a complex phenomenon, especially for milk fat due to its extremely wide FA composition leading to many TAG molecular species. As reviewed in this chapter, extensive research has provided considerable insight into the crystallization properties of milk fat in the anhydrous state, in emulsion (natural milk fat globules, processed lipid droplets, recombined cream) and in complex dairy products (butter, cheeses). This book chapter highlights the recent research demonstrating from scientific points of views that the crystallization properties of milk fat are affected by (1) its FA and TAG compositions, (2) cooling rates and tempering, (3) shear, (4) presence of minor lipid compounds (FFAs, MAGs, DAGs, phospholipids), (5) its dispersion state, i.e. anhydrous bulk *versus* emulsified in numerous droplets. Polymorphic evolutions have been characterized as a function of temperature on heating and in isothermal conditions, e.g. after rapid cooling from the melt. Understanding the functional properties of milk TAG crystals networks requires investigations at several scale levels (microscopic level, nanoscale, molecular scale) performed as a function of temperature or as a function of time in isothermal conditions and then the combination of complementary techniques (rheology, polarised light microscopy, electron microscopy, NMR, DSC, XRD including USAXS, SAXS and WAXS). Undoubtedly, the pursuit of fundamental knowledge in the area of TAG crystallization can yield fascinating new insights that will increase further the value of milk fat in food applications.

References

Bayard, M., Leal-Calderon, F., & Cansell, M. (2017). Free fatty acids and their esters modulate isothermal crystallization of anhydrous milk fat. *Food Chemistry, 218*, 22–29.

Ben Amara-Dali, W., Lesieur, P., Artzner, F., Karray, N., Attia, H., & Ollivon, M. (2007). Anhydrous goat's milk fat: Thermal and structural behaviors studied by coupled DSC and X-ray diffraction. 2. Influence of cooling rate. *Journal of Agricultural and Food Chemistry, 55*, 4741–4751.

Ben Amara-Dali, W., Lopez, C., Lesieur, P., & Ollivon, M. (2008). Crystallization properties and polymorphism of triacylglycerols in goat's milk fat globules. *Journal of Agricultural and Food Chemistry, 56*, 4511–4522.

Bugeat, S., Briard-Bion, V., Pérez, J., Pradel, P., Martin, B., Lesieur, S., et al. (2011). Enrichment in unsaturated fatty acids and emulsion droplet size affect the crystallization behaviour of milk triacylglycerols upon storage at 4°C. *Food Research International, 44*, 1314–1330.

Bugeat, S., Perez, J., Briard-Bion, V., Pradel, P., Ferlay, A., Bourgaux, C., et al. (2015). Unsaturated fatty acid enriched vs. control milk triacylglycerols: Solid and liquid TAG phases examined by synchrotron radiation X-ray diffraction coupled with DSC. *Food Research International, 67*, 91–101.

Buldo, P., Kirkensgaard, J. J. K., & Wiking, L. (2013). Crystallization mechanisms in cream during ripening and initial butter churning. *Journal of Dairy Science, 96*, 6782–6791.

Campos, R., Narine, S. S., & Marangoni, A. G. (2002). Effect of cooling rate on the structure and mechanical properties of milk fat and lard. *Food Research International, 35*, 971–982.

Couvreur, S., Hurtaud, C., Lopez, C., Delaby, L., & Peyraud, J. L. (2006). The linear relationship between the proportion of fresh grass in the cow diet, milk fatty acid composition, and butter properties. *Journal of Dairy Science, 89*, 1956–1969.

Foubert, I., Vanhoutte, B., & Dewettinck, K. (2004). Temperature and concentration dependent effect of partial glycerides on milk fat crystallization. *European Journal of Lipid Science and Technology, 106*, 531–539.

Fredrick, E., Moens, K., Heyman, B., Fischer, S., Van der Meeren, P., & Dewettinck, K. (2013). Monoacylglycerols in dairy recombined cream: I. The effect on milk fat crystallization. *Food Research International, 51*, 892–898.

Fredrick, E., Van de Walle, D., Walstra, P., Zijtveld, J. H., Fischer, S., Van der Meeren, P., et al. (2011). Isothermal crystallization behavior of milk fat in bulk and emulsified state. *International Dairy Journal, 21*, 685–695.

Gliguem, H., Ghorbel, D., Lopez, C., Michon, C., Ollivon, M., & Lesieur, P. (2009). Crystallization and polymorphism of triacylglycerols contribute to the rheological properties of processed cheese. *Journal of Agricultural and Food Chemistry, 57*, 3195–3203.

Gliguem, H., Lopez, C., Michon, C., Lesieur, P., & Ollivon, M. (2011). The viscoelastic properties of processed cheeses depend on their thermal history and fat polymorphism. *Journal of Agricultural and Food Chemistry, 59*, 3125–3134.

Grall, D. S., & Hartel, R. W. (1992). Kinetics of butterfat crystallization. *Journal of the American Oil Chemists Society, 69*, 741–747.

Karray, N., Lopez, C., Lesieur, P., & Ollivon, M. (2005). Dromedary milk fat: thermal and structural properties 2. Influence of cooling rate. *Le Lait, 85*(6), 433–451.

Kaylegian, K. E., & Lindsay, R. C. (1995). *Handbook of milk fat fractionation technology and application*. Champaign, IL: AOCS Press.

Lopez, C. (2011). Milk fat globules enveloped by their biological membrane: Unique colloidal assemblies with a specific composition and structure. *Current Opinion in Colloid and Interface Science, 16*, 391–404.

Lopez, C., Bourgaux, C., Lesieur, P., Bernadou, S., Keller, G., & Ollivon, M. (2002). Thermal and structural behavior of milk fat: 3. Influence of cream cooling rate and droplet size. *Journal of Colloid and Interface Science, 254*, 64–78.

Lopez, C., Bourgaux, C., Lesieur, P., & Ollivon, M. (2002). Crystalline structures formed in cream and anhydrous milk fat at 4°C. *Le Lait, 82*, 317–335.

Lopez, C., Bourgaux, C., Lesieur, P., & Ollivon, M. (2007). Coupling of time-resolved synchrotron X-ray diffraction and DSC to elucidate the crystallisation properties and polymorphism of triglycerides in milk fat globules. *Le Lait, 87*, 459–480.

Lopez, C., Bourgaux, C., Lesieur, P., Riaublanc, A., & Ollivon, M. (2006). Milk fat and primary fractions obtained by dry fractionation 1. Chemical composition and crystallisation properties. *Chemistry and Physics of Lipids, 144*, 17–33.

Lopez, C., Briard-Bion, V., Beaucher, E., & Ollivon, M. (2008). Multiscale characterization of the organization of triglycerides and phospholipids in Emmental cheese: From the microscopic to the molecular level. *Journal of Agricultural and Food Chemistry, 56*, 2406–2414.

Lopez, C., Briard-Bion, V., Bourgaux, C., & Perez, J. (2013). Solid triacylglycerols within human fat globules: Crystals with a melting point above in-body temperature formed upon storage of breast milk at low temperature. *Food Research International, 54*, 1541–1552.

Lopez, C., Briard-Bion, V., Camier, B., & Gassi, J.-Y. (2006). Milk fat thermal properties and solid fat content in Emmental cheese: A differential scanning calorimetry study. *Journal of Dairy Science, 89*, 2894–2910.

Lopez, C., Cauty, C., & Guyomarc'h, F. (2015). Organization of lipids in milks, infant milk formulas and various dairy products: Role of technological processes and potential impacts. *Dairy Science & Technology, 95*, 863–893.

Lopez, C., Karray, N., Lesieur, P., & Ollivon, M. (2005). Crystallisation and melting properties of dromedary milk fat globules studied by X-ray diffraction and differential scanning calorimetry. Comparison with anhydrous dromedary milk fat. *European Journal of Lipid Science and Technology, 107*, 673–683.

Lopez, C., Lavigne, F., Lesieur, P., Keller, G., & Ollivon, M. (2001a). Thermal and structural behavior of anhydrous milk fat: 2. Crystalline forms obtained by slow cooling. *Journal of Dairy Science, 84*, 2402–2412.

Lopez, C., Lavigne, F., Lesieur, P., Keller, G., & Ollivon, M. (2001b). Thermal and structural behavior of milk fat: 1. Unstable species of anhydrous milk fat. *Journal of Dairy Science, 84*, 756–766.

Lopez, C., Lesieur, P., Bourgaux, C., Keller, G., & Ollivon, M. (2001). Thermal and structural behavior of milk fat: 2. Crystalline forms obtained by slow cooling of cream. *Journal of Colloid and Interface Science, 240*, 150–161.

Lopez, C., Lesieur, P., Bourgaux, C., & Ollivon, M. (2005). Thermal and structural behavior of anhydrous milk fat. 3. Influence of cooling rate. *Journal of Dairy Science, 88*, 511–526.

Lopez, C., Briard-Bion, V., & Ménard, O. (2014). Polar lipids, sphingomyelin and long-chain unsaturated fatty acids from the milk fat globule membrane are increased in milks produced by cows fed fresh pasture based diet during spring. *Food Research International, 58*, 59–68.

Lopez, C., Lesieur, P., Keller, G., & Ollivon, M. (2000). Thermal and structural behavior of milk fat: 1. Unstable species of cream. *Journal of Colloid and Interface Science, 229*, 62–71.

Lopez, C., & Ollivon, M. (2009). Triglycerides obtained by dry fractionation of milk fat. 2. Thermal properties and polymorphic evolutions on heating. *Chemistry and Physics of Lipids, 159*, 1–12.

Marangoni, A. G., & Lencki, R. W. (1998). Ternary phase behaviour of milk fat fractions. *Journal of Agricultural and Food Chemistry, 46*, 3879–3884.

Mazzanti, G., Marangoni, A. G., & Idziak, S. H. J. (2009). Synchrotron study on crystallization kinetics of milk fat under shear flow. *Food Research International, 42*, 682–694.

Michalski, M.-C., Ollivon, M., Briard, V., Leconte, N., & Lopez, C. (2004). Native fat globules of different sizes selected from raw milk: Thermal and structural behavior. *Chemistry and Physics of Lipids, 132*(2), 247–261.

Ollivon, M., Keller, G., Bourgaux, C., Kalnin, D., Villeneuve, P., & Lesieur, P. (2006). DSC and high resolution X-ray diffraction coupling. *Journal of Thermal Analysis and Calorimetry, 85*, 219–224.

Ramel, P. R., & Marangoni, A. G. (2016). Engineering the microstructure of milk fat by blending binary and ternary mixtures of its melting fractions. *RSC Advances, 6*, 41189–41194.

Ramel, P. R., Peyronel, F., & Marangoni, A. G. (2016). Characterization of the nanoscale structure of milk fat. *Food Chemistry, 203*, 224–230.

Ronholt, S., Kirkensgaard, J. J. K., Mortensen, K., & Knudsen, J. C. (2014). Effect of cream cooling rate and water content on butter microstructure during four weeks of storage. *Food Hydrocolloids, 34*, 169–176.

Smet, K., Coudijzer, K., Fredrick, E., De Campeneere, S., De Block, J., Wouters, J., et al. (2010). Crystallization behaviour of milk fat obtained from linseed-fed cows. *Journal of Dairy Science, 93*, 495–505.

Smiddy, M. A., Huppertz, T., & Van Ruth, S. M. (2012). Triacylglycerol and melting profiles of milk fat from several species. *International Dairy Journal, 24*, 64–69.

Söderberg, I., Hernqvist, L., & Buchheim, W. (1989). Milk fat crystallization in natural milk fat globules. *Milchwissenschaft, 44*, 403–406.

Ten Grotenhuis, E., Van Aken, G. A., Van Malssen, K. F., & Schenk, H. (1999). Polymorphism of milk fat studied by differential scanning calorimetry and real-time X–ray powder diffraction. *Journal of the American Oil Chemists Society, 76*, 1031–1039.

Timms, R. E. (1980). The phase behaviour and polymorphism of milk fat, milk fat fractions and fully hardened milk fat. *Australian Journal of Dairy Technology, 35*, 47–53.

Truong, T., Bansal, N., Sharma, R., Palmer, M., & Bhandari, B. (2014). Effects of emulsion droplet sizes on the crystallisation of milk fat. *Food Chemistry, 145*, 725–735.

Truong, T., Morgan, G. P., Bansal, N., Palmer, M., & Bhandari, B. (2015). Crystal structures and morphologies of fractionated milk fat in nanoemulsions. *Food Chemistry, 171*, 157–167.

Tzompa-Sosa, D. A., Ramel, P. R., Van Valenberg, H. J. F., & Van Aken, G. A. (2016). Formation of b polymorphs in milk fats with large differences in triacylglycerol profiles. *Journal of Agricultural and Food Chemistry, 64*, 4152–4157.

Van Aken, G. A., Ten Grotenhuis, E., Van Langevelde, A. J., & Schenck, H. (1999). Composition and crystallization of milk fat fractions. *Journal of the American Oil Chemists Society, 76*, 1323–1331.

Van Aken, G. A., & Visser, K. A. (2000). Firmness and crystallization of milk fat in relation to processing conditions. *Journal of Dairy Science, 83*, 1919–1932.

Vanhoutte, B., Dewettinck, K., Foubert, I., Vanlerberghe, B., & Huyghebaert, A. (2002). The effect of phospholipids and water on the isothermal crystallisation of milk fat. *European Journal of Lipid Science and Technology, 104*, 490–495.

Wright, A. J., Hartel, R. W., Narine, S. S., & Marangoni, A. G. (2000). The effect of minor components on milk fat crystallization. *Journal of the American Oil Chemists Society, 77*, 463–475.

Wright, A. J., & Marangoni, A. G. (2003). The effect of minor components on milk fat microstructure and mechanical properties. *Journal of Food Science, 68*, 182–186.

Wright, A. J., & Marangoni, A. G. (2006). Crystallization and rheological properties of milk fat. In P. F. Fox & P. L. H. McSweeney (Eds.), *Advanced dairy chemistry, volume 2 Lipids* (3rd ed., pp. 245–332). New York: Springer.

Chapter 10
Rheology and Texture of Cream, Milk Fat, Butter and Dairy Fat Spreads

Braulio A. Macias-Rodriguez and Alejandro G. Marangoni

1 Introduction

'Rheology' is a branch of physics concerned with deformation and flow experienced by complex fluids and soft materials such as foods when acted on by forces. Such forces may be 'naturally' exerted (e.g. gravitational or interaction forces holding a structure) or deliberately applied during their industrial process, use or consumption. Without exception, rheological phenomena occur in cream, milk fat, butter and dairy blends where it plays essential roles in fundamental, technological and sensorial aspects. Specifically, rheological properties provide information about interaction forces and reversible/irreversible flow of the structural elements of the mesoscopic network. It also relates to the application, "in-use" textural and sensorial properties (e.g. incorrect blending of milk fat fractions leads to macroscopic softening attributed to eutectic formation). Furthermore, it contributes to understanding the effects of formulation and processing. This information is used to establish rheology-structure relationship (e.g. develop models linking shear modulus and microstructure), rheology-texture relationships (e.g. describe firmness in terms of shear compliance), and rheology-formulation-processing relationships (e.g. assess the effect of cooling on firmness), all equally important to understand, control and improve product quality and process performance.

In this chapter, we highlight the above-mentioned aspects, while focusing largely on the characterization of rheology and texture of cream, milkfat, butter and dairy spreads and in some cases in their relationship. At this point, it is important to make a distinction that rheological properties are inherent to the tested material, while textural attributes are not. Texture, earlier coined 'psycho-rheology' by Sir Scott

B. A. Macias-Rodriguez (✉) · A. G. Marangoni
Department of Food Science, University of Guelph, Guelph, ON, Canada
e-mail: amarango@uoguelph.ca

T. Truong et al. (eds.), *Dairy Fat Products and Functionality*,
https://doi.org/10.1007/978-3-030-41661-4_10

Blair (considered the founding father of food rheology by many researchers), is a multifaceted field of study (Scott Blair, 1947). In the words of Szczesniak (2002), "*Texture is the sensory and functional manifestation of the structural, mechanical and surface properties of foods detected through the senses of vision, hearing, touch and kinetics*". Although it is more reasonable to measure rheological material functions, these properties are frequently unable to describe features only captured in the realm of texture. Such complexity arises due to the highly heterogeneous and multiscale structure of foods dictated by their formulation and processing regimes, and their deformations that they undergo during processing, use and oral processing (this latter aspect concerns thin-film rheology or tribology as covered at length in Chap. 11).

Considering the contribution of both rheology and texture to the understanding of cream, milk fat, butter and dairy spreads, it is the aim of this chapter to offer an integrated account of both fields. To achieve this, we summarize major empirical and fundamentals methods to measure rheological properties and textural attributes, detail some relations to calculate relevant parameters, and provide and contrast previous studies when possible. We take this approach due to the high heterogeneity of the literature. Making clear and meaningful connections among all studies is an elusive task due to arbitrary testing conditions and deformations applied to the materials under investigation. We also describe the structures of cream, butter and milk fat, and their proposed link to rheology and texture, with particular emphasis on milk fat crystal networks. In milk fat, these properties and attributes depend in a complex manner on at least three factors: volume fraction (or solid fat content), crystal microstructure and crystal interactions (Narine & Marangoni, 1999a). Finally, we summarize some of the major formulation and processing approaches taken to tailor the rheology of these products.

2 Technological Implications of Rheology and Texture of Cream, Milkfat and Derived Products

The study of rheology and texture in cream, milk fat and their derived products: butter, recombined butter and milkfat blends has born out of the necessity for assessing, controlling and improving their industrial process, quality, consumer acceptability and even for preventing fraudulence.

Cream, the starting point of manufacture of milk fat and butter, is an extremely complex system from a rheological standpoint, far more complicated than milk due to their higher solid fat content, especially at low temperatures. The quality of cream is judged based on sensory perceptions such as "body", a property that the consumer falsely associates with "richness", is a term that is difficult to be precisely reconciled with rheological properties, though it shows strong correlation with viscosity (Scott Blair, 1958).

Milk fat, a natural product obtained from cream, is the major constituent of butter. To broaden the range of functionality of milk fat, it can be separated into fractions with different chemical make-up and melting ranges that affect rheological and textural properties of butter and dairy spreads (Van Aken & Visser, 2000). In these products, quality is largely dependent on firmness or hardness, i.e. resistance to deformation or penetration, and spreadability, i.e. the ease with which the material spreads on bread or another substrate (Prentice, 1993). For example, a "good" butter should not be too firm otherwise, it would tear the bread, nor too spreadable, otherwise it would not remain as a continuous layer on the surface of the bread and appear "oily" or "sticky". It should be neither tough nor crumbly or brittle. A "good" butter should appear "vivid" (i.e. *show some elasticity*) when spread on bread (Scott Blair, 1953). The terms 'hardness' and 'firmness' are used interchangeably throughout the chapter, though some authors suggest the use of 'firmness' for recoverable viscoelastic deformations and 'hardness' for non-recoverable plastic deformations (Faber, Jaishankar, & McKinley, 2017a). The notion of plastic deformation is applicable to butter since it is not completely deprived of elasticity when sheared during spreading. Textural attributes can be correlated to rheological properties (e.g. modulus, compliance, viscosity).

Due to its practical importance, the texture and rheology of butter has been a subfield of research which has laid the foundations of fat rheology. Much work has been published on sensory panels involving craftsmen, *ad hoc* methods and development or application of empirical and fundamental methods aimed to grade or measure firmness and spreadability and determine moduli and viscosities (Prentice, 1993; Scott Blair, 1954, 1958; Wright, Scanlon, Hartel, & Marangoni, 2001). These reports have also investigated the influence of processing and crystallization conditions on textural attributes and rheological properties, supporting their relevance in product quality (Mulder, 1953; Prentice, 1984a, 1993; Scott Blair, 1954, 1958).

3 Structure of Cream, Milk Fat and Butter

Structurally, cream resembles milk with the major difference being the higher content of fat globules in cream. It has been suggested that cream contains globules displaying a bimodal size distribution, with larger globules being more abundant (Prentice, 1993). The distance between the globules, whose average size ≈3 μm, is marginal ≈0.35 μm for creams with fat content of 48% (Prentice, 1993; van Vliet & Walstra, 1979). Increasing and reducing the fat content, reduces and increases the distance between globules, respectively, which alters the viscosity of the cream (Prentice, 1993). At the molecular level, milk fat is composed of an extremely heterogenous triacylglycerols (TAG) mixture. At the nanoscale, TAGs crystallize into lamellae and then into platelets—the fundamental crystal unit. At the submicron and microscopic scale, platelets aggregate into clusters or flocs that make up the microscopic crystal network in which liquid oil is embedded. The microstructure has a tremendous bearing on rheological properties (DeMan & Beers, 1987; Heertje, 1993; Narine & Marangoni, 1999a; Tang & Marangoni, 2007). Within the microstructure, crystal aggregates differing in size (0.1–140 μm) and morphology and being held together by a wide spectrum of bonds of variable degrees of strength and reversibility (van den Tempel, 1961; DeMan & Beers, 1987; Narine and Marangoni 1999e; Shama & Sherman, 1970). Furthermore, butter comprises a water phase dispersed in a continuous oil phase containing crystalline aggregates and fat globules (partly broken or intact). This structure is imparted by churning, physical working

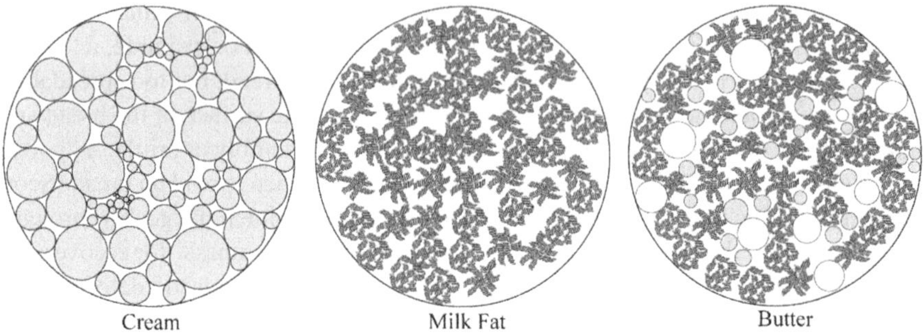

Fig. 10.1 Schematic structure of cream, milk fat and butter. Gray globules correspond to oil droplets, white globules represent water droplets and clusters correspond to crystalline fat

and crystallization, where the membrane of fat globules collapses, globules coalesce, oil leaks out and crystal aggregates form. Figure 10.1 shows a schematic representation of the microstructures associated with cream, milk fat and butter respectively.

4 Rheological and Textural Characterization

Evaluation tests can be broadly classified into three main categories: imitative, empirical, and fundamental tests-based on its foundations and the information obtained from them (Scott Blair, 1958). Each type of test has its own merits and limitations. Imitative tests attempt to resemble the conditions the material will be subjected to during their use, such as machines that imitate spreading of butter on bread by measuring shear imposed by a knife edge. Although they show some correlation with sensory scores or empirical methods, they lack a solid foundation and methodology, control of deformations and quantitative measures. Empirical tests imitate more closely the basic motions of deformations applied during processing and product usage. Their measures (typically textural attributes such as firmness, spreadability) depend on instrument configuration and correlate well with sensory assessment of texture. They are useful for quality control and product development routines such as for adjusting milk fat blending or butter-making. Fundamental tests are rigorously defined in physical and mathematical terms, and aid the measure of true or apparent (for nonhomogeneous flows) bulk properties. They are used for research and development purposes and require a certain degree of expertise. In the following section, we describe the principles of the main empirical and fundamental methods developed early and still in use to measure the rheology of cream, milk fat, butter and dairy fat blends. We also briefly discussed emerging rheological approaches and techniques. It would be impractical to review the many discontinued empirical methods (at least in research grounds). For the interested readers, an extensive account of such methods appears in (Deman, 1983; Mulder, 1953; Prentice, 1984a, 1984b, 1993; Scott Blair, 1954, 1958).

4.1 Imitative and Empirical Tests

4.1.1 Indentation and Penetration Tests

Indentation and penetration have been by far the most popular methods to evaluate the texture of butter, milk fat and edible fats in general. They offer simple and inexpensive characterization which is useful for investigating the effect of processing conditions on hardness and correlate well with sensory panels (Van Aken & Visser, 2000). They are based on the resistance of a material to be "pierced" or indentated by a test body: rod, cone, sphere, needle, etc. The resistance is measured as load, depths or rates of indentation/penetration of the testing body. Some of the early used approaches can be more or less described in chronologically order as the "*oleogrammeter*" developed by Brullet (a vertical rod applied on hardened surfaces of fat loaded with increasing weights aimed to detect butter-margarine adulteration), compression of cylinders of butter (under constant loads and fixed times to establish correlations between hardness and melting point), penetration studies with vertical rods of different dimensions to estimate butter temperature relationships, Kruisheer's penetrometer, Mohr and Wellm's cone- and sphere-yield tests (to assess used for in-container butter) (Scott Blair, 1954). To date, cone penetrometry remains the most popular method to measure the texture of butter or dairy spreads (Wright et al., 2001). A schematic representation of a typical cone is shown in Fig. 10.2.

Conical configurations varying in loads and angles allows to cover a range of textures. Most studies included constant load experiments and to less extent constant rate experiments. Both methods show good agreement with one another (Haighton, 1959; Hayakawa & DeMan, 1982). Results are reported as penetration depths or converted into yield values, hardness or spreadability indexes, using various equations dependent on the testing body and test conditions (Haighton, 1959).

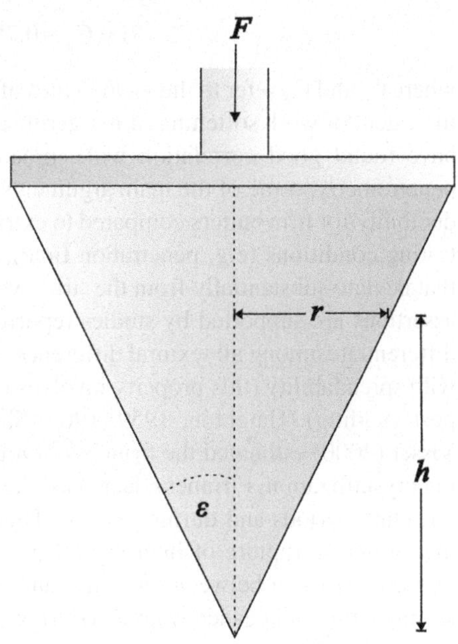

Fig. 10.2 Schematic representation of a cone penetrometer. Applied force *F*, cone radius *r*, arbitrary penetration height *h*

Considering that most of the force is used to overcome the yield point (the stress at which deformations are a combination of elastic 'reversible' and plastic 'irreversible' deformations) and provided that the motion is slow, an empirical yield value is defined as the force load per unit cross-sectional area of the cone as given by,

$$\text{yield value} = \mathrm{K'W} / \mathrm{p}^{\mathrm{n}} \tag{10.1}$$

where K' is a constant dependent on the cone angle, W is the weight of the cone (g), p is the penetration depth raised to a fractional power $1.4 \leq n \leq 2$ as found empirically. The values of these empirical coefficients appear to depend on hardness and type of fat (Haighton, 1959; Tanaka, de Man, & Voisey, 1971) suggesting a complex relationship between the cone and yield value (e.g. for butter $n \approx 1.6$). The yield value may be affected by frictional forces between the fat and the tested material, though these are negligible for truncated cones (Wright et al., 2001). Since yield value and viscosity show some linear proportionality, measuring either of these parameters or a combination of both is necessary for defining the rheology of a system (Mulder, 1953). Using this relationship, yield values have been assigned to margarines and shortenings according to their usability (assessed by 'thumb' tests), though values are not universal but rather specific to product type and country (Haighton, 1959). Alternative definition of apparent yield stress (*AYS*) has been proposed (International Dairy Federation):

$$\mathrm{AYS} = \mathrm{P} / A_{proj} = gw / \pi d^2 tan^2(\varepsilon) \tag{10.2}$$

where g is acceleration due to gravity, w is the weight of the cone, ε is the angle of the cone and d is the penetration depth (Wright et al., 2001). Penetration values can be also converted to hardness defined as force divided by penetration area (similar to *AYS* but divided by the impression area A_{imp}) (Wright et al., 2001). Alternatively, they can be converted to spreadability index (SI) using the following relation:

$$\mathrm{SI} = \mathrm{C_u} - 0.75(\mathrm{C_u} - \mathrm{C_w}) \tag{10.3}$$

where $\mathrm{C_u}$ and $\mathrm{C_w}$ refer to the yield values of unworked and worked fats, to determine the extent of work softening of margarine and butter (Haighton, 1965). Yield values have found good correlation with spreadability. Despite the advantage of cone penetrometry, some of the main arguments made against its use include poor reproducibility for firm butters compared to extrusion and 'sectility' tests, use of arbitrary testing conditions (e.g. penetration time), and ill-defined measures of yield value that deviate substantially from the 'true' yield stress (Atkins & Tabor, 1965). These assertions are supported by studies reporting the inability of cone penetrometry to differentiate among all textural differences of butter and correlations of yield values with spreadability (this property involves extensive shear and structural breakdown post yielding) (Haighton, 1959, 1965; Shama & Sherman, 1970). Van Aken and Visser (2000) estimated the firmness of milk fat (expressed as the yield value) during crystallization. Firmness decreased during kneading but it increased in between kneading periods and during storage. This behavior was attributed to softening of the fat due to rupture of inter-crystal bonds during kneading, reformation of new primary bonds in between kneading and recrystallization of fewer primary bonds (compared to unkneaded samples) during storage.

4.1.2 Extrusion Test

Extrusion has been used to a lesser extent to evaluate the texture of butter. Its main advantage is that it mimics the flow of butter during spreading (though in an empirical manner). It involves measuring the thrust of a piston required to extrude a soft material such as butter through an orifice or nozzle. In principle, it resembles a sectility test, in that in extrusion flow is confined to one side, whereas in cutting flow occurs past both sides of the cutting edges (Prentice, 1984a). Previous tests on butter have shown that the force of extrusion correlates well with spreadability as determined by subjective assessment (Prentice, 1993). The force of extrusion results from two major components: one associated with force to induce extrusion through the orifice, which is constant if properties and rate of extrusion remain constant; and the other with the force to overcome friction at the wall, which varies as the test progresses, e.g. frictional contribution diminishes as the barrel empties (Prentice, 1984a). Some criticisms to this test include time-consuming mounting of the sample and poor control of measuring temperature (Mortensen, 1983).

4.1.3 Wire Cutting Tests (Sectility)

Wire cutting consists of driving a standard wire through a block of sample and measuring the force required to cut the sample (at constant or variable speed) (Scott Blair, 1954). Their main advantage is that it is simple, accessible and reproducible and that the cutting force or 'sectility' is simply related to the diameter of the wire. Since wire cutting involves fracture, deformation, and friction in general, it offers a viable method to measure these properties using some simple assumptions described (Kamyab, Chakrabarti, & Williams, 1998) for cheese cutting. A schematic diagram of wire cutting perpendicularly through a surface is depicted in Fig. 10.3.

For a block of width B, cut by a wire of diameter d, where fractures are assumed to arise from elastic splitting and from yielding and friction, the fracture grows by length dx when the cutting force F moves dx, and F bears the fracture toughness G_c (for elastic splitting), yield stress σ_y and frictional force $\mu\sigma_y$:

$$Fdx = G_C B dx \quad \left(\text{force associated with elastic splitting}\right) \tag{10.4}$$

$$F = 2B \int_0^{\pi/2} \frac{d}{2}\left(\sigma_y cos\theta + \mu\sigma_y sin\theta\right) d\theta \left(\text{force associated with yielding and friction}\right) \tag{10.5}$$

The total force is then

$$F = BG_C + B\left(1+\mu\right)\sigma_y d \qquad i.e., \quad \frac{F}{B} = G_C + \left(1+\mu\right)\sigma_y d \tag{10.6}$$

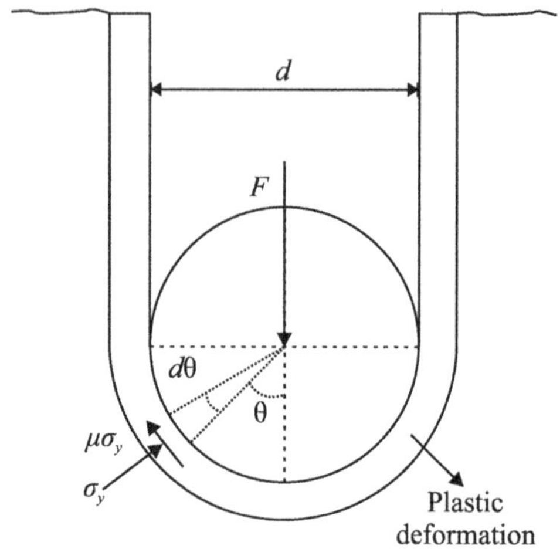

Fig. 10.3 Wire cutter of a block showing frictional and plastic deformation

This means that for steady-state cutting, *F/B* is proportional to *d* with slope of $(1+\mu)\sigma_y$ and intercept of G_c. A more elaborate equation than Eq. (10.6) has been proposed to account for plastic deformations (Kamyab et al., 1998). To the best of our knowledge, G_c has not been estimated for butter or milk fat blends but it should be somewhere around 10 J/m^2 as reported by (Kloek, van Vliet, & Walstra, 2005) using a more general analysis for hydrogenated palm oil-sunflower oil blends. Previous studies have shown excellent correlations of 'sectility' with firmness and viscosity of butter determined by compression of cylindrical specimens and with spreadability assessed subjectively (Scott Blair, 1954). A common argument made against the estimation of rheological properties with a wire cutter in butter is its anisotropic crystalline structure, which may lead to varying resistance in different cutting planes (e.g. for laminated structures, less resistance will be expected in the parallel direction to the lamination).

4.2 Fundamental Tests

4.2.1 Viscometry

Viscometric flows consists in applying steady shear to fluids or semifluid materials. These tests are typically performed with rotational viscometers or rheometers using Couette or cup-and-bob geometries. Results are commonly reported as steady shear viscosity η as a function of shear rate $\dot{\gamma}_0$. Much effort has been devoted to study the viscosity of cream given that it is the initial point of the manufacture of butter and it is a product by itself (Prentice, 1993). Cream is a structured fluid, an emulsion of milk fat in milk plasma (Huppertz & Kelly, 2006). Several factors influence its viscosity such as the volume fraction of fat globules and solid fat, their size and the rheology of continuous and dispersed phases, among others. These, in turn, are affected by origin, method of preparation and history of the cream. The rheology of cream is mainly non-Newtonian, i.e. pronounced shear thinning occurs. However, at

T ≈ 40–80 °C conceiving that all triglycerides contained within the fat globules are molten, it behaves nearly Newtonian and resembles somewhat the flow of milk, such as for creams containing ≤50% fat (or $\phi_{fat} \leq 0.4$) at approximate shear rates of $\dot{\gamma}_0 \geq 10\ s^{-1}$ (Phipps, 1969). For cream with similar fat content, a decrease of 3% of the apparent viscosity η_{app} (i.e. dependent on shear rate) per 1 °C rise of temperature has been estimated (Lyons & Pyne, 1933). In general, decreasing the fat content of the cream increases the encountering distance of the fat globules during shearing and thus reduces the resistance to mutual rotations (and thus of viscosity), whereas increasing the fat content, promotes the 'hindering' effect, partial coalescence during shear, which altogether increases viscous resistance. For example, η_{app} of cream with 54% fat compares to that of 48% fat only if the primer is subjected to a stress three times greater than the latter, to achieve the same flow rate.

4.2.2 Pressure-Driven Flow: Capillary Rheometry (Orifice Die Extrusion)

Orifice die extrusion flows are useful for determining viscosities at controlled shear rates closed to those encountered during manufacturing and usability of pastes such as butter (e.g. during spreading $\dot{\gamma} \approx 300\ s^{-1}$ and cutting $\dot{\gamma} \approx 16\ s^{-1}$) (Castro, Giles, Macosko, & Moaddel, 2010; Prentice, 1993). They can also be used for the determination of yield stress. For this, a capillary rheometer consists in driving a tool (e.g. a piston, ram) at a linear speed S to force material through a barrel with diameter D_0 and a die of diameter D and length L (see Fig. 10.4) (Macosko, 1994).

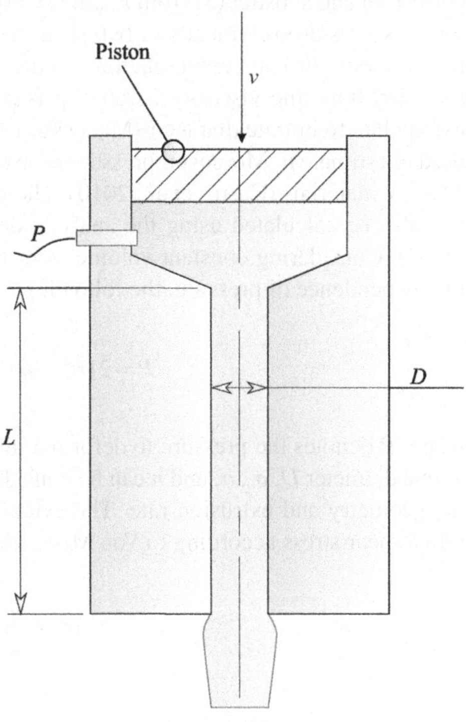

Fig. 10.4 Schematic view of a material being extruded through an orifice die with diameter D and length L in a capillary rheometer. Pressure P is applied by a piston with velocity v, and measured during the test

During this operation, the total pressure drops (P_{tot}) are measured which are made up of two major contributions: pressured drops at the entrance (P_{en}), associated with extensional viscosities μ, and shear pressure drops at the capillary wall (P_{shear}), associated with shear viscosities η. To interpret the results, common corrections that need to be applied include Rabinowitch, Bagley, Mooney and wall slip corrections to account for non-Newtonian behavior, entrance pressure losses, and wall slip at the die wall respectively. The shear stress at the wall (σ_w) and the corrected non-Newtonian shear rate ($\dot{\gamma}_w$) at the wall are calculated with standard formulas to determine apparent viscosities (η_{app}):

$$\sigma_w = \frac{(P_{tot}) \cdot D}{4L} \tag{10.7}$$

$$\dot{\gamma}_w = \frac{8D_0^2 \cdot S}{D^3}\left(\frac{3}{4} + \frac{1}{4}\frac{d\,lnQ}{d\ln\sigma_w}\right) \tag{10.8}$$

and

$$\eta_{app} = \frac{\sigma_w}{\dot{\gamma}_w} \tag{10.9}$$

where Q is the flow rate. The term in the parenthesis corresponds to the Rabinowitch correction. To calculate true shear stress σ, P_{en} can be calculated using the Bagley correction and substracted from P_{tot} in Eq. (10.7). The correction consists in obtaining pressures drops typically at two shear rates (using capillaries with the same D_0 but different D/L) and extrapolation to a die of zero length. To determine true shear rate (and thus true viscosity), wall slip is corrected using the Mooney relation to extrapolate to infinite diameter (Macosko, 1994). The entrance pressure can also be used to estimate μ. Moreover, pressure drops can be utilized to calculate yield stress of pasty materials (Castro et al., 2010). The yield stress σ_y using capillary extrusion can also be calculated using the analysis developed by Benbow and Bridgewater (1993). Considering constant volume, zero-length orifice die, plastic behavior and rate-dependence of pressure, the following relationship was proposed:

$$P = 2\left(\sigma_0 + \alpha V^n\right)\ln\frac{D_0}{D} \tag{10.10}$$

where P denotes the pressure to deform a material from its original diameter D_0 to a final diameter D, σ_0, α, and n can be considered material constants independent of die geometry and extrusion rate. The extensional yield stress σ_0 can be converted into a shear stress according to Von Mises yield criterion:

$$\sigma_0 = \frac{\sigma_y}{\sqrt{3}} \tag{10.11}$$

Castro et al. (2010) recently applied this method to the characterization of the yield stress of soft solids and found good correlation with results obtained from rotational steady shear measurements. Pressure drops and rheological properties of butter such as yield stress σ_y and apparent viscosity η_{app} have been reported as a function of measuring temperature, storage temperature and time ('ageing') using various relations such as the Hagen-Poiseuille (assuming Newtonian flow), Casson and power-law equations have been used to describe yield stress (Hanck & Wall, 1966), estimated as $\sigma_y \approx 2$–12 kPa at T = 5–15 °C (Kawanari, Hamann, Swartzel, & Hansen, 1981). Values of σ_y and η_{app} were affected by test temperature and 'ageing' temperature, although the latter factor had little effect on properties determined by empirical compression and penetrometry suggesting capillary extrusion was more sensitive to such changes. Temperature is inversely correlated to apparent viscosity (i.e. higher temperature, less resistance to flow). This effect is less pronounced for butter containing high melting triglycerides possibly due to a greater degree of crystallinity leading to firmer products (Shukla, Rizvi, & Bartsch, 1995). Flow behavior of cream has also been studied but with a capillary 'consistometer', in which the material is forced by compressed air at known pressure through a series of standard glass capillary tubes (Scott Blair, Hening, & Wagstaff, 1939). Flow of natural, homogenized and reconstituted cream revealed five types of characteristic behavior: (1) truly fluid; constant viscosity, (2) viscosity is dependent of stress and independent of dimensions of the capillary—in most cases viscosity falls with increasing stress (structural viscosity), (3) viscosity is independent of stress but dependent of capillary dimensions—in the majority of cases narrower or longer capillaries show lower viscosity (structural breakdown), and (4) viscosity dependent on stress and dimensions of capillary. Such complex behavior might be attributed to the fat content of the cream and varying amount of crystalline fat within the globules. A major instrumental concern when conducting capillary rheometry at high shear rates is the likelihood of wall slippage, especially due to the self-lubricating nature of butter, milk fat and dairy fat blends. Thus, it is customary practice to correct for wall slip as shown above. Other corrections and equation (such as those described above) to get reliable estimation of material properties can make the tests laborious.

4.2.3 Compression

Compression tests are one of the most popular tests for determining fundamental rheological, fracture properties and empirical textural attributes due to their practicability and easiness of interpretation. They involve deforming a specimen of known dimensions (typically a cylinder) at constant force (creep) or at constant crosshead speed (uniaxial compression if only the upper plate is mobile) for a standardized time. For creep compression, deformation is measured as a function of time, whereas for the latter, force is recorded as a function of time. Davis (1937) and DeMan, Gupta, Kloek, and Timbers (1985) conducted compressive creep (force loading) and recovery (force unloading) on cylindrical samples of butter. Davis (1937) determined modulus of elasticity G for the recovered deformations and viscosity defined as the ratio of the compressive stress to the strain rate ($\eta = \sigma/\dot{\gamma}$) from the plastic non-recovered deformations, whereas DeMan et al. (1985) differentiated among instantaneous elasticity and retarded elasticity. Accounts on how to calculate similar materials measures albeit obtained from creep shear are given subsequently. A combination of apparent shear modulus G and viscosities η provided a measure of 'firmness' (the resistance to creep deformation typically measured over long time), and a ratio of viscosity to elasticity η/G offered a measure of 'springiness' (i.e. the extent of instantaneous recovered strain evaluated at short times), though not justification of such relationships were provided (Davis, 1937). Recently, clear relationships between such properties have been developed for shear creep and recovery compliance (Faber, Jaishankar, & McKinley, 2017b). Increases in loading time, magnitude of force, work softening and temperature all lowered elastic and viscous properties (DeMan et al., 1985; Scott Blair, 1938). Scott Blair conducted successive loading experiments similar to those by Davis (1937) but measured the time required for achieving a strain deformation of 50% to estimate a pseudo-viscosity which he linked with spreadability. For uniaxial compression, force-time curves are reported or converted into normal stress σ_n versus Hencky strain $\in$ or strain rate $\dot{\epsilon}$. Previous studies on butter and milkfat have also interpreted firmness or hardness as the maximum load of the force-time curve, and other textural attributes have been determined by double compression imitative tests referred as to "Texture Analysis". To convert force-time curves, treating butter or milk fat as an incompressible material, the following equations have been applied (Kloek et al., 2005):

$$\sigma_n = F / A \tag{10.12}$$

$$\epsilon_h = \ln\left(h_t / h_0\right) \tag{10.13}$$

$$\dot{\epsilon} = \left(d\epsilon / dt\right) = \dot{h} / h \tag{10.14}$$

where A is the circular area of a cylinder with diameter D, h_0 and h_t are the specimen heights at the beginning and during the test. Apparent rheological measures (i.e. measures that vary according specimen height) of the Young's modulus E_{app} (measured as $d\sigma_n/d\in$ where $\in \to 0$), yield point $\sigma_{\text{y\_app}}$ (defined as the maximum stress) and viscosity η_{app} (determined from the slope $d\sigma_n/d$). Some limitations of compressive test include their limited range of accuracy, e.g. true elastic modulus cannot be measured for yield strains below ≈2% (paradoxically this falls somewhat within the yielding region of fats), manifestation of cracks or shear bands (regions of 'strain localization') due to large strains, e.g. in firm and brittle butters compressed at low temperature, buckling or bulging (cylindrical sample becomes barrel-shaped) and strong frictional effects at the boundary of the sample. Adequate specimen sizes ($h_0/D < 1.5$ but not too low to avoid friction) and lubricated plates (e.g. coated with oil or Teflon) can be used to circumvent the last two effects (Vliet, 2013).

4.2.4 Squeeze Flow

Squeeze flow are relatively simple to implement and allow a wide range of shear rates. These flows are encountered namely during rolling compression. Squeeze flows entails compressing a highly viscous or soft solid typically between two parallel circular plates to squeeze it out radially. The use of lubricated sample-wall interfaces can lead to principal planes (of vanishing shear stress) that achieve frictionless or perfect slip, so pure elongational strain (or biextensional flow) occurs during compression (see Fig. 10.5).

Note that unlike other rheometric tests where slip makes measurements more intricate and lengthy, the attainment of perfect slip rather simplifies the calculations involved in this test. Calculations are based on initial theory developed by (Chatraei, Macosko, & Winter, 1981) for homogeneous deformation (perfect slip) under constant compressive deformation (creep), and adapted to constant displacement rate experiments (Campanella & Peleg, 2002). Assuming a perfect slip boundary condition, for a material being squeezed for a time *t* between circular plates of radius *R*, the relationship of the squeezing force *F*(t) and time dependent sample thickness is given by:

$$F(t) = \frac{3\pi\eta_b R^2}{h(t)} \dot{h}(t) \tag{10.15}$$

where η_b is the biaxial extensional viscosity, and $\dot{h}$ is the steady squeeze rate [or −d*h*(*t*)/d*t* or −d*H*(*t*)/d*t*]. Depending on the type of experiment, *F*(t) or $\dot{h}(t)$ remains constant. For this relationship to hold, experiments must be conducted at large aspect ratios $h \ll R$ ($h/R \approx$ 10–30) to reduce end effects and increase signal (this latter aspect is applicable to semi-fluids). The first aspect imposes some limitation in the consistency of the measured sample, e.g. preparing pristine solid-like butter disks with appropriate *h/R* is somewhat difficult (ratios of $h/R \approx$ 2–4 have been used previously in the literature of butter). Researchers have conducted squeeze flow of butter with constant area (i.e. sample fills entirely the plates). Assuming perfect slip for butter (based on its self-lubrication), steady (at $\dot{\gamma}_0 \approx 10^{-3}\ s^{-1}$) and non-steady (not stress nor shear rate were constant) viscosities of butter have been estimated as $\eta_b \approx 10^8$ Pa s at 15–17 °C. Butters made of high melting triglycerides have shown higher η_b (~1 log order higher at 17 °C) than those formulated with anhydrous milk

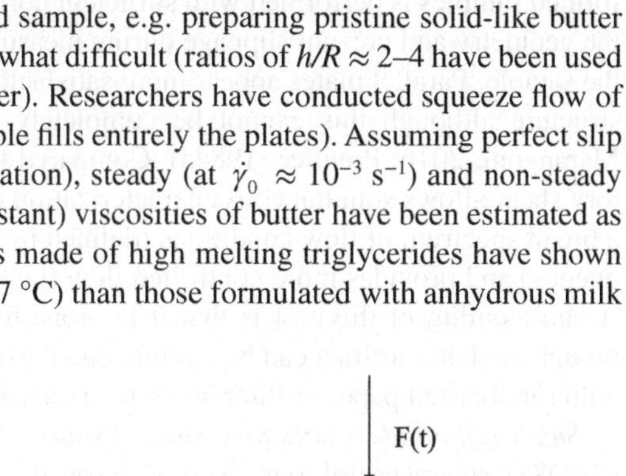

Fig. 10.5 Schematic view of a material moving radially outward during lubricated squeeze flow between parallel plates

fat, and their viscosity increases for both types of butter during storage. These findings support that solid-like high melting triglycerides increase resistance of crystal aggregates to flow and that sintering occurs during storage. Some disadvantages of squeeze flow include that at very large Hencky strains ($\in \approx 3$) the lubricated condition may be violated. In such a case, complex flow fields occur where both slip and shear boundaries coexist, something that is not always easy to detect experimentally and requires more elaborate analysis of the data.

4.2.5 Drag Flow: Oscillatory Shear

Shear rheometry can be performed at small or large strains or stress with rotational rheometers. In this respect, oscillatory shear is the most popular fundamental test employed in butter, milk fat and dairy spreads. Oscillatory shear consists of applying a sinusoidal input function (strain or stress) and measuring the associated response comprising in-phase (stored energy or elastic modulus) and out-of-phase (loss energy or loss modulus) components with respect to the input function. Depending on the amplitude of the input function, oscillatory shear can be divided into two regimes: small amplitude and large amplitude, which probe linear and nonlinear viscoelastic regions, respectively (Hyun et al., 2011). An intermediate regime between small and large amplitude has been referred as to the medium amplitude regime where the nonlinear response grows asymptotically. We do not make a distinction of such regime, and any deformation beyond the linear viscoelastic region is considered large amplitude or nonlinear. Two types of geometries: cone and plate and parallel plate geometries are commonly used for oscillatory shear. In cone and plate, geometries are brought together slowly to minimize sample breakdown or residual stresses affecting the measurement. In parallel plates, loading of stiffer preformed samples is performed with sufficient normal force to allow full contact with the geometry and prevent slippage during measurements but not too high to disturb the sample. Parallel plates appear more satisfactory to reduce damage to the microstructure although this cannot be completely eliminated (Macias-Rodriguez & Marangoni, 2016; Prentice, 1984a). Compared to other fundamental tests, oscillatory shear allows simultaneous characterization of elastic and viscous properties in a broad spectrum of flow conditions (defined by stress or strain and oscillatory frequency) and provides more controlled flow (i.e. gradual increase of deformations). A shortcoming of this test is that it is prone to edge fracture and wall slippage, though the latter artifact can be circumvented using modified surfaces (e.g. covered with sandblasted paper or filter paper to enhance adhesion to the plates).

Small amplitude oscillatory shear (SAOS). Small amplitude oscillatory shear (SAOS) tests are useful to measure viscoelastic properties of the underlying microstructure. SAOS imposes relatively small strains or stresses in the linear viscoelastic region (LVR) typical of materials interacting via short-range van der Waals forces such as butter and edible fats in general (van den Tempel, 1961). As mentioned, to avoid disturbance of the original network, careful loading (preferably with normal force control) must be performed (Macias-Rodriguez & Marangoni, 2016; Thareja et al., 2011). During SAOS tests, strain and stress maintain their linear proportionality and the crystal network exhibits viscoelastic solid-like behavior ($G' > G''$) characterized by high modulus and weak frequency (ω) dependence (van den Tempel, 1961; Narine & Marangoni, 1999b; Macias-Rodriguez & Marangoni, 2016; Rohm & Weidinger, 1993; Thareja et al., 2013). For a sinusoidal strain excitation $\gamma(t) = \gamma_0 \sin(\omega t)$, a sinusoidal stress response $\sigma(t) = \gamma_0 \sin(\omega t + \delta)$ is obtained at the same input frequency ω, and with phase angle δ. The response can be decomposed as

$$\sigma(t) = \gamma_0 G'(\omega)\sin(\omega t) + \gamma_0 G''(\omega)\cos(\omega t) \quad (10.16)$$

in which G', the in-phase elastic modulus or stored energy represents the real component, and G'', the out-of-phase viscous modulus or dissipated energy represents the imaginary component of the complex modulus G^* at a given frequency (ω) (Ferry, 1980; Macosko, 1994; Tschoegl, 1989). A Fourier analysis (which converts the time function into frequency domain) of the stress response reveals in the LVR region, only the first or fundamental harmonic ($n = 1$) associated with ω occurs. Figure 10.6 depicts a frequency sweep and strain amplitude sweep of butter illustrating its weak frequency dependence and values of viscoelastic moduli of $G' \approx 10^6$ Pa and $G'' \approx 10^6$ Pa. This is in good agreement with previous studies on butter obtained from high-melting triglyceride (HMT) and anhydrous milk fat (AMF) (Rohm & Weidinger, 1993; Shukla & Rizvi, 1995).

Values of elastic modulus, complex viscosities and critical strains of $G' \approx 10^6$–10^7 Pa, $\eta^* \approx 10^5$ Pa s and $\gamma_c \approx 10^{-4}$–10^{-3} at T = 15–17 °C respectively were reported (Rohm & Weidinger, 1993; Shukla & Rizvi, 1995). HMT butter showed less temperature dependence and higher magnitudes of the rheological functions due to higher solid fat content and arguably due to microstructure. Both butters have shown increases of G' during storage attribute to 'setting' (increased in firmness due to continuous crystallization). From a practical perspective, the elastic modulus G' correlates well with material hardness (Suresh & Marangoni, 2001).

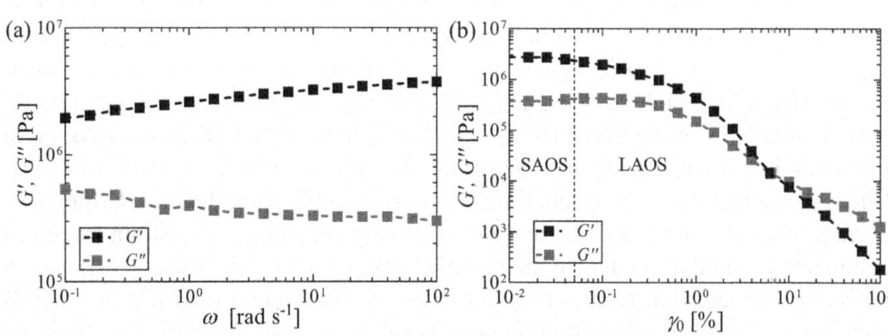

Fig. 10.6 Storage G' and loss G'' moduli of butter (T = 18 °C) measured during (**a**) frequency sweep ($\gamma_0 = 0.01\%$) and (**b**) strain amplitude sweep ($\omega = 3$ rad/s). Dotted line in (**b**) indicates approximate transition from linear (SAOS) to nonlinear regime (LAOS)

Large amplitude oscillatory shear (LAOS). Above a critical strain or critical stress, the material plastically deforms as it starts to yield and 'flow'. As alluded to previously, the yield stress σ_y is an important property that correlates with the usability and sensory perceptions of butter. Despite much controversy about the existence of a 'true' yield stress, the past notion that flow occurs at extremely long timescales below a critical value can be considered negligible at the timescales of applications of most materials (Coussot, 2007). In butter, the confusion rather arose due to the absence of a stress overshoot in steady shear experiments. Nevertheless, this is a common feature of other yield stress materials (Dinkgreve, Paredes, Denn, & Bonn, 2016). From oscillatory experiments, σ_y can be best defined in two main ways: the first requires intersecting stress and strain curves using power law equations to obtain a dynamic yield stress, and the second determining the stress at the intersection of G' and G'' curves as a function of strain to estimate a static yield stress. The primer gives the lowest value of σ_y and γ_y, whereas the latter provides the highest values of σ_y and γ_y since the material have already experience some yielding. Dinkgreve et al. (2016) concluded that for thixotropic materials, both static and dynamic yield stresses must be considered. At large amplitudes (LAOS), complex and nonlinear responses arise in viscoelastic materials such as butter and milk fat. LAOS offers many advantages over more traditional nonlinear rheological methods (e.g. steady shear, capillary rheometry) such as simultaneous and full viscoelastic characterization in deformation and timescale domains, controlled flow (i.e. gradual increase of oscillatory shear minimizes slip) ample operational window (i.e. deformations go well beyond the LVR and thus it can differentiate among microstructures insensitive to SAOS), superior sensitivity (Hyun et al., 2011). LAOS allows independent variation of two parameters: loading strain amplitude γ_0 (for strain-control tests) and frequency ω deformation, yielding viscoelastic responses in a 2D regime map termed the Pipkin space (Pipkin, 1972). In this, nonlinear viscoelastic measures $G'(\omega, \gamma_0)$ and $G''(\omega, \gamma_0)$ are seamlessly linked with linear viscoelastic moduli $G'(\omega)$ and $G''(\omega)$, and with the steady flow viscosity $\eta(\dot{\gamma})$ (Dealy & Wissbrun, 1999; Ewoldt & Bharadwaj, 2013). For a strain input test, the stress material response is in the LAOS or nonlinear regime when the viscoelastic moduli are either variant to changes in γ_0, e.g. $G'(\omega, \gamma_0)$ or $G''(\omega, \gamma_0)$ or the stress response is no longer sinusoidal. As seen in Fig. 10.6, the onset of nonlinear behavior occurs at $\gamma_y \approx 0.06\%$ ($\sigma_y \approx 1360$ Pa). There are several approaches to analyze the nonlinear LAOS data, which include investigating the behavior of the first-harmonic moduli, time-domain raw waveforms τ(t) or two-coordinate axes figures referred as to Lissajous-Bowditch curves and analyzing the raw waveforms via FT rheology, Chebyshev stress decomposition, time dependent moduli, etc. The first-harmonic or average viscoelastic moduli G' and G'' (normally the output of a strain sweep in commercial rheometers) denote the global ('full' or 'intercycle') stress response in the nonlinear behavior. In Fig. 10.6, it can be seen that butter (and fats in general) undergoes average elastic softening coupled with increase dissipation or thinning due to disruption of the crystal network as strain increases. Time-domain raw signals σ(t) and Lissajous-Bowditch curves qualitatively distinguish among material response and capture the onset of nonlinear behavior. Lissajous-Bowditch curves are closed loop plots of γ_0 on the abscissa and σ(t) on the ordinate (elastic representation) or $\dot{\gamma}_0$ on the abscissa and $\sigma(t)$ on the ordinate (viscous representation). Figure 10.7 shows raw elastic Lissajous-Bowditch plots of butter within and outside the LVR for butter at T = 18 °C.

Within the LVR region ($\gamma_0 < 0.01\%$), the plots mirror nearly 'perfect' ellipses where the tangent slope corresponds to G' and the area enclosed by the ellipse represents G''. Beyond LVR at amplitudes where yield stress is exceeded ($\gamma_0 > 0.01\%$), the plot become gradually distorted and acquire square-like shapes enclosing increasingly larger areas (Ewoldt, Hosoi, & McKinley, 2008; Hyun et al., 2011). Typical features, e.g. global strain softening, and additional local features, e.g. intracycle stiffening, masked by the average viscoelastic moduli can be visualized. Global or average elastic softening, is manifested as inter-cycle clockwise rotation in the slope of the stress-strain curve at strain minima $\gamma_0 = 0$ (i.e. at the 'origin' where strain rate $\dot{\gamma}$ is at maxima) toward the strain-axis. Local strain stiffening is clearly visible as the intracycle upturn of the shear stress at strain maxima $\gamma_0 = \text{max}$ (i.e. at the 'extreme' where $\dot{\gamma} = 0$). Stress overshoots, akin to those observed during in start-up shear, appear in the upper left quadrant of the Lissajous-Bowditch and indicate yielding. Such a stress overshoot is rather 'smooth' (where the tangent to the nearly flat part of the curve cuts the stress axis) for butter and is closely related (though lower) to the stress associated with spreading (Prentice, 1984a). The reversibility of the observed behavior during flow reversal evokes microstructure "healing" or thixotropy (Renou, Stellbrink, & Petekidis, 2010; Kim, Merger, Wilhelm, & Helgeson, 2014; Ewoldt & McKinley, 2010) as previously reported for butter (Sone, 1961; Macias-Rodriguez & Marangoni, 2016). A similar analysis was early described by Elliot and Ganz (1971) and Prentice (1984b) to compare yielding during steady shear with that of oscillatory shear for butter. Application of oscillatory strain amplitude exceeding the yield stress resulted in similar behavior to start-up shear, as demonstrated by nearly square waves similar to those depicted in Fig. 10.7 at $\gamma_0 = 40\%$. Prentice also suggested that Lissajous-Bowditch curves of

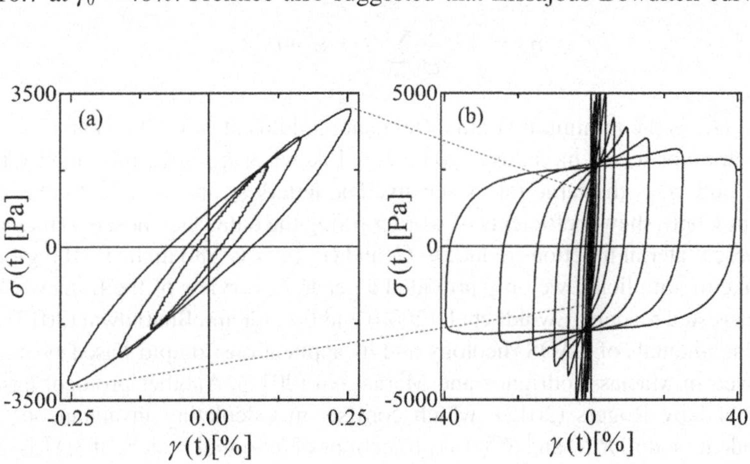

Fig. 10.7 Raw Lissajous-Bowditch plots (elastic perspective) of butter (T = 18 °C) obtained from an oscillatory shear test ($\omega = 3.6$ rad/s) (**a**) linear to mildly nonlinear transition (SAOS to LAOS), (**b**) fully nonlinear response (LAOS)

cream reflected a type of 'fugitive elasticity' (or fluid viscosity) established under the action of shearing forces but rapidly relaxed as shearing ceases (Prentice, 1984c). As mentioned earlier, the first-harmonic viscoelastic moduli are the common output of most rheometers, whereas the locally defined measures should be calculated using any of the frameworks proposed to analyze the nonlinear response such as FT, Chebyshev stress decomposition method. The FT analysis, converts time-domain periodic signals into frequency-domain signals that make up a spectrum encompassing the fundamental frequency and its higher order odd integers ($n = 1, 3, 5\ldots$) in the case of nonlinear response. The leading higher-order harmonic (i.e. the third harmonic) signals the onset of nonlinear behavior. Despite its utility, this "elegant" approach provides little if any physical insight into the nonlinear response. Therefore, other frameworks that overcome this weakness have been proposed such as the Chebyshev stress decomposition. According this, the stress response can be decomposed into elastic and viscous stresses, described by Chebyshev polynomial series of the first kind, which are interrelated to the Fourier series. Similar to the third order Fourier harmonic, Chebyshev coefficients indicate departure from linearity. Some of the metrics developed to capture the nonlinear response, verbatim (Ewoldt et al., 2008):

$$G_M^{'} \equiv \frac{d\tau}{d\gamma} = \sum_{n:odd} nG_n^{'} = e_1 - 3e_3 + \ldots, \tag{10.17}$$

$$G_L^{'} \equiv \frac{\tau}{\gamma} = \sum_{n:odd} G_n^{'} \left(-1\right)^{(n-1)/2} = e_1 + e_3 + \ldots, \tag{10.18}$$

$$\eta_M^{'} \equiv \frac{d\tau}{d\dot{\gamma}} = \frac{1}{\omega} \sum_{n:odd} nG_n^{''} \left(-1\right)^{\frac{n-1}{2}} = \upsilon_1 - 3e_3 + \ldots, \tag{10.19}$$

$$\eta_L^{'} \equiv \frac{\tau}{\dot{\gamma}} = \frac{1}{\omega} \sum_{n:odd} G_n^{''} = \upsilon_1 + \upsilon_3 + \ldots, \tag{10.20}$$

where $G_M^{'}$ is the minimum-strain or tangent modulus at $\gamma(t) = 0$ and and $G_L^{'}$ is the large-strain or secant modulus at $\gamma(t) = \gamma_{max}$. Likewise, $\eta_M^{'}$ is the minimum-rate viscosity and $\eta_L^{'}$ is the large-rate viscosity. The letters e_n and v_n refer to elastic and viscous Chebyshev coefficients of n order fitting the data, and chosen as they allow. All these material functions reduce to G' and G'' ($\eta' = G''/\omega$) in the LVR region. For the sake of simplicity, we only provided a general overview of the framework. For the interested readers, Ewoldt et al. (2008) and Ewoldt and Bharadwaj (2013) cover the fundamentals of LAOS rheology and its applications to lipid-based systems are reviewed in Macias-Rodriguez and Marangoni (2017). Another protocol has been developed by Rogers (2012), which consists in calculating instantaneous time-dependent moduli $R'(t)$ and $R''(t)$ as projections of binomial vectors of 3D Lissajous-Bowditch plot (stress vs. strain vs. strain rate) onto the strain-stress and shear rate-stress plane. This analysis provides a 'complete' picture of yielding that is not revealed by the stress-decomposition Chebyshev framework or any measures at 'fixed' points. It must be noted that this area of research is ongoing progress and so far there is not a definite answer as to whether a protocol is more suitable or appropriate than the other. Each protocol has its own merits and weaknesses that shall be considered prior to their implementation.

4.2.6 Drag Flow: Step Shear Stress (Creep) and Step Shear Strain (Stress Relaxation)

Creep tests consist in applying constant step load (small or large stresses σ_0) for some extended time and measuring displacement (strain) before and after removal (recovery) of the load. The time dependence of creep tests makes them particularly suitable for evaluating mechanical responses both for a short time (material compliance and firmness) and for long times (such as creeping of stacked butter in-store). The load and the duration of the time period should be high and long enough respectively, to induce sufficient creep motion and quantify the Newtonian viscosity, but not too high or long to trigger formation of cracks or induce irreversible plastic deformations as observed in butter (DeMan et al., 1985). Another use of creep tests is for the determination of the yield stress σ_y, where for $\sigma_0 < \sigma_y$ results in $\gamma(t)$ curves characterized by an increase in strain and then a plateau or saturation (a hallmark of the solid regime of pastes), and $\sigma_0 > \sigma_y$ the same curves tend to a straight line with slope 1 in logarithmic scale, indicating infinite deformation at constant rate (a hallmark of the liquid regime of pastes) (Coussot, Tabuteau, Chateau, Tocquer, & Ovarlez, 2006). Experimentally, for stresses $\sigma_0 < \sigma_y$, creep compliance curves overlap or nearly overlap onto each other, whereas for stresses $\sigma_0 > \sigma_y$, deviations from this behavior occur and indicate the onset of nonlinear behavior. However, calculation of σ_y by this method may prove inefficient in stiff pastes (e.g. butter at low temperature) due to heterogeneous flow imposed by sudden stress jumps and higher sensitivity of this method to structural changes over long periods at constant stress (Dinkgreve et al., 2016). For stress inputs $\sigma_0 < \sigma_y$, the rheological behavior of butter and milk fat bears some resemblance with a Burgers body (see Fig. 10.8a), comprising a Maxwell model (spring and dashpot in series) coupled to a Kelvin-Voigt model (spring and dashpot in parallel) (Steffe, 1996) (Fig. 10.8b).

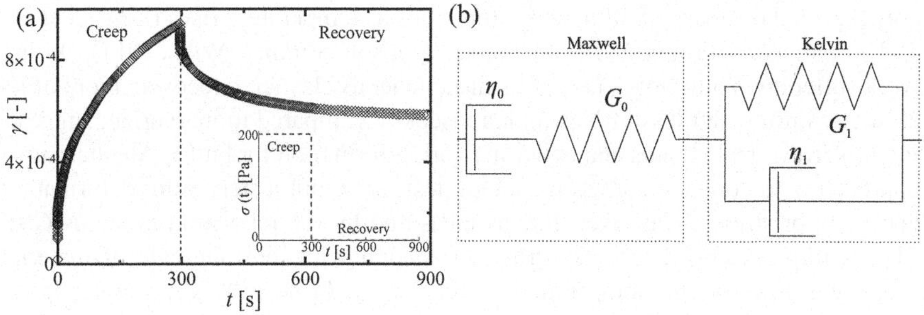

Fig. 10.8 (**a**) Strain deformation of butter (T = 18 °C) measured during (**a**) creep (σ_0 = 200 Pa) and recovery (σ_0 = 00 Pa). (**b**) Burgers model comprising Maxwell and Kelvin elements in series

The compliance function $J(t) = \gamma(t)/\sigma_0$ during the creep (a) and recovery (b) is described by the following equations:

$$J(t) = \frac{t_c}{\eta_0} + \frac{1}{G_0} + \frac{1}{G_1}\left[1 - \exp\left(-\frac{t_c}{\lambda_1}\right)\right] \text{Creep} \quad (10.21)$$

$$J(t) = \frac{t_c}{\eta_0} + \frac{1}{G_1}\left[1 - \exp\left(-\frac{t_c}{\lambda_1}\right)\right] \exp\left(-\frac{t_r - t_c}{\lambda_1}\right); t_r > t_c \text{ Recovery} \quad (10.22)$$

where t_c and t_r are the creep and recovery times, respectively, G_0 and μ_0 are the elastic modulus and viscosity of the Maxwell element, G_1 and μ_1 are similar parametersin the Kelvin-Voigt element and $\lambda_1 = \frac{\eta_1}{G_1}$ is the characteristic retardation time. Notethat G_0 and G_1 are equivalent to instantaneous J_0 and retarded compliances J_1 as $J = 1/G$ in the LVR. Increase the number of linear elements (e.g. adding an extra Kelvin body in series) is customary to describe the complex relaxation behavior, this, however, increases the phenomenology of the model (Scott Blair & Burnett, 1959). It has been suggested that firmness is a time-dependent texture attribute related to the time shear creep compliance $J(t)$:

$$F(t) = \frac{1}{J(t)} \quad (10.23)$$

Previous studies have shown that $J(t)$ can differentiate between soft and firm materials (e.g. cheese) (Brown, Foegeding, Daubert, Drake, & Gumpertz, 2003; Ewoldt, 2014). Figure 10.9 shows an illustration of a creep and recovery compliance obtained for butter T = 18 °C. At first glance, butter shares some resemblance with a Burgers body; however, materials properties calculated from the creep phase are not necessarily the same with those obtained from the recovery phase even for stress inputs within the LVR. Scott Blair and Burnett (1959) attributed this behavior to marked thixotropic behavior at the beginning of straining and recovery of renneted milk gels, and offered some additions to the Burgers model. Past literature on butter investigating their viscoelastic properties as a function of temperature and expectedly reported decrease of all material functions as temperature rises (DeMan et al., 1985; Shama & Sherman, 1970; Vithanage, Grimson, Smith, & Wills, 2011). During work softening, butter loses less of its instantaneous elasticity, recovers more of its elasticity during and less of Newtonian viscosity compared to margarine, suggesting less irreversible bonds and better network reformation for butter. An alternative approach is to conduct a stress relaxation test, i.e. input a step strain deformation and monitor stress decay over time as measured by the relaxation modulus $G(t)$. This is important to determine stress dissipation with the timescale of material usage, e.g. lowering the temperature of butter from 17 °C to 8 °C promotes faster stress relaxation.

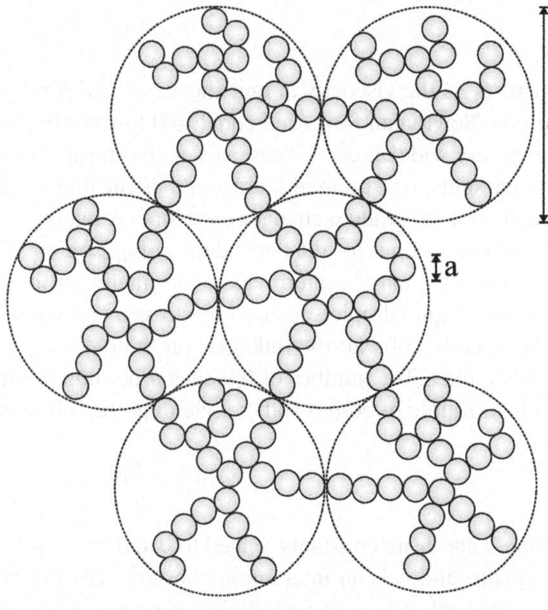

Fig. 10.9 Putative microstructure of a fractal aggregate in a fat crystal network. Particles of size *a* aggregate in flocs with limits ξ represented by dotted lines, which form a close packing

5 Microscopic Modeling: Structure and Rheology Relationships

5.1 Cream

Some attempts have been made to provide some quantitative relationships between solid fat content and steady state viscosity of cream. For Newtonian flow, e.g. in creams containing ≤50% fat at T = 40–80 °C, hydrodynamic interaction determines the viscosity of the cream η as a function of volume fraction of dispersed particles ϕ, as described by Euler's equation:

$$\eta = \eta_0 \left(1 + \frac{1.25\phi}{1 - \phi / \phi_{max}}\right)^2 \tag{10.24}$$

where η_0 is the viscosity of the continuous phase and ϕ_{max} is the maximum volume fraction depending on shape and size distribution of the dispersed particles (spherical for cream). For a correct description of the data, ϕ must take into account the sum of volume fraction of fat globules, casein micelles, protein molecules and lactose molecules, and $\eta_0 \approx 1.02\eta_{water}$ (considering the continuous phase is roughly a 1% solution of salts) and $\phi_{max} \approx 0.9$. This relationship holds for $\eta/\eta_0 > 10$ at $\dot{\gamma}_0 \leq 10\ s^{-1}$. For non-Newtonian shear-thinning behavior, e.g. in creams containing 60% fat at T = 15–80 °C, η_{app} has been described using a power law equation of the following type:

$$\eta'_{app} = \eta'_1 \dot{\gamma}^{-\beta} \quad (10.25)$$

where η'_1 is the viscosity at unit shear rate and β is a constant having a finite value for non-Newtonian flow (and equal to 0 for Newtonian flow). The power law holds over a very wide range of shear rates, including at very low shear rates ($\dot{\gamma}_0 \approx 10^{-4}\,s^{-1}$) though such measurements are impractical due to the biological nature of milk which undergo gradual changes over time. A major weakness of this equation is that it does not account for the dependence of η_{app} on $\dot{\gamma}_0$. In this regard, it has been suggested that η_{app} may be interpreted as originating from disruption of linkages amongst dispersed liquid droplets, caused by stress σ and shear rate $\dot{\gamma}_0$ (Blair, 1965). On one side, σ tends to break the linkages, on the other $\dot{\gamma}_0$ hinders droplets to become in close contact. The number of linkages n per unit volume is linearly proportional to the logarithmic of both σ and $\dot{\gamma}_0$, yielding the following relationship:

$$\ln \dot{\gamma} = \frac{a}{b} ln\tau + C \quad (10.26)$$

where a and b are constants related to σ and $\dot{\gamma}_0$ ($a > b$ for shear-thinning fluids such as cream) and C is an integration constant. The Eq. (10.26) holds in a wider range of $\dot{\gamma}_0$ though it fails at very high shear rates at which all linkages may be broken (Blair, 1965). The picture of physical linkages appears more suitable for partially-coalesced high-fat creams displaying 'network-like' formation; otherwise such concept can be envisaged as averages in time and space of the forces acting on the globules (Prentice, 1993; van Vliet & Walstra, 1979).

5.2 *Butter, Milk Fat and Dairy Spreads*

Constitutive equations have been proposed to model the rheology of butter, milk fat and dairy spreads. Some of these efforts include the visco-plastic model (Tanaka et al., 1971), the modified Bingham model (Elliot & Green, 1972) and the viscous Maxwell-Bingham model (Diener & Heldman, 1968). Tanaka et al. (1971) proposed to express stress (obtained from penetration) as the sum of stress caused by plastic and viscous deformations where the primer is associated with yield value and the latter with apparent viscosities post yielding. Elliot and Green (1972) proposed the use of a modified Bingham model after observation of the stress response of butter subjected to steady shear and dynamic oscillations. The model consisted of a viscous element and yield stress element comprising static and dynamic yield stresses, and an elastic element of modulus connected in series. Diener and Heldman (1968) assigned the Maxwell element of his model to the fat globule theorized as being purely viscous internally and elastic externally or at the boundaries, the yield element to the failure of crystal grain boundary and viscous elements of the free fat in the crystal granules. Despite the utility of these models as "visual aids" to describe empirical equations, the use of spring and dashpots do not represent the actual structure being subjected to deformations. Therefore, we will turn our attention to mesoscopic models that aim to establish structure-rheology relationships, particularly those applicable to fat crystals networks present in milk fat and dairy spreads. Several microscopic models to link the structure and rheology of fat crystal networks have been proposed starting from the early proposed linear chain model (van den Tempel, 1979) to the fractal model (Marangoni & Rousseau, 1996; Narine & Marangoni, 1999c; Tang & Marangoni, 2007; Vreeker, Hoekstra, den Boer, & Agterof, 1992). The discussion presented here will be restricted to latter given its relevance and sound basis. The fractal model operates under the assumption that fat particles (crystalline nanoplatelets) assemble into fractal (self-similar) flocs or clusters (Fig. 10.9). Above a critical particle fraction ($\phi_c \approx 0.05$–0.10), flocs grow enough and overlap ('gelation' occurs) forming a continuous microscopic crystal network with solid-like or elastic behavior.

The elasticity of the system scales as a function of particle fraction (ϕ = SFC/100), according to which two discrete regimes may be distinguished (Marangoni & Rousseau, 1996; Narine & Marangoni, 1999b). At low particle concentration ($\phi < 0.1$), the strong-link regime occurs, i.e. individual clusters grow large and their elasticity (not that of the cluster links) dictates the elastic modulus G' of the system, where each aggregate has a backbone which bears the force applied to it:

$$G' \sim \phi^{\left[(d+x)/(d-D)\right]} \tag{10.27}$$

where d is the Euclidean dimension of the embedding space (usually 3), D is the fractal dimension, which describes spatial distribution and morphology of the network, x is the backbone fractal dimension (~1 to 1.3). The fractal dimension also provides information on the aggregation mechanism of the network, e.g. $D = 1.75$ for DLCA and $D = 2.1$ for RLCA. At high particle concentration ($\phi > 0.1$ such as in most model and complex fat systems), the weak-link regime occurs, i.e. small clusters behave as rigid springs and the links among clusters or microstructures govern the elasticity of the system:

$$G' \sim \phi^{1/(d-D)} \tag{10.28}$$

An additional pre-exponential term $\dfrac{A}{\pi a \gamma d_0^2}$ (for crystal aggregates with spherical-like morphology) added to the weak-link relationship helps to discern among crystal networks showing similar D and ϕ but different G' (Marangoni, 2000). In the pre-exponential term, A corresponds to the Hamaker's constant, a to the size of the primary crystal unit and d_0 to the intercluster distance (Narine & Marangoni, 1999b). A modified fractal model has also been proposed for less "ideal" systems that do not obey exact power law dependence, due to highly heterogeneous crystal networks.

In such systems, stress localization happens in a small fraction of the interconnected network which serves as "weakest bonds" (Tang & Marangoni, 2007). The separate effects of the network microstructure, stress distribution and interconnectivity are determined as follows:

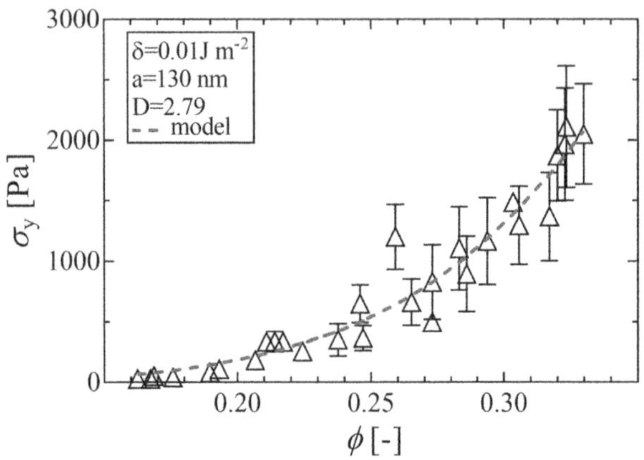

Fig. 10.10 Yield stress σ_y as a function of volume fraction ϕ for blends of milk fat with canola oil. Samples were crystallized statically from 40 to 5 °C at a cooling rate of 1 °C/min, and annealed at 5 °C for 2 h. Symbols represent the average and standard deviation of 2–6 samples fitted to the model in Eq. (10.30). Estimates of the model parameters are indicated. The surface free energy δ was fixed as a constant

$$G' \sim \phi_e^{1/(d-D)} \tag{10.29}$$

where the effective volume fraction of solids responsible for stress-bearing is $\phi_e = 1 - e^{-k\phi^b}$ and k and b are constants. Furthermore, a rough estimation of σ_y of milk fat crystal networks (Fig. 10.10) can be obtained applying thermodynamic considerations as follows:

$$\sigma_y \sim \frac{6\delta}{a}\phi^{1/(d-D)} \tag{10.30}$$

where δ is equivalent to the crystal-melt interfacial tension ($\delta \approx 0.01$ J m^{-2}). Dividing both sides by γ_y yields an estimation of the Young modulus $\in$:

$$\in = \frac{\sigma_y}{\gamma_y} \sim \frac{6\delta}{a\varepsilon^*}\phi^{1/d-D} \tag{10.31}$$

where the shear modulus G can be estimated from $\in$ assuming a Poisson ratio $\mu = 0.5$ for incompressible viscoelastic materials, through the following relation:

$$\epsilon = 2(1+\mu)G \tag{10.32}$$

Overall, despite the elegance of these approaches, there still seems to be some degree of uncertainty on the estimation of G', since milk fat or butter set relatively slowly compared to other type of fats due to its highly heterogeneous composition that hinders crystal packing. The exact variation of G' as a function of ϕ, and pre-exponential affecting G' also requires accurate estimation of Hamaker's constant for two interacting crystal nanoplatelets and structural information of both nano and microstructure. Obtaining such information is by no means trivial. For a comprehensive description of network mesoscopic models, the readers are referred to (Narine & Marangoni, 1999a; Tang & Marangoni, 2007).

6 Comparison Among Methods for Rheological and Texture Characterization

Direct comparisons amongst different methods and research reports cannot be drawn prior to careful assessment of the measuring principles of the test, testing conditions and relationships between load, time, deformation during measurement. Such relationships are strictly defined in fundamental tests (e.g. shear oscillatory, shear creep, etc) for homogeneous deformations. For heterogeneous deformations, rheological functions become variant to testing conditions and thus become apparent. In empirical tests, flows are poorly defined, properties depend on measuring specifications and comparisons among tests are made at the practitioner's own risk. While correlations among empirical tests such as penetrometry, indentation, empirical extrusion tests have been widely found, these do not warrant that similar properties had been measured as suggested in the literature. Strong correlations have been found between quite different properties, e.g. yield value and viscosity (using cone penetrometry). Dolby found excellent agreement between his 'sectility' measurements and Scott Blair's apparent viscosities determined by Scott Blair's parallel plate plastometer, and conceptualized that similar properties may have been measured (Scott Blair, 1954). Scott Blair's viscosities were determined at arbitrary conditions and thus comparisons appear somewhat dubious (Mulder, 1953). Drawing correlations between yield value and viscosity seems strictly dependent on the specific rheology of the butter and measurement conditions (e.g. extent of deformation). For example, a firm and plastic butter may show a high yield stress to initiate "flow" and high viscosities due to 'gradual' yielding (i.e. proportionality between yield stress and viscosity holds true). A firm and brittle butter will not follow the same trend, yield stress may be high but viscosity post-yielding will drop substantially as it breaks catastrophically. Some experimental techniques such as oscillatory shear will be more sensitive to capture such differences than empirical instruments such as cone penetrometry. Several authors have correctly asserted that penetration cannot discriminate among all butter or fat blends rheology. Direct comparison of small deformation and large deformation tests seem troublesome since the primer are attributed with viscoelastic properties of the pristine butter or milk fat microstructure, whereas the latter with the same properties or viscoelastic textural attributes of the 'broken' microstructure. In some cases, sensory correlation with instrumental parameters or material properties (e.g. viscosity) is remarkably successful (Scott Blair, 1958).

7 Properties and Attributes of Most Interest

The main fundamental properties of interest in butter, milk fat and milk fat blends include their elastic and viscous properties, yield stress and thixotropy. These influence widely measured attributes of butter such as firmness, spreadability, 'setting', 'work softening'. In general, the elastic modulus is large not major differences among butters may be expected provided they are formulated, processed and stored at similar conditions. Butters do not show marked elasticity (i.e. they have a narrow linear regime) and yield stress differences are typically marginal. This does not imply that elastic properties are unimportant since they contribute to solid-like behavior and provide "a vivid appearance" (Mulder, 1953). On the other hand, viscous properties determine whether the material displays a brittle-like behavior or a ductile-like behavior. In butter and milk fat the yield stress refers as the critical value beyond which the material transitions from purely elastic deformations to plastic deformations. The estimation of this value appears to be obtained by oscillatory shear experiments in butter and milk fat despite certain limitations. Firmness and butter are two major important viscoelastic texture attributes affecting the acceptability of butter, which are inversely related. Firmness is a time-dependent attribute and as earlier mentioned, it can be best estimated by the shear creep time compliance. Spreadability involves large shear deformations such as those obtained by extrusion experiment. Thus, apparent viscosities at arbitrary shear rates determined by extrusion seems to provide a fair assessment of this property. It has been argued that viscosity and spreadability are inversely related (Davis, 1937), i.e. spreadability is directly related to the extent of shear-thinning of butter. 'Setting' refers to the increase of firmness over storage time due to continuous crystallization and crystal aggregation, which causes an increase in the viscoelastic moduli during storage. On the opposite case, 'work softening' has been used to describe a decrease in the consistency of butter when worked or kneaded (Van Aken & Visser, 2000). This decrease is due to strong strain softening of the elastic modulus and shear thinning of the viscosity and thus these more general terms appear more appropriate. After application of shear, the modulus and viscosity increase partially or fully during 'rest' as network restructuring occurs, a property referred as to thixotropy. A distinction must be made between the decrease and subsequent recovery of viscoelastic properties or textural attributes (i.e. viscoelastic moduli, firmness) that occur from the action of strong shear (e.g. butter working) and those from 'mild' shear (e.g. spreading) (Prentice, 1993; Sone, 1961). In the first case, mechanical behavior results mainly from melting and re-crystallization of the network, whereas in the second case, it arises due to thixotropic behavior.

8 Control of Rheological Properties of Milk Fat and Butter

There are several formulation and processing schemes aimed to tailor the rheology of milk fat, butter and dairy fat blends. Here, we briefly cover those approaches which have been extensively reviewed (Wright et al., 2001). Original FA and TAG composition of milk fat varies according source, season, animal breed, feed, among other factors. This can be altered using various processes including blending, fractionation and interesterification. In addition to composition, pretreatment of milk cream and crystallization conditions such as cooling, shear or mechanical working all affect crystallization and formation of mesoscopic crystal networks that determine rheological properties.

A commonly used approach to modify the original composition of milk fat is to modify the feed of the cows such as with the addition of vegetable or fish oils high in unsaturation or whole oilseeds (e.g. canola seeds). This typically increases oleic acid content while reducing saturated fatty acids content and leading to softer butters. Supplementation of the feed with sufficiently high levels of stearic acid can also promote desaturation of the mammary gland and improve butter spreadability.

Blending with vegetables oil increases the level of unsaturation of milk fat and affects crystallization as observed in melting profiles and hardness index (Rousseau, Hill, & Marangoni, 1996). This operation needs to comply with standards of identity for butter and dairy fat blends. Fractionation separates milk fat into various fractions with distinct TAG chemical makeup and physical properties, which are recombined in various proportion to improve the spreadability of butter. For example, very high melting fractions melt at $T_m > 50$ °C and provide structural integrity, while low melting fraction melt at T = 10–25 °C and reduce hardness, of recombined butters (Wright & Marangoni, 2006). Interesterification randomizes FAs along the TAG backbone, altering crystallization and thus physical properties. Chemical and enzymatic interesterification improve cold spreadability of butter though this comes to reduction of butter flavor (Wright & Marangoni, 2006).

Aging or ripening of cream involves holding it at specific temperatures typically overnight to alter crystallization of the fat contained within the globules. This cost-effective and successful approach has been a widespread practice to reduce firmness and increase spreadability of butter. Churning of cream induces partial phase inversion, i.e. rupture of fat globule and release of fat crystal for their further aggregation and crystallization. Depending on the process, butters with varying rheology can be obtained. For example, continuously-churned butter is firmer than batch-churned butter, since in the primer fat globules are fully destroyed and hence more crystals are available to form the mesoscopic network. Mechanical treatment can also be applied during manufacture of butter, e.g. continuous agitation and shearing during crystallization enhances secondary nucleation, removal of heat of crystallization, which lead to creation of discrete crystal aggregates, whereas batch crystallization promotes formation of larger crystals. Crystal morphology affects the texture of butter, e.g. adequate mechanical working and cooling during crystallization leads to firm yet spreadable butter, whereas excessive working and poor heat dissipation, causes melting that leads 'sticky' butter. Storage conditions influence 'setting' of butters and their final texture, e.g. mild increases in storage temperature can double the firmness of butter (Mortensen & Danmark, 1981).

9 Conclusions

Major developments on the rheology and texture of butter, milk fat and dairy spreads took place in the past as the subfield of food rheology emerged. Early studies were mainly empirical and consisted largely of the measurement of attributes such as firmness and spreadability. With the advent of modern rheometric techniques, more fundamental methods have been introduced to characterize rheological properties of these products. With increasing consumption of dairy products forecasted for the upcoming years, rheology and texture will continue to provide valuable insights into the quality, usability and consumer acceptability of butter and dairy spreads.

References

Atkins, A. G., & Tabor, D. (1965). Plastic indentation in metals with cones. *Journal of the Mechanics and Physics of Solids, 13*, 149–164.

Benbow, J., & Bridgewater, J. (1993). *Paste flow and extrusion*. Oxford, UK: Clarendon Press.

Blair, G. W. S. (1965). On the use of power equations to relate shear-rate to stress in non-Newtonian liquids. *Rheologica Acta, 4*, 53–55.

Brown, J. A., Foegeding, E. A., Daubert, C. R., Drake, M. A., & Gumpertz, M. (2003). Relationships among rheological and sensorial properties of young cheeses. *Journal of Dairy Science, 86*, 3054–3067.

Campanella, O. H., & Peleg, M. (2002). Squeezing flow viscometry for nonelastic semiliquid foods — theory and applications. *Critical Reviews in Food Science and Nutrition, 42*, 241–264.

Castro, M., Giles, D. W., Macosko, C. W., & Moaddel, T. (2010). Comparison of methods to measure yield stress of soft solids. *Journal of Rheology, 54*, 81–94.

Chatraei, S., Macosko, C. W., & Winter, H. H. (1981). Lubricated squeezing flow: A new biaxial extensional rheometer. *Journal of Rheology, 25*, 433.

Coussot, P. (2007). Rheophysics of pastes: A review of microscopic modelling approaches. *Soft Matter, 3*, 528–540.

Coussot, P., Tabuteau, H., Chateau, X., Tocquer, L., & Ovarlez, G. (2006). Aging and solid or liquid behavior in pastes. *Journal of Rheology, 50*, 975.

Davis, J. G. (1937). The rheology of cheese, butter and other milk products. *The Journal of Dairy Research, 8*, 245–264.

Dealy, J. M., & Wissbrun, K. F. (1999). *Melt rheology and its role in plastic processing: Theory and applications*. New York, NY: Van Nostrand Reinhold.

Deman, J. M. (1983). Consistency of fats: A review. *Journal of the American Oil Chemists' Society, 60*, 82–87.

DeMan, J. M., & Beers, A. M. (1987). Fat crystal networks: Structure and rheological properties. *Journal of Texture Studies, 18*, 303–318.

DeMan, J. M., Gupta, S., Kloek, M., & Timbers, G. E. (1985). Viscoelastic properties of plastic fat products. *Journal of the American Oil Chemists' Society, 62*, 1672–1675.

Diener, R. G., & Heldman, D. R. (1968). Methods of determining rheological properties of butter. *Transactions of ASAE, 11*, 444–0447.

Dinkgreve, M., Paredes, J., Denn, M. M., & Bonn, D. (2016). On different ways of measuring "the" yield stress. *Journal of Non-Newtonian Fluid Mechanics, 238*, 233–241.

Elliot, J. H., & Ganz, A. J. (1971). Modification of food characteristics with cellulose hydrocolloids I: Rheological characterization of an organoleptic property (unctuousness). *Journal of Texture Studies, 2*, 220–229.

Elliot, J. H., & Green, C. E. (1972). Modification of food characteristics with cellulose hydrocolloids II: The modified bingham body-a useful rheological model. *Journal of Texture Studies, 3*, 194–205.

Ewoldt, R. H. (2014). Extremely soft: Design with rheologically complex fluids. *Soft Robotics, 1*, 12–20.

Ewoldt, R. H., & Bharadwaj, N. A. (2013). Low-dimensional intrinsic material functions for nonlinear viscoelasticity. *Rheologica Acta, 52*, 201–219.

Ewoldt, R. H., Hosoi, A. E., & McKinley, G. H. (2008). New measures for characterizing nonlinear viscoelasticity in large amplitude oscillatory shear. *Journal of Rheology, 52*, 1427–1458.

Ewoldt, R. H., & McKinley, G. H. (2010). On secondary loops in LAOS via self-intersection of Lissajous–Bowditch curves. *Rheologica Acta, 49*, 213–219.

Faber, T. J., Jaishankar, A., & McKinley, G. H. (2017a). Describing the firmness, springiness and rubberiness of food gels using fractional calculus. Part II: Measurements on semi-hard cheese. *Food Hydrocolloids, 62*, 325–339.

Faber, T. J., Jaishankar, A., & McKinley, G. H. (2017b). Describing the firmness, springiness and rubberiness of food gels using fractional calculus. Part I: Theoretical framework. *Food Hydrocolloids, 62*, 311–324.

Ferry, J. D. (1980). *Viscoelastic properties of polymers*. New York: Wiley.

Haighton, A. J. (1959). The measurement of the hardness of margarine and fats with cone penetrometers. *Journal of the American Oil Chemists' Society, 36*, 345–348.

Haighton, A. J. (1965). Worksoftening of margarine and shortening. *Journal of the American Oil Chemists' Society, 42*, 27–30.

Hanck, R. C., & Wall, C. W. (1966). Pressure losses and rheological properties of flowing butter. *Journal of Food Science, 31*, 534–541.

Hayakawa, M., & DeMan, J. M. (1982). Interpretation of cone penetrometer consistency measurements of fats. *Journal of Texture Studies, 13*, 201–210.

Heertje, I. (1993). Microstructural studies in fat research. *Food Structure, 12*, 77–94.

Huppertz, T., & Kelly, A. L. (2006). Physical chemistry of milk fat globules. In P. L. McSweeney & P. F. Fox (Eds.), *Advanced dairy chemistry. Volume 2: Lipids* (pp. 173–212). New York, NY: Springer.

Hyun, K., Wilhelm, M., Klein, C. O., Cho, K. S., Nam, J. G., Ahn, K. H., et al. (2011). A review of nonlinear oscillatory shear tests: Analysis and application of large amplitude oscillatory shear (LAOS). *Progress in Polymer Science, 36*(12), 1697–1753.

Kamyab, I., Chakrabarti, S., & Williams, J. G. (1998). Cutting cheese with wire. *Journal of Materials Science, 33*, 2763–2770.

Kawanari, M., Hamann, D. D., Swartzel, K. R., & Hansen, A. P. (1981). Rheological and texture studies of butter. *Journal of Texture Studies, 12*, 483–505.

Kim, J., Merger, D., Wilhelm, M., & Helgeson, M.E. (2014). Microstructure and nonlinear signatures of yielding in a heterogeneous colloidal gel under large amplitude oscillatory shear. *Journal of Rheology, 58*, 1359–1390.

Kloek, W., van Vliet, T., & Walstra, P. (2005). Large deformation behavior of fat crystal networks. *Journal of Texture Studies, 36*, 516–543.

Lyons, J., & Pyne, G. T. (1933). Factors affecting the body or viscosity of cream and related matters. *The Economic Proceedings of the Royal Dublin Society, 2*, 461–500.

Macias-Rodriguez, B., & Marangoni, A. G. (2016). Rheological characterization of triglyceride shortenings. *Rheologica Acta, 55*, 767–779.

Macias-Rodriguez, B. A., & Marangoni, A. A. (2017). Linear and nonlinear rheological behavior of fat crystal networks. *Critical Reviews in Food Science and Nutrition, 58*(14), 2398–2415.
Macosko, C. W. (1994). *Shear rheometry: Pressure driven flows*. New York, NY: Wiley-VCH, Inc.
Marangoni, A. G. (2000). Elasticity of high-volume-fraction fractal aggregate networks: A thermodynamic approach. *Physical Review B: Condensed Matter and Materials Physics, 62*, 13951–13955.
Marangoni, A. G., & Rousseau, D. (1996). Is plastic fat rheology governed by the fractal nature of the fat crystal network? *Journal of the American Oil Chemists' Society, 73*, 991–994.
Mortensen, B. K. (1983). Physical properties and modification of milk fat. In P. F. Fox (Ed.), *Developments in dairy chemistry, volume 2: Lipids* (pp. 159–194). Essex, UK: Applied Science Publishers Ltd.
Mortensen, B. K., & Danmark, H. (1981). Firmness of butter measured with a cone penetrometer. *Milchwissenschaft, 36*, 393–395.
Mulder, H. (1953). The consistency of butter. In G. W. Scott Blair (Ed.), *Foodstuffs their plasticity, fluidity and consistency* (pp. 91–123). New York, NY: Interscience Publishers, Inc..
Narine, S. S., & Marangoni, A. G. (1999a). Relating structure of fat crystal networks to mechanical properties: A review. *Food Research International, 32*, 227–248.
Narine, S. S., & Marangoni, A. G. (1999b). Mechanical and structural model of fractal networks of fat crystals at low deformations. *Physical Review. E, Statistical Physics, Plasmas, Fluids, and Related Interdisciplinary Topics, 60*, 6991–7000.
Narine, S. S., & Marangoni, A. G. (1999c). Microscopic and rheological studies of fat crystal networks. *Journal of Crystal Growth, 198–199*, 1315–1319.
Phipps, L. W. (1969). The interrelationship of the viscosity, fat content and temperature of cream between 40° and 80°C. *The Journal of Dairy Research, 36*, 417–246.
Pipkin, A. C. (1972). *Lectures on viscoelasticity theory* (2nd ed.). New York, NY: Springer-Verlag.
Prentice, J. H. (1984a). Plastic fats. In J. H. Prentice (Ed.), *Measurements in the rheology of foodstuffs* (pp. 140–152). Essex, UK: Elsevier Ltd.
Prentice, J. H. (1984b). Measurements on some fluids and their interpretation. In *Measurements in the rheology of foodstuffs* (pp. 108–129). Essex, UK: Elsevier.
Prentice, J. H. (1984c). Measurements on some fluids and their intepretation. In J. H. Prentice (Ed.), *Measurements in the rheology of foodstuffs* (pp. 108–129). Essex, UK: Elsevier Ltd.
Prentice, J. H. (1993). Rheology and texture of dairy products. *Journal of Texture Studies, 3*, 415–458.
Renou, F., Stellbrink, J., & Petekidis, G. (2010). Yielding processes in a colloidal glass of soft star-like micelles under large amplitude oscillatory shear (LAOS). *Journal of Rheology, 54*, 1219.
Rogers, S. A. (2012). A sequence of physical processes determined and quantified in (LAOS): An instantaneous local 2D/3D approach. *Journal of Rheology, 56*, 1129–1151.
Rohm, H., & Weidinger, K. H. (1993). Rheological behaviour of butter at small deformations. *Journal of Texture Studies, 24*, 157–172.
Rousseau, D., Hill, A. R., & Marangoni, A. G. (1996). Restructuring butterfat through blending and chemical interesterification. 3. Rheology. *Journal of the American Oil Chemists' Society, 73*, 983–989.
Scott Blair, G. W. (1938). The spreading capacity of butter. I. *The Journal of Dairy Research, 9*, 208–214.
Scott Blair, G. W. (1947). The role of psychophysics in rheology. *Journal of Colloid Science, 2*, 21–32.
Scott Blair, G. W. (1953). *Foodstuffs: Their plasticity, fluidity and consistency*. Amsterdam, The Netherlands: North-Holland.
Scott Blair, G. W. (1954). The rheology of fats: A review. *Journal of the Science of Food and Agriculture, 5*, 401–405.
Scott Blair, G. W. (1958). Rheology in food research. *Advances in Food Research, 8*, 1–61.
Scott Blair, G. W., & Burnett, J. (1959). On the creep, recovery, relaxation and elastic "memory" of some renneted milk gels. *British Journal of Applied Physics, 10*, 15–20.

Scott Blair, G. W., Hening, J. C., & Wagstaff, A. (1939). The flow of cream through narrow glass tubes. *The Journal of Physical Chemistry, 43*, 853.

Shama, F., & Sherman, P. (1970). The influence of work softening on the viscoelastic properties of butter and margarine. *Journal of Texture Studies, 1*, 196–205.

Shukla, A., & Rizvi, S. S. H. (1995). Viscoelastic properties of butter. *Journal of Food Science, 60*, 902–905.

Shukla, A., Rizvi, S. S. H., & Bartsch, J. A. (1995). Rheological Characterization of butter using lubricated squeezing flow. *Journal of Texture Studies, 26*, 313–323.

Sone, T. (1961). The rheological behavior and thixotropy of a fatty plastic body. *Journal of the Physical Society of Japan, 16*, 961–971.

Steffe, J. F. (1996). *Rheological Methods in Food Processing Engineering*. East Lansing, MI: Freeman Press.

Suresh, N., & Marangoni, A. G. (2001). Elastic modulus as and indicator of macroscopic hardness of fat crystal networks. *LWT- Food Science and Technology, 34*, 33–40.

Szczesniak, A. S. (2002). Texture is a sensory property. *Food Quality and Preference, 13*, 215–225.

Tanaka, M., de Man, J. M., & Voisey, P. W. (1971). Measurement of textural properties of foods with a constant speed cone penetrometer. *Journal of Texture Studies, 2*, 306–315.

Tang, D., & Marangoni, A. G. (2007). Modeling the rheological properties and structure of colloidal fat crystal networks. *Trends in Food Science and Technology, 18*, 474–483.

Thareja, P., Golematis, A., Street, C. B., Wagner, N. J., Vethamuthu, M. S., Hermanson, K. D., et al. (2013). Influence of surfactants on the rheology and stability of crystallizing fatty acid pastes. *Journal of the American Oil Chemists' Society, 90*, 273–283.

Thareja, P., Street, C. B., Wagner, N. J., Vethamuthu, M. S., Hermanson, K. D., & Ananthapadmanabhan, K. P. (2011). Development of an in situ rheological method to characterize fatty acid crystallization in complex fluids. *Colloids and Surfaces A: Physicochemical and Engineering Aspects, 388*, 12–20.

Tschoegl, N. W. (1989). *The phenomenological theory of linear viscoelastic behavior: An introduction*. Berlin, Germany: Springer-Verlag.

Van Aken, G. A., & Visser, K. A. (2000). Firmness and crystallization of milk fat in relation to processing conditions. *Journal of Dairy Science, 83*, 1919–1932.

van den Tempel, M. (1961). Mechanical properties of plastic-disperse systems at very small deformations. *Journal of Colloid Science, 16*, 284–296.

van den Tempel, M. (1979). Rheology of concentrated suspensions. *Journal of Colloid and Interface Science, 71*, 18–20.

van Vliet, T., & Walstra, P. (1979). Relationship between viscosity and fat content of milk and cream. *Journal of Texture Studies, 11*, 65–68.

Vithanage, C. R., Grimson, M. J., Smith, B. G., & Wills, P. R. (2011). Creep test observation of viscoelastic failure of edible fats. *Journal of Physics Conference Series, 286*, 12008.

Vliet, T. (2013). *Rheology and fracture mechanics of foods*. New York, NY: CRC Press.

Vreeker, R., Hoekstra, L. L., den Boer, D. C., & Agterof, W. G. M. (1992). The fractal nature of fat crystal networks. *Colloids and Surfaces, 65*, 185–189.

Wright, A. J., & Marangoni, A. G. (2006). Crystallization and rheological properties of milk fat. In P. F. Fox & P. L. H. McSweeney (Eds.), *Advanced dairy chemistry. Volume 2: Lipids* (pp. 245–291). New York, NY: Springer.

Wright, a. J., Scanlon, M. G., Hartel, R. W., & Marangoni, A. G. (2001). Rheological properties of milkfat and butter concise reviews in food science. *Journal of Food Science, 66*, 1056–1071.

Chapter 11
Tribological Properties of Liquid Milks and Dairy Fat Structured Products

Phuong Nguyen, Yang Zhu, and Sangeeta Prakash

1 Introduction

Milk lipids in bovine and other animal milk exist naturally in the form of colloidal suspension of emulsified globules called milk fat globules (Truong, Palmer, Bansal, & Bhandari, 2016). The size, composition, and structure of native milk fat globules play an important role in the chemical & physical properties, sensory characteristics, and even nutritional profile of milk (Lopez et al., 2011; Michalski, Ollivon, Briard, Leconte, & Lopez, 2004) as well as many fat-containing dairy products such as cheese, cream cheese, ice cream, yoghurt and butter (Nguyen, Kravchuk, Bhandari, & Prakash, 2017; Ningtyas, Bhandari, Bansal, & Prakash, 2017; Truong et al., 2016).

In dairy products, milk fat presents in different forms, e.g. native globules (unhomogenised milk, cream), complex emulsions (homogenised milk, cream), membrane-disrupted free fat in a gel matrix (cheese, yoghurt), agglomerated fat in aerated systems (ice cream, whipped cream) or a continuous, free fat phase (butter, ghee, milk chocolate) (Truong et al., 2016).

The oral processing of dairy products is a complex process, where the dairy food is transported, manipulated, broken down and finally swallowed. During this process, milk fat could be retained in its globule form, released as free fat or a mixture with other food ingredients. This process varies greatly with different food products, oral processing styles and the emotion of the consumers (Engelen & De Wijk, 2012). Food residence time in mouth depends on the form of food product: the

P. Nguyen · Y. Zhu
School of Agriculture and Food Sciences, The University of Queensland, St Lucia, QLD, Australia

S. Prakash (✉)
ARC Dairy Innovation Hub, The University of Queensland, St Lucia, QLD, Australia
e-mail: s.prakash@uq.edu.au

T. Truong et al. (eds.), *Dairy Fat Products and Functionality*,
https://doi.org/10.1007/978-3-030-41661-4_11

residence time in mouth is generally longer for solid food (in the order of minute), then shorter for semi-solid food (up to 8 s) and shortest for liquid food (few seconds) (Chen & Lolivret, 2011; Engelen & De Wijk, 2012). The type of food also determines the variety of oral movement and contact between food and oral surfaces: oral contact with solid food can occur through the lips, tongue, palate, cheeks and teeth, all of which provide textural information as the food is broken down from the first bite to the stage the food bolus is swallowed; on the other hand, for many liquid and semi-solid food, there are limited oral activities like oral manipulation, tongue rolling and swallowing with little or even no chewing requirement. In this case, the fluid only has a quick contact with the oral surfaces as it flows through the oral cavity, thus the mouthfeel is dominated by the flow behaviour of the food and what remains after the food is swallowed, i.e. after-feel sensations. As a result, the texture of liquid and semi-solid has to be assessed rapidly and is mainly governed by their flow and lubricating properties during retention in the mouth (Engelen & De Wijk, 2012).

Traditionally, the mouthfeel of a fluid is assessed by using a rheometer to measure viscosity as a function of shear rate usually between 10^{-1} and 1000 s^{-1} (Shama & Sherman, 1973). It has been proposed that for a wide range of food products, such as Newtonian fluids, true solutions, weak gels, flocculated emulsions and lemon pie filling, the dynamic small deformation measurements at oscillatory frequency of 50 s^{-1} had a good correlation with perceived thickness, stickiness and sliminess (Shama & Sherman, 1973). Kokini (1987) suggested a mathematical relation between creaminess, smoothness and viscosity. However, this was not supported by other research work that showed smoothness and creaminess sensation are independent of viscosity and rheological behaviour of the food (Baier et al., 2009; Kokini, 1987). In addition, it has been found that the way the brain responds to fat texture is independent of viscosity (Verhagen, Rolls, & Kadohisa, 2003). Physically, in oral processing, the food is not only subjected to simple shear as applied in standard rheological equipment but also to a complex shear flow pattern, which involves compression of tongue against other oral surfaces during which the food no longer behaves like a bulk fluid but rather acts as a thin film between tongue and palate. It is not the bulk behaviour of an emulsion as studied with standard rheology that predicts the creamy sensation of a product, but the lubrication property of this thin film that contributes to different types of mouthfeel sensory perception especially those related to fat content, usually described by measuring friction between two surfaces (Baier et al., 2009; Giasson, Israelachvili, & Yoshizawa, 1997; Joyner, Pernell, & Daubert, 2014; Malone, Appelqvist, & Norton, 2003; Sonne, Busch-Stockfisch, Weiss, & Hinrichs, 2014). Furthermore, due to the short residence time of the liquid or semi-solid food in the mouth, the afterfeel sensation becomes more important than other mouthfeel sensations, however, this obviously cannot be simply described by viscosity of the bulk fluid. Besides, it has been found that non-rheological cues such as surface tension, lubrication and deposition of the food onto oral surfaces also influence the perception of texture and mouthfeel (Chen, 2009). For example, smoothness was found to not correlate with rheological parameter but lubrication properties (Kokini, 1987; Kokini & Cussler, 1983; Malone et al., 2003).

Creaminess was also observed to be enhanced with better lubrication properties for the dairy fluids having the same viscosity (Baier et al., 2009).

Due to the above reasons, in order to understand food texture and mouthfeel sensations especially for dairy product with varying fat contents, the tribometer has attracted a lot of interest. Tribology (or thin film rheology) provides an important approach to determine properties of food materials in the form of thin film that cannot be deduced from bulk properties (Baier et al., 2009; Chen, 2009; Kokini, 1987; Kokini & Cussler, 1983; Malone et al., 2003). By changing the relative speeds between the two rubbing surfaces, lubrication properties of a food material could be analyzed not only in the form of a bulk fluid (when the food enters the mouth), but also as a layer being squeezed between oral surfaces and finally as a thin film that remains on the tongue surface which is responsible for after-feel sensation (Dresselhuis, De Hoog, Cohen Stuart, & Van Aken, 2007).

To sum up, for dairy products especially those in liquid and semi-solid form, the sensory behaviour can be predicted by assessing its tribology behavior since it provides better discrimination for sensory attribute related to fat globule and fat content. The aim of this chapter is to summarize the tribological properties of milk fat globules and dairy products (milk, yoghurt and cream cheese) and investigate the effect of milk fat globule (its size, distribution and population) on the lubrication properties and sensory of the final product.

2 Tribology Method for Milk Fat Globules and Dairy Products

In food tribology, friction between two rubbing surfaces is usually presented as a Stribeck curve that shows the relation between the friction coefficient and relative speed between the two surfaces v_s (m/s) or a combined parameter of fluid viscosity η (Pa s), relative speed v_s (m/s) and the surface load F_L (Pa) (Prakash, Tan, & Chen, 2013). The Stribeck curve is usually divided into three different regimes including the boundary, mixed and hydrodynamic regimes. Hydrodynamic regime is observed when the food enters the contact zone between the two rubbing surfaces and the rate of entrainment of the food into the contact zone is due to surface motion, and the sufficiently fluid pressure produced by the food fluid forces the surfaces apart and sustains the load on the surfaces. The thickness of the fluid film between the two contacts and the friction generated depends on velocity, load, lubricant viscosity and relation between pressure and viscosity. In an opposite scenario, when the hydrodynamic fluid pressure in the contact zone is insufficient to separate the surfaces, then the lubrication properties of the fluid is characterized by the presence of an "immobile" layer that does not participate in the hydrodynamic flow of the bulk food. Under the lubricating effect of the thin film, the maximum contact region between the two surfaces can be achieved. This effect is called boundary lubrication, and the regime is called boundary regime (Williams, 2005). Between these two regimes is the mixed regime, where the two surfaces are in partial contact and the contact area is less compared to boundary regime. The rate of fluid flow into the contact zone of this regime is sufficiently large to partly separate the rubbing surfaces but the lubricant film thickness and the length of the asperities (Bhushan, 1998) of the test surfaces are of similar sizes (Cassin, Heinrich, & Spikes, 2001).

The instrument used to measure tribological characteristics or lubrication of food materials is called tribometer. The machine is generally comprised of two interacting surfaces in relative movement at a controlled speed, and the friction force (F_R) and friction coefficient (μ) between them can be measured. The friction coefficient is an indication of lubrication properties of a measured fluid between the interaction surfaces. It is no longer a constant value, depends on contact surface properties, surface load, relative moving speed and lubricant physical properties of that fluid. In food tribology, lubrication properties of food fluid between the two moving surfaces can be related to its sensory characteristics as the food is squeezed and sheared between oral surfaces (such as tongue and palate) during oral processing (Prakash et al., 2013).

For most data presented in this chapter, lubrication properties of dairy fluid were measured on a Discovery Hybrid Rheometer, using ring-on-plate tribo-rheometry (TA Instrument, USA). The upper contact surface was steel surface of a half-ring geometry, which is a ring interrupted in three sections that only half of the ring is in contact with the substrate. The lower contact surface was 3M Transpore Surgical Tape 1527-2 (3M Health Care, USA), which has similar hydrophobicity and surface roughness with human tongue (Nguyen, Nguyen, Bhandari, & Prakash, 2016). The tape was easily applied and replaced after every measurement. The measurement condition was controlled to be similar to sensory analysis: the temperature was maintained at 35 °C and normal force at 2 *N* during the experiment.

3 Tribology Properties of Dairy Liquid

The investigated dairy liquids include pasteurized and ultra-high temperature treatment (UHT) cow milk; pasteurized buffalo milk and chocolate milk. Pasteurized and UHT cow milk are commercial milk from the same manufacturer, while buffalo and chocolate milk were developed in our food processing laboratory.

3.1 *Effect of Fat Level*

Figure 11.1 shows tribology behaviour of three types of dairy liquid with varying fat levels:

(a) Commercial pasteurized homogenized cow milk purchased from local supermarket was from the same brand name with fat content varying from 0.1% to 4.9%;

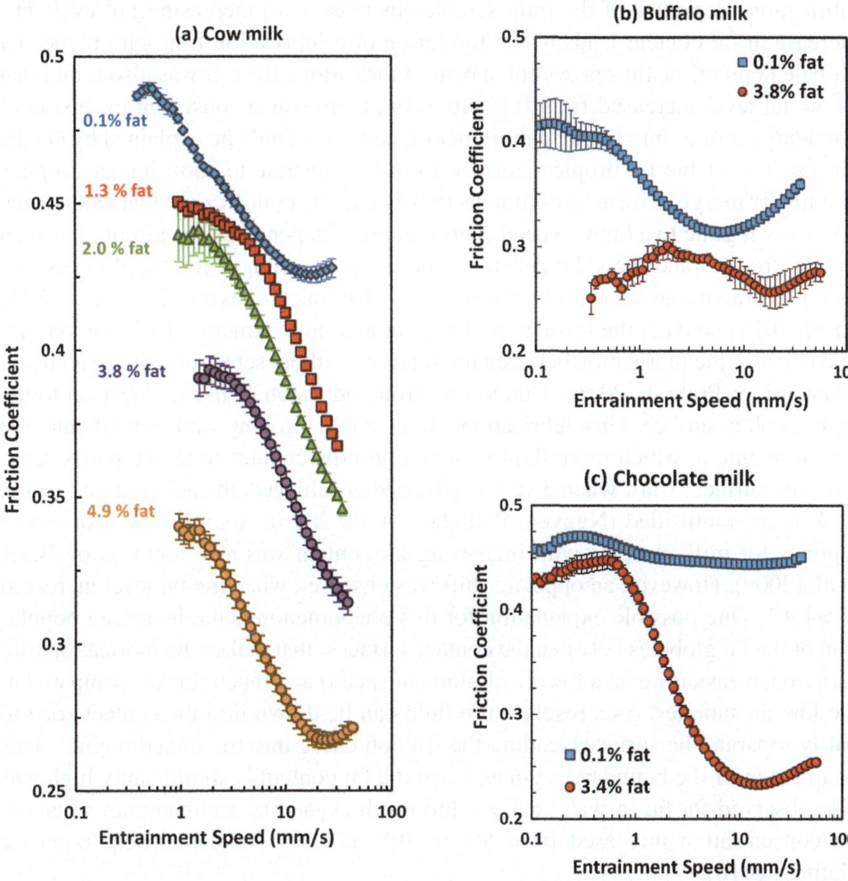

Fig. 11.1 Tribology properties of (**a**) commercial cow milk, (**b**) buffalo milk and (**c**) chocolate cow milk with different fat levels

(b) Unhomogenized raw buffalo milk obtained from local suppliers were pasteurized and processed in the food processing laboratory have 0.1% and 3.8% fat.
(c) Chocolate cow milk with 0.1% and 3.4% fat was reconstituted from skim milk and whole milk powder with 1.5% cocoa powder and 7% caster sugar. In order to achieve a well-mixed solution, each ingredient is added slowly into the mixture in the order of sugar, milk powder and cocoa powder. The samples were then passed through a bench-top UHT plant following the process described by Prakash, Huppertz, Karvchuk, and Deeth (2010).

For commercial cow milk as presented in Fig. 11.1a, friction curves for the five samples showed a typical Stribeck shape, except for samples with 1.3%, 2% and 3.8% fat wherein the hydrodynamic regimes were not observed due to limitation of the experiment set up. A clear discrimination between samples with different fat contents at all investigated entrainment speeds can be observed, indicating the lubrication properties of the milk samples increase with increasing fat level. The increase in fat content leads to the formation of a lubrication film with higher fat globule population, thus providing a better lubrication effect. It was also found that as the fat level increased from 0.1% to 3.8%, there was a consistent extension of boundary regimes into higher entrainment speed. This could be explained by *(1)* the coalescence of the fat droplets adsorbed on the substrate to form bigger droplets that finally merge to form a continuous thin film in the contact zone that extends the boundary regime to a higher speed. This extension depends on the affinity between the fat droplets and the solid substrate and the population of the fat globules that could be drawn into the liquid-solid interface (Chojnicka-Paszun, De Jongh, & De Kruif, 2012); and *(2)* the formation of thicker and more structured lubrication film that enhance the lubrication between the surfaces without separating them (Nguyen, Bhandari, & Prakash, 2016). Due to the strong adhesion of the fat layer on to the hydrophobic surface, this lubrication effect could prolong and extend into the mixed regime at which more fluid is drawn into the contact zone and partly separates the surface. Until when a very high speed is achieved, the adhered fat layer is broken and emulsified (Nguyen, Bhandari, et al., 2016). A similar shift of mixed regimes for milk samples with increasing fat content was also reported by Baier et al. (2009). However, an opposite shift was observed when the fat level increased to >4.9%. One possible explanation for this phenomenon is the increased population of the fat globules between the contact surfaces, that makes the lubrication film form much easier (i.e. at a lower entrainment speed) and much thicker compared to the low fat samples. As a result, more fluid can be drawn into the contact zone to partly separate the surfaces leading the friction curve into the mixed regime. This contraction of the boundary regimes when the fat content is significantly high was also observed for fat-in-water dairy solutions thickened by xanthan gum when the fat concentration increased from 5% to 20% (Debon, Vanhemelrijck, Baier, & Guthrie, 2010).

For other dairy liquids like buffalo or chocolate milk (Fig. 11.1b, c), a similar behaviour was observed:

1. The friction curves resembled typical Stribeck curves for most samples;
2. There was a clear discrimination between the friction curves for the investigation levels for all regimes: friction value decreased as the fat level increased;
3. A consistent shift of mixed regimes (or extension of boundary regime) to higher speeds when the fat level increased was observed.

However, fat is not the sole factor that affects the lubrication of dairy liquid. This is demonstrated by the clear difference in tribology behaviour of unhomogenised pasteurized cow milk and buffalo milk with the same fat content (3.8%) presented in Fig. 11.2. The other factors that may contribute to the difference in lubrication properties of the two milk samples could be their protein content, fat globule size or other chemical compounds in dairy liquid such as lactose, calcium, etc., or the combination of them. Protein has been found to increase the friction between two rubbing surfaces as compared to water due to its ability to adsorb onto the hydrophilic lower surface (i.e. surgical tape) and reduce the contact angle (Nguyen, Nguyen, et al., 2016). The presence of calcium or any other solid particles in the dairy liquid could significantly change the sample lubrication behaviour and increase friction between the two rubbing surfaces (Prakash et al., 2013).

3.2 *Effect of Fat Globule Size*

Besides the fat percentage, the size of fat globule could also contribute to the lubrication properties of dairy liquid. Cow milk with different particle sizes were achieved by subjecting unhomogenised cow milk (3.8% fat) to different pressures using Avestin Emulsiflex C5 (ATA Scientific, Taren Point, NSW, Australia). The different levels of homogenization generated three samples with different particle size (3.99, 2.93 and 1.15 μm) as presented in Fig. 11.3. The particle size D_{43} of the milk samples (A, B and C) decrease with increasing homogenization pressure. Particle size distribution of samples A and B have only one peak at around 4 μm while sample C has two peaks with higher volume fraction for the smaller size particles (Fig. 11.3b).

Figure 11.3a shows that at high entrainment speed, the friction curves of all three different cow milks were overlapping, which indicates they had similar viscosity. However, at low entrainment speed, the cow milk with large sized of fat globules had lower friction value compared to sample B and C that contained smaller sized fat globules. With the relative motion of the two contact surfaces, the fat will enter the contact zone to form a lubrication film. As larger size of fat globules enters into the contact zone, the lubrication film formed can separate the two surfaces better whereby lubrication is enhanced. This result indicates that fat globule size influences the tribological property of cow milk, especially at low entrainment speed.

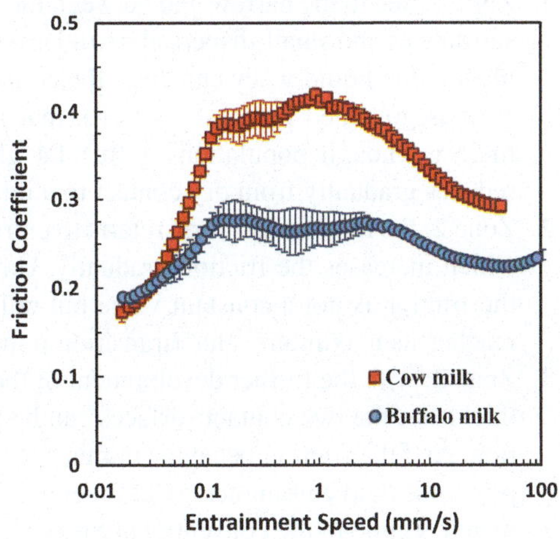

Fig. 11.2 Tribology properties of unhomogenized pasteurized cow milk and buffalo milk with the same fat level of 3.8%

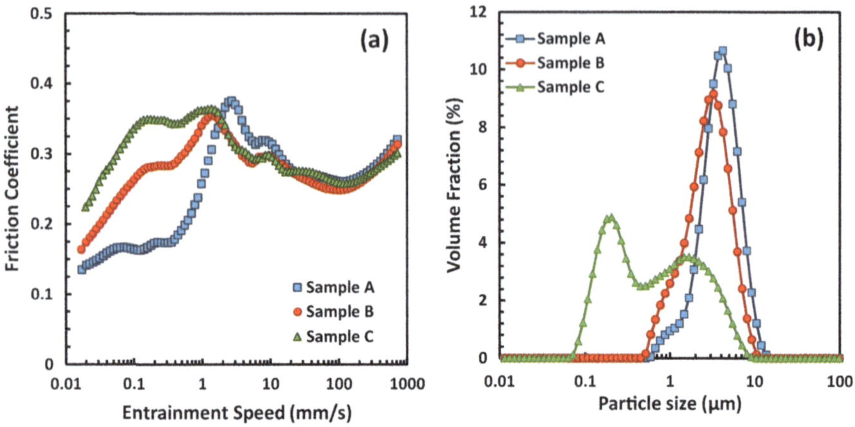

Fig. 11.3 Tribology properties (**a**) and particle size distribution (**b**) of cow milk samples of 3.8% fat with different fat globule size (D_{43} for samples A, B and C are 3.99, 2.93 and 1.15 μm, respectively)

4 Tribology Properties of Yoghurt

The investigated yoghurt includes skim, trim and full fat yoghurt developed in our food processing laboratories.

As shown in Fig. 11.4, the friction curves of yoghurt samples did not resemble the traditional Stribeck curve, but rather followed the schematic diagram in Fig. 11.5. This is because the internal structure of yoghurt is complex with liquid whey and cross-linked gel network formed by denatured whey proteins associated with the casein micelles, in which fat globules are entrapped. The friction curve can be generally divided into five zones:

1. Zone 1: due to the narrow gap between the two contact surfaces, only the soluble substances and small-dispersed particles (such as whey protein and free fat globules) in the liquid whey can enter the contact zone. The friction behavior in this zone is governed by these particles that migrated from the gel matrix. When there is enough population of free fat globules in the contact zone, friction reduces gradually from dry contact to a minimum value at transition point T_1.
2. Zone 2: the fluid (in gel form) starts to entrain into the contact zone in this stage, which increases the friction gradually. Unlike the conventional Stribeck curve, the friction is not a constant value but rather increases linearly with speed and reaches its maximum value (transition point T_2).
3. Zone 3: with the further development of the lubrication film and the increases in thickness, the two contact surfaces can be partly separated that reduces the friction. And the friction behavior in this zone is governed by sample viscosity that promotes fluid entrainment (Cassin et al., 2001). This regime corresponds to the mixed regime in the conventional Stribeck curve.

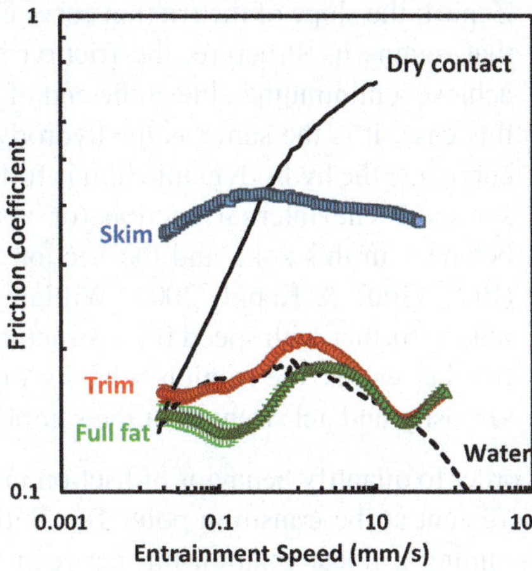

Fig. 11.4 Tribology properties of pot-set yoghurt with different fat levels (adopted from Nguyen et al., 2017)

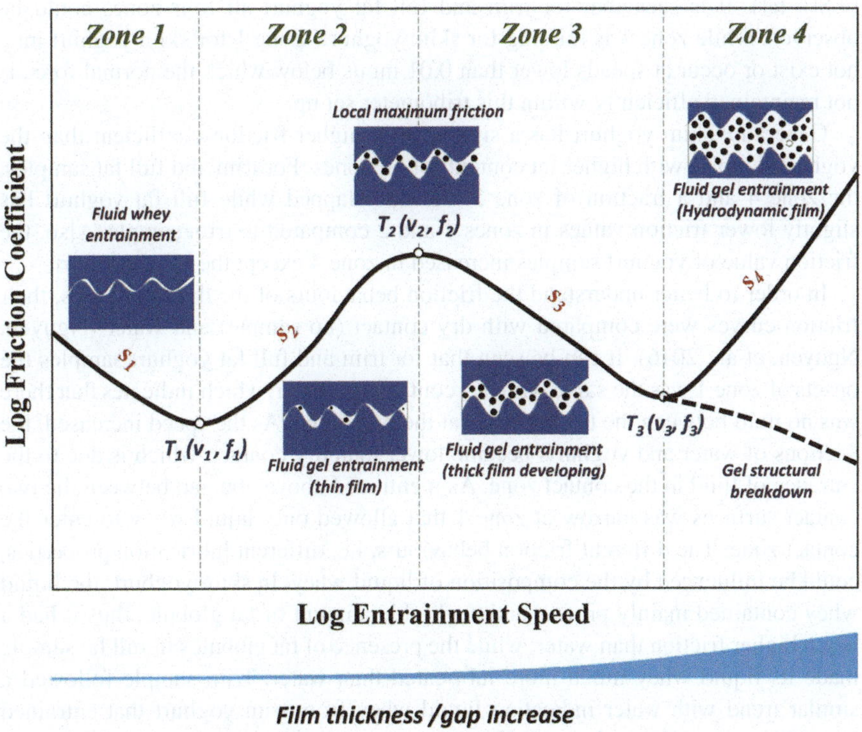

Fig. 11.5 Schematic diagram of fluid entrainment in the contact zone for yoghurt (adopted from Nguyen et al., 2017)

4. Zone 4: the slope of the friction curve changes at transition point T_3. For the fluid that retains its structure, the friction increases again with increasing speed to achieve a minimum value at the end of the mixed regime (transition point T_3). In this case, it is the same as the hydrodynamic regime in the traditional Stribeck curve, i.e. the hydrodynamic film is fully developed and completely separates the surfaces. The internal friction (or viscosity of fluid) determines the friction behavior in this zone and the friction coefficient increases linearly with speed (Butt, Graf, & Kappl, 2004; Williams, 2005). Oppositely, the friction might reduce further with speed if the structure of the fluid breaks down at high speeds. In other words, the friction behavior of fluid within this zone is governed by the viscosity and gel strength of the sample together.

In order to quantify behavior of friction in each zone, the speed (mm/s) and friction coefficient at the transition point T_i is defined as v_i, f_i, and s_i is the slope of zone i (assuming a linear relationship between logarithm of friction and logarithmic of entrainment speed in each zone). Figure 11.4 shows the friction curves of yoghurt samples with different levels of fat: skim (0.1% fat), trim (1.3% fat) and full fat (3.8% fat). It is seen that for trim and full fat yoghurt all four zones could be observed, while zone 1 is missing for skim yoghurt. Zone 1 for skim yoghurt may not exist or occur at speeds lower than 0.01 mm/s below which the normal force is not maintained efficiently within this tribometer set up.

Generally, skim yoghurt has a significantly higher friction coefficient than the yoghurt samples with higher fat content for all zones. For trim and full fat samples, the zone 4 and a fraction of zone 3 were overlapped while full fat yoghurt has slightly lower friction values in zones 1 and 2 compared to trim sample. Also, the friction value of yoghurt samples increased in zone 4 except the skim yoghurt.

In order to better understand the friction behaviours of the three yoghurts, their friction curves were compared with dry contact (no sample) and water (Nguyen, Nguyen, et al., 2016). It can be seen that for trim and full fat yoghurt samples the onsets of zone 1 was the same with dry contact and water, which indicates that there was no fluid between the two surfaces at the beginning. As the speed increased, the frictions of water and yoghurts became lower than dry contact, which is due to the entrance of fluid in the contact zone. As mentioned above, the gap between the two contact surfaces was narrow at zone 1 that allowed only liquid whey to enter the contact zone. The different friction behaviours, i.e. different lubrication properties, could be influenced by the composition of liquid whey. In skim yoghurt, the liquid whey contained mainly protein and negligible amount of fat globule, thus it had a much higher friction than water; while the presence of fat globules in full fat sample made its liquid whey much more lubricated than water. Trim sample followed a similar trend with water indicating liquid whey from trim yoghurt that entrained into the contact zone had similar friction properties like water. The mechanism of friction change with different protein/fat contents could be explained by the adsorption of protein or/and fat on the contact surfaces that has been discussed elsewhere (Nguyen, Bhandari, et al., 2016).

At the start of zone 2 (Fig. 11.5), the coefficient of friction of all the three yoghurts increased due to the entrainment of yoghurt sample (in the gel form) into the contact zone. The friction values decreased in the order from skim, trim to full fat yoghurt due to an increase in fat level in the samples present inside the protein network, and the fat could entrain into the contact zone leading to an increase in the population of fat globules between the two surfaces that decreases the friction value.

After reaching the maximum friction (at transition point T_2), the curve enters zone 3 wherein the friction decreases with speed due to entrainment of more yoghurt gel that develops a thicker lubrication film, partially separating separation the two surfaces. The friction behavior in this zone is governed by both thin film and hydrodynamic lubrication. As fat level increases, more fat embed inside the protein network that improve the gel strength and viscosity of yoghurt, leading to the yoghurt samples having better lubrication property. Therefore, the slope of skim yoghurt was gentler than for trim and full fat yoghurt in zone 3 (s_3).

From transition point T_3, the friction curves for both trim and full fat yoghurt increased linearly with speed, showing hydrodynamic lubrication. Since the friction in hydrodynamic regime is governed by fluid viscosity (Butt et al., 2004), the overlapping of trim and full fat samples in this zone indicates their similar gel structure and viscosity. Compared to the yoghurt samples containing high fat, the friction curve of skim yoghurt decreased further as the speed increases, indicating a structural breakdown of the skim yoghurt at higher speed.

5 Tribology Properties of Cream Cheese

The investigated cream cheese samples include low (0.5% fat), medium (5.5% fat) and high (11.6% fat) fat samples that were developed in our food processing laboratories.

The tribological behaviour of cream cheese (Fig. 11.6) can be presented as a friction curve, which also followed the schematic diagram in Fig. 11.5, except that zone 1 was missing for low and high fat cream cheese. Zone 1 for these two samples may not exist or occur at speeds lower than 0.01 mm/s below which the normal force is not maintained efficiently within this tribometer set up. Additionally, in medium and high fat cream cheese, the regimes that differentiate the frictional behaviour of cream cheese were clearly observed, while low-fat cream cheese showed a decrease in friction coefficient with the increase of sliding speed (Fig. 11.6).

Low-fat cream cheese had a significantly higher friction coefficient with a very distinctive stick and slide pattern when compared to medium-fat and high-fat cream cheese samples. The improvement of the lubrication property of cream cheese with increasing fat level is due to the formation of an interfacial film of fat, which acts as a spacer between the tribo-pair (Nguyen, Bhandari, et al., 2016). For low-fat cream cheese, the friction did not change considerably with sliding speed. While for medium-fat and high-fat samples, although during the medium speed (~1–80 mm/s) and high speed (~> 80 mm/s) regimes the friction curves overlapped each other, the friction coefficient of medium-fat was slightly higher than the high-fat sample at low- speed regime. The overlapping of medium and high-fat samples reflects their similar gel structure and viscosity.

Cream cheese is an emulsion system of fat droplets within an aqueous phase (Coutouly, Riaublanc, Axelos, & Gaucher, 2014), therefore the presence of fat allows the entrainment of the bulk emulsion to be suppressed, enabling an extended low regime with very low friction coefficient. In addition, Dresselhuis, De Hoog, Cohen Stuart, Vingerhoeds, and Van Aken (2008) found that the large particle of fat globules give rise to an increased hydrostatic pressure, which keeps the tongue and palate separated, thus lowering the friction.

As seen in Fig. 11.7, the images clearly indicate a change in the cheese structure as a function of fat content. The microstructure of cream cheese containing 11.6% fat was distinctly different from that of cream cheese with 5.5% and 0.5% fat. The high-fat cream cheese was characterized by a protein gel particle matrix (red) interspersed with fat globules (green) of various sizes and shapes, and it had more intact

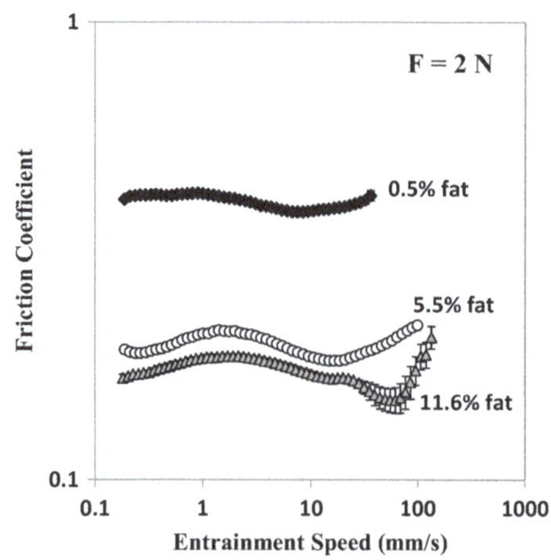

Fig. 11.6 Tribology behaviour of cream cheese with different fat levels. The symbols for samples with 0.5% fat, 5.5% fat and 11.6% are presented in black, grey and white symbols, respectively (adopted from Nguyen, Bhandari, et al., 2016)

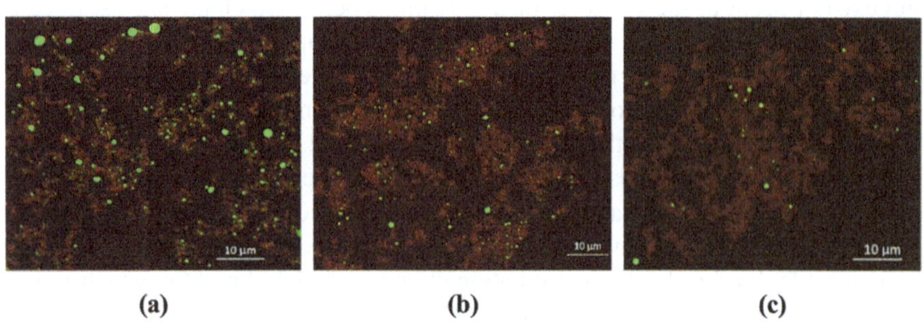

Fig. 11.7 Confocal micrographs of cream cheese samples having different fat levels (**a**) high-fat (**b**) medium-fat, (**c**) low-fat; fat (green) and protein (red) were stained with Nile red and Rhodamine B, respectively (adopted from Ningtyas et al., 2017)

milk fat globules that also aggregated, resulting in large fat globules (Fig. 11.7a). While for low (Fig. 11.7c) and medium-fat (Fig. 11.7b) cream cheese, fat occurred as uniformly dispersed discrete globules with little evidence of clumping.

A reduction in fat content resulted in a decrease in the number of fat globules, therefore low-fat cream cheese had a more compact protein matrix with less open spaces that would have been occupied by milk fat globules. Figure 11.7c clearly confirms the dense and compact protein network around the fat globules in low-fat cream cheese compared with that of medium (Fig. 11.7b) and high-fat (Fig. 11.7a) cream cheese. Thus, the low-fat cheese had higher friction coefficient and cause the astringency feeling in mouth. In contrary, as the amount and size of fat increased, the clumping and coalescence of fat globules became more evident. This will make the texture of high-fat cream cheese soft since the fat droplets disrupt the protein matrix, as explained by Romeih, Michaelidou, Biliaderis, and Zerfiridis (2002). And more fat will penetrate from the cheese during chewing, which acts like a lubricant to decreasing the friction coefficient.

6 Role of Fat Globule in Sensory Perception of Dairy Food

Fat plays an important role in oral sensations perceived from dairy fat structured products like smoothness, creaminess, astringency, etc. Pasteurized cow milk was investigated by a trained panel for studying the influence of fat globule on sensory perception of dairy food. Ten trained panelists (5 males, 5 female, age 20–40 years, healthy subjects, lactose tolerant) participated in the sensory evaluation test through Quantitative Descriptive Analysis (QDA®). They had been selected after completing scaling exercise and basic tastes & texture exercise as per ISO 22935 (ISO, 2009). During the test, a randomized complete block design was used to compare the sensory attributes of milk samples. Equal amounts of each sample were placed into 60 mL cups labeled with randomly selected 3-digit codes and equilibrated at room temperature (22–25 °C) for at least 1 h before each testing. Deionized water also served to panelists for palate cleaning. Panelists evaluated samples for pre-defined textural attributes (Table 11.1) using a linear scale with increasing score from 0 to

Table 11.1 Definitions of the textural attributes

Attributes	Definitions
Thickness/viscosity	Resistance to flow in the mouth before saliva modifies the sample
Smoothness	Perceived smoothness of the sample squeezed between palate and tongue
Chalkiness	Powdery sensation
Astringency	Dry feeling
Fatty feel	Intensity of fat
Oily coating	Oily coating in the mouth after swallowing
Residual coating	Intensity of residues in the mouth after swallowing

15 with anchors marked at 1.25 cm from either end. The evaluation was conducted in individual sensory booths with red light to prevent any interference from the product's appearance.

Test procedure and data collection were programmed using Compusense® software. Since the panelists are allowed to use different parts of the scale to determine the sensory score by themselves, the difference in scores among products is always relative. However, in a properly designed study with correctly trained panelists, this difference between panelists is taken out of the analysis. Besides, panelists' internal consistency and the reproducibility of the panel are assessed on the basis of repeated evaluations of the same products.

Principal component analysis was performed on sensory scores for five pasteurized cow milk samples and the seven attributes (smoothness, viscosity, residual coating, astringency, chalkiness, fatty feel and oily coating) that were best separating the individual samples (Fig. 11.8). The first two principal components accounted for 91.4% of the variability in the results. It is seen that fatty feel and oily coating are negatively correlated with chalkiness and astringency as expected. This is due to the presence of chalkiness and astringency that reduces fatty and oil perception. In the first dimension, the fatty feel (or less astringency) texture increases from low-fat milk to high-fat milk. Attribute vectors for viscosity and smoothness form the second principle dimension. In term of viscosity and smoothness, milk samples could be divided into two groups: lower viscosity and smoothness (cow milk contains 0.1% and 2.0% fat) and higher viscosity and smoothness (cow milk contains 4.9% fat). Besides, cow milk contains 2.0% and 4.9% fat also appear to have higher oily coating feeling than other samples.

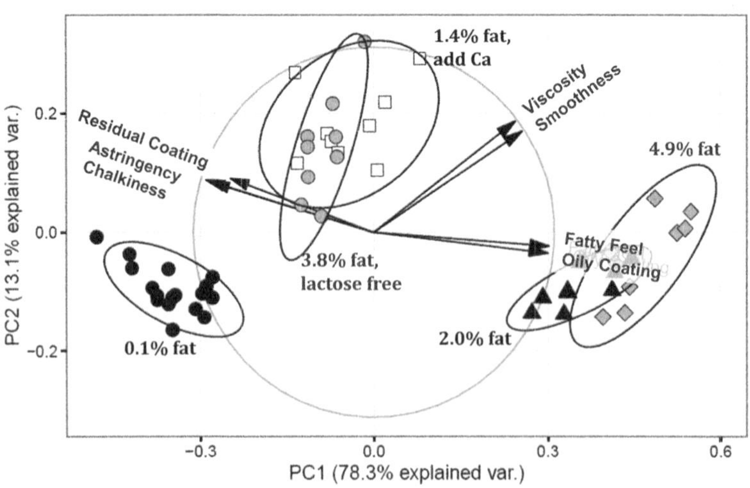

Fig. 11.8 Sensory results from Quantitative Descriptive Analysis for pasteurized cow milk with different fat levels

7 Conclusions

This chapter highlights the importance of milk fat in determining the tribological behaviour of liquid milks and dairy fat-containing products. For dairy liquid, the milk fat globules can have a direct impact on the tribology behavior of products since the milk fat can deposit on the contact surfaces and govern the friction behavior. In addition, both milk fat level and milk fat globule size affect the lubrication property of dairy fat structured products. As milk fat intersperses within the protein network of semi-solid dairy products, such as yoghurt and cream cheese, it can change the product structure, in turn, influencing the entrainment of fluid into the contact zone and resulting in the alteration of the tribology behavior of dairy products.

References

Baier, S., Elmore, D., Guthrie, B., Lindgren, T., Smith, S., & Steinbach, A. (2009). *A new tribology device for assessing mouthfeel attributes of foods*. In 5th International Symposium on Food Structure and Rheology, ETH Zurich, Zurich, Switzerland, 2009.

Bhushan, B. (1998). Contact mechanics of rough surfaces in tribology: Multiple asperity contact. *Tribology Letters, 4*, 1–35.

Butt, H.-J., Graf, K., & Kappl, M. (2004). *Friction, lubrication, and Wear. Physics and chemistry of interfaces*. Wiley-VCH Verlag GmbH & Co. KGaA.

Cassin, G., Heinrich, E., & Spikes, H. A. (2001). The influence of surface roughness on the lubrication properties of adsorbing and non-adsorbing biopolymers. *Tribology Letters, 11*, 95–102.

Chen, J. (2009). Food oral processing: A review. *Food Hydrocolloids, 23*, 1–25.

Chen, J., & Lolivret, L. (2011). The determining role of bolus rheology in triggering a swallowing. *Food Hydrocolloids, 25*, 325–332.

Chojnicka-Paszun, A., De Jongh, H. H. J., & De Kruif, C. G. (2012). Sensory perception and lubrication properties of milk: Influence of fat content. *International Dairy Journal, 26*, 15–22.

Coutouly, A., Riaublanc, A., Axelos, M., & Gaucher, I. (2014). Effect of heat treatment, final pH of acidification, and homogenization pressure on the texture properties of cream cheese. *Dairy Science & Technology, 94*, 125–144.

Debon, S. J. J., Vanhemelrijck, J. G. R., Baier, S. K., & Guthrie, B. D. (2010). Tribology device for assessing mouthfeel attributes of foods. Google Patents.

Dresselhuis, D. M., De Hoog, E. H. A., Cohen Stuart, M. A., & Van Aken, G. A. (2007). Chapter 31: Tribology as a tool to study emulsion behaviour in the mouth. In *Food colloids: Self-assembly and material science*. The Royal Society of Chemistry.

Dresselhuis, D. M., De Hoog, E. H. A., Cohen Stuart, M. A., Vingerhoeds, M. H., & Van Aken, G. A. (2008). The occurrence of in-mouth coalescence of emulsion droplets in relation to perception of fat. *Food Hydrocolloids, 22*, 1170–1183.

Engelen, L., & De Wijk, R. A. (2012). Oral processing and texture perception. In *Food oral processing*. Wiley-Blackwell.

Giasson, S., Israelachvili, J., & Yoshizawa, H. (1997). Thin film morphology and tribology study of mayonnaise. *Journal of Food Science, 62*, 640–652.

ISO. (2009). ISO 22935-2 milk and milk products — Sensory analysis. *Part 2: Recommended methods for sensory evaluation* (p. 23): ISO and IDF 2009.

Joyner, H. S., Pernell, C. W., & Daubert, C. R. (2014). Impact of formulation and saliva on acid milk gel friction behavior. *Journal of Food Science, 79*, E867–E880.

Kokini, J. L. (1987). The physical basis of liquid food texture and texture-taste interactions. *Journal of Food Engineering, 6*, 51–81.
Kokini, J. L., & Cussler, E. L. (1983). Predicting the texture of liquid and melting semi-solid foods. *Journal of Food Science, 48*, 1221–1225.
Lopez, C., Briard-Bion, V., Ménard, O., Beaucher, E., Rousseau, F., Fauquant, J., et al. (2011). Fat globules selected from whole milk according to their size: Different compositions and structure of the biomembrane, revealing sphingomyelin-rich domains. *Food Chemistry, 125*, 355–368.
Malone, M. E., Appelqvist, I. A. M., & Norton, I. T. (2003). Oral behaviour of food hydrocolloids and emulsions. Part 1. Lubrication and deposition considerations. *Food Hydrocolloids, 17*, 763–773.
Michalski, M.-C., Ollivon, M., Briard, V., Leconte, N., & Lopez, C. (2004). Native fat globules of different sizes selected from raw milk: Thermal and structural behavior. *Chemistry and Physics of Lipids, 132*, 247–261.
Nguyen, P. T. M., Bhandari, B., & Prakash, S. (2016). Tribological method to measure lubricating properties of dairy products. *Journal of Food Engineering, 168*, 27–34.
Nguyen, P. T. M., Kravchuk, O., Bhandari, B., & Prakash, S. (2017). Effect of different hydrocolloids on texture, rheology, tribology and sensory perception of texture and mouthfeel of low-fat pot-set yoghurt. *Food Hydrocolloids, 72*, 90–104.
Nguyen, P. T. M., Nguyen, T. A. H., Bhandari, B., & Prakash, S. (2016). Comparison of solid substrates to differentiate the lubrication property of dairy fluids by tribological measurement. *Journal of Food Engineering, 185*, 1–8.
Ningtyas, D. W., Bhandari, B., Bansal, N., & Prakash, S. (2017). A tribological analysis of cream cheeses manufactured with different fat content. *International Dairy Journal, 73*, 155–165.
Prakash, S., Huppertz, T., Karvchuk, O., & Deeth, H. (2010). Ultra-high-temperature processing of chocolate flavoured milk. *Journal of Food Engineering, 96*, 179–184.
Prakash, S., Tan, D. D. Y., & Chen, J. (2013). Applications of tribology in studying food oral processing and texture perception. *Food Research International, 54*, 1627–1635.
Romeih, E. A., Michaelidou, A., Biliaderis, C. G., & Zerfiridis, G. K. (2002). Low-fat white-brined cheese made from bovine milk and two commercial fat mimetics: Chemical, physical and sensory attributes. *International Dairy Journal, 12*, 525–540.
Shama, F., & Sherman, P. (1973). Identification of stimuli controlling the sensory evaluation of viscosity II. Oral methods. *Journal of Texture Studies, 4*, 111–118.
Sonne, A., Busch-Stockfisch, M., Weiss, J., & Hinrichs, J. (2014). Improved mapping of in-mouth creaminess of semi-solid dairy products by combining rheology, particle size, and tribology data. *LWT – Food Science and Technology, 59*, 342–347.
Truong, T., Palmer, M., Bansal, N., & Bhandari, B. (2016). *Effect of milk fat globule size on the physical functionality of dairy products*. Cham, Switzerland: Springer.
Verhagen, J. V., Rolls, E. T., & Kadohisa, M. (2003). Neurons in the primate orbitofrontal cortex respond to fat texture independently of viscosity. *Journal of Neurophysiology, 90*, 1514–1525.
Williams, J. A. (2005). *Engineering tribology*. New York: Cambridge University Press.

Chapter 12
Microstructural Engineering of Milk Fat and Related Products

Pere R. Ramel and Alejandro G. Marangoni

1 Introduction

Milk fat is widely used as a key ingredient in many food products because of its importance in the development of the structural, mechanical, and sensory properties of food products. The functionality of milk fat is greatly dependent on its thermal properties (i.e. crystallization and melting behavior). Therefore, the characterization of the crystal structure of milk fat as affected by thermal processes, composition, and other various processing conditions has been extensively studied in order to understand structure–function relationships as well as to gain opportunities for microstructural engineering of the crystal network formed by milk fat (Campos, Narine, & Marangoni, 2002; Herrera & Hartel, 2000; Lopez, Lesieur, Bourgaux, & Ollivon, 2005; Marangoni et al., 2012; Ramel & Marangoni, 2016; Shi, Smith, & Hartel, 2001).

Milk fat is composed mainly of triacylglycerols (TAGs), which are molecules composed of a glycerol backbone with three fatty acid moieties esterified onto it. The composition of milk fat is very complex because it is comprised of various fatty acids and TAGs (Jensen, 2002; Jensen, Ferris, & Lammi-Keefe, 1991). During crystallization, TAGs adapt a chair or tuning fork conformation, and organize to form ordered domains (Acevedo & Marangoni, 2015; Marangoni et al., 2012; Marangoni & Wesdorp, 2013a; Timms, 1984). This process is called nucleation, which is essential for the formation of a fat crystal network (Marangoni et al., 2012; Martini, Herrera, & Hartel, 2001; Sato, 2001). Once stable nuclei are formed, crystal growth rapidly occurs and the number of nuclei formed initially, determines the amount and size of the resulting milk fat crystal network (Campos et al., 2002; Marangoni & Wesdorp, 2013b; Martini et al., 2001). These crystallization steps are affected by

---

P. R. Ramel · A. G. Marangoni (✉)
Department of Food Science, University of Guelph, Guelph, ON, Canada
e-mail: amarango@uoguelph.ca

T. Truong et al. (eds.), *Dairy Fat Products and Functionality*,
https://doi.org/10.1007/978-3-030-41661-4_12

cooling rate, temperature, and other factors such as composition and shear during crystallization.

The structure of milk fat has been described at different length scales such as molecular level (polymorphism), nanoscale, and micro or mesoscale. At the micro or mesoscale structural level, milk fat structure is characterized by the size, shape, and mass distribution of the fat crystals. Using polarized light microscopy (PLM), the native morphology and the effect of various factors on these fat crystals can be directly observed (Marangoni et al., 2012; Marangoni, Tang, & Singh, 2006; Singh, Bertoli, Rousset, & Marangoni, 2004). Since fat crystals are birefringent, they appear bright while the background and liquid oil appear dark under PLM (Fig. 12.1). Other techniques for characterizing the microstructure of milk fat include scanning electron microscopy (SEM) and the analysis of crystallization kinetic data using the Avrami model for describing the crystallization kinetics and morphology of fat crystals (Marangoni & Mcgauley, 2003; Mazzanti, Guthrie, Sirota, Marangoni, & Idziak, 2004; Ramel & Marangoni, 2016; Wright, Batte, & Marangoni, 2005). The macro-properties of milk fat such as hardness and spreadability are largely dependent on its microstructure (Wright & Marangoni, 2003; Wright, Scanlon, Hartel, & Marangoni, 2001). Many studies have therefore focused on tailoring the microstructure of milk fat with the aim of improving specific finished food product functionalities. This has been shown in work by Singh et al. (2004), where matching microstructures of palm oil-based fats resulted in similar hardness of the crystal network.

In the engineering of the microstructure of milk fat, it is important to note that the crystal network formed by milk fat, or fats in general, is fractal in nature (Marangoni, 2002; Wright et al., 2001; Wright & Marangoni, 2003). That is, large

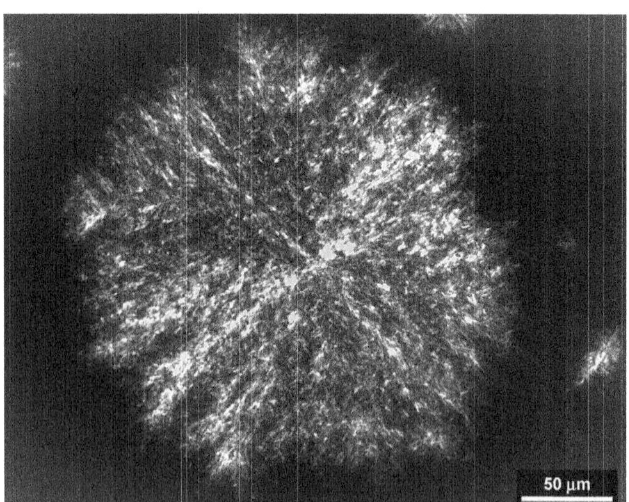

Fig. 12.1 Spherulitic crystal growth of milk fat observed under polarized light microscope (PLM). Adapted from Marangoni, Narine, Acevedo, and Tang (2013)

crystals observed under PLM are far larger aggregates of repeating units of primary crystals called crystalline nanoplatelets (CNPs). Therefore, the properties of the CNPs and the changes that the CNPs incur upon processing could affect the larger microstructures of milk fat (Acevedo & Marangoni, 2010a, 2015; Marangoni et al., 2012; Ramel, Co, Acevedo, & Marangoni, 2016). However, relating the properties at the nanoscale structural level to the microstructure of milk fat is still a growing area of research as most studies have focused on characterizing the microstructure (i.e., changes in size and shape of the large poly-crystal aggregates).

In this chapter, therefore, the effect of various internal and external factors on the microstructure of milk fat is described with some correlations with effects on the nanostructure. Furthermore, opportunities for the engineering of the microstructure of milk fat in relation to the delivery of certain functionalities are discussed.

2 Effect of Various Factors on the Microstructure of Milk Fat

In order to gain insights into strategies that can be used to alter milk fat microstructure, the effect of various processing conditions and composition on crystal size, shape, and mass distribution is discussed in the following sections.

2.1 Effect of Processing Conditions on Milk Fat Microstructure

2.1.1 Crystallization Temperature

Crystallization temperature is very critical in the formation of fat crystals as the supersaturation of TAGs and formation of stable nuclei are dependent on the degree of undercooling of the melt (Dibildox-Alvarado, Marangoni, & Toro-Vazquez, 2010; Marangoni et al., 2012; Sangwal & Sato, 2012). Crystallization temperature has been shown to affect the type of TAGs that crystallize (i.e., solid to liquid fat ratio) (Campos, Litwinenko, & Marangoni, 2003; van Aken, Grotenhuis, Langevelde, & Schenk, 1999; Wright et al., 2001). Since milk fat is composed of various melting fractions such as low melting (LMF), middle melting (MMF) and high melting (HMF) fractions, various microstructures can be formed at different temperatures. This has been shown by Ramel and Marangoni (2016) where at high undercooling conditions (5 °C), one-dimensional growth (e.g., rod or needle-like crystals) is favored (Fig. 12.2). On the other hand, at 15 and 20 °C, various microstructures can be observed depending on the composition of the milk fat fractions present (Fig. 12.3). The stability of the crystals formed during the crystallization process and during storage is also dependent on crystallization temperature. Temperature fluctuations during crystallization was found to affect the resulting microstructure

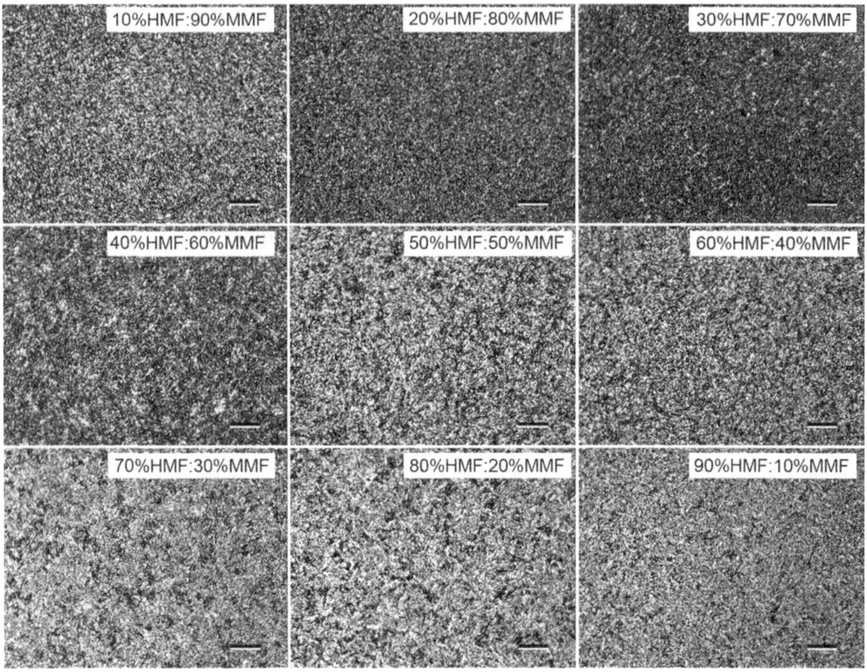

Fig. 12.2 Polarized light (PLM) micrographs of binary mixtures of high melting (HMF) and middle melting (MMF) fractions of milk fat, crystallized isothermally at 5 °C. Scale bar corresponds to 100 μm. Reproduced from Ramel and Marangoni (2016) with permission from the Royal Society of Chemistry

of butter (Buldo, Andersen, & Wiking, 2013). Due to temperature cycling, a harder, more stable and denser crystal network is obtained. This is possibly brought about by flocculation of fat globules and sintering wherein TAGs melt and migrate during increase of temperature to fill gaps in the network, thus resulting in a more compact crystal structure.

2.1.2 Cooling Rate

Another important parameter during crystallization, and possibly one of the most studied factors that affect milk fat microstructure, is cooling rate. Cooling rate mainly affects the number of nuclei formed, which therefore affects the size and number of crystals formed upon crystal growth. Microscopically, at fast cooling rates, smaller crystals are formed while at slow cooling rates, larger crystals develop (Campos et al., 2002; Martini & Marangoni, 2007; Wiking, De Graef, Rasmussen, & Dewettinck, 2009). Considering the same number of TAGs in a system, during crystallization at fast cooling rates, numerous nuclei are formed because the temperature at which high melting TAGs start to crystallize is achieved much faster.

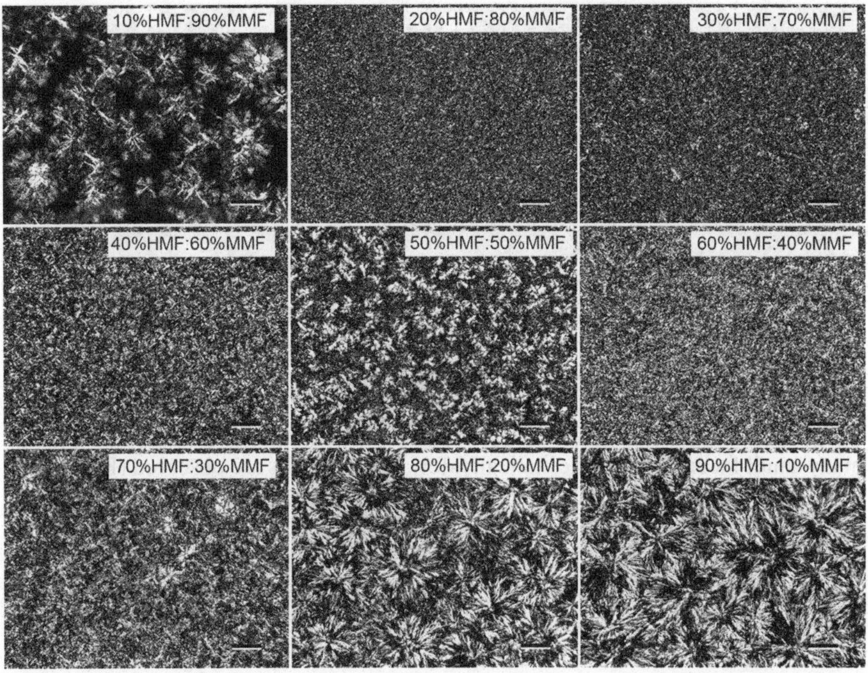

Fig. 12.3 Polarized light (PLM) micrographs of binary mixtures of high melting (HMF) and middle melting (MMF) fractions of milk fat, crystallized isothermally at 20 °C. Scale bar corresponds to 100 μm. Reproduced from Ramel and Marangoni (2016) with permission from the Royal Society of Chemistry

These nuclei serve as starting point for crystal growth, therefore TAGs are distributed throughout these nuclei forming many small crystals (Fig. 12.4a). On the other hand, at slow cooling rates, few nuclei are formed, therefore fewer starting points are available resulting in few large crystals (Fig. 12.4b). As mentioned earlier, since large crystal aggregates observed under PLM are fractal aggregates of CNPs, Acevedo and Marangoni (2010b), showed that at the nanoscale, subjecting blends of fully hydrogenated canola oil and high oleic sunflower oil to slow and fast cooling rates resulted in CNPs that are large and small, respectively. This observation can be possibly extended to milk fat as milk fat is also a TAG crystal network.

2.1.3 Agitation, Ultrasound, and Shear-Induced Crystallization

Controlled agitation, ultrasound, and shear are often applied in processing to improve heat transfer and uniform mixing of fats, generally increase solid fat content, and induce phase transitions in products such as chocolate, margarine, fat spreads, and butter. Microscopically, the application of critical agitation speeds, shear rates, or ultrasound intensities during crystallization results in the formation of

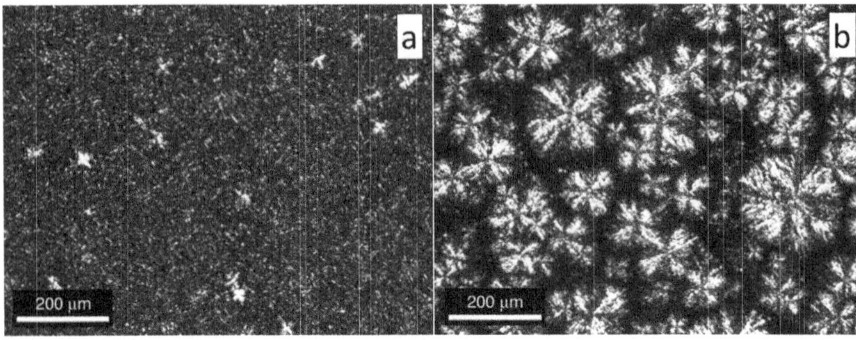

Fig. 12.4 Polarized light (PLM) micrographs of anhydrous milk fat crystallized (**a**) rapidly (1.0 °C/min) and (**b**) slowly (0.1 °C/min). Adapted from Martini and Marangoni (2007)

many small fat crystals, while statically, few large crystals are observed (Kaufmann, De Graef, Dewettinck, & Wiking, 2012; Martini, Suzuki, & Hartel, 2008; Wagh, Birkin, & Martini, 2016). Nevertheless, the size of crystals is affected by other crystallization parameters such as degree of undercooling and cooling rate. The mechanisms by which these processes affect milk fat crystallization are through the introduction of a nucleation seed (i.e., catalytic impurities) and the increase in heat and mass transfer, which allow the global crystallization of high melting TAGs, resulting in numerous nuclei formation and thereby many small crystals. Above a critical shear rate or ultrasound intensity, detrimental effects are usually observed, such as decrease in solid fat content due to melting and breakage of the crystal network (Wagh, Walsh, & Martini, 2013). At the nanoscale, although not studied for milk fat, other TAG systems showed that above critical shear rates, a good correlation between the nanoscale and microscale of fat crystals is observed. That is, smaller CNPs result in smaller crystal aggregates. However, below this critical shear rate, larger CNPs than those formed under static crystallization are observed (Acevedo, Block, & Marangoni, 2012; Acevedo & Marangoni, 2010b).

2.2 Effect of Composition on the Microstructure of Milk Fat

2.2.1 Milk Fat Composition and Minor Components

Milk fat is mainly composed of TAGs, with mono- and diacylglycerols, free fatty acids, cholesterols, glycolipids and other polar lipids as minor components. The presence of minor components in native milk fat was found to delay the onset of crystallization of milk fat. However, equilibrium SFC and overall mechanical and structural properties of milk fat remained unaffected (Wright, Hartel, Narine, & Marangoni, 2000; Wright & Marangoni, 2002, 2003).

2.2.2 Milk Fat Fractions

As mentioned earlier, TAG families in milk fat can be categorized according to their melting points—HMF, MMF, and LMF (Dimick, Reddy, & Ziegler, 1996; Kaylegian & Lindsay, 1995; Timms, 1980; van Aken et al., 1999). Due to presence of these fractions with different melting points, three physical phases may be present during crystallization of milk fat, namely, solid, liquid, and semi-solid at certain temperatures (Campos et al., 2003; Marangoni & Lencki, 1998; Timms, 1984). The solid phase at higher temperatures is mainly composed of HMF and some MMF, while LMF is expected to crystallize at very low temperatures (Danthine, 2012; Dimick et al., 1996; Kaylegian & Lindsay, 1995; Timms, 1984). This is then the basis for fractionation techniques (dry or with the aid of a solvent) developed for separation of these melting fractions. HMF, MMF, LMF, and their subsequent mixtures have been extensively studied for their properties and applications. At high concentrations of HMF in binary and ternary mixtures of milk fat fractions result in the formation of many small rod or needle-like crystals (i.e., one-dimensional growth) (Fig. 12.5). On the other hand, relatively higher concentrations of MMF and LMF result in large spherulites (i.e., multidimensional growth) (Fig. 12.6). This can be

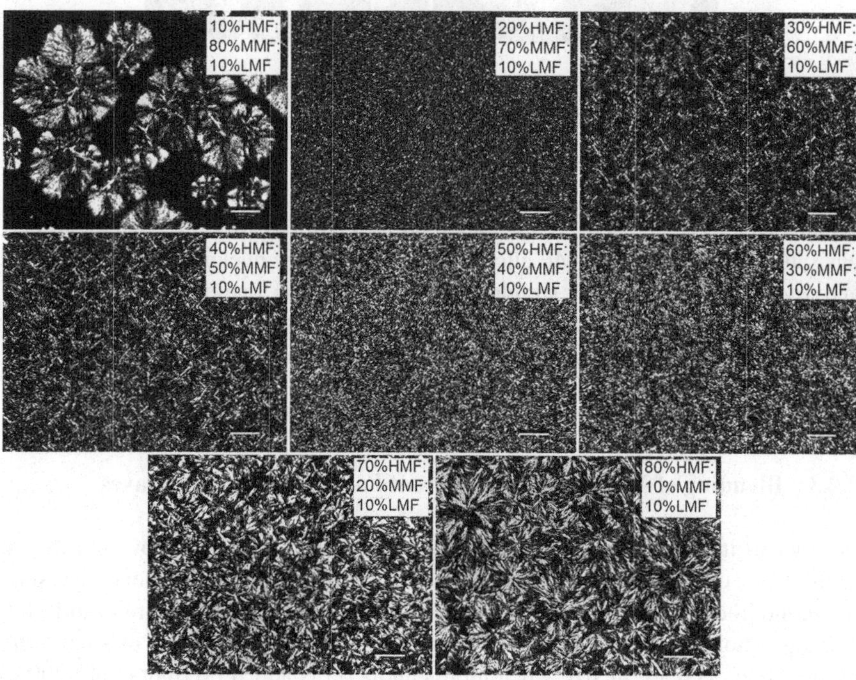

Fig. 12.5 Polarized light (PLM) micrographs of ternary mixtures of high melting (HMF), middle melting (MMF), and low melting (LMF) fractions of milk fat, showing increasing amounts of HMF, crystallized isothermally at 20 °C. Scale bar corresponds to 100 μm. Reproduced from Ramel and Marangoni (2016) with permission from the Royal Society of Chemistry

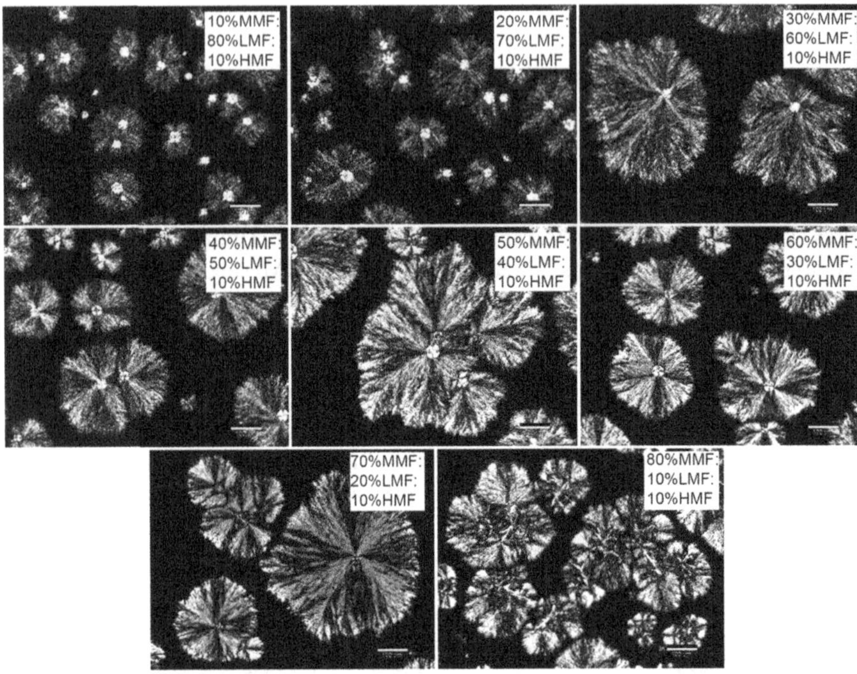

Fig. 12.6 Polarized light (PLM) micrographs of ternary mixtures of high melting (HMF), middle melting (MMF), and low melting (LMF) fractions of milk fat, showing increasing amounts of MMF, crystallized isothermally at 20 °C. Scale bar corresponds to 100 μm. Reproduced from Ramel and Marangoni (2016) with permission from the Royal Society of Chemistry

explained by the fact that at certain undercooling conditions, the presence of numerous saturated TAGs from HMF allow formation of many nuclei, and for nucleation and crystal growth to occur simultaneously resulting in one-dimensional growth. On the other hand, with high amounts of LMF, higher ratios of liquid fat are present which provides space for multidimensional growth and decreased supersaturation (inherent slow crystallization) (Ramel & Marangoni, 2016).

2.2.3 Blending of Milk Fat with Different Fats And Oils, and Waxes

In order to increase unsaturated fat contents of milk fat (*i.e.*, improve nutritional profile), it is usually blended with other fats and oils. However, such blending results in the modification of the crystallization behavior and crystal structure of milk fat. Adding canola oil to milk fat, for example, leads to the formation of spherulitic clusters that decrease in size when more canola oil is added (Wright et al., 2005). This could be explained by the dissolution of milk fat solids and dispersion of the crystalline clusters as more oil is added. In contrast, waxes are usually added to increase the melting point and accelerate hardening of milk fat. Addition of sunflower oil wax to milk fat, for example, results in faster crystallization and formation of smaller crystals (Kerr, Tombokan, Ghosh, & Martini, 2011; Martini, Carelli, & Lee, 2008). This is because waxes have high melting points, which allow them to crystallize at a lower supercooling and therefore serve as seeding nuclei for milk fat crystallization.

2.2.4 Emulsifiers

Milk fat is usually present in foods as an emulsion. The addition of emulsifiers in food is critical for maintaining the stability of emulsions. During milk fat crystallization, emulsifiers were found to play a role mainly for accelerating or delaying nucleation (Cerdeira et al., 2005; Wagh et al., 2013). Emulsifiers that have similar molecular structure with TAGs in milk fat and therefore serve as starting nuclei were found to enhance crystallization while those that are structurally dissimilar delays crystallization. Accelerated nucleation, in turn, results in the formation of many small crystals while delayed crystallization results in the formation of a few crystals. At the nanoscale, using other TAG systems, it was shown that the addition of emulsifiers could potentially affect the surface properties of CNPs, thereby affecting inter-crystalline interactions between CNPs (Acevedo & Marangoni, 2014). This, in turn, affects the mechanical strength and other mechanical properties of CNPs and its fractal aggregates.

3 Effect of Resulting Microstructure on the Properties of Milk Fat and Related Products

Considering the same amount of solid fat content, the size, shape, and spatial distribution of milk fat crystals have been shown to affect the macro-properties of milk fat and milk-fat products. In this section, the impact of the changes in microstructure described above on the properties of milk fat and related products is described.

3.1 Textural Properties of Milk Fat and Related Products

The physical properties of milk fat have been shown to affect the hardness and mouthfeel of milk fat and related products. Crystal size or particle size has been found to be inversely proportional to the hardness or firmness of milk fat (Wright et al., 2001). That is, as crystal size or particle size decreases, the hardness of the milk fat crystal network increases, while hardness decreases when crystal size increases. This can be explained by the fact that at similar solid fat contents, crystal size affects inter-crystalline interactions due to surface contact. When smaller fat crystals are present, the surface area for interaction of crystals is large and therefore the crystal network is harder or firmer. This is because as surface area increases, the number of inter-crystalline interactions also increases. On the other hand, when larger crystals are present, less surface area is available for inter-crystalline interactions, resulting in a soft or weak crystal network (Campos et al., 2002; deMan & Beers, 1987; Kerr et al., 2011). Relating to the processes described above, enhanced nucleation (from the addition of emulsifiers and waxes) or fast cooling and shear application result in smaller crystals and therefore harder or firmer crystal network. On the other hand, low undercooling (e.g., from the addition of canola oil, LMF) and increased crystal growth result in larger, and dispersed crystals resulting in a softer crystal network. Other processes such as temperature cycling resulting in sintering of the fat crystal network results in a stronger crystal network due to the formation of solid bridges among fat crystals.

In butter, rapid cooling leads to the formation of smaller crystals, resulting in a firmer butter product. On the other hand, addition of canola oil results in softening. Working of butter after churning results in the breaking of the crystal network; however, upon setting, hardness or firmness of butter increases. Furthermore, working allows for the expulsion of liquid oil, and reduction of the size of water droplets increasing contact between crystals and consequently increasing the firmness of the product (Wright et al., 2001).

3.2 Rheology

The rheological properties of milk fat are largely dependent on solid fat content or the ratio between solid and liquid fat at specific temperatures. The viscoelastic or plastic properties of milk fat and related products are affected by the balance between solid and liquid fat, or the containment of the liquid fat in the solid crystal network. Too much liquid fat could result in a runny or very soft product while very high solid fat contents could result in extensive hardening or brittleness. The spatial distribution of crystal mass, as measured by image analysis (e.g., box-counting method) is a good indicator of the ratio between solid and liquid fat. It can also be used to determine the fractal dimension of the fat crystal network. As mentioned earlier, fat crystal networks are fractal in nature, wherein the large polycrystalline aggregates are formed by the continuous stacking and aggregation of repeating units of primary crystals—CNPs (Acevedo & Marangoni, 2015; Marangoni, 2002; Ramel et al., 2016). The fractal dimension (i.e., measure of the order in which solid mass is spatially distributed in the network of fat crystal networks) is correlated with elastic modulus (Narine & Marangoni, 1999). Increasing fractal dimensions results in increased hardness of a material (Wright et al., 2001). Furthermore, higher fractal dimensions indicate more ordered crystal distribution and lower porosity of the crystal network, which in turn, increases oil binding (Blake, Co, & Marangoni, 2014; Marangoni et al., 2012).

4 Conclusions

The effect of different factors on the microstructure of milk fat and subsequent effects of the resulting microstructure on the rheological properties of milk fat and related products were described. Besides manipulation of the processes and composition for modifying the microstructure of milk fat, a growing area of research is the engineering of the nanostructure of fats. The size, surface properties, and intercrystalline interactions of CNPs have been shown to be effectively modified by processing conditions and composition. As the larger poly-crystals are fractal aggregates of the CNPs, the properties of the CNPs and its aggregation behavior could have large effects on the microstructure or macro-properties of milk fat. The microstructural level can be manipulated effectively using basic unit operations, thus opening the door to fine-tuning of structure and mechanical properties.

References

Acevedo, N. C., Block, J. M., & Marangoni, A. G. (2012). Critical laminar shear-temperature effects on the nano- and mesoscale structure of a model fat and its relationship to oil binding and rheological properties. *Faraday Discussions, 158*, 171. https://doi.org/10.1039/c2fd20008b

Acevedo, N. C., & Marangoni, A. G. (2010a). Characterization of the nanoscale in triacylglycerol crystal networks. *Crystal Growth & Design, 10*, 3327–3333. https://doi.org/10.1021/cg100468e

Acevedo, N. C., & Marangoni, A. G. (2010b). Toward nanoscale engineering of triacylglycerol crystal networks. *Crystal Growth & Design, 10*, 3334–3339. https://doi.org/10.1021/cg100469x

Acevedo, N. C., & Marangoni, A. G. (2014). Engineering the functionality of blends of fully hydrogenated and non-hydrogenated soybean oil by addition of emulsifiers. *Food Biophysics, 9*, 368–379. https://doi.org/10.1007/s11483-014-9340-9

Acevedo, N. C., & Marangoni, A. G. (2015). Nanostructured fat crystal systems. *Annu Rev Food Sci Technol, 6*, 71–96. https://doi.org/10.1146/annurev-food-030713-092400

Blake, A. I., Co, E. D., & Marangoni, A. G. (2014). Structure and physical properties of plant wax crystal networks and their relationship to oil binding capacity. *Journal of the American Oil Chemists' Society, 91*, 885–903. https://doi.org/10.1007/s11746-014-2435-0

Buldo, P., Andersen, U., & Wiking, L. (2013). Microstructure and material properties of milk fat systems during temperature fluctuations. *Food Biophysics, 8*, 262–272. https://doi.org/10.1007/s11483-013-9299-y

Campos, R., Narine, S., & Marangoni, A. G. (2002). Effect of cooling rate on the structure and mechanical properties of milk fat and lard. *Food Research International, 35*, 971–981. https://doi.org/10.1016/S0963-9969(02)00159-X

Campos, R. J., Litwinenko, J. W., & Marangoni, A. G. (2003). Fractionation of milk fat by short-path distillation. *Journal of Dairy Science, 86*, 735–745. https://doi.org/10.3168/jds.S0022-0302(03)73654-6

Cerdeira, M., Pastore, V., Vera, L. V., Martini, S., Candal, R. J., & Herrera, M. L. (2005). Nucleation behavior of blended high-melting fractions of milk fat as affected by emulsifiers. *European Journal of Lipid Science and Technology, 107*, 877–885. https://doi.org/10.1002/ejlt.200500257

Danthine, S. (2012). Physicochemical and structural properties of compound dairy fat blends. *Food Research International, 48*, 187–195. https://doi.org/10.1016/j.foodres.2012.03.004

deMan, J. M., & Beers, A. M. (1987). Review fat crystal networks: structure and rheological properties. *Journal of Texture Studies, 18*, 303–318.

Dibildox-Alvarado, E., Marangoni, A. G., & Toro-Vazquez, J. F. (2010). Pre-nucleation structuring of triacylglycerols and its effect on the activation energy of nucleation. *Food Biophysics, 5*, 218–226. https://doi.org/10.1007/s11483-010-9163-2

Dimick, P. S., Reddy, S. Y., & Ziegler, G. R. (1996). Chemical and thermal characteristics of milk-fat fractions isolated by a melt crystallization. *Journal of the American Oil Chemists' Society, 73*, 1647–1652. https://doi.org/10.1007/bf02517966

Herrera, M. L., & Hartel, R. W. (2000). Effect of processing conditions on crystallization kinetics of a milk fat model system. *Journal of the American Oil Chemists' Society, 77*, 1177–1188. https://doi.org/10.1007/s11746-000-0184-4

Jensen, R. G. (2002). The composition of bovine milk lipids: January 1995 to December 2000. *Journal of Dairy Science, 85*, 295–350. https://doi.org/10.3168/jds.S0022-0302(02)74079-4

Jensen, R. G., Ferris, A. M., & Lammi-Keefe, C. J. (1991). The composition of milk fat. *Journal of Dairy Science, 74*, 3228–3243. https://doi.org/10.3168/jds.S0022-0302(91)78509-3

Kaufmann, N., De Graef, V., Dewettinck, K., & Wiking, L. (2012). Shear-induced crystal structure formation in milk fat and blends with rapeseed oil. *Food Biophysics, 7*, 308–316. https://doi.org/10.1007/s11483-012-9269-9

Kaylegian, K. E., & Lindsay, R. C. (1995). In K. E. Kaylegian & R. C. Lindsay (Eds.), *Handbook of milkfat fractionation and technology application*. Champaign: AOCS Press.

Kerr, R. M., Tombokan, X., Ghosh, S., & Martini, S. (2011). Crystallization behavior of anhydrous milk fat-sunflower oil wax blends. *Journal of Agricultural and Food Chemistry, 59*, 2689–2695. https://doi.org/10.1021/jf1046046

Lopez, C., Lesieur, P., Bourgaux, C., & Ollivon, M. (2005). Thermal and structural behavior of anhydrous milk fat. 3. Influence of cooling rate. *J Dairy Sci, 88*, 511–526. https://doi.org/10.3168/jds.S0022-0302(05)72713-2

Marangoni, A. G. (2002). The nature of fractality in fat crystal networks. *Trends in Food Science and Technology, 13*, 37–47. https://doi.org/10.1016/S0924-2244(02)00029-8

Marangoni, A. G., Acevedo, N., Maleky, F., Co, E., Peyronel, F., Mazzanti, G., et al. (2012). Structure and functionality of edible fats. *Soft Matter, 8*, 1275. https://doi.org/10.1039/c1sm06234d

Marangoni, A. G., & Lencki, R. W. (1998). Ternary phase behavior of milk fat fractions. *Journal of Agricultural and Food Chemistry, 46*, 3879–3884. https://doi.org/10.1021/jf9801668

Marangoni, A. G., & Mcgauley, S. E. (2003). Relationship between crystallization behavior and structure in cocoa butter. *Crystal Growth & Design, 3*, 95–108. https://doi.org/10.1021/cg025580l

Marangoni, A. G., Narine, S. S., Acevedo, N. C., & Tang, D. (2013). Nanostructure and microstructure of fats. In A. G. Marangoni & L. H. Wesdorp (Eds.), *Structure and properties of fat crystal networks* (2nd ed.). Boca Raton: CRC Press.

Marangoni, A. G., Tang, D., & Singh, A. P. (2006). Non-isothermal nucleation of triacylglycerol melts. *Chemical Physics Letters, 419*, 259–264. https://doi.org/10.1016/j.cplett.2005.11.085

Marangoni, A. G., & Wesdorp, L. H. (2013a). Crystallography and polymorphism. In A. G. Marangoni & L. H. Wesdorp (Eds.), *Structure and properties of fat crystal networks* (2nd ed.). Boca Raton: CRC Press.

Marangoni, A. G., & Wesdorp, L. H. (2013b). Nucleation and crystalline growth kinetics. In A. G. Marangoni & L. Wesdorp (Eds.), *Structure and properties of fat crystal networks* (2nd ed.). Boca Raton: CRC Press.

Martini, S., Carelli, A. A., & Lee, J. (2008). Effect of the addition of waxes on the crystallization behavior of anhydrous milk fat. *Journal of the American Oil Chemists' Society, 85*, 1097–1104. https://doi.org/10.1007/s11746-008-1310-2

Martini, S., Herrera, M. L., & Hartel, R. W. (2001). Effect of cooling rate on nucleation behavior of milk fat - sunflower oil blends. *Journal of Agricultural and Food Chemistry, 49*, 3223–3229. https://doi.org/10.1021/jf001101j

Martini, S., & Marangoni, A. G. (2007). Microstructure of dairy fat products. In A. Tamime (Ed.), *Structure of dairy products*. Oxford: Blackwell Publishing.

Martini, S., Suzuki, A. H., & Hartel, R. W. (2008). Effect of high intensity ultrasound on crystallization behavior of anhydrous milk fat. *Journal of the American Oil Chemists' Society, 85*, 621–628. https://doi.org/10.1007/s11746-008-1247-5

Mazzanti, G., Guthrie, S. E., Sirota, E. B., Marangoni, A. G., & Idziak, S. H. J. (2004). Effect of minor components and temperature profiles on polymorphism in milk fat. *Crystal Growth & Design, 4*, 1303–1309. https://doi.org/10.1021/cg0497602

Narine, S. S., & Marangoni, A. G. (1999). Fractal nature of fat crystal networks. *Physical Review E, 59*, 1908–1920. https://doi.org/10.1103/PhysRevE.59.1908

Ramel, P. R., Co, E. D., Acevedo, N. C., & Marangoni, A. G. (2016). Structure and functionality of nanostructured triacylglycerol crystal networks. *Progress in Lipid Research, 64*, 231–242. https://doi.org/10.1016/j.plipres.2016.09.004

Ramel, P. R., & Marangoni, A. G. (2016). Engineering the microstructure of milk fat by blending binary and ternary mixtures of its fractions. *RSC Advances, 6*, 41189–41194. https://doi.org/10.1039/C6RA07114G

Sangwal, K., & Sato, K. (2012). Nucleation and crystallization kinetics of milk fat. In A. G. Marangoni (Ed.), *Structure-function analysis of edible fats* (1st ed.). Urbana: AOCS Press.

Sato, K. (2001). Crystallization behaviour of fats and lipids—a review. *Chemical Engineering Science, 56*, 2255–2265. https://doi.org/10.1016/S0009-2509(00)00458-9

Shi, Y., Smith, C. M., & Hartel, R. W. (2001). Compositional effects on milk fat crystallization. *Journal of Dairy Science, 84*, 2392–2401. https://doi.org/10.3168/jds.S0022-0302(01)74688-7

Singh, A. P., Bertoli, C., Rousset, P. R., & Marangoni, A. G. (2004). Matching avrami indices achieves similar hardnesses in palm oil-based fats. *Journal of Agricultural and Food Chemistry, 52*, 1551–1557. https://doi.org/10.1021/jf0346531

Timms, R. E. (1980). The phase behaviour and polymorphism of milk fat, milk fat fractions and fully hardened milk fat. *Australian Journal of Dairy Technology, 35*, 47–53.

Timms, R. E. (1984). Phase behaviour of fats and their mixtures. *Progress in Lipid Research, 23*, 1–38. https://doi.org/10.1016/0163-7827(84)90004-3

van Aken, G. A., Grotenhuis, E., Langevelde, A. J., & Schenk, H. (1999). Composition and crystallization of milk fat fractions. *Journal of the American Oil Chemists' Society, 76*, 1323–1331. https://doi.org/10.1007/s11746-999-0146-8

Wagh, A., Birkin, P., & Martini, S. (2016). High-intensity ultrasound to improve physical and functional properties of lipids. *Annual Review of Food Science and Technology, 7*, 23–41. https://doi.org/10.1146/annurev-food-041715-033112

Wagh, A., Walsh, M. K., & Martini, S. (2013). Effect of lactose monolaurate and high intensity ultrasound on crystallization behavior of anhydrous milk fat. *Journal of the American Oil Chemists' Society, 90*, 977–987. https://doi.org/10.1007/s11746-013-2244-x

Wiking, L., De Graef, V., Rasmussen, M., & Dewettinck, K. (2009). Relations between crystallisation mechanisms and microstructure of milk fat. *International Dairy Journal, 19*, 424–430. https://doi.org/10.1016/j.idairyj.2009.03.003

Wright, A. J., Batte, H. D., & Marangoni, A. G. (2005). Effects of canola oil dilution on anhydrous milk fat crystallization and fractionation behavior. *Journal of Dairy Science, 88*, 1955–1965. https://doi.org/10.3168/jds.S0022-0302(05)72871-X

Wright, A. J., Hartel, R. W., Narine, S. S., & Marangoni, A. G. (2000). The effect of minor components on milk fat crystallization. *Journal of the American Oil Chemists' Society, 77*, 463–475. https://doi.org/10.1007/s11746-000-0075-8

Wright, A. J., & Marangoni, A. G. (2002). Effect of DAG on milk fat TAG crystallization. *Journal of the American Oil Chemists' Society, 79*, 395–402. https://doi.org/10.1007/s11746-002-0495-5

Wright, A. J., & Marangoni, A. G. (2003). The effect of minor components on milk fat microstructure and mechanical properties. *Journal of Food Science, 68*, 182–186. https://doi.org/10.1111/j.1365-2621.2003.tb14137.x

Wright, A. J., Scanlon, M. G., Hartel, R. W., & Marangoni, A. G. (2001). Rheological properties of milkfat and butter. *Journal of Food Science, 66*, 1056–1071.

Chapter 13
Oil Structuring in Dairy Fat Products

Ashok R. Patel

1 Introduction

A range of dairy products are consumed on a regular basis as sources of macronutrients (fats and proteins) and micronutrients such as minerals and fat-soluble vitamins. Among these, there are products such as butter and cheese that form a part of daily diets and then there are others such as whipped cream and ice-creams that are usually considered as indulgence products. From colloid science point of view, these products could be broadly classified as structured emulsions (butter and butter spreads), coagulated gels (various cheese types) and foamed emulsions (ice-creams and whipped creams). These products have different microstructures (Fig. 13.1), all of them containing a significant proportion of milk fat distributed either in the bulk or dispersed phases. Milk fat is composed primarily of triglycerides (TAGs) with a significantly high proportion of saturated fatty acids (Table 13.1). Palmitic acid, the main fatty acid in milk fat, is known to increase the risk of cardiovascular disease (CVD) (Wang et al., 2017). And it has been consistently suggested by health agencies that replacing dairy fats with vegetable oils rich in polyunsaturated fatty acids reduces the risk of CVD (Chen et al., 2016; Nettleton, Brouwer, Geleijnse, & Hornstra, 2017). However, the high melting fraction of milk fat (composed of TAGs rich in long-chain fatty acids) is responsible for providing the underlying colloidal network of crystalline particles, which in turn governs the macrostructure and organoleptic properties of dairy fat products. Such properties include spreadability of butter and cheese spreads, plasticity of baking butter, hardness of cooking butter, voluminous body of whipped cream, texture of cheese, creaminess of ice-cream and melt-in-mouth effect of most dairy fat products. In addition, the stabilizing effect provided by bulk crystallization of milk fat in butter and interfacial stabilization of partially coalesced fat globules in whipped cream and ice cream is also dependent

A. R. Patel (✉)
Guangdong Technion Israel Institute of Technology, Shantou, China
e-mail: ashok.patel@gtiit.edu.cn

T. Truong et al. (eds.), *Dairy Fat Products and Functionality*,
https://doi.org/10.1007/978-3-030-41661-4_13

on the high melting TAGs in milk fat. Due to this broad range of functionality provided by milk fat, it is a challenging prospect to replace high melting milk fat with liquid vegetable oils rich in polyunsaturated fatty acids without compromising on the product attributes of reformulated dairy fat products.

From a general perspective, there has been a growing trend in the lipid community to identify alternative routes for structuring liquid vegetable oils in order to reformulate a broad range of lipid-based products (including dairy-related products) owing to a combination of factors including (1) recent policy changes in United States with regards to removal of *trans*-fats from food products, (2) raising concerns among consumers about the negative effect of saturated fat consumption and (3) the ecological damage caused by palm oil usage. These have together provided new motivations for food material scientists to focus on innovations in

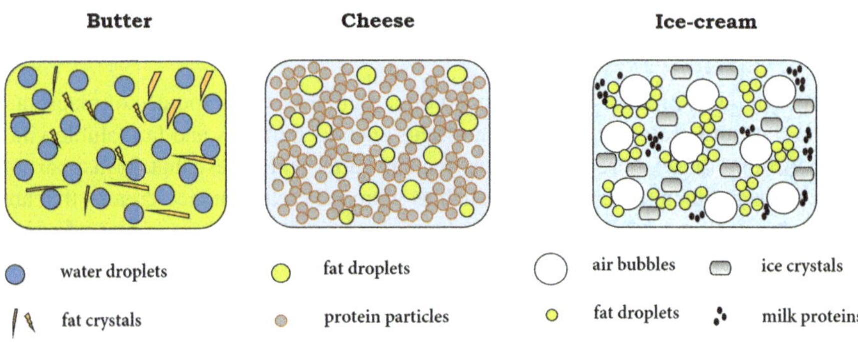

Fig. 13.1 Representative microstructures of dairy fat products such as butter (water droplets dispersed in oil continuous phase containing fat crystals), cheese (fat droplets embedded in protein matrix), and ice-cream (air bubbles stabilized by partially coalesced fat droplets dispersed in water continuous phase containing ice crystals)

Table 13.1 Composition of the major fatty acids in milk fat

Fatty acid	Average range (wt%)
Butyric acid (4:0)	2–5
Caproic acid (6:0)	1–5
Caprylic acid (8:0)	1–3
Capric acid (10:0)	2–4
Lauric acid (12:0)	2–5
Myristic acid (14:0)	8–14
Pentadecanoic acid (15:0)	1–2
Palmitic acid (16:0)	22–35
Palmitoleic acid (16:1)	1–3
Margaric acid (17:0)	0.5–1.5
Stearic acid (18:0)	9–14
Oleic acid (18:1)	20–30
Linoleic acid (18:2)	1–3
Linolenic acid (18:3)	0.5–2

formulating lipid-based products with the aim of improving their nutritional profiles (*trans* fat- free, low in saturated fats, and high in unsaturated fats) and decreasing the over-reliance on palm oil (Patel & Edible, 2017).

Over the last decade, the field of oil structuring has seen a rapid advancement. Different approaches and structuring agents identified so far, may find applications in commercial settings very soon. In the following section, first, the basic concept of fat crystallization is introduced followed by a general overview of the field of oil structuring. The section concludes with a brief review of some of the examples of the applications of structured oil in dairy fat and related products.

2 Crystallization Behaviour of Milk TAGs

As discussed in the previous section, the network of crystalline particles affect a range of macrostructure and organoleptic properties of dairy fat products; hence, it is very important to first understand the crystallization behaviour of TAGs in milk fat. This knowledge can lay a foundation for exploring alternative ways for oil structuring.

Due to the broad TAGs profile, milk fat displays a broad melting range instead of a discrete melting point. Being a mixture of high and low melting TAGs, the crystallization behaviour is a combination of both melt crystallization as well as crystallization from a supersaturated solution where low melting TAGs can be considered to act as solvents for high melting TAGs. In general, fat crystallization behaviour can be broadly defined by three phenomena: nucleation, crystal growth and polymorphism. In addition to these basic phenomenon, the final development of bulk fat also involves a range of structural reorganization at different length scales as shown in Fig. 13.2 (Patel & Dewettinck, 2015).

2.1 Nucleation

Nucleation is described as the formation of the crystalline phase from the liquid phase through the organization of TAGs into a crystal lattice. Three types of nucleation that can generally occur in fats include:

Primary homogeneous nucleation which occurs in the absence of any foreign materials. This process is very rare in milk fat because of the requirement of a very deep supercooling which is practically difficult to attain.

Primary heterogeneous nucleation which is initiated at the surfaces of catalytic impurities such as dust, seed particles or templates of high melting lipid components such as partial glycerides etc. In milk fat, micelles formed by monoacylglycerides may act as catalytic impurities for heterogeneous nucleation (Huppertz, Kelly, & Fox, 2009).

Secondary nucleation occurs when small pieces break from existing crystals and act as nuclei and is also important in milk fat.

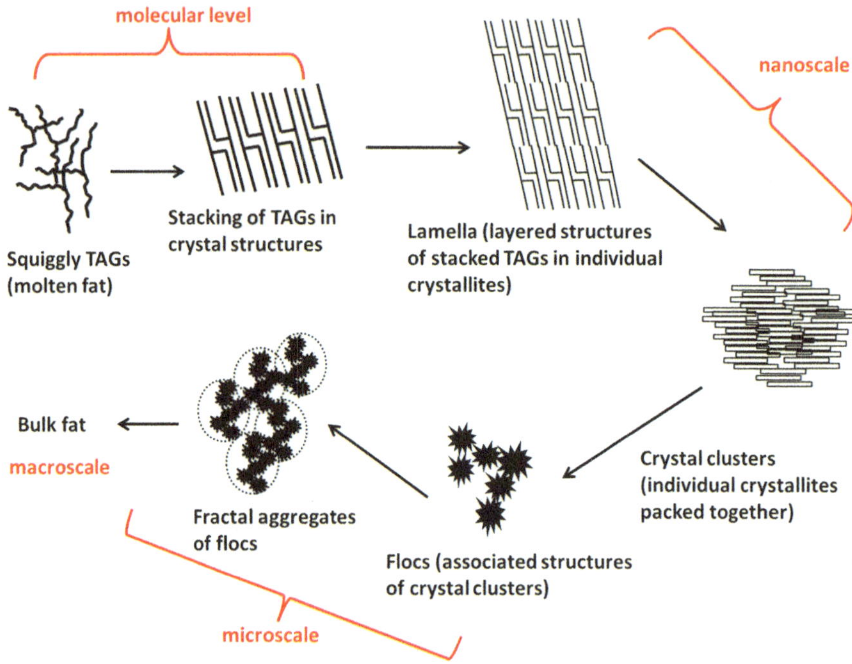

Fig. 13.2 Different structural levels (from molecular to macroscale level) involved in bulk fat development. Reproduced with permission from Patel & Dewettinck, Current update on the influence of minor lipid components, shear and presence of interfaces on fat crystallization. Current Opinion in Food Science 2015, 3:65–70

2.2 *Crystal Growth*

Crystal growth is caused by the inclusion of other TAGs from bulk liquid in the existing crystal lattice. The TAG molecules can incorporate on either of the flat, kinked or stepped surfaces. The crystal growth involves firstly the diffusion of TAG molecules from the bulk phase to the crystal site and secondly the proper orientation to align with the existing packing of the molecules.

2.3 *Polymorphism*

Polymorphism governs the order of the packing of TAG molecules in the crystals. Polymorphic forms are solid phases of the same chemical composition that differ among themselves in crystalline structures. Milk fats like other fats are known to display monotropic polymorphism wherein one phase is stable, and others are metastable under all conditions, regardless of the external conditions such as temperature and pressure. The three polymorphic forms of milk fat crystals include α and β'

(metastable) and β (stable) forms. Fat crystallization follows the Ostwald's step rule which states that the least stable polymorphs (α, in this case) is first to crystallize on cooling followed by step-wise transition to the most stable form. In milk fat, the majority of crystals are known to remain in the metastable β′ form, even after prolonged storage (Wright & Marangoni, 2006).

Fat crystals can grow into spherulitic or needle-shaped structures. In milk fat spherulitic crystal structures are predominant, these crystals have a very dense centre with decreasing density as the distance from the crystal centre increases. Spherulites continue to grow and aggregate into a three-dimensional network to take the form of bulk fat. Since the whole crystallization process of fats is rather 'slow', processing parameters like rapid (quenching) and deep cooling (supercooling) along with application of shear and incorporation of surface-active species is commonly employed to obtain the desired product (Patel, 2015). Generally, a faster cooling rate with deep cooling ensures that a large number of formed nuclei do not get enough time to grow, hence resulting in the formation of larger population of smaller crystals. Smaller size of crystals provide a higher overall effective surface area leading to stronger crystal-crystal interactions in the network. The crystalline network with immobilized liquid oil constitutes conventional route of oil structuring.

For a colloid scientist, this structured system is more akin to a particle-filled gel where a sufficiently high mass fraction of crystalline particles is required to obtain physical gels of liquid oil. Typically, a mass fraction of at least 0.2–0.25 is need for firm gels. Alternatively, such physical gels could be prepared using oleogelators instead of solid fats at relatively lower mass fraction (as low as 0.005 in some case). In the following section, different approaches of oil structuring is briefly discussed with specific emphasis on oleogelation through direct dispersion of structuring materials.

3 Different Approaches for Oil Structuring

There are different ways of categorizing oil structuring approaches, e.g., based on the molecular characteristics of structuring agents (low molecular weight organic compounds, polymeric compounds and inorganic compounds), their chemical types (lipid and non-lipid based oleogelators), and the molecular assemblies formed by the agents (crystalline particles, self-assembled structures, agglomerated inorganic particles and polymeric strands) (Patel & Dewettinck, 2016). The above mentioned ways of categorizing oil structuring approaches suitably covers oleogels which are fabricated by direct dispersion of gelling agent(s) but leaves out many unconventional approaches which may be considered more practical for use in commercial settings. Alternatively, a more comprehensive way of classifying structuring approaches is based on the processing steps involved in creating the structured oil systems as shown in Fig. 13.3.

3.1 Direct Dispersion of Structuring Agents in Oil Phase

Almost all molecular gels are created by direct dispersion of structuring agent at elevated temperature followed by cooling down to ambient or cool temperature to trigger precipitation/crystallization/self-assembly of molecules into building blocks such as fine crystalline units, fibres, micelles etc. that can further associate *via* non-covalent interactions to form a space-filling continuous network. The solvent is then physically trapped into this network leading to a 'gel-like' state.

This is the most common approach used for preparing edible gels, lipid-based structuring agents such as partial glycerides, fatty acids, fatty alcohols, sorbitan esters and waxes are first dispersed above their melting points in liquid oils to create a clear molecular (supersaturated) solution followed by lowering of temperature to trigger crystallization (undercooling). Steps of nucleation, crystal growth, aggregation and network formation (as seen with high melting TAGs) are involved in oil structuring with lipid-based gelators as this approach is similar to fat-based structuring. However, the type and the morphology of crystals formed, the tendency of their unidirectional growth and limited aggregation of the formed crystals, differentiates them from the conventional crystals of TAGs, resulting in a continuous network (that shows high oil binding capacity) at much lower mass fraction of crystalline

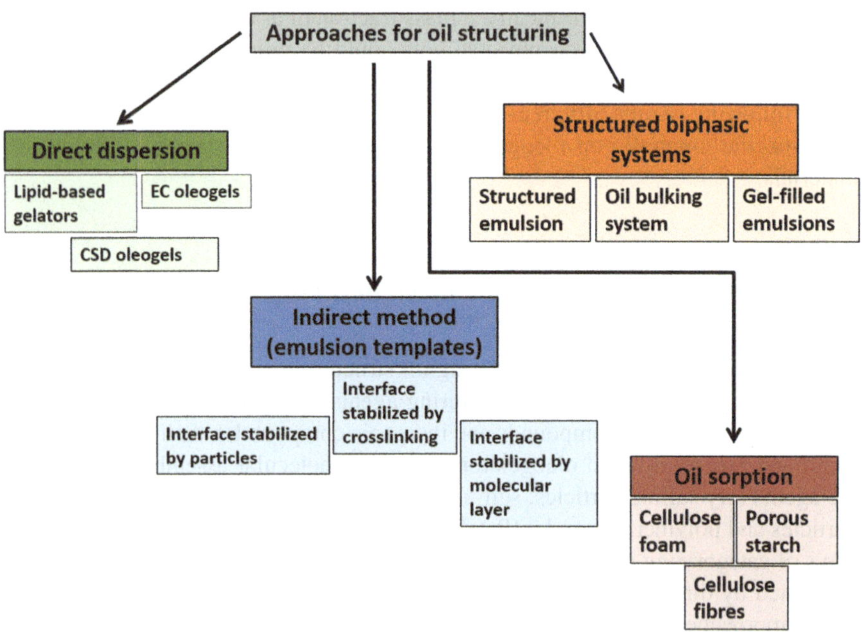

Fig. 13.3 Various approaches used for edible oil structuring. EC and CSD = Ethyl cellulose and colloidal silicon dioxide respectively. Reproduced with permission from Patel & Dewettinck, Edible oil structuring: an overview and recent updates. Food & Function, 2016, 7, 20–29. Published by The Royal Society of Chemistry

phase (as low as 0.5 wt%). The formed network from these lipid-based gelators display thermoreversible properties as seen with the conventional fat. This ability of lipid-based gelators to structure oil reversibly at very low concentrations is quite appealing to the food industry. Accordingly, this category of structuring agents has been studied extensively for edible applications (discussed further in the text).

Waxes Among lipid-based gelators that form crystalline particles, natural waxes are by far the most studied group of materials for structuring edible oils. Chemically, waxes are composed of multiple chemical entities, the four major components are wax esters (WEs), fatty acids (FAs), fatty alcohols (Falcs) and linear hydrocarbons (HCs). Because of the presence of molecular units with linear structures, crystals formed by waxes show predominantly higher one-dimensional (1D) and two-dimensional (2D) growth resulting in needle-like to plate-like structures which are more efficient at gelling liquid oil at a concentration as low as 0.5 wt% (Patel, Babaahmadi, Lesaffer, & Dewettinck, 2015a). In addition to oil gelation, waxes are also known to stabilize oil continuous emulsions by a combination of bulk and interfacial crystallization (Fig. 13.4) (Patel, 2015). Owing to their oil structuring and colloid stabilization properties, a wide range of natural waxes have been explored so far for oil structuring. However, the results obtained with respect to critical gelling concentrations and rheological behaviour differ widely in different studies which makes it difficult to select suitable waxes for food applications. Recently, a comprehensive study was conducted to first elucidate the compositional characterization

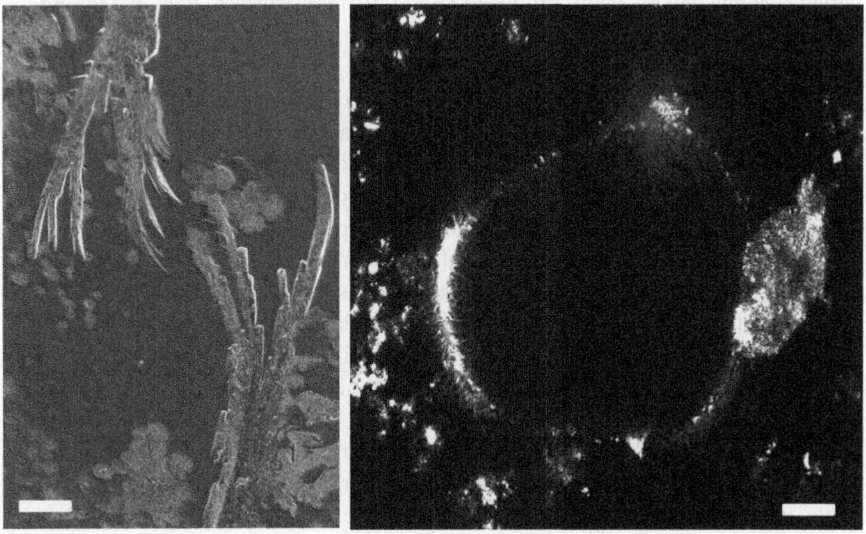

Fig. 13.4 (On left) Needle-like crystals of wax (sunflower wax) seen under cryo-SEM (scale bar = 20 μm); (on right) Confocal image (taken in reflectance mode) of oleogel-based emulsion clearly showing the accumulation of fine wax crystals (shellac wax) at water-oil interface (Pickering stabilization; scale bar = 30 μm). Reproduced from Patel A. R., Alternative Routes to Oil Structuring, Springer International Publishing, 2015, UK

(percentage and chain length of each chemical class) of a set of waxes and further understand their role in the structuring of liquid oil (Doan et al., 2017). Using a set of waxes and a common solvent (rice bran oil), it was found that WEs result in strong and brittle gels with high modulus and yield stress, while HCs and long-chain FAs (above C16) contribute to stability and consistency of gels (high critical stress and consistency index) (Doan et al., 2017).

Partial Glycerides Partial glycerides- mono and di- glycerides (MAGs and DAGs) are comparatively more polar than TAGs due to their structural features. Their crystallization behaviour is thus quite different from TAGs although they all belong to the acylglycerol family (i.e. fatty acid esters of glycerol). The advantage of using these partial glycerides (particularly, DAGs) over TAGs is that, while, they can mimic the crystal network formation and temperature induced texture reversibility shown by TAGs, their digestion in the gastro intestinal tract is quite different from that of TAGs and does not follow the usual mucosal resynthesize of TAGs, resulting in decreased body fat accumulation (Osaki et al., 2005). Both MAGs and DAGs are known to form a crystalline network in liquid oils which results in gel formation. In particular, MAGs with saturated long-chain fatty acids have been extensively studied for oil gelation (Ojijo, Neeman, Eger, & Shimoni, 2004). Since these partial glycerides are already being used in food industry as direct additives without any specified maximum limits (*quantum satis*), they can be considered to have a better acceptance compared to the other lipidic structurants. In addition, due to the interesting phase behaviour displayed by MAGs in oil-water mixtures, a range of oil-water gelled systems can be easily produced (Batte, Wright, Rush, Idziak, & Marangoni, 2007a, 2007b).

Sphingolipids Sphingolipids are a class of biologically relevant lipids containing a sphingosine (a type of aliphatic amino alcohol) backbone with either one hydroxyl group esterified to a fatty acid side chain (ceramides) or both hydroxyl group substituted with a fatty acid and a non-fatty acid group (sphingomyelin, cerebroside and ganglioside). Ceramides, which are the simplest of sphingolipids have been studied for their oil gelling properties, it has been reported that certain short-chain ceramides are effective oleogelators. The increase in the fatty acid chain length adversely affects the success of ceramide as a gelator. The shorter fatty acid chain lengths are known to promote fibril growth, while longer chain lengths promote spherulite or platelet structures (Wang & Rogers, 2015).

Shellac Shellac is a resin purified from the secretion of lac insects, *Laccifer Lacca*. It is not a single compound, but a complex mixture of polar and non-polar components consisting of polyhydroxy polycarboxylic esters, acids and alkanes. Being a complex mixture of fatty esters and alcohols, shellac has a lipophilic nature and tends to self-assemble into colloidal structures based on the solvent properties (Patel, Schatteman, De Vos, & Dewettinck, 2013a).

Shellac shows good miscibility with oil at temperatures above its melting range and forms an oleogel when the dispersion of shellac in liquid oil is cooled to room temperature. These oleogel samples could be obtained even after several cycles of heating and cooling, indicating their thermo-reversible nature. Figure 13.5 shows a comparative picture of samples prepared by heating the shellac dispersions at different concentrations in rapeseed oil followed by cooling to room temperature. As seen from Fig. 13.5, a concentration of 2 wt% is sufficient to create an oleogel. The microscopy studies revealed the formation of crystalline particles which forms a continuous network that entraps the liquid oil into a gel-like phase.

Phytosterol-Oryzanol Mixtures Mixed component oleogels of phytosterols and oryzanol are one of the most studied structured oil systems. When mixed at certain proportions, they co-assemble to form nanoscale tubular structures, and depending on the concentration, these nanoscale tubules aggregate to form a 3-D network that can physically immobilize liquid oil through capillary forces to form a viscoelastic gel. The building blocks in these oleogels are self-assembled tubules of nanoscale dimensions (Bot et al., 2012). These tubules have a complex helical, ribbon-like structure which is mediated by the intermolecular hydrogen bonding between the hydroxyl group of phytosterol and carbonyl group of oryzanol. Beta-sitosterol and other related phytosterols such as dihydrocholesterol, cholesterol and stigmasterol can also produce gels with oryzanol. Structurally, the presence and position of hydroxyl was found to be critical for gel formation while the composition of alkyl group did not appear to have any major impact on the gelling properties of phytosterols (Bot & Agterof, 2006). Although the combination is able to form clear (or slightly hazy) oil gels, the effectiveness of structuring in the presence of water becomes less because of the formation of thicker sitosterol monohydrate fibres (microscale dimensions) (den Adel, Heussen, & Bot, 2010).

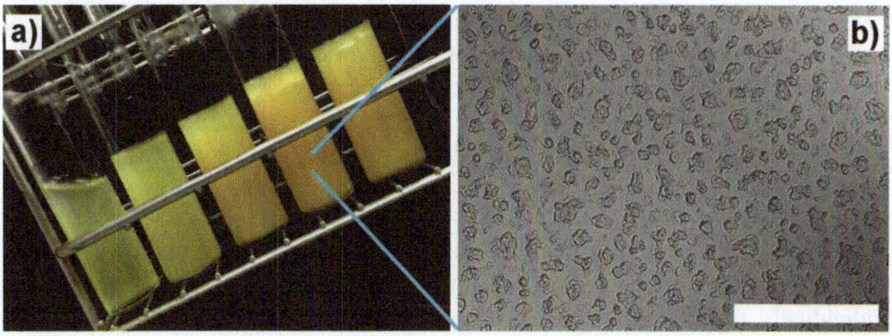

Fig. 13.5 (**a**) Photograph of oleogels prepared at varying concentrations of shellac. (**a**) From left to right: 1, 2, 4, 5 and 6 wt% shellac in rapeseed oil and (**b**) optical microscopy image of an oleogel with 5 wt% shellac (scale bar = 200 mm). Reproduced with permission from Patel et al., Shellac as a natural material to structure a liquid oil based thermo reversible soft matter system. RSC Adv., 2013, 3, 5324–5326. Published by The Royal Society of Chemistry

Fatty Acid-Fatty Alcohol Mixtures Fatty acid and fatty alcohol systems (stearic acid/stearyl alcohol, in particular) were among the first mixed component oleogels studied for potential use in edible fields (Gandolfo, Bot, & Flöter, 2004). The synergistic effect in these systems was explained based on mixed crystal formation. It was hypothesized that the mixed crystals displayed finer crystal sizes as well as an altered crystal morphology (needle-like for combination versus platelet-like for mono components) (Schaink, van Malssen, Morgado-Alves, Kalnin, & van der Linden, 2007). However, it has recently been reported that the synergistic effect is rather linked to an increase in the crystalline mass as well as the spatial distribution of the crystalline mass (Blach et al., 2016).

Unlike phytosterol-oryzanol system, oleogels can be formed individually by both fatty acids and fatty alcohols (Schaink et al., 2007). However, the mixed oleogels display relatively higher firmness and also provides the possibility of tuning the rheological properties by altering the proportion of individual components.

Phytosterol-Monoglyceride Mixtures Both phytosterols (PS) and saturated monoglycerides (MAGs) are able to gel liquid oil on their own. However, these monocomponent oleogels suffer from certain stability issues on storage. For instance, crystals of PS tend to agglomerate and settle to the bottom resulting in a contraction of the crystalline phase (Vaikousi, Lazaridou, Biliaderis, & Zawistowski, 2007). Oleogels prepared with MAGs also suffer from storage stability issues owing to the slow polymorphic transition that leads to the formation of gritty β-crystals on aging (Da Pieve, Calligaris, Panozzo, Arrighetti, & Nicoli, 2011). In addition, mono component gels made using MAGs are sensitive to processing conditions especially shear (Da Pieve, Calligaris, Co, Nicoli, & Marangoni, 2010). Moreover, the high concentration of MAGs required for the formation of mono-component gels may also affect the sensorial quality of the oleogels due to the surface-active nature of MAGs. Surprisingly by combining MAGs with PS at certain proportions, a synergistic enhancement in the rheological properties of oleogels was observed due to the formation of finer crystalline particles with completely different morphology compared to MAG and PS crystals (Bin Sintang, Rimaux, Van de Walle, Dewettinck, & Patel, 2017a). In addition, based on a four-week storage study, it was confirmed that the contraction of crystalline phase and consequent phase separation were also prevented. This was a first such demonstration of identifying synergistic combination that can overcome drawbacks associated with individual structurants (Bin Sintang, Rimaux, et al., 2017a; Patel, 2015).

Mixed Systems Containing Lecithin as an Additive The above mentioned example of mixed gel systems are all based on a combination of components that can individually form structured material phases or combine together to form such material in-situ in the solvent (Patel, 2017a). In addition to these two classes, oleogel systems can also be created by combining a structured material forming component with a non-crystallizing additive (that can impact either the self-assembling properties of gelator or promote an effective spatial distribution of building blocks formed by gelator or strengthen the network linkages among the formed building blocks) (Buerkle & Rowan, 2012; Patel, 2017a). Lecithin is the most common surfactant used in these type of oleogels. Lecithin on its own cannot gel oils unless a small amount of polar solvent (such as water or glycerol) is introduced into the mix. But due to its amphiphilic properties, it can be combined together with other components such as sorbitan tristearate, α-tocopherol and sucrose esters to form oleogel where lecithin can contribute as crystal modifier, network strengthener (formation of weak junctions between crystals) and stabilization of complex micellar structure (Bin Sintang et al., 2017b; Han et al., 2014; Nikiforidis & Scholten, 2014; Patel, 2017b; Pernetti, van Malssen, Kalnin, & Flöter, 2007).

Ethylcellulose Oleogels Ethyl cellulose (EC) is a hydrophobic cellulose derivative synthesize chemically from cellulose through substitution of hydroxyl group on monomeric glucose units which forms the backbone of cellulose. Like other cellulose derivatives, the properties of EC is dependent on the degree of polymerization (DP) and degree of substitution (DS). DP gives a measure of the molecular weight of the polymer whereas DS provides information about the average number of hydroxyl groups substituted on the cellulose backbone. Since glucose monomers have 3-hydroxyl groups that can be substituted, the DS for cellulose derivatives vary between zero to three. A DS of approximately 2.4–2.5 (with approximately 47–48% ethoxy content) results in a hydrophobic polymer that is immiscible in water and miscible in various apolar solvents (Koch, 1937). Commercially, EC polymers are supplied in specific viscosity grades which can be correlated to their molecular weight (Davidovich-Pinhas, Barbut, & Marangoni, 2014).

EC-based oleogels are usually prepared by dispersing polymer powder in liquid oil at temperatures above the glass transition temperature of polymer ($T_g \approx 130$ °C) followed by cooling to lower temperatures. At high temperatures, the polymer chains are unfolded and on subsequent cooling, the unfolded polymer chains form a backbone for cross-linked gel network. The gel is stabilized by hydrogen bonding among polymer chains along with some hydrophobic interactions between acyl chains of oil and the side chains of EC. The minimum gelling concentration (C_g) of EC required for gelation is strongly affected by the molecular weight of polymer and the type of oil used. For instance, the C_g of EC 45 cP is 4 and 6 wt% in soybean and canola oil respectively (Zetzl, Marangoni, & Barbut, 2012).

Colloidal Silicon Dioxide (CSD) Oleogels The use of inorganic particles such as clay as structuring agents to from organogels is well-documented in the cosmetics field where bentonite clays have been extensively studied for their gelation functionalities. In foods domain, CSD has been studied for oil structuring as mono component gelator as well as in combination with fat crystals (Chauhan, Dullens, Velikov, & Aarts, 2017; Patel, Mankoc, Bin Sintang, Lesaffer, & Dewettinck, 2015b). Gelation mechanism in colloidal silicon dioxide oleogels involves a long-range organization of colloidal particles into mesoscopic agglomerated structures. When used together with fat, a composite network of fat crystals and aggregated silica particles is obtained. As recently demonstrated by Chauhan et al., this could be a useful strategy to create reduced solid fat alternatives with similar rheological behaviour and thermal properties as the full-fat systems (Chauhan et al., 2017).

3.2 Indirect Methods of Using Hydrophilic Polymers for Oil Structuring (Emulsion Templates)

Using polymers for oil structuring is a promising strategy because there are many polymers that are approved for use in foods and most of them have been well-characterized. However, as most of the food polymers are inherently hydrophilic in nature, they show limited dispersion in oil which makes them ineffective in structuring oils. In order to use hydrophilic polymers for oil structuring, it is important to first pre-hydrate them in aqueous phase and arrest these hydrated conformations in dehydrated form such that they can be used for physical entrapment of liquid oils. Food polymers such as proteins and modified polysaccharides are surface-active and a conformational framework of these components can be created from their aqueous dispersions by first inducing their adsorption at the oil-water interfaces followed by removal of water to obtain dried microstructures with entrapped/embedded oil phase. As demonstrated in various studies, using this technique a range of structured oil systems such as oil powders, gels and soft solids can be created from emulsion templates (Adelmann, Binks, & Mezzenga, 2012; Liu, Chen, Guo, Yin, & Yang, 2016; Mezzenga & Ulrich, 2010; Patel et al., 2015c; Romoscanu & Mezzenga, 2006; Tavernier, Patel, Van der Meeren, & Dewettinck, 2017). Success of this approach is dependent on the formation of a stable interfacial layer that is elastic enough to withstand the dehydration during drying. This has been achieved through three different ways: (a) by stabilizing interfaces using surface active protein and subsequently crosslinking it through thermal or chemical means, (b) using protein particles followed by complexation with lipid additive, and (c) using a combination of surface and non-surface active food polymers that can increase the viscoelasticity of the interface (Patel & Dewettinck, 2016).

3.3 Structuring Aided by Physical Sorption of Liquid Oil

In the last few years, there has been a rising interest in developing porous materials that can take up a large quantity of hydrophobic oil in the presence of water. Such porous structures with high oil absorption functionality can find important applications in the management of oil spill accidents. Related concept could also be used to create porous materials from food-grade components such as cellulose derivatives (hydroxyl propyl methyl cellulose and methylcellulose), these porous cryogel show exceptional oil sorption properties and could be used for creating structured oleogels containing as high as 98 wt% of liquid oil (Patel, Schatteman, Lesaffer, & Dewettinck, 2013b).

Other materials such as carrageenan aerogels, cellulose fibres and porous starch have also been explored for their oil-absorbing properties (Manzocco et al., 2017; Patel & Dewettinck, 2016).

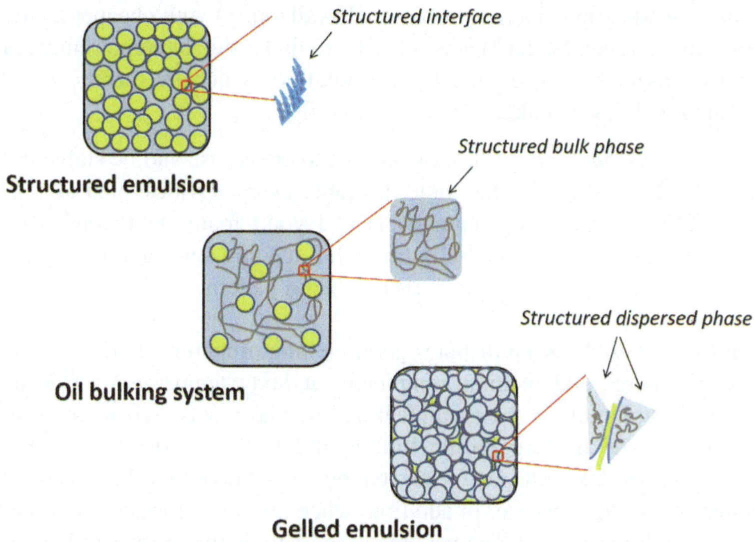

Fig. 13.6 Schematic representation of three different types of structured biphasic systems: structured emulsions, oil bulking system and gelled emulsion with structured interface, bulk phase and continuous phase, respectively. Structured emulsion and oil bulking system have oil droplets dispersed in water continuous phase whereas the gelled emulsion is composed of tightly packed gelled water droplets dispersed in oil continuous phase. Reproduced with permission from Patel & Dewettinck, Edible oil structuring: an overview and recent updates. Food & Function, 2016, 7, 20–29. Published by The Royal Society of Chemistry

3.4 Structured Biphasic Systems

The three different types of structured biphasic systems used in food structuring are depicted using simple schematics in Fig. 13.6. In the case of structured emulsions, the interfaces around the dispersed oil droplets are structured with the help of water-swollen, crystallized monoglyceride layers. In the oil bulking systems, the dispersed oil droplets are embedded in the matrix of water phase which is gelled using network of polymers, and the gelled emulsions are formed by tightly packed dispersed water droplets gelled using synergistic combination of hydrocolloids such as locust bean gum and carrageenan.

4 Functionality of Milk Fat and Possible Applications of Structured Oil in Dairy Fat and Related Products

Partial or complete replacement of milk fat has always been a focus for many food manufacturing industries dealing with dairy products. Traditionally, the main reason was to reduce the cost of products by replacing expensive milk fat (due to competitive market and changes in seasonal availability) with cheaper alternatives like tailor-made vegetable fat blends. In order to find a suitable replacement, a thorough consideration has to be given to functionalities which are specific to milk fat. These functionalities include:

Crystallization Behaviour Milk fat is known to crystallise and be stable in the β′ polymorph. This gives milk fat some desirable characteristics such as relatively high crystallisation rate, smaller crystals and good foam stabilising properties. Hence, when partially replacing milk fat with β-tending vegetable fats, a due consideration needs to be given to their crystallization pattern.

Melting Properties Milk fat displays good melting properties, being quite hard at lower temperatures and melting completely at body temperature. This melting behaviour is important for the mouthfeel and flavour release when eating the finished product. The melting profile further affects the consistency of milk fat, influencing, for example, the texture and eating properties of products such as butter and butter spreads and cheese. In addition, when air is incorporated into the products, such as whipped cream and ice-cream, the ratio between the solid fat and liquid oil (measured through solid fat content) becomes very important for the stability, texture and creaminess of the product.

Unique Taste Due to its unique fatty acid composition (high amount of short-chain fatty acids), milk fat has a strong but desirable creamy/buttery flavour. The compounds responsible for this typical flavour are formed when the short-chain fatty acids undergo reactions such as lactonization and oxidation. In addition, butter also lacks a strong off-flavour which is generally associated with other animal fats.

Colloidal Instability Partial coalescence is one of the colloidal instability displayed by milk fat when used in emulsified products such as whipped cream and ice-cream. Partially coalesced fat droplets form a three dimensional network which is responsible for stabilizing air bubbles and in turn providing the creamy texture of the final product. This property is dependent on both temperature (suggesting a reliance on the solid fat content) and the aeration conditions.

Based on the above mentioned functionalities, it is quite clear that a complete replacement of milk fat will result in a severe loss of product attributes. Hence, the best case scenario is to replace only a part of milk fat with non-dairy fat to obtain products with enhanced nutritional profile. However, these non-dairy fat-containing analogue end up having a relatively higher amount of saturated fats. Structured oil could thus be used in these products to lower the amount of solid fat while simultaneously improving the nutritional profile of the products. As this field is still in its infancy, the reported literature on applications of structured oil in dairy related products are rather scarce, few examples of such published results are discussed below.

Wax-Based Oleogels in Structuring of Reduced-Fat Ice-Cream Wax-based oleogels were evaluated in ice-cream formulations in two separate studies done by the same group. The initial study compared the properties of ice-cream formulations prepared using rice bran wax (RBW) oleogel, liquid oil and butter as fat sources in presence of a commercial emulsifier which was a mixture of mono diglyceride and polysorbate 80 (Zulim Botega, Marangoni, Smith, & Goff, 2013a). Authors observed that 10 wt% RBW oleogel performed comparatively better than liquid oil in terms of overrun, fat globule and air bubble sizes. Further, the fat droplet aggregation at the air cell interface seen in oleogel sample in contrast to oil spreading seen with HOSO suggested that RBW oleogel behaved more as crystallized fat droplets than like liquid oil. However, the overrun of oleogel samples were comparatively lower than milk fat and the fat structure that was formed by the oleogel droplets was not sufficient to cause the expected delay in structural collapse (Zulim Botega et al., 2013a).

In the follow-up study, candelilla wax (CDW) and carnauba wax (CBW) were included in addition to RBW along with glycerol mono oleate as emulsifier and formulations were prepared using continuous and batch freezing (Zulim Botega, Marangoni, Smith, & Goff, 2013b). The authors found that the RBW performed better than CDW and CBW oleogels in terms of higher meltdown stabilization and fat aggregation. The inclusion of GMO as an emulsifier was found to have an influence on wax crystal morphology resulting in the formation of large protruding crystals that facilitated fat droplet agglomeration. In addition, they found that continuous freezing and using a relatively higher concentration of oleogel (15 wt%) showed a better shape retention (Zulim Botega et al., 2013b).

Results from these two studies strongly suggest that there is a potential for replacing milk fat with wax structured vegetable oils.

Oleogels in Cream Cheese Spreads RBW and EC oleogels were used in the formulation of reduced fat cream cheese spreads with an aim of nutritional enhancement of the fatty acid profile (Bemer, Limbaugh, Cramer, Harper, & Maleky, 2016). Oleogels at 10 wt% gelator concentration using high oleic soybean oil and soybean oil as the solvents were simultaneously prepared during the cream cheese manufacturing process instead of preparing them beforehand. Results from microstructure analysis suggested a successful incorporation of oleogels in the formulation matrix resulting in a network structure and fat globule size which were similar to the control sample. Textural evaluation revealed that oleogel cream cheese samples (OCC) showed comparable hardness, spreadability, and stickiness with the full-fat cream cheese control. The samples were also subjected to sensorial evaluation to compare the organoleptic properties of OCC with that of full-fat control and the results suggested that the palatability of OCC could be improved to reduce its strong flavour and bitterness (Bemer et al., 2016).

Oleogels Based on Hydrophilic Cellulose Derivatives in Sandwich Cookie Creams Tanti et al. recently evaluated the possibility of replacing icing shortening with structured oil (based on hydroxyl propyl methylcellulose and methylcellulose) in recipes of sandwich cookie creams (Tanti, Barbut, & Marangoni, 2016). Formulations were prepared by combining 40% fat (icing shortening or structured oil or in several combinations) with 60% icing sugar, the icing shortening was replaced at 50, 75 and 100% levels. A complete replacement of icing shortening resulted in a hard, unacceptable product but 50 and 75% replacement of icing shortening led to products which had comparable textural properties to commercial benchmark (Tanti et al., 2016).

Although the fat replaced in this study was non-dairy in nature, the strategy used in this study could benefit replacement of conventional icing shortening which is based on butter.

Oleogels as Spreadable Fat and Butter Alternatives Yilmaz and Ogutcu carried out sensory and consumer tests to evaluate the possibility of using wax-based oleogel of hazelnut oil (aromatized with 0.5 wt% diacetyl as butter flavor) as butter alternative. Based on the Hedonic attributes (such as appearance, odor, flavour and spreadability) tested by the consumers, oleogel products showed potential as butter alternatives. When tested with consumers (among 120 consumers) for their buying decisions, around 43% and 57% indicated that they would definitely buy the product while 24–29% indicated that they would try once and then decide.options to be explored for future commercial applications in reduced fat products.

Anhydrous Milk Fat–Sunflower Oil Wax Blends Hybrid systems which are structured oil systems that rely on combining fat hardstocks with oleogelators are considered as one of the options to be explored for future commercial applications in reduced-fat products (Patel, 2017c; Patel & Edible 2017).

Combining of one of the several oleogelators with anhydrous milk fat (AMF) could be useful in developing hybrid systems which could be used in the formulation of dairy fat products.

Kerr et al. studied the effect of sunflower oil wax on the crystallization behaviour and functional properties of anhydrous milk fat (Kerr, Tombokan, Ghosh, & Martini, 2011). The incorporation of wax resulted in a decrease in the induction times of nucleation and smaller crystal formation of AMF. The smaller size of crystal ensured a stronger network formation and an increase in the hardness. These results indicate that there is a potential to develop hybrid systems using AMF as the fat hardstock.

5 Conclusion

Although a lot of progress has been made in the recent few years, the field of oleogelation is still in its growing phase. Different structuring agents including lipidic materials (waxes, partial glycerides, fatty acids/alcohols etc.), polymers and inorganic particles have been extensively investigated for their role in oil structuring both for fundamental exploration and application-oriented work. They have also been studied in a range of food products mainly including chocolates, spreads & bakery products. As demonstrated in few reported studies, incorporation of structured oil systems could also be beneficial in the reformulation of a broad category of dairy fat and related products. Milk fat has many unique properties which makes it quite challenging to completely replace it with structured oil systems. However, the hybrid system created by combining milk fat and oleogelators could pave way for a new range of dairy products with healthier lipid profiles while still maintaining the characteristics of conventional products.

References

Adelmann, H., Binks, B. P., & Mezzenga, R. (2012). Oil powders and gels from particle-stabilized emulsions. *Langmuir, 28*, 1694–1697.

Batte, H. D., Wright, A. J., Rush, J. W., Idziak, S. H. J., & Marangoni, A. G. (2007a). Phase behavior, stability, and mesomorphism of monostearin–oil–water gels. *Food Biophysics, 2*, 29–37.

Batte, H. D., Wright, A. J., Rush, J. W., Idziak, S. H. J., & Marangoni, A. G. (2007b). Effect of processing conditions on the structure of monostearin–oil–water gels. *Food Research International, 40*, 982–988.

Bemer, H. L., Limbaugh, M., Cramer, E. D., Harper, W. J., & Maleky, F. (2016). Vegetable organogels incorporation in cream cheese products. *Food Research International, 85*, 67–75.

Bin Sintang, M. D., Rimaux, T., Van de Walle, D., Dewettinck, K., & Patel, A. R. (2017a). Oil structuring properties of monoglycerides and phytosterols mixtures. *European Journal of Lipid Science and Technology, 119*, 1500517.

Bin Sintang, M. D., et al. (2017b). Mixed surfactant systems of sucrose esters and lecithin as a synergistic approach for oil structuring. *Journal of Colloid and Interface Science, 504*, 387–396.

Blach, C., et al. (2016). Revisiting the crystallization behavior of stearyl alcohol: Stearic acid (SO: SA) mixtures in edible oil. *RSC Advances, 6*, 81151–81163.

Bot, A., & Agterof, W. G. M. (2006). Structuring of edible oils by mixtures of γ-oryzanol with β-sitosterol or related phytosterols. *Journal of the American Oil Chemists' Society, 83*, 513–521.

Bot, A., et al. (2012). Elucidation of density profile of self-assembled sitosterol + oryzanol tubules with small-angle neutron scattering. *Faraday Discussions, 158*, 223–238.

Buerkle, L. E., & Rowan, S. J. (2012). Supramolecular gels formed from multi-component low molecular weight species. *Chemical Society Reviews, 41*, 6089–6102.

Chauhan, R. R., Dullens, R. P. A., Velikov, K. P., & Aarts, D. G. A. L. (2017). The effect of colloidal aggregates on fat crystal networks. *Food & Function, 8*, 352–359.

Chen, M., et al. (2016). Dairy fat and risk of cardiovascular disease in 3 cohorts of US adults. *The American Journal of Clinical Nutrition, 104*, 1209–1217.

Da Pieve, S., Calligaris, S., Co, E., Nicoli, M. C., & Marangoni, A. G. (2010). Shear nanostructuring of monoglyceride organogels. *Food Biophysics, 5*, 211–217.

Da Pieve, S., Calligaris, S., Panozzo, A., Arrighetti, G., & Nicoli, M. C. (2011). Effect of monoglyceride organogel structure on cod liver oil stability. *Food Research International, 44*, 2978–2983.

Davidovich-Pinhas, M., Barbut, S., & Marangoni, A. G. (2014). Physical structure and thermal behavior of ethylcellulose. *Cellulose, 21*, 3243–3255.

den Adel, R., Heussen, P. C., & Bot, A. (2010). Effect of water on self-assembled tubules in β-sitosterol + γ-oryzanol-based organogels. *Journal of Physics: Conference Series, 247*, 12025.

Doan, C. D., et al. (2017). Chemical profiling of the major components in natural waxes to elucidate their role in liquid oil structuring. *Food Chemistry, 214*, 717–725.

Gandolfo, F. G., Bot, A., & Flöter, E. (2004). Structuring of edible oils by long-chain FA, fatty alcohols, and their mixtures. *Journal of the American Oil Chemists' Society, 81*, 1–6.

Han, L., et al. (2014). Structure and physical properties of organogels developed by sitosterol and lecithin with sunflower oil. *Journal of the American Oil Chemists' Society, 91*, 1783–1792.

Huppertz, T., Kelly, A. L., & Fox, P. F. (2009). *Dairy fats and related products* (pp. 1–27). Chichester: Wiley-Blackwell. https://doi.org/10.1002/9781444316223.ch1
Kerr, R. M., Tombokan, X., Ghosh, S., & Martini, S. (2011). Crystallization behavior of anhydrous milk fat–sunflower oil wax blends. *Journal of Agricultural and Food Chemistry, 59*, 2689–2695.
Koch, W. (1937). Properties and uses of ethylcellulose. *Industrial and Engineering Chemistry, 29*, 687–690.
Liu, X., Chen, X.-W., Guo, J., Yin, S.-W., & Yang, X.-Q. (2016). Wheat gluten based percolating emulsion gels as simple strategy for structuring liquid oil. *Food Hydrocolloids, 61*, 747–755.
Manzocco, L., et al. (2017). Exploitation of κ-carrageenan aerogels as template for edible oleogel preparation. *Food Hydrocolloids, 71*, 68–75.
Mezzenga, R., & Ulrich, S. (2010). Spray-dried oil powder with ultrahigh oil content. *Langmuir, 26*, 16658–16661.
Nettleton, J. A., Brouwer, I. A., Geleijnse, J. M., & Hornstra, G. (2017). Saturated fat consumption and risk of coronary heart disease and ischemic stroke: A science update. *Annals of Nutrition & Metabolism, 70*, 26–33.
Nikiforidis, C. V., & Scholten, E. (2014). Self-assemblies of lecithin and [small alpha]-tocopherol as gelators of lipid material. *RSC Adv, 4*, 2466–2473.
Ojijo, N. K. O., Neeman, I., Eger, S., & Shimoni, E. (2004). Effects of monoglyceride content, cooling rate and shear on the rheological properties of olive oil/monoglyceride gel networks. *Journal of the Science of Food and Agriculture, 84*, 1585–1593.
Osaki, N., et al. (2005). Metabolities of dietary triacylglycerol and diacylglycerol during the digestion process in rats. *Lipids, 40*, 281.
Patel, A. R. (2015). In A. R. Patel (Ed.), *Alternative routes to oil structuring* (pp. 1–14). Cham: Springer. https://doi.org/10.1007/978-3-319-19138-6_1
Patel, A. R. (2017a). A colloidal gel perspective for understanding oleogelation. *Current Opinion in Food Science, 15*, 1. https://doi.org/10.1016/j.cofs.2017.02.013
Patel, A. R. (2017b). Stable 'arrested' non-aqueous edible foams based on food emulsifiers. *Food & Function, 8*, 2115. https://doi.org/10.1039/C7FO00187H
Patel, A. R. (2017c). Methylcellulose-coated microcapsules of Palm stearine as structuring templates for creating hybrid oleogels. *Materials Chemistry and Physics, 195*, 268–274.
Patel, A. R., & Dewettinck, K. (2015). Current update on the influence of minor lipid components, shear and presence of interfaces on fat crystallization. *Current Opinion in Food Science, 3*, 65–70.
Patel, A. R., & Dewettinck, K. (2016). Edible oil structuring: An overview and recent updates. *Food & Function, 7*, 20–29.
Patel, A., & Edible, R. (2017). 'Oleocolloids': The final frontier in food innovation? *Journal of Agricultural and Food Chemistry, 65*, 3432–3433.
Patel, A. R., Schatteman, D., De Vos, W. H., & Dewettinck, K. (2013a). Shellac as a natural material to structure a liquid oil-based thermo reversible soft matter system. *RSC Advances, 3*, 5324–5327.
Patel, A. R., Schatteman, D., Lesaffer, A., & Dewettinck, K. (2013b). A foam-templated approach for fabricating organogels using a water-soluble polymer. *RSC Advances, 3*, 22900–22903.
Patel, A. R., Babaahmadi, M., Lesaffer, A., & Dewettinck, K. (2015a). Rheological profiling of organogels prepared at critical gelling concentrations of natural waxes in a triacylglycerol solvent. *Journal of Agricultural and Food Chemistry, 63*, 4862–4869.
Patel, A. R., Mankoc, B., Bin Sintang, M. D., Lesaffer, A., & Dewettinck, K. (2015b). Fumed silica-based organogels and 'aqueous-organic' bigels. *RSC Advances, 5*, 9703–9708.
Patel, A. R., et al. (2015c). Biopolymer-based structuring of liquid oil into soft solids and oleogels using water-continuous emulsions as templates. *Langmuir, 31*, 2065–2073.
Pernetti, M., van Malssen, K., Kalnin, D., & Flöter, E. (2007). Structuring edible oil with lecithin and sorbitan tri-stearate. *Food Hydrocolloids, 21*, 855–861.

Romoscanu, A. I., & Mezzenga, R. (2006). Emulsion-templated fully reversible protein-in-oil gels. *Langmuir, 22*, 7812–7818.

Schaink, H. M., van Malssen, K. F., Morgado-Alves, S., Kalnin, D., & van der Linden, E. (2007). Crystal network for edible oil organogels: Possibilities and limitations of the fatty acid and fatty alcohol systems. *Food Research International, 40*, 1185–1193.

Tanti, R., Barbut, S., & Marangoni, A. G. (2016). Hydroxypropyl methylcellulose and methylcellulose structured oil as a replacement for shortening in sandwich cookie creams. *Food Hydrocolloids, 61*, 329–337.

Tavernier, I., Patel, A. R., Van der Meeren, P., & Dewettinck, K. (2017). Emulsion-templated liquid oil structuring with soy protein and soy protein: κ-carrageenan complexes. *Food Hydrocolloids, 65*, 107–120.

Vaikousi, H., Lazaridou, A., Biliaderis, C. G., & Zawistowski, J. (2007). Phase transitions, solubility, and crystallization kinetics of phytosterols and phytosterol–oil blends. *Journal of Agricultural and Food Chemistry, 55*, 1790–1798.

Wang, T.-M., & Rogers, M. A. (2015). Biomimicry – An approach to engineering oils into solid fats. *Lipid Technology, 27*, 175–178.

Wang, Y., et al. (2017). Saturated palmitic acid induces myocardial inflammatory injuries through direct binding to TLR4 accessory protein MD2. *Nature Communications, 8*, 13997.

Wright, A. J., & Marangoni, A. G. (2006). In P. F. Fox & P. L. H. McSweeney (Eds.), *Advanced dairy chemistry volume 2 lipids* (pp. 245–291). Boston: Springer. https://doi.org/10.1007/0-387-28813-9_7

Zetzl, A. K., Marangoni, A. G., & Barbut, S. (2012). Mechanical properties of ethylcellulose oleogels and their potential for saturated fat reduction in frankfurters. *Food & Function, 3*, 327–337.

Zulim Botega, D. C., Marangoni, A. G., Smith, A. K., & Goff, H. D. (2013a). The potential application of rice bran wax oleogel to replace solid fat and enhance unsaturated fat content in ice cream. *Journal of Food Science, 78*, C1334–C1339.

Zulim Botega, D. C., Marangoni, A. G., Smith, A. K., & Goff, H. D. (2013b). Development of formulations and processes to incorporate wax oleogels in ice cream. *Journal of Food Science, 78*, C1845–C1851.

Chapter 14
Role of Differentiated-Size Milk Fat Globules on the Physical Functionality of Dairy-Fat Structured Products

Tuyen Truong and Bhesh Bhandari

1 Introduction

Bovine milk fat is one of the most important lipids in the human diet and originally exists as globules dispersed in the milk plasma. The typical size of milk fat globules (MFG) ranges from 0.1 to 15 μm with an average diameter of 4 μm (Walstra, 1995) as illustrated in Fig. 14.1a. The MFG size can be classified into three size fractions, i.e. small (<1 μm), intermediate (1–8 μm) and large (>8 μm) MFG sizes (Michalski, Briard, & Michel, 2001; Walstra & Oortwijn, 1969) with their volume-based percentages of 5%, 80% and 1–2%, respectively. The small size fraction accounts for 80% of the MFG size distribution on the basis of number of globules. Formation of various MFG sizes is governed by processes of assembly, growth and secretion of fat globules in the milk-secreting cells of the mammary gland of mammals (Timmen & Patton, 1988). Generally, MFG origins from the endoplasmic reticulum membranes where tiny intracellular lipid droplets (<0.5 μm) having a triacylglycerol (TAG) core enveloped by a single layer of proteins and polar lipids are generated. These lipid micro-droplets fuse to form bigger droplets, regarded as cytoplasmic lipid droplets, whose droplet-droplet fusion is regulated by specific calcium and protein complexes and fusion-promoting agents, e.g. gangliosides (Valivullah, Bevan, Peat, & Keenan, 1988). The intermediate size MFGs are progressively coated by the plasma membrane when being transported to the apical plasma membrane, resulting in the final tri-layer structure of intact milk fat globule membrane

T. Truong
ARC Dairy Innovation Hub, The University of Queensland, St Lucia, QLD, Australia

School of Science, RMIT University, Melbourne, VIC, Australia

B. Bhandari (✉)
ARC Dairy Innovation Hub, School of Agriculture and Food Sciences,
The University of Queensland, St Lucia, QLD, Australia
e-mail: b.bhandari@uq.edu.au

T. Truong et al. (eds.), *Dairy Fat Products and Functionality*,
https://doi.org/10.1007/978-3-030-41661-4_14

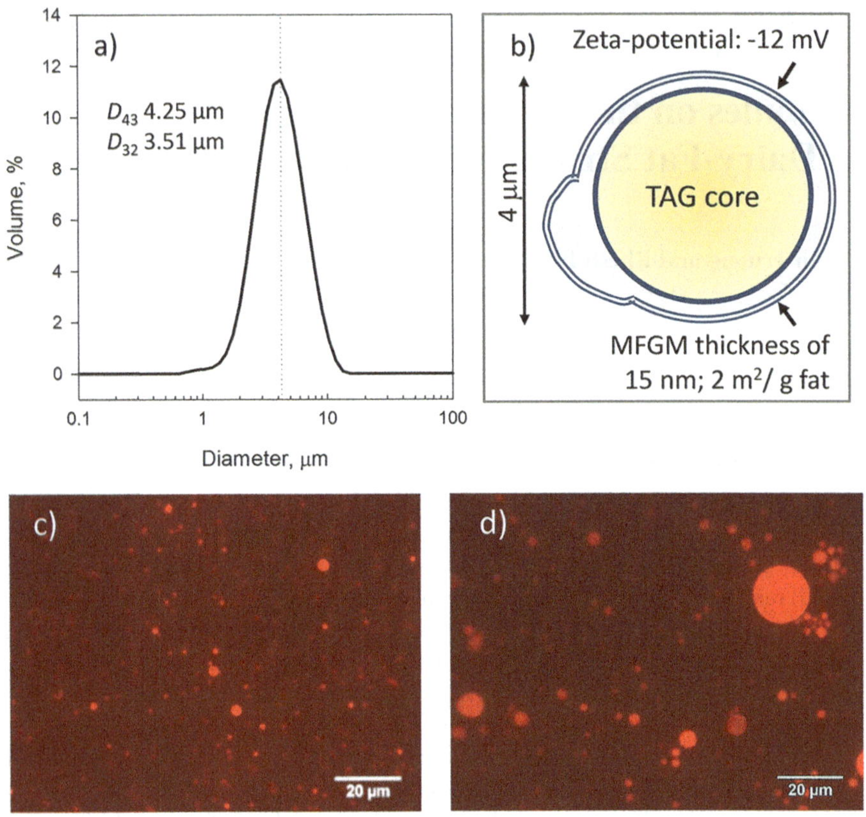

Fig. 14.1 Illustrations of MFG size distribution measured at 25 °C by using laser light scattering technique (**a**) and structure of bovine MFG with TAG core being enveloped by intact MFGM (**b**). Fluorescence microscopic images of small (**c**) and large (**d**) MFG size fractions obtained by two-stage centrifugal separation (scale bar: 20 µm)

(MFGM) (Fig. 14.1b) (Heid & Keenan, 2005). Mean values of specific surface area and zeta-potential of MFG are about 2.2 m^2 g^{-1} fat and −13.5 mV, respectively (Huppertz & Kelly, 2006; Michalski, Michel, Sainmont, & Briard, 2002). The structure, composition and biochemical functions of MFGM are also MFG-size dependent as discussed elsewhere (Lopez, 2011). It is postulated that the large MFG size is due to post-secretion fusion between large and smaller globules (Timmen & Patton, 1988). Hence, the wide diversity of MFG size in secreted milk is a result of extensive growth of MFG size when being transported from the origins to the secretion sites. Measurement of MFG size can be done using numerous techniques such as microscopy (Ong, Dagastine, Kentish, & Gras, 2010; Precht, 1988; Truong, Morgan, Bansal, Palmer, & Bhandari, 2015), Coulter counting (Cornell & Pallansch, 1966; Walstra & Oortwijn, 1969), laser diffraction, static and dynamic light scattering (McCrae & Lepoetre, 1996; Michalski et al., 2001; Robin & Paquin, 1991), spectroscopy, ultrasound (Miles, Shore, & Langley, 1990), scanning flow cytometry

(Konokhova et al., 2014) and electroacoustics (Wade & Beattie, 1997). Among these measurement methods, laser light scattering techniques is widely adopted to analyse MFG size as well as size distribution. Common expression of mean diameters of MFG size includes number mean (d_n, $D_{1,0}$), volume mean (d_v, $D_{3,0}$), volume surface-weighted mean (d_{vs}, $D_{3,2}$), and volume moment-weighted mean (d_{vm}, $D_{4,3}$) (Fig. 14.1a).

Apart from its physiological role in delivering energy and nutrition to the suckling calf, the wide diversity of size is of industrial interest since each size class may have additional functions. In fact, MFG sizes can be varied among breeds, seasons and lactation stages (Carroll et al., 2006; Mesilati-Stahy & Argov-Argaman, 2014; Wiking, Stagsted, Lennart, & Nielsen, 2004). MFG size distribution can also be altered through milking times and feeding strategy (Avramis, Wang, McBride, Wright, & Hill, 2003; Couvreur, Hurtaud, Marnet, Faverdin, & Peyraud, 2007; Wiking, Nielsen, Bavius, Edvardsson, & Svennersten-Sjaunja, 2006). The effect of breeds and feeding strategy on MFG size has been discussed in **C** of this book. Manipulation of MFG size in post-farm can be achieved by using conventional dairy processing methods such as gravity separation, homogenisation, microfiltration and centrifugation (Dhungana, Truong, Palmer, Bansal, & Bhandari, 2017; Ma & Barbano, 2000; Michalski et al., 2006; Panchal, Truong, Prakash, Bansal, & Bhandari, 2017). It has been reported that differentiated-size MFG possess different chemical composition and physical properties (Lopez et al., 2011; Michalski, Ollivon, Briard, Leconte, & Lopez, 2004). These discrepancies suggest potential industrial strategy of manipulating MFG size in structuring of fat-structured products and developing improved functionalities for dairy fat-containing products such as cream, butter, whipped cream, cheese, yoghurt etc.

The current chapter discusses the importance of MFG size in processing of dairy-fat structured products with the view of potential applications to the production of innovative dairy ingredients and products. It will provide a comprehensive overview of size-dependent variations in physical and chemical properties as well as methodologies to alter the size of both native and emulsified MFGs. Recent studies on utilisation of size-differentiated MFG in dairy-fat structured products will also be highlighted.

2 Variations in Chemical Properties of Differentiated-Size Milk Fat Globules

Bovine milk fat is highly enriched in TAGs, having the average amount of TAGs more than 98% on weight basis. Other primary components accounting for the remaining 2% include monoacylglycerols, diacylglycerols, free fatty acids, and phospholipids (MacGibbon & Taylor, 2006). The bovine milk fat has also been regarded as one of the most chemically complex natural fats existed with numerous types of fatty acids (>400) and TAG species (>200) (Gresti, Bugaut, Maniongui, & Bezard, 1993). As listed in Table 14.1, complex fatty acids such as short-chain

Table 14.1 The amount (%w/w) of principal fatty acids in milk fat (MF) and compositional difference between small and large MFG fractions

Fatty acid	Fatty acid common name	Amount (%w/w)			
		MF[a]	MF-TAGs[b]	Small MFG	Large MFG
$C_{4:0}$	Butyric	2–5	3.6		
$C_{6:0}$	Caproic	1–5	2.4		
$C_{8:0}$	Caprylic	1–3	1.2		
$C_{10:0}$	Capric	2–4	2.9		
$C_{12:0}$	Lauric	2–5	3.5	+4.1%	
$C_{14:0}$	Myristic	8–14	11.2	+5%	
$C_{14:1}$	Myristoleic	0.8	2.0		
$C_{15:0}$	Pentadecanoic	1–2	1.4		
$C_{16:0}$	Palmitic	22–35	29.4	+3.3%	
$C_{16:1}$	Palmitoleic	1–3	3.0	+20.2% (Fauquant et al., 2005; Michalski et al., 2005)	
$C_{17:0}$	Margaric	0.5–1.5	0.8		
$C_{18:0}$	Stearic	9–14	10.6		Enriched (Wiking et al., 2004)
$C_{18:1}$ *cis*	Oleic	20–30	24.2	TAG core: ↑ (Fauquant et al., 2005; Lopez et al., 2011)	↓ (Martini et al., 2006; Timmen & Patton, 1988)
$C_{18:1}$ *trans*		3.9			
$C_{18:2}$	Linoleic	1–3	3.0	TAG core: ↑ (Fauquant et al., 2005; Lopez et al., 2011)	
$C_{18:3}$	Linolenic	0.5–2	0.7[c]		
Conjugated linoleic acids (CLA)					
$C_{18:2}$,cis-9 trans-11	Rumenic acid	75–90 (Bauman, Corl, & Peterson, 2003)		87%; MFG 2.9 μm) (Michalski et al., 2005) ↑; MFG 1.6 μm (Lopez et al., 2011)	82–85% (MFG 4.9–5.7 μm)

[a]Data combined from Kaylegian and Lindsay (1995) and MacGibbon and Taylor (2006)
[b]Data compiled from Wright and Marangoni (2002)
[c]Includes $C_{20:0}$

(C_4–C_8), medium-chain (C_{10}–C_{12}), and long-chain fatty acids (C_{14}–C_{18}) are found in milk lipids with the long-chain fatty acids being abundant (81.9%). The most abundant long-chain fatty acids in milk fat are myristic (8–14%), palmitic (22–35%), stearic (9-14%), and oleic (20–30%) acid (Table 14.1).

Variations in chemical composition of differentiated MFG size fractions are of nutritional interests as they can be used towards the development of nutrient-fortified dairy products. Few attempts have been made to examine the compositional differences across the size range of bovine MFG. Given the differences from sources of milk fat and sampling/analytical methods, a clear tendency in compositional variations between small and large MFGs has not been established. The difficulty is also partly due to overlapping of MFG size ranges being fractionated. Table 14.1 also represents the size-dependent variations in fatty acid composition between small and large MFG size fractions in selected studies.

Regarding individual fatty acids, two separate studies performed by Briard, Leconte, Michel, and Michalski (2003) (small MFG: 1.0–3.3 μm; large MFG: 5.9–7.3 μm) and Lopez et al. (2011) (small MFG: 1.6 μm; large MFG: 6.6 μm) reported that the amount of short-chain fatty acids tended to be unchanged between the small and large MFG fractions. Various investigations on MFG-size dependent changes of saturated, medium-chain fatty acids in secreted cow milk showed that their proportions increased with bigger MFGs (Martini, Cecchi, & Scolozzi, 2006; Mesilati-Stahy, Mida, & Argov-Argaman, 2011; Wiking et al., 2004). Contrast to this finding on varying MFG size by herd management strategy, small and large MFG sizes fractionated by microfiltration technique exhibited a different trend. Fauquant, Briard, Leconte, and Michalski (2005) reported that higher concentrations of lauric, myristic and palmitic acids were found in the small size fraction (2.3–3.7 μm) concerning its large size counterpart (5.2–8.0 μm). Similar tendency, e.g. increasing medium-chain fatty acids in small MFG size fractions, were also reported by Lopez et al. (2011) and Michalski, Briard, and Juaneda (2005). With respect to long-chain fatty acids, the same trend was found across number of studies (Briard et al., 2003; Briard-Bion, Juaneda, Richoux, Guichard, & Lopez, 2008; Fauquant et al., 2005; Michalski et al., 2005; Timmen & Patton, 1988; Wiking et al., 2004) that large MFG size fractions had significantly higher proportion of stearic acid.

Regarding unsaturated fatty acids, the reported differences in the relative amount of palmitoleic (C16:1), oleic (C18:1) and linoleic (C18:2) acids between small and large MFG were contradictory. The small MFG fraction had greater proportion of palmitoleic (Fauquant et al., 2005, Michalski et al., 2005 but less amount of oleic and linoleic acids (Fauquant et al., 2005; Lopez et al., 2011; Wiking et al., 2004). However, other authors (Martini et al., 2006; Timmen & Patton, 1988) reported that the amount of oleic acid was greater in the small MFG fraction. Wiking et al. (2004) also found that palmitoleic acid was enriched with large MFG fraction. Bovine milk fat also contains conjugated linoleic acids (CLA), which have been known to have positive effects on human health such as anti-obesity, anti-carcinogenicity, and anti-diabesity (Belury, 2002). Rumenic acid (C18:2, *cis*-9 *trans*-11), which is a primary component of CLA, was found to be enriched in small MFG fraction (2.9 μm) as compared to its large MFG counterpart (4.9–5.7 μm). Its concentrations in both fractions were 87% and 82–85%, respectively (Michalski et al., 2005). Lopez et al. (2011) also reported similar observation. On the other hand, a few CLA isomers (*trans*-8, *cis*-10, *trans*-11, and *trans*-13) are more concentrated in the larger MFG fraction (Michalski et al., 2005).

3 Dependence of Physical Properties on Milk Fat Globule Sizes

As previously described, milk fat globules have a wide size range (0.1–15 μm) in which, technically, the smallest size class (i.e. 0.1 μm) is about 100 times smaller than its largest counterpart. Considering the small and large MFG fractions at the same bulk volume, the small MFG fraction will have not only a higher total number of globules but also a larger ratio of surface area to volume. Furthermore, there is an increase in curvature with decreasing globule/droplet size. As a result, one can expect no*table* discrepancies in physical properties between the small and large MFG fractions.

Physical Stability Bovine milk has been regarded as an oil-in-water emulsion or a colloidal suspension since its microstructure composes of milk fat globules dispersed in the milk plasma, which contains serum proteins, casein micelles, sugars and minerals (Walstra, Geurts, Noomen, Jellama, & Van Boekel, 1999). Thus, milk fat globules are readily subjected to physical instabilities such as droplet aggregation, creaming, flocculation and partial coalescence, causing alteration in their structural organisation and spatial distribution. Milk is homogenised to reduce the MFG size below 1 μm to enhance the physical stability for shelf-life extension of drinking milk. Typical size of homogenized milk fat globules is about 0.4 μm, which renders adequate physical stability to homogenized milk against creaming phenomenon. Recent work performed on recombined and standardised commercial cream (23–28% w/w fat) covering three MFG size ranges of 0.13, 0.6 and 3.9 μm showed that the sub-micron- (0.6 μm) and nano-sized (0.13 μm) droplets were relatively stable after 1 month of storage at 4 °C (Hussain, Truong, Bansal, & Bhandari, 2017). Since small MFG size is less prone to partial coalescence and creaming, the small MFGs are undesirable for butter making. This is due to the large MFG size that facilitates partial coalescence, improving the efficiency of the churning process (Walstra, Wouters, & Geurts, 2005). The MFG size also affects cold agglutination in raw milk. When raw milk is subjected to cooling, agglutinin causes a precipitation of cryoglobulins in raw milk onto the MFG. The cold agglutination induces aggregation of MFGs, causing formation of large floccules and a subsequent creaming layer. Thus, raw milk contains small MFGs will be more stable against cold agglutination because it needs more agglutinin covering greater surface area with smaller MFG size (Walstra et al., 2005).

Viscosity In general, there is a slight increase in viscosity of milk and dairy emulsions with smaller MFG size regardless of fat content. This is due to smaller droplet size and/or narrow size distribution causing greater colloidal repulsion and monodispersed close packing (Pal, 1996). As such inter-droplet resistance increases, leading to a corresponding increase in the bulk viscosity (Long, Zhao, Zhao, Yang, & Liu, 2012). Kietczewska, Kruk, Czerniewicz, Warminska, and Haponiuk (2003) reported that reduction in MFG size from 2.7 to 1.0 μm in 3.3% fat milk resulted in higher viscosities (1.8–1.96 mPa s). Regarding dairy emulsion systems, the viscosities of both low- (10%) and high- (36%) fat containing emulsions having various MFG size ranges also increased with smaller emulsion droplet sizes (Lopez, 2005; Truong, Bansal, & Bhandari, 2014). Measurement of apparent viscosities in recombined dairy cream (23% fat) also showed that the apparent viscosities increased from 0.026 to 3.14 Pa s when the droplet size was reduced from micron- (3.9 μm) to sub-micron size range (0.24–0.59 μm) (Hussain et al., 2017).

Crystallisation Properties Crystallisation behaviour and crystalline structure of milk fat in milk, natural/recombined cream and milk fat emulsions are dependent on MFG size (Bugeat et al., 2011; Lopez et al., 2002; Michalski, Ollivon, et al., 2004; Truong et al., 2015; Truong, Bansal, Sharma, Palmer, & Bhandari, 2014). Regarding native MFG, it was reported that the small MFG size fraction (0.93 μm) obtained by microfiltration exhibited delayed crystallisation as compared to its larger MFG size fraction (7.15 μm). The longitudinal packing of milk fat crystals (e.g. double chain length) tended to be more in the large MFG size fraction (Michalski, Ollivon, et al., 2004). In this work, it is also pointed out the difference in crystallisation properties of differentiated-size native MFGs within the size range of 0.93–7.14 μm may be governed by thermal history and cooling rate rather than the direct influence of MFG size (Michalski, Ollivon, et al., 2004). The dependence of crystallisation properties on MFG size is more evident in milk fat emulsion systems where anhydrous or fractionated milk fat is fabricated in the form of dispersed droplets surrounding by aqueous phase containing dairy-based emulsifiers (whey proteins and/or caseins). In these milk fat emulsion systems, crystallisation temperature, solid fat content and melting enthalpy were found to be lower with smaller droplet size (Bugeat et al., 2011; Lopez et al., 2002; Truong, Bansal, Sharma, et al., 2014). These differences are partly attributed to lack of impurities to catalyse numerous amount of smaller droplet size, resulting in limiting the rate of crystallisation. Confinement of milk fat into tiny droplet boundary at nano-size scale seemed to alter the crystalline structure of milk fat. For instance, triple chain length structure of crystalline milk fat was absent in milk fat nanoemulsions (~200 nm) enriched in unsaturated fatty acids as compared to that of micron-sized emulsions contained the same composition (Bugeat et al., 2011; Truong et al., 2015). More information on the effects of MFG size on crystallisation of milk fat can be found in **C** of this book.

Structural Properties of Milk Fat Crystals When milk fat is crystallised into solid-state, MFG size had an impact on arrangement (crystalline structure) and shape (crystal morphology) of milk fat crystals. Depending on the microscopic techniques used and the MFG size range investigated, features of milk fat crystals have been described differently. Classification of milk fat crystals can be based on birefringence of the crystals under polarised light microscopy (Walstra, 1967). As illustrated in Fig. 14.2a, four main types of crystals were visualised in cream, namely O (no birefringence), N (needle-type), L (layer-type) and M (mixed type, e.g. combination of L and N types). Goff (1997) reported that MFG contained needle-type fat crystals within the interior part (Fig. 14.2b). Lopez et al. (2002) used this classification to

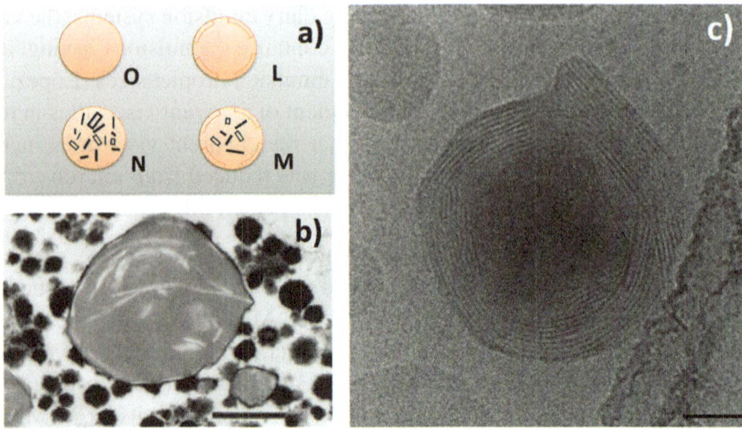

Fig. 14.2 Four main types of milk fat crystals (O, L, M and N) as observed under polarised light microscope (redrawn from Walstra (1967)) (**a**); visualisation of crystal morphologies in MFG as captured by electron microscopic techniques showing needle-type crystals within interior part of MFG (adapted from Goff, 1997); (**b**) and straight orientation of lamellar TAG layers in nano-sized dairy emulsion droplets (adapted from Truong et al., 2015). Scale bars in (**b**) and (**c**) represents 0.5 μm and 50 nm, respectively

describe the size dependence of structural properties in native milk fat globules. Accordingly, the largest MFG exhibited N-type whilst fat crystals were tiny in the smallest MFG fraction which can be deemed as type O. Nevertheless, under slow cooling regime (0.5 °C min^{-1}), the largest and smallest MFG was found to contain type M and spherulite-shaped crystals, respectively (Lopez et al., 2002). Observation of milk fat globules upon crystallisation using freeze-fracturing and electronic microscopy revealed the location of crystal shell as well as the presence of concentric layers of 5 mm in thickness surrounding the globules (Precht, 1988). Usage of high-resolution electron microscopy permits observation of milk fat emulsions in nano-sized range. As shown in Fig. 14.2c, arrangement of TAG lamellar layers at the outer part were of a straight orientation when the milk fat nanoemulsion (200 nm) was cooled at a very slow cooling rate (0.1 °C min^{-1}) (Truong et al., 2015). Truong et al. (2015) also reported the impact of droplet size in milk fat emulsions on morphologies of milk fat crystals. Owing to the physical confinement of nano-sized droplets, it is likely that typical crystals could not bend along the extreme curvature of such tiny droplets. Therefore, straight orientation of TAG lamellar layers is favourable, leading to protrude of fat crystals at the outer part (Truong et al., 2015).

Optical Properties Dispersed particles in milk, particularly fat globules and casein micelles, influences the colour and opacity of milk. Since milk fat has a wide distribution of globule size, it tends to have stronger light scattering than that of casein micelles (Walstra et al., 2005). Thus, estimation of milk fat globule size has been undertaken using spectroturbidimetry in early days (Ashworth, 1951; Goulden,

1958), static light scattering (Michalski et al., 2001), dynamic light scattering (Truong, Bansal, & Bhandari, 2014) and visible and near-infrared (Vis/NIR) spectroscopy (Aernoutsa et al., 2015) in recent days. The latest technique revealed that there is a reduction of the visible and near-infrared bulk scattering coefficient and scattering anisotropy factor with decreasing milk fat globule size obtained by ultrasonic homogenization of raw milk (Aernoutsa et al., 2015). This implies that milk fat globule size influences the optical properties of milk. Since smaller fat globules cause an increase in light scattering, the appearance of milk appears to be whiter with smaller MFGs (Fox & McSweeney, 1998). For example, it is reported that luminosity of sonicated milk (MFG size <1 μm; $L*$: 92.37) is remarkably greater that of raw milk ($L*$: 87.820 (Fox & McSweeney, 1998).

Electrical Conductivity Electrical conductivity of milk varies between 4.0 and 5.5 mS cm^{-1} at ambient temperature. Major components of milk such as casein micelles, milk fat globules, lactose and salts contribute to the electrical conductivity in which contribution from soluble salts is the greatest. Lactose and proteins have an indirect influence on the electrical conductivity by their impact on the viscosity of milk. Fat itself has poor conductivity and the presence of milk fat globules also immobilises the charge-carrying ions; thus, it can be deduced that the electrical resistance of milk will be higher with increasing milk fat content. Few attempts have been made to measure electrical conductivity of milk having smaller fat globule size (Banach, Żywica, & Kiełczewska, 2008; Mabrook & Petty, 2003). It was reported that smaller MFG size caused a slight increase in conductance (5.05 ± 0.03 mS versus 4.85 ± 0.03 mS) in commercial full-fat milk (Mabrook & Petty, 2003). Regarding homogenised milk, larger droplet size within 1.5–5 μm did not exhibit any significant difference in conductance properties. Nevertheless, when milk was homogenised at higher pressure (20 MPa) to obtain smaller MFG size (1.07 μm), the impedance remarkably decreased. The difference in conductivity is attribu*table* to the effect of homogenisation on disintegration of casein micelles rather than any direct influence of MFG size. The shearing force exerted during homogenisation process may cause dissociation and solubilisation of the colloidal calcium phosphate from the micelles to a certain extent, resulting in the imbalance of mineral salts in milk serum. As such, the electrical conductivity is altered.

4 Main Approaches to Manipulate MFG Size

There have been many attempts to manipulate MFG size distribution since different MFG size fractions might have potential benefits in nutritional properties, processability and physical functionality of dairy fat-based products and ingredients. In general, methodologies to vary MFG size can be categorised into three main strategies, e.g. herd management, fractionation on MFG size basis and shear processing, as summarised below. The MFG size reported in this book chapter is based on volume-weighted mean diameter, unless otherwise specified.

4.1 Herd Management

Strategies of herd management involve in selection of different breed of cows, modification of cow feed inputs and milking practice such as milking frequency and milking at various stages of lactation.

It has been known that variation in MFG size naturally exists between individual cows (Couvreur et al., 2007; Logan, Auldist, Greenwood, & Day, 2014) with the span in MFG size can be up to 1 μm (Mulder & Walstra, 1974). The discrepancy in mean MFG size among a single herd of Holstein-Friesian cows (n = 78) was noted to be as wide as 2.5–5.7 μm (Logan, Auldist, et al., 2014). There is also a correlation between breed of cows and MFG size. For example, measurement of the average diameter of MFG produced by Italian Friesians, German Friesians and Jersey cows using florescent microscopy showed that their MFG sizes were different (5.3, 4.93 and 4.97 μm, respectively) (Martini, Cecchi, Scolozzi, Leotta, & Verita, 2003). Similar observation was reported by Banks, Clapperton, Muir, and Girdler (1986) and Carroll et al. (2006). That is, MFG size in Jersey milks tended to have a greater number of large fat globules (>5 μm) and wider size distribution that those of Friesians milks.

Alteration of MFG size can also be achieved by modifying dietary supplementation for cow diets. Previous studies on influence of lipid dietary supplements on bovine milk composition found that cows on diets enriched in unsaturated lipids secreted smaller fat globules (fish oil: 1.84 versus 2.31 μm; linseed oil: 4.56 versus 4.73 μm, whole soybean: 4.07 versus 4.18 μm; and fresh grass: 3.65 versus 3.94 μm) with narrower size distribution in their milks (Avramis et al., 2003; Couvreur et al., 2007; Hurtaud, Faucon, Couvreur, & Peyraud, 2010; Lopez et al., 2008). In contrast, MFG size seemed to increase with the addition of saturated fatty acids into the dietary supplements (Wiking, Bjorck, & Nielsen, 2003). The modification of cow diets in this way is thought to be associated with synthesis and secretion of milk fat (Wiking et al., 2003, 2004). Given that MFGM material is available, the addition of a greater amount of dietary fat into feeding inputs induces higher lipid content, resulting in the formation of larger MFG. Nevertheless, it was found that enzyme activity of γ-glutamyl transpeptidase, which is an indicator of production of membrane material, decreased with resultant large MFG size obtained from high lipid diets. This implies that the supply of polar lipids, the main component of membrane material, is limited in secretory cells, in this case, causing the preferential production of greater MFG size (Wiking et al., 2004).

It appears that MFG size changes along lactation stage (Mesilati-Stahy & Argov-Argaman, 2014; Wiking et al., 2004) in response to alteration of energy balance, which is positive towards late lactation. This results in greater availability of MFGM material that can sufficiently cover greater proportion of micro-lipid droplets whereby a larger amount of small fat globules can be formed (Martini, Altomonte, Pesi, Tozzi, & Salari, 2013; Wiking et al., 2006). Walstra et al. (2005) also found that there is a decrease in MFG size (from 4.4 to 2.9 μm) with advancing lactation stage. Few studies reported the influence of milking times and frequency on MFG size. The increase in milking times facilitates the growth of MFG size with possible shift from medium- to large-sized fat globules (Wiking et al., 2006). Fat globules also tended to grow larger (4.28 ± 0.06 to 4.39 ± 0.07 μm) with increased daily milking frequency from two to four (Wiking et al., 2006). On the other hand, milking system and milk interval did not have a significant influence on MFG size (Abeni, Degano, Calza, Giangiacomo, & Pirlo, 2005).

Taken together, the herd management strategies can alter the native MFG size and size distribution to a range of 3.0–5.3 μm. Given that discrete size fractions are hardly yielded whereas complex supply chain management is required, this approach limits practical application from an industry perspective. Thus, manipulation of MFG size by adaptation of conventional dairy processing technologies may be more feasible. The post-farm strategies such as gravity separation, centrifugation, microfiltration and homogenisation are discussed as followings.

4.2 Fractionation of MFG on Size Basis

Gravity Separation The separation of MFG size fractions on gravity basis resembles the natural creaming process of milk. Owing to the difference in density of serum and fat phases in milk, lower density material (i.e. fat) rises over higher counterpart (i.e. serum/water) according to Stokes' law. This phenomenon also depends on the size of the milk fat globules as well as relative difference in composition between the small and large size fractions of MFG (Ma & Barbano, 2000). Since large MFG contain less mass of MFGM than the volume of fat, their density is lower than that of smaller MFG. Thus, large MFG tended to be risen over smaller MFG. For example, largest MFG size fraction (3.6 μm) was found on the top layer of aged (2–48 h at 4 or 15 °C) milk (3.75%w/w fat content) contained in a vertical column container. This was followed by smaller size fractions of 2.8, 2.3 and 1.2 μm toward to the bottom (Ma & Barbano, 2000). Using similar gravity separation method on standardised cream having 10% w/w fat, it was shown that the gradient of MFG size spans from 2.8 to 4.8 μm (Eden, Dejmek, Lofgren, Paulsson, & Glantz, 2016). The rising speed of MFG is also a function of temperature and aging time. The MFG size (4.5 ± 0.06 μm) remained the same when whole milk was kept at 5 (4.4 ± 0.03 μm) and 40 °C (4.3 ± 0.07 μm). It became larger (4.62 ± 0.05 μm) at the top layer as well as creamed faster when the temperature was at 25 °C. A study performed on gravity-based fractionation of MFG shows that extended aging time (48 h at 4 °C) facilitates greater fractionation efficiency with achievable small MFG size fraction of 1.2 μm (Ma & Barbano, 2000). The gravity separation method can also be done in a two-stage procedure in which the control milk (3.58 μm) was fractionated at 4 °C for 6 h to yield semi-skim milk in the first stage. The second step of separation of semi-skim milk obtained skim and cream portions. MFG size of the latter fraction (3.45 μm) was slightly different to that of control milk (O'Mahony, Auty, & McSweeney, 2005).

Ultrasonic-Assisted Separation Ultrasonication technique utilises primary acoustic radiation force to induce physical destabilisation of milk fat globules whereby the creaming process is facilitated due to formation of floccules and clusters of MFG (Juliano et al., 2011; Leong et al., 2014). There are numerous factors controlling the separation efficacy of MFG using ultrasonication such as physical properties of the MFG (size, solid/liquid proportion of fat) and ultrasonic operating parameters (frequency of ultrasound, processing time, mode of operation, specific energy density input etc.) (Juliano et al., 2011; Leong et al., 2014). A previous study performed on recombined emulsion (3.5% fat) showed that smaller fat droplets (2.7 μm) were more resistant to creaming in comparison to larger droplets in raw milk (4.9 μm) and coarser emulsion (9.3 μm) upon sonication (400 kHz or 1.6 MHz) for 5 min at 35 °C. Based on the discrepancy in MFG size in top and bottom fractions obtained in sonicated natural whole milk (4.3–4.5 μm), it was found that the milk pre-cooled to 5 °C had the lowest differentiation (4.39 and 4.44 μm). The separation efficacy improved with broadest discrepancy in MFG size pre-heated at 25 °C (4.0–4.9 μm) (Leong, Juliano, et al., 2014). Possible explanations for this observation could be due to alteration of immunoglobulins in sonicated milks and/or associated change in the solid/liquid ratio of the milk fat, impacting the formation of floccules of MFGs. Use of higher frequency ultrasound results in greater separation efficacy of MFG. Sonication of natural whole milk at 1 MHz appears to yield greater differentiation in MFG size fractions at the top (4.9 μm) and bottom (4.0 μm) than those obtained with 600 kHz sonication (4.7 and 4.4 μm, respectively) (Leong, Juliano, et al., 2014). An attempt has also been made to further improve ultrasonic separation efficacy with multi-stage ultrasonic fractionation. The smallest achievable size was about 3.38 μm, which was 0.9 μm different to the MFG size of the original milk (4.28 μm) (Leong et al., 2016). The ultrasound-assisted separation of MFG is fully covered in Chap. 18.

Centrifugation Centrifugation is a well-established method to concentrate MFG in commercial manufacture of various dairy-fat based products such as skim milk, cream and butter. Similar to the naturally occurring gravitation separation method, the centrifugation method is also based on a density basis but with 6500-fold faster sedimentation velocity (TetraPak, 2009). As compared to the gravity separation method, a larger differentiation in small (2.5–3.0 μm) and large (5.0 μm) MFG size fractions can be achieved with mild centrifugation at 150 × *g* (Logan et al., 2014). A two-step centrifugal method has been developed aimed at fractionating more discrete MFG size fractions (Timmen & Patton, 1988). Timmen and Patton (1988) reported that skim and cream portions obtained from the two-step centrifugal method enriched in small MFG (1.02–1.77 μm) and large MFG (2.76–3.33 μm). There are few attempts to modify the cream separators to obtain more discrete MFG size fractions. Eden et al. (2016) modified the geometry of the bowl disc and used only four discs to centrifuge standardised milk (4% w/w fat) at 1200 × *g*, obtaining the skim and cream portions. The fractionated cream was further concentrated using an unmodified cream separator at 5300 × *g*. The processing temperature of both fractionation and concentration stages was 55 °C. The MFG size of the resultant creams (35.8–44.1% w/w fat) was varied about 0.5 μm from the original milk (4.3 μm) in the range of 4.1–4.9 ± 0.1 μm. Dhungana et al. (2017) reported that larger differentiation (2.9 μm) in MFG size fractions (1.35–4.28 μm) of fractionated creams (33–65% w/w fat) can be achieved when using a novel two-stage centrifugal fractionation method. This method utilised a commercial cream separator as a continuous centrifuge in the first stage by removing all the cones from separating disc. Normal set up of commercial cream separator (no removal of the cones) was employed for the second stage to concentrate the skim and cream fractions obtained in the previous step. Figure 14.1c and d represent microscopic images of small and large MFG fractions in native dairy creams obtained by this two-stage centrifugal fractionation method (unpublished data).

Microfiltration Microfiltration is a membrane processing technique that can fractionate MFG size effectively (Michalski et al., 2006). Utilisation of different membrane pore sizes (2–12 μm) and operating parameters of membrane processing (permeate flux, tangential shear stress, volume reduction factor etc.), one can obtain discrete MFG size classes as small as 0.9–3.3 μm in permeate and much larger MFG size in retentate (5–7.5 μm) from the original milk of 4.2 μm (Michalski et al., 2006). The microfiltration technique was found to maintain the integrity of fat globules without damage of MFG owing to shear or cavitation as reflected by unchanged zeta-potential values of the resultant MFG size fractions (Michalski, Michel, et al., 2002). However, this process is susceptible to membrane fouling (Michalski et al., 2006).

Up to date, among available methods for fractionating native MFG on size basis without shear, centrifugation method seems to be commercially viable since it has the potential to be applicable and effective in terms of cost-efficiency due to use of conventional processing equipment, high throughput and size differentiation. It is thought that with further modification, the two-stage centrifugal separation method could provide an efficient mean of producing differentiate-sized creams for improved dairy-fat based products.

4.3 Downsize of Milk Fat Globules with Mechanical Shear

Reduction of MFG size based on mechanical shear processing is a common practice in industrial processing of milks, aiming at improving their physical stability. Depending on homogenisation conditions, the fat globule size can be reduced to the sub-micron range. Owing to rupture forces of cavitation, shear and high turbulence, the MFGs are divided into finer droplets in accompany with disruption of milk fat globule membrane into fragments. Since the globule surface area increases with reduction of MFG size, stabilisation of newly generated droplets in homogenised milks is governed by dairy proteins (whey protein and adsorbed caseins) and the MFGM fragments (Michalski, Cariou, Michel, & Garnier, 2002). As such, the shear-processing can lead to alteration of physicochemical properties of homogenised milks regarding their native counterparts. This is due to differences in the fat globule size as well as properties of emulsified layers in the reformed MFGs. Beyond that, use of high pressure homogenisation can cause dissociation of casein micelles into casein micellar fragments and modification of whey proteins (Lee, Lefèvre, Subirade, & Paquin, 2009). Thus, the reduced MFG size, the emulsified MFG properties and the modified structures might have either positive or negative impacts on functionalities of dairy products made from homogenised milks.

The three common sizing methods to reduce fat globule size in dairy processing are homogenisation, microfluidisation and ultrasonication. Among them, the homogenisation of milk is a conventional practice in dairy industry in which pressure differences (caused by homogenisation valve and collision with the impact ring) result in rupture forces (i.e. shear, cavitation and high turbulence). Generally, milk is pre-heated to 50–60 °C and subjected to the first- and second-stage valves at 10–30 MPa and 3–5 MPa, respectively. The former aims to break up the fat globules, whereas the latter prevents formation of agglomerates (Walstra et al., 1999). Homogenisation of milks at high pressure in the range of 50–350 MPa has also been investigated. With microfluidisation technique, the fat globule size is reduced by intensive disruption forces generated in an interaction chamber where two streams of product (e.g. milk, cream) collide with each other, causing breaking up of fat globules. It is reported that microfluidisation technique has peak shear rate as high as 10^8–10^9 s^{-1} (Mason, Wilking, Meleson, Chang, & Graves, 2006). In ultrasonication technique, homogenisation of milk using ultrasonic waves generated by acoustic power (typically 20 kHz frequency) creates intense mechanical vibrations. This generates cavitation forces whereby milk fat globules can be disrupted.

Table 14.2 presents selected operating conditions of downsizing of milk fat globules using the mechanical shear. Conventional homogenisation of milks at 18–20 MPa can reduce the MFG size about tenfold with average MFG size (Hayes & Kelly, 2003; Thiebaud, Dumay, Picart, Guiraud, & Cheftel, 2003). However, MFG size seems not to be significantly further reduced with high pressure homogenisation. As shown in Table 14.2, comparable MFG size range (0.4–0.9 μm) was obtained when milks were homogenised at 200 MPa as against conventional homogenisation (0.7–0.8 μm). It is reported that usage of higher pressure up to 300 MPa can reduce the MFG size as small as 0.15–0.16 μm and 0.30–0.43 μm in milk (Serra, Trujillo, Quevedo, Guamis, & Ferragut, 2007) and cream (Rodarte, Zamora, Trujillo, & Juan, 2018), respectively. The ineffectiveness of high pressure homogenisation in further reduction of MFG size may be related to shortage of the available emulsifiers to stabilise the newly generated surface area resulting from numerous amount of smaller fat globules. Thus, the addition of extraneous emulsifiers may facilitate greater reduction of MFG size. A recent study performed on dairy cream (38% fat) showed that nano-sized range of MFG (0.13–0.14 μm) can be achieved with the addition of 4–5% sodium caseinate into cream prior to microfluidisation (Panchal et al., 2017). When pressure is applied at a moderate level, microfluidisation method appears to be more effective than conventional homogenisation in reduction of MFG size. A previous study showed that microfluidisation and homogenisation of milk at the same pressure (40 MPa) resulted in different MFG size range, i.e. 280 and 400 nm, respectively (Dalgleish, Tosh, & West, 1996). Major factors influence the efficiency of microfluidisation in reduction of MFG size include microfluidising pressure and fat content of the sample (Hardham, Imison, & French, 2000; Olson, White, & Richter, 2004). Downsizing of MFG using ultrasonication is a function of ultrasonic power and treatment time (Table 14.2). In general, application of higher ultrasonic power with prolonged treatment duration caused greater reduction of MFG size in both milk and cream (Bermudez-Aguirre, Mawson,

Table 14.2 Selected studies on downsizing of milk fat globules using mechanical shear via conventional/high pressure homogenisation, microfluidisation and ultrasonication

Sample	Operating conditions	MFG size obtained	References
Conventional and high pressure homogenisation (HPH)			
Milk	18 MPa	D_{43} 0.7 μm	Hayes & Kelly, 2003)
Milk	1–2 stages: 50–200 MPa	D_{43} 0.62–3.20 μm	
Pre-warmed milk prior to 2-stage HPH		D_{43} 0.4–0.5 μm	
Milk (4, 14, and 24 °C)	200 MPa	D_{43} 0.90, 0.65 and 0.37 μm, respectively	Thiebaud et al. (2003)
Milk	Conventional	D_{43} 0.88 μm	Hayes, Fox, and Kelly (2005)
	Pasteurized and HPH	D_{43} 0.48–0.86 μm	
Milk (40 °C)	200 MPa	0.15 μm	Serra et al. (2007)
Milk (30 °C)	300 MPa	0.16 μm	
Cream (20% fat)	300 MPa; single stage	D_{43} 0.30–0.43 μm	Rodarte et al. (2018)
Microfluidisation			
Milk	40 MPa	280 nm	Dalgleish et al. (1996)
Milk	50 and 100 MPa	460 and 304 nm	Olson et al. (2004)
	150 and 200 MPa	361 and 383 nm	
Milk (42 and 54 °C)	75, 125 and 170 MPa	0.39–0.50 μm	Bucci, Van Hekken, Tunick, Renye, and Tomasula (2018)
Cream (38% fat) with extraneous protein added)	62 MPa	D_{32} 0.13 μm	Panchal et al. (2017)
Ultrasonication			
Milk	20–40 W for 1–10 min	2–5 μm	Wu et al. (2000), Ertugay et al. (2004), Bermudez-Aguirre et al. (2008)
Milk	400–450 W	0.5–0.7 μm	
Milk (70–75 °C)		0.57–0.95 μm	Villamiel and de Jong (2000)
Milk	31 W for 30 min; 50 °C	1.49 μm	Chandrapala et al. (2016)
Cream (42% fat)	50 W for 10 min; 10 and 50 °C	3.35 and 1.63 μm, respectively	

& Barbosa-Canovas, 2008; Chandrapala et al., 2016; Ertugay, Sengul, & Sengul, 2004; Villamiel & de Jong, 2000; Wu, Hulbert, & Mount, 2000).

There is an effect of temperature on size reduction of MFG in milks and creams regardless of homogenisation techniques employed (Chandrapala et al., 2016; Serra et al., 2007; Thiebaud et al., 2003). As summarised in Table 14.2, smaller MFG size was obtained with a higher inlet temperature of milks and cream. Thiebaud et al. (2003) reported that MFG size of milk being homogenised at 4, 14 and 24 °C with 200 MPa was in the range of 0.9, 0.65 and 0.37 μm, respectively. Ultrasonication of cream at 50 °C yielded smaller MFG size (1.63 μm) as compared to that of lower temperature (10 °C; 3.35 μm) (Chandrapala et al., 2016).

5 Influence of MFG Size on Properties and Functionalities of Dairy-Fat Structured Products

Milk fat, casein micelles and whey proteins are major components of various dairy products (Fig. 14.3) in which the fundamental characteristics of each component will affect the product properties. As previously discussed, there is a dependence of chemical composition and physical characteristics on MFG size. Thus, variation of MFG size in dairy products can be expected to influence texture, flavour, sensory perception and physical functionalities. Baes on previous findings in the dairy-related field, a summary of possible influence of differentiated-size MFG on fundamental characteristics (e.g. interfacial properties, crystallisation properties, physical stability and optical properties) and resultant alterations on physical functionalities of milk, yoghurt, cheese, butter, whipped cream and ice cream is presented in Fig. 14.3.

5.1 *Milk and Dairy Cream*

Variation in MFG size was found to influence various physical properties and functionalities of fluid milk such as thermal stability, gelation, foaming and sensory properties.

Thermal Stability It is known that reduction of MFG size by homogenisation affects heat stability of milk. The homogenisation needs to be carried out prior to the pre-warming or concentration step to maintain the heat stability of milk. The reverse order will cause a reduction of the heat stability. It was demonstrated that heat coagulation time at 120 °C increased with homogenised milk having MFG size

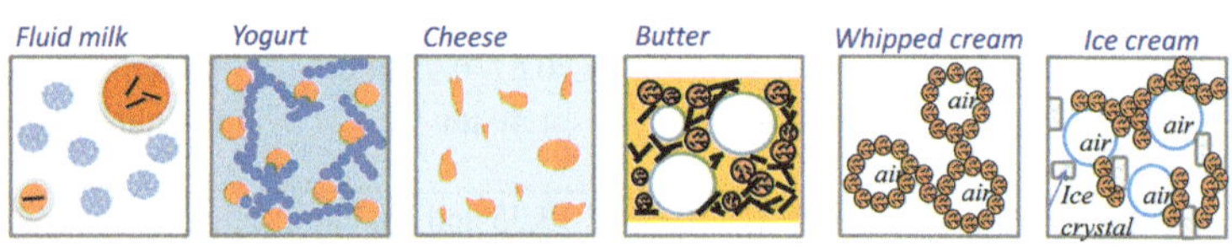

Physical state	Liquid	Gel	Plastic solid	Solid	Foam	Solid foam
Compositional difference → interfacial properties	Heat stability Foamability Foam stability	Syneresis Firmness	Water retention Free oil leakage Syneresis Firmness Cheese yield	Water retention	Foamability Foam stability	
Crystallisation properties			Rheology	Churning of cream	Foaming	
Physical stability	Creaming		Creaming			
Optical properties	Colour		Colour			

Fig. 14.3 Illustrations of role of milk fat in structuring wide variety of dairy products and associated influence of MFG size on physical functionality of the products

of less than 0.4 μm (Whiteley & Muir, 1996). Since no changes in heat stability was in association with MFG size above 0.4 μm, it was suggested that the varied MFG size did not have any direct effect on the heat stability. Instead, alteration of protein conformation upon homogenisation may govern the heat stability of homogenised milk. A recent study performed on heat stability of native creams (Dhungana, Truong, Bansal, & Bhandari, 2019) provides insights into the role of MFG size as well as the connection between heat treatment with complexity of serum phase in the context of MFG size reduction. In this study, five fractions of MFG size (1.45, 2.45, 3.85 and 4.50 μm) were obtained from fractionation of native creams (18 and 36% fat content) using the two-stage centrifugal method (Dhungana et al., 2017). Testing of heat stability of all cream samples at 140 °C showed that the heat stability depended on both the MFG size and the fat content. The smaller MFG size fraction was more heat s*table* with low fat content. In contrast, the larger MFG size fraction having high fat content tended to maintain heat stability better than that of smaller size fraction. It was proposed that apart from MFG size, possible influence of the heat stability may include changes in the composition of serum phase, fat content and MFGM properties.

Gelation Gelation of milk can be induced by rennet addition and/or acidification in which the protein network is built up by fusion and/or linkage of caseins, providing pores (typically 4 and 15 μm in rennetted gels and 1–30 μm in acid gels) for fat globules to be embedded within (Cho, Lucey, & Singh, 1999; Mellema, Heesakkers, van Opheusden, & van Vliet, 2000; Michalski, Cariou, et al., 2002). Depending on the relative size between fat globules and pore size, the fat globules can be fitted into or excluded from the void spaces, resulting in enhancing the gel firmness or weakening the gel integrity, respectively (Logan et al., 2015; Michalski, Cariou, et al., 2002). It was postulated that native MFGs act as inert fillers since their intact MFGM do not interact with the casein network. In homogenised and recombined milks, the surface of MFGM is generally modified whereby adsorption of whey proteins and caseins permits interaction with the gel structure. Thus, the effect of MFG size on gelation of milk not only involves in optimum packing between fat globule and gel pores size but also presence and volume fractions of other interacting particles forming the gel network. For both rennet- and acid-induced gels, Michalski, Cariou, et al. (2002) found that G' increased with smaller MFG size fraction obtained from microfiltrated milk (2.8 vs. 4.6 μm), homogenised milk (1.6–2 μm) and reconstituted skim milk (1.9–3.9 μm). In another study performed on fresh milk obtained from selection of cow breeding, it was reported that rennet-induced gel contained large MFG (4.6–5.15 μm) tended to firmer than that of small MFG (3.5–3.6 μm) at the similar size of casein micelles (Logan et al., 2015). Regarding acid-induced gel, measurement of yield stress and strain of homogenised skim milk (50–850 bar) showed that the measured rheological values were higher with smaller MFG size (0.2 versus 2.0 μm) (Ji, Lee, & Anema, 2011). In this case, the reinforcement of gel structure might be partly related to changes in protein adsorption behaviour and characteristics of other interacting particles caused by homogenisation process (Ji et al., 2011).

Foamability and Foam Stability Foaming characteristics of whole milk homogenised at different pressures to obtain different particle sizes (0.39, 0.44, 0.76, and 1.50 μm) have been investigated (Borcherding, Hoffmann, Lorenzen, & Schrader, 2008). It was shown that the foaming properties at 50 °C were governed by milk protein fractions contained in the skim milk phase rather than a direct influence of MFG size. However, the influence of fat globules was more evident when foaming was carried out at lower temperature 20–30 °C (Borcherding et al., 2008). At this point, it is not clear whether the physical state of milk fat across a temperature range of 20–50 °C have had any impact on the foaming properties observed in this temperature region. Dependence of foamability and foam stability on MFG size (0.2, 0.6 and 1.2 μm) of various milk fat emulsions (10% anhydrous milk fat or milk fat fractions) has also been studied (Truong, Bansal, & Bhandari, 2014). In general, the extremely small droplets (i.e. 0.2 μm) could not retain in the liquid films, causing thinning of the films and subsequent foam collapsing. Further investigation is needed to elucidate the role of MFG size in foamability and foam stability of milks where associated alterations of protein conformation is minimized.

Sensory Properties Perception of dietary fat has been known to have multimodal stimulus in which there is an involvement of sensory modalities (in-mouth tactile sensations, vision, taste, olfaction) and product-related factors (structure of food matrix, viscosity, aroma, tasting temperature) (Le Calve et al., 2015; Mattes, 2009; Schoumacker et al., 2017). Few attempts have been made to relate the perception of creaminess in milk with different MFG sizes (Goudedranche, Fauquant, & Maubois, 2000; Richardson & Booth, 1993). For instance, Richardson and Booth (1993) reported that the smaller MFG obtained by homogenisation in thickened milk (18.6 Pa s) contributed to greater creaminess. This can be partly attributed to the greater surface area and numerous amount of smaller MFG in combination with adequate thickness, providing overall perception of creaminess in the thickened milk. From milk fat emulsion (4% fat) perspective, it was found that fat aggregates having average size above 5 μm contributed to powdery feeling and textural perception such as thickness and chalkiness (Fibrianto, 2013).

5.2 *Yoghurt*

As previously discussed, MFGs play an important role in determining the gelation properties of milk. In yoghurt system, depending on their interactions with the protein matrix in either active or inert modes, milk fat globules can promote or disrupt the microstructure of yoghurts, respectively. The influence of MFG on water retention and associated gel strength as well as syneresis of yoghurts have been investigated. It was reported that yoghurt made from thermo-sonicated milk with smaller MFG size (0.4 μm) had two-fold water retention than that of made from the conventional method (Riener, Noci, Cronin, Morgan, & Lyng, 2009). Reduction of MFG size using microfluidisation technique showed that larger gel particles can be formed with smaller MFG. Nevertheless, associated changes in texture and the amount of water retained in the matrix of yoghurt made from microfluidised milk had only a slight improvement (Ciron, Gee, Kelly, & Auty, 2010). Usage of high pressure homogenisation was able to further reduce MFG size as small as D_{32} 0.12–0.16 μm. Such small MFG was found to have positive effects on gel strength, gel firmness and syneresis of yoghurt as compared with the sample made from skim milk powder using conventional manufacturing method (Serra et al., 2007). In those studies, reduction of MFG size was done using mechanical shear processing that has altered the globule membrane reactivity with inclusions of whey proteins and casein. Thus, it will be interesting to examine the effect of native MFG size on the physical functionality of yoghurt products.

5.3 Cheese

The primary structure of cheese is a protein matrix composed of aggregated and linked casein micelles in which water, salts and MFG are dispersed. MFG can act either as inert fillers or interacting particles within the cheese protein network depending on the integrity of MFG. Generally, native MFGM is non-interactive whereas homogenised MFGM being partly replaced with caseins and/or whey proteins is prone to interacting with the cheese protein network through hydrophobic interactions (Everett & Olson, 2000; Lucey, Johnson, & Horne, 2003). Regarding cheese prepared from native MFG, small MFG fraction obtained by microfiltration was found to improve water binding capacity in Emmental and Camembert cheeses. This enhancement with the cheeses containing small MFG fraction was due to greater surface area to volume ratio and associated changes in MFGM material, leading to the higher moisture content and softer texture in the cheeses (Michalski et al., 2003, 2004). Proteolysis and lipolysis seem to be enhanced with smaller MFG in Camembert, Emmental, and Italian cheese (Jana & Upadhyay, 1992; Michalski et al., 2003, 2007; Michalski, Camier, et al., 2004). This may be partly explained by more specific sites are available for enzyme activity with a greater proportion of MFGM in smaller MFG fraction. A contrasting observation was reported in Emmental cheese and miniature Cheddar cheese that large MFG (5.6 and 4.68 μm, respectively) caused greater lipolysis upon ripening with remarkable increase in free fatty acid (O'Mahony et al., 2005). It was postulated that the differences in liberation of free fat in these cheese matrices may be caused by different crystallisation behaviour of fat within small MFG. Since the larger MFG may contain crystalline structures that were able to disrupt the MFGM (O'Mahony et al., 2005), making it more prone to the enzyme activity.

Native MFG size also influences product properties and functionalities in Camembert, Emmental, mini Swiss and fresh cheeses. The cheeses made from smaller MFG appeared to be softer with lower rheological values (Goudedranche et al., 2000; Michalski et al., 2003, 2007; Michalski, Camier, et al., 2004). Improved stretching and elasticity was also found in Emmental cheese containing small

MFG. With regarding to sensory properties, the cheeses made from smaller MFG was perceived as smoother and more elastic texture (Goudedranche et al., 2000; Michalski et al., 2003). St-Gelais, Passey, Hache, and Roy (1997) also reported effect of MFG size on low-fat Cheddar cheese manufactured from low mineral retentate powder in which large MFG fraction improved colour, flavour and texture of the cheese.

Reduction of MFG size via homogenisation approach has been shown to affect enzyme activity (lipolysis, proteolysis), processability (fat loss to whey, free oil release, cheese yield), physicochemical properties (moisture, viscosity, firmness, whiteness, rennet time, syneresis), and functionality (flowability, stretchability) (Green, Marshall, & Glover, 1983; Lemay, Paquin, & Lacroix, 1994; Michalski, Camier, et al., 2004; Rowney, Hickey, Roupas, & Everett, 2003; Rudan, Barbano, Gu, & Kindstedt, 1998; Schenkel, Samudrala, & Hinrichs, 2013). These influences are attributed to MFG size reduction and modification of MFG surface upon homogenisation (Green et al., 1983). For instance, casein may be less available to build up the cheese protein network since it is incorporated into the homogenised MFGM, resulting in weaker structure of the cheese protein matrix (Green et al., 1983).

5.4 Butter

Butter is made from churning of dairy cream whereby a phase inversion of oil-in-water to water-in-oil emulsions can be occurred. Thus, the physical properties of initial dairy cream significantly influence the processability of butter making and physical functionality of resultant butter. It has been demonstrated that churning time of cream is MFG-size dependent. A previous study performed on secreted milks having large (2.3 μm) and small MFGs (1.84 μm), which was obtained by modifying diet of cow with fish meal, showed that the small MFG induced longer churning time (Avramis et al., 2003). When cow diets were modified with extruded linseed, it appears that small MFGs (3.49 μm) had shorter churning time as compared to that of larger ones (4.18 μm). In these studies, the discrepancy in churning time may be related to possible changes in MFG composition resulting from the different cow diets. In another study, Goudedranche et al. (2000) found no difference in churning abilities between small (below 2 μm) and large (above 2 μm) MFGs fraction obtained by microfiltration of native MFG in cream. The churnability of differentiated-sized MFG prepared from recombined cream (38% fat; emulsifying anhydrous milk fat with sodium caseinate) within the range of 0.17–3.50 μm has also been studied (Panchal et al., 2017). Since the recombined cream having smaller size tended to be less prone to destabilisation, churning time was extended with reduction of MFG size. Addition of low molecular surfactant (Tween 80) into the sodium caseinate stabilised cream samples promotes the "competitive destabilisation" at the interface of water and oil, leading to a significant reduction of churning time (Panchal et al., 2017). Apart from churning time, few studies showed that the small MFG had a negative impact on process and product such as higher fat loss during churning and greater water retention in the resultant butter (Goudedranche et al., 2000; Hurtaud et al., 2010).

Few attempts have been made to prepare butter made from differentiated-size dairy creams. Goudedranche et al. (2000) reported that the butter made from smaller fat globules obtained by microfiltration method tended to be greasy, oily than that of prepared from control butter. In contrast, Avramis et al. (2003) and Hurtaud et al. (2010) found that small MFG improved physical functionality of the resultant butter, i.e. more spreadable, softer texture and better mouthfeel. Since the small MFG was obtained from modifying cow diet with meal enriched in unsaturated fatty acid, it is difficult to interpret whether the associated changes in physical functionality related to MFG size or the compositional differences with varied MFG size. Thus, it would be interesting to further explore the feasibility of manipulating MFG size to improve the physical functionality of butter.

5.5 Whipped Cream and Ice Cream

Whipped cream and ice cream are common dairy aerated products in which partial coalescence of MFG is essential to stabilise air cells embedded in the dairy matrices (Goff, 1997). Formation of foam and foam stability in those products are governed by several factors such as whipping conditions (time, intensity and temperature), ice crystal size, serum viscosity and interactions between fat, emulsifiers/stabilisers at the air-water interface (Goff, 1997). Apart from those influential factors, MFG size has been found to impact on foamability and foam stability of whipped cream and ice cream. It has been demonstrated in native dairy creams having different MFG size (3.8–4.9 μm) obtained by gravity separation and subsequent bowl disk centrifugal separation that an increase in MFG size reduced whipping time by 22 ± 7% with greater overrun. Similar tendency was noted with the homogenised creams having similar MFG size range, indicating that the foamability is dependent on true MFG size (Eden et al., 2016). In this study, no difference in serum drainage (based on MFG size) in whipped cream was found. In another study, it has been reported that dairy foam containing smaller MFG fraction, which was prepared from microfiltration technique, was more *stable* with less foam collapse (12%) as against the foam made from large MFG fraction (29%) and control (25%) whipped cream (Michalski et al., 2006). Regarding ice cream system, influence of MFG size in homogenised cream has been evaluated whereas there is little information on manipulation of native MFG size in ice cream manufacturing. Koxholt, Eisenmann, and Hinrichs (2001) used homogenised cream within size range of 0.44–3.33 μm to study melt-down of ice cream. It was found that foam structures containing smaller fat particles (0.44–0.85 μm) had faster melt-down that their larger counterpart (0.85–3.33 μm). The contrasting observations on the effect of MFG size in whipped cream and dairy cream systems suggest that these different influences may be also attributed from the discrepancy in dairy processing and associated MFG preparation.

6 Conclusion

This book chapter underscores the importance of manipulating MFG size, which spans from 0.1 to 15 μm in naturally bovine MFG, to improve processability and physical functionality of a wide variety of dairy-fat structured products. Beyond the impact of MFG size, there is a compositional difference among MFG size fractions, giving rise to discrepancies in the fundamental properties of MFG. Herd management, physical separation and microfiltration are effective methods to preserve the integrity of MFGM in native size-classified milks. However, further improvement needs to be done to obtain more discrete MFG size fractions with these methods. Mechanical processing is a common practice to reduce the MFG to nano-sized scale (i.e. approximately 0.2 μm). When being subjected to the mechanical shear, the MFGM of emulsified globules is disrupted whereby caseins and whey proteins are incorporated into the reformed MFGM. This alteration makes the emulsified MFG become interacting particles with the protein network whereas non-interacting native MFG acts as inert fillers within the dairy matrices. As such, the inactive and active modes of MFG also contribute to particle interactions within the dairy matrices, apart from the influence of true MFG size. It has been demonstrated that differentiated-sized MFG fractions participate in structuring dairy-fat containing products where the associated differences in composition, structure and crystallisation behaviour of milk fat can be used as a means of enhancing physical functionality and processability of fluid milks, dairy creams, cheeses, yoghurts, butter, whipped cream and ice cream. With availability of advanced techniques for dairy processing and characterisation of dairy products, it is suggested that further research to be undertaken to obtain more discrete MFG size fraction with wider size ranges. The next advance can be manipulation of the complex interactions between differentiated-sized fat globules and other particles in dairy matrices to develop innovative, low-fat, health-promoting dairy products.

Acknowledgements This research was supported under Australian Research Council's Industrial Transformation Research Hub (ITRH) funding scheme (IH120100005). The ARC Dairy Innovation Hub is a collaboration between the University of Melbourne, the University of Queensland and Dairy Innovation Australia Ltd.

References

Abeni, F., Degano, L., Calza, F., Giangiacomo, R., & Pirlo, G. (2005). Milk quality and automatic milking: Fat globule size, natural creaming, and lipolysis. *Journal of Dairy Science, 88*(10), 3519–3529.

Aernoutsa, B., Beersa, R. V., Wattéa, R., Huybrechtsa, T., Jordensb, J., Vermeulenc, D., et al. (2015). Effect of ultrasonic homogenization on the Vis/NIR bulk optical properties of milk. *Colloids and Surfaces B: Biointerfaces, 126*, 510–519.

Ashworth, U. S. (1951). Turbidity as a means for determining the efficiency of homogenization. *Journal of Dairy Science, 34*(4), 317–320.

Avramis, C. A., Wang, H., McBride, B. W., Wright, T. C., & Hill, A. R. (2003). Physical and processing properties of milk, butter, and Cheddar cheese from cows fed supplemental fish meal. *Journal of Dairy Science, 86*(8), 2568–2576.

Banach, J. K., Żywica, R., & Kiełczewska, K. (2008). Effect of homogenization on milk conductance properties. *Polish Journal of Food and Nutrition Sciences, 58*(1), 107–111.

Banks, J. M., Clapperton, J. L., Muir, D. D., & Girdler, A. K. (1986). The influence of diet and breed of cow on the efficiency of conversion of milk constituents to curd in cheese manufacture. *Journal of the Science of Food and Agriculture, 37*(5), 461–468.

Bauman, D. E., Corl, B. A., & Peterson, G. P. (2003). *The biology of conjugated linoleic acids in ruminants*. Champaign, IL: AOCS Press.

Belury, M. A. (2002). Dietary conjugated linoleic acid in health: Physiological effects and mechanisms of action. *Annual Review of Nutrition, 22*, 505–531.

Bermudez-Aguirre, D., Mawson, R., & Barbosa-Canovas, G. V. (2008). Microstructure of fat globules in whole milk after thermosonication treatment. *Journal of Food Science, 73*(7), E325–E332.

Borcherding, K., Hoffmann, W., Lorenzen, P. C., & Schrader, K. (2008). Effect of milk homogenisation and foaming temperature on properties and microstructure of foams from pasteurised whole milk. *LWT-Food Science and Technology, 41*(10), 2036–2043.

Briard, V., Leconte, N., Michel, F., & Michalski, M. C. (2003). The fatty acid composition of small and large naturally occurring milk fat globules. *European Journal of Lipid Science and Technology, 105*(11), 677–682.

Briard-Bion, V., Juaneda, P., Richoux, R., Guichard, E., & Lopez, C. (2008). trans-C18: 1 isomers in cheeses enriched in unsaturated fatty acids and manufactured with different milk fat globule sizes. *Journal of Agricultural and Food Chemistry, 56*(20), 9374–9382.

Bucci, A. J., Van Hekken, D. L., Tunick, M. H., Renye, J. A., & Tomasula, P. M. (2018). The effects of microfluidization on the physical, microbial, chemical, and coagulation properties of milk. *Journal of Dairy Science, 101*, 6990–7001.

Bugeat, S., Briard-Bion, V., Perez, J., Pradel, P., Martin, B., Lesieur, S., et al. (2011). Enrichment in unsaturated fatty acids and emulsion droplet size affect the crystallization behaviour of milk triacylglycerols upon storage at 4 degrees C. *Food Research International, 44*(5), 1314–1330.

Carroll, S. M., DePeters, E. J., Taylor, S. J., Rosenberg, M., Perez-Monti, H., & Capps, V. (2006). Milk composition of Holstein, Jersey, and Brown Swiss cows in response to increasing levels of dietary fat. *Animal Feed Science and Technology, 131*(3-4), 451–473.

Chandrapala, J., Ong, L., Zisu, B., Gras, S. L., Ashokkumar, M., & Kentish, S. E. (2016). The effect of sonication and high pressure homogenisation on the properties of pure cream. *Innovative Food Science & Emerging Technologies, 33*, 298–307.

Cho, Y. H., Lucey, J. A., & Singh, H. (1999). Rheological properties of acid milk gels as affected by the nature of the fat globule surface material and heat treatment of milk. *International Dairy Journal, 9*(8), 537–545.

Ciron, C. I. E., Gee, V. L., Kelly, A. L., & Auty, M. A. E. (2010). Comparison of the effects of high-pressure microfluidization and conventional homogenization of milk on particle size, water retention and texture of non-fat and low-fat yoghurts. *International Dairy Journal, 20*(5), 314–320.

Cornell, D. G., & Pallansch, M. J. (1966). Counting and sizing fat globules electronically. *Journal of Dairy Science, 49*(11), 1371.

Couvreur, S., Hurtaud, C., Marnet, P. G., Faverdin, P., & Peyraud, J. L. (2007). Composition of milk fat from cows selected for milk fat globule size and offered either fresh pasture or a corn silage-based diet. *Journal of Dairy Science, 90*(1), 392–403.

Dalgleish, D. G., Tosh, S. M., & West, S. (1996). Beyond homogenization: The formation of very small emulsion droplets during the processing of milk by a Microfluidizer. *Netherlands Milk and Dairy Journal, 50*(2), 135–148.

Dhungana, P., Truong, T., Bansal, N., & Bhandari, B. (2019). Thermal and UHT stability of native, homogenized and recombined creams with different average fat globule sizes. *Food Research International, 123*, 153–165.

Dhungana, P., Truong, T., Palmer, M., Bansal, N., & Bhandari, B. (2017). Size-based fractionation of native milk fat globules by two-stage centrifugal separation. *Innovative Food Science & Emerging Technologies, 41*, 235–243.
Eden, J., Dejmek, P., Lofgren, R., Paulsson, M., & Glantz, M. (2016). Native milk fat globule size and its influence on whipping properties. *International Dairy Journal, 61*, 176–181.
Ertugay, M. F., Sengul, M., & Sengul, M. (2004). Effect of ultrasound treatment on milk homogenisation and particle size distribution of fat. *Turkish Journal of Veterinary & Animal Sciences, 28*(2), 303–308.
Everett, D. W., & Olson, N. F. (2000). Dynamic rheology of renneted milk gels containing fat globules stabilized with different surfactants. *Journal of Dairy Science, 83*(6), 1203–1209.
Fauquant, C., Briard, V., Leconte, N., & Michalski, M. C. (2005). Differently sized native milk fat globules separated by microfiltration: Fatty acid composition of the milk fat globule membrane and triglyceride core. *European Journal of Lipid Science and Technology, 107*(2), 80–86.
Fibrianto, K. (2013). *Contribution of anhydrous milk fat to oral processing and sensory perception of liquid milks*. PhD Thesis, The University of Queensland.
Fox, P. F., & McSweeney, P. L. (1998). *Dairy chemistry and biochemistry*. London: Blackie Academic & Professional.
Goff, H. D. (1997). Instability and partial coalescence in whippable dairy emulsions. *Journal of Dairy Science, 80*(10), 2620–2630.
Goudedranche, H., Fauquant, J., & Maubois, J. L. (2000). Fractionation of globular milk fat by membrane microfiltration. *Le Lait, 80*(1), 93–98.
Goulden, J. D. S. (1958). Some factors affecting turbimetric methods for the determination of fat in milk. *Journal of Dairy Research, 25*(2), 228–235.
Green, M. L., Marshall, R. J., & Glover, F. A. (1983). Influence of homogenization of concentrated milks on the structure and properties of rennet curds. *Journal of Dairy Research, 50*(3), 341–348.
Gresti, J., Bugaut, M., Maniongui, C., & Bezard, J. (1993). Composition of molecular-species of triacylglycerols in bovine-milk fat. *Journal of Dairy Science, 76*(7), 1850–1869.
Hardham, J. F., Imison, B. W., & French, H. M. (2000). Effect of homogenisation and microfluidisation on the extent of fat separation during storage of UHT milk. *Australian Journal of Dairy Technology, 55*(1), 16–22.
Hayes, M. G., Fox, P. F., & Kelly, A. L. (2005). Potential applications of high pressure homogenisation in processing of liquid milk. *Journal of Dairy Research, 72*(1), 25–33.
Hayes, M. G., & Kelly, A. L. (2003). High pressure homogenisation of raw whole bovine milk (a) effects on fat globule size and other properties. *Journal of Dairy Research, 70*(3), 297–305.
Heid, H. W., & Keenan, T. W. (2005). Intracellular origin and secretion of milk fat globules. *European Journal of Cell Biology, 84*(2-3), 245–258.
Huppertz, T., & Kelly, A. L. (2006). Physical chemistry of milk fat globules. In P. F. Fox & P. L. H. McSweeney (Eds.), *Advanced dairy chemistry volume 2: Lipids*. New York: Springer.
Hurtaud, C., Faucon, F., Couvreur, S., & Peyraud, J. L. (2010). Linear relationship between increasing amounts of extruded linseed in dairy cow diet and milk fatty acid composition and butter properties. *Journal of Dairy Science, 93*(4), 1429–1443.
Hussain, H., Truong, T., Bansal, N., & Bhandari, B. (2017). The effect of manipulating fat globule size on the stability and rheological properties of dairy creams. *Food Biophysics, 12*(1), 1–10.
Jana, A. H., & Upadhyay, K. G. (1992). Homogenisation of milk for cheesemaking. *Australian Journal of Dairy Technology, 47*, 72–79.
Ji, Y. R., Lee, S. K., & Anema, S. G. (2011). Effect of heat treatments and homogenisation pressure on the acid gelation properties of recombined whole milk. *Food Chemistry, 129*(2), 463–471.
Juliano, P., Kutter, A., Cheng, L. J., Swiergon, P., Mawson, R., & Augustin, M. A. (2011). Enhanced creaming of milk fat globules in milk emulsions by the application of ultrasound and detection by means of optical methods. *Ultrasonics Sonochemistry, 18*(5), 963–973.
Kaylegian, K. E., & Lindsay, R. C. (1995). *Handbook of milkfat fractionation technology and applications*. Champaign: AOCS Press.

Kietczewska, K., Kruk, A., Czerniewicz, M., Warminska, M., & Haponiuk, E. (2003). The effect of high-pressure homogenization on changes in milk colloidal and emulsifying systems. *Polish Journal of Food and Nutrition Sciences, 12*(1), 43–46.

Konokhova, A. I., Rodionov, A. A., Gilev, K. V., Mikhaelis, I. M., Strokotov, D. I., Moskalensky, A. E., et al. (2014). Enhanced characterisation of milk fat globules by their size, shape and refractive index with scanning flow cytometry. *International Dairy Journal, 39*(2), 316–323.

Koxholt, M. M. R., Eisenmann, B., & Hinrichs, J. (2001). Effect of the fat globule sizes on the meltdown of ice cream. *Journal of Dairy Science, 84*(1), 31–37.

Le Calve, B., Saint-Leger, C., Babas, R., Gelin, J. L., Parker, A., Erni, P., et al. (2015). Fat perception: How sensitive are we? *J Texture Stud, 46*(3), 200–211.

Lee, S.-H., Lefèvre, T., Subirade, M., & Paquin, P. (2009). Effects of ultra-high pressure homogenization on the properties and structure of interfacial protein layer in whey protein-stabilized emulsion. *Food Chemistry, 113*(1), 191–195.

Lemay, A., Paquin, P., & Lacroix, C. (1994). Influence of microfluidization of milk on cheddar cheese composition, color, texture, and yield. *Journal of Dairy Science, 77*(10), 2870–2879.

Leong, T., Johansson, L., Juliano, P., Mawson, R., McArthur, S., & Manasseh, R. (2014). Design parameters for the separation of fat from natural whole milk in an ultrasonic litre-scale vessel. *Ultrasonics Sonochemistry, 21*(4), 1289–1298.

Leong, T., Johansson, L., Mawson, R., McArthur, S. L., Manasseh, R., & Juliano, P. (2016). Ultrasonically enhanced fractionation of milk fat in a litre-scale prototype vessel. *Ultrasonics Sonochemistry, 28*, 118–129.

Leong, T., Juliano, P., Johansson, L., Mawson, R., McArthur, S. L., & Manasseh, R. (2014). Temperature effects on the ultrasonic separation of fat from natural whole milk. *Ultrasonics Sonochemistry, 21*(6), 2092–2098.

Logan, A., Auldist, M., Greenwood, J., & Day, L. (2014). Natural variation of bovine milk fat globule size within a herd. *Journal of Dairy Science, 97*(7), 4072–4082.

Logan, A., Day, L., Pin, A., Auldist, M., Leis, A., Puvanenthiran, A., et al. (2014). Interactive effects of milk fat globule and casein micelle size on the renneting properties of milk. *Food and Bioprocess Technology, 7*(3), 3175–3185.

Logan, A., Leis, A., Day, L., Øiseth, S. K., Puvanenthiran, A., & Augustin, M. A. (2015). Rennet gelation properties of milk: Influence of natural variation in milk fat globule size and casein micelle size. *International Dairy Journal, 46*, 71–77.

Long, Z., Zhao, M. M., Zhao, Q. Z., Yang, B., & Liu, L. Y. (2012). Effect of homogenisation and storage time on surface and rheology properties of whipping cream. *Food Chemistry, 131*(3), 748–753.

Lopez, C. (2005). Focus on the supramolecular structure of milk fat in dairy products. *Reproduction Nutrition Development, 45*(4), 497–511.

Lopez, C. (2011). Milk fat globules enveloped by their biological membrane: Unique colloidal assemblies with a specific composition and structure. *Current Opinion in Colloid & Interface Science, 16*(5), 391–404.

Lopez, C., Bourgaux, C., Lesieur, P., Bernadou, S., Keller, G., & Ollivon, M. (2002). Thermal and structural behavior of milk fat - 3. Influence of cooling rate and droplet size on cream crystallization. *Journal of Colloid and Interface Science, 254*(1), 64–78.

Lopez, C., Briard-Bion, V., Menard, O., Beaucher, E., Rousseau, F., Fauquant, J., et al. (2011). Fat globules selected from whole milk according to their size: Different compositions and structure of the biomembrane, revealing sphingomyelin-rich domains. *Food Chemistry, 125*(2), 355–368.

Lopez, C., Briard-Bion, V., Menard, O., Rousseau, F., Pradel, P., & Besle, J. M. (2008). Phospholipid, sphingolipid, and fatty acid compositions of the milk fat globule membrane are modified by diet. *Journal of Agricultural and Food Chemistry, 56*(13), 5226–5236.

Lucey, J. A., Johnson, M. E., & Horne, D. S. (2003). Invited review: Perspectives on the basis of the rheology and texture properties of cheese. *Journal of Dairy Science, 86*(9), 2725–2743.

Ma, Y., & Barbano, D. M. (2000). Gravity separation of raw bovine milk: Fat globule size distribution and fat content of milk fractions. *Journal of Dairy Science, 83*(8), 1719–1727.
Mabrook, M. F., & Petty, M. C. (2003). Effect of composition on the electrical conductance of milk. *Journal of Food Engineering, 60*(3), 321–325.
MacGibbon, A. K. H., & Taylor, M. W. (2006). Composition and structure of bovine milk lipids. In P. F. Fox & P. L. McSweeney (Eds.), *Advanced dairy chemistry 2: Lipids*. New York: Springer.
Martini, M., Altomonte, I., Pesi, R., Tozzi, M. G., & Salari, F. (2013). Fat globule membranes in ewes' milk: The main enzyme activities during lactation. *International Dairy Journal, 28*(1), 36–39.
Martini, M., Cecchi, F., & Scolozzi, C. (2006). Relationship between fat globule size and chemical and fatty acid composition of cow's milk in mid lactation. *Italian Journal of Animal Science, 5*(4), 349–358.
Martini, M., Cecchi, F., Scolozzi, C., Leotta, R., & Verita, P. (2003). Milk fat globules in different dairy cattle breeds part I: Morphometric analysis. *Italian Journal of Animal Science, 2*, 272–274.
Mason, T. G., Wilking, J. N., Meleson, K., Chang, C. B., & Graves, S. M. (2006). Nanoemulsions: Formation, structure, and physical properties. *Journal of Physics: Condensed Matter, 18*(41), R635–R666.
Mattes, R. D. (2009). Is there a fatty acid taste? *Annual Review of Nutrition, 29*, 305–327.
McCrae, C. H., & Lepoetre, A. (1996). Characterization of dairy emulsions by forward lobe laser light scattering - Application to milk and cream. *International Dairy Journal, 6*(3), 247–256.
Mellema, M., Heesakkers, J. W. M., van Opheusden, J. H. J., & van Vliet, T. (2000). Structure and scaling behavior of aging rennet-induced casein gels examined by confocal microscopy and permeametry. *Langmuir, 16*(17), 6847–6854.
Mesilati-Stahy, R., & Argov-Argaman, N. (2014). The relationship between size and lipid composition of the bovine milk fat globule is modulated by lactation stage. *Food Chemistry, 145*, 562–570.
Mesilati-Stahy, R., Mida, K., & Argov-Argaman, N. (2011). Size-dependent lipid content of bovine milk fat globule and membrane phospholipids. *Journal of Agricultural and Food Chemistry, 59*(13), 7427–7435.
Michalski, M. C., Briard, V., & Juaneda, P. (2005). CLA profile in native fat globules of different sizes selected from raw milk. *International Dairy Journal, 15*(11), 1089–1094.
Michalski, M. C., Briard, V., & Michel, F. (2001). Optical parameters of milk fat globules for laser light scattering measurements. *Le Lait, 81*(6), 787–796.
Michalski, M. C., Camier, B., Briard, V., Leconte, N., Gassi, J. Y., Goudedranche, H., et al. (2004). The size of native milk fat globules affects physico-chemical and functional properties of Emmental cheese. *Le Lait, 84*(4), 343–358.
Michalski, M. C., Camier, B., Gassi, J. Y., Briard-Bion, V., Leconte, N., Famelart, M. H., et al. (2007). Functionality of smaller vs control native milkfat globules in emmental cheeses manufactured with adapted technologies. *Food Research International, 40*(1), 191–202.
Michalski, M. C., Cariou, R., Michel, F., & Garnier, C. (2002). Native vs. damaged milk fat globules: Membrane properties affect the viscoelasticity of milk gels. *Journal of Dairy Science, 85*(10), 2451–2461.
Michalski, M. C., Gassi, J. Y., Famelart, M. H., Leconte, N., Camier, B., Michel, F., et al. (2003). The size of native milk fat globules affects physico-chemical and sensory properties of Camembert cheese. *Le Lait, 83*(2), 131–143.
Michalski, M. C., Leconte, N., Briard-Bion, V., Fauquant, J., Maubois, J. L., & Goudedranche, H. (2006). Microfiltration of raw whole milk to select fractions with different fat globule size distributions: Process optimization and analysis. *Journal of Dairy Science, 89*(10), 3778–3790.
Michalski, M. C., Michel, F., Sainmont, D., & Briard, V. (2002). Apparent zeta-potential as a tool to assess mechanical damages to the milk fat globule membrane. *Colloids and Surfaces B: Biointerfaces, 23*(1), 23–30.

Michalski, M. C., Ollivon, M., Briard, V., Leconte, N., & Lopez, C. (2004). Native fat globules of different sizes selected from raw milk: Thermal and structural behavior. *Chemistry and Physics of Lipids, 132*(2), 247–261.

Miles, C. A., Shore, D., & Langley, K. R. (1990). Attenuation of ultrasound in milks and Creams. *Ultrasonics, 28*(6), 394–400.

Mulder, H., & Walstra, P. (1974). The fat dispersion. In H. Mulder & P. Walstra (Eds.), *The milk fat globule. Emulsion science as applied to milk products and comparable foods* (pp. 54–66). Wageningen: The Netherlands Center for Agricultural Publishing and Documentation.

O'Mahony, J. A., Auty, M. A. E., & McSweeney, P. L. H. (2005). The manufacture of miniature Cheddar-type cheeses from milks with different fat globule size distributions. *Journal of Dairy Research, 72*(3), 338–348.

Olson, D. W., White, C. H., & Richter, R. L. (2004). Effect of pressure and fat content on particle sizes in microfluidized milk. *Journal of Dairy Science, 87*(10), 3217–3223.

Ong, L., Dagastine, R. R., Kentish, S. E., & Gras, S. L. (2010). The effect of milk processing on the microstructure of the milk fat globule and rennet induced gel observed using confocal laser scanning microscopy. *Journal of Food Science, 75*(3), E135–E145.

Pal, R. (1996). Effect of droplet size on the rheology of emulsions. *AICHE J, 42*(11), 3181–3190.

Panchal, B. R., Truong, T., Prakash, S., Bansal, N., & Bhandari, B. (2017). Effect of fat globule size on the churnability of dairy cream. *Food Research International, 99*(Pt 1), 229–238.

Precht, D. (1988). In N. Garti & K. Sato (Eds.), *Fat crystal structure in cream and butter. Crystalllization and polymorphism of fats and fatty acids* (pp. 305–361). New York: Marcel Dekker.

Richardson, N. J., & Booth, D. A. (1993). Effect of homogenization and fat content on oral perception of low and high viscosity model creams. *Journal of Sensory Studies, 8*, 133–143.

Riener, J., Noci, F., Cronin, D. A., Morgan, D. J., & Lyng, J. G. (2009). The effect of thermosonication of milk on selected physicochemical and microstructural properties of yoghurt gels during fermentation. *Food Chemistry, 114*(3), 905–911.

Robin, O., & Paquin, P. (1991). Evaluation of the particle-size of fat globules in a milk model emulsion by photon-correlation spectroscopy. *Journal of Dairy Science, 74*(8), 2440–2447.

Rodarte, D., Zamora, A., Trujillo, A.-J., & Juan, B. (2018). Effect of ultra-high pressure homogenization on cream: Shelf life and physicochemical characteristics. *LWT-Food Science and Technology, 92*, 108–115.

Rowney, M. K., Hickey, M. W., Roupas, P., & Everett, D. W. (2003). The effect of homogenization and milk fat fractions on the functionality of Mozzarella cheese. *Journal of Dairy Science, 86*(3), 712–718.

Rudan, M. A., Barbano, D. M., Gu, M. R., & Kindstedt, P. S. (1998). Effect of the modification of fat particle size by homogenization on composition, proteolysis, functionality, and appearance of reduced fat Mozzarella cheese. *Journal of Dairy Science, 81*(8), 2065–2076.

Schenkel, P., Samudrala, R., & Hinrichs, J. (2013). Thermo-physical properties of semi-hard cheese made with different fat fractions: Influence of melting point and fat globule size. *International Dairy Journal, 30*(2), 79–87.

Schoumacker, R., Martin, C., Thomas-Danguin, T., Guichard, E., Le Quéré, J. L., & Labouré, H. (2017). Fat perception in cottage cheese: The contribution of aroma and tasting temperature. *Food Quality and Preference, 56*, 241–246.

Serra, M., Trujillo, A. J., Quevedo, J. M., Guamis, B., & Ferragut, V. (2007). Acid coagulation properties and suitability for yogurt production of cows' milk treated by high-pressure homogenisation. *International Dairy Journal, 17*(7), 782–790.

St-Gelais, D., Passey, C. A., Hache, S., & Roy, P. (1997). Production of low-fat cheddar cheese from low and high mineral retentate powders and different fractions of milkfat globules. *International Dairy Journal, 7*(11), 733–741.

TetraPak. (2009). Centrifugal separators and milk fat standardisation. In *Dairy processing handbook* (pp. 91–113). Lund: TetraPak Processing Systems AB.

Thiebaud, M., Dumay, E., Picart, L., Guiraud, J. P., & Cheftel, J. C. (2003). High-pressure homogenisation of raw bovine milk. Effects on fat globule size distribution and microbial inactivation. *International Dairy Journal, 13*(6), 427–439.
Timmen, H., & Patton, S. (1988). Milk-fat globules - Fatty-acid composition, size and invivo regulation of fat liquidity. *Lipids, 23*(7), 685–689.
Truong, T., Bansal, N., & Bhandari, B. (2014). Effect of emulsion droplet size on foaming properties of milk fat emulsions. *Food and Bioprocess Technology, 7*(12), 3416–3428.
Truong, T., Bansal, N., Sharma, R., Palmer, M., & Bhandari, B. (2014). Effects of emulsion droplet sizes on the crystallisation of milk fat. *Food Chemistry, 145*, 725–735.
Truong, T., Morgan, G. P., Bansal, N., Palmer, M., & Bhandari, B. (2015). Crystal structures and morphologies of fractionated milk fat in nanoemulsions. *Food Chemistry, 171*, 157–167.
Valivullah, H. M., Bevan, D. R., Peat, A., & Keenan, T. W. (1988). Milk lipid globules - control of their size distribution. *Proc Natl Acad Sci U S A, 85*(23), 8775–8779.
Villamiel, M., & de Jong, P. (2000). Influence of high-intensity ultrasound and heat treatment in continuous flow on fat, proteins, and native enzymes of milk. *Journal of Agricultural and Food Chemistry, 48*(7), 3068–3068.
Wade, T., & Beattie, J. K. (1997). Electroacoustic determination of size and zeta potential of fat globules in milk and cream emulsions. *Colloids and Surf B:Biointerfaces, 10*(2), 73–85.
Walstra, P. (1967). On the crystallization habit in fat globules. *Netherlands Milk and Dairy Journal, 21*(3/4), 166–191.
Walstra, P. (1995). Physical chemistry of milk fat globules. In P. F. Fox (Ed.), *Advanced dairy chemistry vol. 2: Lipids* (pp. 131–178). London: Chapman & Hall.
Walstra, P., Geurts, T. J., Noomen, A., Jellama, A., & Van Boekel, M. A. J. S. (1999). *Dairy technology: Principles of milk properties and processes*. New York: Marcel Dekker, Inc..
Walstra, P., & Oortwijn, H. (1969). Estimating globule-size distribution of oil-in-water emulsions by coulter counter. *J Colloid Interface Sci, 29*(3), 424.
Walstra, P., Wouters, J. T. M., & Geurts, T. J. (2005). *Dairy Science and Technology*. CRC: Press.
Whiteley, A. J., & Muir, D. D. (1996). Heat stability of homogenised concentrated milk. 1. Comparison of microfluidiser with a valve homogeniser. *Milchwissenschaft-Milk Science International, 51*(6), 320–323.
Wiking, L., Bjorck, L., & Nielsen, J. H. (2003). Influence of feed composition on stability of fat globules during pumping of raw milk. *International Dairy Journal, 13*(10), 797–803.
Wiking, L., Nielsen, J. H., Bavius, A. K., Edvardsson, A., & Svennersten-Sjaunja, K. (2006). Impact of milking frequencies on the level of free fatty acids in milk, fat globule size, and fatty acid composition. *Journal of Dairy Science, 89*(3), 1004–1009.
Wiking, L., Stagsted, J., Lennart, B., & Nielsen, J. H. (2004). Milk fat globule size is affected by fat production in dairy cows. *International Dairy Journal, 14*(10), 909–913.
Wright, A. J., & Marangoni, A. G. (2002). The effect of minor components on milkfat crystallization, microstructure, and rheological properties. In A. G. Marangoni & S. S. Narine (Eds.), *Physical properties of lipids* (p. 589). New York: Marcel Dekker.
Wu, H., Hulbert, G. J., & Mount, J. R. (2000). Effects of ultrasound on milk homogenization and fermentation with yogurt starter. *Innovative Food Science & Emerging Technologies, 1*(3), 211–218.

Chapter 15
Dairy Lipids in Infant Formulae to Reduce the Gap with Breast Milk Fat Globules: Nutritional and Health Benefits Associated to Opportunities

Christelle Lopez

1 Introduction

Dietary lipids play a major role in infant nutrition, development and health. The lipids are quantitatively important as providers of energy during the early months of life, qualitatively important as providers of bioactive molecules and of major importance as structured molecules organized in the form of complex supramolecular assemblies. The biological fluid secreted by lactating mothers, i.e. human breast milk, is a complex and unique fluid that evolution adapted to satisfy neonatal needs. Lipids contribute the major portion (45–55%) of the energy contained in human milk (metabolizable energy content of human milk is about 65–70 kcal/100 mL), with a total fat intake of approximately 25 g/day in a fully breast-fed infant during the 6 months of life (Grote et al., 2016; Innis, 2011). The other sources of energy provided by breast milk are lactose and proteins. Breast milk lipids also constitute a source of bioactive molecules (e.g. the polyunsaturated fatty acids of the *n*-6 and *n*-3 series) and structural components (phospholipids, sphingolipids, cholesterol) of functional importance. Breast milk lipids act as vehicles for transport and absorption of lipid-soluble compounds such as vitamins (A, D, E and K). Lipids in human milk are extremely complex and include a wide diversity of molecular species. Although the precise physiological roles of milk lipids are not yet fully understood, they are known to modulate gastrointestinal function, lipoprotein metabolism, membrane composition and function, signaling pathways, that greatly affect infant growth, development and health.

Breast milk is the gold standard for neonatal nutrition, highly recommended as the exclusive component of the infant's diet up to 6 months of life (AAP, 2012;

C. Lopez (✉)
INRAE, UMR STLO, Rennes, France

INRAE, UR BIA, Nantes, France
e-mail: christelle.lopez@inrae.fr

T. Truong et al. (eds.), *Dairy Fat Products and Functionality*,
https://doi.org/10.1007/978-3-030-41661-4_15

WHO, 2011). However, when breast-feeding is not possible, infant formulae (IF) are the best alternative. The best way to improve IF quality and functions in infants is to mimic both the composition and structure of human breast milk. Much interest has been developed in the last decades on the qualitative importance of dietary lipids given to infants, for example, the amount of polyunsaturated fatty acids (PUFA) and the relative intake of PUFA from *n*-3 and *n*-6 series. More recently, the importance of the positional distribution of fatty acids on the 3 *sn*-positions of the triacylglycerol (TAG) molecules has raised since it is involved in the digestion mechanisms and has nutritional impacts for the infants. Very recently, the supramolecular organization of lipids in emulsion droplets has gained much attention due to differences existing between the large fat globules covered by a biological membrane rich in polar lipids, glycoproteins and cholesterol naturally secreted in breast milk, and the tiny processed lipid droplets covered by milk proteins and devoid of biological membrane components produced in IF. Most of IF are manufactured with blends of vegetable oils and DHA-rich oils to provide the essential fatty acids. However, the use of bovine milk lipids in IF is currently increasing.

This chapter provides information about the lipids in breast milk and in IF. The health benefits and opportunities provided by the introduction of milk lipids in IF are developed.

2 Breast Milk Fat Globules: Unique Lipid Assemblies Covered by a Biological Membrane

Human breast milk from healthy mothers is a timely-adapted and balanced source of nutrients and bioactive compounds ensuring proper growth and development of infants. Breast-feeding is, therefore, the gold standard for early nutrition of full-term infants up to 6 months of age and is highly recommended to provide the benefits of human milk components (WHO, 2011). Among all breast milk components, lipids are of high importance.

2.1 Composition of Breast Milk Lipids

Lipids are the most variable components of human milk, including quantitative and qualitative variations. The fat content of breast milk is markedly influenced by the stage of lactation (lower fat content in colostrum during early lactation compared to mature milk), the mother's diet and genetic background (Demmelmair & Koletzko, 2018). During weaning, the fat content increases from the fore-milk (1–3 g/100 mL) to the hind-milk (5–8 g/100 mL) secreted by mothers of term infants (Saarela, Kokkonen, & Koivisto, 2005). Variations in breast milk fat content also occur over 24 h (Khan et al., 2013). Breast milk contains 2–8 g of lipids per 100 mL, with an average around 3.5–4.5 g/100 mL. About 98% of milk lipids correspond to hydrophobic molecules called triacylglycerols (TAG; tri-esters of fatty acids and glycerol). Other quantitatively minor lipids of high nutritional importance include the polar lipids that represent 0.2–1% of milk lipids and the sterols (mainly cholesterol) that account for about 0.2–0.4%. The lipids in human milk have been the focus of reviews (Demmelmair & Koletzko, 2018; Jensen, 1999; Koletzko, 2016).

2.1.1 From the Fatty Acid Composition to the Structure of Triacylglycerols

Fatty Acid Composition Fatty acids from both colostrum and mature human milks exhibit variations. Colostrum fatty acid profile was reported to be mainly dependent on maternal nationality and age rather than mode of delivery and maternal body mass index (Fidler & Koletzko, 2000; Sinanoglou et al., 2017). Mature breast milk fatty acid profile varies greatly depending on geographical location and dietary habits of the lactating mothers and also during lactation. However, major trends can be defined regarding fatty acid composition of breast milk.

Table 15.1 shows the fatty acid composition of human breast milk. Breast milk contains fatty acids with 10–24 carbon atoms. It is rich in long-chain saturated fatty acids (34–47%), such as palmitic acid (C16:0; 15–25%), stearic acid (C18:0; 5–11%), myristic acid (C14:0; 5–9%), which constitute important sources of energy for the infants. Breast milk contains a high amount of monounsaturated fatty acids (31–43%), mainly the oleic acid (C18:1 *n*-9; 24–37%) and substantial amounts of polyunsaturated fatty acids (PUFA) of the *n*-6 series (*n*-6 PUFA; 12–26%) and *n*-3 series (*n*-3 PUFA; 0.8–3.6%) that are of utmost importance for infant health. The PUFA linoleic acid (LA: C18:2 *n*-6; 6–24%) and alpha-linolenic acid (ALA: C18:3 *n*-3; 0.25–3.4%) are recognized as dietary essential fatty acids because of the inability of animal tissues to introduce the necessary double bonds in the carbon chain before carbon 9. The increase in the maternal diet of LA in the last 60 years is reflected in a significant increase in the LA content of breast milk. The level of ALA has remained fairly stable during this same period, resulting in a marked increase in the LA/ALA ratio from approximately 6–8% before 1970 to 14–16% since 1980 (Ailhaud et al., 2006). LA (*n*-6 PUFA) and ALA (*n*-3 PUFA) are precursors of long-chain PUFA. The negative impact of *n*-6 PUFA in excess has been evidenced. Studies showed that high LA intake inhibits *n*-3 fatty acid synthesis in humans and DHA incorporation in tissues. The Eden Study showed that a high LA content in the colostrum could limit the benefit of colostrum DHA on cognition in children (Bernard et al., 2017).

Breast milk contains long-chain PUFA that are essential in infants for brain and retina development, i.e. the docosahexaenoic acid (DHA: C22:6*n*-3; traces—0.8%) and eicosapentaenoic acid (EPA: C20:5*n*-3; traces—0.25%), for biological membranes and nervous system, i.e. the arachidonic acid (ARA: C20:4*n*-6; traces —0.9%). Long-chain PUFA serve as indispensable structural components of

Table 15.1 Fatty acid composition (g per 100 g of total fatty acids) of human milk and infant formulae

Fatty acids	Human milk	Infant formulae		
		(1)VO	(2)VO	(3)MF+VO
C4:0	0—traces	0	0	2.4
C6:0	0—traces	0.2	0	1.3
C8:0	0—traces	1.85	1.2 (0.4–2.1)	1.7
C10:0	1.4 (1–2)	1.55	1.1 (0.1–1.7)	2.2
C12:0	5.4 (3–7)	11.15	5.4 (0.2–13.6)	6.3
C14:0	7.3 (5–9)	5.0	4.6 (0.9–7.0)	7.2
C16:0	26 (15–27)	23.5	26.3 (15.9–31.7)	18.9
C16:1*n*-7	3.5 (3.5–4.2)	0	0.6 (0.2–1.1)	1.1
C18:0	9.5 (5–11)	3.95	5.3 (3.2–7.7)	6.7
C18:1*n*-9	33.6 (24–37)	28.5	37.6 (31.6–42.3)	28.1
C18:2*n*-6 (LA)	14 (6–24)	19.0	14.0 (10.0–19.0)	16.7
C18:3*n*-3 (ALA)	1.7 (0.25–3.4)	1.85	1.6 (1.2–2.0)	1.5
C20:4*n*-6 (ARA)	0.5 (0.25–0.90)	0	0.3 (0.1–0.6)[a]	0
C22:6*n*-3 (DHA)	0.32 (0.10–0.80)	0	0.2[a]	0

Mean value and, between brackets, minimum—maximum values reported in literature
(1) Infant formula containing a mixture of vegetable oils (VO): 48% palm oil, 25% coconut oil, 27% soya oil
(2) Infant formulae containing mixtures of vegetable oils (VO): palm oil, rapeseed oil, soybean oil and coconut oil as major oils. Adapted from Straarup et al. (2006)
(3) Infant formula containing milk fat, corn oil and other non-specified vegetable oils (MF+VO)
[a]Added using fish oils or/and unicellular microorganisms
Abbreviations: LA = linoleic acid, ALA = alpha-linolenic acid, ARA = arachidonic acid, DHA = docosahexanoic acid

cellular membranes and are deposited to a considerable extent in the growing brain and the retina during perinatal development. The supply of preformed long-chain PUFA with human milk lipids has been related to functional outcomes of the recipient infants such as visual acuity and development of cognitive functions during the first year of life. Breast milk in many Western countries has a ARA/DHA ratio of approximately 2/1. Many Asian and Scandinavian breast milks have much lower ARA/DHA ratios because of a higher consumption of DHA-rich fish. The content in PUFA varies during lactation. During the course of the first year of lactation, the contents of both LA (200 mg/dL) and ALA (20 mg/dL) in human milk increase by 8–38% whereas the contents of long-chain PUFA ARA (15–16 mg/dL) and DHA (7–8 mg/dL) decrease by 32–52% (Koletzko, 2016).

TAG Composition and Intramolecular Structure Human milk fat consists of many TAG molecular species (Figs. 15.1 and 15.2). More than 100 different TAG molecules were detected in human milk fat samples (Kallio, Nylund, Bostrom, & Yang, 2017; Kim, Park, & Shim, 2015; Linderborg et al., 2014; Zhang et al., 2019). The composition of TAG in human milk can vary depending on the region, lactation and diet. The same fatty acid can be in different *sn*-position on the glycerol backbone,

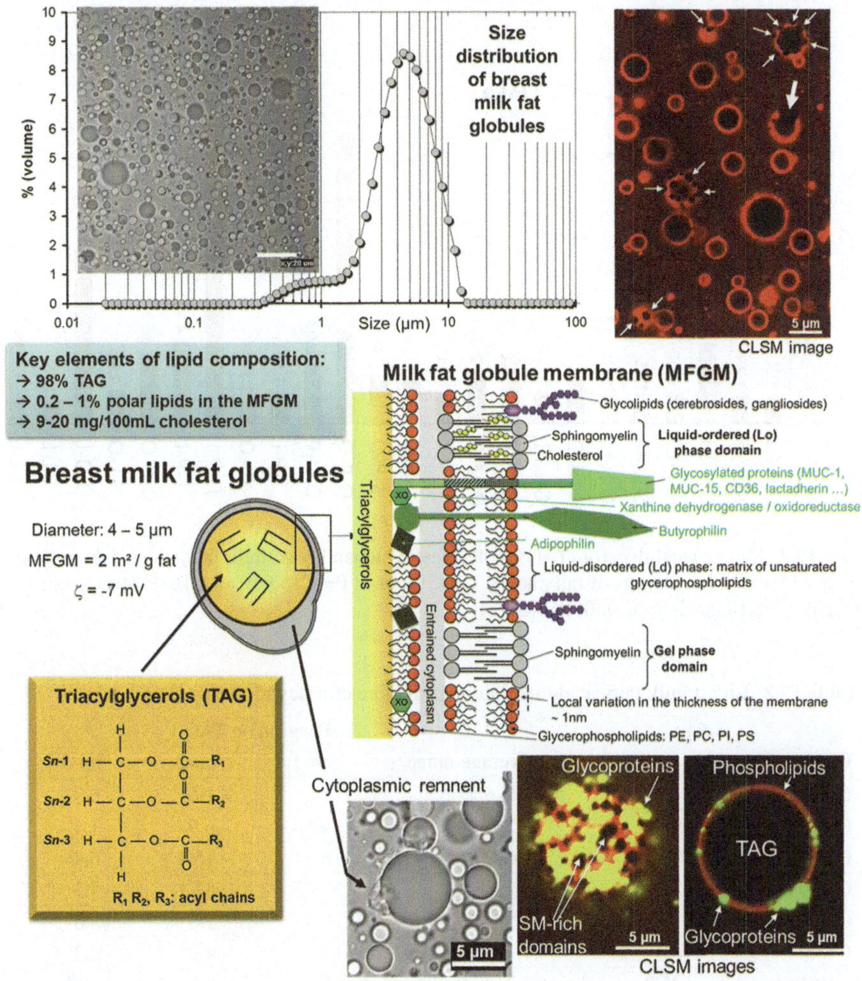

Fig. 15.1 Breast milk fat globules: lipid assemblies with a core of triacylglycerols covered by a biological membrane. Confocal laser scanning microscopy (CLSM) images show in red the fluid matrix of polar lipids, in black the micro-domains rich in sphingomyelin and cholesterol, and in yellow/green the glycoproteins in the milk fat globule membrane (MFGM). A schematic representation of the human MFGM is proposed. Adapted from Lopez and Ménard (2011)

e.g. O-P-O and O-O-P (*sn*1-*sn*2-*sn*3; O = oleic acid, C18:1*n*-9; P = palmitic acid, C16:0). However, the breast milk fatty acids are not randomly distributed on the TAG molecules: a regio-specificity exists for each fatty acid. The saturated fatty acids, such as C16:0, are largely located in the *sn*-2 position (C16:0 >70% at *sn*-2 position in human milk; Table 15.2). The unsaturated fatty acids, such as C18:1*n*-9 and C18:2*n*-6 (LA), are mainly located in the *sn*-1 and *sn*-3 positions of human milk TAG (Innis, 2011; Jensen, 1999). The “Unsaturated-Saturated-Unsaturated”

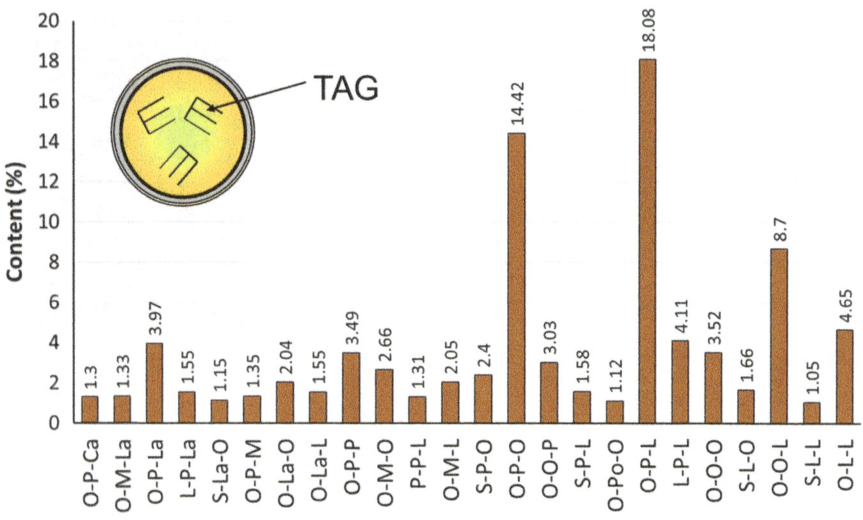

Fig. 15.2 Main triacylglycerol (TAG) molecules in human breast milk fat. Adapted from Zhang et al. (2019). Abbreviations of fatty acids: O= C18:1*n*-9; P= C16:0; Ca= C10:0; M= C14:0; La= C12:0; L= C18:2*n*-6; S= C18:0; Po= C16:1

Table 15.2 Breast milk fatty acids and internal structure of triacylglycerols (TAG)

Fatty acids	Average	Average range	Position on TAG		
			Sn-1	*Sn*-2	*Sn*-3
C10:0	3.0	0.6–6.2	1.1	1.6	5.9
C12:0	9.8	2.2–34.9	4.5	6.9	1.4
C14:0	8.8	1.6–27.6	6.5	15.4	6.4
C16:0	21.9	12.7–29.2	18.7	57.1	5.3
C16:1	3.5	3.5–4.2	3.4	1.6	5.8
C18:0	6.4	1.1–9.7	14.2	4.9	2.2
C18:1*n*-9	25.2	7.2–40.0	44.0	8.1	50.5
C18:2*n*-6	12.7	3.5–30.0	7.2	3.7	12.7
C18:3*n*-3	0.92	0.27–2.71	ND	ND	ND

structure of TAG accounts for about 49% of the total TAG content in breast milk from China (Zhang et al., 2019).

The regio-distribution of the fatty acids on the three *sn*- positions of the glycerol backbone in TAG molecules, and more particularly the "Unsaturated-Saturated-Unsaturated" structure of dietary TAG, are of importance regarding TAG digestion and absorption in the gastrointestinal tract of infants given breast milk (Fig. 15.3) (Innis, 2011; Mu & Hoy, 2004). Indeed, gastric lipase hydrolyses most specifically the fatty acids located on the *sn*-3 position while pancreatic lipase has a specificity for external positions (*sn*-1 and *sn*-3). Lipids are absorbed by infants in the form of *sn*-2 monoacylglycerols (*sn*-2 MAG; Fig. 15.3) and in the form of lipid micelles containing non-esterified fatty acids from the hydrolysis of *sn*-1 and *sn*-3 positions of TAG. The esterification of palmitic acid (C16:0) on the *sn*-2 position of TAG in breast milk favors its intestinal absorption as *sn*-2 monoacylglycerols that is utilized for resynthesis of chylomicron TAG for providing energy to the infants. The non-esterified (free) saturated fatty acids released upon enzymatic hydrolysis, that were located in the *sn*-1 and *sn*-3 positions of TAG molecules, can interact in the intestine with calcium ions and form insoluble soaps that will be lost in the feces (Fig. 15.3). This would cause a considerable loss of energy for the infants. The advantages of C16:0 at the *sn*-2 position in human milk for fat and mineral absorption, behavior of the infants have recently been summarized in systematic and narrative reviews (Bar-Yoseph, Lifshitz, & Cohen, 2013; Miles & Calder, 2017; Petit, Sandoz, & Garcia-Rodenas, 2017). Table 15.3 shows the amount of C16:0 esterified in the *sn*-2 position of various sources of dietary fats (human milk fat, bovine milk fat, vegetable oils) and structured TAG.

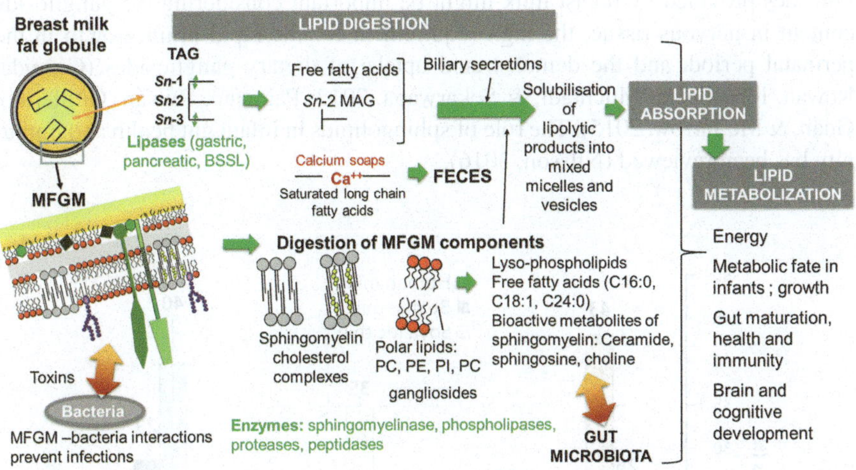

Fig. 15.3 Fate of breast milk fat globules upon digestion in the gastro-intestinal tract of infants and main roles played by lipids

Table 15.3 Stereospecific distribution of palmitic acid, C16:0, in human breast milk, bovine milk and vegetable oils

	%C16:0 at *sn*-2 position
Human breast milk fat	70–88
Bovine milk fat	40–45
Blends of vegetable fats commonly used in infant formulae	10–20[a]
Structured TAG	39–47[a]

[a]Adapted from Sun et al. (2018)

The results are expressed in % at *sn*-2 position of total C16:0. Adapted from Hageman, Danielsen, et al. (2019)

2.1.2 Polar Lipids

Breast milk contains polar lipids, at levels accounting for 0.2–1% of total lipids (10–40 mg/100 mL) (Jensen, 1999), that are mainly located in the biological membrane surrounding the TAG core of milk fat globules (Fig. 15.1). The main polar lipids found in the human MFGM are the glycerophospholipids (phosphatidylcholine, PC; phosphatidylethanolamine, PE; phosphatidylinositol, PI; phosphatidylserine, PS) and the sphingolipids, mainly milk sphingomyelin (SM), but also cerebrosides and gangliosides (Fig. 15.4) (Lopez & Ménard, 2011; Yao et al., 2016).

Sphingolipids are based on a sphingoid backbone. In MFGM, the dominating sphingolipid is sphingomyelin (phosphocholine as head group) and in much smaller amounts glucosylceramides (glucose), lactosylceramides (lactose), and with more complex glycosyl residues gangliosides (monosaccharides, N-acetylgalactoseamine, sialic acid and others) occur. Gangliosides are exclusively located in the MFGM and act as decoy receptors for pathogens, which may prevent infections of infants, and they can modulate the behavior of immune cells (Rueda, 2007). The dietary gangliosides provided by breast milk might be important considering the ganglioside content in nervous tissue, the high requirement for the rapid brain growth in the perinatal period, and the demonstrated uptake of dietary gangliosides (Gurnida, Rowan, Idjradinata, Muchtadi, & Sekarwana, 2012; Palmano, Rowan, Guillermo, Guan, & Mc Jarrow, 2015). The role of sphingolipids in infant gut health and immunity has been reviewed (Nilsson, 2016).

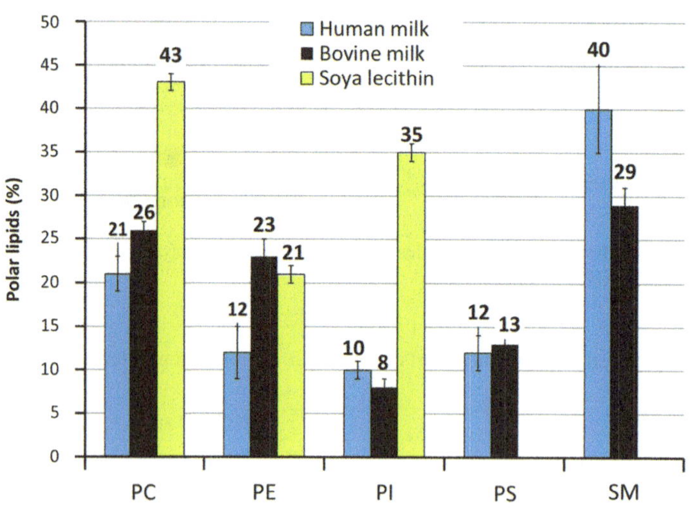

Fig. 15.4 Main polar lipids found in the biological membrane surrounding fat globules in human milk and bovine milk. Comparison with polar lipids in soya lecithin, that is commonly used as emulsifier in infant formulae. Adapted from Lopez et al. (2015)

The MFGM contributes to the dietary intake of breastfed infants, and provides sialic acid (gangliosides), and choline (10% provided by PC and SM) (Leermakers et al., 2015; Wang, 2012).

The individual fatty acid profile of each class of human milk polar lipids is specific (Benoit et al., 2010; Yao et al., 2016). Milk-SM contains high amount of long-chain saturated FA (C16:0, C22:0, C23:0, C24:0); PC is dominated by C16:0 and C18:1 *n*-9; PE by C18:1 *n*-9; PI and PS are rich in C18:1 *n*-9.

Comparing the fatty acid composition of milk total lipids and polar lipids reveals some general tendencies although there are huge differences in the fatty acid composition between the individual polar lipids than in total lipids, while C18:0 tends to be higher in polar lipids (Cilla, Quintaes, Barberá, & Alegría, 2016; Jensen, 1999; Wang et al., 2000). About 85% of the milk long-chain PUFA are contributed by TAG and only 15% are esterified on polar lipids from the MFGM (Harzer, Haug, Dieterich, & Gentner, 1983). ARA contributes up to 12% to phosphatidylethanolamine or phosphatidylinositol, but the percentage in sphingomyelin is about 0.4% and thus similar to total lipids, while DHA contributes up to 5% and 3% in phosphatidylethanolamine and phosphatidylserine, respectively (Cilla et al., 2016). The MFGM is, therefore, an interesting source of PUFA.

The nutritional importance of the MFGM lipids is not primarily based on their long-chain PUFA content, but they provide a variety of specific lipids, such as the high amount of sphingolipids, and some bioactivities in the gastro-intestinal tract that could be important and then further elucidated.

2.1.3 Sterols in Human Milk

Human milk contains 9–20 mg/100 mL cholesterol (Jensen, 1999). Various other minor sterol species exist, such as desmosterol and phytosterols (β-sitosterol, stigmasterol). However, cholesterol largely dominates in sterol composition and accounts for 94–96% of total sterol in human milk (Benoit et al., 2010; Jensen, 1999; Yao et al., 2016). Sterols are mainly located in the biological membrane surrounding fat globules in milk where they play important structural roles, mainly in the formation of ordered lipid domains (Fig. 15.1) (Lopez, Madec, & Jiménez-Flores, 2010). The amount of sterols in milk can, therefore, depend on total fat content and on the size of fat globules since it is proportional to the amount of membrane. Cholesterol is the substrate for the synthesis of bile acids, lipoproteins, vitamin D and hormones. It also acts by stabilizing the structure of cellular membranes and is incorporated into brain lipids mainly during the first months of life (Kinney, Karthigasan, Borenshteyn, Flax, & Kirschner, 1994). During the neonatal period, cholesterol found in breast milk was very early suspected to contribute to cholesterol homeostasis in the adult. It was demonstrated later by the high cholesterol content in breast milk leads to transient high total serum cholesterol concentration in infancy but low in adulthood (Owen et al., 2008).

2.2 Organization of Lipids in Breast Milk

The Milk Fat Globules Breast milk is a natural oil-in-water (O/W) emulsion secreted by the mammary epithelial cells of the lactating mothers. Intracellular lipid droplets formed from the endoplasmic reticulum grow in size as they move toward the apical cell membrane. They are then enveloped by the apical plasma membrane and are voided into the lumen of the alveoli of the milk secreting gland (Heid & Keenan, 2005). The mechanism of milk lipid secretion leads to the formation of droplets called the milk fat globules. Sometimes, cytoplasmic remnants remain attached to fat globules (Fig. 15.1).

The size distribution of breast milk fat globules ranges from 0.3 to 15 μm with a mean diameter of about 4–5 μm (Fig. 15.1) (Lopez & Ménard, 2011). The variation in size of human milk fat globules raises unsolved questions about their specific function and metabolic fate when ingested by neonates. Nanometer-sized lipid-protein particles termed "lactosomes" have also been characterized in breast milk and the question of their potential metabolic, immune regulatory and protective role is still open (Argov-Argaman et al., 2010).

Milk fat globules are unique natural delivery systems of lipids and bioactive molecules in the gastrointestinal tract of newborns and efficient conveyers of energy. They are also involved in the protection of the neonate toward intestinal infections (Hamosh et al., 1999; Peterson, Patton, & Hamosh, 1998). Milk lipids present a very specific and complex structure at several levels of scale. Breast milk fat globules have a core rich in TAG that can be in the liquid or solid state depending on temperature, and are enveloped by a biological membrane called the milk fat globule membrane (MFGM).

Crystallization Properties of Human Milk TAG in the Core of Fat Globules Human milk can be transferred directly from the mother to the infant during breast-feeding or expressed by the mother's breast to be stored refrigerated or in a freezer for later feeding to her infant. Also, human milk banks collect milk from lactating mothers in order to help vulnerable and high-risk infants in neonatal intensive care units of hospitals. In this latter case, milk is frozen and heat-treated to ensure its microbiological safety. The impact of cooling breast milk has raised the attention of scientists.

Due to their high amount of long-chain saturated fatty acids (C16:0, C18:0; Table 5.2), human milk TAG can crystallize at low temperature. For example, storage of breast milk in the fridge induce the crystallization of TAG within fat globules (Lopez, Briard-Bion, Bourgaux, & Pérez, 2013). After storage in the fridge, the TAG crystals have a melting temperature above the physiological temperature of digestion in infants. Since solid TAG are not hydrolyzed by the digestive enzymes, the crystallization of milk fat that occurs upon storage of breast milk in the fridge could negatively impact the digestibility of high melting temperature TAG and the bioavailability of C16:0. It has therefore been recommended to warm breast milk around 50 °C before consumption by the infants.

The Milk Fat Globule Membrane: A Biological Membrane with a Unique Organization The MFGM acts to stabilize physically the milk emulsion, facilitates the digestion of the fat by the infant and may play other functional and physiological roles that are not fully elucidated. The average MFGM thickness is typically 10–50 nm. The MFGM contains specific membrane proteins that are integral or peripheral (e.g. xanthine oxidase, adipophilin, fatty acid-binding protein, the mucins MUC1, MUC 4, MUC 15, lactadherin (PAS 6/7), CD36, butyrophilin), enzymes, polar lipids and cholesterol. The MFGM is a trilayer of polar lipids, where the first monolayer originates from the endoplasmic reticulum and the outer bilayer comes from the apical plasma membrane of the mammary epithelial cells (Heid & Keenan, 2005). The outer bilayer of the MFGM exhibits a heterogeneous lateral packing with a phase separation of ordered lipid domains in the gel or liquid-ordered phase that are rich in sphingomyelin and cholesterol (Fig. 15.1) (Lopez et al., 2010; Lopez & Ménard, 2011; Zou et al., 2012). The micro-domains are surrounded by a fluid matrix composed of the low melting temperature unsaturated polar lipids PC, PE, PI and PS. These ordered lipid micro-domains revealed in the MFGM were called lipid rafts by Christelle Lopez's group (INRAE, France) by analogy with the rafts found in biological membranes (Lopez et al., 2010; Simons & Ikonen, 1997). Figure 15.1 shows images of the MFGM taken by confocal microscopy in situ in milk and present a schematic representation of the MFGM. The membrane-specific proteins are heterogeneously distributed in the MFGM, not located in the ordered lipid micro-domains, and protrude in the aqueous phase surrounding fat globules to form the glycocalyx (Hamosh et al., 1999; Lopez & Ménard, 2011).

Although the MFGM is only some nanometers thick, while the diameters of the milk fat globules are in the micrometer range, the MFGM could be very important for the physiological properties of human milk. There are good indications from in vitro studies, animal studies and observational studies that several MFGM compounds are important for infant development, including the development of cognitive functions.

2.3 *Fate of Human Milk Fat Globules in the Gastrointestinal Tract of Infants: From Digestion to Absorption*

The unique assembly of lipids in breast milk, in the form of fat globules of 5 μm diameter surrounded by a biological membrane (Fig. 15.1), provides bioactive components from the MFGM, a protection toward infections and an efficient digestion and absorption of lipids in the gastro-intestinal tract of infants. Milk fat globules are therefore of major significance for the growth, development and health of breast-fed infants.

The efficient absorption of milk lipids requires the digestion of milk fat globules in the gastro-intestinal tract of newborns (Fig. 15.3). The hydrolysis of the milk fat globules begins in the stomach of infants by the gastric lipase and then continues in the duodenum, where the pancreatic lipase in conjunction with colipase and the bile salt-stimulated lipase (BSSL) endogenous to human milk complete the process initiated in the stomach (Bernback, Blackberg, & Hernell, 1990; Lindquist & Hernell, 2010). The persistence of milk fat globules in the upper intestine allows the discharge of bioactive components in the distal part of the intestine and contributes to their physiological impact in breast-fed infants. In human milk, the external layers of the MFGM contain a group of glycoprotein filaments which are believed to enhance digestion by helping to bind lipase (Jensen, Ferris, & Lammikeefe, 1992). The fate of milk fat globules upon digestion in the gastrointestinal tract of newborns and the role played by milk fat globule and MFGM components (i.e. glycoproteins, free fatty acids) has raised attention of scientists for many years (Hamosh et al., 1999). However, the mechanisms of milk fat globule digestion and the identification of all the bioactives released in the gut during digestion are not yet fully known and will continue to be the focus of research studies in a near future.

3 Lipids in Infant Formulae: Composition and Structure

When mothers cannot or do not want to breast-feed their infant nor express their milk, infant formulae (IF) are the best alternative. According to the regulation, IF are the manufactured foods ingested by healthy infants during the first months of life (children under the age of 12 months), designed to satisfy by themselves the specific nutritional requirements of such infants until the introduction of appropriate complementary feeding. The composition of IF is complex and roughly based on a human mother's milk composition at approximately 1–3 months *post-partum*. IF can be produced as powder or liquid and permit the bottle-feeding of babies.

Compositional and structural differences exist between human breast milk lipids and those found in IF. The composition (fatty acids, molecular species and structure of TAG) of breast milk fat globules result from the maternal diet and characteristics and to biological mechanisms while the composition of lipids in IF results from formulation. The structure and interfacial properties of breast milk fat globules result from the biological mechanisms of their secretion, while the structure of the emulsion and the interfacial properties of lipid droplets in IF result from technological processing, mainly the homogenization step that governs the size of the lipid droplets.

3.1 Manufacture of IF

IF are food products manufactured at the industrial scale by the combination of technological steps of blending, heat treatment, concentration, homogenization, spray-drying and packaging. The objective is to produce stable and reproducible end-products IF (Fig. 15.5). They are processed either as ready-to-feed (liquid) or

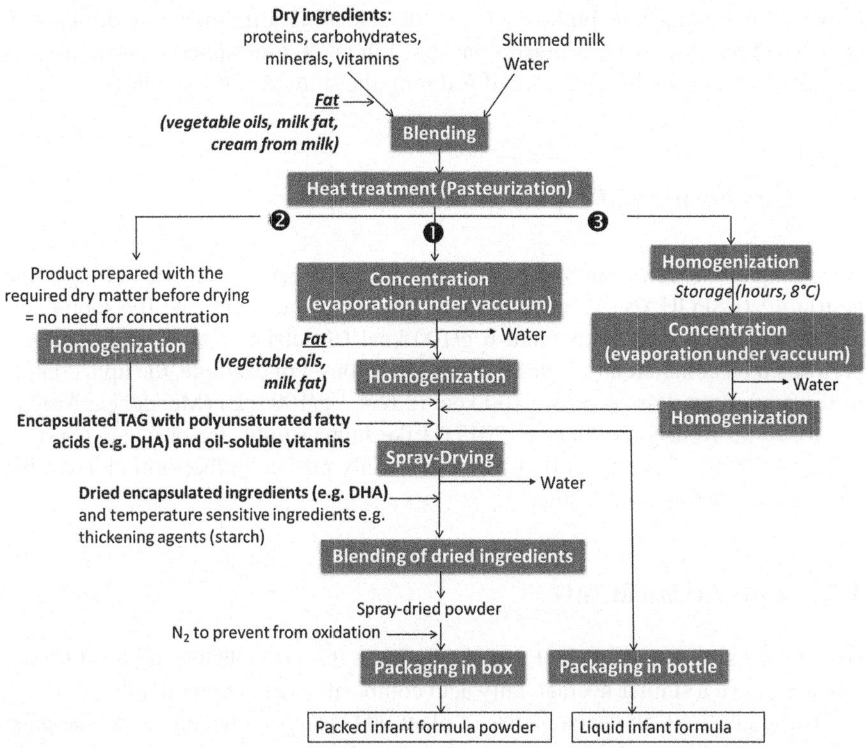

Fig. 15.5 Flow chart of infant formula production

as powdered milk. IF are oil-in-water (O/W) emulsions containing proteins, free amino acids, carbohydrates and other nutrients (minerals, vitamins, oligoelements, nutritional factors) dispersed in water.

In the manufacture of IF, homogenization is an essential technological step as it creates the O/W emulsion and prevents coalescence and formation of free fat during storage. Homogenization also prevents creaming and separation of the oil phase from water in the reconstituted IF prepared to bottle feed the infants. However, during the mechanical process of homogenization, the ingredients are blended together and submitted to pressure (from 15 to 25 MPa), that can alter their structure and functional properties. Hence, some ingredients may not be added before the homogenization step. It is, for example, the case for oils rich in *n*-3 and *n*-6 long-chain PUFA (DHA, ARA) that are highly sensitive to oxidation. In order to ensure their chemical protection towards oxidation, the long-chain PUFA-rich oils can be encapsulated in a matrix composed of modified starch or caseins, together with antioxidant molecules (e.g. vitamin C, tocopherols), and used as spray-dried powder ingredients. These encapsulated long-chain PUFA-rich oils can be added (1) in the concentrated and homogenized product just before the spray-drying step in order to limit the duration of their contact with pro-oxidant molecules (e.g. iron, copper) dispersed in the aqueous phase, or (2) by dry blending after the spray-drying process. The final IF powder is always packaged in an N_2 atmosphere in order to prevent oxidation of the long-chain PUFA during the storage of the product.

3.2 Composition of Lipids in IF

According to the European Society for Pediatric Gastroenterology Hepatology And Nutrition (ESPGHAN), the minimum total lipid content in IF should be 4.4 g/100 kcal and the maximum 6 g/100 kcal (Koletzko et al., 2005). Analysis performed on commercial IF can highlight variations. For example, the lipid content in IF varies from 3.46 to 6.33 g/100 kcal (2.6–4.3 g/100 mL) (Mendonça, Araújo, Borgo, & de Rodrigues Alencar, 2017). Like human milk, fat provides approximately 40–50% of energy in IF. Lipids are mainly present in the form of TAG, but also in low amount of emulsifiers.

3.2.1 Fatty Acids and TAG

As in breast milk, the lipids in IF are mainly TAG. It is possible to combine different oils to achieve a similar average fatty acid composition as in human milk.

Over the last 40–50 years, blends of plant oils have been widely used. The most commonly used plant oils are palm oil, palm olein oil, coconut oil, soybean oil, rapeseed oil, corn oil, safflower oil, and (high oleic) sunflower oil. Each of these plant oils has a specific fatty acid composition (Table 15.4). Blends of these plant oils are the fat source for IF to provide targeted fatty acid profile (Table 15.1), mainly the saturated fatty acids such as C16:0, the monounsaturated C18:1*n*-9, and the polyunsaturated C18:2*n*-6 (LA) and C18:3*n*-3 (ALA). Depending on the different plant oils blended, the FA composition and TAG molecular species present in IF can vary. For example, palm olein and palm oils are sources of long-chain saturated and monounsaturated fatty acids (43% C16:0; 37% C18:1*n*-9) but do not contain short- or medium-chain fatty acids. On another hand, coconut oil is an excellent source of saturated medium-chain fatty acids (Table 15.4).

IF formulated without palm oil or alternative source of C16:0 have palmitate levels as low as 8% of the total fatty acids *versus* 26% in breast milk, whereas lauric acid (C12:0) amounts to 12–13.4% *versus* 5% in breast milk (Hageman, Danielsen, Nieuwenhuizen, Feitsma, & Dalsgaard, 2019). In IF with dairy fat, C16:0 levels reach 16–20% (Table 15.1) (Hageman, Danielsen, et al., 2019). In preterm infants, because of possible intestinal immaturity, facilitation of fat absorption through the inclusion of medium-chain fatty acids in the diet may be useful. Hence, most preterm and some term IF contain medium-chain TAG derived from coconut oil.

In accordance with existing regulations, all IF contain LA and ALA but in variable amounts, depending on the blends of plant oils (Table 15.5). For example, soy oil provides these PUFA. Regulation from Europe stipulates that the levels of LA

Table 15.4 Composition of lipids in vegetable oils commonly used in the formulation of infant formulae and in anhydrous dairy fat

Lipid composition	Vegetable oils Palm oil	Cononut oil	Soya oil	Rapeseed oil	Sunflower oil	Oleic sunflower oil	Dairy fat
Fatty acids wt%							
C4:0	0	0	0	0	0	0	(1.4–3.3)
C6:0	0	0.5	0	0	0	0	2.1 (1.6–2.2)
C8:0	0	8	0	0	0	0	1.7 (1.3–3.0)
C10:0	0	6	0	0	0	0	3.5 (2.0–4.0)
C12:0	0	47	0	0	0	0	(2.0–5.0)
C14:0	1.25	19.5	0	0	0	0	(8–14)
C16:0	43.5	8.75	10.75	4.75	5.95	4.0	(22–35)
C16:1	0	0	0.09	0.18	0.12	0	(1–3)
C18:0	4.5	3	3.7	1.9	4.0	4.0	(9–14)
C18:1*n*-9	40.0	7.5	23.5	60.6	25.0	77.0	(20–30)
C18:2*n*-6 (LA)	9.85	1.72	53.23	18.7	61.0	15.0	(1–3)
C18:3*n*-3 (ALA)	0.15	0.01	6.50	0.08	0.08	0.5	(0.5–2)
LA/ALA	65.7	172	8.2	763	763	30	

and ALA should cover a ratio between 5 and 15 (EFSA, 2014). The analysis of commercial IF revealed that the LA/ALA ratio can vary from approximately 5–13 (Table 15.5). However, high LA intake have been revealed to be deleterious, and a LA/ALA ratio above 10 could induce a default in the conversion toward long-chain *n*-3 PUFA that is of high importance in the development of infant brain. Intake of LA levels could be reduced with a concomitant *n*-3 PUFA increase: 1% LA intake and ratio of 2–3 for LA/ALA should be enough to maintain a proper equilibrium for an optimal bioconversion to long-chain *n*-3 PUFA.

Despite long-chain PUFA such as DHA and ARA are present in breast milk and play important roles in early infant development, some, but not all, IF contain appreciable amounts of added long-chain PUFA (Table 15.5). A supply of these long-chain PUFA in IF is thought to be important since endogenous synthesis is insufficient to maintain tissue levels equivalent to breast-fed infants (Lien, Richard, & Hoffman, 2018). These long-chain PUFA, particularly DHA, are needed for optimal brain development and visual function in infants. In order to consider advances in knowledge to improve infant nutrition, the composition of IF has evolved during the last years, as shown in Table 15.5 with IF collected in supermarkets or pharmacy. An obligation of DHA supplementation at levels of 20–50 mg/100 kcal is mandatory since 2020 by European regulation for IF (EFSA, 2014). A supplementation in ARA is not imposed by Europe. Long-chain PUFA added to IF are typically

Table 15.5 Lipid composition of infant formulae for infants from birth to 6 months of age

IF name	Fats	Per 100 mL[a]	Emulsifiers
Enfamil® premium lipil DHA	Vegetable oils (palm olein, coconut, soy, high oleic sunflower oils); oils from unicellular organisms: ARA from *Moetierella Alpina* oil and DHA from *Crypthecodinium cohnii*	Lipids: 3.7 g C18:2*n*-6: 610 mg C18:3*n*-3: 47 mg LA/ALA: 13 DHA: 11.4 mg ARA: 23 mg	Soy lecithin
Enfamil® Enspire™ (since 2016)	Vegetable oils (palm olein, coconut, soy, high oleic sunflower oils); DHA and ARA blends; MFGM 6g/L	ND	
Biostime SN-2 Bio Plus (2019)	Vegetable oils (sunflower, rapeseed, coconut oils); Cream (from milk); ARA from *Moetierella Alpina* oil and DHA from fish oil "Absence of palm oil"	Lipids: 3.4 g Saturated FA: 1.3 g C18:2*n*-6: 589 mg C18:3*n*-3: 74 mg LA/ALA: 8 DHA: 7.4 mg ARA: 11 mg	Soy lecithin
Picot (2013)	Vegetable oils (palm, rapeseed, soy oils)	Lipids: 3.7 g Saturated FA: 1.3 g C18:2*n*-6: 608 mg C18:3*n*-3: 101 mg LA/ALA: 6	Soy lecithin
Picot (2017)	Cream (from milk); vegetable oils (rapeseed, sunflower, high oleic sunflower oils)	Lipids: 3.1 g Saturated FA: 1.3 g C18:2*n*-6: 439 mg C18:3*n*-3: 73 mg LA/ALA: 6	Soy lecithin
Picot (2019)	Cream (from milk); vegetable oils (rapeseed, high oleic sunflower, sunflower oils); fish oil as a source of DHA; *Moetierella Alpina* oil as a source of ARA "Absence of palm oil"	Lipids: 3.6 g Saturated FA: 1.3 g C18:2*n*-6: 486 mg C18:3*n*-3: 67.5 mg LA/ALA: 7.2 DHA: 11.5 mg ARA: 11.5 mg	Soy lecithin
Modilac Doucéa (2013)	Vegetable oils (palm, high oleic sunflower, soya, coconut, rapeseed oils); fish oil	Lipids: 3.5 g C18:2*n*-6: 557 mg C18:3*n*-3: 42 mg LA/ALA: 13.3 DHA: 7 mg ARA: 12.6 mg	Soya lecithin
Modilac Doucéa (2019)	Vegetable oils (high oleic sunflower, sunflower, rapeseed oils); anhydrous milk fat; fish oil; *Moetierella Alpina* oil "Absence of palm oil"	Lipids: 3.4 g Saturated FA: 0.9 g C18:2*n*-6: 608 mg C18:3*n*-3: 58 mg LA/ALA: 10.5 DHA: 17 mg ARA: 19 mg	Soya lecithin Sunflower lecithin

(continued)

Table 15.5 (continued)

IF name	Fats	Per 100 mL[a]	Emulsifiers
Gallia calisma (2013)	Vegetable oils (palm, coconut, rapeseed, sunflower oils); *Moetierella Alpina* oil; fish oil	Lipids: 3.1 g Saturated FA: 1.5 g C18:2*n*-6: 537 mg C18:3*n*-3: 52 mg LA/ALA: 10.3 DHA: 6.3 mg ARA: 12.6 mg	Soy lecithin, mono and diglycerides
Gallia (2013)	Vegetable oils (palm, rapeseed, coconut, sunflower oils); *Moetierella Alpina* oil; fish oil	Lipids: 3.5 g Saturated FA: 1.5 g C18:2*n*-6: 461 mg C18:3*n*-3: 85 mg LA/ALA: 5.4 DHA: 6.6 mg ARA: 12 mg	Soy lecithin
Gallia calisma (2017–2019)	Vegetable oils (palm, rapeseed, coconut, sunflower oils); *Moetierella Alpina* oil; fish oil	Lipids: 3.5 g Saturated FA: 1.5 g C18:2*n*-6: 458 mg C18:3*n*-3: 85 mg LA/ALA: 5.4 DHA: 11–6.5 mg ARA: 12–6.5 mg	Soy lecithin
Nidal (2013)	Vegetable oils	Lipids: 3.4 g Saturated FA: 1.4 g C18:2*n*-6: 511 mg C18:3*n*-3: 63 mg LA/ALA: 8.1	Soya lecithin
Nidal omega3 (2013)	Vegetable oils; fish oil; ARA from *Moetierella Alpina* oil	Lipids: 3.6 g Saturated FA: 1.5 g C18:2*n*-6: 529 mg C18:3*n*-3: 65 mg LA/ALA: 8.1 DHA: 7.9 mg ARA: 7.9 mg EPA: 2.3 mg	Soya lecithin
Nidal omega3 (2019)	Vegetable oils (high oleic sunflower, coconut, rapeseed, sunflower oils); fish oil "Absence of palm oil"	Lipids: 3.2 g Saturated FA: 0.8 g C18:2*n*-6: 475 mg C18:3*n*-3: 41 mg LA/ALA: 11.6 DHA: 17 mg	Soya lecithin
Blédilait (2019)	Vegetable oils (palm, coconut, rapeseed, sunflower oils); *Moetierella Alpina* oil; fish oil	Lipids: 3.4 g Saturated FA: 1.5 g C18:2*n*-6: 445 mg C18:3*n*-3: 82 mg LA/ALA: 5.4 DHA: 6.4 mg ARA: 6.4 mg	Soy lecithin, mono and diglycerides

(continued)

Table 15.5 (continued)

IF name	Fats	Per 100 mL[a]	Emulsifiers
Guigoz	Vegetable oils	Lipids: 3.6 g Saturated FA: 1.4 g C18:2*n*-6: 541 mg C18:3*n*-3: 66 mg LA/ALA: 8.2	Soy lecithin
Guigoz (2019)	Vegetable oils (high oleic sunflower, coconut, rapeseed sunflower oils); fish oil "Absence of palm oil"	Lipids: 3.6 g Saturated FA: 0.9 g C18:2*n*-6: 555 mg C18:3*n*-3: 49 mg LA/ALA: 11.3 DHA: 17.4 mg ARA: ND	Soy lecithin
Physiolac bio (2013)	Vegetable oils (high oleic sunflower and rapeseed oils)	Lipids: 3.46 g C18:2*n*-6: 378 mg C18:3*n*-3: 47.25 mg LA/ALA: 8	Sunflower lecithin
Physiolac (2013)	Vegetable oils	Lipids: 3.59 g C18:2*n*-6: 690 mg C18:3*n*-3: 69 mg LA/ALA: 10	Soya lecithin
Physiolac precision (2019)	Vegetable oils (INFAT®: palm, palmist, rapeseed, sunflower oils); fish oil; *Moetierella Alpina* oil	Lipids: 3.4 g Saturated FA: 1.3 g C18:2*n*-6: 533 mg C18:3*n*-3: 63 mg LA/ALA: 8.5 DHA: 6.8 mg ARA: 10.8 mg EPA: 1.4 mg	Soya lecithin
Novalac	Vegetable oils (palm, rapeseed, palmist, sunflower oils)	Lipids: 3.5 g Saturated FA: 1.4 g C18:2*n*-6: 400 mg C18:3*n*-3: 62.4 mg LA/ALA: 6.4	Sunflower lecithin
Babylait	Vegetable oils (rapeseed, coconut, high oleic sunflower, maize oils)	Lipids: 3.3 g C18:2*n*-6: 660 mg C18:3*n*-3: 92 mg LA/ALA: 7.2	Soya lecithin
Milumel	Vegetable oils (rapeseed, sunflower, high oleic sunflower oils)	Lipids: 3.1 g Saturated FA: 1.1 g C18:2*n*-6: 526 mg C18:3*n*-3: 71 mg LA/ALA: 7.4	Soy lecithin

[a]LA = C18:2*n*-6; ALA = C18:3*n*-3; LA/ALA

from algae, fungal oils, marine oils, or egg-yolk derived lipids (Table 15.5). The addition of DHA can be performed with the addition of oils from fish (for example tuna oil) or alguae (for example *Cryptheconium Cohnii oil*; *Schizochytrium sp.* oil) (Table 15.5). The addition of arachidonic acid (ARA; C20:4 *n*-6) is generally performed with the addition of, for example, oil from the fungus *Moetierella Alpina*

(Table 15.5). Irrespective of the fat blend used (plant oils, milk fat), DHA and ARA are added as separate ingredients to IF.

TAG from vegetable oils show stereospecific positioning of the fatty acids, with the saturated fatty acids typically positioned on the outer *sn*-1 and *sn*-3 positions, but less than 20% of C16:0 is esterified in the *sn*-2 position of TAG (Mu & Hoy, 2004). The structure of TAG in vegetable oils is therefore different to the structure of TAG in breast milk. Saturated vegetable fats such as palm oil contain high amount of C16:0 which is almost exclusively esterified at the *sn*-1 and *sn*-3 positions of TAG. The main TAG molecular species in palm oil are P-O-P, P-O-O, P-O-LA (P: palmitic acid; O: oleic acid; LA: linoleic acid). On the other hand, highly unsaturated vegetable oils not only contain unsaturated fatty acids at the *sn*-2 position of TAG but also an abundance of TAG with 2 or 3 unsaturated fatty acids. For example, soybean oil has about 70% of C18:2*n-6* (LA) in the *sn*-2 position of TAG. The main TAG molecular species in soybean oil are LA-LA-LA, LA-LA-O, LA-LA-P (LA: linoleic acid; O: oleic acid).

Examination of the fatty acid profile and regio-distribution of fatty acids on TAG in IF revealed considerable differences with breast milk, mainly a lower proportion of C16:0 in the *sn*-2 position in IF containing only vegetable oils (Straarup, Lauritzen, Faerk, Hoy, & Michaelsen, 2006). Significant differences in TAG composition were found between human milk from Chinese mothers and IF, such as much higher medium-chain TAG and saturated TAG in IF, indicating that the IF developed by foreign manufacturers were not suitable for Chinese babies (Tu, Ma, Bai, & Du, 2017).

These TAG compositional and structural differences can impact the digestion of lipids and absorption by the infants. This is why recent research studies in the field of IF focus on the structuration of TAG, by positioning fatty acids of nutritional interest in the *sn*-2 position of the glycerol backbone, either by synthesis or by interesterification. Recently, TAG generated through an enzymatic process from vegetable oils or combination of vegetable oils and other oils (for example fish oil) have become available. The most common product is *sn*-2 palmitate also called beta-palmitate, a structured TAG with C16:0 esterified preferentially in the *sn*-2 position, which is used in IF currently on the market (for example INFAT® OPO, SN-2 palmitate by Advanced Lipids; Table 15.5). Beta-palmitate is the resulting product of the enzymatic interesterification of palm oil and high oleic sunflower oil, where the TAG P-O-P is transformed to O-P-O. These structured TAG make it possible to produce IF with TAG structures higher in *sn*-2 palmitate, often above 40% (ranging from 39 to 47%) of the total palmitic acid content (17–25%) (Bar-Yoseph et al., 2013; Miles & Calder, 2017; Sun, Wei, Su, Zou, & Wang, 2018).

3.2.2 Sterols in IF

Cholesterol is not present in the IF formulated with the blending of plant oils. The analysis of sterol contents in IF showed that those based on vegetable fats contained, on average, 0.185 mg/L of cholesterol (Claumarchirant, Matencio, Sanchez-Siles, Alegría, & Lagarda, 2015), while human breast milk contains 90–150 mg/L of cholesterol (Koletzko, 2016). IF containing a blend of vegetable oils and bovine milk fat contain higher levels of sterols, on average 0.927 mg/L (Claumarchirant et al., 2015), which is still low compared to human breast milk. However, dairy lipids (milk fat globules and MFGM) are a promising source of cholesterol in IF.

3.2.3 Lipid emulsifiers in IF

Non-dairy lipid emulsifiers and stabilizers are added to IF to ensure emulsion stability (Table 15.5). Lecithin from vegetable origin such as soy lecithin or sunflower lecithin, from fish and krill or egg yolk phospholipids, are the common emulsifiers used in IF generally recognised as safe. Monoglycerides and diglycerides (citric acid esters of mono and diglycerides; CITREM) can also be used. The emulsifiers will coat the surface of oil droplets, together with proteins (from milk: caseins, whey proteins; from rice or other vegetable origin), and effectively decrease the surface tension conferring at least short-term stability to the O/W emulsion. Moreover, the lipid emulsifiers make the emulsion more heat-resistant compared to emulsions containing lipid droplets only stabilized by proteins. Monoglycerides may also be added to IF as anti-foaming agents since excessive foaming of reconstituted milks is considered undesirable.

The nature of the emulsifier can impact the susceptibility of the lipid droplets to coalescence and break up within the gastro-intestinal tract of infants, thereby altering the total surface area of lipid exposed to the digestive lipases. The characteristics of the interfacial layer between the TAG core of the lipid droplets and the surrounding aqueous phase, including the kind of emulsifier, can impact the adsorption and activity of the digestive lipases at the oil/water interface. In vitro studies reported that the fatty acids contained in CITREM compounds are released to a large extent by gastric lipase, pancreatic lipase, pancreatic-lipase-related protein 2 and carboxyl ester hydrolase (Amara et al., 2014).

In IF formulated with plant oils, the specific components of the MFGM are generally absent. This relates to MFGM proteins, but also MFGM complex lipids (phospholipids, sphingolipids, ceramides, gangliosides) and cholesterol. However, residual milk polar lipids from the MFGM can be found in skimmed milk powder used in IF.

The metabolic impact of MFGM polar lipids surrounding breast milk TAG vs. non-dairy emulsifiers such as soy or sunflower lecithins in IF is not known.

3.3 Microstructure of Fat in IF: Processed Lipid Droplets Coated by Proteins and Non-Dairy Lipid Emulsifiers

Figure 15.6 shows the organization of fat in IF powders and after rehydration. The lipid droplets are tailored during the manufacture process, mainly during the technological process of homogenization. The objectives of homogenization are (1) to

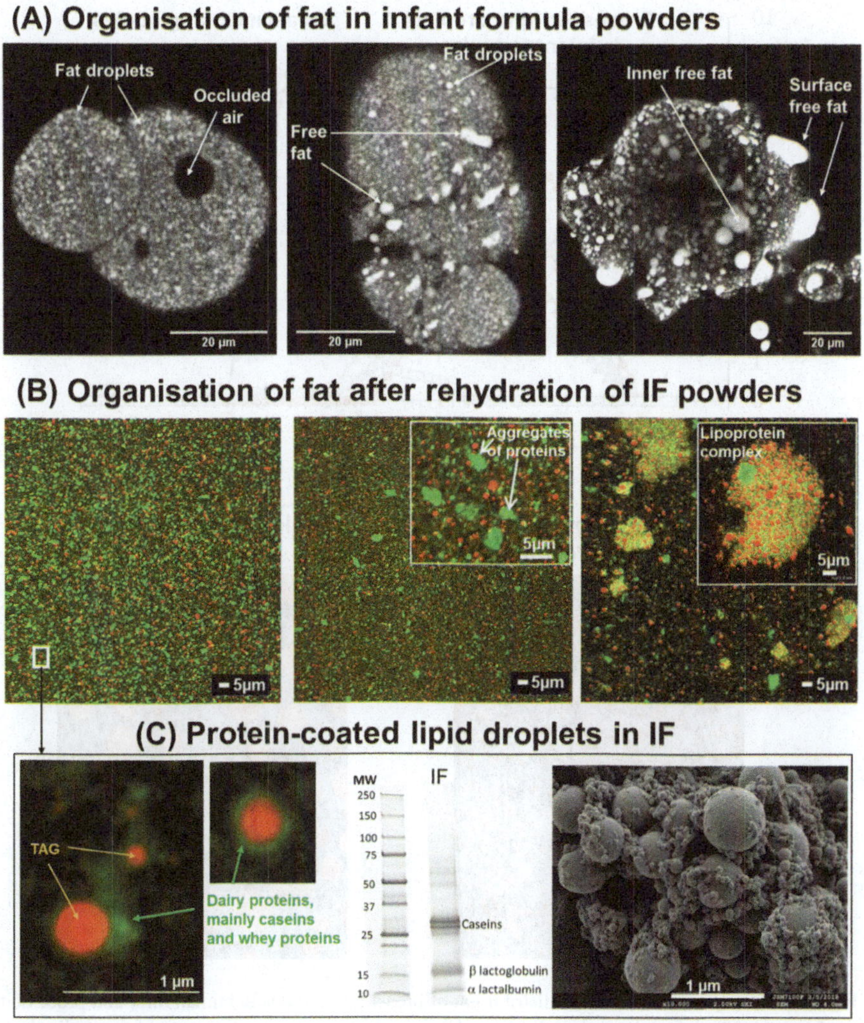

Fig. 15.6 Microstructural observations of lipids (**A**) in infant formula powder particles (fat is in white), (**B**) after rehydration (fat in red, proteins in green), (**C**) focus on the surface composition and structure showing the adsorption of dairy proteins. Adapted from Lopez et al. (2015)

create an oil-in-water emulsion by mixing the blend of fats and the other components (proteins, minerals, emulsifiers), and (2) to form small lipid droplets that will be physically stable upon the drying process and the storage of IF powders. *In-situ* structural observations of IF powders performed by confocal microscopy revealed that fat can be dispersed in small spherical droplets homogeneously distributed in the powder particle (Fig. 15.6A, left). Fat can also be present in non-spherical large fat droplets and in the non-emulsified organization also called free fat (Fig. 15.6A, right). The formation of high free fat content in IF mainly occurs in the case of IF

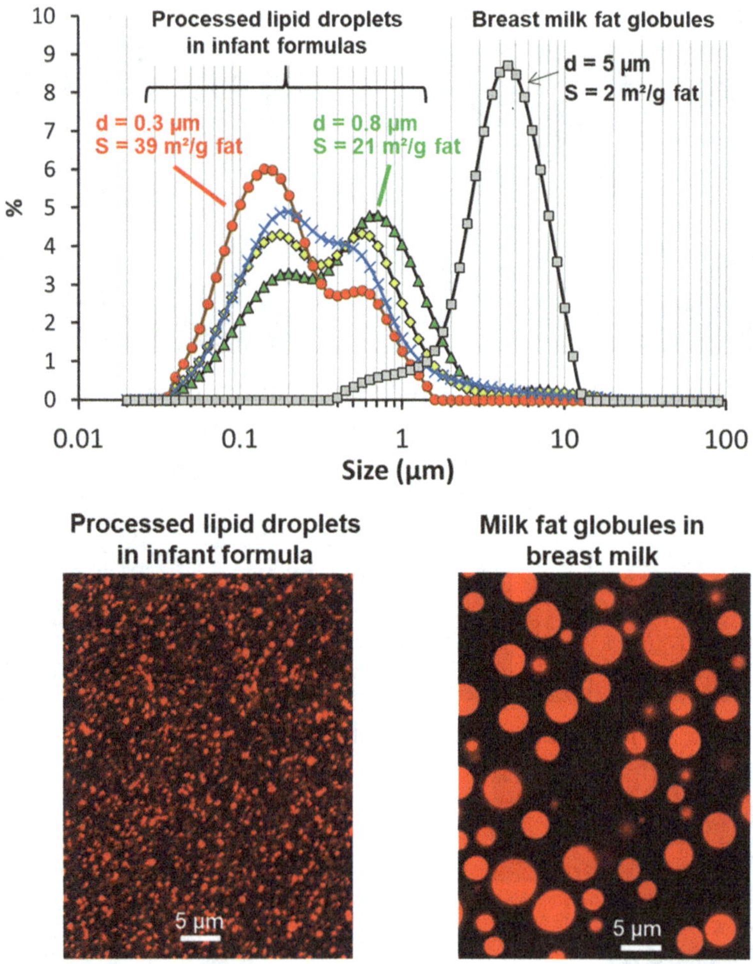

Fig. 15.7 Comparison of the size distribution of breast fat globules in human milk and processed lipid droplets in infant formulae. Bottom: CLSM images showing the core of lipid droplets. Adapted from Lopez et al. (2015); Lopez, Cauty and Guyomarc'h (2019)

containing hydrolyzed proteins that have a low ability to stabilize the lipid droplets. The physical destabilization of the lipid droplets in the powder induces the formation of non-emulsified or free fat inside the grain powder (inner free fat) or at its periphery (surface free fat) (Fig. 15.6A, right). Free fat favors the oxidation of lipids, alters the functional properties and the quality of IF powders (e.g. flow and wetting properties, poor rehydration).

After rehydration of the IF powder, most of the standard IF exhibit tiny lipid droplets with a size distribution ranging from 0.05 to 2 µm and exhibit a large TAG/water interface, from about 20–40 m^2/g fat (Lopez, Cauty, & Guyomarc'h, 2015). The size of processed lipid droplets in IF is smaller than the size of breast fat globules in human milk (Fig. 15.7). The small size of IF lipid droplets results from the pressure applied during the homogenization step. Differences in the size distribution of the lipid droplets exist between different commercial IF. They result from variations in the technological parameters used by different manufacturers or for different markets. The lipid droplets are mainly coated by dairy proteins, the casein micelles and whey proteins (Fig. 15.6). These structural parameters and interfacial composition are different from the size of fat globules in breast milk and have consequences on the mechanisms of lipid digestion in IF (Armand et al., 1996, 1999).

In some cases, IF can contain aggregates of proteins or complexes formed between lipid droplets and proteins, that may result from the thermo-induced denaturation of whey proteins occurring during the heat treatments performed for the microbial safety of IF (Fig. 15.6B; Lopez et al., 2015). Such lipoprotein complexes induced by the industrial process raise questions about the accessibility of TAG and proteins by the digestive enzymes in the gastro-intestinal tract of newborns and then about the nutritional and health impacts.

4 Dairy Lipids in Infant Formulae: Health Benefits for Infants and Potentialities

The underlying aim in the design and development of IF is to achieve a similar digestive outcome and nutritional yield to that of human milk. Scientific evidence continues to accumulate that the quality and structure of dietary lipids provided to infants has a marketed impact on health outcomes. As such, bovine milk itself is not an appropriate substitute of breast milk for infant nutrition. However, some of its components (e.g. milk fat globules, TAG, medium-chain fatty acids, C16:0, LA, ALA, MFGM, cholesterol) can be reformulated together with, for example, plant oils and structured TAG to achieve an equivalent balance to human breast milk. Bovine milk was the source of fat in IF until the 1970s and is still used in some parts of the world (Delplanque, Gibson, Koletzko, Lapillonne, & Strandvik, 2015). Since 1970s, vegetable oils have been extensively used in IF for 2 main reasons: (1) to provide high levels of mono unsaturated fatty acids and PUFA, and (2) the lower cost of vegetable oils compared to dairy fat. Palm oil has been widely used in IF to provide C16:0 to infants. However, palm oil free IF is becoming increasingly popular because of unsustainable production methods and environmental reasons (Table 15.5). On the opposite, the use of dairy lipids in IF is currently increasing (Table 15.5). Recent research studies showed that the use of dairy lipids in combination with plant oils enables a lipid profile closer to breast milk in terms of fatty acid composition, TAG structure, polar lipids and cholesterol contents with health benefits in infants.

This part of the chapter will develop the advantages provided by dairy lipids composition, TAG and MFGM, and the interest in developing lipid droplets biomimetic of breast milk fat globules. The potentialities of the addition of dairy lipids in IF are described.

4.1 Dairy FA and TAG: Specificity and Health Benefits

Bovine milk fat contains about 98% of TAG molecules, which is similar to breast milk fat. The bovine fatty acid composition comprises short-chain (C4:0–C6:0), medium-chain (C8:0–C14:0) and long-chain (>C14:0) fatty acids (Table 15.4). Bovine TAGs contain higher amount of saturated fatty acids than breast milk TAG (70% vs. 40%, respectively), including short-chain and medium-chain fatty acids. Bovine milk fat contains a lower amount of monounsaturated fatty acids (24% vs. 36%) and PUFA (2% vs. 18%) than breast milk fat.

Dairy C16:0 Content and sn-2 Position on TAG Compared to plant oils that are widely formulated in IF, dairy fat provides (1) short- and medium-chain fatty acids, such as coconut oil and (2) high C16:0 content (about 25–30% in dairy fat), such as palm oil. Interestingly, dairy TAG have a higher percentage of C16:0 esterification at the *sn*-2 position compared with palm oil. Dairy TAG contain about 25–30% C16:0 and 45% of C16:0 is esterified at the *sn*-2 position of TAG, with 47% distributed at the *sn*-1 position and 8% at the *sn*-3 position (Table 15.3). The formulation of IF with bovine milk fat, therefore, increases the proportion of C16:0 in the *sn*-2 position of TAG compared to palm oil.

The fatty acid composition (i.e. medium-chain fatty acids, C16:0) and the TAG stereospecific structure (i.e. C16:0 in *sn*-2 position) of bovine milk fat might be beneficial for digestion and absorption of fatty acids and further outcomes. Hageman et al. (2019) performed in vitro studies to investigate the release of fatty acids from IF containing different fat blends, e.g. 100% vegetable fat *versus* 67% bovine milk fat and 33% vegetable fat, and compared this to human milk. These authors found that the total FA release was not affected by the composition of fat. However, different time-dependent release of individual fatty acids were observed, which might result in differences in absorption and other health effects in vivo.

The high proportion of C16:0 in *sn*-2 position of TAG provided by dairy TAG in IF decreases the formation of calcium soaps in the infant intestine and then contributes in a higher absorption of calcium leading to an improved bone strength.

It is important to note that, despite the supportive research studies showing the beneficial impact of *sn*-2 palmitate (C16:0 in the *sn*-2 position of TAG), guidelines on the composition of IF do not support its inclusion. This demonstrates that much more research studies in infants are needed in this area.

Dairy LA and ALA in Infant Formulae: DHA Levels in Infants In infants, it has been shown that DHA is essential for brain development and functions. Some studies showed that infants receiving IF containing dairy lipids had an intermediate status in *n*-3 long-chain PUFA between breastfed infants and infants fed IF containing a mixture of vegetable oils (Courage et al., 1998). An increased number of research studies have focused on the impact of the partial replacement of plant oils by dairy fat in IF, on the DHA levels in infants. Bovine milk contains lower amount of LA and ALA than breast milk but the partial incorporation of dairy fat in IF could improve in infants the conversion pathway from ALA to *n*-3 long-chain PUFA.

A study performed in healthy Italian newborns showed that an IF containing a mix of dairy lipids and plant oils increased the endogenous conversion of *n*-3 long-chain FA from precursor ALA, leading to higher total *n*-3 fatty acids, DPA and DHA status in red blood cells than a plant oil-based IF. Interestingly, the DHA values were closer to those obtained with breast-feeding (Gianni et al., 2018). From inclusion at age below 3 weeks to the 4 subsequent months, the newborns exhibited a normal growth without any significant impact of dairy lipids on gastrointestinal symptoms or infant behaviour. Modifying lipid quality in IF by adding dairy lipids should, therefore, be considered as an interesting method to improve *n*-3 FA status in infants.

Benefits of the addition of dairy lipids in IF were observed in animal models. In young mice, it has been reported that dairy fat blend improves brain DHA level and neuroplasticity (Dinel et al., 2016). In young rats, Prof Delplanque's group and collaborators focused on the role played by dairy fat on the DHA level of brain (Delplanque, Du, Martin, & Guesnet, 2018), which is an important goal in neonates. This group studied the impact on blood and brain DHA levels of dairy fat included in different blends of vegetable oils complying with the lipids recommendations for IF (LA: 16%; ALA: 1.6–2.5%; LA/ALA ratio: 5–10). They also evaluated the impact of pure dairy fat presenting very low levels of LA and ALA (1.9% and 0.8%, respectively), but with a proper LA/ALA ratio of 2.6 which they compared to previous ones and to rapeseed oil rich in ALA (8%). The three main findings of these studies are as follows:

- dairy-fat-based diet (50% dairy, 50% vegetable oils) with 1.5% ALA content is more efficient than a pure vegetable oil blend with as much ALA (1.5%) and the same LA/ALA ratio of 10 to increase the brain DHA in the growing rat. Specific and complex component of dairy fat could be an explanation, such as the short and medium-chain fatty acids that are highly oxidizable after absorption and may thereby spare ALA from oxidation and favor ALA partitioning towards the desaturation and elongation pathways, increasing the long-chain *n*-3 PUFA levels such as those of DHA (Drouin et al., 2018).
- dairy-fat-based diet (50% dairy, 50% vegetable oils) enriched with 2.3% ALA is even more efficient. This could be attributed to both the increased level of dietary ALA and the concomitant decrease in the LA/ALA ratio from 10 to 5 which has been recognized as an important factor driving the bioconversion of ALA into DHA, because of the competition between the parent *n*-3 and *n*-6 fatty acids for desaturation and elongation pathways.
- dairy-fat-based diet containing pure dairy fat (100%) with only 0.8% of ALA and 1.9% of LA is as efficient as an 8% ALA rapeseed diet (22% LA) to increase the brain DHA in the growing rat, both presenting a similar very low LA/ALA ratio (less than 3) and present results comparable to the 2.3% ALA dairy/vegetable blend.

The role of the delta6-desaturase enzyme could be involved in this process and is crucial to explain these last results: ALA is the precursor of DHA but also its competitor for the last delta6-desaturase step and is regulated by substrate levels. An excess of ALA could represent the first substrate, producing increasing quantities of some n-3 long-chain fatty acids (EPA, docosa-pentaenoic acid: DPAn-3) and secondarily could limit the implication of delta6-desaturase in the second control point for DHA conversion. Explanation for rapeseed is exactly reverse and could represent an excess of precursor, which could limit the bioconversion to DHA, even reducing its level when intake is above the optimal intake (around 2.5–3% of total FA). The proof of this intra-cascade n-3 competition for delta6-desaturase has been validated. Finally, pure dairy fat, despite very low levels of PUFA (1.5–3% LA and 0.5–0.8% ALA) but associated with a very favourable LA/ALA ratio similar to rapeseed oil (maximum 3/1), was able to provide the proper conditions for a bioconversion of ALA to n-3 long-chain fatty acids and DHA necessary for the brain of young animals. Together, these observations clearly demonstrated that brain DHA levels can be substantially improved by dairy fat based-diets. The use of fats that are low in PUFA such as dairy may confer some metabolic advantages in that they allow better endogenous conversion of ALA to DHA.

4.2 MFGM: Source of Bioactive Molecules

Human milk contains complex lipids (glycerophospholipids, sphingolipids, cholesterol) and other MFGM components such as membrane-specific highly glycosylated proteins which are not provided with conventional IF based on plant oils as the lipid source.

Growing interest of supplementing IF with bovine MFGM comes from the scientific proofs of bioactivities and benefits obtained thanks to randomized infant trials, with reported impact for example on brain development and cognitive functions, immunity and gut physiology, and reduction of infections in the neonates. Several reviews have been dedicated to the supplementation of IF with bovine MFGM (Hernell, Timby, Domellof, & Lonnerdal, 2016; Timby, Domellöf, Lönnerdal, & Hernell, 2017). Although milk polar lipids and more widely MFGM components have key functions in infants, there is currently no requirement for the introduction of MFGM and individual components in IF. MFGM supplementation to IF has been proven to be safe and well-tolerated (Billeaud et al., 2014). No effects were reported on growth or long-term health outcomes. These studies stimulated further research on the preparation of MFGM-enriched ingredients. This supplementation of bovine MFGM in IF is possible thanks to the production of food-grade ingredients at the industrial scale. These two aspects are developed in the next paragraphs of the chapter.

Benefits Provided by the Supplementation of IF with Bovine MFGM During the past 2 decades, in vitro and animal studies as well as clinical trials in infants provided scientific proofs that dietary exposure to IF containing MFGM provides beneficial effects. Although controversies exist, several of MFGM components have been related to nutritional and health-enhancing functions and the bovine MFGM has been considered as a potential nutraceutical (Spitsberg, 2005). The benefits of MFGM for infants have been reviewed (Demmelmair, Prell, Timby, & Lönnerdal, 2017). In the following section, an overview of the most recent research studies showing health benefits provided by the MFGM is proposed.

MFGM Improves Brain and Cognitive Development The benefits provided by the supplementation in MFGM or in bioactive components from the MFGM (sialic acid, gangliosides, sphingomyelin, cholesterol, proteins) have been reported. In pre-school children consuming MFGM concentrate in chocolate formula milk for 4 months, beneficial effects on behavioral and emotional regulation were reported (Veereman-Wauters et al., 2012). In a randomized trial including infants below 2 months of age, the consumption of a MFGM-supplemented, low-energy, low-protein experimental IF showed a positive association with neurocognitive development. At 12 months of age, the cognitive score was significantly higher in the MFGM-supplemented experimental IF group than in the standard IF group but was not significantly different from that of breastfed infant group (Timby, Domellof, Hernell, Loennerdal, & Domellof, 2014). Rats receiving a diet supplemented with cholesterol displayed better performances in memory and learning than rats receiving a normal diet (Ya et al., 2013). In piglets, dietary supplementation with glycomacropeptide as a provider of sialic acid was reported to provide faster learning of difficult tasks and to improve memory (Wang et al., 2007). With regard to neurodevelopment, the ganglioside content of the MFGM might be highly relevant, considering the high ganglioside content in nervous tissue, the high requirement in the perinatal period due to the rapid brain growth, and the demonstrated uptake of dietary gangliosides (Palmano et al., 2015). Human infants fed IF supplemented with gangliosides or milk-SM displayed improved cognitive, neurobehavioral and motor development (Gurnida et al., 2012; Tanaka et al., 2013). However, whether MFGM supplementation benefits result from the action of a single or a combination of bioactive components is not yet known.

MFGM Prevents Infection In 6–12-month-old infants in Peru, daily supplementation with MFGM enriched protein significantly decreased the duration of diarrhea episodes and the incidence of bloody diarrhea by almost 50% considering confounding factors (Zavaleta et al., 2011). In agreement with these findings, a study performed with young European pre-school infants aged 2.5–6 years, consuming daily MFGM-enriched complimentary food reported a significant decreased number of days with fever without any impact on diarrhea, constipation and cough during the four-month intervention period, compared to a corresponding supplement without MFGM components (Veereman-Wauters et al., 2012). In Swedish children recruited before 2 months of age and fed with MFGM-supplemented IF until 6 months of age, the preventive effect of MFGM on infections was also reported (Timby et al., 2015). Among infants fed the MFGM-supplemented IF, fewer episodes of acute otitis media and lower antipyretic use during the intervention were reported compared with infants fed a control IF. Glycoproteins from the MFGM (e.g. butyrophilin, mucins, lactadherin) and glycolipids (gangliosides) may participate in the defense against infections by preventing pathogen adhesion to epithelium (Fuller, Kuhlenschmidt, Kuhlenschmidt, Jimenez-Flores, & Donovan, 2013; Sprong, Hulstein, Lambers, & van der Meer, 2012) (Fig. 15.3). For some of the MFGM proteins, gastric stability has been demonstrated and they have been suggested to contribute to the protection against bacteria and viruses in the neonatal gastrointestinal tract and to affect the immune system (Hamosh et al., 1999; Peterson et al., 1998). As the proteome of bovine MFGM has been found similar to the human MFGM proteome, these findings suggest the possibility of beneficial effects of supplementing IF with bovine MFGM.

Milk sphingolipids (gangliosides) could play a protective role against bacterial toxins and bacterial development via a competition for bacteria binding sites as many bacteria adhere to epithelial cells via glycosphingolipids. Sphingolipids from the MFGM transiting in the gut may decrease pathogen adherence to the intestinal mucosa facilitating pathogen elimination (Sprong et al., 2012) and may influence the gut microbiota composition (Fig. 15.3). Sprong et al. (2012) also evidenced in vitro the bactericidal activity of digestion products of sphingolipids and more specifically of lyso-SM.

Altogether, research studies showed that the MFGM contributes to the protection of the infants from pathogens. In contrast, in a multicenter trial with French and Italian full-term neonates below 14 days of age, the safety evaluation of standard IF and IF enriched with a protein-rich MFGM fraction or a lipid-rich MFGM fraction did not reveal differences between groups in terms of diarrhea and intestinal discomfort nor in ear, respiratory and gastrointestinal infections (Billeaud et al., 2014).

MFGM and Sphingolipids Are Involved in Gut Health and Immunity Appropriate intestinal barrier maturation is essential for absorbing nutrients and preventing pathogens and toxins from entering the body. Beneficial effects of the MFGM and MFGM components, mainly the sphingolipids, on gut epithelial barrier have been reported.

In challenging conditions against pathogenic bacteria such as *Clostridium difficile* or *Listeria monocytogenes* involving rats, authors reported that MFGM provides protection, probably by stimulating mucin secretion and preventing adherence of pathogens to the intestinal mucosa (Bhinder et al., 2017; Sprong et al., 2012). In weaned mice, providing a MFGM-rich fat feed decreased the inflammatory response to a systemic lipopolysaccharide (LPS) challenge and was associated with decreased gut permeability (Snow, Ward, Olsen, Jimenez-Flores, & Hintze, 2011). These effects may be partly related to gangliosides that inhibited degradation of tight junctions occurring during LPS-induced acute inflammation (Park, Thomson, & Clandinin, 2010). Moreover, authors reported that in mice fed a high-fat diet for 4 weeks, milk-SM had significantly lowered serum LPS compared to control, which may have been due to altered distal gut microbiota (lower fecal Gram-negative bacteria; higher *Bifidobacterium*) (Norris, Jiang, Ryan, Porter, & Blesso, 2016). In rat pups, 10 days feeding with MFGM-supplemented formula normalized delayed intestinal growth induced by standard IF feeding compared to suckling rats (Bhinder et al., 2017). Using neonatal piglets, authors reported that MFGM in IF accelerated the maturation of the intestinal immune system that was closer to the one observed in mother-fed piglets, induced mucosal growth without any impact on epithelial permeability (Le Huerou-Luron et al., 2018; Lemaire et al., 2018).

For further information about the role of sphingolipids in infant gut health and immunity, a very interesting review of scientific knowledge is recommended (Nilsson, 2016). Sphingolipids are important polar lipids in the MFGM but are not found in standard IF prepared with vegetable oils. Digestion of milk-SM in the gastrointestinal tract of infants generates the bioactive metabolites ceramide, sphingosine and sphingosine-1-phosphate (Fig. 15.3). The distal small intestine and colon are exposed to milk-SM and its bioactive metabolites (Nilsson, 2016; Ohlsson et al., 2010). Humans digest and absorb most of the SM in normal diets and the level of sphingolipid metabolites to which the colon is exposed can be influenced by realistic amounts of dietary SM such as milk-SM (Ohlsson et al., 2010). These compounds are both metabolic intermediates during synthesis and degradation of sphingolipids, and bioactive compounds with numerous signaling functions. After digestion of milk-SM by enzymes, its fatty acids (e.g. C24:0) are absorbed and transferred to tissues. Milk-SM is also a source of choline, known to be important in neonates for phospholipid synthesis during organ growth and for acetylcholine formation. Hence, milk-SM digestion must also be viewed in relation to choline production (Claumarchirant et al., 2016). In adult rats fed 3H-choline labeled SM, 30% of the radioactivity was in liver PC after 4 h, indicating that choline released during SM digestion is extensively reutilized for hepatic PC synthesis (Nilsson & Duan, 2006).

Milk-SM, which is a major polar lipid in the MFGM, is believed to play an important role in neonatal gut maturation during the suckling period. Milk-SM was reported to accelerate enzymatic and morphological maturation of the intestine in artificially reared rats (Motouri et al., 2003). However, it is unknown whether the effects were caused by milk-SM itself or to its metabolites. Rat studies suggest systemic effects of dietary milk-SM, including an increase of central nervous system myelination in a deficit model (Oshida et al., 2003).

Gangliosides, which are exclusively located in the MFGM, are able to modulate the behavior of immune cells. Intact gangliosides containing sialic acid have beneficial effects in the gut, e.g. reduce pro-inflammatory signaling, influence on gut bacterial flora, protective functions by their interactions with pathogens and bacterial toxins, effects on mucosal epithelial and immune functions (Miklavcic, Schnabl, Mazurak, Thomson, & Clandinin, 2012; Rueda, 2007). Hence, sphingolipids provided by the MFGM are important for mucosal functions, gut integrity and immune maturation in the neonate.

In infant pup rats the addition of bovine MFGM, whose lipids differ from human milk MFGM, to the formula made intestinal development and microbiome of formula fed rats more similar to that of breastfed rats compared to non-supplemented formulae (Bhinder et al., 2017).

Milk Polar Lipids and Sphingomyelin Affect TAG lipolysis upon Digestion The impact of milk polar lipids, including their specificity to form sphingomyelin-cholesterol complexes (Lopez & Ménard, 2011), on the mechanisms involved in lipolysis occurring in the gastro-intestinal tract of infants is not well-known and discrepancies exist between authors. Nilsson and Duan established that milk-SM and its metabolites may influence TAG hydrolysis, cholesterol absorption, lipoprotein formation, and mucosal growth in the gut (Nilsson & Duan, 2006). The role played by the milk-SM in on the activity of the gastric lipase at the surface of lipid droplets has been demonstrated (Favé et al., 2007). The specificity of milk polar lipids rich in milk-SM in modulating the mechanisms of milk lipid digestion is of growing interest, mainly in comparison with soy lecithin that is widely used in IF (Table 15.5). Mathiassen et al. showed that exchanging soy lecithin with dairy polar lipids increased gastric lipase activity by 2.5-fold (Mathiassen et al., 2015). In mice, Lecomte et al showed that milk polar lipids introduced in an emulsion together with dairy proteins as emulsifiers resulted in a quicker elevation and clearance of plasma TAG compared to soy phospholipids (Lecomte et al., 2015). As regards to the different phases in which milk-SM can occur (gel phase below Tm, fluid phase above Tm, liquid-ordered phase in presence of cholesterol; Lopez, Cheng, & Perez, 2018), nutritional studies should further consider the biophysical properties of milk-SM to better understand its role in the mechanisms of TAG hydrolysis.

MFGM and Dairy Fat Modulate Gut Microbiota The interaction between dietary lipids and gut microbiota is of increasing interest since it is well-known that the gut microbiome plays a crucial role in the maturation of the gastrointestinal immune defence. IF-feeding is usually associated with a higher bacterial richness and diversity and with different taxonomic composition compared with breast-feeding. Not much is known on the effect of dairy lipids on gut microbiota composition. In piglets, the composition of fecal microbiota has been reported to differ between piglets fed IF containing dairy lipids and MFGM or IF containing plant oils or mother-reared (Le Huerou-Luron et al., 2018). Supplementing IF with dairy lipids and MFGM increased *Proteobacteria* and *Bacteroidetes* while decreasing *Firmicutes phyla* compared with piglets receiving IF exclusively based on plant oils. The effects of milk fat and MFGM containing IF on the gut microbiota composition, and the underlying mechanisms, need to be further elucidated in infants.

4.3 Structure of the Emulsion and Interfacial Properties: Benefits of Large Droplets Coated with Milk Polar Lipids and MFGM in Infants

The organizations of lipids in human milk and in IF are different, i.e. the structure of TAG molecules, the size of lipid droplets and of the composition of their surface (MFGM vs. dairy proteins and non-dairy lipid emulsifiers, respectively) (Lopez et al., 2015; Michalski, Briard, Michel, Tasson, & Poulain, 2005) (Figs. 15.1, 15.6, and 15.7).

Pioneer studies clearly showed that dietary lipid structure affects lipolysis and the metabolic fate of fatty acids due to distinct differences in lipid digestion and absorption kinetics as well as post-prandial response (Armand et al., 1996, 1999). Any difference in postprandial lipid handling can impact lipid availability for the development of metabolic organs. This, in turn, could program metabolic homoeostasis, energy balance and metabolic response with potential impact on later life health. In recent years, studies have shown that the macrostructure of lipids in IF (i.e. the size of processed lipid droplets) and the interfacial composition (i.e. presence of MFGM and/or milk polar lipids at the surface of processed lipid droplets) are involved in metabolic programming. Animal studies highlighted the importance of the structure of lipid droplets on later adiposity and metabolism. Providing to mice pups (from post-natal day 16–42) IF with large droplets (i.e. modal diameter 6.25 μm) containing TAG from plant oils origin and covered by polar lipids from the MFGM (concept IMF Nuturis®; Nutricia Research; patent WO2010027258A1) reduced fat accumulation as well as fasting plasma leptin, resistin, glucose and lipids (TAG and total cholesterol) in adults fed a western diet, compared to the group fed standard IF (Oosting et al., 2012, 2014). The same group investigated the potential role played by large lipid droplets only, of polar lipid-coating only or the combination of both (i.e. the two parameters considered in the concept IF Nuturis®). IF with both large lipid droplet size and MFGM coating administrated during the neonatal period contributed to the observed protective effect against obesity in later life (Baars et al., 2016). Although the mechanisms remain unclear, these animal studies showed that early nutrition is associated with sustained effects on later life obesity. A clinical trial is ongoing to test the Nuturis® (NCT01609634; New Infant Formula Trial in Healthy Term Subjects on Growth, Body Composition, Tolerance and Safety; estimated study completion date: December 2019). An IF containing large, phospholipid coated lipid droplets was recently found to support adequate growth in healthy Asian infants during the first 4 months of life as compared with a standard IF (Shek, Yu, Wu, Zhu, & Chan, 2017).

Several mechanisms might explain the different responses observed following consumption of the Concept IF Nuturis®, such as access of digestive lipases to the TAG core of the fat droplets resulting in differences in fat digestion and absorption. Also, a difference in clotting behavior of casein in the acidic gastric milieu due to the different coating of the two formulae tested, that is, only proteins vs. partly protein and phospholipids, might have affected gastric emptying and thereby postprandial responses. However, the exact underlying mechanism remains to be determined. Further, it is unknown to what extent these findings can be extrapolated to infants, and if these postprandial changes translate into longer-term health effects.

Scientific proofs of the fact that the structure of the lipid droplets ingested during neonatal period programs body composition and metabolism in adulthood must be confirmed in clinical trials.

4.4 Opportunities to Produce Food-Grade Ingredients Enriched in Bovine MFGM

Technologies have become available to obtain MFGM from cow's milk and the bovine compounds also exhibit bioactivities in human infants and adults as reported by clinical studies (Hernell et al., 2016; Timby et al., 2017).

Various dairy streams permit the recovery of MFGM and are available in sufficient amount to produce ingredients at the industrial scale (Fig. 15.8) (Dewettinck et al., 2008; Jiménez-Flores & Brisson, 2008; Vanderghem et al., 2010). MFGM can be isolated from bovine milk fat globules, e.g. after concentration in creams, shear at low temperature to break fat globules, melting and centrifugation to separate anhydrous fat from the aqueous phase, called beta serum, containing MFGM fragments and polar lipids. MFGM can also be recovered from by-products of the cow's dairy industry, e.g. buttermilk obtained from butter making, butter serum recovered during the preparation of anhydrous milk fat, acid or neutral whey obtained from cheese production (Fig. 15.8). Treatments during dairy processing (e.g. churning, agitation, pressing of cheese curd) break the MFGM into fragments and release them in the aqueous phase.

Buttermilk, butter serum, beta serum and cheese whey are suitable sources because of their low cost and their relatively high content in MFGM components. A weak point concerning these by-products is that they have been treated at high

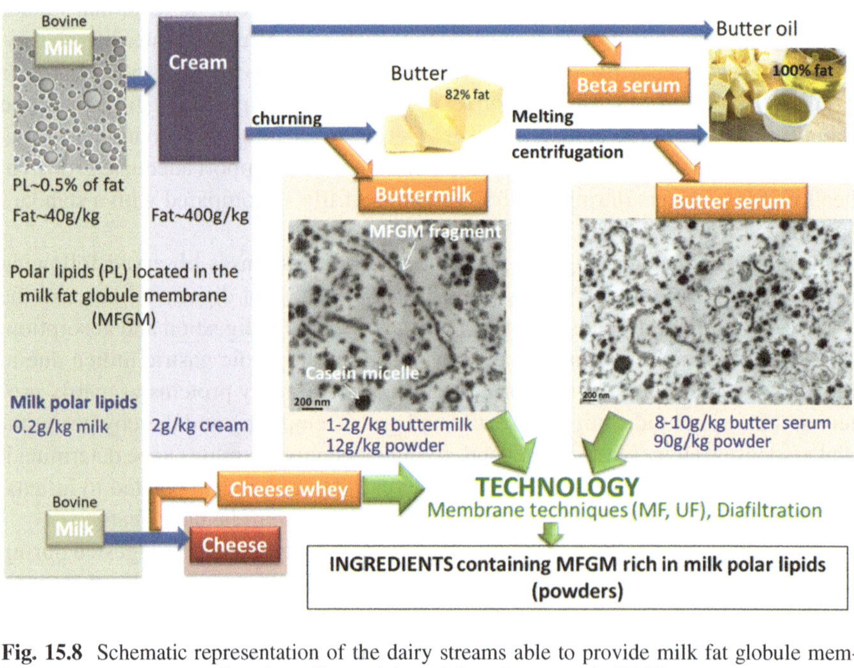

Fig. 15.8 Schematic representation of the dairy streams able to provide milk fat globule membrane components to produce ingredients. Adapted from Lopez et al. (2019)

temperatures several times for bacterial and safety reasons, inducing a denaturation of proteins, interactions between MFGM proteins and whey proteins, and interactions between sugars and proteins. These dairy streams are not produced with the same volumes (e.g. buttermilk is produced with higher volumes compared to butter serum), and do not have the same composition, in terms of dry matter, amount of MFGM and milk polar lipids, proteins. For example, butter serums exhibit a higher polar lipid content than buttermilks: 9.5 vs. 1.6%. As a result, they are not equivalent sources of MFGM and polar lipids. Moreover, the relative proportion of milk polar lipids can be altered by processing. The relative proportion of milk-SM is 23% in milk and cream, while butter serums contain a higher proportion of milk-SM (34%) and buttermilks contain a lower proportion of milk-SM (19%) (Lopez, Blot, Briard-Bion, Cirié, & Graulet, 2017). A similar relative amount of each polar lipid class was reported for milk and cheese whey. The lipid composition of industrial buttermilks and butter serums has been determined with a special emphasis on sphingolipid and ceramide isoforms (Bourlieu et al., 2018). Another variation in polar lipid composition concerns the fatty acids. It has been reported that cow diet affects the fatty acid composition of milk polar lipids recovered in by-products of the dairy industry such as buttermilks and butter serums (Lopez, Blot, et al., 2017).

The economic valorization of dairy streams containing MFGM, mainly buttermilk and butter serum, is improved by the development of technological processes able to concentrate and purify MFGM and milk polar lipids. For the last 20 years, industrial and academic research teams have focused on the development of processes able to selectively fractionate components and particularly to recover and concentrate MFGM fragments and milk polar lipids from the dairy streams (buttermilk, butter serum, cheese whey). The main challenge for these technological studies is to separate MFGM fragments and milk polar lipids from the other components (TAG, proteins, lactose and minerals). The different steps involved in the recovery of MFGM materials have been reviewed (Dewettinck et al., 2008; Gassi et al., 2016; Holzmueller & Kulozik, 2016; Singh, 2006). The objective is to selectively remove proteins (caseins and whey proteins), lactose and minerals and to isolate and concentrate MFGM and/or specific fractions such as the polar lipids. The main technological steps are e.g. acid or rennet-induced precipitation of proteins followed by centrifugation to recover the aqueous phase, filtration using membrane techniques such as microfiltration and ultrafiltration, diafiltration to remove lactose and minerals. The final products are dehydrated using spray-drying to allow their conservation. The ingredients containing MFGM and milk polar lipids are mainly found in the form of powders. These technological processes lead to the production of high added-value MFGM-enriched ingredients compared to dried by-products, e.g. buttermilk powder.

Further work is required to optimize food-grade down-stream processes for MFGM components that can be applied in the food industry.

Table 15.6 provides an overview of MFGM-containing ingredients that have been used in research studies and/or that are currently commercially available. The

Table 15.6 Examples of ingredients containing milk fat globule membrane (MFGM) components that are commercially available and have been used in research studies

Ingredient name	Company	Composition Total fat (%)	PL (%)	Proteins (%)	Lactose (%)	Studies
Lacprodan® MFGM-10	Arla foods	16–22	6–8	73–80	3	Timby et al. (2015), Timby, Domellof, et al. (2014), Timby, Loennerdal, Hernell, & Domellof (2014), and Zavaleta et al. (2011)
BAEF	Corman	16–23	6	26–31	35–40	Oosting et al. (2012)
SM2	Corman	12–16	6–8	26.5–30.5	45–49	Baars et al. (2016)
SM3	Corman	20–26	11–16	12–24	45–53	
SureStart™ Lipid100	Fonterra	19–21	7.6–7.9	30–31	39–42	Anderson, MacGibbon, Haggarty, Armstrong, and Roy (2018), Gallier et al. (2015), and Oosting et al. (2014)
SureStart™ Lipid70[a]	Fonterra	17	6.3	72	1	
INPULSE	Lactalis	25–35	11–16	55–60	12–17	Veereman-Wauters et al. (2012)

[a]Ingredient produced from whey derived from the manufacture of cheese products

ingredients have a complex composition. They contain TAG, proteins, i.e. skim milk derived proteins (caseins and whey proteins) and MFGM-specific proteins, phospholipids, lactose and minerals. The relative proportions of these components depend on the technological steps used to prepare the ingredients. The total amount of proteins and the relative proportion of the different kinds of proteins can affect the functional properties of the ingredient. In these ingredients prepared mainly from by-products of the cream and cheese industry, the polar lipids and membrane-specific proteins are closely associated in the MFGM fragments released during processing of cream (Fig. 15.8) (Gallier et al., 2015; Lopez et al., 2015, 2017). Ingredients enriched in MFGM have recently become available in sufficient quantity and quality to be added in IF. Randomized clinical studies with experimental IF incorporating MFGM from bovine milk have already provided some evidence of clinical benefits.

The preparation of pure MFGM or milk polar lipids is possible from a technological point of view. Ingredients are commercially available (e.g. Lacprodan PL-75 produced by Arla Foods, PC700 produced by Fonterra). However, the utilization of these functional ingredients is difficult from an economic point of view since they are expensive. Bovine milk-SM purified from milk polar lipids is also available commercially (e.g. provided by Avanti Polar Lipids) but with a high quotation.

4.5 Opportunities to Prepare MFGM-Coated Lipid Droplets Bio-Inspired by Breast Milk Fat Globules

In recent years, IF containing MFGM-coated lipid droplets have been prepared and showed health benefits compared to protein-coated lipid droplets (Baars et al., 2016; Baumgartner, van de Heijning, Acton, & Mensink, 2017; Oosting et al., 2012, 2014). These O/W emulsions were prepared by adding ingredients containing MFGM fragments (Table 15.6) together with TAG (e.g. from vegetable oils, milk fat or a mixture of both) before the homogenization process. Such emulsions are bioinspired from milk fat globules. The biological and health functions of these emulsions are governed by the TAG/water interface, in terms of amount, composition and structure. To fully benefit the advantages provided by the MFGM, the composition and the structure of the surface of these lipid droplets are of primary importance. However, to date, information remains scarce.

The structure of the concept IF Nuturis®, that comprises processed lipid droplets of large size coated by MFGM components reported to be involved in early lipid programming, has been investigated (Gallier et al., 2015; Oosting et al., 2012, 2014). The authors used confocal microscopy with a dye having affinity for the polar lipids (i.e. Annexin-V Alexa Fluor 488) to show their presence at the TAG/water interface as a sign for the localization of MFGM components (Oosting et al., 2012, 2014). Further characterizations of the emulsion have been performed by the combination of confocal and electron microscopy (Gallier et al., 2015). The authors reported the presence of MFGM components i.e. polar lipids, glycoproteins, glycolipids and cholesterol adsorbed at the surface of the lipid droplets. Gallier and coworkers hypothesized that a thin monolayer membrane (5–10 nm) was present at the surface of the lipid droplets. However, the resolution of electron microscopy does not allow knowing if the components form a monolayer or if the trilayer organization of the MFGM can be preserved. In this study, the interactions between casein micelles and MFGM fragments in the ingredient have been observed. Moreover, the presence of casein micelles adsorbed at the surface of the lipid droplets in the concept IMF Nuturis® has been reported. The potential role played by the casein micelles adsorbed at the TAG/water interface has not been discussed, mainly as regards to the mechanisms of lipid digestion in the gastrointestinal tract of infants. By comparing the structure of the emulsion in human milk, concept IF and control IF, the authors concluded that the processed MFGM-coated lipid droplets in the concept IF are closer to the human milk fat globules than the processed protein-coated lipid droplets in the control IF.

In a more recent study, the preparation of MFGM-coated lipid droplets biomimetic of milk fat globules, the physical stability of this emulsion and its behavior as a function of pH were reported (Lopez, Cauty, et al., 2017) (Fig. 15.9). The ingredient prepared from industrial butter serum (Gassi et al., 2016) contains linear MFGM fragments of various lengths that have been released from the surface of bovine milk fat globules during phase inversion of concentrated cream. The ingredient also contains vesicles formed by MFGM and/or by MFGM polar lipids. TEM images of

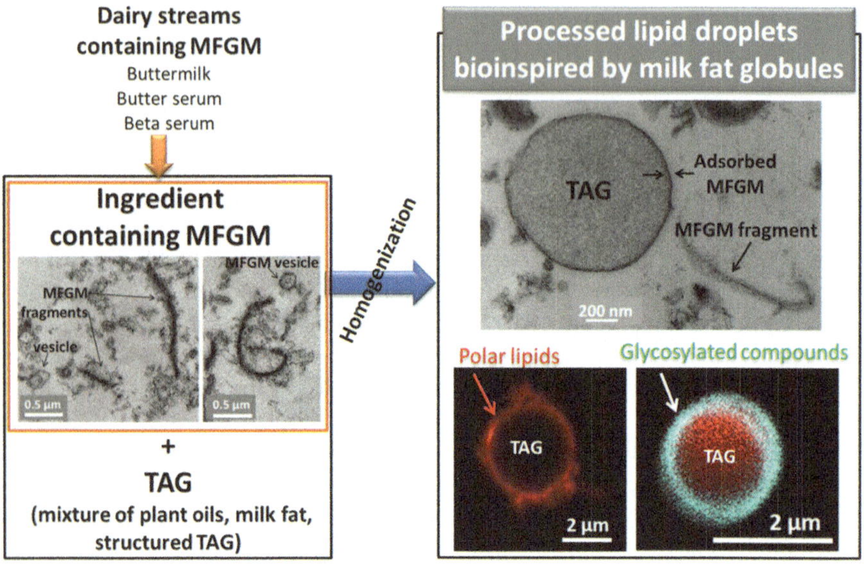

Fig. 15.9 Tailoring of milk fat globule membrane (MFGM)-coated processed lipid droplets bioinspired by milk fat globules. Preparation of MFGM-containing ingredient from dairy streams and homogenization with triacylglycerols (TAG) to form the processed lipid droplets covered by MFGM fragments containing polar lipids and glycosylated compounds (glycoproteins and glycolipids) as observed using transmission electron microscopy (top) and confocal laser scanning microscopy (bottom). Adapted from Lopez, Cauty, et al. (2017)

the emulsion showed a thin layer of MFGM components present at the surface of the lipid droplets with some MFGM fragments adsorbed at the surface of the lipid droplets and protruding in the aqueous phase (Fig. 15.9). Also, Some MFGM fragments were present in the aqueous phase of the emulsion, showing a partitioning of MFGM components between the TAG/water interface and the aqueous phase. CLSM images with specific fluorescent probes permitted to identify polar lipids and glycosylated molecules (i.e. glycoproteins and glycolipids) at the surface of the lipid droplets. The polar lipids were organized as a thin layer and as small spherical vesicles. This study also showed that the glycocalyx present around milk fat globules can be formed around processed lipid droplets coated with bovine MFGM. The description of the biomimetic emulsions and of the surface of MFGM-coated lipid droplets performed in independent studies using different ingredients leads to convergent information (Gallier et al., 2015; Lopez, Cauty, et al., 2017).

As a conclusion, it is possible to prepare MFGM-coated lipid droplets in IF to mimic the structure of breast milk fat globules covered by the MFGM. The size of lipid droplets can be modulated by the shear applied during emulsion formation (i.e. the homogenization pressure) and both the amount and the type of tension-active molecules able to stabilize the interface formed upon homogenization. The biophysical studies performed using the combination of microscopy techniques showed that the MFGM fragments and MFGM components (e.g. polar lipids, proteins) added during the preparation of the emulsion are able to move to and adsorb at the surface of the processed lipid droplets during homogenization. The presence of MFGM fragments in the emulsions shows that they are not disrupted under pressure during homogenization under the range of pressures used. Other components such as the proteins present in the ingredient (i.e. caseins, whey proteins, membrane-specific proteins), can also be adsorbed at the surface of the lipid droplets. An important aspect to consider is the order of introduction of the ingredients during the preparation of the emulsion since it can strongly affect the composition and the structure of the lipid droplet surface as well as the functional properties of the MFGM-coated lipid droplets.

4.6 Manufacture of IF Containing Dairy Lipids

Dairy lipids can be introduced in IF in three different ways:

1. anhydrous milk fat or butter oil that provides dairy TAG. Table 15.5 shows an example of IF powder, Modilac Doucéa, that contains anhydrous milk fat. Dairy TAGs need to be melted above their final melting temperature (around 40 °C) and can be blended with other oils such as plant oils. The oils are then homogenized to form the oil-in-water emulsion. See Fig. 15.5 for the preparation of IF.
2. ingredients enriched in MFGM or milk polar lipids that are obtained from by-products of dairy industries. Table 15.5 shows an example of IF powder, Enfamil® Enspire™, that contains MFGM. The ingredients rich in MFGM are generally introduced in the aqueous phase before the homogenization step. MFGM can then be located at the TAG/water interface, depending on their competition with dairy proteins to stabilize the interface.
3. full fat milk or cream that corresponds to the concentration of fat globules from bovine milk, and contains dairy TAG and all components of the MFGM. Table 15.5 shows examples of IF powders containing cream from bovine milk in their formulation (Biostime, Picot). Bovine milk fat globules have a large size (mean diameter around 4 μm) and, as a consequence, can phase separate during the process and be physically destabilized in the IF powder (formation of free fat highly sensitive to oxidation). To assure the physical stability of lipid droplets and the quality of IF powders towards oxidation and functional properties (e.g. wettability), milk fat globules are homogenized during the manufacture of IF together with the other oils. The processed lipid droplets formed in these IF are a mixture of TAG from milk and plants in their core, surrounded by dairy proteins and possibly MFGM components at the TAG/water interface. When cream is included as an ingredient in IF, polar lipid content is usually higher with a wide range of molecular species from different polar lipid classes specific to the bovine MFGM (SM, PC, PE, PI, PS).

In the future, processing techniques used in the manufacture of IF (homogenization, heat treatment, spray-drying) will need to be adapted to preserve the structure of milk fat globules enveloped by the MFGM and all the health benefits provided by the variety of dairy lipid components.

5 Conclusions

The partial replacement of plant oils by dairy lipids in IF, associated with the formation of MFGM coated lipid droplets, will contribute in reducing the gap with breast milk fat globules by considering both the composition and the structure. The technological opportunities exist to improve IF and the nutritional and health benefits in infants. Dairy lipids have large similarities with breast milk lipids in terms of variety of fatty acids, presence of short and medium-chain fatty acids, low levels of LA, a proper level of ALA and LA/ALA ratio which is presently better than the human breast milk, *sn*-2 position of palmitic acid on TAG, MFGM, cholesterol. Consequently, the use of dairy fat in IF, as well as the absolute amount of polyunsaturated LA and ALA, should be considered in the future to improve the quality of IF.

References

Ailhaud, G., Massiera, F., Weill, P., Legrand, P., Alessandri, J.-M., & Guesnet, P. (2006). Temporal changes in dietary fats: Role of n-6 polyunsaturated fatty acids in excessive adipose tissue development and relationship to obesity. *Progress in Lipid Research, 45*, 203–236.

Amara, S., Patin, A., Giuffrida, F., Wooster, T. J., Thakkar, S. K., Bénarouche, A., et al. (2014). In vitro digestion of citric acid esters of mono- and diglycerides (CITREM) and CITREM-containing infant formula/emulsions. *Food & Function, 5*, 1409–1421.

American Academy of Pediatrics (AAP). (2012). Section on breastfeeding. Breastfeeding and the use of human milk. *Pediatrics, 129*, e827–e841.

Anderson, R. C., MacGibbon, A. K. H., Haggarty, N., Armstrong, K. M., & Roy, N. C. (2018). Bovine dairy complex lipids improve in vitro measures of small intestinal epithelial barrier integrity. *PLoS One, 13*, e0190839.

Argov-Argaman, N., Smilowitz, J. T., Bricarello, D. A., Barboza, M., Lerno, L., Froehlich, J. W., et al. (2010). Lactosomes: Structural and compositional classification of unique nanometer-sized protein lipid particles of human milk. *Journal of Agricultural and Food Chemistry, 58*, 11234–11242.

Armand, M., Hamosh, M., Mehta, N. R., Angelus, P. A., Philpott, J. R., Henderson, T. R., et al. (1996). Effect of human milk or formula on gastric function and fat digestion in the premature infant. *Pediatric Research, 40*, 429–437.

Armand, M., Pasquier, B., André, M., Borel, P., Senft, M., Peyrot, J., et al. (1999). Digestion and absorption of 2 fat emulsions with different droplet sizes in the human digestive tract. *The American Journal of Clinical Nutrition, 70*, 1096–1106.

Baars, A., Oosting, A., Engels, E., Kegler, D., Kodde, A., Schipper, L., et al. (2016). Milk fat globule membrane coating of large lipid droplets in the diet of young mice prevents body fat accumulation in adulthood. *The British Journal of Nutrition, 115*, 1930–1937.

Bar-Yoseph, F., Lifshitz, Y., & Cohen, T. (2013). Review of sn-2 palmitate oil implications for infant health. *Prostaglandins, Leukotrienes and Essential Fatty Acids, 89*, 139–143.
Baumgartner, S., van de Heijning, B. J. M., Acton, D., & Mensink, R. P. (2017). Infant milk fat droplet size and coating affect postprandial responses in healthy adult men: A proof-of-concept study. *European Journal of Clinical Nutrition, 71*, 1108–1113.
Benoit, B., Fauquant, C., Daira, P., Peretti, N., Guichardant, M., & Michalski, M.-C. (2010). Phospholipid species and minor sterols in French human milks. *Food Chemistry, 120*, 684–691.
Bernard, J. Y., Armand, M., Peyre, H., Garcia, C., Forhan, A., De Agostini, M., et al. (2017). Breastfeeding, polyunsaturated fatty acid levels in colostrum and child intelligence quotient at age 5-6 years. *The Journal of Pediatrics, 183*, 43–50.e3.
Bernback, S., Blackberg, L., & Hernell, O. (1990). The complete digestion of human-milk triacylglycerol in vitro requires gastric lipase, pancreatic colipase-dependent lipase, and bile-salt stimulated lipase. *The Journal of Clinical Investigation, 85*, 1221–1226.
Bhinder, G., Allaire, J. M., Garcia, C., Lau, J. T., Chan, J. M., Ryz, N. R., et al. (2017). Milk fat globule membrane supplementation in formula modulates the neonatal gut microbiome and normalizes intestinal development. *Scientific Reports, 7*, 45274.
Billeaud, C., Puccio, G., Saliba, E., Guillois, B., Vaysse, C., Pecquet, S., et al. (2014). Safety and tolerance evaluation of milk fat globule membrane-enriched infant formulas: A randomized controlled multicenter non-inferiority trial in healthy term infants. *Clinical Medicine Insights: Pediatrics, 8*, CMPed.S16962.
Bourlieu, C., Cheillan, D., Blot, M., Daira, P., Trauchessec, M., Ruet, S., et al. (2018). Polar lipid composition of bioactive dairy co-products buttermilk and butterserum: Emphasis on sphingolipid and ceramide isoforms. *Food Chemistry, 240*, 67–74.
Cilla, A., Quintaes, K. D., Barberá, R., & Alegría, A. (2016). Phospholipids in human milk and infant formulas: Benefits and needs for correct infant nutrition. *Critical Reviews in Food Science and Nutrition, 56*, 1880–1892.
Claumarchirant, L., Cilla, A., Matencio, E., Sanchez-Siles, L. M., Castro-Gomez, P., Fontecha, J., et al. (2016). Addition of milk fat globule membrane as an ingredient of infant formulas for resembling the polar lipids of human milk. *International Dairy Journal, 61*, 228–238.
Claumarchirant, L., Matencio, E., Sanchez-Siles, L. M., Alegría, A., & Lagarda, M. J. (2015). Sterol composition in infant formulas and estimated intake. *Journal of Agricultural and Food Chemistry, 63*, 7245–7251.
Courage, M., Mccloy, U., Herzberg, G., Andrews, W., Simmons, B., Mcdonald, A., et al. (1998). Visual acuity development and fatty acid composition of erythrocytes in full-term infants fed breast milk, commercial formula, or evaporated milk. *Journal of Developmental & Behavioral Pediatrics, 19*, 9–17.
Delplanque, B., Du, Q., Martin, J.-C., & Guesnet, P. (2018). Lipids for infant formulas. *OCL, 25*, D305.
Delplanque, B., Gibson, R., Koletzko, B., Lapillonne, A., & Strandvik, B. (2015). Lipid quality in infant nutrition: Current knowledge and future opportunities. *Journal of Pediatric Gastroenterology and Nutrition, 61*, 8–17.
Demmelmair, H., & Koletzko, B. (2018). Lipids in human milk. *Best Practice & Research Clinical Endocrinology & Metabolism, 32*, 57–68.
Demmelmair, H., Prell, C., Timby, N., & Lönnerdal, B. (2017). Benefits of lactoferrin, osteopontin and milk fat globule membranes for infants. *Nutrients, 9*, 817.
Dewettinck, K., Rombaut, R., Thienpont, N., Le, T. T., Messens, K., & Van Camp, J. (2008). Nutritional and technological aspects of milk fat globule membrane material. *International Dairy Journal, 18*, 436–457.
Dinel, A. L., Rey, C., Bonhomme, C., Le Ruyet, P., Joffre, C., & Layé, S. (2016). Dairy fat blend improves brain DHA and neuroplasticity and regulates corticosterone in mice. Prostaglandins. *Leukotrienes and Essential Fatty Acids, 109*, 29–38.
Drouin, G., Catheline, D., Sinquin, A., Baudry, C., Le Ruyet, P., Rioux, V., et al. (2018). Incorporation of dairy lipids in the diet increased long-chain omega-3 fatty acids status in post-weaning rats. *Frontiers in Nutrition, 5*, 42.

EFSA. (2014). Scientific opinion on the essential composition of infant and follow-on formulae. *EFSA Journal, 12*, 24–32.
Favé, G., Leveque, C., Peyrot, J., Pieroni, G., Coste, T. C., & Armand, M. (2007). Modulation of gastric lipolysis by the phospholipid specie: Link to specific lipase-phospholipid interaction at the lipid/water interface? *The FASEB Journal, 21*, A1010.
Fidler, N., & Koletzko, B. (2000). The fatty acid composition of human colostrum. *European Journal of Nutrition, 39*, 31–37.
Fuller, K. L., Kuhlenschmidt, T. B., Kuhlenschmidt, M. S., Jimenez-Flores, R., & Donovan, S. M. (2013). Milk fat globule membrane isolated from buttermilk or whey cream and their lipid components inhibit infectivity of rotavirus in vitro. *Journal of Dairy Science, 96*, 3488–3497.
Gallier, S., Vocking, K., Post, J. A., Van de Heijning, B., Acton, D., Van der Beek, E. M., et al. (2015). A novel infant milk formula concept: Mimicking the human milk fat globule structure. *Colloids and Surfaces B: Biointerfaces, 136*, 329–339.
Gassi, J. Y., Blot, M., Beaucher, E., Robert, B., Leconte, N., Camier, B., et al. (2016). Preparation and characterisation of a milk polar lipids enriched ingredient from fresh industrial liquid butter serum: Combination of physico-chemical modifications and technological treatments. *International Dairy Journal, 52*, 26–34.
Gianni, M. L., Roggero, P., Baudry, C., Fressange-Mazda, C., Galli, C., Agostoni, C., et al. (2018). An infant formula containing dairy lipids increased red blood cell membrane Omega 3 fatty acids in 4 month-old healthy newborns: A randomized controlled trial. *BMC Pediatrics, 18*, 53.
Grote, V., Verduci, E., Scaglioni, S., Vecchi, F., Contarini, G., Giovannini, M., et al. (2016). Breast milk composition and infant nutrient intakes during the first 12 months of life. *European Journal of Clinical Nutrition, 70*, 250–256.
Gurnida, D. A., Rowan, A. M., Idjradinata, P., Muchtadi, D., & Sekarwana, N. (2012). Association of complex lipids containing gangliosides with cognitive development of 6-month-old infants. *Early Human Development, 88*, 595–601.
Hageman, J. H. J., Danielsen, M., Nieuwenhuizen, A. G., Feitsma, A. L., & Dalsgaard, T. K. (2019). Comparison of bovine milk fat and vegetable fat for infant formula: Implications for infant health. *International Dairy Journal, 92*, 37–49.
Hageman, J. H. J., Keijer, J., Dalsgaard, T. K., Zeper, L. W., Carriere, F., Feitsma, A. L., et al. (2019). Free fatty acid release from vegetable and bovine milk fat-based infant formulas and human milk during two-phase in vitro digestion. *Food & Function, 10*, 2102–2113.
Hamosh, M., Peterson, J. A., Henderson, T. R., Scallan, C. D., Kiwan, R., Ceriani, R. L., et al. (1999). Protective function of human milk: The milk fat globule. *Seminars in Perinatology, 23*, 242–249.
Harzer, G., Haug, M., Dieterich, I., & Gentner, P. R. (1983). Changing patterns of human milk lipids in the course of the lactation and during the day. *The American Journal of Clinical Nutrition, 37*, 612–621.
Heid, H. W., & Keenan, T. W. (2005). Intracellular origin and secretion of milk fat globules. *European Journal of Cell Biology, 84*, 245–258.
Hernell, O., Timby, N., Domellof, M., & Lonnerdal, B. (2016). Clinical benefits of milk fat globule membranes for infants and children. *The Journal of Pediatrics, 173*, S60–S65.
Holzmueller, W., & Kulozik, U. (2016). Isolation of milk fat globule membrane (MFGM) material by coagulation and diafiltration of buttermilk. *International Dairy Journal, 63*, 88–91.
Innis, S. M. (2011). Dietary triacylglycerol structure and its role in infant nutrition. *Advances in Nutrition, 2*, 275–283.
Jensen, R., Ferris, A., & Lammikeefe, C. (1992). Lipids in human-milk and infant formulas. *Annual Review of Nutrition, 12*, 417–441.
Jensen, R. G. (1999). Lipids in human milk. *Lipids, 34*, 1243–1271.
Jiménez-Flores, R., & Brisson, G. (2008). The milk fat globule membrane as an ingredient: Why, how, when? *Dairy Science & Technology, 88*, 5–18.
Kallio, H., Nylund, M., Bostrom, P., & Yang, B. (2017). Triacylglycerol regioisomers in human milk resolved with an algorithmic novel electrospray ionization tandem mass spectrometry method. *Food Chemistry, 233*, 351–360.

Khan, S., Hepworth, A. R., Prime, D. K., Lai, C. T., Trengove, N. J., & Hartmann, P. E. (2013). Variation in fat, lactose, and protein composition in breast milk over 24 hours: Associations with infant feeding patterns. *Journal of Human Lactation, 29*, 81–89.

Kim, K.-M., Park, T.-S., & Shim, S.-M. (2015). Optimization and validation of HRLC-MS method to identify and quantify triacylglycerol molecular species in human milk. *Analytical Methods, 7*, 4362–4370.

Kinney, H., Karthigasan, J., Borenshteyn, N., Flax, J., & Kirschner, D. (1994). Myelination in the developing human brain - Biochemical correlates. *Neurochemical Research, 19*, 983–996.

Koletzko, B. (2016). Human milk lipids. *Annals of Nutrition & Metabolism, 69*(Suppl 2), 28–40.

Koletzko, B., Baker, S., Cleghorn, G., Neto, U. F., Gopalan, S., Hernell, O., et al. (2005). Global standard for the composition of infant formula: Recommendations of an ESPGHAN coordinated international expert group. *Journal of Pediatric Gastroenterology and Nutrition, 41*, 584–599.

Le Huerou-Luron, I., Bouzerzour, K., Ferret-Bernard, S., Menard, O., Le Normand, L., Perrier, C., et al. (2018). A mixture of milk and vegetable lipids in infant formula changes gut digestion, mucosal immunity and microbiota composition in neonatal piglets. *European Journal of Nutrition, 57*, 463–476.

Lecomte, M., Bourlieu, C., Meugnier, E., Penhoat, A., Cheillan, D., Pineau, G., et al. (2015). Milk polar lipids affect in vitro digestive lipolysis and postprandial lipid metabolism in mice–3. *The Journal of Nutrition, 145*, 1770–1777.

Leermakers, E. T. M., Moreira, E. M., Kiefte-de Jong, J. C., Darweesh, S. K. L., Visser, T., Voortman, T., et al. (2015). Effects of choline on health across the life course: A systematic review. *Nutrition Reviews, 73*, 500–522.

Lemaire, M., Dou, S., Cahu, A., Formal, M., Le Normand, L., Rome, V., et al. (2018). Addition of dairy lipids and probiotic Lactobacillus fermentum in infant formula programs gut microbiota and entero-insular axis in adult minipigs. *Scientific Reports, 8*, 11656.

Lien, E. L., Richard, C., & Hoffman, D. R. (2018). DHA and ARA addition to infant formula: Current status and future research directions. *Prostaglandins, Leukotrienes and Essential Fatty Acids, 128*, 26–40.

Linderborg, K. M., Kalpio, M., Makela, J., Niinikoski, H., Kallio, H. P., & Lagstrom, H. (2014). Tandem mass spectrometric analysis of human milk triacylglycerols from normal weight and overweight mothers on different diets. *Food Chemistry, 146*, 583–590.

Lindquist, S., & Hernell, O. (2010). Lipid digestion and absorption in early life: An update. *Current Opinion in Clinical Nutrition and Metabolic Care, 13*, 314–320.

Lopez, C., Blot, M., Briard-Bion, V., Cirié, C., & Graulet, B. (2017). Butter serums and buttermilks as sources of bioactive lipids from the milk fat globule membrane: Differences in their lipid composition and potentialities of cow diet to increase n-3 PUFA. *Food Research International, 100*, 864–872.

Lopez, C., Briard-Bion, V., Bourgaux, C., & Pérez, J. (2013). Solid triacylglycerols within human fat globules: β crystals with a melting point above in-body temperature of infants, formed upon storage of breast milk at low temperature. *Food Research International, 54*, 1541–1552.

Lopez, C., Cauty, C., & Guyomarc'h, F. (2015). Organization of lipids in milks, infant milk formulas and various dairy products: Role of technological processes and potential impacts. *Dairy Science & Technology, 95*, 863–893.

Lopez, C., Cauty, C., & Guyomarc'h, F. (2019). Unraveling the complexity of milk fat globules to tailor bioinspired emulsions providing health benefits: The key role played by the biological membrane. *European Journal of Lipid Science and Technology, 121*, 1800201.

Lopez, C., Cauty, C., Rousseau, F., Blot, M., Margolis, A., & Famelart, M.-H. (2017). Lipid droplets coated with milk fat globule membrane fragments: Microstructure and functional properties as a function of pH. *Food Research International, 91*, 26–37.

Lopez, C., Cheng, K., & Perez, J. (2018). Thermotropic phase behavior of milk sphingomyelin and role of cholesterol in the formation of the liquid ordered phase examined using SR-XRD and DSC. *Chemistry and Physics of Lipids, 215*, 46–55.

Lopez, C., Madec, M.-N., & Jiménez-Flores, R. (2010). Lipid rafts in the bovine milk fat globule membrane revealed by the lateral segregation of phospholipids and heterogeneous distribution of glycoproteins. *Food Chemistry, 120*, 22–33.

Lopez, C., & Ménard, O. (2011). Human milk fat globules: Polar lipid composition and in situ structural investigations revealing the heterogeneous distribution of proteins and the lateral segregation of sphingomyelin in the biological membrane. *Colloids and Surfaces B: Biointerfaces, 83*, 29–41.

Mathiassen, J. H., Nejrup, R. G., Frøkiær, H., Nilsson, Å., Ohlsson, L., & Hellgren, L. I. (2015). Emulsifying triglycerides with dairy phospholipids instead of soy lecithin modulates gut lipase activity. *European Journal of Lipid Science and Technology, 117*, 1522–1539.

Mendonça, M. A., Araújo, W. M. C., Borgo, L. A., & de Rodrigues Alencar, E. (2017). Lipid profile of different infant formulas for infants. *PLoS One, 12*, e0177812.

Michalski, M. C., Briard, V., Michel, F., Tasson, F., & Poulain, P. (2005). Size distribution of fat globules in human colostrum, breast milk, and infant formula. *Journal of Dairy Science, 88*, 1927–1940.

Miklavcic, J. J., Schnabl, K. L., Mazurak, V. C., Thomson, A. B. R., & Clandinin, M. T. (2012). Dietary ganglioside reduces proinflammatory signaling in the intestine. *Journal of Nutrition and Metabolism, 2012*, 1.

Miles, E. A., & Calder, P. C. (2017). The influence of the position of palmitate in infant formula triacylglycerols on health outcomes. *Nutrition Research, 44*, 1–8.

Motouri, M., Matsuyama, H., Yamamura, J., Tanaka, M., Aoe, S., Iwanaga, T., et al. (2003). Milk sphingomyelin accelerates enzymatic and morphological maturation of the intestine in artificially reared rats. *Journal of Pediatric Gastroenterology and Nutrition, 36*, 241.

Mu, H. L., & Hoy, C. E. (2004). The digestion of dietary triacylglycerols. *Progress in Lipid Research, 43*, 105–133.

Nilsson, A. (2016). Role of sphingolipids in infant gut health and immunity. *The Journal of Pediatrics, 173*, S53–S59.

Nilsson, Å., & Duan, R.-D. (2006). Absorption and lipoprotein transport of sphingomyelin. *Journal of Lipid Research, 47*, 154–171.

Norris, G. H., Jiang, C., Ryan, J., Porter, C. M., & Blesso, C. N. (2016). Milk sphingomyelin improves lipid metabolism and alters gut microbiota in high fat diet-fed mice. *The Journal of Nutritional Biochemistry, 30*, 93–101.

Ohlsson, L., Hertervig, E., Jönsson, B. A., Duan, R.-D., Nyberg, L., Svernlöv, R., et al. (2010). Sphingolipids in human ileostomy content after meals containing milk sphingomyelin. *The American Journal of Clinical Nutrition, 91*, 672–678.

Oosting, A., Kegler, D., Wopereis, H. J., Teller, I. C., van de Heijning, B. J. M., Verkade, H. J., et al. (2012). Size and phospholipid coating of lipid droplets in the diet of young mice modify body fat accumulation in adulthood. *Pediatric Research, 72*, 362–369.

Oosting, A., van Vlies, N., Kegler, D., Schipper, L., Abrahamse-Berkeveld, M., Ringler, S., et al. (2014). Effect of dietary lipid structure in early postnatal life on mouse adipose tissue development and function in adulthood. *British Journal of Nutrition, 111*, 215–226.

Oshida, K., Shimizu, T., Takase, M., Tamura, Y., Shimizu, T., & Yamashiro, Y. (2003). Effects of dietary sphingomyelin on central nervous system myelination in developing rats. *Pediatric Research, 53*, 589–593.

Owen, C. G., Whincup, P. H., Kaye, S. J., Martin, R. M., Davey Smith, G., Cook, D. G., et al. (2008). Does initial breastfeeding lead to lower blood cholesterol in adult life? A quantitative review of the evidence. *The American Journal of Clinical Nutrition, 88*, 305–314.

Palmano, K., Rowan, A., Guillermo, R., Guan, J., & Mc Jarrow, P. (2015). The role of gangliosides in neurodevelopment. *Nutrients, 7*, 3891–3913.

Park, E. J., Thomson, A. B., & Clandinin, M. T. (2010). Protection of intestinal occludin tight junction protein by dietary gangliosides in lipopolysaccharide-induced acute inflammation. *Journal of Pediatric Gastroenterology and Nutrition, 50*, 321–328.

Peterson, J. A., Patton, S., & Hamosh, M. (1998). Glycoproteins of the human milk fat globule in the protection of the breast-fed infant against infections. *Neonatology, 74*, 143–162.

Petit, V., Sandoz, L., & Garcia-Rodenas, C. L. (2017). Importance of the regiospecific distribution of long-chain saturated fatty acids on gut comfort, fat and calcium absorption in infants. *Prostaglandins, Leukotrienes and Essential Fatty Acids, 121*, 40–51.

Rueda, R. (2007). The role of dietary gangliosides on immunity and the prevention of infection. *British Journal of Nutrition, 98*, S68–S73.

Saarela, T., Kokkonen, J., & Koivisto, M. (2005). Macronutrient and energy contents of human milk fractions during the first six months of lactation. *Acta Paediatrica, 94*, 1176–1181.

Shek, D. T. L., Yu, L., Wu, F. K. Y., Zhu, X., & Chan, K. H. Y. (2017). A 4-year longitudinal study of well-being of Chinese university students in Hong Kong. *Applied Research in Quality of Life, 12*, 867–884.

Simons, K., & Ikonen, E. (1997). Functional rafts in cell membranes. *Nature, 387*, 569–572.

Sinanoglou, V. J., Cavouras, D., Boutsikou, T., Briana, D. D., Lantzouraki, D. Z., Paliatsiou, S., et al. (2017). Factors affecting human colostrum fatty acid profile: A case study. *PLoS One, 12*, e0175817.

Singh, H. (2006). The milk fat globule membrane—A biophysical system for food applications. *Current Opinion in Colloid & Interface Science, 11*, 154–163.

Snow, D. R., Ward, R. E., Olsen, A., Jimenez-Flores, R., & Hintze, K. J. (2011). Membrane-rich milk fat diet provides protection against gastrointestinal leakiness in mice treated with lipopolysaccharide. *Journal of Dairy Science, 94*, 2201–2212.

Spitsberg, V. L. (2005). Invited review: Bovine milk fat globule membrane as a potential nutraceutical. *Journal of Dairy Science, 88*, 2289–2294.

Sprong, R. C., Hulstein, M. F. E., Lambers, T. T., & van der Meer, R. (2012). Sweet buttermilk intake reduces colonisation and translocation of Listeria monocytogenes in rats by inhibiting mucosal pathogen adherence. *British Journal of Nutrition, 108*, 2026–2033.

Straarup, E. M., Lauritzen, L., Faerk, J., Hoy, C. E., & Michaelsen, K. F. (2006). The stereospecific triacylglycerol structures and fatty acid profiles of human milk and infant formulas. *Journal of Pediatric Gastroenterology and Nutrition, 42*, 293–299.

Sun, C., Wei, W., Su, H., Zou, X., & Wang, X. (2018). Evaluation of sn-2 fatty acid composition in commercial infant formulas on the Chinese market: A comparative study based on fat source and stage. *Food Chemistry, 242*, 29–36.

Tanaka, K., Hosozawa, M., Kudo, N., Yoshikawa, N., Hisata, K., Shoji, H., et al. (2013). The pilot study: Sphingomyelin-fortified milk has a positive association with the neurobehavioural development of very low birth weight infants during infancy, randomized control trial. *Brain and Development, 35*, 45–52.

Timby, N., Domellof, E., Hernell, O., Loennerdal, B., & Domellof, M. (2014). Neurodevelopment, nutrition, and growth until 12 mo of age in infants fed a low-energy, low-protein formula supplemented with bovine milk fat globule membranes: A randomized controlled trial. *The American Journal of Clinical Nutrition, 99*, 860–868.

Timby, N., Domellöf, M., Lönnerdal, B., & Hernell, O. (2017). Supplementation of infant formula with bovine milk fat globule membranes. *Advances in Nutrition, 8*, 351–355.

Timby, N., Hernell, O., Vaarala, O., Melin, M., Lönnerdal, B., & Domellöf, M. (2015). Infections in infants fed formula supplemented with bovine milk fat globule membranes. *Journal of Pediatric Gastroenterology and Nutrition, 60*, 384.

Timby, N., Loennerdal, B., Hernell, O., & Domellof, M. (2014). Cardiovascular risk markers until 12 mo of age in infants fed a formula supplemented with bovine milk fat globule membranes. *Pediatric Research, 76*, 394–400.

Tu, A., Ma, Q., Bai, H., & Du, Z. (2017). A comparative study of triacylglycerol composition in Chinese human milk within different lactation stages and imported infant formula by SFC coupled with Q-TOF-MS. *Food Chemistry, 221*, 555–567.

Vanderghem, C., Bodson, P., Danthine, S., Paquot, M., Deroanne, C., & Blecker, C. (2010). Milk fat globule membrane and buttermilks: From composition to valorization. *Biotechnologie, Agronomie, Société et Environnement, 14*, 485–500.

Veereman-Wauters, G., Staelens, S., Rombaut, R., Dewettinck, K., Deboutte, D., Brummer, R.-J., et al. (2012). Milk fat globule membrane (INPULSE) enriched formula milk decreases febrile episodes and may improve behavioral regulation in young children. *Nutrition, 28*, 749–752.

Wang, B. (2012). Molecular mechanism underlying sialic acid as an essential nutrient for brain development and cognition. *Advances in Nutrition, 3*, 465S–472S.

Wang, B., Yu, B., Karim, M., Hu, H., Sun, Y., McGreevy, P., et al. (2007). Dietary sialic acid supplementation improves learning and memory in piglets. *The American Journal of Clinical Nutrition, 85*, 561–569.

Wang, L., Shimizu, Y., Kaneko, S., Hanaka, S., Abe, T., Shimasaki, H., et al. (2000). Comparison of the fatty acid composition of total lipids and phospholipids in breast milk from Japanese women. *Pediatrics International, 42*, 14–20.

WHO. (2011). *Exclusive breastfeeding for six months best for babies everywhere*. WHO: World Health Organization. Retrieved from http://www.who.int/mediacentre/news/statements/2011/breastfeeding_20110115/en/

Ya, B., Liu, W., Ge, F., Zhang, Y., Zhu, B., & Bai, B. (2013). Dietary cholesterol alters memory and synaptic structural plasticity in young rat brain. *Neurological Sciences, 34*, 1355–1365.

Yao, Y., Zhao, G., Xiang, J., Zou, X., Jin, Q., & Wang, X. (2016). Lipid composition and structural characteristics of bovine, caprine and human milk fat globules. *International Dairy Journal, 56*, 64–73.

Zavaleta, N., Kvistgaard, A. S., Graverholt, G., Respicio, G., Guija, H., Valencia, N., et al. (2011). Efficacy of an MFGM-enriched complementary food in diarrhea, anemia, and micronutrient status in infants. *Journal of Pediatric Gastroenterology and Nutrition, 53*, 561–568.

Zhang, X., Qi, C., Zhang, Y., Wei, W., Jin, Q., Xu, Z., et al. (2019). Identification and quantification of triacylglycerols in human milk fat using ultra-performance convergence chromatography and quadrupole time-of-flight mass spectrometery with supercritical carbon dioxide as a mobile phase. *Food Chemistry, 275*, 712–720.

Zou, X.-Q., Guo, Z., Huang, J.-H., Jin, Q.-Z., Cheong, L.-Z., Wang, X.-G., et al. (2012). Human milk fat globules from different stages of lactation: A lipid composition analysis and microstructure characterization. *Journal of Agricultural and Food Chemistry, 60*, 7158–7167.

Chapter 16
Ghee, Anhydrous Milk Fat and Butteroil

Surendra Nath Battula, N. Laxmana Naik, Rajan Sharma, and Bimlesh Mann

1 Introduction

Oils and fats have been constituents of human nutrition from ancient times; they contain the highest level of energy of all components of food and supply essential elements for the body. Dietary lipids provide essential fatty acids and facilitate the absorption of lipid-soluble vitamins (FAO-WHO, 2010). They also largely determine the texture and flavour of foods and hence, enhance their taste and acceptability. In addition, fats slow gastric emptying and intestinal motility, thereby prolonging satiety. Milk fat is the third main source of lipids for human nutrition (Aguedo et al., 2008); it has been harvested for human use for thousands of years (Gnanasambandam et al., 2017). Historically, milk fat and milk fat-based products have found a special place in food habits of all cultures as people all over the world have consumed them. It is widely acknowledged that milk fat imparts excellent flavour and superior mouthfeel to milk products (Aguedo et al., 2008; Omar et al., 2017; Reddy, 2010). Milk fat in the form of butter or cream has limited shelf life owing to lipolysis and microbiological deterioration. Since these types of spoilage take place in the water phase or at the interface between the water and the fat phases, it is well known that removal of water from cream and butter by converting them to anhydrous milk fat, butteroil and ghee extends the keeping quality of milk fat (Illingworth et al., 2009; Mortensen, 2011). The latest is the major form of utilization of milk fat in the Indian sub-continent. In this Chapter, the product description, method of manufacturing, composition, physico-chemical properties and functionality of ghee, anhydrous milk fat and butteroil are described.

S. N. Battula (✉) · N. Laxmana Naik
Southern Regional Station, ICAR-National Dairy Research Institute, Bengaluru, India

R. Sharma · B. Mann
ICAR-National Dairy Research Institute, Karnal, India

T. Truong et al. (eds.), *Dairy Fat Products and Functionality*,
https://doi.org/10.1007/978-3-030-41661-4_16

1.1 Production Statistics and Market Trend

The world milk output in 2017 reached to 811 million tons, which is 1.4% higher than in 2016; however, the world butter trade declined in 2017 by nearly 11% (FAO, 2017). The production figures for anhydrous milk fat and ghee are often converted into butter equivalents and included in the butter data. In general, the production figures for all milk fat product types are pooled under the names anhydrous milk fat or butteroil independent of their composition, and without consideration for the specifications established by the Codex Alimentarius Commission (Mortensen, 2011). As per the FAO (2017), data on Milk and Milk Products: Price and Trade Update, World trade for butter in 2017 was around 929,000 tons. India is the largest milk producing country in the world (165,612,000 tons in 2017) (FAO, 2017). In India, about 27.5% of the total milk production is converted into ghee and 6.5% as butter. A report published by IMARC Group (https://www.imarcgroup.com/anhydrous-milk-market), entitled "Anhydrous Milk fat (Butteroil) Market: Global industry trends, share, size, growth, opportunity and forecast 2017–2022" finds that the global anhydrous milk fat market reached a volume of 465.5000 tons in 2016, grew at a CAGR of 4.7% during 2009–2016. According to another report on the "Ghee Market: Global industry trends, share, size, growth, opportunity and forecast 2018–2023", the global ghee market grew at a CAGR of around 3.8% during 2010–2017 reaching a volume of 5.6 million tons in 2017 (https://www.imarcgroup.com/ghee-market). As per the last 5 years data (FAO, 2017), New Zealand, European Union, Belarus, Australia, India and the United States of America are the major butter exporting countries. China, Russia, Egypt, Mexico, the United States of America are the major butter importing countries. Apart from these, many Asian and European countries, which are not self-sufficient in their dairy production, often import anhydrous milk fat.

1.2 Standards and Specifications

The Codex Alimentarius Commission (CAC) under the joint Food and Agricultural Organization (FAO) and World Health Organization (WHO) has established standards with revisions and amendments, for milk fat products intended for further processing or culinary use. The standard CODEX STAN: 280-1973 applies to products designated for anhydrous milk fat, milk fat, anhydrous butteroil, butteroil, and ghee, which are defined as fatty products derived exclusively from milk and/or products obtained from milk by means of processes that result in almost total removal of water and non-fat solids. Further, ghee is defined as a product exclusively obtained from milk, cream or butter by means of processes that result in almost total removal of water and non-fat solids, with an especially developed flavour and physical structure (Codex, 2018).

Table 16.1 is derived from these above specifications/standards and shows the main criteria by which these products are classified with respect to their fat/

Table 16.1 Codex standards and other quality parameters for anhydrous milk fat, butteroil and ghee

Composition and quality parameters	Anhydrous milk fat/ Anhydrous butteroil	Milk fat	Butteroil	Ghee
Minimum milk fat (% m/m)	99.8	99.6	99.6	99.6
Maximum moisture (% m/m)	0.1			
Maximum free fatty acids as oleic acid (g/100 g)	0.3	0.4	0.4	0.4
Maximum peroxide value (milli-equivalents of oxygen/kg fat)	0.3	0.6	0.6	0.6
Taste and odour	Acceptable for market requirements after heating a sample to 40–45 °C			
Texture	Smooth and fine granules to liquid, depending on temperature			
Copper; maximum level (mg/kg)	0.05			
Iron; maximum level (mg/kg)	0.2			

Table 16.2 Food additives permitted in butteroil, anhydrous milk fat and ghee as per Codex standards

Food additives	Maximum level
Ascorbyl esters	500 mg/kg
Butylated hydroxy anisole	175 mg/kg
Butylated hydroxy toluene	75 mg/kg
Citric acid	GMP
Propyl gallate	100 mg/kg
Sodium dihydrogen citrate	GMP
Tocopherols	500 mg/kg
Trisodium citrate	GMP

moisture content, and also the main quality parameters, such as free fatty acids (FFA) and peroxide value (PV), and other contaminants that are used to classify them into one of the aforementioned products. Starter cultures of harmless lactic acid producing bacteria are the permitted ingredients in the manufacture of butter oil, anhydrous milk fat, and ghee. Food additives permitted as per the General Standard for Food Additives (CODEX STAN 192-1995) in Food Category 02.1.1 (butteroil, anhydrous milk fat, ghee) are shown in Table 16.2. The products covered by this standard shall comply with the maximum limits for contaminants that are specified for the product in the General Standard for Contaminants and Toxins in Food and Feed (CODEX STAN 193-1995). The milk used in the manufacture of the products covered by this standard shall comply with the maximum limits for contaminants and toxins specified for milk by the General Standard for Contaminants and Toxins in Food and Feed (CODEX STAN 193-1995) and with the maximum residue limits for veterinary drug residues and pesticide residues established for milk by the CAC.

Food Safety and Standards Act (FSSA): In India, the standards for ghee (Table 16.3) are laid down by the Food Safety and Standards Act (2006), (FSSR, 2011), which are mandatory in nature. As per this Act, milk fat, ghee, butteroil,

Table 16.3 Standards for milk fat products as per Food Safety and Standards Act (2006), India

Parameter	Milk fat, Butteroil	Anhydrous milk fat, Anhydrous butteroil	Ghee
Moisture, maximum (%, m/m)	0.4	0.1	0.5
Milk fat, minimum (%, m/m)	99.6	99.8	99.5
Butyro-refractometer reading at 40°C	40–44	40–44	40.0–45.0[a]
Reichert Meissl value, minimum	28	28	Not less than 24.0[b]
Polenske value	1.0–2.0	1.0–2.0	–
Free fatty acids as oleic acid, maximum (%)	0.4	0.3	3.0
Peroxide value (milli equivalent of oxygen/kg fat), maximum	0.6	0.3	–
Baudouin test	Negative	Negative	Negative

[a]Butyro-refractometer reading
[b]Reichert Meissl value differ in different states of India and are also different for cotton tract and non-cotton tract areas.

anhydrous milk fat and anhydrous butteroil are fatty products derived exclusively from milk or products obtained from milk, or both, by means of processes, which result in almost total removal of water and milk solids-not-fat. Ghee has specially developed flavour and physical structure as a result of its method of manufacturing. The raw materials (milk or milk products) used shall be free from added flavour, colour or preservative. In the case of anhydrous milk fat and butteroil, the permitted additives are ascorbyl palmitate: 500 mg/kg; propyl gallate: 100 mg/kg; ethyl gallate: 100 mg/kg; octyl gallate: 100 mg/kg; dodecyl gallate: 100 mg/kg; butylated hydroxyl anisole: 175 mg/kg and citric acid: Good Manufacturing Practice (GMP). As per the Act, no antioxidant or any additive is permitted to be added to ghee.

These standards mainly focus on the manufacturing practices followed (moisture and free fatty acid levels) and purity (Reichert value or the Reichert-Meissl-Wollny value or Reichert-Meissl-Wollny number; RM value, Polesnke value, Butyro-Refractometer; BR reading, Baudouin test). Since RM value and BR reading are affected by several factors such as breed, species and feed, which vary from state to state in the country, the FSSA standards for these parameters vary accordingly. Further, since, cottonseed/cake feeding affects the RM value and BR reading considerably, cotton tract areas are identified, and the ghee produced in such areas have different standards so as not to discriminate the genuine ghee as adulterated ghee.

2 Product Description

2.1 Ghee

Ghee is distinctly different from the other milk fat products in terms of flavour and texture and therefore, several definitions of ghee have been proposed. Ghee may be defined as a pure clarified fat exclusively obtained from milk, cream or butter, by means of processes involving the application of heat at atmospheric pressure, which result in the almost total removal of moisture and solids-not-fat and which gives the product a characteristic flavour and texture. Codex Alimentarius defines ghee as a product exclusively obtained from milk, cream or butter, by means of processes, which result in almost total removal of water and not-fat-solids, with an especially developed flavour and physical structure (Codex Alimentarius, 2018). Heat induced changes in milk proteins/lactose during the clarification process impart a distinctive, pleasant cooked, nutty flavour to ghee. The ghee prepared from *makkan*, the *desi* butter (obtained by churning the fermented milk, *dahi*), contains many of the volatile compounds formed during the fermentation of milk and their subsequent fractionation into *desi* butter. These impart a distinctive characteristic flavour when the butter is clarified into ghee. Ghee of cow milk has a distinct golden yellow colour, attributable to carotene. On the other hand, the ghee of buffalo milk is noted for its whitish appearance with a greenish tinge, attributed to bilirubin and biliverdin. During the slow cooling process after heat desiccation and subsequent storage, ghee develops typical granular crystals, which remain partly dispersed in the middle layer and partly in the bottom layer of clear liquid fat. Ghee is also popular though with some other names in many other tropical and semi-tropical countries. *Samna* in Egypt (Abou-Donia & El-Agamy, 1993), *Samin* in Sudan (Dirar, 1993), *Maslee* or *Samn* in Middle East, *Rogan* in Iran (Urbach & Gordon, 1994), and *Samuli* in Uganda (Sserunjogi et al., 1998).

2.2 Butteroil

Butteroil is the oily product obtained from butter, also known as clarified butter. It is a popular product and widely consumed food commodity. One reason of popularity is its long shelf-life (Kaya, 2000). It is made from cream or white butter of varying age, and alkali is permitted to neutralize the free fatty acids. Butteroil has an even texture and a slightly crystalline yellow colour. Though, there are slight variations in the taste and aroma depending on regional preferences around the world, it has a light taste and a subtle sweet aroma.

2.3 Anhydrous Milk Fat

Anhydrous milk fat possesses a light yellow colour and has a clean bland taste, free from sour, bitter, rancid, oxidized or other objectionable flavours. It should be made from fresh cream or white butter and alkali is not permitted to neutralize the free fatty acids. Intense heat treatment is not used in the manufacturing process. Consequently, anhydrous milk fat lacks heated butter flavours. In the following text the term 'Anhydrous milk fat' is used for both anhydrous milk fat/anhydrous butteroil, milk fat, and butteroil; because both the production technology and the applications of these products are quite similar. Moreover, anhydrous milk fat is the term commonly used in both industry and international trade. Raw material and processing specifications are the same as for anhydrous butteroil except the use of alkali to neutralize free fatty acids as mentioned above. If the neutralization option is used, the resulting product then falls outside the definition of anhydrous milk fat and into butteroil categories (Illingworth et al., 2009).

3 Ghee

3.1 Method of Production

Ghee is produced by clarifying butter at temperatures ranging from 110 to 120°C for a flash to a few minutes of continuous heating. The oily phase is then separated from the residue through clarification and/or filtration. There are three methods of ghee preparation viz., indigenous, creamery-butter and direct cream (Fig. 16.1). Relative merits of these methods of ghee manufacture are summarized in Table 16.4.

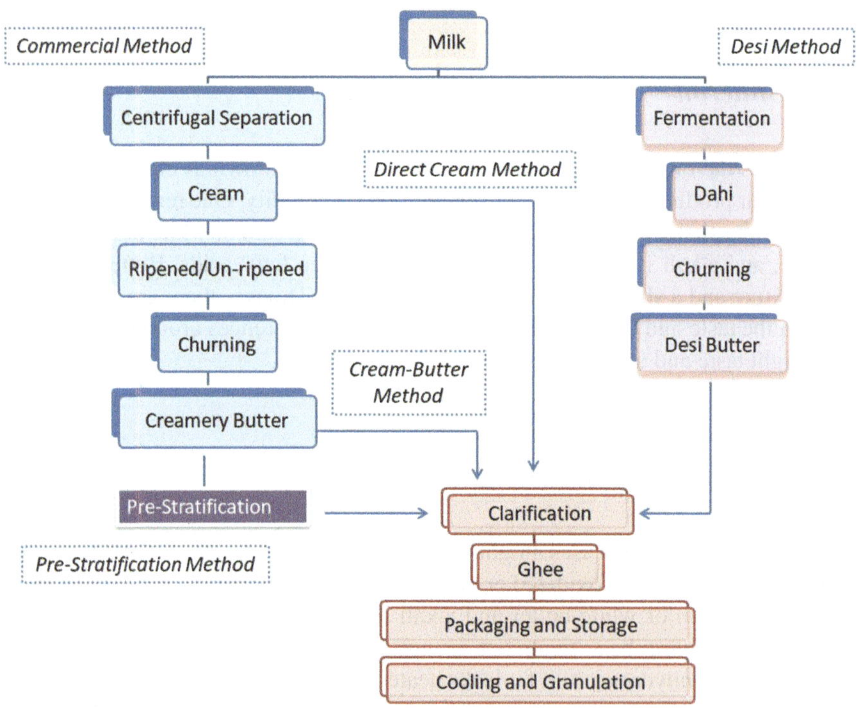

Fig. 16.1 Block diagram of ghee manufacturing methods

Table 16.4 Relative merits of different methods of making ghee

Parameter	Desi method	Cream butter method	Direct cream method	Prestratification method
Fat recovery (%)	88–90	88–92	92	>93
Aroma	Strong nutty	Pleasantly rich	Pleasantly rich	Pleasant
Flavour	Acid	Normal	Normal	Normal
Texture	Packed coarse grains	Slushy fine grains (cow) or packed fine grains (buffalo)	Fine grains	Fine grains
Clarification using heat	Easy, economical and prestratification possible	Easy, economical and prestratification possible	Easy and economical	Easy and economical
By-product(s)	Buttermilk and ghee residue	Skimmed milk, buttermilk and ghee residue	Skimmed milk, and ghee residue	Skimmed milk, buttermilk, butter serum and ghee residue
Keeping quality	Moderate	Good	Good	Good
Adaptability	Small scale	Large scale	Large scale	Large scale

Data adopted from Rajorhia (1993) and Aneja et al. (2002)

In the indigenous method, practiced usually in house-hold and unorganized sector, boiled and cooled milk is fermented with a starter culture to *dahi*, which is churned with an indigenous corrugated wooden beater or in a domestic mixer-grinder, separating *makkan*, the *desi* butter, and clarifying it into ghee in an open pan with direct heat. Under commercial conditions, cream separated from whole milk is separated into butter using a churn (batch process) or in a continuous butter making machine. The butter is then heated at 110–120°C for no hold to a few minutes in a stainless steel steam kettle to ghee. To save on energy and time, butter is usually subjected to a process called pre-stratification, in which butter is melted at about 50–60°C, and the separated serum is drained off; and the oily phase is then clarified to ghee. In the pre-stratification method, advantages such as economy in fuel consumption and production of ghee with low acidity and comparatively longer shelf life have been claimed (Bahadur et al., 1950; Ganguli & Jain, 1973). In the direct cream process, cream (50–70% fat) is directly heat clarified to ghee. Warner (1976) described the stages of heat clarification process of ghee. To convert cream or butter to ghee, heat is applied at a controlled temperature at the various stages of processing. Initially, the temperature is gradually raised to about the boiling point of water, while stirring to control frothing. In the second stage, most of the free water evaporates, which requires a considerable amount of heat. As most of the water evaporates, the rate of heating is controlled and maintained at about 103°C to prevent the charring of solids-not-fat so as not to develop flavours and/or a brown colour. Overheating could drive off desirable volatile flavour materials and also impair the formation of suitable grains upon cooling. The impairment of crystal formation appears to be associated with the possible volatilization of some short-chain FFA, which changes the normal composition of the fat in ghee. Finally, the temperature is raised to between 105 and 118°C with constant agitation in order to remove the water bound to the solids-not-fat and to develop the characteristic flavour. Whatever is the process, ghee is separated from the residue, which is generally referred to as ghee residue through clarification and or filtration, filled into pouches, bottles or tins and the filled containers are stored at 22–23°C for 12–24 h for granulation.

3.2 Ghee Composition

Ghee is mainly isolated milk fat and has almost all the constituents of the lipid component of milk. The chief difference between anhydrous milk fat and ghee pertains with regard to the level of phospholipids and flavour components. General composition of cow and buffalo ghee as compiled by Sharma (1981) is shown in Table 16.5. A general composition of bovine milk lipids collected from literature is also shown in Table 16.6.

Triglycerides account for nearly 96–98% of the total milk lipids. More than 200 species of triglycerides alone have been quantified in milk fat (Gresti et al., 1993). The distribution of fatty acids within milk fat is not random (Jensen et al., 1991).

Table 16.5 Major and minor constituents of cow and buffalo ghee/milk fat

Constituents	Cow ghee	Buffalo ghee
Saponifiable constituents		
Triglycerides		
Short chain (%)	37.6	45.3
Long chain (%)	62.4	54.7
Trisatutated (%)	39.0	40.7
High melting (%)	4.9	8.7
Partial glycerides		
Diglycerides (%)	4.3	4.5
Monoglycerides (%)	0.7	0.6
Phoshpolipids (mg %)	38.0	42.5
Unsaponifiable constituents		
Cholesterol (mg %)	330	275
Lansterol (mg %)	9.32	8.27
Lutein (μg/g)	4.2	3.1
Squalene (μg/g)	59.2	62.4
Carotene (μg/g)	7.2	–
Vitamin A (μg/g)	9.2	9.5
Vitamin E (μg/g)	30.5	26.4
Ubiquinone	5.0	6.5
Flavour compounds		
Free fatty acids (% oleic acid)	0.2	0.2
Total carbonyls (μM/g)	7.2	8.64
Volatile carbonyls (μM/g)	0.33	0.26
Head space carbonyls (μM/g)	0.035	0.027

Table 16.6 Composition of bovine milk lipids

Class	A	B	C
Triglycerides (%)	97.5	95.8	97–98
Diglycerides (%)	2.5	4.2	2.9
Monoglycerides (%)	0.36	2.25	0.28–0.59
Cholesterol esters (%)	Trace	0.02	NR
Cholesterol (%)	0.31	0.46	0.42
Phospholipids (%)	0.596	1.11	0.20–1.00
Free fatty acids (%)	NR	0.28	0.10–0.44

A: Christie (1988); B: Bitman and Wood (1990); C: Jensen (2002). *NR*: not reported

The acyltransferase enzymes involved in milk fat synthesis preferentially synthesize triglycerides with short-chain fatty acids in the *sn*-3 position and longer chain fatty acids in the *sn*-1 and *sn*-2 positions. Nearly 85% of the butyric acid and 58% of the caproic acid in milk fat are found at *sn*-3 position (Dimick et al., 1970). Although milk fat contains hundreds of triglyceride species, it is often discussed in terms of groups or fractions of triglycerides, which are chemically and physically distinct (Bornaz et al., 1993; Marangoni & Lencki, 1998; Timms, 1980). Triglyceride fractions are typically distinguished from each other on the basis of melting behavior. Accordingly, there are three main fractions of triglycerides, the high-melting, medium-melting and low-melting fractions (HMF, MMF and LMF, respectively).

Minor lipids account for 2–4% of the lipids (Bitman & Wood, 1990; Christie, 2003). These lipids include compounds such as partial glycerides, free fatty acids, phospholipids and cholesterol. The minor lipids present in the highest proportions in milk fat tend to be the phospholipids. Nearly 70% of the phospholipids in milk are associated with fat globule membrane (Swaisgood, 1996). The major species of phospholipids in milk fat are sphingomyelin (35.3%), phosphatidylcholine (33.6%) and phosphatidylethanolamine (22.3%) (Swaisgood, 1996). The phospholipids in milk represent a source of long-chain polyunsaturated fatty acids (Jensen et al., 1991). Buffalo ghee contains slightly higher levels of phospholipids than cow ghee. Partial acylglycerols are present in milk fat as a result of incomplete triglyceride synthesis and lipolytic activity. Lipolytic enzymes preferentially liberate fatty acids from the sn-1 and sn-3 positions of triglycerides (Walstra et al., 1999). Lipolysis results in an increase in the proportion of diacylglycerols and monoacylglycerols as well as free fatty acids. The remaining minor lipids include sterols, carotenoids, glycolipids, squalene and hydrocarbons, which are present in trace amounts (Jensen & Newburg, 1995). Cholesterol is the predominant sterol present in milk fat and represents most of the unsaponifiable lipid material in milk. A small fraction of the cholesterol (~10%) is esterified to a fatty acid (Jensen & Newburg, 1995). Nearly 85% of the cholesterol in milk is found in the fat globule membrane (Swaisgood, 1996). Carotenoids are responsible for the yellow colour of cow milk fat and originate from plant material in the diet (Gunstone et al., 1986). Milk is considered to be a good source of Vitamin A, but contains only small amounts of Vitamin D, E and K (Buss et al., 1984). The total Vitamin A bioactivity can be obtained by summation of the concentrations and activities of the different forms of Vitamin A. In milk fat, this gives an average value of approximately 12 retinol equivalents/g fat or 40 IU/g (MacGibbon & Taylor, 2006). In milk, α-tocopherol accounts for virtually all of Vitamin E, although very small amounts of β-tocopherols and ɣ-tocopherols are present. Concentration of α-tocopherol in milk fat varies widely, ranging from 18 to 35 μg/g fat (Wearne, 1999).

Various workers identified and determined the fatty acid composition of milk fat by using gas liquid chromatography (GLC) (Glass & Jenness, 1971; Hawke, 1957; Jensen et al., 1962; Murthy & Narayanan, 1971; Patton et al., 1960; Smith & Ronning, 1961; Thompson et al., 1959). Although over four hundred and six different fatty acids have been identified in bovine milk fat, only a few (~12) of these are present at concentrations exceeding 1.0% (Jensen et al., 1991; Jensen & Newburg, 1995). Levels of important fatty acids in milk fat as reported by various workers are shown in Table 16.7. Fats from ruminant species are characterized by a relatively high proportion of short- and medium-chain fatty acids (4:0–14:0) and trans unsaturated acids (Jensen et al., 1991). These acids are synthesized in the mammary gland, while polyunsaturated fatty acids are partially hydrogenated by microorganisms in rumen, resulting in a number of geometric and positional unsaturated acid isomers (Gunstone et al., 1986). The presence of 8.3% short-chain fatty acids (4:0–8:0) and 6.6% medium-chain (10:0–12:0) in milk fat distinguishes it from all other fats, in which those types of fatty acids are not present (Jensen & Newburg, 1995). They have a large effect on the melting and solidification properties of milk fat and thus, their role is comparable to the role of unsaturated fatty acids in vegetable fats. Milk fat is a relatively high saturated fat containing about 65% saturated fatty acids of the total fatty acids. Intake of saturated fatty acids is considered as a strong indicator of

Table 16.7 Fatty acid profiles of milk fats derived from cow and buffalo

Fatty acids	Cow milk fat[a]	Buffalo milk fat[a]	Cow milk fat[b]
4:0	3.2	4.4	3.49
6:0	2.1	1.5	2.36
8:0	1.2	0.8	1.36
10:0	2.6	1.3	2.89
12:0	2.8	1.8	4.03
14:0	11.9	10.8	10.79
14:1	2.1	1.3	1.10
16:0	29.9	33.1	26.91
16:1	2.2	2.0	1.50
18:0	10.1	12.0	10.51
18:1	27.4	27.2	24.82
18:2	1.5	1.5	2.68
18:3	0.6	0.5	0.51
20:0	NR	NR	0.10

NR: not reported

[a]Murthy and Narayanan (1971)

[b]Maniongui, Gresti, Bugaut, Gauthier, and Bezard (1991)

serum total, low-density lipoprotein (LDL) and high-density lipoprotein (HDL) cholesterol levels (Woodside & Kromhout, 2005). However, all the saturated fatty acids do not increase cholesterol levels. Lauric, myristic and palmitic acids raise blood total and LDL cholesterol levels, whereas stearic and short and medium-chain fatty acids have little or no effect on blood cholesterol levels (German & Dillard, 1998). Further, short and medium-chain fatty acids are directly absorbed from the intestine into portal circulation and are not transported through the bloodstream to the liver by chylomicrons; consequently, these fatty acids would be expected to have different effects on blood lipids than long-chain fatty acids (McNamara, 1991). The unsaturated fatty acids constitute about 30% contributed mainly by oleic acid (MacGibbon & Taylor, 2006; Walstra et al., 1999). The proportion of mono-enes, di-enes and poly-enes is respectively about 27, 2.5, and 0.8% (Walstra et al., 1999). Some minor fatty acids have odd-numbered chain lengths or have branched chains (iso and ante-iso). Some fatty acids contain keto or hydroxyl groups and even a cyclohexyl group is detected in some fatty acids (Jensen, 1992). Fatty composition of milk fat is influenced by factors like species, breed, feed (Grummer, 1991; Hawke & Taylor, 1983) diet (Christie, 1995; Grummer, 1991; Jensen et al., 1991; Nielsen, 1971), region (Dolby, 1949), season (MacGibbon & McLennan, 1987; Wolff et al., 1995) and stage of lactation (Christie, 1995; Kleyn, 1992; MacGibbon & McLennan, 1987). In temperate countries, winter butter, for example, typically has a higher palmitic acid content and lower oleic acid content than summer butter and, overall, a higher level of saturation, which makes it notoriously harder than summer butter (Prentice, 1972). Seasonal variation is related to the fact that the polyunsaturated fatty acid content of fresh green fodder consumed by ruminants during the summer is higher than that of winter feed. Buffalo milk fat is reported to have higher levels of palmitic and stearic acids and lower levels of unsaturated fatty acids than cow milk fat (Lakshminarayana & Rama Murthy, 1985; Murthy & Narayanan, 1971).

3.3 Sensory Properties of Ghee

Ghee is prepared by a number of methods, each one giving ghee of different sensory attributes. Traditional criteria of quality in ghee centre around its sensory properties namely, flavour texture and sometimes colour. These attributes of ghee have long served to characterize and identify the product. Regional preferences (Rajorhia, 1993) occur for flavour and texture in ghee used in India. The flavour preferences vary from undercooked to overcooked, while the texture preferences vary from fine to coarse grains. Indian Standards Institution (1966) provides a scorecard (Table 16.8) for assessing the quality of ghee by considering the sensory properties of ghee.

Flavour The flavour of ghee cannot be surpassed, a fact known and taken to heart since the eighth century BC when it was mentioned in Indian texts (Rangappa & Achaya, 1974). Ghee has a rich aroma, which is described as pleasant, nutty, lightly cooked or caramelized aroma. The aroma is generally evaluated by slightly rubbing

Table 16.8 Score card for evaluation of sensory scores of ghee

Characteristics	Maximum score	Sample score
Flavour	50	
Texture	20	
Colour	10	
Freedom from suspended impurities	15	
Packaging	5	

Note: Grading: excellent (A): >90; good (B): 80–90; fair (C): 60–79; poor (D): <59

a little ghee on the back of one hand with the index finger of the other, when frictional and body heat releases the aromatic constituents, which can then be inhaled. Taste of ghee can best be described as lack of oiliness or of blandness, sweetly rather than sharply acid. Flavour is a composite attribute, comprising mainly taste but with overtones of smell, feel and temperature perception. The highly prized characteristic flavour of ghee is due to a complex mixture of compounds produced during the various stages of processing as well as those naturally present in milk (Achaya, 1997; Joseph & Appachar, 1980; Yadav & Srinivasan, 1992). It is not due to a single flavour component or single class of flavour compounds. Rather, an optimum blend of various flavour compounds (n > 100), viz., fatty acids, carbonyls, lactones etc., is responsible (Wadhwa & Jain, 1990). Flavour of ghee is subjected to variations depending on the method of preparation, ripening of cream, temperature and time of clarification of cream/butter etc. Flavour constituents of ghee are discussed separately in the subsequent section.

Colour A bright yellow colour, caused by the presence of carotenoid pigments, is associated with cow ghee and constitutes a desirable criterion in areas where the latter is preferred. Indeed in such regions, buffalo ghee, which lacks carotene and is white or pale-cream in colour, is sometimes coloured with natural pigments like turmeric, curcumin, annatto or synthetic fat-soluble dyes to raise its price. Such an addition is not permitted in India and is considered as illegal. The colour of fats always appears deeper to the eye when melted than when in solid form, a general phenomenon connected with viewing by transmitted and reflected light. Buffalo ghee made using the indigenous route appears greenish due to the presence of bile pigments, bilirubin and biliveridin (Kumar et al., 1985).

Texture Completely melted ghee, on cooling to room temperature, can assume the form of large, coarse grains suspended discretely or in clusters in a liquid phase. Grain formation is a unique property of ghee as it contains a large variety of triglycerides having wide range of melting points. Granularity is considered by the common buyer as an important criterion of quality and even purity. Therefore, under market conditions, great importance is attached to the grainy structure in ghee when judging its quality. Granulation in ghee is due to the crystallization phenomenon of the high melting triglycerides. The fatty acid profile of milk fat, temperature at which ghee is clarified, followed by the rate at which it is cooled, seeded and held under quiescent storage before packing, and temperature of ghee storage in the marketing network are various factors that contribute towards grainy texture of ghee.

The partly granular form assumed by ghee is primarily due to certain content of glycerides of higher melting saturated fatty acids, especially palmitic and stearic. As the fatty acid composition is influenced by several factors, it is expected that these factors also would have an effect on the appearance of the ghee. Arumughan and Narayan (1982) reported that buffalo ghee produced bigger grains than cow ghee and such a difference was due to the presence of higher levels of myristic, palmitic and stearic acids, and trisaturated glycerides in the former than in the latter. Feed is another factor that influences the fatty acid composition significantly and therefore affects the texture of ghee. For instance, a higher proportion of cotton seed in the diet gives rise to augmented stearic acid content in ghee, which looks tallowy in appearance and can be quite hard (Rangappa & Achaya, 1974).

The desi method produces large crystals in ghee. On the other hand, ghee obtained by creamery butter or direct cream processes need not necessarily have large crystals (De, 1980). Nagesh (1981) reported that size and quantity of grains were more in case of ghee made from butter than cream. He also reported that the quantity and size of the grains formed were more in cow ghee prepared from fresh than ripened butter / cream. The temperature of clarification has also been found to influence the texture of ghee. Ghee prepared employing higher temperatures of clarification gives better grain size and quantity. As the temperature of clarification increased from 100 to 120°C, the size and the quantity of grains increased. At higher clarification temperature, phospholipids and some of the substances from solid-not-fat portion of the cream or butter may get transferred to milk fat phase and such substances while entering the fat phase may act as nuclei for the formation of crystals and also affect the size and quantity of grains formed during the crystallization of milk fat (Lakshminarayana, 1983; Nagesh, 1981). Rate of cooling is another important factor affecting the grain formation in ghee; rapid cooling causes solidification of large number of nuclei simultaneously. Therefore, the crystals that grow are very small in size but large in number. On the other hand, in slow cooled ghee, the grain size increased but the quantity decreased. Rachana et al. (2013) reported that rapid cooling gave crystals of only about 0.1–0.136 mm in size and the quantity of crystals found was as high as 80–85%. In slow cooled ghee, the crystal size grew up to 0.206–0.223 mm and the quantity of crystals found was 30–35%. Storage temperature was the main factor influencing the size and quantity of grains in ghee. The temperature at which the maximum grain size obtained for cow and buffalo ghee was 28 and 30 °C, respectively (Nagesh, 1981). The storage period of ghee also has a considerable influence on grain size and liquid portion. Arumughan and Narayanan (1982) studied granulation of buffalo ghee and cow ghee at 29°C. The grain size was observed to increase from first to third day in both buffalo and cow ghee. It is often difficult to get consistency in the granulation of ghee under commercial conditions. Seeding of melted ghee with grains or crystals of high melting triglycerides was found to have a significant influence on the grain formation in ghee. The size of the grains in such seeded ghee is small, but the quantity of grains is higher. This was because of large number of nuclei provided by the seeded crystals for the formation of grains simultaneously. Shape of the crystal is needle like in the case of seeded ghee, while it is spherical in the case of unseeded ghee. This difference in the crystal shape might be due to the sudden formation of the crystals in the case of seeded ghee. Various workers have reported that addition of emulsifiers like monoglycerides (Kumar & Negi, 2004) or surfactants (Marangoni, 2005) could improve granulation in ghee/milk fat.

3.4 Flavour Constituents of Ghee

Flavour characteristics of ghee develop at almost all the stages of processing concerned with ghee making. However, fermentation of milk (as in indigenous process) or cream and heat clarification process of cream/butter influence the development of flavour in ghee to a high degree. Further, natural constituents present in milk, cream and butter such as indole, skatole, dimethylsulphide, diacetyl, carbonyl compounds and free acids are transferred to fat phase during heat clarification. These constituents contribute to the flavour of ghee. Wadhwa and Jain (1990) extensively reviewed the chemistry of ghee flavours and variations in the level of flavour components as affected by various technological parameters. Flavour of ghee analyzed through GLC has revealed a wide spectrum consisting of more than 100 flavour compounds. A number of fermentation constituents (e.g., lactic-, butyric-, caproic- and caprylic acids, and carbonyls etc.) produced during souring are partly transferred to ghee and impart flavour to ghee. During heat clarification, the browning constituents produced, as a result of reaction between carbohydrates and proteins; lactones and acrolein produced from fat, carbonyls and fatty acids produced due to decomposition of fat, lactose degradation constituents, volatile constituents from cooked proteins etc., are partly transferred to the fat phase and influence the flavour of ghee to a great extent.

Free Fatty Acids Ghee owes its pleasing flavour to several fatty acids. These compounds are produced from fatty acid glycerides by lipolysis during fermentation of milk or cream and processing treatments during the preparation of ghee. The free fatty acid level of fresh ghee is about 6–12 mg/g. The lower fatty acids (C6–C12), though present in low concentration (0.4–1 mg/g) accounting only 5–10% of total free fatty acids, contribute significantly to the ghee flavour (Gaba & Jain, 1973).

Carbonyls Carbonyls play an important role in flavours and off-flavours of fat rich dairy products like ghee. This class of compounds includes, broadly, monocarbonyls and dircarbonyls. Monocarbonyls further constitute alkan-2-ones, alkanals, alk-2-enals and alka-2,4-dienals. Alken-2-ones or methyl ketones are reported in several dairy products, especially heat processed. They impart cooked flavour in foods, in which butterfat is used as shortening. Alkan-2-ones are produced by the hydrolysis of ketogenic glycerides (β-ketoglycerides) followed by decarboxylation of β-ketocarboxylic acids during various processing treatments involved in the preparation of ghee. Also, lipolysis of triglycerides through Penicillium moulds during fermentation of milk or cream may be the source of alkan-2-ones. Penicillium lipases liberate free saturated fatty acids and further oxidize them at β position to form β-ketoacids, which on decarboxylation yield methyl ketones. Aldehydes namely n-alkanals, alk-2-enals and alka-2,4-dienals are formed by the oxidation of unsaturated fatty acids during fermentation of milk or cream and heat clarification of cream or butter (Yadav & Srinivasan, 1992). Polar carbonyls, which include dicarbonyls, α-ketoacids, and glyoxals and furfurals are produced during fermentation and heat clarification processes of ghee making. It was reported that the level of polar carbonyls increased by ripening of cream and heat clarification temperature of butter (Rao & Ramamurthy, 1984).

Lactones Lactones contribute significantly to the flavour of fat based dairy products especially the heat processed ones. These compounds are produced by the hydrolysis of lactogenic (hydroxyl) glycerides followed by dehydration (lactonization) of hydroxy acids. Addition of microbial lipase accelerates the lactone production and thus, flavour is enhanced. The lipolysis of delta-hydroxy acid glycerides yields free delta-hydroxy acids, which undergo ring closure to form lactones. Alternatively, delta-ketoacid glycerides undergo lipolysis to yield delta-keto acids, which undergo reduction to form hydroxyl acids and then are converted to lactones. The other compounds expected to play a considerable role in ghee flavour are: hydrocarbons, esters, phenols, benzaldehyde, furans, alkyl pyrazines, sulphur compounds like trithiolane, dithione, thiazoline, thiazole, oxyzole, imidazone etc. (Wadodkar et al., 1996).

3.5 Flavour Development in Ghee Clarification

A good part of the normal development of rich aroma takes place during the rendering process. The most important factor controlling the intensity of flavour in ghee is the temperature of clarification. Ghee prepared at 120°C or above has an intense flavour, which is usually referred to as cooked or burnt. In contrast, ghee prepared at about 110°C has a somewhat mild flavour, often referred to as curdy. The desi method generally produces ghee with the most desirable flavour (Ganguli & Jain, 1973). The heating process generates flavour compounds through the interaction of protein degradation products, lactose and minerals (Achaya, 1997), and possibly through the degradation of FFA and lactose (Fennema, 1985). Carbonyls (aldehydes and ketones) and lactones are believed to play a significant role in the typical flavour of ghee (Yadav & Srinivasan, 1992). The heating process does not only result in production of compounds, which positively improve the ghee flavour, but also seems to drive off putrefactive odours that may have developed in the butter. Some of the constituents produced during clarification along with other flavour constituents produced during fermentation are partly transferred to ghee phase. As clarification is an important step in the transfer of flavour constituents, the sensory attributes of the ghee are almost a direct measure of the temperature and duration of heating. If the clarification is inadequate, the aroma of the resulting ghee would be buttery and on the other extreme, it would be burnt if the butter is overheated. As serum constituents are drained off in the pre-stratification process, there is less chance for the production of flavour compounds from proteins and lactose during the subsequent clarification. Therefore, flavour of ghee made by this process is usually mild.

Fermentation Desi ghee has an intense flavour when compared to ghee made from sweet cream butter. This is due to the production of flavour compounds during fermentation process involved in the manufacture of ghee. Ghee made from sweet cream butter has a mild aroma due to the omission of fermentation process. If cream is selectively ripened in the ghee making process, the flavour could be considerably improved. A number of flavour compounds are produced during fermentation of cream, part of which is transferred to butter. During boiling of butter to ghee, flavour compounds migrate to fat phase resulting in ghee with a good aroma. The flavour improvement in ghee obtained from ripened materials is ascribed to the metabolic activity of the starter bacteria on various cream/butter constituents like lactose, citrate and glucose. The flavour compounds produced during ripening process by starter activity get incorporated in the final product on boiling of butter to ghee resulting in good aroma. The incorporation is facilitated by the acidity in the ripened cream/ butter. In addition, it is likely that the lower pH produced by the starter activity also helps in increasing the intensity of chemical reactions (Yadav & Srinivasan, 1987).

4 Anhydrous Milk Fat and Butteroil Production Process

4.1 Method of Production

The principle behind the manufacture of milk fat products that include anhydrous milk fat and all its variants is the removal of the water and water-soluble components of milk to leave only pure fat or oil. Most important general requirement is purity of milk fat and its stability to autoxidation. To secure this, good-quality fresh milk should be used. Contamination by traces of copper is highly detrimental. There are numerous manufacturing processes as presented in Fig. 16.2. As a rule, one starts from butter. Alternatively, one can make from pasteurized cream (40% fat) that is concentrated to high-fat cream (70–80% fat), followed by phase inversion (from an oil-in-water emulsion to a water-in-oil emulsion). To achieve this, cream can be passed through an agitator, a special pump, or even a homogenizer; often, the phase inversion occurs easier if the cream first is subjected to "washing," i.e., diluting it with water and re-separating. If high fat cream is passed through a scraped surface heat exchanger, while being cooled sufficiently for fat crystallization to occur, then butter is formed. Further, the product will be concentrated by heating to 99.6% fat. It is then vacuum-dried to remove residual moisture. The moisture content should not exceed 0.1%, because otherwise moisture droplets may form at low temperature. If the moisture content is higher (up to 0.4%), the product is usually designated "butter oil."

Industrialized process for manufacturing of anhydrous milk fat from cream is shown in Fig. 16.3 and from butter method is shown in Fig. 16.4. In the cream method, the pasteurized cream is concentrated in a special separator (4) to a fat content of 70–80% fat after which the highly concentrated fat emulsion is

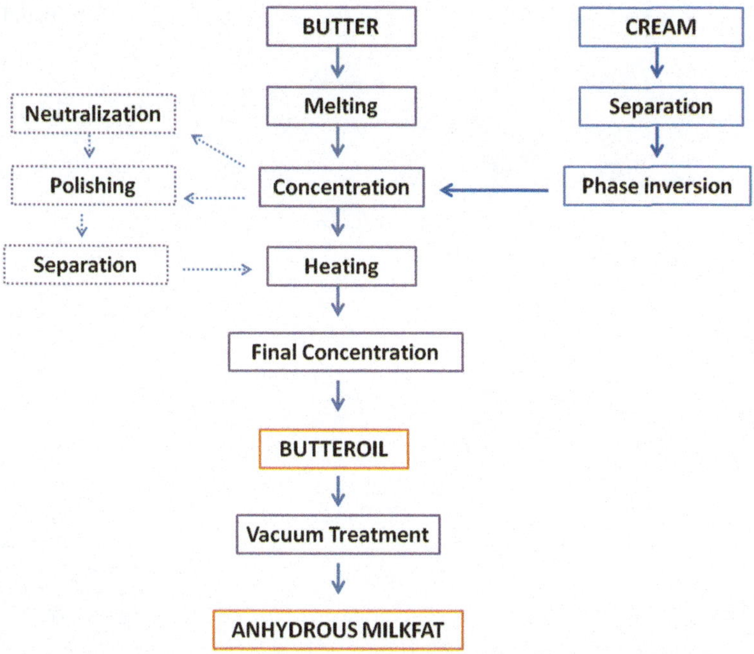

Fig. 16.2 Block diagram of butteroil and anhydrous milk fat production from cream or butter

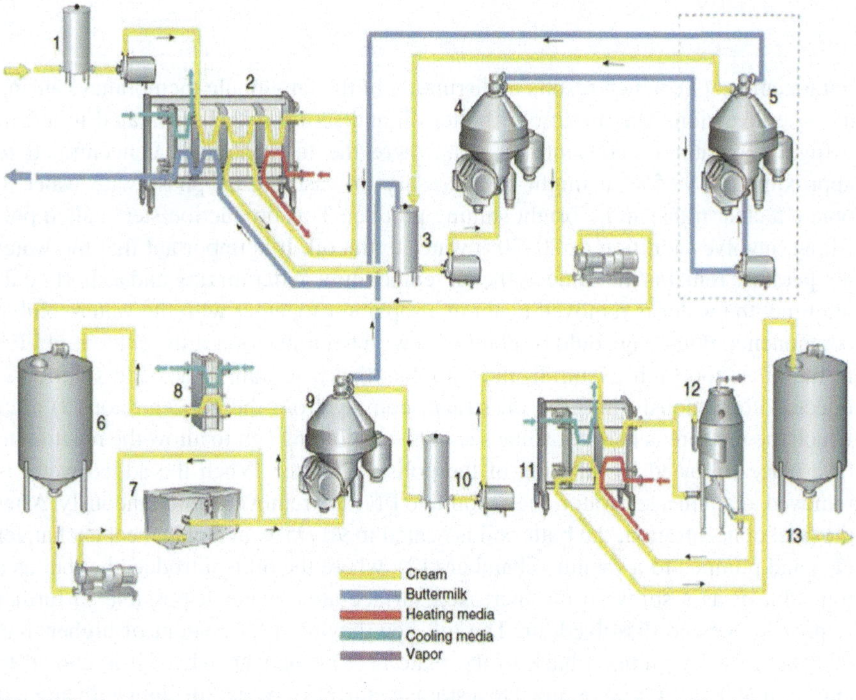

Fig. 16.3 Production lines for anhydrous milk fat from cream. *Note*: 1, 3 and 10 = balance tanks; 2, 8 and 11 = plate heat exchangers; 4 = pre-concentrator; 5 = separator for 'buttermilk' from the pre-concentrator-4 (optional); 6 = buffer tank; 7 = homogenizer for phase inversion; 9 = final concentrator; 12 = vacuum chamber; 13 = storage tank. Reproduced with permission from Dairy Processing Handbook, Tetra Pak A/B, Lund, Sweden (Bylund, 2003)

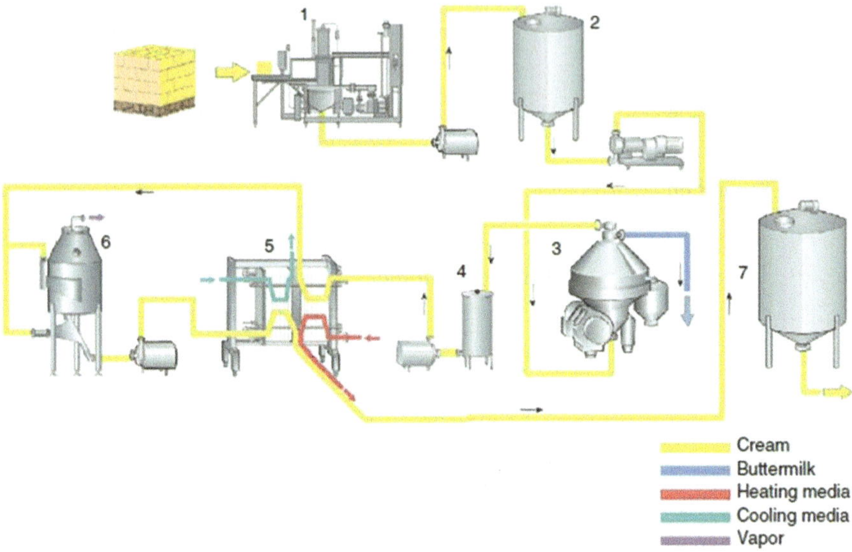

Fig. 16.4 Production lines for anhydrous milk fat from butter. *Note:* 1 = melter and heater for butter; 2 = holding tank; 3 = concentrator; 4 = balance tank; 5 = plate heat exchanger for heating/cooling; 6 = vacuum chamber; 7 = storage tank. Reproduced with permission from Dairy Processing Handbook, Tetra Pak A/B, Lund, Sweden (Bylund, 2003)

homogenized (7), which results in disruption of the fat globule membranes causing a phase inversion. The mixture of butter oil and serum is then separated in a centrifugal separator or concentrator (9), where the fat content is concentrated to approximately 99.5%. It might be necessary to wash the oil phase with water to obtain a clear transparent, bright shining product. This production step, called polishing, involves addition of 20–30% water to the oil. It is important that the water temperature remains the same as the oil temperature. After mixing and a short holding time, the water is removed again in a separator together with the water-soluble components. Polishing could be combined with neutralization if the content of FFA in the oil is too high. Neutralization implies that a sodium hydroxide solution is injected continuously into the oil stream coming from the concentrator (9) after which the mixture is held for some seconds in a holding cell to allow the reaction to take place followed by addition of the polishing water. When the added water is removed again in a separator, the saponified FFA are removed simultaneously. After the final concentration, the butter oil is heated to 90–95°C in a plate heat exchanger (11) and pumped to a vacuum chamber (12), where the oil is introduced either as a thin film or as a spray so the increased surface area makes it possible to further remove water and dissolved air. The oil, now having a fat content of higher than 99.8%, is finally pumped back to the plate heat exchanger, where it is cooled to approximately 40°C and pumped to a storage tank (13), where flushing with an inert gas and addition of antioxidants is a possibility before packaging (Anonymous, 2003; Illingworth et al., 2009; Mortensen, 2015).

Anhydrous milk fat (AMF) manufacturing from butter method starts by melting of the butter. Both sweet cream and cultured butter, as well as salted and unsalted butter, can be used, often in the form of cold-stored or frozen butter stored for some time. The melting process often starts by heating the butter in a temperature-controlled room, but microwave thawing can also be used. After warming, further heating follows in a special melting equipment (1) by indirect heating with steam or hot water to a temperature well above the melting point of the fat. If the production of AMF takes place immediately after the butter manufacture without any intermediate storage, the butter is pumped directly to the melting equipment. After melting, the liquid blend of milk fat and butter serum is pumped to a holding tank (2), where it is held for about 30 min to ensure complete melting of the fat and aggregation of the proteins in the aqueous phase. From the holding tank, the blend is pumped to a special separator or concentrator (3). The separation is a critical stage in the process as it is very dependent on the type of butter used. The more or less un-denaturated proteins in the serum phase of sweet cream butter often cause in a certain degree of emulsification of the fat, which complicates the separation. This can be counteracted by lowering the pH-value of the serum phase to about 4.5, for example, by adding citric acid during melting. This denatures the proteins and consequently reduces their emulsifying ability. After separation, the fat phase is further treated in a similar process as described when using cream as starting material (Anonymous, 2003; Illingworth et al., 2009; Mortensen, 2015).

During manufacture and, most importantly, during the packing and storage of AMF products, it is essential that the fat is protected as far as possible from the ravages of oxidation. Nitrogen sparging into the liquid fat to scavenge any dissolved oxygen is preferable. Measurement of dissolved oxygen at the time of packing is important, and to ensure a long shelf-life, it should be limited to a maximum level of 3%. During packing, any air, for example, in drums, should be replaced with nitrogen before filling is commenced, and packaging should ideally be filled from the bottom to prevent any incorporation of air during filling. Enzymatic hydrolysis of fat in milk or cream produces FFA. The solubility of these acids in water depends on their chain lengths with short-chain acids being water soluble and long-chain ones more soluble in oil. Alkali refining, originally used by the vegetable oil industry (Anderson, 1962), is applied in this process. Sodium hydroxide solution (6–10%) is injected and mixed into the fat stream using a static mixer or other device to ensure good mixing. The butter oil and caustic mixture is held in the holding tube for 10–20 s to allow the reaction to take place. Hot water (60–70°C) is then added with a further 5–10 s holding time before the stream passes on to the polishing separator. Polishing is usually sufficient to reduce any residual soaps to 35–90 mg/kg, but there is little margin for overcoming process fluctuations. A second washing and polishing step would reduce alkalinity to perhaps 10–50 mg/kg consistently. If the neutralization option is used, the resultant fat then falls outside the definition of AMF and into one of the butteroil categories. Vacuum drying is generally applied, e.g., at 40°C and 2 kPa. It causes a decrease of the moisture content to below

0.1% and also the oxygen content decreases significantly. A product made in this way may keep for some years if made from milk without any initiation of autoxidation, if stored in isolation from air and light, and if copper contamination has been rigorously prevented.

4.2 Composition

Anhydrous milk fat is one of the most complex fats found in nature. This complexity originates from its triacylglycerol composition and the extreme diversity of its fatty acids (e.g., chain length, position, geometry and number of double bonds, and branching), more than 400 of which have been identified (Jensen & Newburg, 1995). More than 200 individual molecular species of even-numbered triglycerides have also been quantified (Gresti et al., 1993). The consequence of this diversity of triglycerides and fatty acid chain length is that anhydrous milk fat has a broad melting range (approximately −40 to +40 °C) and no true melting point. Moreover, this fat-composition related complexity is dramatically enhanced by the existence of a polymorphism of monotropic type for each triglyceride (Minato, Yano, Ueno, Smith, & Sato, 1997; Ollivon & Perron, 1992; Small, 1986). Typical composition and nutritional values of anhydrous milk fat and butteroil are presented in Table 16.9.

5 Physico-Chemical Characteristics of Ghee, Anhydrous Milk Fat and Butteroil

The unique fatty acid composition of milk fat such as the presence of short and medium-chain fatty acids and a high proportion of saturated fatty acids is reflected in unique physico-chemical characteristics. These characteristics help to some extent in the detection of adulteration of milk fat/ghee with other cheaper oils and fats. Hence, some of these form part of specifications of FSSA and AG Mark standards. The physico- chemical characteristics of cow and buffalo ghee, anhydrous milk fat and butter oil are shown in Table 16.10.

6 Packaging and Shelf Life of Ghee, Anhydrous Milk Fat and Butteroil

Butter has been the traditional form of storage for milk fat, but in some cases ghee, anhydrous milk fat and butteroil are more preferable, because it requires less storage space. These products do not need to be refrigerated in any of its presentations in order to be preserved, facilitating usage flexibility, lower storage costs, and guaranteeing higher shelf life than the butter, ghee is rich in saturated fatty acids (>65%)

Table 16.9 Composition and nutritional values of anhydrous milk fat and butteroil

Parameters	Anhydrous milk fat	Butter oil
Calories (kcal/100 g)	880	876
Total Fat (milk fat) (g/100 g)	99.9	99.48
Moisture (g/100 g)	<0.1	0.24
Protein (g/100 g)	<0.01	0.28
Carbohydrate (g/100 g)	<0.01	Nil
Sugars (Lactose) (g/100 g)	<0.01	Nil
Dietary Fibre (g/100 g)	Nil	Nil
Cholesterol (mg/100 g)	240	256
Fatty Acids (g/100 g Product)		
Saturated fatty acids	66.2	61.92
Mono unsaturated fatty acids	20.2	28.73
Poly unsaturated fatty acids	1.3	3.69
Vitamin A (mg retinol/100 g)	1.0	0.7
Vitamin D (mg/100 g)	Trace	Trace
Vitamin E (mg/100 g)	3.8	2.8
Vitamin K (mg/100 g)	Trace	8.6
Vitamin C (mg/100 g)	<0.1	Trace

Adapted from: USDA National Nutrient Database for Standard Reference; National Agricultural Library; United States Department of Agriculture (USDA) (2015); 1400 Independence Ave., S.W.; Washington, DC 20250 USA., Basic Report: 01003, Butter oil, anhydrous

Table 16.10 Physico-chemical characteristics of ghee, anhydrous milk fat and butteroil

Parameters	Anhydrous milk fat	Cow ghee[a]	Buffalo ghee[a]	Butteroil[b]
Melting point, (°C)	31–35	28–41	32–43	32
Refractive index	1.4534–1.4549	NR	NR	1.4632
Specific gravity	0.909–0.925	0.887	0.887	0.87
Specific heat at 40°C (kJ/kg)	2.1	NR	NR	NR
Smoke point (°C)	193.33	192	198	252
Flash point (°C)	287.77	NR	NR	NR
Fire/Ignition point (°C)	471.11	NR	NR	NR
Solubility of water in AMF (40°C)	0.20	NR	NR	NR
Viscosity at 40°C (mPa. s)	31	NR	NR	NR
Free fatty acids as oleic acid (%m/m)	0.2	0.2	0.2	0.18
Peroxide Value (meq O_2/kg)	0.2	NR	NR	0.25
Iodine value (mL 0.1N $Na_2S_2O_3$/g)	29–39	33.7	29.4	35.1
Reichert Meissl value	22.4–31.8	26.7	32.3	NR
Polenske value	1.5–3.4	1.76	1.41	NR
Kirchner value	17–27	22.1	28.4	NR
Saponification value	225–235	227.3	230	224
Butyro-refractometer reading (at 40°C)	NR	42.3	42.0	NR

NR: not reported/not available
[a]Aneja, Mathur, Chandan, and Banerje (2002)
[b]Kaya (2000)

than many other polyunsaturated fatty acid (PUFA) rich vegetables oils, its shelf life is relatively lower (3–6 months at room temperature). This is mainly due to a low level of natural antioxidants, which are present at higher levels in vegetable oils. Under commercial practices, ghee is usually packed in flexible multilayer packages, glass/polyethylene terephthalate (PET) jars and lacquered tin cans. Different volumes such as 5–10, 100, 200, 500 and 1000 mL to 15 kg are packed depending on consumer/market requirements. For bulk buyers, products are also supplied in road tankers. Ghee is stored at ambient temperature and it keeps well for about 6 months. Shelf life of butteroil is 12 months from the date of manufacture at room temperature. Anhydrous milk fat typically packed in 200 L drums in nitrogen headspace can be stored for several months at +4 °C and 12–24 months at room temperature. The major spoilage that occurs in these products is the development of off-flavours during storage due to autoxidation, which is a self-catalytic process. Several factors such as content of unsaturated fatty acids and antioxidants, processing and storage conditions, and packaging affect the rate of autoxidation. Cockerell et al., (1971) reported that the main factors responsible for increasing the rate of oxidation in fatty foods are contact with atmospheric oxygen, dissolved oxygen, presence of peroxides, water activity (aw,), light, temperature of storage, presence of metal catalysts and antioxidants. During autoxidation, primary and secondary oxidation products are formed, which besides affecting flavour and colour also may impart toxicity to the products.

Autoxidation Autoxidation also results in degradation of components such as essential fatty acids, vitamins, flavour compounds, pigments etc. There are numerous evidences pointing out that oxidized lipids could have negative health implications (Pukalskas et al., 2005). Thus, oxidation of lipids not only produces unpleasant flavours and discolouration, but can also decrease the nutritional quality and safety. The oxidative stability of the frying oils is vital for the shelf life and nutritional value of oil itself and for the foods, which are prepared using the oil/fat. The high temperature at which the foods are fried leads to greater deterioration of the oil and the foods which are fried using the oil due to thermal oxidation. Considering the high value of ghee, measures to delay the onset of autoxidation, so that it can have an extended shelf life, are of great importance to the dairy industry. One of the best ways to protect lipids against autoxidation and thermal oxidation is the incorporation of antioxidants during processing. Antioxidants may act as free radical quenchers, reducing compounds, singlet oxygen scavengers or as pro-oxidant metal suppressors. The use of synthetic antioxidant application is restricted in several countries because of their possible toxicity and carcinogenic effects. Therefore, use of these compounds is the target of questions on their safety, which necessitated the search for natural antioxidants that can act individually or synergistically with other additives. Plants contain many types of effective antioxidants such as phytosterols, phenolic acids, flavonoids and carotenoids. Oxidative stability of ghee during accelerated oxidation conditions was assessed (Pawar et al., 2014) by comparing the antioxidant activities of *vidarikand* (*Pueraria tuberosa*), *shatavari* (*Asparagus racemosus*) and *ashwagandha* (*Withania somnifera*) herb extracts (aqueous and ethanolic) with reference to butylated hydroxy anisole (BHA). The ethanolic extracts of the herbs were more effective in preventing the development of the peroxides and conjugated dienes in ghee compared to their aqueous extracts and showed higher induction period as compared to their aqueous counterparts as anlaysed in Rancimat. It was reported that the *vidarikand* had the maximum antioxidant activity among all the herbs. Antioxidant potential of orange peel extract in ghee at 0.5 and 1% was evaluated at different storage temperatures including 60 ± 2°C. The results showed that the ghee samples incorporated with 1% orange peel extract followed by 0.5% showed higher DPPH (2,2-diphenyl-1-picrylhydrazyl) radical scavenging activity and lower development of PV, thiobarbituric acid (TBA) and FFA than ghee incorporated with 0.02% BHA (Asha et al., 2015). Similarly, addition of 1% pomegranate peel extract reduced peroxide, TBA and carbonyl values in the ghee samples stored at 60°C for 21 days than BHA (0.02%) treated ghee samples and control (Arpita, 2015). However, plant extracts containing natural antioxidants may adversely affect the sensory properties of ghee and have poor thermal stability and hence are not suitable for use in ghee. Addition of rosemary extract (RME), a source of natural antioxidants, at 0.1%, increased the antioxidant potential of ghee in terms of radical scavenging activity (DPPH assay) without affecting sensory and physicochemical properties. Rancimat analysis at 130°C revealed that ghee added with RME showed a significantly higher induction period (4.4 h) than control (1.7 h). Based on peroxide and TBA value determinations, it was observed that shelf-life of ghee added with RME was significantly higher than that of control and ghee added with BHA (Rahila et al., 2018). Addition of RME has also imparted thermal stability to ghee and delayed the onset of cholesterol oxidation, when used as a frying medium. Findings suggested that RME could be used as a source of natural antioxidants for imparting autoxidative and thermal stability to ghee.

Thermal Oxidation The chemistry of lipids oxidized at high temperature is more complex since both thermolytic and oxidative reactions are simultaneously involved. Both saturated and unsaturated fatty acids undergo chemical decomposition when exposed to heat in the presence of oxygen. During the oxidation of fat at elevated temperatures, hydroperoxide decomposition and secondary oxidation can occur at extremely rapid rates. During frying, ghee is exposed continuously, or repeatedly, to elevated temperature in the presence of air and moisture. A number of chemical reactions including oxidation, hydrolysis and polymerization of frying fat take place, leading to the formation of many undesirable secondary products. As these reactions proceed, the functional, sensory, and nutritional quality of oil deteriorates and may reach a point where it has to be ultimately discarded (Stevenson & Esjin, 1984). Fats used for culinary purposes are almost always heated in air. Therefore, oxidative changes usually accompany and probably precede the thermal changes (Artman, 1969). Various kinds of fats and fatty materials have been treated under a wide range of temperatures, times and degree of aeration (Crnjar et al., 1981). From these studies, it was observed that there were considerable changes in the physical and chemical properties of oils and fats during thermal oxidation. When a fat is heated, it first shows a gain in weight as oxygen is absorbed, and its peroxide value may increase. But as heating continues, the peroxides decompose and the breakdown products start to distill off leading to a net loss in weight. The refractive index and UV absorption of the fat increase owing to the conjugation of double bonds and accumulation of oxygenated products, which include carbonyls, hydroxyl and epoxy derivatives. Eventually, the fat turns brown in colour and the colour is attributed to αβ and αα′ unsaturated carbonyls (Mukai et al., 1968). Various factors like temperature and duration of heating, extent of aeration (including the surface volume ratio of the fat), degree of unsaturation of the fat and presence of catalysts influence the extent to which these changes take place and their sequence (Rai & Narayanan, 1984). It has been shown that fatty acids on the 1st and 3rd positions of triglyceride molecules are more susceptible to oxidative attack than those on the 2nd position (Endres et al., 1962). It is generally proposed that the changes brought about in a fat during thermal oxidation result from the formation and rapid decomposition of hydroperoxides. Similar changes result when the fat is first oxidized at low temperature and then heated in inert atmosphere (Kumazawa & Oyama, 1965). However, there is no clear evidence that all of the changes reported to occur during thermal oxidation take place through the intermediacy of hydroperoxides. It seems quite possible that the free radicals might form more directly and react with each other or with oxygen to form the final products. The above works have observed that for a given total duration of heating at a specified temperature, continuous heating is less destructive than the intermittent heating. Obviously, during the intermittent heating substantial concentration of hydroperoxides is formed and these hydroperoxides decompose to give all at once, a higher concentration of radicals than is achieved by the slow and steady attack of oxygen at a fixed high temperature.

Presence of cholesterol oxidation products (COP) in foods has evoked much interest due to health concerns. Nath and Murthy (1988) showed that the ghee produced and stored under normal conditions did not contain COP. Kumar and Singhal (1992) reported that 25-hydroxycholesterol and cholestantriol were formed in intermittently heated ghee. Nath et al., (1996) observed that when ghee used for frying for a short period did not result in the formation of cholesterol oxides, but the same culinary process lasting for 15 min generated COP. Further, the addition of ghee residue as a source of natural antioxidants delayed the formation of COPs up to 60 min of frying. Rahila (2016) and Rahila et al. (2018) reported that γ-oryzanol (0.5%), sesamol (0.5%) and RME (0.1%) prevented the formation of COP in ghee during one hour of deep-frying.

7 Variants of Ghee and Anhydrous Milk Fat

Milk fat contributes to unique functionality and flavor; therefore, they are used in a wide variety of foods, pharmaceuticals and Ayurvedic applications. Accordingly, based on the applications, a variety of value added ghee and anhydrous milk fat products are available in market and these include fractionated ghee and anhydrous milk fat, organic ghee, medicated ghee, herbal ghee, species and breed specific ghee and so on. Some of these product variants are sold at premium prices.

Medicated and Herbal Ghee About 55–60 medicated ghee types are reported in literature. When ghee is cooked with various substances like paste of drugs, and liquids like decoction, juice or milk etc., in specific proportions over mild heat till the appearance of certain pharmaceutical parameters is known as medicated ghee (Sujatha & Sarashetti, 2015). These substances may also impart intense flavour, colour, and distinctive fragrance to the developed products. The herbal ghee is prepared by incorporating the active principles of the herbs, spices and condiments added at different stages of processing of ghee without essentially altering the sensory status of the final product. *Arjuna* ghee, *Shatavari* ghee, *Brahmi* ghee, *Panchgavya* ghee, *Panchcoal* ghee and *Ashwagandha* ghee are the popular herbal ghee variants. When herbs are mixed with ghee, their activity and utility is potentiated many times. Herbs impart the final product with improved physico-chemical and sensory quality, and nutritional benefits making it a value added product. The bioactive compounds of the herbs extend the shelf life of the developed product and also some herbs improve the textural and grain quantity and grain size of the ghee. A large variety of naturally occurring plant materials, such as soya bean, safflower, amla (*Phylanthus amblica*) fruits, curry (*Murraya koenigi*) leaves and betel (*Piper betel*) leaves have been shown to improve the shelf-life of ghee (Rao & Singh, 1990).

Fractions of Ghee, Anhydrous Milk Fat and Butteroil Milk fat is often discussed in terms of two to three fractions: the high-, middle- and low-melting fractions. In reality, however, numerous fractions of unique composition can be obtained by selective fractionation techniques. Cooling rate, agitation, solvent properties and the presence of additives also influence crystallization and properties of the fractions obtained (Illingworth, 2002; Kaylegian & Lindsay, 1995).

Low-Cholesterol Ghee Low-cholesterol ghee with 90% less cholesterol was prepared (Kumar et al., 2010) using β-cyclodextrin. The physico-chemical properties such as RM value, Polenske value, BR reading at 40°C, iodine value and free fatty acid level as oleic acid in the low-cholesterol ghee remained almost unaltered. A similar trend was also observed in buffalo ghee. Fat soluble Vitamins (β-carotene, A and E) in both cow and buffalo low-cholesterol ghee were very similar to that of the respective ghee samples. However, 65–70% loss of Vitamin D was observed in low-cholesterol ghee.

Low Cholesterol Anhydrous Milk Fat Removal of cholesterol is not a routine operation in the manufacture of anhydrous milk fat. Two processes have been cited. One, which does not seem to have any effect on the flavour of anhydrous milk fat, is to mix the fat with β-cyclodextrin, a modified starch. This molecule surrounds the cholesterol molecule, allowing it to be filtered from the fat. The second process was developed by S.A. Fractionnement Tirtiaux of Belgium; it involves steam distillation which employs a deodorizer as used in the physical refining of edible oils. The high temperatures required for this process also destroys the natural carotene colour of the anhydrous milk fat and removes the natural antioxidants; what is more critical is that the natural flavour for which anhydrous milk fat is prized is also distilled off.

Species and Breed Specific Ghee Milk produced by some of the species such as goat, camel and donkey are gaining attention in promoting health due to the presence of several bioactive substances in them. In general, it is believed that milk from indigenous cattle has better therapeutic properties than that from exotic or crossbreds. Therefore, ghee from such breeds and species is used in special Ayurvedic applications.

8 Functionality and Use of Ghee, Anhydrous Milk Fat and Butteroil

Milk fat ingredients contribute unique functionality and flavor that enable their use in a wide variety of food applications. Ghee, anhydrous milk fat and butteroil are convenient to use in liquid form because it is easy to mix with and meter into other products. Milk fat could be used in almost any food application in which a fat source is needed. An essential characteristic of the products is their prolonged keeping quality achieved by the very low moisture content, which makes the products suitable for serving as a buffer during fluctuations in milk production over the year. It also makes the products suitable for shipment to faraway destinations to help leveling the unequal milk production in the world. Customized fat products for various applications can be obtained by fractionation of ghee, anhydrous milk fat and butteroil.

Ghee is the most widely used milk product in the Indian sub-continent and primarily used for cooking and frying and as dressing or toppings for various foods. In India, 80% of ghee produced is used for culinary purposes. The remaining 20% is used for confectionery, including small amount consumed on auspicious occasions like religious ceremonies (Rajorhia, 1993). Ghee is used for table/garnishing purpose, mainly as flavouring agent on hot rice, rotis (Indian bread), and shallow and deep-frying of traditional Indian foods. It is also a preferred medium of cooking for the preparation of several sweet-meats. Ghee is considered as supreme cooking/frying medium for the preparation of several traditional foods as it imparts the desired flavour and texture to them. Further, it is used for medicinal (Ayurvedic formulations) and religious (lighting of divine lamps and brightening of *homa*, the sacred fire) purposes (Mortensen, 2015; Rahila et al., 2018).

Butteroil helps to bring out a nutty flavor in fried foods, which makes them taste much better than with most other oils. Butteroil has got a remarkably high smoke point of 252°C, and hence, it is one of the best oils for deep frying and also it is much easier to fry with butteroil as it does not splatter much. An important application of anhydrous milk fat is the production of recombined liquid milk, but it can also be used in the production of blends, low-fat dairy spreads, blended spreads, ice cream, processed cheese, sauces, frying, grilling, roasting medium as well as in the bakery and confectionary industry. Anhydrous milk fat has the potential to be used as a replacement of hydrogenated fats due to its high content of palmitic and low content of trans-fatty acids. It is also known to be high in stearic acid, which has been shown to have a neutral effect on coronary heart disease (Aro et al., 1997; Tarrago-Trani et al., 2006). Besides, anhydrous milk fat has good sensory attributes such as flavor and mouthfeel (Kaylegian et al., 1993).

The choice of milk fat depends on its composition, functionality and cost. Cream and butter contain high levels of moisture, which can limit their use because of shelf-life concerns or the moisture requirements of the desired application. Ingredients such as ghee, anhydrous milk fat, butteroil have additional applications due to their lower moisture content and desired sensory and physico-chemical propeties (Gnanasambandam et al., 2017).

9 Conclusion

All over the world, surplus milk fat is conserved in the form of ghee, anhydrous milk fat and butteroil, rather than butter, because of savings in cold storage capacity and the anticipated long shelf-life at ambient temperature. Current consumer trends related to clean label are driving food companies to proactively move away from ingredients such as hydrogenated oils to "cleaner" ingredients such as ghee, butteroil anhydrous milk fat, butter and cream for their food products. Moreover, emerging research is changing nutritional perspectives related to milk fat. Ghee, anhydrous milk fat, butteroil, and similar products have many applications, and in many parts of the world, they are of vital importance for the economy of the dairy industry.

References

Abou-Donia, S. A., & El-Agamy, S. I. (1993). Ghee. In R. Macrae, R. K. Robinson, & M. J. Sadler (Eds.), *Encyclopaedia of food science, food technology and nutrition* (Vol. 6, pp. 3992–3994). London, UK: Academic Press Ltd.

Achaya, K. T. (1997). Ghee, vanaspati and special fats in India. In F. D. Gunstone & F. B. Padley (Eds.), *Lipid technologies and applications* (pp. 369–390). New York, NY: Marcel Dekker Inc..

Aguedo, M., Hanon, E., Danthine, S., Paquot, M., Lognay, G., Thomas, A., et al. (2008). Enrichment of anhydrous milk fat in polyunsaturated fatty acid residues from linseed and rapeseed oils through enzymatic interesterification. *Journal of Agricultural and Food Chemistry, 56*(5), 1757–1765.

Anderson, A. J. C. (1962). Section 2: Neutralisation or de-acidification. In P. N. Williams (Ed.), *Refining oils and fats for edible purposes* (2nd ed., pp. 28–213). Oxford, UK: Pergamon Press.

Aneja, R. P., Mathur, B. N., Chandan, R. C., & Banerjee, A. K. (2002). Fat rich products. In *Technology of Indian milk products* (pp. 183–198). New Delhi, India: Dairy India Publication.

Anonymous. (2003). Anhydrous milk fat (AMF) and butteroil. In *Dairy processing handbook* (2nd ed., pp. 293–299). Lund, Sweden: Tetra Pak Processing Systems AB.

Aro, A., Jauhiainen, M., Partanen, R., Salminen, I., & Mutanen, M. (1997). Stearic acid, trans fatty acids, and dairy fat: Effects on serum and lipoprotein lipids, apolipoproteins, lipoprotein (a), and lipid transfer proteins in healthy subjects. *The American Journal of Clinical Nutrition, 65*(5), 1419–1426.

Arpita, R. M. (2015). *Modeling the antioxidant potential of pomegranate peel extract in ghee*. PhD Thesis. Southern Regional Station, ICAR- National Dairy Research Institute (Deemed University), India.

Artman, N. R. (1969). The chemical and biological properties of heated and oxidized fats. In *Advances in lipid research* (Vol. 7, pp. 245–330). Amsterdam, The Netherlands: Elsevier.

Arumughan, C., & Narayan, K. M. (1982). Triacylglycerol composition of buffalo milk fat. *Journal of Dairy Research, 49*, 81–85.

Asha, A., Manjunatha, M., Rekha, R. M., Surendranath, B., Heartwin, P., Rao, J., et al. (2015). Antioxidant activities of orange peel extract in ghee (butter oil) stored at different storage temperatures. *Journal of Food Science and Technology, 52*(12), 8220–8227.

Bahadur, B., Srinivasan, M. R., & Ray, S. C. (1950). Utilisation of induced stratification of butter melt in ghee making. *Indian Journal of Dairy Science, 3*(3), 94–106.

Bitman, J., & Wood, D. L. (1990). Changes in milk fat phospholipids during lactation. *Journal of Dairy Science, 73*(5), 1208–1216.

Bornaz, S., Fanni, J., & Parmentier, M. (1993). Butter texture: The prevalent triglycerides. *Journal of the American Oil Chemists Society, 70*(11), 1075–1079.

Buss, D. H., Jackson, P. A., & Scuffam, D. (1984). Composition of butters on sale in Britain. *Journal of Dairy Research, 51*(4), 637–641.

Bylund, G. (2003). *Dairy processing handbook*. Lund, Sweden: Tetra Pak Processing Systems AB.

Christie, W. W. (1988). *Lipid Analysis: Isolation, separation, identification and structural analysis of lipids* (pp. 93–96). Bridgwater, UK: Pergamon Press.

Christie, W. W. (1995). Composition and structure of milk lipids. In P. F. Fox (Ed.), *Advanced dairy chemistry-2* (pp. 1–56). London, UK: Chapman and Hall.

Christie, W. W. (2003). *Lipid Analysis: Isolation, separation, identification and structural analysis of lipids*. Bridgwater, UK: Elsevier.

Cockerell, I., Francis, B., & Halliday, D. (1971). Changes in the nutritive value of concentrate feedstuffs during storage. In *Proceedings of Conference of Feed Resources and improvement of Animal feeding methods in the CENTO Region countries* (pp. 181–192). London, UK: Tropical Products Institute.

Codex Alimentarius. (2018). *Milk and milk products* (2nd ed.). Codex Standard for Milk fat Products Stan 280-1973.

Crnjar, E. D., Witchwoot, A., & Nawar, W. W. (1981). Thermal oxidation of a series of saturated triacylglycerols. *Journal of Agricultural and Food Chemistry, 29*(1), 39–42.

De, S. (1980). Indian dairy products. In *Outlines of dairy technology* (pp. 433–462). New Delhi, India: Rekha Printers Pvt Ltd.

Dimick, P. S., McCarthy, R. D., & Patton, S. (1970). Milk fat synthesis. In *Physiology of digestions and metabolism in the ruminant* (pp. 29–54). Newcastle-upon-Tyne, UK: Oriel Press.

Dirar, H. A. (1993). *The indigenous fermented foods of the Sudan: A study in African food and nutrition*. Wallingford, UK: CAB International.

Dolby, R. M. (1949). The properties of New Zealand butters and butterfats: II. The relation of hardness of New Zealand commercial butter to composition of the butterfat. *Journal of Dairy Research, 16*(3), 336–347.

Endres, J. G., Bhalerao, V. R., & Kummerow, F. A. (1962). Thermal oxidation of synthetic triglycerides. I. Composition of oxidized triglycerides. *Journal of the American Oil Chemists Society, 39*(2), 118–121.

FAO. (2017). *Milk and milk products: Price and trade update, trade and markets division (EST)* (pp. 1–7). Rome, Italy: Food and Agriculture Organization of the United Nations (FAO).

FAO-WHO. (2010). Fat and fatty acid requirements for Adults. In *FAO food and nutrition paper-91, fats and fatty acids in human nutrition: Report of an expert consultation* (pp. 55–62). Rome, Italy: FAO-WHO.
Fennema, O. R. (1985). *Food chemistry* (2nd ed.). New York, NY: Marcel Dekker Inc..
FSSR. (2011). *Food products standards*. In Food safety and standards (Food products standards and food additives) regulations. http://www.fssai.gov.in
Gaba, K. L., & Jain, M. K. (1973). Note on the flavour changes in ghee on storage: Their sensory and chemical assessment. *The Indian Journal of Animal Sciences, 43*(1), 67–70.
Ganguli, N. C., & Jain, M. K. (1973). Ghee: Its chemistry, processing and technology. *Journal of Dairy Science, 56*(1), 19–25.
German, J. B., & Dillard, C. J. (1998). Fractionated milk fat: Composition, structure, and functional properties. *Food Technology, 52*(2), 38.
Glass, R. L., & Jenness, R. (1971). Comparative biochemical studies of milk. 6. Constituent fatty acids of milk fats of additional species. *Comparative Biochemistry and Physiology, 28*, 353–359.
Gnanasambandam, R., Torres-Gonzalez, M., Burrington, K. J., & Kapoor, R. (2017). *Milk fat and related ingredients serving today's marketplace* (pp. 1–16).
Gresti, J., Bugaut, M., Maniongui, C., & Bezard, J. (1993). Composition of molecular species of triacylglycerols in bovine milk fat. *Journal of Dairy Science, 76*(7), 1850–1869.
Grummer, R. R. (1991). Effect of feed on the composition of milk fat. *Journal of Dairy Science, 74*(9), 3244–3257.
Gunstone, F. D., Harwood, J. L., & Padley, F. B. (1986). *The lipid hand book* (pp. 1–169). London, UK: Chapman and Hall.
Hawke, J. C. (1957). The fatty acids of butterfat and the volatile acids formed on oxidation. *Journal of Dairy Research, 24*(3), 366–371.
Hawke, J. C., & Taylor, M. W. (1983). Influence of nutritional factors on the yield, composition and physical properties of milk fat. In *Developments in dairy chemistry* (Vol. 2, pp. 37–81). Dordrecht, The Netherlands: Springer.
Illingworth, D. (2002). Fractionation of fats. In A. G. Marangoni & S. S. Narine (Eds.), *Physical properties of lipids* (pp. 411–448). New York, NY: Marcel Dekker.
Illingworth, D., Patil, G. R., & Tamime, A. Y. (2009). Anhydrous milk fat manufacture and fractionation. In *Dairy fats and related products* (pp. 108–166). Oxford, UK: Wiley-Blackwell.
Indian Standards Institution. (1966). *IS 3508 (1966) reaffirmed- 2003: Method of sampling and test for ghee [FAD19: Dairy products and equipment]*. New Delhi, India: Bureau of Indian Standards.
Jensen, R. G. (1992). Fatty acids in milk and dairy products. In C. K. Chowk (Ed.), *Fatty acid in foods and their health implications* (pp. 95–135). New York, NY: Marcel Dekker.
Jensen, R. G. (2002). The composition of bovine milk lipids: January 1995 to december 2000. *Journal of Dairy Science, 85*(2), 295–350.
Jensen, R. G., Ferris, A. M., & Lammi-Keefe, C. J. (1991). The composition of milk fat. *Journal of Dairy Science, 74*(9), 3228–3243.
Jensen, R. G., Gander, G. W., & Sampugna, J. (1962). Fatty acid composition of the lipids from pooled, raw milk. *Journal of Dairy Science, 45*(3), 329–331.
Jensen, R. G., & Newburg, D. S. (1995). Milk lipids. In R. G. Jensen (Ed.), *Handbook of milk composition* (pp. 545–575). New York, NY: Academic Press.
Joseph, A. M., & Appachar, S. R. (1980). Effect of ripening cream with selected lactic acid bacteria on the quality of ghee. *Journal of Dairy Research, 47*(3), 411–415.
Kaya, A. (2000). Properties and stability of butter oil obtained from milk and yoghurt. *Food, 44*(2), 126–129.
Kaylegian, K. E., Hartel, R. W., & Lindsay, R. C. (1993). Applications of modified milk fat in food products. *Journal of Dairy Science, 76*(6), 1782–1796.
Kaylegian, K. E., & Lindsay, R. C. (1995). *Handbook of milk fat fractionation technology and application*. Champaign, IL: American Oil Chemists' Society Press.
Kleyn, D. H. (1992). Textural aspects of butter. *Food Technology, 46*, 118–121.

Kumar, M., Sharma, V., Lal, D., Kumar, A., & Seth, R. (2010). A comparison of the physico-chemical properties of low cholesterol ghee with standard ghee from cow and buffalo creams. *International Journal of Dairy Technology, 63*(2), 252–255.

Kumar, N., & Singhal, O. P. (1992). Effect of processing conditions on the oxidation of cholesterol in ghee. *Journal of the Science of Food and Agriculture, 58*(2), 267–273.

Kumar, R., & Negi, S. (2004). Granulation of ghee using emulsifiers. In *Presentation, XXXIII dairy industry conference, New Delhi* (p. 106).

Kumar, M.A., Dastur, N. N., & Bhatt, M. V. (1985). Studies on the pigments of buffalo milk ii. identification of bilirubin ix alpha isomer in sour buffalo milk. *Milchwissenschaft, 40*(12), 724–726.

Kumazawa, H., & Oyama, T. (1965). Estimation of total carbonyl content in oxidized oil by 2,4-dinitrophenylhydrazine. *Journal of Japan Oil Chemists' Society, 14*(4), 167–171.

Lakshminarayana, M.(1983). *Fractionation of buffalo milk fat and studies on physico-chemical properties of fractions of buffalo milk fat.* Ph.D. Thesis. N.D.R.I (Deemed University), Bangalore.

Lakshminarayana, M., & Rama Murthy, M. K. (1985). Cow and buffalo milk fat fractions. Yield physico-chemical characteristics and fatty acid composition. *Indian Journal of Dairy Science, 38*, 256–264.

MacGibbon, A. K. H., & McLennan, W. D. (1987). Hardness of New Zealand patted butter: Seasonal and regional variations. *New Zealand Journal of Dairy Science and Technology, 22*(2), 143–156.

MacGibbon, A. K. H., & Taylor, M. W. (2006). Composition and structure of bovine milk lipids. In *Advanced dairy chemistry volume 2 lipids* (pp. 1–42). Boston, MA: Springer.

Maniongui, C., Gresti, J., Bugaut, M., Gauthier, S., & Bezard, J. (1991). Determination of bovine butterfat triacylglycerols by reversed-phase liquid chromatography and gas chromatography. *Journal of Chromatography. A, 543*, 81–103.

Marangoni, A. G. (2005). Crystallization kinetics in fat crystal networks. In *Fat crystal networks* (pp. 21–82). New York, NY: Marcel Dekker.

Marangoni, A. G., & Lencki, R. W. (1998). Ternary phase behavior of milk fat fractions. *Journal of Agricultural and Food Chemistry, 46*(10), 3879–3884.

McNamara, J. P. (1991). Regulation of adipose tissue metabolism in support of lactation. *Journal of Dairy Science, 74*, 706–719.

Minato, A., Yano, J., Ueno, S., Smith, K., & Sato, K. (1997). FT–IR study on microscopic structures and conformations of POP–PPO and POP–OPO molecular compounds. *Chemistry and Physics of Lipids, 88*(1), 63–71.

Mortensen, B. K. (2011). Butter and other milk fat products. In J. W. Fuquay, P. F. Fox, & P. L. H. McSweeney (Eds.), *Encyclopedia of dairy sciences* (2nd ed., p. 502). Boston, MA: Elsevier/Academic Press.

Mortensen, B. K. (2015). *Anhydrous milk fat/butter oil and ghee.*

Mukai, M., Thomas, J. F., & Tebbens, B. D. (1968). Fate of airborne benzo [a] pyrene. *Environmental Science & Technology, 2*(1), 33–39.

Murthy R. M. K., & Narayana, K. M. (1971). Fatty acid composition of buffalo and cow milk fat by gas liquid chromatography. *Milchwisenschaft., 26*, 693–697.

Nagesh, K. (1981). *Studies on some physico-chemical aspects of grain formation in ghee.* Msc. Thesis. University of Agricultural Sciences, Bangalore.

Nath, B. S., & Murthy, M. R. (1988). Cholesterol in Indian ghee. *The Lancet, 332*(8601), 39.

Nath, B. S., Usha, M. A., & Rama Murthy, M. K. (1996). Effect of deep-frying on cholesterol oxidation in ghee. *Journal of Food Science and Technology, 33*(5), 425–426.

Nielsen, V. H. (1971). Curing texture defects in butter. *American Dairy Review, 33*, 28–29.

Ollivon, M., & Perron, R. (1992). Chapter 5, physical properties of fats. In A. Karleskind & J. P. Wolff (Eds.), *Oils and fats manual* (Vol. 1, pp. 433–529). Paris, France: Lavoisier.

Omar, K. A., Gounga, M. E., Liu, R., Mwinyi, W., Aboshora, W., Ramadhan, A. H., et al. (2017). Triacylglycerol composition, melting and crystallization profiles of lipase catalysed anhydrous milk fats hydrolysed. *International Journal of Food Properties, 20*(suppl 2), 1230–1245.

Patton, S., Evans, L., & McCarthy, R. D. (1960). The action of pancreatic lipase on milk fat. *Journal of Dairy Science, 43*(1), 95–96.

Pawar, N., Gandhi, K., Purohit, A., Arora, S., & Singh, R. R. B. (2014). Effect of added herb extracts on oxidative stability of ghee (butter oil) during accelerated oxidation condition. *Journal of Food Science and Technology, 51*(10), 2727–2733.

Prentice, J. H. (1972). Rheology and texture of dairy products. *Journal of Texture Studies, 3*(4), 415–458.

Pukalskas, A., van Beek, T. A., & de Waard, P. (2005). Development of a triple hyphenated HPLC–radical scavenging detection–DAD–SPE–NMR system for the rapid identification of antioxidants in complex plant extracts. *Journal of Chromatography A, 1074*(1-2), 81–88.

Rachana, C. R., Nath, B. S., Reshma, M. V., & Armughan, C. (2013). Variation in grainy texture of commercial ghee in relation to laboratory ghee and its blends. *International Journal of Dairy Technology, 66*(1), 90–97.

Rahila, M. P. (2016). *Effect of natural antioxidants on the autoxidative and thermal stability of ghee (clarified butterfat)*. PhD Thesis. Southern Regional Station, ICAR-National Dairy Research Institute (Deemed University), India.

Rahila, M. P., Surendra Nath, B., Laxmana Naik, N., Pushpadass, H. A., Manjunatha, M., & Franklin, M. E. E. (2018). Rosemary (Rosmarinus officinalis Linn.) extract: A source of natural antioxidants for imparting autoxidative and thermal stability to ghee. *Journal of Food Processing and Preservation, 42*(2), e13443.

Rai, T., & Narayanan, K. M. (1984). Development of carbonyls in ghee and refined groundnut oil during intermittent frying. *Indian Journal of Dairy Science, 37*(3), 230–233.

Rajorhia, G. S. (1993). Ghee. In R. Macrae, R. K. Robinson, & M. J. Sadler (Eds.), *Encyclopaedia of food science, food technology and nutrition* (Vol. 4, pp. 2186–2192). London, UK: Academic Press Ltd.

Rangappa, K. S., & Achaya, K. T. (1974). *Indian dairy products*. Bombay, India: Asia Publishing House.

Rao, D. V., & Ramamurthy, M. K. (1984). Polar carbonyls in cow and buffalo ghee. *Indian Journal of Dairy Science, 37*, 98–102.

Rao, D. V., & Singh, H. (1990). Use of antioxidants in ghee-a review. *Indian Journal of Dairy Science, 43*(3), 359–363.

Reddy, Y. S. (2010). Improving plasticity of milk fat for use in baking by fractionation. *Journal of the American Oil Chemists' Society, 87*(5), 493–497.

Sharma, R. S. (1981). Ghee: A resume of recent researches. *Journal of Food Science and Technology, 18*, 70–77.

Small, D. M. (1986). The physical chemistry of lipids. *Handbook of Lipid Research, 4*, 285–343.

Smith, L. M., & Ronning, M. (1961). Comparison of fatty acid composition of milk fats produced by cows fed alfalfa, oat, or ground, pelleted alfalfa hay. *Journal of Dairy Science, 44*, 1170–1174.

Sserunjogi, M. L., Abrahamsen, R. K., & Narvhus, J. (1998). A review paper: Current knowledge of ghee and related products. *International Dairy Journal, 8*(8), 677–688.

Stevenson, S. G., & Mand Esjin, N. M. (1984). Quality control in the use of deep frying oils. *Journal of the American Oil Chemists' Society, 61*, 1102–1108.

Sujatha, K., & Sarashetti, R. S. (2015). Vision and ghee. *Journal of Biological and Scientific Opinion, 3*, 143–146.

Swaisgood, H. E. (1996). Characteristics of milk. In O. W. Fennema (Ed.), *Food chemistry* (3rd ed., pp. 841–878). New York, NY: Marcel Dekker. Inc..

Tarrago-Trani, M. T., Phillips, K. M., Lemar, L. E., & Holden, J. M. (2006). New and existing oils and fats used in products with reduced trans-fatty acid content. *Journal of the American Dietetic Association, 106*(6), 867–880.

Thompson, M. P., Brunner, J. R., & Stine, C. M. (1959). Characteristics of high-melting triglyceride fractions from the fat-globule membrane and butter oil of bovine milk. *Journal of Dairy Science, 42*(10), 1651–1658.

Timms, R. E. (1980). The phase behaviour and polymorphism of milk fat, milk fat fractions and fully hardened milk fat. *Australian Journal of Dairy Technology, 35*(2), 47–53.

Urbach, G., & Gordon, M. H. (1994). Flavours derived from fats. In D. P. J. Moran & K. K. Rajah (Eds.), *Fats in food products* (pp. 347–405). London, UK: Blackie Academic and Professional.

USDA National Nutrient Database for Standard Reference; National Agricultural Library; United States Department of Agriculture (USDA). (2015). *Basic report: 01003, butter oil, anhydrous.*

Wadhwa, B. K., & Jain, M. K. (1990). Chemistry of ghee flavour-a review. *Indian Journal of Dairy Science, 43*(4), 601–607.

Wadodkar, U. R., Murthi, T. N., & Punjrath, J. S. (1996). Isolation of ghee volatiles by vacuum degassing, their separation and identification using gas chromatography/mass spectrometry. *Indian Journal of Dairy Science, 49*(3), 185–198.

Walstra, P., Geurts, T. J., Noomen, A., Jellema, A., & van Boekel, M. A. J. S. (1999). *Dairy technology: Principles of milk properties and processes* (Vol. 50-70, pp. 501–510). New York, NY: Marcel Dekker/CRC Press.

Warner, J. N. (1976). Principles of dairy processing. *Chaos, 13*, 253–259.

Wearne, S. (1999). *Nutrient analysis of other milks and creams*. Food Surveillance Information Sheet (United Kingdom). Number 178.

Wolff, R. L., Bayard, C. C., & Fabien, R. J. (1995). Evaluation of sequential methods for the determination of butterfat fatty acid composition with emphasis ontrans-18: 1 acids. Application to the study of seasonal variations in french butters. *Journal of the American Oil Chemists' Society, 72*(12), 1471–1483.

Woodside, J. V., & Kromhout, D. (2005). Fatty acids and CHD. *Proceedings of the Nutrition Society, 64*(4), 554–564.

Yadav, J. S., & Srinivasan, R. A. (1987). Role of starter culture in enhancing ghee flavour: A review. *IJDS, 40*(2), 153–157.

Yadav, J. S., & Srinivasan, R. A. (1992). Advances in ghee flavour research. *Indian Journal of Dairy Science, 45*, 338–338.

Chapter 17
Dairy Creams and Related Products

Pramesh Dhungana and Bhesh Bhandari

1 Introduction

Milk fat globules, which are the building blocks of cream, are naturally secreted in the bovine mammary gland. The secretion process involves accumulation of small triacylglycerol (TAG) microdroplets followed by covering up from three-layered membrane and eventual release as fat globules in bulk milk. Since the fat globule contains a larger proportion of TAG at their core and a membrane with numerous health-promoting substances, it is considered as a food ingredient with a source of both energy and functional compounds.

According to Codex (2014), a cream is the fluid milk product comparatively rich in fat, in the form of an emulsion of fat-in-skimmed milk, obtained by physical separation from milk and containing milkfat not less than 10% (w/w). In Australia, a regular dairy cream sold as "cream" must be cream with not less than 350 g/kg of milkfat (FSANZ, 2016). A cream may contains approved food stabilisers and thickener as additives singly or in combination if the use of such ingredients is justifiable (Codex, 2014). There are numerous varieties of cream in the market with varying in fat content, acidity, and texture. The common varieties of creams available in the Australian market are tabulated in Table 17.1. Depending upon types of cream, the manufacturing steps may involve standardization, heat treatment, homogenization, fermentation and packaging.

Creams can be broadly divided into two types based on the nature of origin: natural and recombined cream. Natural cream is concentrated form of naturally synthesised milk fat globules, whereas recombined cream is concentrated form of mechanically created fat globules from a mixture of anhydrous milk fat and suitable emulsifier(s). A recombined cream might contain a natural or artificial emulsifier.

P. Dhungana · B. Bhandari (✉)
ARC Dairy Innovation Hub, School of Agriculture and Food Sciences,
The University of Queensland, St Lucia, QLD, Australia
e-mail: b.bhandari@uq.edu.au

T. Truong et al. (eds.), *Dairy Fat Products and Functionality*,
https://doi.org/10.1007/978-3-030-41661-4_17

Table 17.1 Various types of market cream in Australia (data from self-survey)

Cream type	Fat content (% w/w)	Food additives	Possible functionalities of the respective additives	References
Sour cream	38.2	Culture	Fermentation	Born (2013)
Light sour cream	19.3	Milk solids	Body of cream	UOG (2018)
		Gelatine	Melt-in-mouth gel	Williams, Phillips, and Vries (2004)
Light thickened cream	18	Potassium alginate	Gelling agent	Williams et al. (2004)
		Gaur gum	Thickener and stabilizer	Mudgil, Barak, and Khatkar (2014)
Double cream	45	Pectin	Short texture	Williams et al. (2004)
		Carrageenan	Low concentration gelling agent	Williams et al. (2004)
Thickened cream	35	Carrageenan	Forms gel at low concentration	Williams et al. (2004)
		Gaur gum	Thickener and stabilizer	Mudgil et al. (2014)
Whipped cream (Canister)	27.5	Sodium alginate	Rapid setting cold gelling agent	Williams et al. (2004)
		Carrageenan	Low concentration gelling agent	Williams et al. (2004)
		Mono- or di-glycerides of fatty acids	Induce partial coalescence	Fredrick et al. (2013)
		Nitrous oxide	Water soluble, odourless, tasteless, non-toxic and no after taste	Getz, Smith, Tracy, and Prucha (1937)
Cream powder	≥42	Modified starch	Emulsifier-filler	Keogh (2004)
		Gum acacia	Emulsifier-filler	

2 Natural Cream

2.1 Gravity Separation

Natural cream can be prepared by either gravity separation or centrifugal separation method. Gravity separation method follows the Stokes' equation for settling velocity. If milk is allowed to stand; fat globules being lighter than milk serum (skim-milk), start to rise with upward gravitational force (f_u) given by equation (Towler, 1994):

$$f_u = 4\pi r^3 g\left(\rho_s - \rho_f\right)/3 \tag{17.1}$$

where r = radius of globule; g = acceleration due to gravity; ρ_s = density of milk serum; and ρ_f = density of fat globules.

In the case of equilibrium, the upward force is equal to the frictional force experienced by fat globules. The frictional force given by Stokes' Law is

$$f_r = 6\pi\eta r v \tag{17.2}$$

where η = viscosity of milk serum; and ν = velocity of the fat globule.

By combining two Eqs. (17.1) and (17.2), the upward velocity of rising fat globules can be written as:

$$v = D^2 g\left(\rho_s - \rho_f\right)/18\eta \tag{17.3}$$

where D = diameter of fat globules (Fig. 17.1).

The Eq. (17.3) indicates that the upward velocity of fat globule is directly proportional to the square of its radius and density difference between milk serum and fat globules and is inversely proportional to the viscosity of milk serum. The viscosity of milk serum and densities of fat globules and milk serum can be varied by changing the temperature of milk (Towler, 1994). The gravity separation method is not time efficient and requires a great deal of attention to assure food safety. Therefore, it is not a common practice at the industrial level. Artisanal milk product manufacturers often use this method.

Beside fat globule size and density, the other important factor that promotes gravity separation of fat globules in raw milk at a lower temperature is agglutinin. Agglutinin promotes attachment of cryoglobulins (lipoproteins and immunoglobulins) on milk fat

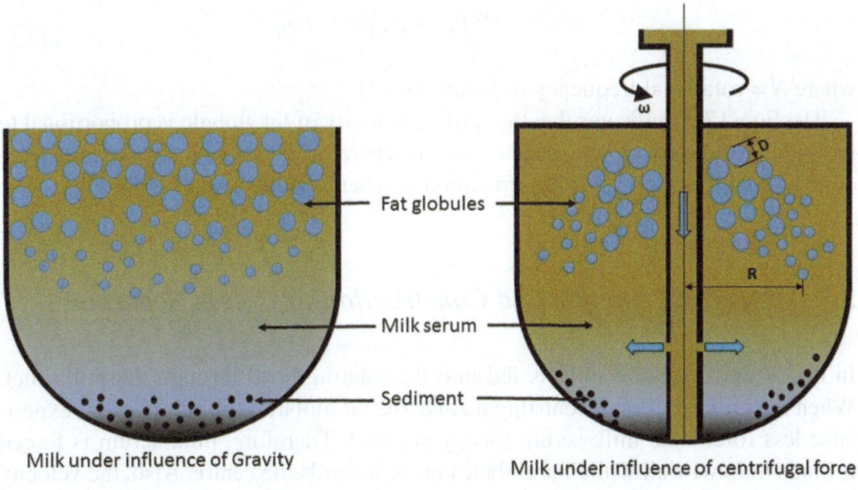

Fig. 17.1 Difference between the path of fat globules during gravity (left) and centrifugal (right) separation; D diameter of fat globules, R radial of fat globules from the axis of rotation; and ω angular velocity

globule surface and causes flocculation. However, warming up of milk above 37 °C reduces such flocculation because of the detachment of cryoglobulins (Everett, 2007). In addition, pasteurisation of milk at relatively high temperature denatures agglutinin causing a significant slowdown of gravity separation (creaming) (Wilbey, 1996).

2.2 Centrifugal Separation

A widely used method for manufacturing of commercially natural dairy cream is centrifugal separation, which enables rapid separation of a fat fraction of milk from milk serum. Technically, the centrifugal separation method can be considered as an accelerated gravity separation method where the upward velocity of fat globules increased by many folds using external force.

Centrifugal separation of cream is based on the use of centrifugal force to separate particles and liquids of different density and size (Saravacos & Kostaropoulos, 2016). Use of external force dramatically increases the settling velocity of fat globules resulting in rapid separation of fat globules from milk serum. The resultant fat globule velocity can be expressed as following by replacing "*g*" with centrifugal parameters:

$$v = D^2 \left(\rho_s - \rho_f\right) R\omega^2 / 18\eta \tag{17.4}$$

where R = radial of fat globules from the axis of rotation (Fig. 17.1); ω = angular velocity (radians s^{-1})

Equation (17.4) can be rewritten as:

$$v = 2\pi^2 D^2 \left(\rho_s - \rho_f\right) RN^2 / 9\eta \tag{17.5}$$

where N = rotational frequency (revolutions s^{-1}).

The Eq. (17.5) indicates that the settling velocity of fat globule is proportional to the square of the rotational speed of cream separator. Difference between the principle of centrifugal and gravity separation is schematically given in Fig. 17.1.

2.3 Working Principle and Construction of Cream Separator

In the cream separator, milk is fed into the rotating bowl through the milk inlet. When the milk reaches to centrifugal zone, the fat globules (lighter portion) experience less force than milk serum (heavy portion). Therefore, milk serum is forced towards to bowl wall while fat globules move towards the centre. Also, the velocity of the lighter dispersed fat phase depends on their size as larger size move relatively faster than the smaller size globules (Eq. (17.5)). Such action results in separate phases of cream and skim milk coming out through the different outlets. In the case of commercial cream separator (Fig. 17.2), numerous separating cones are integral parts and these discs form numerous narrow channels that substantially increase the efficiency of the separation process.

The fat separation efficiency of cream separator depends upon various factors such as disc configuration, the rotational frequency of separator, temperature of milk, the feed rate of milk, fat globule size and rotational frequency of separator. Temperature not only decreases the viscosity of milk but also increase the density difference between milk serum and fat phase that cumulatively increases the efficiency of the separation process. The bowl diameter, number of separating cones, the position of holes in separating cones and gap between adjacent separating cones play a vital role in centrifugal separation process (Towler, 1994). At a fixed rotational frequency, wider bowl results in better separation and so is true for a condition with a small gap of separating cones. Wider bowl (1) increases residence time of fat globules under the influence of centrifugal time; (2) ensures complete migration of fat globules towards centre; (3) narrow gap of separating cones which dramatically decrease the distance to be travelled by fat globules and (4) ensures no remixing by providing laminar flow of fluid within the narrow channel. The feed rate of milk controls the residence time of fat globules within separating zone. The lower the feed rate, the higher the separating efficiency; however, a very low feed rate may decrease the separation efficiency as thick cream may hinder outflow of the cream itself. Similarly, centrifugal speed of separating bowl plays a very crucial role in determining the efficiency of a cream separator. Provided all the other parameters

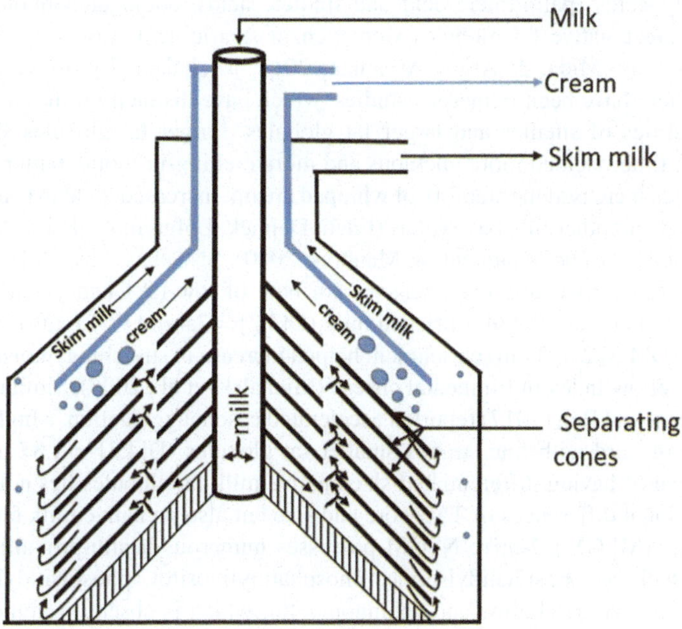

Fig. 17.2 Schematic diagram of a centrifugal cream separator

constant in Eq. (17.5), i.e., at a fixed dimension of separating bowl and fixed cream separation temperature, higher centrifugal speed increases the efficiency of the cream separator by increasing separation velocity of fat globules. This highlights that a reasonable degree of cream separation efficiency can be achieved by decreasing bowl size and increasing rotational speed or vice versa.

3 Fractionation of Natural Cream Based on Milk Fat Globule Sizes

Natural cream is concentrated form of native milk fat globules. Size of milk fat globules varies from 0.1 to 20 μm (Pieter Walstra, 1999). It is now well established that there is a considerable variation in physicochemical and nutritional properties of fat globules depending upon their size. These variations are associated with compositional differences in TAG core, milk fat globule membrane (MFGM) and size itself.

Size of milk fat globule significantly affects physical properties such as creaming stability, viscosity etc. Creaming stability, viscosity, and whiteness of milk/cream increase with a decrease in milk fat globule size. In contrast, milk fat crystallisation temperature and melting enthalpy, and electrical conductivity of globules decrease with decrease in size of fat globules (Truong, Palmer, Bansal, & Bhandari, 2016a). Native smaller sized fat globules were rich in medium-chain and unsaturated fatty acids (palmitoleic acid and linoleic acid), conjugated linoleic acids whereas lager native fat globules were rich in stearic acid (Lopez et al., 2011; Mesilati-Stahy, Mida, & Argov-Argaman, 2011; Michalski, Briard, & Juaneda, 2005). There have been numerous studies, which have distinctly differentiated the functionalities of smaller and larger fat globules. Larger fat globules shortened whipping time, yielded more unctuous and more melting-in-mouth butter. Smaller fat globules increased the stability of whipped cream, increased yield in Camembert cheese and smoother full-fat yoghurt (Edén, Dejmek, Löfgren, Paulsson, & Glantz, 2016; Goudédranche, Fauquant, & Maubois, 2000). St-Gelais, Passey, Haché, and Roy (1997) reported improved sensory attribute of low-fat Cheddar cheese produced from lager fat globule enriched milk (D[3,2] = 2.4 μm vs 1.6 μm). Small fat globules (D[4,3] 2.5–3 μm) enrichment helped increase in stretching, moisture content and yellow index in Emmental cheese (Michalski et al., 2006). Similarly, Luo, Wang, Guo, and Ren (2017) reported accelerated casein aggregation, a high storage modulus of curd with fine strands smaller fat globules (D[4,3] = 1.87 μm). The importance of having differentiated sized native milk fat globules lies not only on compositional differences of TAG core and size but also on native milk fat globule membrane (MFGM). Native NFGM possesses numerous health-promoting compounds such as phosphatidylserine, phosphatidylinositol, phosphatidylethanolamine, phosphatidylcholine and sphingomyelin, which is absent in homogenised and recombined MFGM (Deeth, 1997; Lopez et al., 2011). Besides, presence of intact native MFGM makes fat globules as an inert filler (structure breaker) in milk acid gel since native MFGM interact with protein matrix poorly; however, homogenised milk fat globules act as active filler because of the interaction of MFGM with protein matrix (Vliet & Dentener-Kikkert, 1982). Native milk fat globules have also been found easily digestible than milk fat globules created after homogenization. Berton et al., 2012reported that although the surface area of homogenised fat globules (D[4.3] 0.18–0.29 μm) increases 25-folds, the homogenised fat globules is only twice the enzymatic activity (human pancreatic lipase) as compared to the native fat globules (D[4.3] 4.11–4.31 μm). Moreover, the catalytic activity of human pancreatic lipase was 4.6 times higher on native milk fat globule of D[4,3] 1.8 μm than 6.7 μm (Berton et al., 2012).

In recent days, few methods are developed to fractionate native milk fat globules on size basis for accessing the possibilities of the industrial significance of the heterogeneity in size-based milk fat globules properties. Most of the methods/processes for the production of size based fractionated natural cream comprises two stages: (1) fractionation of milk into various streams having different average fat globule size and (2) normal cream separation of the respective streams. In all size-based fractionated cream production processes stage 1 is different depending up-on nature of process; however stage 2 can be applied to all to obtain cream with different average fat globule sizes (Dhungana, Truong, Palmer, Bansal, & Bhandari, 2017; Edén et al., 2016; Goudédranche et al., 2000; Luo et al., 2017; Olsson & Mamic, 2015).

3.1 Gravity Separation

Gravity separation method utilises the size variation and size-dependent variation in creaming velocity of milk fat globules due to the difference in density between dispersed fat and serum phases.

It is apparent from the Eq. (17.3) that large fat globules in milk at constant temperature and serum composition rise at the top at higher velocity resulting in more substantial proportion of larger fat globules in top layers of milk at a fixed time interval.

Figure 17.3 depicts the theoretical distribution of fat globules in a gravity separation vessel. Each of the layers is then separated in typical cream separator to get creams having different average fat globule size. In early days, Zannoni (1981) reported a common practice of gravity separation method typically used in various part of Italy involved storage of raw milk in a shallow basin at 10–16 °C for 6–12 h to obtain desired fat content and average fat globule size. Rate of creaming or fat content and size of a certain fraction of milk during gravity separation can be varied by manipulating the milk storage temperature (Ma & Barbano, 2000). The second stage of the gravity separation method can be done by carrying out normal centrifugal cream separation of the semi-skimmed milk layers such as A, B and C in Fig. 17.3. A similar approach has been used to get size based fractions of cream by Olsson and Mamic (2015).

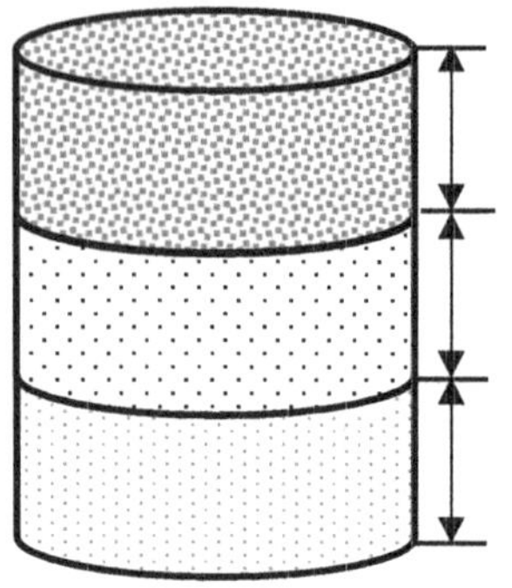

Fig. 17.3 A schematic diagram showing size-based layers of milk formed during gravity separation at a point of storage time; The proposed order of the average size of fat globules in each is A > B > C

3.2 Ultrasonically Assisted Gravitation Separation

Ultrasonically assisted separation occurs when acoustic radiation forces generated from standing wave sound field displace the fat globules towards pressure nodes and antinode plane (Leong, Johansson, Juliano, McArthur, & Manasseh, 2013). The underlying mechanism of milk fat separation on size basis using ultrasonication has been fully covered in Chap. 18. In general, facilitation of milk fat globule separation by ultrasonication depends upon many factors viz. size, surface properties of fat globules and acoustic force. The ultrasonic treatment has been found more effective for larger fat globules than small fat globules. Rate of creaming at low temperature (5 °C) is much efficient with high-frequency ultrasound treatment than that of low frequency as high frequency produces more significant acoustic force than low frequency (Leong et al., 2016).

3.3 Separation with Modified Cream Separator

An emerging method to produce cream with different average fat globules size is the use of modified cream separator. Manufacture of creams with this method follows two-stage centrifugal separation. The first stage is a "fractionation" process, whereas the second stage is a concentration (normal) process. Modification of cream separator involves either partial (Edén et al., 2016) or complete removal (Dhungana et al., 2017) of separating cones from separating disc. It is suggested that complete removal of separating cones increases the efficiency of size-based fractionation of fat globules. Removal of separating cones still provides free space to the milk fat globules inside the disc to experience centripetal force. When milk fat globules in the milk are released inside the rotating disc, the amount of the centripetal force experienced by the fat globules differs depending upon their size. Large fat globules, being less dense and bigger in size, acquire less speed than denser small fat globules. Therefore, under the right combination of milk feed temperature and feed rate, the small milk fat globules reach the inner wall of the separating disc much earlier than larger milk fat globules (Dhungana et al., 2017). Small fat globules follow the route of skim milk whereas large fat globules exit through the cream outlet. In the second stage, called concentration, each of the two streams from the first stage is subjected separately to a normal cream separation process to get creams with entirely different average milk fat globule size. A schematic view of two-stage separation is given in Fig. 17.4.

4 Recombined Cream

A dairy cream can be recreated by mixing milk fat concentrates (cream or anhydrous milk fat) and suitable emulsifier(s) and/or stabiliser(s); and followed by application of heat and suitable mechanical treatment (Fredrick et al., 2013; Towler, 1994). There are numerous types of emulsifiers and stabilizers, which could be of either protein (dairy and non-dairy) or phospholipids or surfactants or carbohydrate-

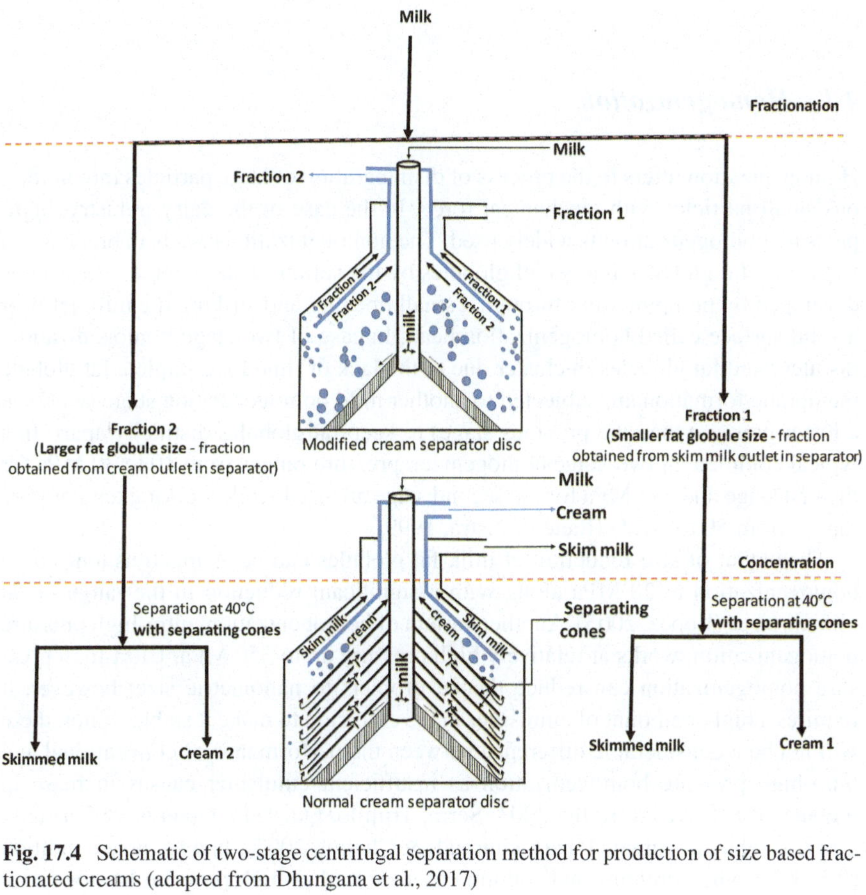

Fig. 17.4 Schematic of two-stage centrifugal separation method for production of size based fractionated creams (adapted from Dhungana et al., 2017)

based compounds. An emulsifier might be called as a stabilizer or texture modifier depending upon their intended functionalities (Chen, 2015; Ozturk & McClements, 2016). Casein is the most widely used protein-based emulsifier because of its milk origin, excellent heat stability, interfacial properties and nutritional value (Hu et al., 2015; van Lent, Le, Vanlerberghe, & Van der Meeren, 2008). Casein stabilised recombined cream had lower crystallisation temperature, and better stability towards coalescence than that of whey protein stabilised cream (Relkin, Sourdet, & Fosseux, 2003). The reported mechanical means to prepare recombined dairy cream are homogenisation, high-pressure homogenisation, microfluidization and ultrasonication (Truong, Palmer, Bansal, & Bhandari, 2016b).

The use of recombined dairy cream is increasing steadily because of some distinct advantages such as lower storage cost of raw materials, and ease of making a cream with desired characteristics, and independent of milking season (van Lent et al., 2008). A major change that is consistently being observed in recombined cream is the creation of an entirely new type of membrane with characteristic composition and microstructure. In addition, the recombination process also enables the creation of fat globules of a desired size range.

4.1 Homogenization

Homogenization refers to the process of disintegration of large particles into numerous small particles with mechanical force. In the case of the dairy industry, high-pressure homogenisation is widely used. The homogenization results in breakage of large milk fat globules into small globules by the action of shear force. Shearing is developed by pumping milk through a small opening and striking the milk jet over a solid surface called homogenization head. In case of two-stage homogenization, disintegrated fat globules in cluster due to the lack of time for complete fat globule membrane formation are subjected to another mild homogenization stage (works at a lower pressure than the previous stage) to keep fat globules distinctly apart. In a typical commercial two-stage homogenizer, pressure ranges from 10 to 30 MPa for the first stage and 3–5 MPa for the second stage are used while working temperature ranges from 50 to 60 °C (Pieter Walstra, 1999).

The extent of size reduction of milk fat globules can be as much as tenfold by homogenization at 20 MPa along with a significant reduction in the range of fat globule sizes (Lopez, 2005). Another variant of homogenization, ultra-high pressure homogenization, works at relatively high pressure up to 350 MPa. Ultra-high pressure homogenization can reduce the fat droplets to nanometric size; however, it requires a higher amount of emulsifier in the emulsion to make it stable. If not, there will not be a considerable difference between the performance of conventional and ultra-high pressure homogenization as insufficient emulsifier causes increase in globule size (Hayes & Kelly, 2003; Serra, Trujillo, Quevedo, Guamis, & Ferragut, 2007; Thiebaud, Dumay, Picart, Guiraud, & Cheftel, 2003). In a suitable condition (0.5–3.0% whey proteins and sodium caseinate), ultra-high pressure homogenizer operated at 87–123 MPa producing recombined emulsion from anhydrous milk fat with mean globule diameter as small as 200 nm (Truong, Palmer, Bansal, & Bhandari, 2013).

4.2 Microfluidisation

Microfluidisation is the process where two high-speed micro streams colloid each other thereby disintegrating the larger particles into small ones, especially to nanometric sizes (Truong et al., 2016b). Although it shares basic steps viz shear, cavitation, turbulence, with homogenization, its effectiveness on getting the same size at constant pressure is much higher than that of high-pressure homogenisation (Dalgleish, Tosh, & West, 1996; Hardam, Imison, & French, 2000). The efficiency of microfluidisation process depends upon various factors such as operating pressure, fat content of the final product and amount of emulsifier (Hardam et al., 2000; Truong et al., 2013); therefore an optimum condition based on prevailing operating condition is necessary to realize better performance (Mahdi, He, & Bhandari, 2006). For example, it was reported that the increase in total protein in native cream from 2.2% (w/w) to 3% (w/w) significantly decreased the size of microfluidised cream from D [3,2] 1.35 μm to 0.26 μm when the cream was microfluidised at 62 MPa and 40 °C (Panchal et al., 2017). However, no significant decrease in average droplet size was observed on further increase in protein content of native cream. Similarly, microfluidisation of recombined mass of anhydrous milk fat, sodium caseinate and water; and commercial cream and sodium caseinate each at fat: protein ratio of 5:1 created creams with D[4,3] 0.37–0.40 μm and 0.16–0.20 μm, respectively, when microfluidisation was done at 45–85 MPa (Hussain, Truong, Bansal, & Bhandari, 2017).

4.3 Ultrasonication

Ultrasonication method has been widely used to homogenise or reduce the droplet size of the emulsion (Hardam et al., 2000; Muthupandian et al., 2010; Villamiel & de Jong, 2000; Wu, Hulbert, & Mount, 2000). Ultrasonication is the process characterised by the acoustic cavitation as a result of mechanical vibration powered by high-frequency sound (~20 kHz) (Truong et al., 2016b). During sonication, a disintegration of larger emulsion droplet happens because of pressure difference created during bubble collapse (Mason, Wilking, Meleson, Chang, & Graves, 2006). The degree of homogenization achieved by ultrasonication depends upon the applied acoustic power, treatment time, and working temperature. In ordinary cases, the use of high acoustic power, longer treatment time coupled with high temperature result in smaller fat globule size (Villamiel & de Jong, 2000; Wu et al., 2000). Sonication of whole milk for 10 min with 180 W and 450 W reduced average fat globule size from 5.5 μm to 2.36 and 0.73 μm, respectively (Ertugay, Sengul, & Sengul, 2004).

5 Major Types of Commercial Cream Products and Their Processing

There are various types of cream products manufactured by following the different technologies with typical characteristics and end-use (Table 17.2). The major types of technologically important methods are discussed below.

5.1 Whipping Cream

Whipping cream is a special variant of cream having an excellent capacity to form foam. Whipping cream (aka light whipping cream in the USA) which is pasteurised or ultra-pasteurised and may be homogenised, contains less than 36% but not less than 30% milkfat in cream (FDA, 2015). Manufacturing of whipping cream involves

Table 17.2 Different cream products and their properties

Cream types	Fat content (%)	Additives	Usage	Specialities	References
Whipping cream	30–40	Carrageenan, alginate, starch, gelatine	Whipped cream, flour confectionery	Forms aerated product	FDA (2015)
Sour cream	≥18	Culture (*Streptococcus lactis* and *Streptococcus cremoris*)	Salad dressing	Acidity not less than 0.5% as lactic acid	http://www.idfa.org/news-views/media-kits/milk/definition
Cream powder	40–70	Lactose, proteins, sorbitol	Dried soups, ice cream	Less than 2% moisture, spray-dried	Tamime (2009)
Cream liqueur	>14	Sucrose, citrate, ethanol	As alcoholic beverage	Alcoholic beverage	Tamime (2009)
Coffee cream	≥18	Phosphates, citrates	Feathering of coffee	Double stage homogenization	FDA (2015), Walstra, Wouters, and Geurts (2005)
Dessert cream	Not less than 18	Phosphates, citrates	Dessert dressing	Single-stage homogenization; thicker than coffee cream, carrageenan, alginate	FDA (2015), Walstra, Walstra, et al. (2005)

standardisation of the fat content of cream and addition of optional ingredients like an emulsifier, stabiliser, sweetener, etc.; followed by pasteurisation, cooling, packaging and storage. Pasteurisation could be done either by High Temperature Short Time (HTST; 85 °C for 30 min) or Ultra High Temperature (UHT) method (140 °C for 2 s). Generally homogenization is not done for whipping cream; however, a UHT treated whipping cream could be homogenised after pasteurization at reasonably low pressure of about 3.5–7 MPa to prevent adverse effects of high heat treatment on creaming stability of cream (Varnam & Sutherland, 2001; Walstra, Walstra, Wouters, & Geurts, 2005). Packaging of whipping cream is done after cooling to 5 °C. The whipping cream should be stored at 4 °C for 24 h in the case of immediate use; otherwise, whippability will be impaired because of insufficient solid fat to trigger partial coalescence during whipping (Early, 1998; Sung & Goff, 2010; Walstra, Wouters, & Geurts, 2005). Functionality and emulsion properties of whipping cream depend also on interfacial serum phase materials. Whey protein isolate stabilised whipping cream demonstrated shorter whipping time than sodium caseinate stabilised cream (Hotrum, Stuart, van Vliet, Avino, & van Aken, 2005). Similarly, presence of a higher amount of protein, and other hydrocolloids (locust bean gum, carrageenan etc.) in serum phase of whipping cream increased the whipping time and decreased overrun (Camacho, Martínez-Navarrete, & Chiralt, 1998; van Lent et al., 2008).

5.2 Sterilised Cream

Sterilised cream is available as both coffee and dessert cream. The major difference between them is the degree of clustering, and so is the methods of achieving such conditions. Coffee cream, also known as light cream in Australia and US, contains milkfat not less than 18% with an average of 20% while acidity remains in the range of 0.14–0.15% as lactic acid (FDA, 2015; Walstra, Wouters, & Geurts, 2005). Coffee cream manufacture starts with standardisation of cream mass by mixing high-quality cream, skim milk and stabilising salt to achieve 20% milkfat and desired pH. Coffee cream is heated relatively at a higher temperature (70–75 °C) before double stage homogenization (first stage at 11–20 MPa and the second stage at 3–5 MPa). Variation in pressure during homogenization depends on the technique employed for sterilisation. In case of bottle sterilisation process which is done at 115 °C for 20 min, homogenization is done at lower to mid-range of pressure while homogenization is done at a higher range of each stage if a cream is UHT sterilised (Lampert, 1965; Spreer, 1998; Varnam & Sutherland, 2001). UHT sterilised cream is packaged online in aseptic condition. Packaged cream is then cooled to 25 °C before storage and/or distribution (Lampert, 1965; Spreer, 1998; Varnam & Sutherland, 2001; Walstra, Wouters, & Geurts, 2005).

Unlike coffee cream, dessert cream, which is thicker than coffee cream, is first UHT treated at 140 °C for 10 s followed by cooling to 50 °C before single-stage aseptic homogenization at 10 MPa. The homogenised cream is aseptically packaged and cooled to 10 °C (Walstra, Wouters, & Geurts, 2005). High heat treatment during sterilisation leads to cooked flavour, irreversible creaming and gelation (Hoffmann & Buchheim, 2006). Addition of just 0.015% of carrageenan improved creaming behaviour and decreased serum loss upon whipping of 30% fat containing UHT cream (Precht, Peters, & Petersen, 1988).

5.3 *Sour Cream*

Sour cream, also known as cultured sour cream, is the pasteurised and lactic acid-producing bacteria fermented cream containing milkfat not less than 18%(w/v) (IDFA, 2018). However, their fat content may range from 10 to 40% depending upon countries (Hoffmann & Buchheim, 2006). Sour cream must have titratable acidity not less than 0.5%, determined as lactic acid. In case of sour cream with added optional ingredients (such as gelatine, starch), the milk fat content should not be less than 18% of dairy portion and not less than 14.4% milkfat of total weight of cream (FDA, 2015; Hoffmann & Buchheim, 2006).

Basic steps of sour cream processing involve standardisation of cream mix, pasteurisation, fermentation and packaging (Born, 2013; Hoffmann & Buchheim, 2006). The cream mix is prepared from high-quality cream and milk solids not fat. Pasteurisation can be done either by vat (73.9–79.4 °C for 30 min) or HTST (82.2–85 °C) method (Born, 2013). The cream is then cooled to 68 °C before homogenization at 13.8–17.2 MPa pressure. A two single-stage homogenization process would result in an excellent body and texture on cream (Born, 2013; Lampert, 1965).

Lactic acid fermentation is done with 1–4% starter culture of *Streptococcus lactis*, *Streptococcus cremoris,* flavour-producing bacteria *Leuconostoc citrovorum* and *L. dextranicum.* The optimum temperature for growth of these bacteria ranges between 21.1 and 23.9 °C. Fermentation is partially stopped after 12–18 h by cooling cream to below 10 °C or preferably to 2–4 °C once either pH or Soxhlet-Henkle (SH) value reaches to 4.6 to 5.1 and 25–35, respectively (Early, 1998; Lampert, 1965; Spreer, 1998). The optional ingredients (e.g. thickener, emulsifier, food flavouring, nutritive sweeteners, salt etc.) which are safe and suitable can be added to sour cream up to permissible limit in order to improve texture, prevent syneresis, increase palatability and attractiveness and extend shelf life (Lampert, 1965; Spreer, 1998).

5.4 *Cream Liqueur*

A cream liqueur is a cream-based alcoholic beverage. One of the world-famous cream liqueur variety is Irish cream liqueur, predominantly sold as Baileys Irish Cream (Mitchell, 2016). Industrial preparation method of traditional Irish cream liqueur (17% by volume alcohol) starts with the mixing of water (85 °C) and tri-sodium citrate followed by addition of sodium caseinate, sugar, caramel and Annatto (Mitchell, 2016). Dairy cream (<10 °C) and spirit are then mixed with a previously prepared base mix in a mix tank, and the final mix is homogenised in a two-stage homogeniser (24.1 MPa and 3.45 MPa) at 55 °C. The homogenised mix is homogenised again as in the previous set of homogenization pressures after addition of flavour. The final liqueur is then cooled to 14 °C followed by packaging in brown colour glass bottles (Hoffmann & Buchheim, 2006; Mitchell, 2016). Brown colour bottle prevents light-induced off-flavour development (Hoffmann & Buchheim, 2006). Stability of cream liqueur largely depends on alcohol content and minerals in the serum phase of liqueur. Increase in alcohol content and the presence of inorganic substances in liquid phase decreased the stability of cream liqueur (Banks & Muir, 1985). These are the major challenges for the production of high alcohol-containing cream liqueur. Another demerit from processors' viewpoint is not being able to produce cream with high alcohol content (Banks & Muir, 1985; Heffernan, Kelly, & Mulvihill, 2009). Banks and Muir (1985) reported that cream liqueur (36–45° proof alcohol and 40% total solids) prepared using washed native cream (reduced inorganic component) where alcohol was mixed before 2 pass homogenization at 31 MPa pressure and 55 °C having better stability than liqueur from whole cream (38 days vs 1 day). In the same report, it was also reported that addition of tri-sodium citrate on washed cream decreased the stability of cream liqueur (Banks & Muir, 1985). However, the addition of tri-sodium citrate enhances the stability of cream liqueur prepared from the whole cream where it acts as calcium sequestrant and protects from casein aggregation. The second approach for the production of high alcohol cream liqueur is the addition of alcohol in two lots. Of the total alcohol to be added, a portion alcohol is added to cream before homogenization and the second portion of alcohol is added to the homogenized mixture of cream and alcohol. Such sequence helps keep emulsion droplet size small (Hoffmann & Buchheim, 2006). Stability of cream liqueur can also be increased by reducing emulsion droplet size, increasing total solid content (as sugar) and addtion of small molecules emulsifier such as monoglycerides. The first two parameters increase viscosity of final product whereas last parameter, when added in optimum amount along with sodium caseinate and tri-sodium citrate, helps reduce phase separaiton (Banks & Muir, 1988; Hoffmann & Buchheim, 2006). Banks and Muir (1988) reported that the cream liqueur would acquire enhanced viscosity, creaminess and whitening strength if 98% of the total fat globules in liqueur were of diameter <0.8 μm. This size range can easily be obtained either by multi-cycle homogenization or by sigle-cycle microfludization (Mitchell, 2016; Panchal et al., 2017).

5.5 *Cream Powder*

A cream powder is a dried form of liquid cream. Dried cream is used in desserts, ice creams, dried soups, packet cake mixes etc. It contains a maximum of 5% by weight moisture, minimum 42% by weight milk fat and a minimum of 34% by weight milk protein in milk solid-not-fat (Codex, 2014). Spray-dried cream powder/ encapsulated fat powder is made from stabilised emulsions. Proteins, modified starches and other suitable hydrocolloids are the major types of emulsifiers used in such products. Another important component of cream powder is fillers. They can be water-soluble carbohydrates, hydrolysed starches, and gums (Keogh, 2004). Fäldt and Bergenståhl (1995) described a method by mixing sodium caseinate and butterfat in a lactose solution at pH 7. The mix is then heated to 70 °C followed by pre-homogenization with a high-speed stirrer. High-pressure homogenization of liquid mass is carried out at 100 MPa pressure for several cycles. Multiple cycle homogenization of emulsion, although being cost-intensive, effectively decreased the proportion of over-sized fat globules and made emulsion droplet size more uniform which may help to improve the powder properties (Hogan, McNamee, O'Riordan, & O'Sullivan, 2001; Muir & Banks, 1986). The homogenized cream is spray-dried with inlet air at 180 °C, and outlet air temperature is maintained at 80–90 °C. Hogan et al. (2001) and Vignolles et al. (2010) also reported similar methods to obtain fat (soy and sunflower oil, respectively) encapsulated powder. In summary, spray-dried fat powder preparation process involves stabilisation of fat-filled emulsion by suitable emulsifier (preferably caseinate) at suitable homogenization pressure, mixing of filler materials (preferably lactose), and followed by spray drying preferably at 180–190 °C as inlet temperature and 80–90 °C as outlet temperature. Sodium caseinate is preferred as emulsifier over other food proteins as it is the most surface tension reducing food protein ever known (Dalgleish, 1989). Food additives that can be added to legal limits to improve powder properties are stabilizers, firming agents, acidity regulators, anticaking agents, and antioxidants (Codex, 2014).

Since the cream powder is a high-fat content product, it is susceptible to lipid oxidation and caking during handling. Caking occurs when surface fat content is high. It also lacks reconstitutability and fat leakage increases with storage time (Fäldt & Bergenståhl, 1995). The melting point of fat core of emulsion before drying plays an important role in the surface fat content of powder (Keogh, 2004). Fäldt and Bergenståhl (1995), reported that the fat powders prepared from emulsion with low melting point fat (soybean oil) and high melting point fat (high melting point rapeseed oil) as core materials had lower amount of surface fat (3% and 15% respectively) than powder prepared with medium melting point fat, butterfat and hardened coconut butter (~34%). Partial crystallinity occurs significantly in medium melting point fat, which leads to partial coalescence of droplets and ultimately facilitate leakage of fat out to the surface (Coupland, 2018; Fäldt & Bergenståhl, 1995). Lactose, a critical factor responsible for fat encapsulated powder stability and functionality, affects cream powder depending upon its concentration and crystallinity (Fäldt & Bergenståhl, 1995; Keogh, 2004). Lactose acts as a continuous phase in the powder system and helps limit to protein-protein interaction that eventually hinders fat globule coalescence and aggregation in both emulsion and drying stage (Keogh, 2004). An increase in lactose concentration along with sodium caseinate reduced surface fat (Fäldt & Bergenståhl, 1996a). However, these authors did not notice such effect of lactose with whey protein. Lactose also affects cream powders adversely upon moisture uptake in poor storage condition (Fäldt & Bergenståhl, 1996b). Recrystallisation of lactose increases powder particle size, porosity and induces coalescence resulting in increased free and surface fat of the powder (Fäldt & Bergenståhl, 1996b; Saito, 1985). Emulsion droplet size before spray drying also plays significance role on fat encapsulated powder. Soottitantawat et al. (2005) reported a significant increase in retention of d-limonene in response to a decrease in emulsion size from 2 to 0.5 μm. In addition, same authors claimed a constant decrease in surface oil content when emulsion droplet size decreased from 4.1 to 0.65 μm. The average droplet size of an emulsion with large average droplet size decreased significantly after atomization, which was not significant in emulsion with smaller average size droplets. Soottitantawat et al. (2005) postulated that the increase in surface oil and decrease in d-limonene flavour in spray-dried powder with the increase in droplet size could be due to the break down of large droplets during atomization. Therefore, a careful selection of emulsifier, drying condition and droplet size that compatible with core material is necessary to obtain fat encapsulated powder with better physical properties.

6 Conclusion and Future Remarks

Dairy cream and associated products have been in the utmost preference of consumers across the globe. These products have reached out to culinary of every ethnicity and race. Dairy cream is a concentrated mass of fat globules obtained from bovine milk. It is separated from milk, based on density differences between milk fat globules and milk serum. The most widely used method for cream separation is centrifugal separation. In addition, centrifugal separator with disc fitted with numerous separating cones is the most commonly used device to separate fat from milk. Milk feed rate and inlet temperature, centrifugal speed, number of separating cones and distance between adjacent cones are the major factors influencing the efficiency of fat separation using such separator. Current development in cream processing is the size-based fractionation of milk fat globules and subsequently getting size-differentiated cream. A great deal of research has shown some marked differences in physicochemical properties of native fat globules depending upon their sizes. These differences have also shown significant impacts on cream-based products. One of the current research trend is focusing on the utilization of such size-dependent differences for improvement and manipulation of product performances. Recombined creams, which are concentrated form of newly created—recreated fat globules either using anhydrous milk fat or natural cream, are also gaining industrial importance because of their easy manoeuvrability and potential applications in new product development. However, a large number of research has shown them incapable of maintaining product characteristics when used as a replacement of natural cream.

Thermal and physical stability of cream are the major properties of a cream, which have a direct influence on processability and product characteristics of almost every kind of cream-based products. In native cream, thermal and physical stability slacken off with time. A conventional remedy is mechanical homogenization of cream, which, if sufficient amount of right emulsifier is provided, effectively increase the physical and thermal stability. However, these achievements come with the expense of native milk fat globule membrane. Since smaller fat globules are more stable towards creaming and thermal stress, selective removal of larger fat globules from milk could improve these properties of resulting creams. However, the lack of cost-effective methodology for size-based fractionation of native fat globules has limited industrial application of such an approach. This could be a future area of research. Some of the specialized cream-based products, especially cream liqueur and cream powder, although having a long history, are still suffering from poor keeping quality. It is postulated that a reduction in cream droplet size in combination with suitable emulsifier could improve this quality. Based on the available literature, future works in the processing of cream-based products could be studied on the effect of nano-sized emulsion droplets on the stability of such specialised creams.

References

Banks, W., & Muir, D. (1988). Stability of alcohol-containing emulsions. In *Advances in food emulsions and foams* (pp. 257–283). London, UK: Elsevier.

Banks, W., & Muir, D. D. (1985). Effect of alcohol content on emulsion stability of cream liqueurs. *Food Chemistry, 18*(2), 139–152. https://doi.org/10.1016/0308-8146(85)90137-2

Berton, A., Rouvellac, S., Robert, B., Rousseau, F., Lopez, C., & Crenon, I. (2012). Effect of the size and interface composition of milk fat globules on their in vitro digestion by the human pancreatic lipase: Native versus homogenized milk fat globules. *Food Hydrocolloids, 29*(1), 123–134.

Born, B. (2013). Cultured/sour cream. In *Manufacturing yogurt and fermented milks* (pp. 381–391). New York, NY: John Wiley & Sons.

Camacho, M. M., Martínez-Navarrete, N., & Chiralt, A. (1998). Influence of locust bean gum/λ--carrageenan mixtures on whipping and mechanical properties and stability of dairy creams. *Food Research International, 31*(9), 653–658. https://doi.org/10.1016/S0963-9969(99)00041-1

Chen, L. (2015). Emulsifiers as food texture modifiers. In J. Chen & A. Rosenthal (Eds.), *Modifying food texture: Novel ingredients and processing techniques* (Vol. 1, pp. 27–49). Cambridge, UK: Woodhead Publishing.

Codex. (2014). Codex standard for milk powders and cream powder. In *Standard for cream powder, half cream powder and high fat milk powder (Vol. A-10-1971)* (pp. 1–6). Geneva: FAO/WHO.

Coupland, J. N. (2018). Crystallization of lipids in oil-in-water emulsion states. In *Crystallization of lipids: Fundamentals and applications in food, cosmetics, and pharmaceuticals* (p. 431). Hoboken, NJ: John Wiley & Sons Ltd.

Dalgleish, D. (1989). *Protein-stabilized emulsions and their properties* (pp. 211–250). London, UK: Elsevier Applied Science.

Dalgleish, D., Tosh, S., & West, S. (1996). Beyond homogenization: The formation of very small emulsion droplets during the processing of milk by a microfluidizer. *Nederlands melk en Zuiveltijdschrift, 50*(2), 135–148.

Deeth, H. (1997). The role of phospholipids in the stability of milk fat globules. *Australian Journal of Dairy Technology, 52*(1), 44.

Dhungana, P., Truong, T., Palmer, M., Bansal, N., & Bhandari, B. (2017). Size-based fractionation of native milk fat globules by two-stage centrifugal separation. *Innovative Food Science & Emerging Technologies, 41*, 235–243.

Early, R. (1998). *Technology of dairy products*. Cham, Switzerland: Springer Science & Business Media.

Edén, J., Dejmek, P., Löfgren, R., Paulsson, M., & Glantz, M. (2016). Native milk fat globule size and its influence on whipping properties. *International Dairy Journal, 61*, 176–181. https://doi.org/10.1016/j.idairyj.2016.06.004

Ertugay, M. F., Sengul, M., & Sengul, M. (2004). Effect of ultrasound treatment on milk homogenisation and particle size distribution of fat. *Turkish Journal of Veterinary and Animal Sciences, 28*(2), 303–308.

Everett, D. W. (2007). Cream products. In Y. H. Hui (Ed.), *Handbook of food products manufacturing* (pp. 723–747). New York, NY: John Wiley & Sons.

Fäldt, P., & Bergenståhl, B. (1995). Fat encapsulation in spray-dried food powders. *Journal of the American Oil Chemists' Society, 72*(2), 171–176. https://doi.org/10.1007/bf02638895

Fäldt, P., & Bergenståhl, B. (1996a). Spray-dried whey protein/lactose/soybean oil emulsions. 1. Surface composition and particle structure. *Food Hydrocolloids, 10*(4), 421–429.

Fäldt, P., & Bergenståhl, B. (1996b). Spray-dried whey protein/lactose/soybean oil emulsions. 2. Redispersability, wettability and particle structure. *Food Hydrocolloids, 10*(4), 431–439.

FDA. (2015). Milk and Cream. Maryland, USA: Department of Health and Human Services, USA Retrieved from https://www.ecfr.gov/cgi-bin/text-idx?SID=9976l6ec95bb95e6efbb038809843f9b&mc=true&node=sp21.2.131.b&rgn=div6

Fredrick, E., Heyman, B., Moens, K., Fischer, S., Verwijlen, T., Moldenaers, P., et al. (2013). Monoacylglycerols in dairy recombined cream: II. The effect on partial coalescence and whipping properties. *Food Research International, 51*(2), 936–945.

FSANZ. (2016). Australia New Zealand Food Standards code- Standard- 2.5.2-Cream. Australia: The Board of Food Standards Australia New Zealand Retrieved from https://www.legislation.gov.au/Details/F2015L00470

Getz, C., Smith, G. F., Tracy, P., & Prucha, M. (1937). Instant whipping of cream by aeration. *Journal of Food Science, 2*(5), 409–428.

Goudédranche, H., Fauquant, J., & Maubois, J.-L. (2000). Fractionation of globular milk fat by membrane microfiltration. *Le Lait, 80*(1), 93–98.

Hardam, J. F., Imison, B. W., & French, H. M. (2000). Effect of homogenisation and microfluidisation on the extent of fat separation during storage of UHT milk. *Australian Journal of Dairy Technology, 55*(1), 16–22.

Hayes, M. G., & Kelly, A. L. (2003). High pressure homogenisation of raw whole bovine milk (a) effects on fat globule size and other properties. *Journal of Dairy Research, 70*(03), 297–305.

Heffernan, S. P., Kelly, A. L., & Mulvihill, D. M. (2009). High-pressure-homogenised cream liqueurs: Emulsification and stabilization efficiency. *Journal of Food Engineering, 95*(3), 525–531. https://doi.org/10.1016/j.jfoodeng.2009.06.018

Hoffmann, W., & Buchheim, W. (2006). Significance of milk fat in cream products. In *Advanced dairy chemistry volume 2: Lipids* (pp. 365–375). New York, NY: Springer.

Hogan, S. A., McNamee, B. F., O'Riordan, E. D., & O'Sullivan, M. (2001). Microencapsulating properties of sodium caseinate. *Journal of Agricultural and Food Chemistry, 49*(4), 1934–1938.

Hotrum, N. E., Stuart, M. A. C., van Vliet, T., Avino, S. F., & van Aken, G. A. (2005). Elucidating the relationship between the spreading coefficient, surface-mediated partial coalescence and the whipping time of artificial cream. *Colloids and Surfaces A: Physicochemical and Engineering Aspects, 260*(1-3), 71–78.

Hu, B., Zhang, L., Liang, R., Chen, F., He, L., Hu, B., et al. (2015). Cross-linking of interfacial casein layer with genipin prevented pH-induced structural instability and lipase digestibility of the fat droplets. *Journal of Agricultural and Food Chemistry, 63*(7), 2033–2040. https://doi.org/10.1021/jf505724c

Hussain, H., Truong, T., Bansal, N., & Bhandari, B. (2017). The effect of manipulating fat globule size on the stability and rheological properties of dairy creams. *Food Biophysics, 12*(1), 1–10. https://doi.org/10.1007/s11483-016-9457-0

IDFA. (2018). *Sour cream. Definitions*. Retrieved from https://www.idfa.org/news-views/media-kits/milk/definition

Keogh, M. K. (2004). Spray-dried microencapsulated fat powders. In O. Charles (Ed.), *Encapsulated and powdered foods* (pp. 477–482). New York, NY: Taylor and Francis.

Lampert, L. M. (1965). *Modern dairy products: Composition, food value, processing, chemistry, bacteriology, testing*. New York, NY: Chemical Publishing Company.

Leong, T., Johansson, L., Juliano, P., McArthur, S. L., & Manasseh, R. (2013). Ultrasonic separation of particulate fluids in small and large scale systems: A review. *Industrial & Engineering Chemistry Research, 52*(47), 16555–16576. https://doi.org/10.1021/ie402295r

Leong, T., Johansson, L., Mawson, R., McArthur, S. L., Manasseh, R., & Juliano, P. (2016). Ultrasonically enhanced fractionation of milk fat in a litre-scale prototype vessel. *Ultrasonics Sonochemistry, 28*, 118–129. https://doi.org/10.1016/j.ultsonch.2015.06.023

Lopez, C. (2005). Focus on the supramolecular structure of milk fat in dairy products. *Reproduction Nutrition Development, 45*(4), 497–511.

Lopez, C., Briard-Bion, V., Ménard, O., Beaucher, E., Rousseau, F., Fauquant, J., et al. (2011). Fat globules selected from whole milk according to their size: Different compositions and structure of the biomembrane, revealing sphingomyelin-rich domains. *Food Chemistry, 125*(2), 355–368.

Luo, J., Wang, Y., Guo, H., & Ren, F. (2017). Effects of size and stability of native fat globules on the formation of milk gel induced by rennet. *Journal of Food Science, 82*(3), 670–678.

Ma, Y., & Barbano, D. M. (2000). Gravity separation of raw bovine milk: Fat globule size distribution and fat content of milk fractions. *Journal of Dairy Science, 83*(8), 1719–1727. https://doi.org/10.3168/jds.S0022-0302(00)75041-7

Mahdi, J. S., He, Y., & Bhandari, B. (2006). Nano-emulsion production by sonication and microfluidization—A comparison. *International Journal of Food Properties, 9*(3), 475–485. https://doi.org/10.1080/10942910600596464

Mason, T. G., Wilking, J., Meleson, K., Chang, C., & Graves, S. (2006). Nanoemulsions: Formation, structure, and physical properties. *Journal of Physics: Condensed Matter, 18*(41), R635.

Mesilati-Stahy, R., Mida, K., & Argov-Argaman, N. (2011). Size-dependent lipid content of bovine milk fat globule and membrane phospholipids. *Journal of Agricultural and Food Chemistry, 59*(13), 7427–7435.

Michalski, M.-C., Briard, V., & Juaneda, P. (2005). CLA profile in native fat globules of different sizes selected from raw milk. *International Dairy Journal, 15*(11), 1089–1094.

Michalski, M.-C., Leconte, N., Briard-Bion, V., Fauquant, J., Maubois, J., & Goudédranche, H. (2006). Microfiltration of raw whole milk to select fractions with different fat globule size distributions: Process optimization and analysis. *Journal of Dairy Science, 89*(10), 3778–3790.

Mitchell, P. C. (2016). Modernization of the traditional Irish cream liqueur production process. In A. McElhatton & M. M. El Idrissi (Eds.), *Modernization of traditional food processes and products* (pp. 31–44). Boston, MA: Springer US.

Mudgil, D., Barak, S., & Khatkar, B. S. (2014). Guar gum: Processing, properties and food applications—A review. *Journal of Food Science and Technology, 51*(3), 409–418. https://doi.org/10.1007/s13197-011-0522-x

Muir, D. D., & Banks, W. (1986). Technical note: Multiple homogenization of cream liqueurs. *International Journal of Food Science & Technology, 21*(2), 229–232. https://doi.org/10.1111/j.1365-2621.1986.tb00444.x

Muthupandian, A., Bhaskaracharya, R., Kentish, S., Lee, J., Palmer, M., & Zisu, B. (2010). The ultrasonic processing of dairy products – An overview. *Dairy Science & Technology, 90*(2-3), 147–168.

Olsson, A., & Mamic, A. (2015). *Method development for fractionation of milk fat globules-for improvement of cream functionality*.

Ozturk, B., & McClements, D. J. (2016). Progress in natural emulsifiers for utilization in food emulsions. *Current Opinion in Food Science, 7*, 1–6.

Panchal, B. R., Truong, T., Prakash, S., Bansal, N., Bhandari, B., & Hub, A. D. I. (2017). Effect of fat globule size on the churnability of dairy cream. *Food Research International, 99*, 229–238.

Precht, D., Peters, K. H., & Petersen, J. (1988). Improvement of storage stability and foaming properties of cream by addition of carrageenan and milk constituents. *Food Hydrocolloids, 2*(6), 491–506. https://doi.org/10.1016/S0268-005X(88)80048-1

Relkin, P., Sourdet, S., & Fosseux, P.-Y. (2003). Fat crystallization in complex food emulsions: Effects of adsorbed milk proteins and of a whipping process. *Journal of Thermal Analysis and Calorimetry, 71*(1), 187–195. https://doi.org/10.1023/A:1022282720983

Saito, Z. (1985). Particle structure in spray-dried whole milk and in instant skim milk powder as related to lactose crystallization. *Food Structure, 4*(2), 16.

Saravacos, G., & Kostaropoulos, A. E. (2016). Mechanical separation equipment. In *Handbook of food processing equipment* (pp. 233–292). New York, NY: Springer.

Serra, M., Trujillo, A., Quevedo, J., Guamis, B., & Ferragut, V. (2007). Acid coagulation properties and suitability for yogurt production of cows' milk treated by high-pressure homogenisation. *International Dairy Journal, 17*(7), 782–790.

Soottitantawat, A., Bigeard, F., Yoshii, H., Furuta, T., Ohkawara, M., & Linko, P. (2005). Influence of emulsion and powder size on the stability of encapsulated d-limonene by spray drying. *Innovative Food Science & Emerging Technologies, 6*(1), 107–114. https://doi.org/10.1016/j.ifset.2004.09.003

Spreer, E. (1998). *Milk and dairy product technology* (Vol. 83). Raton, FL: CRC Press.

St-Gelais, D., Passey, C. A., Haché, S., & Roy, P. (1997). Production of low-fat Cheddar cheese from low and high mineral retentate powders and different fractions of milkfat globules. *International Dairy Journal, 7*(11), 733–741. https://doi.org/10.1016/S0958-6946(97)00084-8

Sung, K. K., & Goff, H. D. (2010). Effect of solid fat content on structure in ice creams containing palm kernel oil and high-oleic sunflower oil. *Journal of Food Science, 75*(3), C274–C279.

Tamime, A. Y. (2009). *Dairy fats and related products*. New York, NY: John Wiley & Sons.

Thiebaud, M., Dumay, E., Picart, L., Guiraud, J., & Cheftel, J. (2003). High-pressure homogenisation of raw bovine milk. Effects on fat globule size distribution and microbial inactivation. *International Dairy Journal, 13*(6), 427–439.

Towler, C. (1994). Developments in cream separation and processing. In R. K. Robinson (Ed.), *Modern dairy technology: Volume 1. Advances in milk processing* (pp. 61–105). Boston, MA: Springer US.

Truong, T., Palmer, M., Bansal, N., & Bhandari, B. (2013). *Physical properties of milk fat in nano-emulsions*. Cham, Switzerland: Springer.

Truong, T., Palmer, M., Bansal, N., & Bhandari, B. (2016a). Effect of milk fat globule size on physical properties of milk. In *Effect of milk fat globule size on the physical functionality of dairy products* (pp. 35–45). Cham, Switzerland: Springer.

Truong, T., Palmer, M., Bansal, N., & Bhandari, B. (2016b). Methodologies to vary milk fat globule size. In *Effect of milk fat globule size on the physical functionality of dairy products* (pp. 15–30). Cham, Switzerland: Springer.

UOG. (2018). *Milk solids-not-fat*. Food Science Department, University of Guelph. Retrieved 27 July, 2018, from https://www.uoguelph.ca/foodscience/book-page/milk-solids-not-fat

van Lent, K., Le, C. T., Vanlerberghe, B., & Van der Meeren, P. (2008). Effect of formulation on the emulsion and whipping properties of recombined dairy cream. *International Dairy Journal, 18*(10–11), 1003–1010. https://doi.org/10.1016/j.idairyj.2008.04.002

Varnam, A., & Sutherland, J. P. (2001). *Milk and milk products: Technology, chemistry and microbiology* (Vol. 1). Heidelberg, Germany: Springer Science & Business Media.

Vignolles, M.-L., Lopez, C., Le Floch-Fouéré, C., Ehrhardt, J.-J., Méjean, S., Jeantet, R., et al. (2010). Fat supramolecular structure in fat-filled dairy powders: A tool to adjust spray-drying temperatures. *Dairy Science & Technology, 90*(2-3), 287–300.

Villamiel, M., & de Jong, P. (2000). Influence of high-intensity ultrasound and heat treatment in continuous flow on fat, proteins, and native enzymes of milk. *Journal of Agricultural and Food Chemistry, 48*(2), 472–478. https://doi.org/10.1021/jf990181s

Vliet, T. v., & Dentener-Kikkert, A. (1982). Influence of the composition of the milk fat globule membrane in the rheological properties of acid milk gels. *Netherlands Milk and Dairy Journal, 36*, 261–265.

Walstra, P. (1999). *Dairy technology: Principles of milk properties and processes*. New York, NY: CRC Press.

Walstra, P., Walstra, P., Wouters, J. T., & Geurts, T. J. (2005). Cream products. In *Dairy science and technology* (2nd ed., pp. 447–466). New York, NY: CRC Press.

Walstra, P., Wouters, J. T., & Geurts, T. J. (2005). *Dairy science and technology*. New York, NY: CRC Press.

Wilbey, R. A. (1996). Estimating the degree of heat treatment given to milk. *International Journal of Dairy Technology, 49*(4), 109–112.

Williams, P., Phillips, G., & de Vries, J. (2004). Hydrocolloid gelling agents and their applications. In *Gums and stabilisers for the food industry* (Vol. 12, pp. 23–31). Cambridge, UK: Royal Society of Chemistry.

Wu, H., Hulbert, G. J., & Mount, J. R. (2000). Effects of ultrasound on milk homogenization and fermentation with yogurt starter. *Innovative Food Science & Emerging Technologies, 1*(3), 211–218. https://doi.org/10.1016/S1466-8564(00)00020-5

Zannoni, M. (1981). [Survey of natural creaming of milk fat in the Parmigiano-Reggiano cheese production area [in Italy]].[Italian]. Scienza e Tecnica Lattiero Casearia.

Chapter 18
Ultrasound-Assisted Cream Separation

Thomas S. H. Leong

1 Introduction

Cream is a dispersion of concentrated butterfat droplets dispersed in an aqueous protein matrix that has a natural tendency to rise. The separation of cream from milk is a process that dates back millennia. The natural tendency for cream to rise is a method that is still used today in the production of some notable traditional cheeses. In Northern Italy, famous hard cheeses such as Parmesan and Romano cheeses, are produced using a semi fat-reduced milk that is created by overnight separation using natural gravity. This process naturally separates the larger cream droplets present in the milk since they rise fastest and tends to retain the smaller cream droplets in the cheese milk. This natural selection of small cream globules from large globules is purported by traditional cheesemakers to contribute to the unique texture and flavour characteristics of these cheeses.

In the context of dairy processing today, cream is a valuable commodity that is routinely separated from whole milk to make butter, cheese and many other products. The invention of the modern-day centrifuge during the nineteenth century, was a major turning point in dairy processing, as it enabled rapid, mass-scale separation of cream from milk. This mass production capability enabled centralised dairy processing facilities, completely transforming the dairy industry.

The modern centrifuge separates cream from milk on the basis of rapidly spinning milk at several thousand times the force of gravity. The fat droplets, which have a lower density than the skim milk phase, are pushed upwards due to the g-forces they are subject to during spinning, and they become rapidly separated from the skim milk phase. The separation is often operated in such a way so that all the fat is pulled from milk and then put back into the milk to carefully defined fat contents.

T. S. H. Leong (✉)
ARC Dairy Innovation Hub, University of Melbourne, Parkville, VIC, Australia
e-mail: thomas.leong@unimelb.edu.au

T. Truong et al. (eds.), *Dairy Fat Products and Functionality*,
https://doi.org/10.1007/978-3-030-41661-4_18

This protocol is also known as standardisation and is the general method by which milks available from the supermarket are produced.

While centrifugation is exceptionally effective, it does bear several potential issues. The rapid spinning required to perform the separation consumes a large amount of energy. Due to the many moving parts and the closely interspaced plates located within the centrifuge, extensive maintenance and long centrifuge cleaning downtime are potential issues that can affect profitability.

Recently, new techniques have become available to provide dairy producers with the ability to control the separation of cream globules more effectively. As mentioned, the selection of small-sized fat droplets from large-sized fat droplets (i.e. size fractionation) has been performed using natural gravity separation in the production of some Northern Italian cheeses. New research following this traditional way of separation has been conducted (Caplan, Melilli, & Barbano, 2013; Ma & Barbano, 2000) that indicate advantages in regard to bacterial quality on top of the natural separation of small cream globules from large globules. Microfiltration (Goudédranche, Fauquant, & Maubois, 2000; Michalski et al., 2006) is another proven method that can be used to selectively separate fat globules at an industrially relevant scale. Size distributions collected from micro-filtered milks indicate excellent capability for microfiltration to specifically recover small and large-sized cream globules from milk (Michalski et al., 2006). The processing of whole milk through membrane pores, however, is prone to fouling, and hence requires extensive cleaning to maintain a suitably high selectivity and flux during operation. Some recent studies have also shown that modified centrifugal separation, can also achieve size-based fractionation (Dhungana et al., 2017; Logan et al., 2014). The modifications require either redesign of existing centrifugal equipment and/or operating using select conditions that favours selective separation of fat globules.

The use of ultrasonic standing waves has recently been reported as an interesting alternative technique by which cream droplets can be rapidly separated from milks. The mechanism for ultrasonic separation consists of no moving parts and promotes separation in a way that mimics natural gravitational separation. This chapter will report on the developments made using this novel separation method that shows great promise as a complementary tool for cream separation in dairy processing.

2 Background of Ultrasound

The use of ultrasound in fluid processing (sonoprocessing) in the classical sense is most well-known to involve the frequency range between 20–100 kHz (Leong, Ashokkumar, & Kentish, 2011). In fluids, the application of ultrasound induces a phenomenon known as acoustic cavitation, which is the nucleation, growth and subsequent collapse of gaseous entities in the fluid.

Ultrasonic separation, also known as megasonic separation, harnesses the effects of ultrasound at a higher frequency range, typically between 400–3000 kHz (Leong et al., 2015). At these frequencies, the transient cavitation collapse events that often occur in the 20–100 kHz range are reduced, such that physical shearing effects become minimal (Leong, Martin, & Ashokkumar, 2016). This section will provide some background theory of ultrasound waves and how they can be applied in the context of cream separation for dairy processing.

2.1 Ultrasound

Sound waves are vibrations that occur through a medium, either air, liquid or solid, and have a characteristic frequency i.e. rate of vibration. For audible sound, the higher the rate of vibration, the more 'high pitched' that the sound will be perceived as. Ultrasonic waves vibrate at a frequency that exceeds human audibility, which are typically above 16 kHz.

Ultrasound can be broadly characterised into several distinct regions that produce different outcomes when applied to fluids, solids or gases. As this chapter focuses on the application of ultrasound in milk and dairy systems, the following theory will focus on the application of ultrasound through fluids.

The 'power ultrasound' region, refers to the application of ultrasound in the frequency range between 20–100 kHz (Leong et al., 2011). At these frequencies, the sound wave influences bubbles present in solution to oscillate. Depending on the amplitude of the sound wave, the bubbles undergo highly non-linear oscillation such that the bubble expands in size dramatically during pressure rarefaction (i.e. the low pressure cycle). In some cases, a tenfold increase in the bubble size occurs during this expansion (Leong et al., 2014). During the compression phase of the oscillation, i.e. high pressure cycle, the bubble compresses and collapses upon itself which, depending on the size of the bubble during expansion, occurs with intense energy. The size of a bubble at which this occurs is known as the resonance size, and there is usually a range of sizes over which a bubble can be subject to this collapse behaviour at a given ultrasound frequency and driving pressure.

The energy that is released is sufficient to increase the temperatures in the centre of the collapsing bubble to ~5000 K and generate pressure shockwaves in the order of several hundred atmospheres. The increase in temperature is sufficient to induce homolysis of water molecules to form highly reactive hydrogen and hydroxide radical species (Weissler, 1959). These physical and chemical effects occur simply from the application of ultrasound to a fluid containing dissolved gaseous material at ambient temperature. Sonochemistry has opened up a whole new capability in fluid processing, whereby ultrasonic vibrations can be harnessed to significantly enhance rates of reaction, as well as accelerate heat and mass transfer, at significantly reduced operating conditions (Ashokkumar & Mason, 2000).

As the frequency of ultrasound vibrations increase, their influence on bubbles change quite dramatically. The resonance size of bubbles is inversely proportional to the ultrasound frequency (Leong, Martin, & Ashokkumar, 2016). The size to which bubbles grow to before they collapse is reduced at elevated frequencies, resulting in reduced physical effects i.e. lower temperature, lower magnitude pressure. The frequency range between 100–1000 kHz, is typically termed the intermediate ultrasound region. Although the physical effects are reduced, this region is characterised by peak formation of chemical radicals (Koda, Kimura, Kondo, & Mitome, 2003; Mason, Cobley, Graves, & Morgan, 2011). The reason is because the active population of bubbles that are at the resonance size are increased, due to the decrease in the resonance size with frequency. Therefore, although the energy of collapse is reduced, the increased number of collapse events that occur in this frequency range makes it quite efficient in the production of radical species suited for enhancing chemical reactions.

By increasing the frequency of ultrasound further (i.e. >1 MHz), their influence on initiating strong bubble collapse events become negligible. These frequencies produce a gentler physical effect on objects within the sound field and is a commonly used mechanism in ultrasonic baths for cleaning sensitive components. Other forces created by the acoustic field such as acoustic streaming and acoustic radiation forces also become stronger with increasing frequencies (Leong, Johansson, Juliano, McArthur, & Manasseh, 2013a). Acoustic radiation forces can become extremely powerful in manipulating the movement of suspended objects within a fluid. If the sound wave inside the fluid can be established as a standing wave, then the suspended objects can be manipulated to move towards specific regions of maximal or minimal pressure within the wave field by the acoustic radiation forces.

2.2 *Acoustic Radiation Forces*

A sound wave reflected upon itself can superimpose to form what is known as an acoustic standing wave (Fig. 18.1). These waves are characterised by alternating regions of high local pressure where constructive superimposition occurs, known as pressure antinodes, and regions of minimum local pressure where destructive interference occurs, known as pressure nodes.

When a standing wave field is sustained in a medium containing a suspension of particulates, the wave will be partially scattered by the particles if there exists an acoustic impedance mismatch between them and the fluid. This produces what is known as the primary acoustic radiation force.

The primary acoustic radiation force influences the suspended particulates, which could be solid particles or immiscible liquid droplets, manipulating them to move towards either the node or antinode of the standing wave. The specific location to which they are transported depends on the material properties of the particulates relative to the surrounding medium, known as the acoustic contrast factor. The contrast factor accounts for the specific gravity/density and the compressibility of the particles, and can be calculated using:

$$\varnothing = \frac{5\rho_p - 2\rho_l}{2\rho_p + \rho_l} - \frac{\beta_p}{\beta_l} \tag{18.1}$$

where ρ is the density, β is the compressibility, and the subscripts l and p refer to the liquid medium and the particles, respectively (Yosioka & Kawasima, 1955).

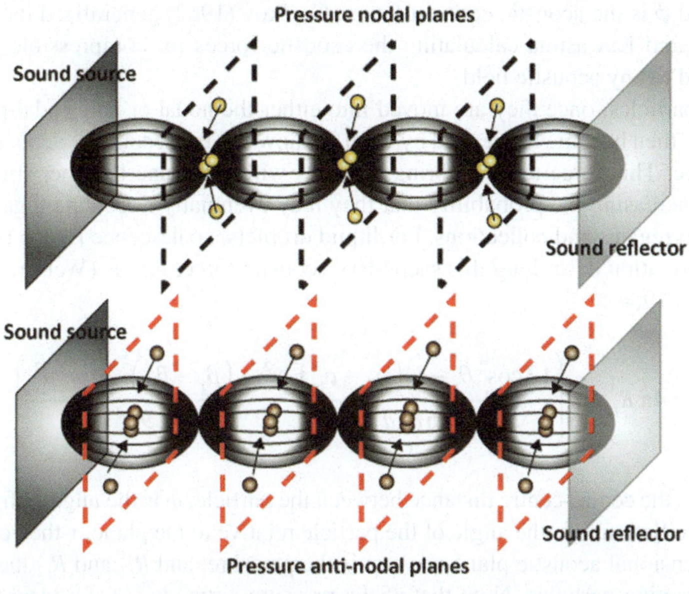

Fig. 18.1 Migration of food materials within an acoustic standing wave field. Food particles/droplets may migrate towards pressure nodal or pressure anti-nodal planes, depending on their material properties. Reprinted with permission from Leong, Johansson, Juliano, McArthur, and Manasseh (2013b). Copyright 2013 American Chemical Society

Particles that have a positive acoustic contrast generally move towards the pressure nodes of the standing wave. Contrarily, those that have a positive contrast move towards the pressure antinodes. Ultrasonic separation works on the principle such that by positioning individual droplets or particles on pressure nodes or antinodes within the standing wave, an enhanced probability to aggregate or coalesce into larger entities occurs. The formation of larger collections of particulates can cause them to separate more quickly out of suspension. Figure 18.1 shows a schematic representation of particles moving towards nodal planes and anti-nodal planes within a standing wave field.

The development of the theoretical understanding of acoustic radiation forces dates back to the work of King (1934), when he first studied the behaviour of spherical incompressible particles in acoustic fields. Yosioka and Kawasima (1955) continued this work, calculating the acoustic forces for compressible particles in planar acoustic waves as:

$$F_{ac} = -\frac{4\pi}{3} r^3 \mathrm{k} \mathrm{E}_{ac} \phi \sin(2kx) \tag{18.2}$$

where r is the particle radius, $k = \frac{2\pi}{\lambda}$ the wavenumber, λ is the wavelength ofsound, E_{ac} the specific energy density, x the distance from a nodal point of the standing wave and ϕ is the acoustic contrast factor. Gor'kov (1962) generalized the work of Yosioka and Kawasima calculating the acoustic forces for compressible particles subjected to any acoustic field.

The particles, once they are moved into either the nodal or anti-nodal planes of pressure, then become influenced by what is known as the secondary acoustic radiation force. This force tends to bring particles within a plane together (Fig. 18.1), further increasing the probability that they may eventually flocculate together into larger aggregates and collections. For liquid droplets, coalescence is also possible.

The equation describing this secondary acoustic force, F_{sec}, is (Weiser, Apfel, & Neppiras, 1984):

$$F_{sec} = 4\pi R_{p_1}^3 R_{p_2}^3 \left(\frac{\left(3\cos^2\theta_r - 1\right)\left(\rho_p - \rho_m\right)^2 v^2}{6\rho_m d^4} - \frac{\left(\beta_p - \beta_m\right)^2 \rho_m \omega^2 p^2}{9d^2} \right) \quad (18.3)$$

Here, d is the centre-centre distance between the particle, ω is the angular frequency of the oscillation, θ is the angle of the particle relative to the plane, v the velocity in a 1-dimensional acoustic plane wave, p is the pressure, and R_{p1} and R_{p2} the radii of the interacting particles. Note that at the pressure antinodes of the ideal standing wave field, the velocity is zero (and vice versa). It has been described that in many situations, the magnitude of this force is 1–2 orders smaller than that of the primary acoustic force (Leong et al., 2013a). The inter-particle distance is an important consideration for the secondary acoustic radiation force. Since the force is inversely proportional to the distance between adjacent particles, the secondary acoustic radiation force becomes strong in magnitude when particles are sufficiently close. The secondary force can be attractive or repulse (Mettin, Akhatov, Parlitz, Ohl, & Lauterborn, 1997), and this is dependent on the driving frequency of the sound source relative to the respective linear resonances of each individual particle (Fig. 18.2). Particles located on an axis approximately perpendicular to the direction of sound propagation will be attractive.

The primary and secondary acoustic radiation forces are fundamental to the ability of an acoustic standing wave to promote enhanced separation of suspended particulates from fluids. There are several other important considerations that influence the ability of a standing wave to manipulate particles and therefore induce rapid separation. The frequency of the applied sound wave is directly proportional to the magnitude of both the primary and secondary radiation forces. This means that the higher the frequency of sound applied, the stronger the forces becomes. While the manipulation and separation of particles in the mm range can be achieved using relatively low frequency vibrational waves, particles of novel interest in the micron size range e.g. fat globules in milk, typically requires ultrasound to be applied at MHz frequencies (1–3 MHz) (Leong et al., 2014). There exists a limitation on using higher ultrasonic frequencies than this approximate range, due to the attenuation of

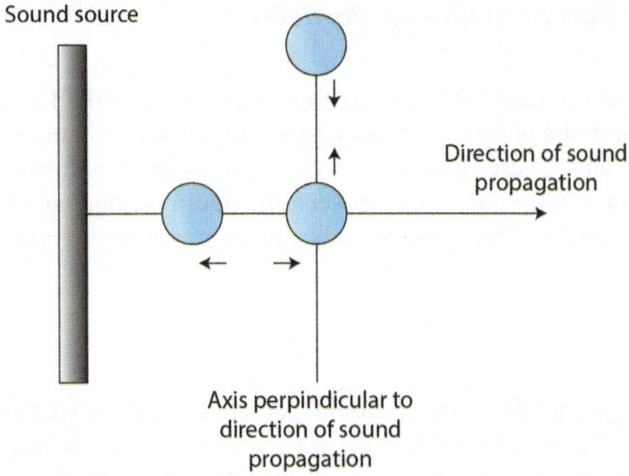

Fig. 18.2 Direction that the secondary radiation force acts on particles when they are parallel to the direction of sound propagation and when they are perpendicular to the direction of sound propagation. Reprinted with permission from Leong et al. (2013b). Copyright 2013 American Chemical Society

high frequency ultrasound that limits its ability to be effectively applied to fluids. The attenuation of the sound wave means that the energy of the vibrations become absorbed by the fluid and will diminish with distance from the sound source (Leong, Coventry, Swiergon, Knoerzer, & Juliano, 2015).

The acoustic energy density of the ultrasonic standing wave established within the fluid is proportional to the magnitude of the primary and secondary acoustic radiation forces. Typically, lower frequency ultrasound can achieve a higher acoustic energy density compared with high frequency ultrasound, which again is influenced by the limitations of acoustic attenuation as a sound wave travels through a fluid.

The size of the particles in question influences the selection of the required ultrasonic frequency and acoustic energy density for the separation. Separation is impossible or ineffective if either the frequency or acoustic intensity is too low to manipulate particles of a very small particle size. As mentioned, it is not possible to use very high frequencies in the context of processing large volumes of material, due to the strong attenuation that high frequency ultrasound experiences with distance. Ultrasonic separation can be highly complementary with flocculation enhanced separation, however. A flocculating agent that can promote smaller particles to stick together and form larger particles will separate more effectively under the influence of an ultrasonic standing wave, that drives these particles closer together to flocculate with higher probability.

2.3 Other Forces Acting on Particles

Ultrasonic separation manipulates particles suspended in a fluid to gather at pressure nodal/anti-nodal planes where they have a high likelihood to form bigger floccules or aggregates. These clusters of particles can cream or sediment at a rapid rate if there exists a density difference between the liquid medium and the material of the particle or droplet. This is due to buoyancy or gravitational force as follows:

$$F_B = \frac{4}{3}\pi r^3 g\left(\rho_l - \rho_p\right) \tag{18.4}$$

where g is the gravity. In a viscous liquid, particles will also experience the Stokes drag force F_D. For small particles, the particles' inertia can be neglected and an assumption that they are always moving in a local steady-state (Bruus, 2012) can be made. For low Reynolds number, the drag force F_D is given by Stokes Law as:

$$F_D = 6\eta\pi r\left(v_l - v_p\right) \tag{18.5}$$

where η is the viscosity of the fluid, v_l is the velocity of the fluid and v_p is the velocity of the particle.

2.4 Acoustic Streaming

Acoustic streaming is the fluid flow generated by the attenuation of an acoustic wave as it propagates through a medium. As the alignment of particles within nodal and anti-nodal planes is a key mechanism for enhanced separation, acoustic streaming may play a detrimental role to the separation as it can exert drag and shear on particles, influencing their behavior and movement within an acoustic field (Spengler & Jekel, 2000). The effects of acoustic streaming on particle manipulation by acoustic radiation force was observed by Spengler, Coakley, and Christensen (2003). Spengler et al. (2003) saw how yeast cells that were manipulated in an ultrasound chamber resulted in changing patterns and velocities over time as they were dragged around the chamber by acoustic streaming. The implication is that if the streaming motion is very strong, it can potentially override the acoustic radiation forces, disturbing or even eliminating the formation of particle bands established at pressure nodal planes.

The origin of acoustic streaming comes from the attenuation of the sound field, which can be caused by viscous attenuation in the medium; scattering on particles, droplets or bubbles; or friction between a vibrating wall and the surrounding fluid. Zarembo (1971) described three main types of acoustic streaming: (1) Eckart streaming, (2) Rayleigh and (3) boundary layer Schlichting streaming. Rayleigh and Schlichting streaming are considered microstreaming events as the length scales over which they propagate is of the order of magnitude of the wavelength of the sound wave for the case of Rayleigh streaming, and orders of magnitude smaller than a wavelength for the case of Schlichting streaming. Eckart streaming is a larger scale streaming condition as the flow patterns are of a length scale significantly greater than the wavelength of sound, and typically can be visually observed as strong fluid motion that occurs within a sound chamber.

Eckart and Rayleigh streaming are the more important streaming events that should be considered in a standing wave field. In a standing wave, the streaming that occurs between pressure nodes and antinodes is called Rayleigh streaming. Eckart streaming is due to the attenuation of the acoustic energy across the bulk of the fluid as the beam of the sound wave propagates through it. When the medium attenuates the wave, a proportion of the acoustic energy is absorbed by the fluid, heating it up. An illustrative diagram of Eckart and Rayleigh streaming is shown in Fig. 18.3.

The viscous acoustic attenuation coefficient for a plane sound wave was derived by Stokes (1848) as:

$$\alpha = \frac{2\nu\left(2\pi f\right)^{2}}{3c^{3}} \tag{18.6}$$

It can be seen that attenuation is proportional to viscosity and to the square of the frequency. This means that streaming effects tend to become more intense with increasing frequency. As mentioned earlier, high frequency ultrasound generates stronger acoustic radiation force. Therefore, consideration needs to be made to balance strong acoustic radiation to manipulate small particles, with the potential formation of strong acoustic streaming forces that may completely disrupt any manipulation that occurs. One of the possible ways in which this can be overcome is to make use of acoustic transparent films that serve to break up the streaming

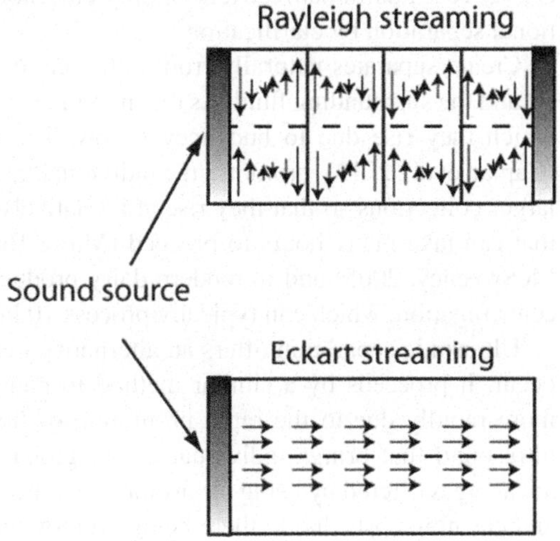

Fig. 18.3 Eckart and Rayleigh acoustic streaming. Eckart streaming propagates through the medium from the sound source and occurs without the requirement for a standing wave to be generated. Rayleigh streaming requires a standing wave consisting of pressure nodal and antinodal regions to develop. Reprinted with permission from Leong et al. (2013b). Copyright 2013 American Chemical Society

forces (Spengler & Jekel, 2000) so that higher frequency ultrasound can be employed more effectively in large separation chambers. These transparent films may not be suitable for heavily fouling fluids such as milks, as milk processing equipment often requires frequent cleaning to maintain hygienic processing conditions.

In high power ultrasonic operations, there is also a formation of a fast jet streaming called "Stuart" streaming. According to Lighthill (1978), at higher powers, Stuart streaming takes the form of an inertially dominated turbulent jet.

3 Cream Separation Using Ultrasound

The concept of ultrasonic separation has been investigated for several decades for various applications (Leong et al., 2013a). Microfluidic devices have been used extensively in the study of ultrasonic separation (Lenshof & Laurell, 2011; Nilsson, Petersson, Jönsson, & Laurell, 2004), as the small, well-controlled geometries that can be achieved within such devices are highly conducive to generating stable, strong standing waves for manipulation of micron-sized particulates in fluids. The rapid separation and removal of lipids from blood cells for blood removed during surgery is an example of the precise control that ultrasonic separation can create (Jonsson et al., 2004; Petersson, Nilsson, Holm, Jonsson, & Laurell, 2004, 2005). Further examples can be found in the review by Leong et al. (2013a).

It is not until recently in the past decade that new ultrasonic transducers capable of generating high frequency ultrasound at high power and hence longer effective working distances, have large-scale applications of ultrasonic separation become viable. An example of large-scale commercial use of ultrasonic separation technology is in the palm oil milling industry of Malaysia (Juliano et al., 2013; Juliano, Swiergon, Mawson, Knoerzer, & Augustin, 2013), where sound waves applied to the mesocarp of the pressed palm oil lead to dramatically enhanced yields of recovery (2–10% additional recovery of oil yield), not previously possible using conventional separation by clarification.

Cream separates naturally from milk due to the density difference between the fat and the surrounding fluid. As the individual droplets of fat are small, the speed at which they rise due to buoyancy is low. The immunoglobulins present in milk (Caplan et al., 2013), promotes the individual fat globules to flocculate together into larger collections so that they rise at a relatively faster rate. This is a slow process that can take many hours to proceed (Ma & Barbano, 2000; O'Mahony, Auty, & McSweeney, 2005) and in modern dairy production, has been largely replaced by centrifugation, which can typically process 10 kL/h of milk.

Ultrasonic separation offers an alternative method by which this separation can occur. It proceeds by a similar method to natural cream separation, albeit much more rapidly due to the rapid promotion of floccule formation from the applied ultrasound that brings individual cream globules together. Some of the possible advantages offered by using ultrasonic separation are potentially lower maintenance and cleaning costs due to there being no moving parts and reduced surface area of contact between the milk and the active processing equipment. It is envisioned that the technology could play a complementary role to centrifugation as it has the potential to influence parameters such as particle size distributions differently. It is known that the distributions of the milk fat globules can play a role in both flavour development and textural attributes of dairy products such as cheeses (Goudédranche et al., 2000). There exists scope for the technology to promote the development of new value-added dairy products.

The following sections will detail the recent developments in cream separation using ultrasonic standing waves for milk systems.

3.1 Proof of Concept

Miles, Morley, Hudson, and Mackey (1995) was one of the first to demonstrate the ability of ultrasonic standing waves to manipulate the fat globules present in milk. Diluted milk samples held within a cuvette were subject to an ultrasonic standing wave (1–3 MHz frequency). Bands of fat globules accumulating at half-wavelength distances, perpendicular to the direction of the sound propagation were observed. These bands of fat droplets collected together and subsequently rose more quickly to the surface of the liquid. Grenvall, Augustsson, Folkenberg, and Laurell (2009); Grenvall, Folkenberg, Augustsson, and Laurell (2012) showed that within a microfluidic flow chamber, the manipulation of the fat globules occurs such that they are positioned towards the pressure antinodes of the standing wave. Because the walls of the microfluidic chamber are typically pressure anti-nodes, the high concentration of fat that accumulated there lead to an eventual clogging up of the flow cell. Nevertheless, from diluted milk, a 3.5-fold enrichment in the fat concentration could be achieved. Although the volume scales demonstrated in these studies were small (30 μl/min), the concept was established that enhanced cream separation could be achieved by manipulation of the fat globules within an acoustic standing wave field.

3.2 Scale-Up Developments and Selective Fractionation

The scale-up potential of ultrasonic separation for milk and dairy systems has more recently been developed and explored. It was successfully demonstrated the viability to scale-up the separation of fat from a recombined milk emulsion held within small tube of 7 mL towards a larger vessel holding 6 L (Juliano et al., 2011, 2013). Different frequencies and transducer configurations in direct contact with the milk emulsion were tested. Runs were performed with one or two transducers placed in vertical (parallel or perpendicular) and horizontal positions (at the reactor base) at 400 kHz, 1 MHz and/or 2 MHz. For these emulsions, it was that the found most efficient creaming after treatment using 400 kHz ultrasound (Juliano, Temmel, Rout, et al., 2013).

The parameters that were found to achieve rapid separation of a recombined milk emulsion, however, were not successful in enhancing cream separation from natural whole milk (Temmel, 2012). This is attributed to several reasons. Firstly, the droplet size distribution and surface composition of natural milk fat are quite different to those in a recombined milk emulsion. The generally smaller sized droplets in whole milk requires a stronger acoustic radiation force to manipulate effectively. The use of 400 kHz ultrasound was not effective in providing a sufficiently strong acoustic radiation force. Secondly, whereas the recombined milk emulsion is stabilised by casein/whey proteins, the milk fat in whole milk is coated by the native milk fat globule membrane. This membrane protects the fat droplets from undergoing coalescence. Mechanistically, this means that the separation is more challenging since coalesced droplets do not redisperse during separation, whereas native fat droplets that are flocculated due to immunoglobulins (Caplan et al., 2013) may become redispersed by excessive acoustic streaming or mixing.

Leong et al. (Leong et al., 2014; Leong, Johansson, et al., 2014) overcame these issues to achieve effective separation of cream from natural whole milk. By using higher frequency ultrasound, in this case, ultrasound >1 MHz, effective manipulation and hence separation of the fat globules present in natural whole milk could be achieved. The caveat in using these higher frequency sound waves, however, is that there is a limited effective distance over which the standing wave can be established. It was found that a short transducer-reflector separation distance (between 30 mm to 85 mm) could enable effective cream separation. Furthermore, the application of two transducers arranged in a parallel set-up could influence more rapid separation of cream due to the ability to achieve a higher energy density per unit volume within the reactor. The same principles can be applied to promote the separation of fat from whey, as demonstrated by Torkamani et al. (2016)

3.3 Selective Ultrasonic Fractionation

The size of the fat droplets determines the rate at which they can be removed by ultrasonic separation. As demonstrated by Leong et al. (2016), it is possible to exploit this to perform selective enrichment and removal of large and small-sized droplets in separated milk streams. In a simple batch process, it was established that the cream became distributed in the separation vessel after sonication, such that the smallest fat globules were retained towards the bottom of the vessel, and large-sized fat globules were enriched within the cream collected near the top of the vessel. By collecting these fractions positioned at the top and bottom of the vessel specifically, it is possible to fractionate the cream in milk according to size. By performing multiple processing passes, the size of these globules can be further refined into specified size distributions. Both the skimmed and concentrated cream fractions can be processed using ultrasonic separation to enrich/concentrate either small or large-sized globules respectively (Leong et al., 2015).

The fractionation of milks into streams with specifically modified particle size distributions may have significant benefit in the production of specialty cheeses and other dairy products. Goudédranche et al. (2000) reported the use of small-sized fat globules in the production of certain cheeses could result in improved sensory qualities. Similarly, Logan et al. (2014) has reported improvement to cheddar cheese yields when using smaller sized fat globules. Large sized fat globules have been reported to produce favourable effects for butter by Goudédranche et al. (2000). One of the possible advantages of ultrasound separation is its ability to achieve fractionation with minimal damage to its natural membrane layer (Leong, Johansson, et al., 2014). One of the next key steps towards practical implementation of ultrasound separation in the dairy industry would be to assess dairy products made using milk streams derived by this method.

3.4 Continuous Flow Processing

In batch processing, the entire milk volume can be subject to a desired processing time to achieve a target separation yield. The milk can also be selectively removed from specific locations within the vessel, which enables fractionation of the milk streams based on how fast they rise within the milk. Continuous flow processing enables the separation of larger volumes of product that can be collected autonomously. Due to the constant flow however, specific fractionation becomes slightly more complex as there is continual movement of milk through the reactor. Fractionation can still be achieved in a continuous flow system by continual removal of bulk streams from the top and bottom. Further refinement of the pooled streams collected at the top and bottom of the vessel, can also be achieved by using multi-pass processing.

By designing a flow-through reactor that minimizes the effective distance over which the standing wave is established (~45 mm), Leong, Juliano, et al. (2015) were able to operate an ultrasonic separation reactor in a continuous flow arrangement for cream separation. When operating with continuous flow, the input and removal rate of the milk into the vessel becomes an important consideration, as this affects the residence time for which the milk is subject to the ultrasonic standing wave. Generally, a longer residence time results in more effective separation. Leong, Juliano, et al. (2015) found that a residence time of 10 min (specific energy = 190 J/g) lead to an effective skimming of 21% fat removal relative to the initial after 1 pass, compared with a significantly lower skimming of 4% relative to the initial for a residence time of 3.3 min (specific energy = 62 J/g). Notably, with continual input of whole milk and removal of cream at the surface and skim at the base, a 'steady-state' could be achieved. This steady state was achieved approximately 10 min after starting of the flow, as milks sampled at the 20 and 30 min sampling points approached a consistent level of cream removal. With multiple passes, a partial skimming of fat down to 50% of the original fat content was achieved after 4 passes. Additionally, there appeared to be a refinement of the cream globules retained in the skim, such that the largest fat globules are removed in the earlier passes of the separation.

In the natural separation of cream from milk, there is an optimum temperature at which cream floccules will form most effectively by the process known as agglutination (Caplan et al., 2013; Mulder & Walstra, 1974). Leong, Juliano, et al. (2014) showed that there also exists a temperature range over which the rate of milk fat separation is highest when influenced by ultrasonic standing waves. Experimental trials indicated that a temperature range between 20 to 60 °C was found to offer fast separation of milk fat.

3.5 Comparison with Other Milk Separation Techniques

Whilst the specificity of fat globule separation has some limitations, i.e. the sharpness of the separation is restricted by the mechanism of the separation, the ultrasonic separation methodology offers an interesting alternative to natural gravity separation and could be a complementary methodology for differential centrifugation (Dhungana et al., 2017; Logan et al., 2014) and microfiltration (Michalski et al., 2006) techniques.

The use of a stage-based processing protocol reported by Leong, Johansson, et al. (2016) follows a similar separation strategy previously presented by O'Mahony et al. (2005) using gravity separation. Whereas the study reported using natural separation required 6–18 h to achieve separation at 4 °C, the use of ultrasonics in the study by Leong, Johansson, et al. (2016) was able to achieve appreciable fractionation after 5–20 min of ultrasonic application. After three processing stages of ultrasonic application, the skimmed fractions resulted in significantly more selective size retainment of small globules (Fig. 18.4) compared with after 6 h of separation without the application of ultrasound in the study by O'Mahony et al. (2005), although the temperature of separation employed by O'Mahony et al. was 4 °C. The separation of fat by both natural gravitational separation and by ultrasound has been reported to be slower at refrigeration temperatures (Leong, Juliano, et al., 2014). Ma and Barbano (2000) reported similar separation at 15 °C and achieved a reduction in the mean particle size in the skimmed fraction of ~1 μm after 2 h. The particle sizes and fat content obtained by Leong et al. after three-stages were comparable to the fractions obtained by Ma and Barbano (2000) at their respective top and bottom, after natural separation for 2 h. The use of differential centrifugation with mild *g*-forces (~150 × *g*) by Logan et al. (2014) has been reported to be useful for fat fractionation. Milks with fractionated distributions similar to Leong, Johansson, et al. (2016) work were reported.

The limitations in fractionation with natural creaming as described by Michalski et al. (2006) are also relevant for ultrasound enhanced separation, since the mechanism for separation proceeds in a similar manner. The main limitation is that smaller

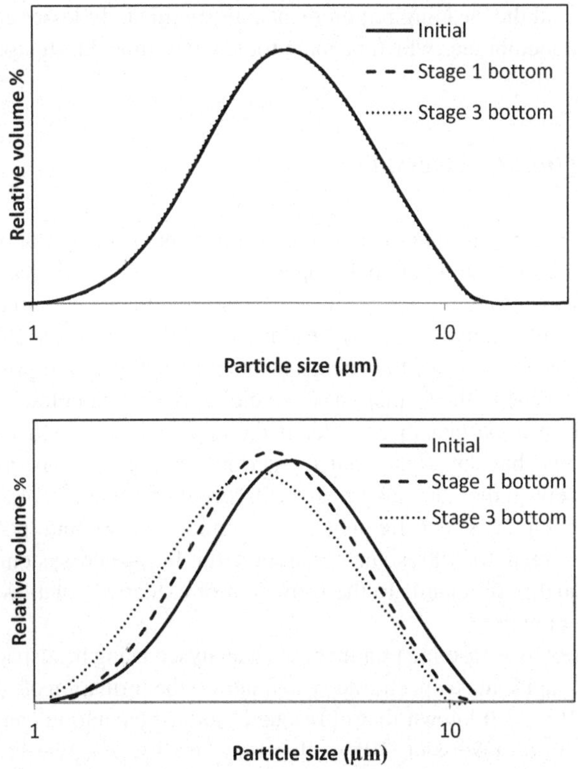

Fig. 18.4 Fractionated skimmed milks obtained from ultrasonic separation (15 min sonication time) compared with control. Reprinted from Leong, Johansson, et al. (2016). Copyright 2016, with permission from Elsevier

fat globules that are already at the top of the holding vessel, will not readily move back down to the bottom of the container and will remain there. Hence, the achievable specificity using ultrasound fractionation tends not to be as high compared with membrane filtration, which separates MFGs based on size exclusion rather than buoyancy and acoustic forces that depend on not just size, but also density and compressibility.

One potential advantage of the ultrasound separation method is that it can be readily tailored to achieve fractionation outcomes (particle size and/or fat concentration) desired for a particular milk product by simple tuning parameters. For example, increasing or decreasing the duration of sonication and controlling the acoustic energy applied to the system can increase the amount and rate of fat removal. Sample collection is also relatively simple and robust and can be designed as a series of under and overflows from a separation reactor to facilitate the removal of fractionated partitions of milk from the bottom, top or other location within the separation reactor. There also exists no moving parts that require frequent maintenance, and the cleaning requirements are significantly lesser compared with microfiltration membranes which become fouled over time during operation.

3.6 Milk Quality Assessment

The application of ultrasound to milks can result in the formation of oxidative radicals that may modify the functional properties and flavour profile. Generally, sonochemical radical formation and strong physical effects, are minimal when using high frequency ultrasound in the MHz range (>1 MHz) for ultrasonic separation. However, in some situations, the energy level employed can still promote the creation of radicals due to the formation of 'stable' cavitation bubbles.

It is therefore important to consider if the application of high frequency, low power ultrasound has any significant impact on the quality of the produced milk streams. In recent studies (Johansson et al., 2016; Juliano et al., 2014), milks subject to ultrasonication at select frequencies (1 and 2 MHz) and power loadings (40–500 J/g) suited for ultrasonic separation for litre-scale systems, have been assessed for quality in regards to the formation of oxidative volatiles and physical damage to fat globules.

Milks subject to ultrasound treatment were analysed using headspace analysis by gas chromatography-mass spectrometry to quantify the formation of oxidative volatiles formed. It is well known that ultrasound-induced cavitation can produce free radicals in solution (Weissler, 1959), that may oxidise sensitive components. In milks, the oxidation of lipids (Juliano et al., 2014) and cholesterols (Sieber, 2005) may have a strong effect on the flavour profile, decreased nutrition and possible increased toxicity.

Johansson et al. (2016) demonstrated that 2 MHz ultrasound (up to 373 kJ/kg) induced minimal sonochemical light emission from reaction with luminol, indicating minimal formation of radicals. By comparison, 1 MHz ultrasound (at 464 kJ/kg) resulted in visibly stronger luminol emission, indicating the formation of a larger amount of radicals. Milks subject to ultrasound at either of these frequencies and energies, however, did not result in a significant increase to the formation of oxidative volatiles relative to controls. Juliano et al. (2014) found that at extended sonication duration when using 1 MHz ultrasound, oxidative volatiles such as nonanal did exceed odour thresholds, but only when temperatures of the milk were maintained above 40 °C during treatment. The reason is because even though radicals are formed, milk contains a number of antioxidants (Cervato & Giovanna, 1999; Lindmark-Månsson & Åkesson, 2000; Taylor & Richardson, 1980) that will buffer the effects of chemical radical species. With prolonged treatment at high temperatures, the buffering capacity will become depleted, enabling attack of lipids to oxidation from the created free radicals due to the ultrasound. A burning 'metallic' off odour, was detected only when milks were subject to high-intensity sonication using high power, low frequency 20 kHz ultrasound confined within an enclosed cell

(Juliano et al., 2014). These results indicated in general that there is minimal degradation of less reactive fat compounds.

The physical effects of ultrasound applied at the frequencies and energy loadings used to separate milk fat globules have also been assessed (Leong, Johansson, et al., 2014; Leong, Juliano, et al., 2014). Zeta-potential (Michalski, Michel, Sainmont, & Briard, 2002) was used to assess milk fat globules subject to ultrasound frequencies at 600 kHz (Leong, Yasui, et al., 2014) and 1 MHz (Leong, Johansson, et al., 2014), with results indicating no significant disruption of the natural milk fat globule membrane. No evidence of coalescence or significant rupture of fat globules was observed by optical microscopy (Leong, Johansson, et al., 2014). The physical damage to the milk fat globule membrane is generally minimal at the frequencies used for ultrasonic separation.

There is a possibility that the applied ultrasound can alter the functionality of the proteins present in milk. Shanmugam, Chandrapala, and Ashokkumar (2012) reported that low frequency ultrasound ~20 kHz caused partial denaturation of the whey proteins to form soluble whey-whey/whey-casein complexes. This denaturation is largely attributed to the physical effects of ultrasound such as strong localised heating and shearing, which are minimal at the higher ultrasonic frequencies employed in ultrasonic separation. At frequencies of ~1 MHz, there is formation of chemical radicals. The amount of radicals needed to cause changes to milk proteins is unknown. The proteins and other components in milk are known to have strong antioxidant capacity (Cervato & Giovanna, 1999; Lindmark-Månsson & Åkesson, 2000) however and may mitigate the impact of oxidation damage to the milk overall. Interestingly, a study by Taylor and Richardson (1980) showed that sonication actually may increase the anti-oxidant capacity of skim milk, possibly due to release of more casein into solution by disruption of casein micelles.

To avoid any physical and/or chemical damage to milk, operation at 2 MHz ultrasound has been recommended by Johansson et al. (2016), as it offers highly effective acoustic separation with minimal radical formation and negligible physical shearing effects. In comparison to the changes in functionality of the proteins created during milk pasteurization from protein denaturation, ultrasound would have minimal or no impact.

3.7 Design Principles for Ultrasonic Separation Reactors

The use of ultrasound to separate fat from milk requires an understanding of several key design parameters that is underlined by the ability of the sound waves to provide uniform treatment of material processed within its confines. As established in recent works (Juliano, Temmel, Rout, et al., 2013; Leong, Johansson, et al., 2014), the reactor geometry and placement of transducers within the reactor are important considerations.

The maximum effective working distance over which a standing wave field can be effectively formed is limited by the attenuation of the sound wave in the fluid medium. The higher the ultrasonic frequency used for the separation, the more strongly attenuated it will be as it propagates through the fluid. What this means is that ultrasound applied in the MHz region is generally limited to shorter effective working distances to achieve a strong standing wave. As shown by Leong, Johansson, et al. (2014), for the separation of cream in milk, the maximum effective working distance over which a standing wave can be established to promote rapid separation using 1 and 2 MHz ultrasound, is ~ 80–100 mm. Leong, Coventry, et al. (2015) also showed that when using high frequency 2 MHz ultrasound, a significant decline of sound pressures occurs beyond an of 55 mm in a large experimental chamber. By contrast, 400 kHz ultrasound, which is less effective for cream separation, was found to penetrate a considerably longer distance (up to 2 m) without a decline in pressure level.

Strong attenuation may also give rise to undesirable acoustic streaming in the form of Eckhart streaming currents, which can cause disruption to the separation of particles aligned within the nodal/anti-nodal planes of the standing wave. In a large ultrasound reactor vessel, acoustic streaming is a natural part of the system and cannot generally be avoided. Provided that a sufficiently strong standing wave is formed, acoustic streaming will likely not be a deterrent to the enhanced separation, and it is possible that it can even promote separation as the streaming can move flocculated cream more rapidly towards the surface (Leong, Juliano, et al., 2015). The use of acoustic transparent films has been reported to reduce the effects of strong acoustic streaming by breaking up the streaming currents before they can propagate through the entire vessel (Spengler & Jekel, 2000). These films have yet to be utilised in milk separation applications, but they may not be suitable for dairy processing applications due to the high cleaning requirements needed.

The alignment and positioning of transducers and the reflector within the reactor to generate standing waves should consider how the material flows inside the reactor and whether or not multiple transducers are being used for the separation. The alignment of multiple transducers relative to the reactor cross-section, can also be positioned either in a parallel or perpendicular arrangement (Juliano, Temmel, Rout, et al., 2013). Perpendicular alignment may create more regions where particle accumulation can occur. Parallel alignment has the advantage of allowing close positioning of the transducers when facing each other. Very short transducer-reflector distances, or in this case transducer-transducer distances, can be realised even with large surface area plates when using a parallel alignment. In most cases, a parallel alignment will offer more optimal reflection and concentration of acoustic energy, although it should be noted that a perfectly parallel alignment between transducer and reflector/transducer is in practice not generally required in large systems to produce effective separation.

In milk separation, enhanced separation due to gravity is a necessary mechanism and for this, it is more optimal for the pressure antinodes to be aligned vertically such that collected material can readily rise or sediment within the sound field. If the particle banding occurs in a horizontal alignment, the natural rise/fall of the cream globules may be hindered since product must pass through multiple aligned bands that can 're-trap' the aggregated material prior to eventual rising/falling beyond the active processing region.

It is also possible to operate ultrasonic transducers in a non-contact set-up to simplify and minimize cleaning requirements. Transmission plates designed with optimal thickness for sound transmission can be used to mount transducers externally (Michaud, Leong, Swiergon, Juliano, & Knoerzer, 2015), which may allow for a more durable system for continuous processing.

4 Outlook and Conclusion

Ultrasonic separation for enhanced cream separation remains a relatively new and understudied technology. While it shows effectiveness in enhancing cream separation via natural creaming mechanisms, the technique is best considered as a complementary rather than a standalone process to replace existing separation techniques such as centrifugation. It could be used, for example, as a pre-treatment to conventional centrifugation, whereby rapid selective removal of large-sized fat globules can be achieved, such that creams formed by subsequent separation from the remaining fat in the milk, will consist of smaller droplets. Further studies to investigate the viability of such an approach, would provide an improved understanding for the future industry application of this technique for enhanced cream separation in the dairy industry. It is likely that the scales at which it can operate at, will be amenable to low to medium volume scale dairy manufacturers, in which the simple yet effective separation of fat droplets by size, would produce a point of difference in the finished dairy products. The method could enable highly selective dairy fat removal and concentration, bringing to market new products for the consumer.

References

Ashokkumar, M., & Mason, T. J. (2000). Sonochemistry. In *Kirk-Othmer encyclopedia of chemical technology*. New York, NY: John Wiley & Sons, Inc..

Bruus, H. (2012). Acoustofluidics 7: The acoustic radiation force on small particles. *Lab on a Chip, 12*, 1014–1021.

Caplan, Z., Melilli, C., & Barbano, D. (2013). Gravity separation of fat, somatic cells, and bacteria in raw and pasteurized milks. *Journal of Dairy Science, 96*(4), 2011–2019.

Cervato, R. C., & Giovanna, B. C. (1999). Studies on the antioxidant activity of milk caseins. *International Journal of Food Sciences and Nutrition, 50*, 291–296.

Dhungana, P., Truong, T., Palmer, M., Bansal, N., Bhandari, B., & Hub, A. D. I. (2017). Size-based fractionation of native milk fat globules by two-stage centrifugal separation. *Innovative Food Science and Emerging Technologies, 41*, 235–243.

Gor'kov, L. P. (1962). On the forces acting on a small particle in an acoustical field in an ideal fluid. *Soviet Physics Doklady, 6*, 773.

Goudédranche, H., Fauquant, J., & Maubois, J.-L. (2000). Fractionation of globular milk fat by membrane microfiltration. *Le Lait, 80*, 93–98.

Grenvall, C., Augustsson, P., Folkenberg, J. R., & Laurell, T. (2009). Harmonic microchip acoustophoresis: A route to online raw milk sample precondition in protein and lipid content quality control. *Analytical Chemistry, 81*, 6195–6200. https://doi.org/10.1021/Ac900723q

Grenvall, C., Folkenberg, J. R., Augustsson, P., & Laurell, T. (2012). Label-free somatic cell cytometry in raw milk using acoustophoresis. *Cytometry Part A, 81*, 1076–1083.

Johansson, L., Singh, T., Leong, T., Mawson, R., McArthur, S., Manasseh, R., et al. (2016). Cavitation and non-cavitation regime for large-scale ultrasonic standing wave particle separation systems–In situ gentle cavitation threshold determination and free radical related oxidation. *Ultrasonics Sonochemistry, 28*, 346–356.

Jonsson, H., Holm, C., Nilsson, A., Petersson, F., Johnsson, P., & Laurell, T. (2004). Particle separation using ultrasound can radically reduce embolic load to brain after cardiac surgery. *The Annals of Thoracic Surgery, 78*, 1572–1578. https://doi.org/10.1016/j.athoracsur.2004.04.071

Juliano, P., Kutter, A., Cheng, L. J., Swiergon, P., Mawson, R., & Augustin, M. A. (2011). Enhanced creaming of milk fat globules in milk emulsions by the application of ultrasound and detection by means of optical methods. *Ultrasonics Sonochemistry, 18*, 963–973. https://doi.org/10.1016/j.ultsonch.2011.03.003

Juliano, P., Swiergon, P., Lee, K. H., Gee, P. T., Clarke, P. T., & Augustin, M. A. (2013). Effects of pilot plant-scale ultrasound on palm oil separation and oil quality. *Journal of the American Oil Chemists' Society, 90*, 1253–1260. https://doi.org/10.1007/s11746-013-2259-3

Juliano, P., Swiergon, P., Mawson, R., Knoerzer, K., & Augustin, M. A. (2013). Application of ultrasound for oil separation and recovery of palm oil. *Journal of the American Oil Chemists' Society, 90*, 579–588. https://doi.org/10.1007/s11746-012-2191-y

Juliano, P., Temmel, S., Rout, M., Swiergon, P., Mawson, R., & Knoerzer, K. (2013). Creaming enhancement in a liter scale ultrasonic reactor at selected transducer configurations and frequencies. *Ultrasonics Sonochemistry, 20*, 52–62. https://doi.org/10.1016/j.ultsonch.2012.07.018

Juliano, P., Torkamani, A. E., Leong, T., Kolb, V., Watkins, P., Ajlouni, S., et al. (2014). Lipid oxidation volatiles absent in milk after selected ultrasound processing. *Ultrasonics Sonochemistry, 21*, 2165–2175.

King, L. V. (1934). On the acoustic radiation pressure on spheres. *Proceedings of the Royal Society of London. Series A, Mathematical and Physical Sciences, 147*, 212–240. https://doi.org/10.1098/rspa.1934.0215

Koda, S., Kimura, T., Kondo, T., & Mitome, H. (2003). A standard method to calibrate sonochemical efficiency of an individual reaction system. *Ultrasonics Sonochemistry, 10*, 149–156.

Lenshof, A., & Laurell, T. (2011). Emerging clinical applications of microchip-based acoustophoresis. *Journal of the Association for Laboratory Automation, 16*, 443–449. https://doi.org/10.1016/j.jala.2011.07.004

Leong, T., Ashokkumar, M., & Kentish, S. (2011). The fundamentals of power ultrasound–a review. *Acoustics Australia, 39*, 54–63.

Leong, T., Coventry, M., Swiergon, P., Knoerzer, K., & Juliano, P. (2015). Ultrasound pressure distributions generated by high frequency transducers in large reactors. *Ultrasonics Sonochemistry, 27*, 22–29.

Leong, T., Johansson, L., Juliano, P., Mawson, R., McArthur, S., & Manasseh, R. (2014). Design parameters for the separation of fat from natural whole milk in an ultrasonic litre-scale vessel. *Ultrasonics Sonochemistry, 21*, 1289–1298.

Leong, T., Johansson, L., Juliano, P., McArthur, S. L., & Manasseh, R. (2013a). Ultrasonic separation of particulate fluids in small and large scale systems: A review. *Industrial and Engineering Chemistry Research, 52*, 16555–16576.

Leong, T., Johansson, L., Juliano, P., McArthur, S. L., & Manasseh, R. (2013b). Ultrasonic separation of particulate fluids in small and large scale systems: A review. *Industrial and Engineering Chemistry Research, 52*, 16555–16576. https://doi.org/10.1021/ie402295r

Leong, T., Johansson, L., Mawson, R., McArthur, S. L., Manasseh, R., & Juliano, P. (2016). Ultrasonically enhanced fractionation of milk fat in a litre-scale prototype vessel. *Ultrasonics Sonochemistry, 28*, 118–129.

Leong, T., Juliano, P., Johansson, L., Mawson, R., McArthur, S., & Manasseh, R. (2015). Continuous flow ultrasonic skimming of whole milk in a liter-scale vessel. *Industrial and Engineering Chemistry Research, 54*, 12671–12681.

Leong, T., Juliano, P., Johansson, L., Mawson, R., McArthur, S. L., & Manasseh, R. (2014). Temperature effects on the ultrasonic separation of fat from natural whole milk. *Ultrasonics Sonochemistry, 21*, 2092–2098.

Leong, T., Knoerzer, K., Trujillo, F. J., Johansson, L., Manasseh, R., Barbosa-Cánovas, G. V., et al. (2015). Megasonic separation of food droplets and particles: Design considerations. *Food Engineering Reviews, 7*, 298–320.

Leong, T., Yasui, K., Kato, K., Harvie, D., Ashokkumar, M., & Kentish, S. (2014). Effect of surfactants on single bubble sonoluminescence behavior and bubble surface stability. *Physical Review E, 89*, 043007.

Leong, T. S., Martin, G. J., & Ashokkumar, M. (2016). Ultrasonic encapsulation–A review. *Ultrasonics Sonochemistry, 35*(Pt B), 605–614.

Lighthill, J. (1978). Acoustic streaming. *Journal of Sound and Vibration, 61*, 391–418.

Lindmark-Månsson, H., & Åkesson, B. (2000). Antioxidative factors in milk. *British Journal of Nutrition, 84*, 103–110.

Logan, A., Day, L., Pin, A., Auldist, M., Leis, A., Puvanenthiran, A., et al. (2014). Interactive effects of milk fat globule and casein micelle size on the renneting properties of milk. *Food and Bioprocess Technology, 7*, 3175–3185.

Ma, Y., & Barbano, D. (2000). Gravity separation of raw bovine milk: Fat globule size distribution and fat content of milk fractions. *Journal of Dairy Science, 83*, 1719–1727.

Mason, T. J., Cobley, A. J., Graves, J. E., & Morgan, D. (2011). New evidence for the inverse dependence of mechanical and chemical effects on the frequency of ultrasound. *Ultrasonics Sonochemistry, 18*, 226–230. https://doi.org/10.1016/j.ultsonch.2010.05.008

Mettin, R., Akhatov, I., Parlitz, U., Ohl, C. D., & Lauterborn, W. (1997). Bjerknes forces between small cavitation bubbles in a strong acoustic field. *Physical Review E, 56*, 2924–2931.

Michalski, M., Leconte, N., Briard-Bion, V., Fauquant, J., Maubois, J., & Goudédranche, H. (2006). Microfiltration of raw whole milk to select fractions with different fat globule size distributions: Process optimization and analysis. *Journal of Dairy Science, 89*, 3778–3790.

Michalski, M.-C., Michel, F., Sainmont, D., & Briard, V. (2002). Apparent ζ-potential as a tool to assess mechanical damages to the milk fat globule membrane. *Colloids and Surfaces B: Biointerfaces, 23*, 23–30.

Michaud, M., Leong, T., Swiergon, P., Juliano, P., & Knoerzer, K. (2015). Design parameters of stainless steel plates for maximizing high frequency ultrasound wave transmission. *Ultrasonics Sonochemistry, 26*, 56–63.

Miles, C. A., Morley, M. J., Hudson, W. R., & Mackey, B. M. (1995). Principles of separating micro-organisms from suspensions using ultrasound. *Journal of Applied Microbiology, 78*, 47–54. https://doi.org/10.1111/j.1365-2672.1995.tb01672.x

Mulder, H., & Walstra, P. (1974). *The milk fat globule: Emulsion science as applied to milk products and comparable foods*. Belfast, Ireland: Commonwealth Agricultural Bureaux.

Nilsson, A., Petersson, F., Jönsson, H., & Laurell, T. (2004). Acoustic control of suspended particles in micro fluidic chips. *Lab on a Chip, 4*, 131–135.

O'Mahony, J. A., Auty, M. A., & McSweeney, P. L. (2005). The manufacture of miniature Cheddar-type cheeses from milks with different fat globule size distributions. *Journal of Dairy Research, 72*, 338–348.

Petersson, F., Nilsson, A., Holm, C., Jonsson, H., & Laurell, T. (2004). Separation of lipids from blood utilizing ultrasonic standing waves in microfluidic channels. *Analyst, 129*, 938–943. https://doi.org/10.1039/B409139f

Petersson, F., Nilsson, A., Holm, C., Jonsson, H., & Laurell, T. (2005). Continuous separation of lipid particles from erythrocytes by means of laminar flow and acoustic standing wave forces. *Lab on a Chip, 5*, 20–22. https://doi.org/10.1039/B405748c

Shanmugam, A., Chandrapala, J., & Ashokkumar, M. (2012). The effect of ultrasound on the physical and functional properties of skim milk. *Innovative Food Science and Emerging Technologies, 16*, 251–258.

Sieber, R. (2005). Oxidised cholesterol in milk and dairy products. *International Dairy Journal, 15*, 191–206.

Spengler, J., & Jekel, M. (2000). Ultrasound conditioning of suspensions - Studies of streaming influence on particle aggregation on a lab- and pilot-plant scale. *Ultrasonics, 38*, 624–628.

Spengler, J. F., Coakley, W. T., & Christensen, K. T. (2003). Microstreaming effects on particle concentration in an ultrasonic standing wave. *AIChE Journal, 49*, 2773–2782.

Stokes, G. G. (1848). On a difficulty in the theory of sound. *Philosophical Magazine, 33*, 349–356. https://doi.org/10.1080/14786444808646119

Taylor, M., & Richardson, T. (1980). Antioxidant activity of skim milk: Effect of sonication. *Journal of Dairy Science, 63*, 1938–1942.

Temmel, S. (2012). *Enhanced separation of creaming in a megasonic cubic reactor*. Erlangen, Germany: Friedrich-Alexander-Universitat Erlangen-Nurnberg.

Torkamani, A. E., Juliano, P., Fagan, P., Jiménez-Flores, R., Ajlouni, S., & Singh, T. K. (2016). Effect of ultrasound-enhanced fat separation on whey powder phospholipid composition and stability. *Journal of Dairy Science, 99*, 4169–4177.

Weiser, M., Apfel, R., & Neppiras, E. (1984). Interparticle forces on red cells in a standing wave field. *Acta Acustica united with Acustica, 56*, 114–119.

Weissler, A. (1959). Formation of hydrogen peroxide by ultrasonic waves: Free radicals. *Journal of the American Chemical Society, 81*, 1077–1081.

Yosioka, K., & Kawasima, Y. (1955). Acoustic radiation pressure on a compressible sphere. *Acustica, 5*, 167.

Zarembo, L. K. (1971). Acoustic streaming. In L. D. Rozenberg (Ed.), *High-intensity ultrasonic fields* (Vol. 85, p. 137). New York, NY: Plenum Press.

Chapter 19
Dairy Fat and Fat-Filled Powders: Process-Product Interactions

Pierre Schuck

1 Introduction

Fat-filled spray-dried dairy powders have achieved great economic, nutritional and functional importance. This includes a wide range of products such as whole milk powder, cream powder, whey derivatives, infant milk formula and dairy ingredients enriched with either milk fat or with vegetable oils.

The most frequently used technique for dehydration of these types of dairy products is spray drying, which involves spraying a concentrated liquid product in the form of droplets into a large drying chamber containing dry, hot steam to eliminate water. Spray drying is commonly used in dairy-based systems (whole and skim) to enhance shelf life by decreasing water activity and minimizing transport costs (Henning, Baer, Hassan, & Dave, 2006; Schuck et al., 2007; Walstra, Geurts, Noomen, Jellema, & van Boekel, 1999).

From an economic point of view, global production of 5.1 million tons of whole milk (WMP) and semi-skimmed milk powder decreased by 1.4% in 2015, following market growth in 2014. This was a consequence of a significant drop (−5.8%) in production in New Zealand, by far the world's biggest WMP producer. This major decrease—the country's first after years of growth (+7.7% per year on average since 2010) was due to reduced deliveries and export demand in 2015. The second-largest producer, China, remained reasonably stable at 1.1 million tons, while the EU, in third position, surpassed its average whole and semi-skimmed milk powder growth rate of 2%, with a 2.9% increase. Brazil's output somewhat stabilized (+0.8%), remaining at 0.6 million tons. In 2014, global production of WMP had increased by a whole 7%, leading to massive stockpiling and market glutting, which gave way to a more modest situation in 2015 (IDF, 2016).

P. Schuck (✉)
INRA Agrocampus Ouest, UMR 1253, STLO, Rennes, France

T. Truong et al. (eds.), *Dairy Fat Products and Functionality*,
https://doi.org/10.1007/978-3-030-41661-4_19

The diversified applications of fat-filled spray-dried dairy powders require a number of specific physical properties. The drying process leads to a range of structural and physicochemical modifications that, in turn, influence the rehydration and handling properties of dairy powders and their shelf life (Westergaard, 2004). Furthermore, increasing the fat content leads to additional quality implications such as poor rehydration and flow properties, and off-flavors attributed to the presence of free and surface fat (Allison, Chang, Randolph, & Carpenter, 1999; Aule & Worstorff, 1975; Buchheim, Samhammer, & Lembke, 1974; Buma, 1971; Crowley, Kelly, Schuck, Jeantet, & O'Mahony, 2016; Vignolles et al., 2009; Vignolles, Jeantet, Lopez, & Schuck, 2007). Except for its special use in chocolate and cake manufacturing, the latter is considered as a quality defect. The drying industry attributes a strong decrease in rehydration properties, flowability and the development of oxidation flavors to free fat.

The recent bibliographic study by Vignolles et al. (2007) highlights two groups of factors that play a major role in the quality of fat-filled dairy powders: the composition and the process. Concerning the composition, several questions arise. For example, what is the content of the fat, the type of fat and its melting point? Does it contain phospholipids? Concerning the encapsulation by proteins, what is the content, the type of protein (casein, caseinates, whey protein (concentrate, isolate), ± denatured)? Finally, is the lactose in a crystal state or an amorphous state? Regarding the manufacturing process, what is the behavior of the fat during homogenization, lactose crystallization, spray drying, storage and rehydration?

The bibliographical study of Vignolles et al. (2007) identified several specific studies on fat-filled powders in an attempt to better understand the interaction of the drying process with the type of fat and the behavior of proteins at the interfaces during the formation of the emulsion during storage and rehydration. It thus revealed that the physical and biochemical characteristics of the emulsions to be dried (droplet size and viscosity) and the powder quality obtained (oxidation, rehydration, flow) are key criteria for research into the functional and nutritional properties of fat-filled dairy powders.

The aim of this chapter is to consider process/product interactions on fat-filled dairy powders from the point of view of the physical properties of the powders, with a focus on free fat content.

2 Dairy Powder Manufacturing Process

2.1 Whole Milk and Semi-Skimmed Milk Powders

The flow chart of whole milk and semi-skimmed milk powder production, consisting of reception, clarification, cooling, standardization, heat treatment, vacuum evaporation, homogenization, spray or roller drying, and storage/packaging is shown in Fig. 19.1 (Schuck, 2011a). Raw full-fat milk used for powder production

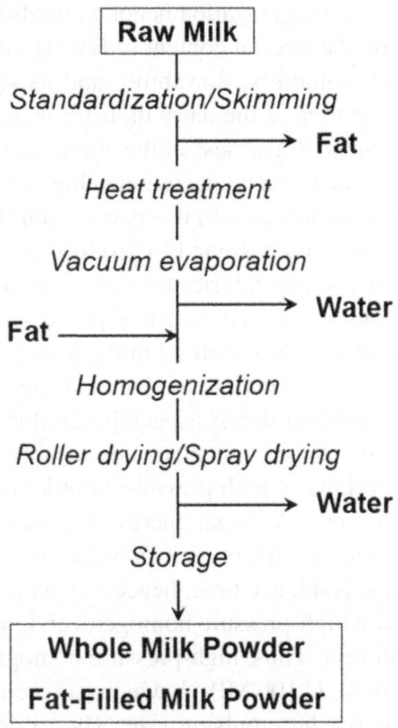

Fig. 19.1 Flowchart of whole, semi-skimmed and fat-filled milk powder production

must be of high chemical, sensory and bacteriological quality, which is regulated by government standards. After reception, milk is clarified, usually by centrifugal separators, and cooled to 4 °C in plate heat exchangers, followed by storage at the same temperature. The next operation is standardization or skimming, which is used to adjust the ratio of milk fat to total solids required in the final product. Even for whole and semi-skimmed milk powders, milk is often totally skimmed to avoid destructuration of the fat globules during vacuum evaporation. The fat is often added just before spray drying during a homogenization step. Heat treatment is commonly performed using the indirect method in a tubular or plate heat exchanger at 88–95 °C for 15–30 s, aiming to destroy pathogenic bacteria and most of the saprophytic microorganisms, to inactivate enzymes (especially lipase), and to activate SH groups in α-lactoglobulin, which results in an antioxidative effect. Vacuum evaporation is used to concentrate milk prior to drying and can be combined with reverse osmosis. Evaporation is performed in multiple-effect vacuum evaporators with mechanical or thermal steam recompression, where energy consumption is approximately 10–30 times lower than for spray drying (Schuck, 2011b). The differences in the concentration of total solids are determined by the drying technique used: 30–35% total solids for roller drying and 45–50% total solids for spray drying. Concentrating milk prior to drying has a positive effect on milk powder quality: milk powder produced from concentrated milk consists of larger powder particles containing less occluded air and, consequently, resulting in better storage stability.

Homogenization is not compulsory but is usually applied with the aim of reducing the free-fat content in full-fat milk powder, which has a negative effect on powder solubility, flowability and its susceptibility to oxidation. The homogenization step used in the dairy industry is mainly designed as an emulsification step (oil-in-water). Regardless of the choice of the emulsification process, the breakdown of the dispersed elements requires high shearing in order to obtain a dispersion of the lipid phase, whose purpose is to obtain the smallest possible oil droplets. Many devices have been designed to produce emulsions. Of all these techniques, some are used only in laboratories (shakers, vibrators, magnetostriction, aerosols, etc.). The most commonly used industrial processes are rotor stators (single agitators, "Ultraturrax" tanks, cutters, colloid mills, high pressure homogenizers (P < 100 MPa) and ultrasonic systems) (Croguennec, Jeantet, & Schuck, 2016). In the case of homogenizers (a method that is especially used in the dairy and food industry), this method consists of forcing a coarse emulsion through an adjustable opening valve by a pump working at high pressure in order to reduce the size of the fat globules as much as possible. At equal energy densities compared to other conventional emulsification systems, high pressure systems are more efficient than the others due to a short average residence time, hence, higher power density. During the expansion in the valve of a high-pressure homogenizer, kinetic energy is dissipated by the viscous stresses in heat. Thus, high-pressure homogenizers produce maximum heating on the order of 20 °C 100 MPa^{-1}, which can then act on the properties of the constituents as well as on the emulsion size. The oil-in-water emulsions are generally insensitive to slight variations in temperature. However, the efficiency of the homogenization is improved by heating, which causes a partial melting of the triglycerides contained in the organic phase. It is well established that the rise in temperature decreases the interfacial tension between the phases as well as the viscosity, according to the Arrhenius law. This action facilitates the work of both dividing the dispersed phase and reducing the mean diameter of the fat globules. On the other hand, the higher the temperature is, the higher the risk of denaturation will be, leading to changes in the interfacial adsorption of the emulsifiers (Croguennec et al., 2016). The correlation between the free fat content and greater fat droplet diameters has already been established: to achieve low free fat content in powders, the fat droplet size in the concentrate must be less than 1 µm before spray drying (Chever et al., 2017; Vignolles et al., 2009).

The homogenization step is often used in the dairy industry to produce dairy ingredients enriched with dairy, animal or vegetable fat. Regardless of the process, the biochemical composition has to be taken into account. Thus, according to Vignolles et al. (2009), the presence of amorphous lactose is essential to emulsify and encapsulate fat: it could act as a hydrophilic protective agent, especially with heat-denatured whey proteins. With amorphous lactose, micellar casein and whey proteins are better emulsifiers than heat-denatured whey proteins in terms of the formation of smaller fat droplets. However, only native whey proteins have proven to be better encapsulating agents with regard to free fat and surface fat. Hence, whey proteins, when weakly heat-denatured, and in the presence of amorphous lactose, could be used as effective ingredients to encapsulate fat in dairy powders.

2.2 *Infant Milk Formula Powder*

Milk is secreted by female mammals after parturition. As an exclusive food of the newborn, it contains all the essential nutrients for baby mammals. Human milk is thus the best food for all human infants. The World Health Organization (WHO) recommends exclusive breastfeeding for infants during the first 6 months of life to achieve optimal growth, development and health, followed by breastfeeding with the addition of complementary foods up to the age of 2 years or beyond (WHO (World Health Organization), 2011). However, there are situations where the mother cannot or does not want to breastfeed. In such situations, an alternative food is necessary: an infant formula. According to the international Codex Alimentarius Commission, an infant formula is defined as a breast-milk substitute that can by itself fulfill all the nutritional requirements of infants from birth up to the introduction of complementary feeding (FAO/WHO, 2007). Spray drying is a process able to retain both the nutritional quality of infant formula and its ease of conservation and use. Moreover, because of the continuous scientific and technical developments of the past century, more and more ingredients have been added to infant formulae, such as vegetable oils instead of cream, adjustment of the casein/whey protein ratio, introduction of polyunsaturated fatty acids, etc. Infant formulae are now formulated to specifically meet the nutritional needs of infants. Considerable progress has been made in recent years in making the composition of infant formulae as close as possible to that of breast milk (Blanchard, Zhu, & Schuck, 2014).

Lipids play an important role in infant formulae. They provide 40–50% of the daily energy intake of an infant. Lipids can also provide essential fatty acids (linoleic acid and α-linolenic acid) and fat-soluble vitamins (A, D, E, K) for infants. Most of the current infant formula preparations use a blend of different vegetable oils (e.g., palm oil, soy oil, coconut oil, etc.) (Blanchard et al., 2014).

Infant formulae are often presented in a sterilized liquid form (ready to feed (RTF)) or in a powder form. Liquid infant formulae are more convenient to use and can avoid possible errors when reconstituting powdered formulae, but the price is relatively higher. Powdered infant formulae are much easier to transport and store. The following section only describes the production of powdered infant formulae. There are two different processes that can be used to manufacture powdered infant formulae: the dry mix process and the wet mix with spray drying process.

2.2.1 Dry Mix Process

In the dry mix process, the ingredients are in powdered form and are mixed together in large batches to produce a uniform powdered infant formula. A base powder may be used for the manufacture of different final powdered infant formulae by adding various dry ingredients. The base powder, however, is often obtained by the dry mix process. Compared to the wet mix process, this process requires a much smaller investment and energy consumption; consequently, the production cost is lower.

However, since there is no heat treatment involved to destroy any eventual bacteria in the mix, post-process contamination may occur during further processing steps (Cordier, 2008). Thus, the microbiological quality of raw materials is critical. Furthermore, different ingredients have different densities, which may result in the inhomogeneity of the composition during long storage and transportation times (Montagne, Van Dael, Skanderby, & Hugelshofer, 2009).

2.2.2 Wet Mix Process

The wet mix process typically involves several steps (Fig. 19.2). Ingredients are first mixed together in a liquid form to obtain a uniform emulsion. After pasteurization, homogenization and evaporation, the concentrated emulsion is dehydrated by spray drying to produce a powdered infant formula. The wet mix process can provide the best uniformity of nutrient distribution in the product. The pasteurization step ensures that any pathogenic bacteria present in the raw materials or contamination during the preparation of the wet mix are eliminated. Montagne et al. (2009) explained that due to the differences in sensitivity to heat, the fat and vitamins are generally mixed with the liquid preparation after heat treatment and vacuum evaporation but before homogenization and spray drying, in order to obtain a low free fat content in powders (Chever et al., 2017; Vignolles et al., 2009). A well-controlled spray drying process can also provide good solubility of the powder during reconstitution. On the other hand, this process requires a bigger investment and higher energy consumption, and the production costs are therefore greater. However, an inappropriate drying operation may lead to non-compliant product quality and

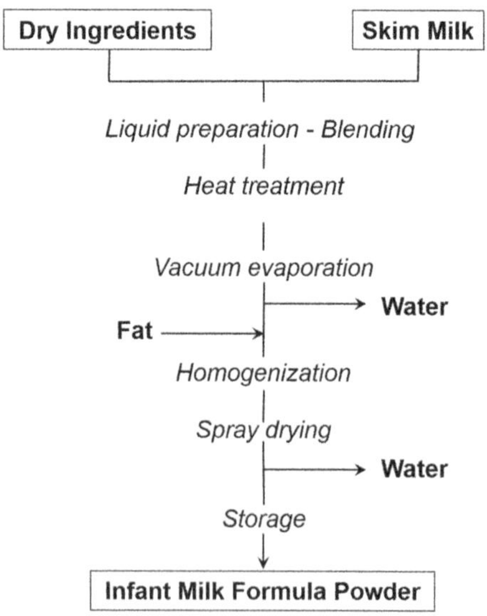

Fig. 19.2 Flowchart of infant milk formula powder production

substantial economic losses (Schuck, 2002), including poor homogenization with big fat globules that increase the free fat content and the poor quality.

As explained in the previous paragraph, free fat content has to be reduced to improve the powder quality obtained (low oxidation, good rehydration and good flowability). Vignolles et al. (2007) explained that free fat decreased with increasing lactose/protein ratios. Aguilar and Ziegler (1994a) found that lactose concentrations of more than 400 g kg^{-1} were necessary to limit the free fat content. However, another question concerns the state of the lactose, amorphous vs. crystal.

Aguilar and Ziegler (1994b) and Fäldt and Bergenståhl (1996) showed that lactose crystals damaged the milk fat globule and the protein coating of fat droplets during processing steps and storage, leading to coalescence. Lactose is present as a continuous phase in milk (King, 1965). It changes from the glassy to the rubbery state above the glass transition temperature (T_g), thus allowing the mobility of the other constituents present in a powder particle and the subsequent release of fat (Shimada, Roos, & Karel, 1991). If the lactose was crystallized to between 85–95% before drying in the α monohydrate state, the crystal could tear the milk fat globule membrane during homogenization, increasing the free fat content (Schuck, 2011c; Vignolles et al., 2007). Once the lactose was crystallized, the solvent could easily diffuse, and the encapsulation efficiency decrease. To avoid this negative effect, a part of the lactose is incorporated into the wet mix and another part into the dry mix in order to obtain a good amorphous lactose ratio to improve fat encapsulation and rehydration and, in crystal form, to improve the dryability and storage conditions (Fäldt & Bergenståhl, 1996; Schuck, 2011c).

2.3 Fat-Filled Whey Powder

The flow chart of whey and permeate powder production, consisting of reception, clarification, cooling, standardization, heat treatment, vacuum evaporation, lactose crystallization, homogenization, spray drying, and storage/packaging is shown in Fig. 19.3 (Schuck, 2011c). Evaporation removes some of the water (85–95%) from whey by boiling under vacuum and increases the dry matter from 6.5 to 50–65% (w/w). At this stage of concentration, whey is a highly concentrated product, but the concentration is not enough to prevent the chemical reactions or changes in state taking place. The water activity (a_w) is close to 0.95–0.99. Therefore, lactose, whose concentration is much higher than its solubility level, changes states during crystallization. This change in state is not instantaneous and depends on complex kinetics related to many factors (product: composition, impurities, additives with their concentrations; process: temperature, stirring conditions, seeding, etc.). The crystallized whey concentrate is then dehydrated more completely by spray drying combined with fluidization. For fat-filled whey and permeate powder, the state of the lactose is a key consideration for infant milk formula powder. As previously explained, to optimize the quality of these powders, a part of the lactose has to be in an amorphous state to improve the fat encapsulation (reduction of the free fat

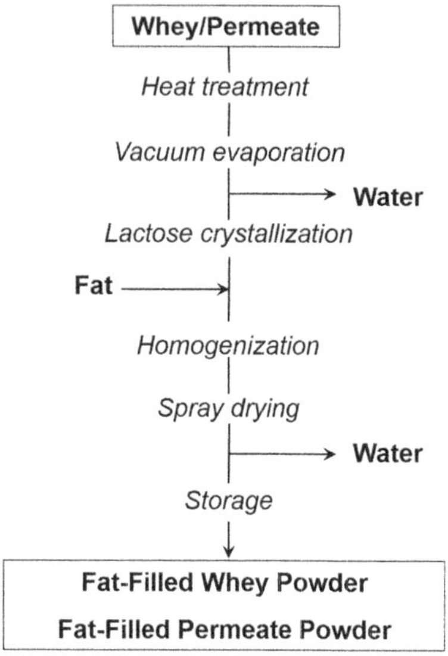

Fig. 19.3 Flowchart of fat-filled whey/permeate powder production

content) and rehydration, and in a crystal state to improve the dryability and storage conditions (Fäldt & Bergenståhl, 1996; Schuck, 2011c).

According to Schuck (2011c), lactose crystallization in whey is a key stage in the manufacture of whey powders. Controlling this stage at an industrial level should increase the prospects of improving the process as well as the physicochemical qualities of the powders. Lactose crystallization occurs in highly supersaturated solutions, indicating that the phenomena of nucleation and crystal growth can take place simultaneously. At the industrial level, lactose crystallization is also performed in a medium with a complex chemical composition. In particular, certain macromolecules such as whey proteins, whose influence on the kinetics of lactose crystallization has received very little attention, are also present. The literature indicates that the kinetics can be modified by the presence of other components at each change of state. In order to fully understand their influence, it is necessary to understand their specific effects on lactose solubility and the laws of the kinetics controlling the stages of mutarotation, nucleation and growth.

Thus, as previously explained, to obtain a high-quality fat-filled whey/permeate powder with a low free fat content, a high rehydration behavior and high dryability without stickiness and cakiness, the optimal crystal/amorphous lactose ratio should be between 0.20–0.25/0.80–0.75. The problem with lactose crystallization is that the process involves a crystal/amorphous lactose ratio at equilibrium between 0.80–0.75/0.20–0.25 (Gernigon et al., 2013). It is for this reason that a heating step is added after the lactose crystallization step, just before fat incorporation and homogenization, to inverse this ratio, in order to solubilize a part of the crystallized lactose. The control of the ratio is mainly done using refractometry measurements according to the study of Gernigon et al. (2013).

3 Quality Control of Fat-Filled Powders

Spray-dried infant formula powders are initially packed in big bags pending the outcome of quality control. Firstly, although infant formula preparation includes heat treatment to eliminate pathogenic bacteria during processing, a microbiological examination is essential and needs to be performed to guarantee the microbiological safety of the product, especially for the detection of *Salmonella*, *Enterobacteria*, and *Enterobacter sakazakii*. Since the final product may be recontaminated at some point during the rest of the process and infant formulae are usually given to infants without further heat treatment. Secondly, the nutrient content of a powdered infant formula must be checked to ensure that it can fully meet the nutritional requirements of infants (e.g., composition profiles of proteins and fats, mineral and vitamin content, etc.). Finally, certain analyses concerning powder solubility, dispersibility, density, granulometry profile and bottle tests should also be performed to assure the physicochemical quality of the powder. Once powder quality is consistent, the powder can then be packed in small metal cans in an aseptic environment. The powder is usually packed in a nitrogen atmosphere to prevent oxidation of the fat component.

According to Schuck (2011b), a dairy powder is characterized not only by its composition (proteins, carbohydrates, fats, minerals and water) but also by its microbiological and physical properties (bulk and particle density, instant characteristics, flowability, floodability, hygroscopicity, degree of caking, whey protein nitrogen index, thermostability, insolubility index, dispersibility index, wettability index, sinkability index, rehydration time, free fat, occluded air, interstitial air, particle size, water activity (a_w), T_g, etc.), which form the basic elements of quality specifications, and well-defined testing methods exist to assess these properties according to international standards. These characteristics depend on drying parameters (type of spray dryer, nozzles/wheels, pressure, agglomeration, as well as the thermodynamic conditions of the air such as temperature, relative humidity, and velocity), the characteristics of the concentrate before spraying (physicochemical characteristics, viscosity, thermosensitivity and availability of water), and storage conditions.

3.1 Free Fat

According to Vignolles et al. (2007), functional properties of dairy powders related to free fat mainly depend on two considerations. Concentrate composition has a great influence through its three major constituents: (1) fat which includes type of fat (milk fat or vegetable oil), melting point, load and surface-active molecules such as phospholipids; (2) proteins, mostly casein or caseinate and whey proteins, e.g. WPI and WPC; and (3) lactose or another carbohydrate in their amorphous or crystallized form. Each constituent is dramatically related to one another by physicochemical conditions, which in turn influences their reactivity and morphology. The process identifies four key-steps: homogenization, pre-crystallization of lactose, drying and storage. The latter has a strong impact on the behavior of lactose and fat. Each step determines fat stability in subsequent ones. Finally, composition and process are highly related but few studies focused on the interaction between both, or only considered a little part of the process. Hence, generalities cannot be extrapolated on how composition and process parameters influence free fat content and fat stability during the overall process, which in turn, interfere in some of the applications of dairy powders: off-flavors from oxidation products, poor rehydration and flowing properties. Moreover, it gradually appears that the measure of surface fat by means of X-ray photoemission spectrometry (XPS) is a powerful tool to predict and assess the functional properties of dairy powders. Further research should focus on the overall process, i.e. at least from the homogenization step to storage. It should also increasingly consider composition consistent with industrial requirements, the relationship between free and surface fat and their consequences on dairy powder properties. This knowledge will ultimately lead to the appraisal of each factor influence, a better understanding of fat migration during drying and storage and finally a better control of the functional properties of dairy powders (Vignolles et al., 2007).

3.2 Density

Densities are classified into three groups: bulk (apparent) density, particle density and the density of the dry milk solids, all of which are very much interrelated. The bulk density of a powder is a complex property that depends on primary factors such as the true or absolute density of the product, the air within each particle (occluded air, OA) and the air between each particle (interstitial air, IA) (Schuck, Dolivet, & Jeantet, 2012). Regarding the occluded air, the content and state of proteins can markedly affect stable foam formation, whereas fat has the opposite effect. High-fat concentrates are much less susceptible to foaming than skim milk. Undenatured whey proteins in skim milk have a greater tendency to foam. Denaturation of whey proteins by high heat treatment reduces foaming. Concentrates with a low total solid content foam more than those with a high total solid content (Pisecky, 1997).

3.3 Rehydration

The hydration ability of a powder in water is an essential property for industrial users of dehydrated ingredients in the liquid phase. It is generally accepted that there are three different stages in the rehydration process: wetting, dispersion and solubilization. The degree of progress in each stage is essentially characterized by three indices: wettability, dispersibility and solubility. Rehydration kinetics depend, on the one hand, on the composition of the powder (notably the surface composition for wetting) as well as the affinity between the components and water and, on the other hand, the accessibility of the powder components to water (influence of the structure [porosity and capillarity] and rehydration conditions [agitation, temperature, solid/liquid concentration] (Jeantet, Schuck, Six, Andre, & Delaplace, 2010; Schuck et al., 2012). The amount and dispersion of fat or free fat in milk powder negatively affect wettability (Gaiani et al., 2006). Since fat is hydrophobic, it inhibits the wetting of milk powder. To improve the wettability of high fat powders, the surface of powder particles is coated with a surface-active agent. Lecithin is one of the most commonly used surface-active agents for the instantization of whole milk powders.

3.4 Glass Transition and Water Activity

The physicochemical properties of free and bound water affect the physical state, transition temperatures, sticking temperature, reaction kinetics and stability of milk products. The emphasis has been on the physical state of non-fatty solids and the effects of water and its physical state on chemical reaction rates, growth of microorganisms and stability. Spray drying, storage and the quality of dairy powders are significantly dependent on both the physical state of lactose (one of the main components of dairy powders) and other carbohydrates, which themselves are dependent on the T_g and a_w.

The water activity of dried milk products is largely correlated with moisture content and temperature. The influence of various processing techniques on the composition and state of individual components also affects water activity.

The water activity of whole milk powders is mainly controlled by the moisture content expressed in non-fat solids because fat has no influence. Thus, differences in a_w of different kinds of dairy powders are due mostly to the state of the proteins and the physical state of the lactose. The ideal moisture content can be determined for the optimal stabilization (at 0.2 a_w and 25 °C) of some dairy powders by using practical or theoretical sorption isotherms (Efstathiou, Feuardent, Méjean, & Schuck, 2002). For example, the corresponding moisture content must be close to 4%, 3% and 2.5%, for skim milk, fat-filled milk at 26% fat and fat-filled milk at 40% fat, respectively (Schuck et al., 2012).

4 Control and Improvement of Powder Properties

Due to the variety and complexity of the fat-filled concentrates to be dried, a rigorous method based on physicochemical and thermodynamic properties is now necessary. Two approaches to the spray drying of fat-filled powder are required in dairy research, one involving the products (availability of water related to the biochemical composition) and the second involving the process (knowledge and improvement of the drying parameters) (Schuck, 2013).

4.1 Availability of the Water

The aim of this section is to explain a recent method (drying by desorption using a thermohygrometer sensor) in order to determine major drying parameters according to food components in relation to their interactions with water (bound and free water) and linked to water transfer kinetics. The studies of Schuck et al. (1998, 2009) have shown that drying by desorption is an excellent tool to determine and optimize the major spray drying parameters in relation to biochemical composition according to water availability and desorption behavior (calculation of extra energy ΔE). Analysis of the desorption curve (measured relative humidity vs. time), combined with knowledge of the temperature, total solids, density and specific heat capacity of the concentrate, air flow rates, theoretical water content in relation to water activity and RH of the outlet air, the current weather conditions, cost per kWh and the percentage of drying in the integrated fluid has made it possible to determine enthalpy, T, RH (including ΔE) for each inlet air, concentrate and powder flow rate, specific energy consumption, energy and mass balance, dryer yield and cost (in € or in $) to remove 1 kg of water or to produce 1 kg of powder. All these results are summarized in Fig. 19.4. This figure is a representation of the software output:

- air characteristics at the dryer/integrated fluid bed inlet and outlet (top),
- flow, energy and cost calculations (bottom) (Schuck et al., 2009).

4.2 Process Improvement

The aim of this section is to explain that the use of a thermohygrometric sensor with some measurements (temperature, absolute (AH) and relative humidity (RH), dry air flow rate, water activity) to calculate the mass and absolute humidity is helpful to prevent sticking in the drying chamber and to optimize powder moisture and water activity in relation to the relative humidity of the outlet air (Schuck, Méjean, Dolivet, & Jeantet, 2005). It was observed that the calculated AH is systematically higher than the measured AH because the calculated AH corresponds to the maximum theoretical value that can be reached. Calculation of AH by means of the mass balance is based on the hypothesis that the air circulating in the spray dryer removes all the water from the concentrate. Thus, if the difference between the calculated and measured absolute humidity of the outlet air is below 2 g of water kg^{-1} dry air (depending on the spray dryer with regard to measurement accuracy), there is no problem of sticking in the spray dryer chambers, regardless of the dairy concentrate used. On the other hand, sticking was observed in this study for differential AH

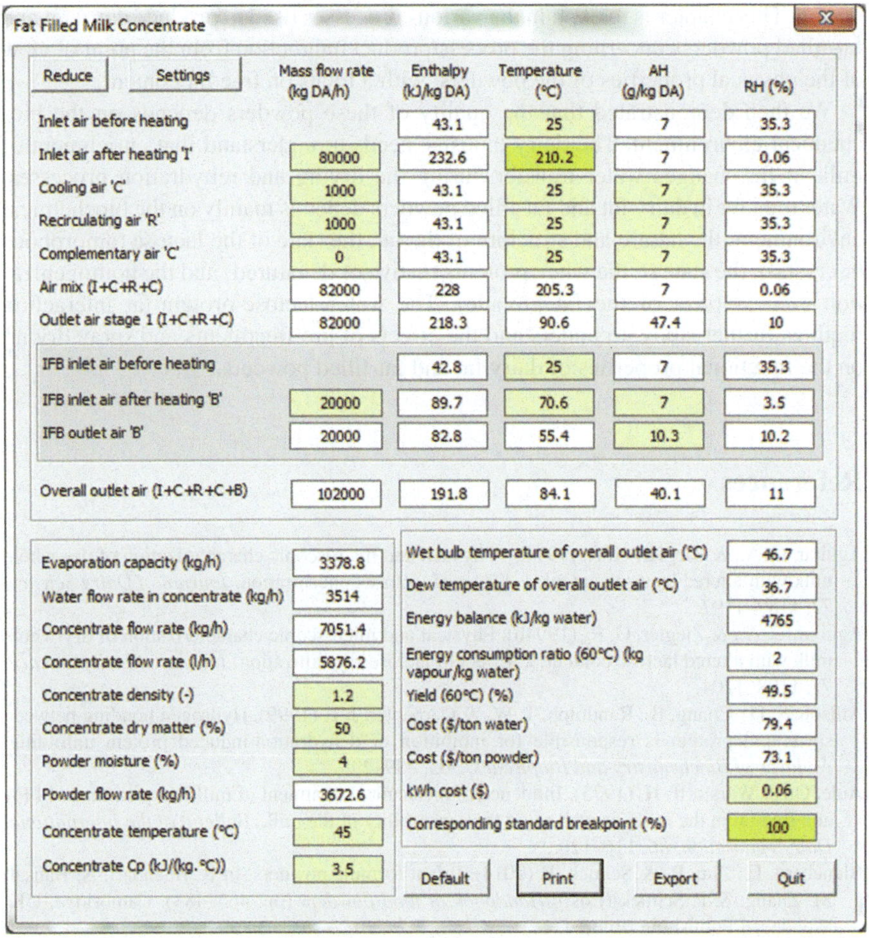

Fig. 19.4 Spray drying parameters of fat-filled milk concentrate calculated by SD2P®

above 2 g water kg^{-1} dry air, corresponding to lower water removal and, consequently, to favorable sticking conditions. The operator can follow the absolute humidity and anticipate a variation in drying parameters according to the differences between calculated and measured absolute humidity.

5 Conclusions

This chapter aimed to suggest different flow charts to produce dairy fat and fat-filled powder and explain the process of dehydration, i.e. spray drying, to understand the effects of spray drying on the quality of dairy fat and fat-filled powders during drying. This chapter is to explain the various processes used to produce dairy fat and fat-filled powders concerning the process/product interaction from the point of view of the physical properties of the powders, with a focus on free fat content.

We then demonstrated that the quality of these powders depends on the biochemical environment. The dairy industry needs to understand that enrichment of milk in fat changes water transfer during the drying and rehydration processes. Water transfer in dairy fat and fat-filled powders depends mainly on the biochemical environment, the nature and structure of the fat, the state of the lactose (amorphous or crystal), the state of the whey proteins (native or denatured) and the homogenization process (size of the fat globule). The water–lactose-protein-fat interaction requires further study, to understand the effects of pre-treatments and spray drying on the functional properties of dairy fat and fat-filled powders.

References

Aguilar, C. A., & Ziegler, G. R. (1994a). Physical and microscopic characterization of dry whole milk with altered lactose content. 1. Effect of lactose concentration. *Journal of Dairy Science, 77*, 1189–1197.

Aguilar, C. A., & Ziegler, G. R. (1994b). Physical and microscopic characterization of dry whole milk with altered lactose content. 2. Effect of lactose crystallization. *Journal of Dairy Science, 77*, 1198–1204.

Allison, S. D., Chang, B., Randolph, T. W., & Carpenter, J. F. (1999). Hydrogen bonding between sugar and protein is responsible for inhibition of dehydration-induced protein unfolding. *Archives of Biochemistry and Biophysics, 365*, 289–298.

Aule, O., & Worstorff, H. (1975). Influence of mechanical treatment of milk on quantities of FFA and free fat in the milk, as well as on the separability of the milk. *Bulletin of the International Dairy Federation, 86*, 116–120.

Blanchard, E., Zhu, P., & Schuck, P. (2014). Infant formula powders. In B. Bhandari, N. Bansal, M. Zhang, & P. Schuck (Eds.), *Handbook of food powders* (pp. 465–483). Cambridge, UK: Woodhead Publishing Limited.

Buchheim, W., Samhammer, E., & Lembke, J. (1974). Light-microscopical visualization of fat on the surface of dried milk particles. *Milchwissenschaft, 29*, 513–517.

Buma, T. J. (1971). Free fat in spray-dried whole milk. 3. Particle size. Its estimation, influence of processing parameters and its relation to free-fat content. *Netherlands Milk and Dairy Journal, 25*, 53–72.

Chever, S., Mejean, S., Dolivet, A., Mei, F., Den Boer, C. M., Le Barzic, G., et al. (2017). Agglomeration during spray drying: Physical and rehydration properties of whole milk/sugar mixture powders. *LWT - Food Science and Technology, 83*, 33–41.

Cordier, J.-J. (2008). Production of powdered infant formulae and microbiological control measures. In J. M. Farber & S. J. Forsythe (Eds.), *Enterobacter sakawakii* (pp. 145–185). Washington, DC: ASM Press.

Croguennec, T., Jeantet, R., & Schuck, P. (2016). From milk to dairy products. In R. Jeantet, T. Croguennec, P. Schuck, & G. Brulé (Eds.), *Handbook of food science and technology: Food biochemistry and technology* (Vol. 3, pp. 3–63). London, UK: ISTE and Wiley.

Crowley, S. V., Kelly, A. L., Schuck, P., Jeantet, R., & O'Mahony, J. A. (2016). Rehydration and solubility characteristics of high-protein dairy powders. In P. L. H. McSweeney & J. A. O'Mahony (Eds.), *Advanced dairy chemistry-1B proteins: Applied aspects* (pp. 99–131). New York, NY: Springer Publishers.

Efstathiou, T., Feuardent, C., Méjean, S., & Schuck, P. (2002). The use of carbonyl analysis to follow the main reactions involved in the process of deterioration of dehydrated dairy products: Prediction of most favourable degree of dehydration. *Le Lait, 82*, 423–439.

Fäldt, P., & Bergenståhl, B. (1996). Spray-dried whey protein/lactose/soybean oil emulsions. 2. Redispersability, wettability and particle structure. *Food Hydrocolloids, 10*, 431–439.

FAO/WHO. (2007). *Standard for infant formulas and formulas for special medical purposes intended for infants, CODEX STAN 72-1981*. Rome, Italy: Food and Agricultural Organization of United Nations.

Gaiani, C., Ehrhardt, J. J., Scher, J., Hardy, J., Desobry, S., & Banon, S. (2006). Surface composition of dairy powders observed by X-ray photoelectronspectroscopy and effects on their rehydration properties. *Colloids and Surfaces. B, Biointerfaces, 49*, 71–78.

Gernigon, G., Baillon, F., Espitalier, F., Le Floch-Fouéré, C., Schuck, P., & Jeantet, R. (2013). Effects of the addition of various minerals, proteins and salts of organic acids on the principal steps of α-lactose monohydrate crystallization. *International Dairy Journal, 30*, 88–95.

Henning, D. R., Baer, R. J., Hassan, A. N., & Dave, R. (2006). Major advances in concentrated and dry milk products, cheese, and milk fat-based spreads. *Journal of Dairy Science, 89*, 1179–1188.

IDF. (2016). The world dairy situation 2016. *Bulletin of the International Dairy Federation, 485*, 264.

Jeantet, R., Schuck, P., Six, T., Andre, C., & Delaplace, G. (2010). The influence of stirring speed, temperature and solid concentration on the rehydration time of micellar casein powder. *Dairy Science and Technology, 90*, 225–236.

King, N. (1965). The physical structure of dried milk. *Dairy Science Abstracts, 27*, 91–104.

Montagne, D. H., Van Dael, P., Skanderby, M., & Hugelshofer, W. (2009). Infant formulae – powders and liquids. In A. Y. Tamime (Ed.), *Dairy powders and concentrated products* (pp. 294–331). Ayr, UK: Wiley-Blackwell.

Pisecky, J. (1997). *Handbook of milk powder manufacture*. Copenhagen, Denmark: Niro A/S.

Schuck, P. (2013). Condensed and powdered milk. In Y. W. Park & G. F. W. Haenlein (Eds.), *Milk and dairy products in human nutrition* (pp. 412–434). Oxford, UK: Wiley-Blackwell.

Schuck, P. (2011a). Dehydrated dairy products/milk powder: Types and manufacture. In J. W. Fuquay, P. F. Fox, & P. L. H. McSweeney (Eds.), *Encyclopedia of dairy sciences* (Vol. 2, 2nd ed., pp. 108–116). San Diego, CA: Academic Press.

Schuck, P. (2011b). Dehydrated dairy products/milk powder: Physical and functional properties of milk powders. In J. W. Fuquay, P. F. Fox, & P. L. H. McSweeney (Eds.), *Encyclopedia of Dairy Sciences* (Vol. 2, 2nd ed., pp. 117–124). San Diego, CA: Academic Press.

Schuck, P. (2011c). Lactose and oligosaccharides/lactose: Crystallization. In J. W. Fuquay, P. F. Fox, & P. L. H. McSweeney (Eds.), *Encyclopedia of dairy sciences* (Vol. 3, 2nd ed., pp. 182–195). San Diego, CA: Academic Press.

Schuck, P. (2002). Spray drying of dairy products: state of the art. Lait 82(4), 375–382.

Schuck, P., Méjean, S., Dolivet, A., Gaiani, C., Banon, S., Scher, J., et al. (2007). Water transfer during rehydration of micellar casein powders. *Le Lait, 87*, 425–432.

Schuck, P., Dolivet, A., & Jeantet, R. (2012). *Analytical methods for food and dairy powders*. Oxford, UK: Wiley-Blackwell.

Schuck, P., Méjean, S., Dolivet, A., & Jeantet, R. (2005). Thermohygrometric sensor: A tool for optimizing the spray drying process. *Innovative Food Science and Emerging Technologies, 6*, 45–50.

Schuck, P., Dolivet, A., Méjean, S., Zhu, P., Blanchard, E., & Jeantet, R. (2009). Drying by desorption: A tool to determine spray drying parameters. *Journal of Food Engineering, 94*, 199–204.

Schuck, P., Roignant, M., Brulé, G., Davenel, A., Famelart, M. H., & Maubois, J. L. (1998). Simulation of water transfer in spray drying. *Drying Technology, 16*, 1371–1393.

Shimada, Y., Roos, Y. H., & Karel, M. (1991). Oxidation of methyl linoleate encapsulated in amorphous lactose-based food model. *Journal of Agricultural and Food Chemistry, 39*, 637–641.

Vignolles, M. L., Jeantet, R., Lopez, C., & Schuck, P. (2007). Free fat, surface fat and dairy powder: Interactions between process and product. A review. *Le Lait, 87*, 187–236.

Vignolles, M. L., Lopez, C., Madec, M. N., Ehrhardt, J. J., Méjean, S., Schuck, P., et al. (2009). Fat properties during homogenization, spray-drying and storage affect the physical properties of dairy powders. *Journal of Dairy Science, 92*, 58–70.

Walstra, P., Geurts, T. J., Noomen, A., Jellema, A., & van Boekel, M. A. J. S. (1999). *Dairy technology: Principles of milk properties and processes*. Basel, Switzerland: Marcel Dekker.

Westergaard, V. (2004). Milk Powder Technology. Evaporation and Spray Drying. Copenhagen: Niro A/S.

WHO (World Health Organization). (2011). *Exclusive breastfeeding for six months best for babies everywhere*. Retrieved 20 January, 2012, from http://www.who.int/mediacentre/news/statements/2011/breastfeeding_20110115/en/

Chapter 20
Effect of Lipids on the Rehydration Behaviours of Milk Protein Concentrate

Dasong Liu, Yun Yu, Qiying Feng, and Peng Zhou

1 Introduction

MPC is a functional dairy protein ingredient with protein content ranging between 40–90%, and it is produced from the pasteurized skim milk sequentially through ultrafiltration, diafiltration, evaporation and spray-drying processes. During the membrane filtration process, the casein micelles and whey proteins are retained in the retentate, while the lactose and soluble salts are removed with the permeate. Therefore, the protein content of MPC is determined by the extent of membrane filtration, and the ratio of caseins to whey proteins (4:1) in MPC is almost the same to that in skim milk. MPC has unique nutritional and functional properties, and it has become increasingly popular in the dairy industry, such as the development and production of yoghurt, cheese, beverage, infant formula and nutritional bar (Singh, 2011).

The rehydration behaviour of MPC is of great importance for fully achieving other functional properties, such as emulsifiability, foamability, viscosity, turbidity, heat stability and gelation property (Bouvier, Collado, Gardiner, Scott, & Schuck, 2013). Therefore, the rehydration of MPC should be fast and complete even at the typical low temperatures and within the time frames for the manufacture of dairy products. However, a major problem associated with MPCs containing increasing amounts of micellar caseins is their poor solubility, especially for those stored at high temperatures for a long period of time. The poor solubility of MPC will limit its application in the dairy industry. Numerous researches have been conducted to investigate the molecular mechanisms that contribute to the solubility loss of MPC

D. Liu · Y. Yu · P. Zhou (✉)
State Key Laboratory of Food Science and Technology, Jiangnan University, Wuxi, Jiangsu Province, China
e-mail: zhoupeng@jiangnan.edu.cn

Q. Feng
Department of Food Science, University of Guelph, Guelph, ON, Canada

T. Truong et al. (eds.), *Dairy Fat Products and Functionality*,
https://doi.org/10.1007/978-3-030-41661-4_20

and also to develop the processing strategies that lead to the increase in solubility of MPC. Some researchers indicated that the casein micelles within MPC can easily associate with each other through non-covalent interactions to form insoluble protein aggregates (Havea, 2006; Mckenna, 2000a). Mimouni, Deeth, Whittaker, Gidley, and Bhandari (2010a) also suggested that a hydrophobic monolayer skin was formed on the surface of MPC powder particle through cross-linking and even fusion of casein micelles during storage, thus resulting in longer dissolution times of MPC (Fyfe et al., 2011).

The presence of intrinsic and extrinsic lipids within MPC will partially affect the association of casein micelles and also the surface hydrophobicity of powder particles, thus further affecting the rehydration behaviours of MPC. Over the past decades, several researches have been done to address these issues from various perspectives. This chapter provides a general review of the relevant literature for three main topics as follows: the rehydration behaviours of MPC, the effect of lipid types and lipid addition methods on the rehydration behaviours of MPC, and the techniques that are commonly applied to characterize the distribution of lipids within MPC powder particles.

2 Rehydration Behaviours of MPC

To be fully functional as a dairy protein ingredient, MPC should readily dissolve in solution upon rehydration. Generally, the rehydration process of MPC can be divided into four phases as follows: wetting, submersing, dispersing, and dissolving (Freudig, Hogekamp, & Schubert, 1999; IDF, 1979). The wetting phase corresponds to the imbibing of water under the influence of capillary forces for the powder particles, while the submersing phase refers to the sinking of powder particles below the surface of water. The dispersing phase corresponds to the disintegrating of powder particles into primary particles, while the dissolving phase refers to the completely disintegrating of powder particles in water. Generally, the delay in any phase will negatively affect the overall rehydration behaviours of MPC. Wetting is the initial phase of powder rehydration and it is mainly affected by the size distribution and the surface composition (e.g., lipid) of powder particles (Fäldt & Bergenståhl, 1994; Gaiani et al., 2010; Granelli, Fäldt, Appelqvist, & Bergenståhl, 1996; Hardas, Danviriyakul, Foley, Nawar, & Chinachoti, 2000; Kim, Chen, & Pearce, 2002). The size distribution affects the porosity of powders and the surface composition determines the contact angle between water and particle surface, which will further affect the rate of water imbibing. Dissolving is the final phase of powder rehydration and it appears to be a more reliable criterion to evaluate the rehydration behaviours of MPC considering that the four rehydration phases are difficult to study independently.

2.1 Deterioration in Rehydration Behaviours

The freshly prepared MPC may present poor solubility and the solubility will continue to decrease during following storage, especially for those stored at high temperatures and also for those containing high levels of proteins (Jimenez-Flores & Kosikowski, 1986; Mistry & Hassan, 1991; Mistry & Pulgar, 1996). Commercial MPC is mainly produced in Oceania, North America and Western Europe, and a large proportion of these products are exported to Asian markets by ship. Therefore, temperature higher than 40 °C is not uncommon during transportation, especially in the holds of ships and on the wharves and in the warehouses in the warm climates. Deterioration reactions in MPC powders occur at a faster rate at higher temperatures. For dehydrated products, a shelf life of 1 month at 40 °C can be equivalent to a shelf life of about 18 months at 23 °C (Carr, Bhaskar, & Ram, 2004). Anema, Pinder, Hunter, and Hemar (2006) reported that the changes in solubility of MPC85 stored at different temperatures as a function of storage time can be collapsed onto a single master curve by using the temperature-time superposition method, suggesting that the same physico-chemical processes affected the solubility of MPC85 stored at different temperatures. Anema et al. (2006) also reported that the half-lives (i.e., the time to reach 60% solubility) for MPC85 stored at 50, 40, 35, 30 and 20 °C were 2, 6, 21, 50 and 210 days, respectively. McKenna (2000a) reported that the solubility of MPC35, MPC56, MPC70, MPC85 and MPC92 was 98%, 95%, 70%, 35% and 27%, respectively. In a survey, 32 commercial MPC samples with protein contents ranging between 56–85% in dry matter were collected fresh from the suppliers in Oceania, North America and Western Europe, and a negative correlation between the protein content and the solubility was observed (Huppertz & Gazi, 2015). A following study by the same group was conducted on a series of MPCs with varying protein contents produced under the controlled conditions from a single batch of skim milk, and the results confirmed that the high protein MPCs were more prone to solubility loss and the solubility loss was accelerated strongly at high storage temperatures (Huppertz & Gazi, 2015).

2.2 Molecular Mechanism

During membrane filtration of skim milk, the concentration of proteins is increased and the lactose is removed, thus leading to the close packing of casein micelles. During the following spray-drying, the casein micelles are forced together by their adsorption to the air-water interface and may even touch at the final air-particle interface. The contact between casein micelles is further promoted as the casein micelles become partially depleted in surface κ-casein during membrane filtration, thus reducing both steric hindrance and electrostatic repulsion (Singh & Creamer, 1991; Singh & Fox, 1987). With such close proximity and the reduction in repulsive forces, further protein-protein interactions may cause cross-linking and even fusion of casein micelles. During the subsequent storage, increased interactions occur between and within casein micelles, resulting in the compaction of casein micelles within powder particles and also the formation of a monolayer skin of closely packed casein micelles on the surface of powder particles. Mimouni et al. (2010a); Mimouni, Deeth, Whittaker, Gidley, and Bhandari (2010b) reported that the combination of different types of interactions (i.e., bridges, direct contact) between casein micelles contributed to the formation of a porous and gel-like structure that restrained the dispersion of individual casein micelles into the surrounding liquid phase without preventing the penetration of water and the solubilization of non-micellar components.

Poor solubility of MPC is mainly due to the delayed rehydration kinetics rather than the formation of insoluble materials. Mimouni et al. (2010b) reported that the slowly solubilising materials consisted almost entirely of casein micelles and the non-micellar components (e.g., whey proteins, lactose and sodium) were readily soluble. The inter-micellar bonds might involve weak forces (e.g., hydrophobic interactions and hydrogen bonds) that would be reversible upon rehydration, rather than the covalent cross-linking such as disulfide bonds, as both the long-term rehydration and the mechanical shearing (e.g., ultrasonication and homogenization) were sufficient to allow almost complete dispersion of powder particles (Chandrapala, Martin, Kentish, & Ashokkumar, 2014; Schokker et al., 2011). Havea (2006) reported that the insoluble materials (sediments) in MPC85 were dissociable under the conditions of non-reduced sodium dodecyl sulfate polyacrylamide gel electrophoresis, suggesting that the insoluble protein aggregates were mainly formed by the non-covalent bonds such as hydrophobic interactions. During membrane filtration of skim milk, the calcium ion concentration increases with the gradual depletion of serum citrate and the calcium ion activity coefficient also increases with the gradual decrease in ionic strength of the serum phase, thus leading to the increase in calcium ion activity (Crowley et al., 2014). The increasing calcium ion activity may further predispose the concentrated system to more significant protein-protein and protein-mineral interactions (e.g., calcium ion bridges) during subsequent processing and storage, which would adversely affect the solubility of MPC.

3 Effect of Lipids on Rehydration Behaviours of MPC

3.1 Intrinsic Lipids

3.1.1 Composition

Bovine milk typically contains 4–5% lipids and the size of fat globules ranges between 0.1–15 μm in diameter. The spherical core of the fat globules is predominantly composed of neutral lipids, and the exterior membrane of the fat globules mainly consists of polar lipids and proteins. The milk fat globule membrane contains the major part of the dairy polar lipids (Evers, 2004; Gaiani et al., 2010;

Rombaut, Camp, & Dewettinck, 2005). Some of the small residue fat globules are neutrally buoyant and cannot be eliminated by centrifugation. Moreover, the physical and mechanical treatments applied during milk processing may cause the partial breakdown of milk fat globule membrane and hence the migration of polar lipids into serum phase. Both the residue fat globules and the polar lipids in the serum phase cannot be removed during the membrane filtration process and will be finally retained within the MPC powder particles as intrinsic lipids. Typically, lipid content as high as 0.15% can be found in skim milk.

Native phosphocaseinate is a similar dairy protein ingredient that is produced from skim milk by microfiltration. Gaiani et al. (2009) reported that native phosphocaseinate contained about 0.4% lipids and the relative concentration of polar lipids was about 66%, while the lipids in milk contained less than 1% polar lipids. The main polar lipids in native phosphocaseinate were sphingomyelin, phosphatidylethanolamine, phosphatidylcholine and phosphatidylinositol, and their relative concentrations were 31–43%, 24–36%, 21–28% and 0–5%, respectively.

3.1.2 Migration of Lipids

The distribution of lipids within powder particles dictates the bulk behaviours in terms of wettability, flowability, and stability to caking and oxidative rancidity. Based on the location within powder particles, the solvent-extractable lipids can be divided into four fractions as follows: the surface lipids, the outer layer lipids from fat globules in the surface layer of the powder particles, the capillary lipids constituted by fat globules that can be reached by the solvent through capillary forces, and the dissolution lipids consisting of lipids reached by solvent through holes left by the already extracted lipids (Vega & Roos, 2006). Some other researchers just simply defined the lipids within powder particles as free lipids and encapsulated lipids (Vignolles et al., 2009; Vignolles, Jeantet, Lopez, & Schuck, 2007). The free lipids, including both surface free lipids and inner free lipids, are not entirely coated by the matrix of proteins and lactose, so they can be easily extracted by solvents. The encapsulated lipids are the fraction that is protected by the matrix combined with proteins and lactose, and they are still present after the extraction of free lipids. It is commonly recognized that the free lipids play a more influential role in determining the rehydration behaviours of dairy protein powders (Vignolles et al., 2007; Vignolles, Lopez, Ehrhardt, et al., 2009). The migration of lipids to the particle surface is regarded as one of the main factors that would deteriorate the wettability of MPC due to the corresponding increase in surface hydrophobicity. The immigration phenomenon of lipids occurs during both the spray-drying process and the following storage, and it could be explained by both the amphiphilic nature and melting points of the intrinsic lipids.

Lipids seem to be the most preferred dairy components to migrate towards the air-water interface during spray-drying and also towards the air-particle interface during storage. Nijdam and Langrish (2006) reported that the surface lipid coverage of dairy powders sharply increased from 0% to 35% as the average lipid content increased from 0% to 5%, while the surface lipid coverage only increased by a further 25% as the average lipid content increased from 5% to 30%. Gaiani et al. (2010) also reported an over-representation of lipids on the particle surface (6%) in comparison with the total lipids (0.4%) for the freshly prepared native phosphocaseinate, and the surface lipid content further increased to 17% when the native phosphocaseinate was stored at both 20 and 50 °C for 60 days, which was strongly correlated to the lengthening of wetting time. The melting points of lipids extracted from dairy powders ranges between −3 and 32 °C (Gaiani et al., 2010). During the typical spray-drying process of dairy powders, the outlet temperature is usually above 70 °C which is higher than the melting points of lipids (Kim, Chen, & Pearce, 2009a, 2009b). Consequently, the lipids should be in the liquid form throughout the spray-drying process, and this liquid form could explain the over-representation of lipids on the particle surface of the freshly prepared dairy protein powders. Kim, Chen, and Pearce (2005) also suggested that the moisture content gradients effectively concentrated the solutes (lipids, proteins and lactose) at the droplet surface, where the moisture content was lower, thus causing the diffusion of these solutes towards the core of the droplets during spray-drying. The lower molecular weight solutes such as lactose would diffuse inwards more rapidly than the higher molecular weight solutes such as lipids and proteins which consequently concentrated at the particle surface. Gaiani et al. (2010) reported that more surface lipids were observed for high protein dairy powders spray-dried at lower outlet temperatures, and this was accompanied by a sharp decrease in the wetting time of high protein dairy powders. With the decrease in spray-drying temperature, the rate of water evaporation was decreased during spray-drying, thus allowing more time for the inward diffusion of lipids (Meerdink & Riet, 1995). Nijdam and Langrish (2006) reported that a higher spray-drying temperature favours the accumulation of lactose over proteins on the particle surface of dairy powders, while the appearance of lipids on the particle surface was not differently affected by the spray-drying temperature. It was suggested that higher spray-drying temperatures hastened the formation of a surface skin that hindered the migration of surface-active proteins towards the surface, while the lipids were completely in a mobile fluid form at the typical spray-drying temperatures (Hassan & Mumford, 1993; Kentish, Davidson, Hassan, & Bloore, 2005). During storage of dairy powders, a portion of the lipids will be in the liquid form, and the balance between fluid and solid phase depends on the specific storage temperature. Gaiani et al. (2009) also reported the appearance of pores on the particle surface of native phosphocaseinate powders after 60 days storage at 50 °C. Both the lipids in liquid form and the presence of pores during storage may allow the release of lipids onto the particle surface of dairy protein powders.

Kim et al. (2005) reported that the surface free lipids in dairy protein powders had a higher proportion of high-melting lipids than the encapsulated lipids. The melting point of the surface lipids determined the wettability of dairy protein powders, and hence it could be directly estimated by measuring the wettability of dairy protein powders in water at a series of temperatures. The dairy protein powders could not be completely wetted at temperatures between 10–37 °C within a reasonable time period of less than 15 min, while the wetting time sharply decreased at and above 38 °C, indicating that the surface lipids melted at about 38 °C. Kim, Chen, and Pearce (2009c) further reported the release of encapsulated low-melting lipids towards the particle surface of dairy protein powders during 6 months of storage at room temperature, thereby reducing the melting points of the surface free lipids. However, the wettability of dairy protein powders was predominantly controlled by the coverage of surface lipids rather than the thickness of surface lipid layers or the melting point of surface lipids.

In a recent research by Nasser et al. (2017), the increases in the migration of surface lipids, the surface crust resistance and the rehydration characteristic times were demonstrated for the micellar casein powders stored at and above 40 °C for 12 months. The whole rehydration process of micellar casein powders was simply divided into two stages, namely water penetration and particle fragmentation. A strong correlation ($R^2 > 0.9$) was demonstrated between the migration of lipids to the surface and the evolution of the wetting times during storage. The ageing-induced increase in the wetting times remained weak compared to the strong increase in the total rehydration times, while the reinforcement of cross-linking between adjacent casein micelles upon aging-induced significant increase of both the fragmentation times and the total rehydration times. Consequently, the delayed casein micelle release that was induced by the cross-linking formation during storage was considered as the rate-limiting stage of rehydration for the aged micellar casein powders.

3.1.3 Oxidation of Lipids

Compared with the encapsulated lipids in dairy protein powder particles, the surface free lipids are usually more susceptible to oxidation due to their direct exposure in air (Nasser et al. 2017). The lipid oxidation occurs even at very low water activities since the migration of large hydrophobic molecules (lipids) does not depend on the water mobility (Thomsen, Lauridsen, Skibsted, & Risbo, 2005a, 2005b). This would help explain why the solubility loss of whole milk powders during storage was not significantly affected by the relative humidity conditions. Singh (1991) reported that both the intermediate products (e.g., free radicals and hydroperoxides) and the end products (e.g., malondialdehyde) produced from the complicated lipid oxidation processes can react with the amino acid residues, such as lysine, cysteine, histidine, methionine, valine and phenylalanine, to form protein cross-links that would prevent the dispersion of dairy protein powders in water. Le, Bhandari, Holland, and Deeth (2011) reported that the solubility of MPC80 decreased rapidly during 12 weeks of storage at 30 °C and 44–84% relative humidity, and this corresponded well with the formation of high-molecular-weight protein complexes that were observed by the reducing two-dimensional electrophoresis as diffuse staining above the casein monomers. It was suggested that the formation of non-disulfide covalent protein cross-linking might be partially accelerated by the lipid oxidation. Moreover, it was also suggested that the protein cross-linking mainly occurred between different casein micelles since the protein cross-linking within casein micelles was unlikely to prevent the release of individual casein micelle from the MPC powder particles during rehydration.

3.2 *Extrinsic Phospholipids*

The presence of intrinsic lipids would negatively affect the rehydration behaviours of dairy protein powders as mentioned above, while the extrinsic phospholipids are usually added during the manufacturing process of dairy protein powders to improve their rehydration behaviours. The commonly used food-grade phospholipids from soybean, egg yolk, sunflower seed and milk fat globule membrane are now commercially available (Thompson & Singh, 2006; Van & Wendel, 2014). Besides, the synthetic phospholipids of plant and animal origin are also available. These phospholipids have been widely used in pharmaceutical, cosmetic and food industries to encapsulate and deliver bioactive components (Guimarães & Ré, 2011; Güney & Kutlu, 2011; Liu, Wang, & Xia, 2012).

3.2.1 Addition of Phospholipids During Fluidized Bed Drying

During the typical rehydration process of dairy protein powders in water, the powder particles firstly overcome the surface tension so that the water can penetrate into the porous network of particles via the capillary forces (Depalo & Santomaso, 2013; Yazdanpanah & Langrish, 2012). The dairy protein powders with remarkable rehydration behaviours usually have large particle size, small contact angle and high porosity with big interstitial void volume (Selomulya, 2007). Lecithin is a kind of food-grade surfactant with amphiphilic properties, and it is commonly added to the particle surface of dairy protein powders during their fluidized bed drying processes. The surface lecithin would not only decrease the surface tension between the powder particles and water but also act as bridges between the primary particles to result in agglomerated granules with larger particle size and higher porosity, thus eventually contributing to the improvement in rehydration behaviours of dairy protein powders (Lin, Mckeigue, & Maldarelli, 1990). Ji, Cronin, Fitzpatrick, and Miao (2017) added 50 g lecithin solution (0.5%, 2% and 5%) to adhere the primary particles of 100 g dairy protein powders by fluidized bed agglomeration, and the resulting granules had larger particle size, lower bulk density, higher porosity and more irregular shapes. This consequently led to a quicker water penetration by the capillary flow as demonstrated by the more rapid decrease in the strength of films formed at the powder/water interfaces during the wetting process of the resulting dairy protein powders. However, the secondary fluidized bed drying is time and energy-consuming, and it may further alter the natural properties of dairy components. Moreover, it is also difficult to achieve a homogeneous coating of the lecithin layer on the surface of powder particles (Fonseca, Bento, Quintero, Gabas, & Oliveira, 2015).

3.2.2 Direct Addition of Phospholipids Before Spray-Drying

The direct addition of extrinsic phospholipids to the feeding solutions before the spray-drying process is a commercially favourable approach to produce instant dairy protein powders in a single step. The commonly added phospholipids possess low melting points and amphiphilic natures, and therefore they will be in the liquid form so as to be able to compete with the intrinsic lipids and proteins for the air-water interface of the atomized droplets during the spray-drying process (Jayasundera, Adhikari, Aldred, & Ghandi, 2009; Miller, Alahverdjieva, & Fainerman, 2008; Young, Sarda, & Rosenberg, 1993). Consequently, the added phospholipids will be preferentially accumulated on the surface of powder particles, thus enhancing the rehydration behaviours of the resulting dairy protein powders. Lallbeeharry et al. (2014) added 1% egg yolk lecithin (w/w, phosphatidylcholine) to the dairy protein feeding solution before spray-drying using a single-droplet drying device, and they found that the co-spray-drying of lecithin (ionic surfactant) with dairy protein feeding solution reduced the wetting time of powder particles from about 35 s to less than 15 s. It was suggested that the added lecithin tended to combine with dairy proteins to form a complex, which then competed with the intrinsic components for the air-water interface of the atomized droplets during spray-drying.

3.2.3 Addition of Phospholipid Nanovesicles Before Spray-Drying

As stated above, the deterioration in rehydration behaviours of MPC during both the manufacturing process and the following storage is mainly caused by the cross-linking between neighboring casein micelles within powder particles (Havea, 2006; Mimouni et al., 2010a, 2010b). Therefore, it is theoretically proposed that any strategy that can slow down or prevent the development of cross-linking between casein micelles within powder particles would effectively improve the rehydration behaviours of MPC. One promising way to reduce the cross-linking of casein micelles is to provide spatial separation between the adjacent casein micelles by introducing smaller-sized inert spacers, such as polysaccharides and phospholipid nanovesicles. In a recent research by Bansal, Truong, and Bhandari (2017), the phospholipid nanovesicles with an average hydrodynamic diameter of about 82 nm were firstly prepared from a soy lecithin dispersion (5%, w/w) by two passes of microfluidization at 103 MPa. Subsequently, these bi-lamellar vesicles were directly added to MPC80 dispersion (11%, w/w) at various levels to achieve 0%, 1%, 5% and 10% (w/w) of total dairy solids, respectively. The mixed suspensions were spray-dried at 160 °C inlet and 75 °C outlet air temperatures, and the powders obtained were immediately stored at 25 or 4 °C and 0.23 water activity for up to 180 days. They found that the addition of nanovesicles at levels as low as 1% and 5% of the total dairy solids could improve the solubility of MPC80 by 13% and 30%, respectively, during storage at 25 °C for 90 days. Even after 180 days of storage at 25 °C, the solubility of MPC80 showed an increasing trend of more than

15% when the addition of nanovesicles reached the levels of both 5% and 10%. It was postulated that the presence of spacer particles (lecithin nanovesicles) that are smaller than the average size of casein micelles (20–500 nm) could lead to a reduction of interactions between the casein micelles within MPC80 powder particles (Bansal et al., 2017).

3.3 Extrinsic Fatty Acids

MPC could be used as a wall material to encapsulate and stabilize the functional poly-unsaturated fatty acids, such as linoleic acid, linolenic acid, arachidonic acid, eicosapentaenoic acid and docosahexaenoic acid. Zhang (2015) reported that the direct addition of extrinsic fatty acids to the MPC85 dispersion before spray-drying significantly improved the rehydration behaviours of the resulting powders. After 60 days of storage at 35 °C, the solubility of MPC85 sharply decreased to less than 40%, while the solubility of MPC85 added with 10% linoleic acid (w/w) remained above 75%. It was suggested that the addition of fatty acid would decrease the pH of the MPC85 dispersion, thus leading to the solubilisation of colloidal calcium phosphate and hence the dissociation of caseins from casein micelles. The smaller-sized non-micellar caseins would adsorb preferentially at the air-water interface during spray-drying, compared with the larger-sized casein micelles (Schokker et al., 2011). This will lead to less accumulation and spreading of casein micelles at the air-particle interface, resulting in a lower susceptibility to both the cross-linking between casein micelles and the consequent formation of a monolayer skin on the surface of powder particles. Besides, the non-micellar caseins would spatially separate the casein micelles within powder particles, thus inhibiting the direct cross-linking of casein molecules in two adjacent casein micelles by acting as inert spacers in a matrix of casein micelles. The reduction in cross-linking between casein micelles both on the surface and in the core of powder particles would improve the rehydration behaviours simultaneously.

4 Techniques to Characterize the Distribution of Lipids in MPC

To fully understand how lipids influence the rehydration behaviours of dairy protein powders, it is of crucial importance to characterize the distribution of lipids within the powder particles both quantitatively and qualitatively (Murrieta-Pazos et al., 2012). The commonly used techniques are introduced in this section.

4.1 Solvent Extraction

Originally, the solvent extraction method is solely applied to classify the intrinsic lipids within dairy protein powders (Vega & Roos, 2006). Nowadays, the solvent extraction method is usually used together with other methods, such as differential scanning calorimetry, high performance liquid chromatography and gas chromatography, to gain more detailed information about the distribution and characteristic of the intrinsic lipids as well as their changes during the processing and storage of dairy protein powders. Based on the typical extraction sequences and methods, the total lipids in dairy protein powders can be divided into the surface free lipids, the inner free lipids and the encapsulated lipids. The free lipids are regarded as the lipids unprotected by protein layers, and they originate from the particle surface as well as the interior of the powder particles. The lipids on the particle surface is mostly free lipids, and the proteins and lactose are located underneath the surface free lipids. Kim et al. (2002) suggested that petroleum ether could extract the free lipids without causing any structural changes, while ethanol could extract almost the total lipids as it caused structural changes that might provide access of solvent to the interior of protein bound fat globules. Murrieta-Pazos, Gaiani, Galet, Calvet, et al. (2012) reported that the melting profiles of total lipids extracted from dairy protein powders showed three endothermic peaks with the typical melting behaviors starting around −40 °C and finishing around 40 °C. These melting peaks might be related to low, intermediate and high melting lipids based on their melting points, respectively. The dairy lipids had many different species in terms of compositions, and hence they would not form pure crystals but tend to crystallize in groups of similar size and structure. Therefore, the polymorphic forms of these mixed crystals might also contribute to the presence of the three melting peaks. The fatty acids with long chains of more than 14 carbons were more present in the encapsulated lipid fractions, while the fatty acids with short and medium chains of less than 14 carbons were more present in the free lipid fractions. Moreover, the fatty acids with long chains were less prone to reach the surface of the powder particles at the classical spray-drying temperatures, resulting in the presence of more fatty acid with short chains at the surface of the powder particles.

4.2 XPS, EDX and ToF-SIMS

XPS (x-ray photoelectron spectroscopy), also named ESCA (electron spectroscopy for chemical analysis), is commonly applied to determine the surface composition of dairy protein powders (Fäldt & Bergenståhl, 1996; Fäldt, Bergenståhl, & Carlsson, 1993). The dairy protein powders are complex systems containing various components such as proteins, lactose, lipids, minerals, vitamins and trace elements. In order to simplify the modeling analysis, it is usually assumed that the dairy protein powders only comprise three major components, i.e. proteins, lactose and lipids, while the presence of minor components such as minerals and vitamins is neglected (Kim et al., 2009a; Nijdam & Langrish, 2006). Besides, it is generally accepted that there is a linear quantitative relationship between the three typical elements (i.e., C, O and N) and the major dairy components (Fäldt et al., 1993). Therefore, the relative surface coverages of proteins, lactose and lipids can be calculated from the percentages of C, O and N at elemental levels measured by XPS. Using XPS, Gaiani et al. (2007, 2009) reported a three-fold increase of surface lipids of native phosphocaseinate after 60 days of storage at both 20 and 50 °C, while the surface proteins correspondingly decreased from 94% to 83%. It is commonly recognized that the surface composition of powder particles is significantly different from the bulk composition of powder particles. During the spray-drying process and the following storage of dairy protein powders, the lipids and proteins would be over-represented on the surface of powder particles at the expense of lactose (Gaiani et al., 2010; Kim et al., 2002). The lipids would be more preferentially accumulated on the particle surface, thus reducing the wettability of dairy protein powders. Nijdam and Langrish (2006) reported that only a small change of average lipid content between 0% and 5% can cause a large increase of the surface lipid coverage, suggesting that a small change in lipid content of the feeding solution before spray-drying would strongly affect the bulk properties of the resulting dairy protein powders.

EDX (energy dispersive x-ray spectroscopy) is usually coupled with SEM (scanning electron microscopy) to analyze the surface composition of dairy protein powders. EDX works in a surface depth of approximately 1 μm, while XPS only focalizes at about the first 10 nm of the particle surface. The composition gradient from the surface to the core of powder particles can be more clearly characterized by the combination of EDX and XPS that allow variable depth of investigation. By coupling EDX and XPS, Murrieta-Pazos, Gaiani, Galet, and Scher (2012) reported that the proteins and lactose were more present at the first 5 nm within the surface layer of the dairy protein powder particles whereas the lipids were more present just under. Moreover, ToF-SIMS (time-of-flight secondary ion mass spectrometry) can provide information on the amount of amino-acids and lipids at the extreme surface (1–3 nm in depth) of dairy protein powder particles. Therefore, the lipid migration towards the particle surface (e.g., extreme surface) could only be measured by ToF-SIMS rather than XPS in some cases (Nasser et al., 2017).

4.3 Microscopy

A series of microscopy techniques have been successfully employed to investigate the microstructure and component distribution of dairy protein powder particles. Two of the most commonly used microscopy techniques, i.e. SEM and CLSM (confocal laser scanning microscopy), are illustrated in this section.

4.3.1 SEM

SEM has been widely used to obtain the structural information (e.g., size, porosity, smoothness and cross-linking) of dairy protein powder particles. Using SEM, it was extensively reported that there were no visible differences between the fresh and aged MPC powder particles, both of which exhibited a smooth surface without the presence of any detectable sub-structures (Fyfe et al., 2011; Mimouni et al., 2010a; Nasser et al., 2017). However, the structural differences between the fresh and aged MPC powder particles became evident during their rehydrating process. The aged MPC powder particles showed stronger micelle-micelle interactions (i.e., cross-linking and fusion) within a surface crust, which acted as a barrier to inhibit the release of casein micelles during the rehydration process of MPC. Fyfe et al. (2011) reported that some small pores were formed on the surface of MPC80 powder particles stored at 25–40 °C and 44–84% relative humidity for 90 days. Gaiani et al. (2009) also observed small pores on the surface of native phosphocaseinate powder particles stored at 50 °C for 60 days. It was believed that the small pores on the surface of dairy protein powder particles would aid the migration of lipids to the particle surface during storage. Moreover, the distribution of lipids on the surface of dairy protein powder particles can also be characterized by comparing the SEM images of power particles before and after the lipid extraction. It was speculated that the surface lipids could act as lubricants to make the surface of dairy protein powder particles appear smoother (Kim et al., 2002; Murrieta-Pazos, Gaiani, Galet, Calvet, et al., 2012; Murrieta-Pazos, Gaiani, Galet, & Scher, 2012; Thomas, Scher, Desobry-Banon, & Desobry, 2004).

4.3.2 CLSM

CLSM is a noninvasive technique that can characterize both the structure and the component distribution within dairy protein powder particles (Mckenna, 2000b; Singh & Gallier, 2017). Using the Nile blue dye and the fluorescently labelled phospholipid, Mckenna (2000a) identified the location of lipids and phospholipids within dairy protein powder particles, respectively. Both the lipids and phospholipids on the surface of powder particles tended to pool at the joining points of agglomerated powder particles during storage. Using CLSM, Vignolles et al. (2009) reported that the dairy protein powders prepared from homogenized emulsions revealed a homogeneous distribution of very small lipid droplets within the powder particles, while the powders prepared from unhomogenized emulsions showed the presence of very large lipid droplets. The large lipid droplets might result in the formation of free lipids, thus reducing the rehydration behaviours of dairy protein powders (Buma, 1971; Fink & Kessler, 1985). With the use of CLSM, Kosasih, Bhandari, Prakash, Bansal, and Gaiani (2016a, 2016b) reported the migration of free lipids into the vacuoles of dairy protein powder particles during the accelerated storage at 37 °C for 18 weeks.

5 Conclusion

To improve the rehydration behaviours of MPC, it is essential to fully understand the mechanism that contributes to the deterioration of the rehydration behaviours of MPC on a molecular level. The effect of lipids on the rehydration behaviours of MPC might be positive or negative, and the specific molecular mechanism depends on both the lipid types and the lipid addition methods during the manufacturing process. The presence of intrinsic lipids will lead to the deterioration of the wetting behaviours of MPC, while the extrinsic polar lipids can be added to improve the wetting and dissolving behaviours of MPC. The combination of various techniques can provide more detailed information about the distribution and characteristic of lipids within MPC powder particles as well as their changes during processing and storage, thus leading to a better understanding about the acting mechanism of lipids on the rehydration behaviours of MPC.

References

Anema, S. G., Pinder, D. N., Hunter, R. J., & Hemar, Y. (2006). Effects of storage temperature on the solubility of milk protein concentrate (MPC85). *Food Hydrocolloids, 20*, 386–393.

Bansal, N., Truong, T., & Bhandari, B. (2017). Feasibility study of lecithin nanovesicles as spacers to improve the solubility of milk protein concentrate powder during storage. *Dairy Science & Technology, 96*, 861–872.

Bouvier, J. M., Collado, M., Gardiner, D., Scott, M., & Schuck, P. (2013). Physical and rehydration properties of milk protein concentrates: Comparison of spray-dried and extrusion porosified powders. *Dairy Science & Technology, 93*, 387–399.

Buma, T. J. (1971). Free fat in spray-dried whole milk. 10. A final report with a physical model for free fat in spray-dried milk. *Netherlands Milk and Dairy Journal, 25*, 159–174.

Carr, A., Bhaskar, V., & Ram, S. (2004). *Monovalent salt enhances solubility of milk protein concentrate*. United States Patent Application US 2004/0208955 A1.

Chandrapala, J., Martin, G. J. O., Kentish, S. E., & Ashokkumar, M. (2014). Dissolution and reconstitution of casein micelle containing dairy powders by high shear using ultrasonic and physical methods. *Ultrasonics Sonochemistry, 21*, 1658–1665.

Crowley, S. V., Megemont, M., Gazi, I., Kelly, A. L., Huppertz, T., & O'Mahony, J. A. (2014). Heat stability of reconstituted milk protein concentrate powders. *International Dairy Journal, 37*, 104–110.

Depalo, A., & Santomaso, A. C. (2013). Wetting dynamics and contact angles of powders studied through capillary rise experiments. *Colloids and Surfaces A: Physicochemical and Engineering Aspects, 436*, 371–379.

Evers, J. M. (2004). The milkfat globule membrane-methodologies for measuring milkfat globule (membrane) damage. *International Dairy Journal, 14*, 747–760.

Fäldt, P., & Bergenståhl, B. (1994). The surface composition of spray-dried protein-lactose powders. *Colloids and Surfaces A: Physicochemical and Engineering Aspects, 90*, 183–190.

Fäldt, P., & Bergenståhl, B. (1996). Spray-dried whey protein/lactose/soybean oil emulsions. 2. Redispersability, wettability and particle structure. *Food Hydrocolloids, 10*, 431–439.

Fäldt, P., Bergenståhl, B., & Carlsson, G. (1993). The surface coverage of fat on food powders analyzed by ESCA (electron spectroscopy for chemical analysis). *Food Structure, 12*, 225–234.

Fink, A., & Kessler, H. G. (1985). The effect of heating on the storage stability of unhomogenized cream (30%). *Milchwissenschaft, 6*, 326–328.

Fonseca, C. R., Bento, M. S. G., Quintero, E. S. M., Gabas, A. L., & Oliveira, C. A. F. (2015). Physical properties of goat milk powder with soy lecithin added before spray drying. *International Journal of Food Science and Technology, 46*, 608–611.

Freudig, B., Hogekamp, S., & Schubert, H. (1999). Dispersion of powders in liquids in a stirred vessel. *Chemical Engineering and Processing, 38*, 525–532.

Fyfe, K. N., Kravchuk, O., Le, T., Deeth, H. C., Nguyen, A. V., & Bhandari, B. (2011). Storage induced changes to high protein powders: Influence on surface properties and solubility. *Journal of the Science of Food and Agriculture, 91*, 2566–2575.

Gaiani, C., Morand, M., Sanchez, C., Tehrany, E. A., Jacquot, M., Schuck, P., et al. (2010). How surface composition of high milk proteins powders is influenced by spray-drying temperature. *Colloids and Surfaces. B, Biointerfaces, 75*, 377–384.

Gaiani, C., Scher, J., Ehrhardt, J. J., Linder, M., Shuck, P., Desobry, S., et al. (2007). Relationships between dairy powder surface composition and wetting properties during storage: Importance of residual lipids. *Journal of Agricultural and Food Chemistry, 55*, 6561–6567.

Gaiani, C., Schuck, P., Scher, J., Ehrhardt, J. J., Arab-Tehrany, E., Jacquot, M., et al. (2009). Native phosphocaseinate powder during storage: Lipids released onto the surface. *Journal of Food Engineering, 94*, 130–134.

Granelli, K., Fäldt, P., Appelqvist, L. Å., & Bergenståhl, B. (1996). Influence of surface structure on cholesterol oxidation in model food powders. *Journal of the Science of Food and Agriculture, 71*, 75–82.

Guimarães, K. L., & Ré, M. I. (2011). Lipid nanoparticles as carriers for cosmetic ingredients: The first (SLN) and the second generation (NLC). In R. Beck, S. Guterres, & A. Pohlmann (Eds.), *Nanocosmetics and nanomedicines* (pp. 101–122). Heidelberg: Springer.

Güney, G., & Kutlu, H. M. (2011). Importance of solid lipid nanoparticles in cancer therapy. *Cancer Nanotechnology, 3*, 400–403.

Hardas, N., Danviriyakul, S., Foley, J. L., Nawar, W. W., & Chinachoti, P. (2000). Accelerated stability studies of microencapsulated anhydrous milk fat. *LWT - Food Science and Technology, 33*, 506–513.

Hassan, H. M., & Mumford, C. J. (1993). Mechanisms of drying of skin-forming materials. III. droplets of natural products. *Drying Technology, 11*, 1765–1782.

Havea, P. (2006). Protein interactions in milk protein concentrate powders. *International Dairy Journal, 16*,15–422.

Huppertz, T., & Gazi, I. (2015). Milk protein concentrate functionality through optimized product-process interactions. *New Food, 18*, 12–17.

IDF. (1979). *Standard 87: Determination of the dispersiblity and wettability of instant dried milk*. Brussels: International Dairy Federation.

Jayasundera, M., Adhikari, B., Aldred, P., & Ghandi, A. (2009). Surface modification of spray dried food and emulsion powders with surface-active proteins: A review. *Journal of Food Engineering, 93*, 266–277.

Ji, J., Cronin, K., Fitzpatrick, J., & Miao, S. (2017). Enhanced wetting behaviours of whey protein isolate powder: The different effects of lecithin addition by fluidised bed agglomeration and coating processes. *Food Hydrocolloids, 71*, 94–101.

Jimenez-Flores, R., & Kosikowski, F. V. (1986). Properties of ultrafiltered skim milk retentate powders. *Journal of Dairy Science, 69*, 329–339.

Kentish, S., Davidson, M., Hassan, H., & Bloore, C. (2005). Milk skin formation during drying. *Chemical Engineering Science, 60*, 635–646.

Kim, E. H. J., Chen, X. D., & Pearce, D. (2002). Surface characterization of four industrial spray-dried dairy powders in relation to chemical composition, structure and wetting property. *Colloids and Surfaces. B, Biointerfaces, 26*, 197–212.

Kim, E. H. J., Chen, X. D., & Pearce, D. (2005). Melting characteristics of fat present on the surface of industrial spray-dried dairy powders. *Colloids and Surfaces. B, Biointerfaces, 42*, 1–8.

Kim, E. H. J., Chen, X. D., & Pearce, D. (2009a). Surface composition of industrial spray-dried milk powders. 1. Development of surface composition during manufacture. *Journal of Food Engineering, 94*, 163–168.

Kim, E. H. J., Chen, X. D., & Pearce, D. (2009b). Surface composition of industrial spray-dried milk powders. 2. Effects of spray drying conditions on the surface composition. *Journal of Food Engineering, 94*, 169–181.

Kim, E. H. J., Chen, X. D., & Pearce, D. (2009c). Surface composition of industrial spray-dried milk powders. 3. Changes in the surface composition during long-term storage. *Journal of Food Engineering, 94*, 182–191.

Kosasih, L., Bhandari, B., Prakash, S., Bansal, N., & Gaiani, C. (2016a). Effect of whole milk concentrate carbonation on functional, physicochemical and structural properties of the resultant spray dried powder during storage. *Journal of Food Engineering, 179*, 68–77.

Kosasih, L., Bhandari, B., Prakash, S., Bansal, N., & Gaiani, C. (2016b). Physical and functional properties of whole milk powders prepared from concentrate partially acidified with CO_2 at two temperatures. *Journal of Food Engineering, 56*, 4–12.

Lallbeeharry, P., Tian, Y., Fu, N., Wu, W. D., Woo, W. M., Selomulya, C., et al. (2014). Effects of ionic and nonionic surfactants on milk shell wettability during co-spray-drying of whole milk particles. *Journal of Dairy Science, 97*, 5303–5314.

Le, T. T., Bhandari, B., Holland, J. W., & Deeth, H. C. (2011). Maillard reaction and protein cross-linking in relation to the solubility of milk powders. *Journal of Agricultural and Food Chemistry, 59*, 12473–12479.

Lin, S. Y., Mckeigue, K., & Maldarelli, C. (1990). Diffusion-controlled surfactant adsorption studied by pendant drop digitization. *AIChE Journal, 36*, 1785–1795.

Liu, G. Y., Wang, J. M., & Xia, Q. (2012). Application of nanostructured lipid carrier in food for the improved bioavailability. *European Food Research and Technology, 234*, 391–398.

Mckenna, A. B. (2000a). *Effect of processing and storage on the reconstitution properties of whole milk and ultrafiltered skim milk powders*. PhD thesis, Massey University, New Zealand.

Mckenna, A. B. (2000b). Examination of whole milk powder by confocal laser scanning microscopy. *The Journal of Dairy Research, 64*, 423–432.

Meerdink, G., & Riet, K. V. (1995). Modeling segregation of solute material during drying of liquid foods. *AIChE Journal, 41*, 732–736.

Miller, R., Alahverdjieva, V. S., & Fainerman, V. B. (2008). Thermodynamics and rheology of mixed protein-surfactant adsorption layers. *Soft Matter, 4*, 1141–1146.

Mimouni, A., Deeth, H. C., Whittaker, A. K., Gidley, M. J., & Bhandari, B. R. (2010a). Investigation of the microstructure of milk protein concentrate powders during rehydration: Alterations during storage. *Journal of Dairy Science, 93*, 463–472.

Mimouni, A., Deeth, H. C., Whittaker, A. K., Gidley, M. J., & Bhandari, B. R. (2010b). Rehydration of high-protein-containing dairy powder: Slow- and fast-dissolving components and storage effects. *Dairy Science & Technology, 90*, 335–344.

Mistry, V. V., & Hassan, H. N. (1991). Delactosed high milk protein powder. 2. Physical and functional properties. *Journal of Dairy Science, 74*, 3716–3723.

Mistry, V. V., & Pulgar, J. B. (1996). Physical and storage properties of high milk protein powder. *International Dairy Journal, 6*, 195–203.

Murrieta-Pazos, I., Gaiani, C., Galet, L., Calvet, R., Cuq, B., & Scher, J. (2012). Food powders: Surface and form characterization revisited. *Journal of Food Engineering, 112*, 1–21.

Murrieta-Pazos, I., Gaiani, C., Galet, L., & Scher, J. (2012). Composition gradient from surface to core in dairy powders: Agglomeration effect. *Food Hydrocolloids, 26*, 149–158.

Nasser, S., Jeantet, R., De-Sa-Peixoto, P., Ronse, G., Nuns, N., Pourpoint, F., et al. (2017). Microstructure evolution of micellar casein powder upon ageing: Consequences on rehydration dynamics. *Journal of Food Engineering, 206*, 57–66.

Nijdam, J. J., & Langrish, T. A. G. (2006). The effect of surface composition on the functional properties of milk powders. *Journal of Food Engineering, 77*, 919–925.

Rombaut, R., Camp, J. V., & Dewettinck, K. (2005). Analysis of phospho- and sphingolipids in dairy products by a new HPLC method. *Journal of Dairy Science, 88*, 482–488.

Schokker, E. P., Church, J. S., Mata, J. P., Gilbert, E. P., Puvanenthiran, A., & Udabage, P. (2011). Reconstitution properties of micellar casein powder: Effects of composition and storage. *International Dairy Journal, 21*, 877–886.

Selomulya, C. (2007). On measurement of food powder reconstitution properties. *Drying Technology, 26*, 3–14.

Singh, H. (1991). Modification of food proteins by covalent crosslinking. *Trends in Food Science and Technology, 2*, 196–200.

Singh, H. (2011). Milk protein products functional properties of milk proteins. In J. W. Fuquay (Ed.), *Encyclopedia of dairy sciences* (pp. 887–893). San Diego: Academic Press.

Singh, H., & Creamer, L. K. (1991). Influence of concentration of milk solids on the dissociation of micellar kappa-casein on heating reconstituted milk at 120 degrees C. *The Journal of Dairy Research, 58*, 99–105.

Singh, H., & Fox, P. F. (1987). Heat stability of milk: Role of β-lactoglobulin in the pH-dependent dissociation of micellar k-casein. *The Journal of Dairy Research, 54*, 509–521.

Singh, H., & Gallier, S. (2017). Nature's complex emulsion: The fat globules of milk. *Food Hydrocolloids, 68*, 81–89.

Thomas, M. E., Scher, J., Desobry-Banon, S., & Desobry, S. (2004). Milk powders ageing: Effect on physical and functional properties. *Critical Reviews in Food Science and Nutrition, 44*, 297–322.

Thompson, A. K., & Singh, H. (2006). Preparation of liposomes from milk fat globule membrane phospholipids using a microfluidizer. *Journal of Dairy Science, 89*, 410–4199.

Thomsen, M. K., Lauridsen, L., Skibsted, L. H., & Risbo, J. (2005a). Temperature effect on lactose crystallization, maillard reactions, and lipid oxidation in whole milk powder. *Journal of Agricultural and Food Chemistry, 53*, 7082–7090.

Thomsen, M. K., Lauridsen, L., Skibsted, L. H., & Risbo, J. (2005b). Two types of radicals in whole milk powder. Effect of lactose crystallization, lipid oxidation, and browning reactions. *Journal of Agricultural and Food Chemistry, 53*, 1805–1811.

Van, H. P., & Wendel, A. (2014). The use of natural and synthetic phospholipids as pharmaceutical excipients. *European Journal of Lipid Science and Technology, 116*, 1088–1107.

Vega, C., & Roos, Y. H. (2006). Invited review: Spray-dried dairy and dairy-like emulsions-compositional considerations. *Journal of Dairy Science, 89*, 383–401.

Vignolles, M. L., Jeantet, R., Lopez, C., & Schuck, P. (2007). Free fat, surface fat and dairy powders: Interactions between process and product. A review. *Le Lait, 87*, 187–236.

Vignolles, M. L., Lopez, C., Ehrhardt, J. J., Lambert, J., Méjean, S., Jeantet, R., et al. (2009). Methods' combination to investigate the suprastructure, composition and properties of fat in fat-filled dairy powders. *Journal of Food Engineering, 94*, 154–162.

Vignolles, M. L., Lopez, C., Madec, M. N., Ehrhardt, J. J., Méjean, S., Schuck, P., et al. (2009). Fat properties during homogenization, spray-drying, and storage affect the physical properties of dairy powders. *Journal of Dairy Science, 92*, 58–70.

Yazdanpanah, N., & Langrish, T. A. G. (2012). Releasing fat in whole milk powder during fluidized bed drying. *Drying Technology, 30*, 1081–1087.

Young, S. L., Sarda, X., & Rosenberg, M. (1993). Microencapsulating properties of whey proteins. 1. Microencapsulation of anhydrous milk fat. *Journal of Dairy Science, 76*, 2868–2877.

Zhang, H. C. (2015). *Effect of lipid on soluble characteristic of milk protein concentrat*e. Master thesis, Jiangnan University, China.

Chapter 21
Butter and Dairy Fat Spreads

Bhavesh Panchal and Bhesh Bhandari

1 Introduction

Oils and fats are important ingredients in a wide variety of manufactured foods, and constitute a significant part of food recipes (Sato & Ueno, 2014). Spreadable fats are one of these which are 'plastic' in nature, able to be spread into a thin layer on food articles such as bread slice. The spreadable products include butter, margarine, and other fat blends and spreads which are usually based on milk fat, other fats (vegetable, animal or marine origin) or blend of milk fat and other fats. Among these, butter is known to be produced since ancient times as a very old way of preserving milk fat and has been known to be used in cooking and even for medical and cosmetic purposes (Early, 1998). Butter remained relatively expensive commodity among fat-based products even after the introduction of commercial production (Varnam & Sutherland, 1994). As a result, 'margarine' was invented as an inexpensive substitute of butter in 1869 by French chemist 'Hippolyte Mege-Mouries' (Bumbalough, 2000; Freeman & Melnikov, 2005; Ghotra, Dyal, & Narine, 2002; Lane, 1998; Varnam & Sutherland, 1994). Margarine was widely accepted as a spread, however, is not recognized considerably similar to butter in terms of either taste, texture or mouth-feel due to its vegetable fat source (Ahmed & Luksas, 1988). Moreover, high-fat content in both margarine and butter was a major health concern for dietary conscious persons irrespective of the kind of fat it contained. As a consequence, a wide range of low fat spreads were introduced (Varnam & Sutherland, 1994).

Few of the early commercially developed spreadable fat products include, 'Bregott'—a fat blend (80 g milk fat and 20 g soybean oil 100 g^{-1} fat) containing 80% fat launched in 1969 (Keogh, 2006; Lane, 1998; Mortensen, 2011; Wilbey, 1994), 'Latt & Lagom'—a low fat spread (60 g milk fat and 40 g soybean oil 100 g^{-1}

B. Panchal · B. Bhandari (✉)
ARC Dairy Innovation Hub, School of Agriculture and Food Sciences,
The University of Queensland, St Lucia, QLD, Australia
e-mail: b.bhandari@uq.edu.au

T. Truong et al. (eds.), *Dairy Fat Products and Functionality*,
https://doi.org/10.1007/978-3-030-41661-4_21

fat) containing 40% fat, launched in 1975 (Keogh, 2006; Lane, 1998; Mortensen, 2011; Wilbey, 1994), 'Clover'—a fat blend similar to 'Bregott' containing 50 g milk fat, 50 g blend of soybean oil and partially hydrogenated soybean oil 100 g^{-1} fat which was launched in 1983 (Keogh, 2006; Lane, 1998; Wilbey, 1994), 'Danelite'—a reduced-fat dairy spread containing 55 g milk fat and 4 g protein 100 g^{-1} was launched in 1986 to improve the potential difficulties of spreadability (Wilbey, 1994), 'St Evel Gold Lowest' and 'Delight Extra Low'—very low fat spreads containing 25% and 20% fat respectively, based on patent (Platt and Gupta, 1989) of spreads with 18–35% fat (Lane, 1998; Wilbey, 1994).

In recent years, the enrichment of polyunsaturated fatty acids, plant sterols, essential fatty acids, minerals and vitamins in product composition and application of emulsifier mixes and water structuring substances has provided wide range of flexibility in fat content, texture, functionality, health benefits and applications of these products (Freeman & Melnikov, 2005; Henning, Baer, Hassan, & Dave, 2006). Butter and other milk fat based products are valuable products for the dairy industry due to their unique taste, textural characteristics, and nutritional value. Though butter has been the target of countless negative nutritional attacks centered on high-fat content, cholesterol, high saturated fat, and calories, per capita consumption of butter in the United States has remained at 2 kg which is similar to butter consumption in early 1980s (Henning et al., 2006). In Australia as well, annual per capita consumption of butter has increased to around 4.7 kg in 2017/18 from 4.0 kg in 2013/14, which indicates consumers continue to remain interested in the 'naturalness' of butter, together with its superior taste and cooking functionality (Dairy Australia, 2018). As per sensory evaluation conducted by Krause, Lopetcharat, and Drake (2007), the key discriminating characteristics found that drive consumers liking of butter over other spreadable fats were a desirable flavor and a natural image. In addition to taste and flavor, the texture of milk fat, as well as butter, are also important considerations in quality evaluations (Wright, Scanlon, Hartel, & Marangoni, 2001). Butter remains a solid at colder temperatures but softens to a spreadable consistency after equilibration to room temperature. Most of the recent efforts have been targeted to improve the consistency of butter in a way to make it more spreadable at low temperature without losing its ability to stand up room temperature (Rønholt, Mortensen, & Knudsen, 2013). Different processing methods and compositions have been used to develop cold-spreadable butter in the past (Wright et al., 2001). However, the key to develop such functional product is to fully understand the structure and physical properties of milk fat along with impact factors. The crystallization behavior of milk fat and rheology of butter are presented in Chaps. 9 and 10, respectively. Hence, this chapter primarily focuses on relative potent manufacturing factors addressing functionality and rheology of the milk fat-based products.

2 Milk Fat-Based Products

2.1 Butter

Butter is a fatty product derived exclusively from milk and/or products obtained from milk, principally in the form of an emulsion of the type water-in-oil, which must contain minimum of 80% (m/m) fat and maximum of 16% (m/m) water and 2% (m/m) non-fat milk solids (CODEX standards 279-1971). Cream is generally used as starting material for the manufacturing of butter which may be fresh (pH around 6.6) or ripened (fermented, pH around 4.6) yielding 'sweet-cream' and 'ripened cream' butter (Fox & McSweeney, 1998). Manufacturing process essentially involves mechanical phase inversion and working of butter grains. It has been known since prehistoric time that if cream is agitated, fat aggregates to form granules and phase inversion take place i.e. conversion of oil in water emulsion (cream) to a water in oil emulsion (butter) as illustrated in Fig. 21.1. The fat content of cream can vary from 30 to 55% for batch method and 38–42% for continuous system (Keogh, 2006). Both traditional batch method and continuous method for the manufacturing of butter involve following major steps:

1. Concentration of the fat phase of milk by gravimetric or mechanical separation
2. Partial crystallization of fat
3. Churning, phase inversion
4. Removal of buttermilk
5. Formation of plasticized water in oil emulsion by working

In batch method, churning is carried out through rotation of batch churn which is a stainless steel barrel of either cylindrical or single/double cone-shaped. While in continuous butter making machine working on Fritz principle, multi-bladed dashers

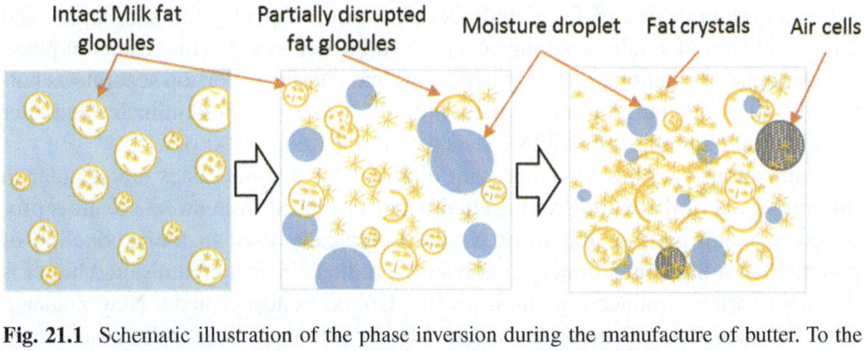

Fig. 21.1 Schematic illustration of the phase inversion during the manufacture of butter. To the left is cream, an oil-in-water emulsion, with the continuous water phase being blue and the milk fat globules dispersed within the water phase. Churning of cream facilitates a phase inversion to a water-in-oil emulsion (middle image). After buttermilk drainage-washing-working moisture is controlled, and upon storage, more dense fat crystal network formed (right image). Butter illustrated by a white continuous fat phase with blue water droplets and milk fat globules dispersed within together with fractions of ruptured milk fat globule membrane

rotating at around 1000 rpm in a churning section destabilizes the cream. In batch churning method, phase inversion of cream happens at relatively slow rate (30 min to 1 h) with large mass, whereas in continuous method, continuously passing small quantity of cream gets destabilized in 1–2 s (Mortensen & Denmark, 2011; Wilbey, 1994). The pre-treated cream is transferred to batch churn or churning section of continuous butter making machine at desired churning temperature. Churning takes place in controlled temperature conditions in both methods. After phase inversion, buttermilk is separated, and butter grains are washed to remove residual non-fat solids. The make-up water, salt and/or colour are then added into the butter according to end product use, and working is carried out at a specified temperature. The homogenous butter is then packed and stored under refrigeration.

As per mechanism of churning, partial crystallization of fat is essential to trigger partial coalescence of fat globules (Walstra, Walstra, Wouters, & Geurts, 2005). Partial coalescence is the process where fat crystal from one partially crystalline droplet penetrates a liquid portion of another partially crystalline droplet resulting into formation of irregularly shaped aggregates sharing part of the membrane. This happens during churning when air is beaten into the cream, and numerous small air bubbles form driving collision of partially crystalline fat globules and their adsorption on air bubbles (McClements, 2004). The fat globules often spread part of their membrane substances and liquid fat over the air-water interface and thereby several globules remain adhered to the bubble, this phenomenon resembles 'floatation'. However, the air bubbles keep colliding with each other during their continuous movement in liquid and thereby tend to grow in size. This exerts a shearing effect on the fat globules trapped between lamellae of the bubbles leading extensive damage to the fat globule membranes. In a way surface area of bubbles diminishes largely driving adhered fat globules towards each other and forming small fat clumps (Walstra et al., 2005).

In our laboratory, an observation of continuously increasing fat globule size during progress of churning (10 °C) before phase inversion is presented in Fig. 21.2. When a certain degree of fat globule destabilization happens, foam collapses and direct collision of small fat clumps occurs to a greater extent. This results in phase inversion, i.e. formation of large butter grains leaving serum portion separate as buttermilk. Thus, in churning mechanism, initially 'floatation' predominates and later 'mechanical clumping' (Fox & McSweeney, 1998; Keogh, 2006).

Mainly four varieties of butter with different flavors and tastes are popular in different parts of the world. An unsalted butter produced from sweet cream is produced in many parts of the world which is typically used in the production of recombined milk, confectionery, and bakery products. Sweet cream salted butter is the main variety produced in countries like United States, Canada, New Zealand, Australia, and Great Britain. An unsalted butter produced from cultured cream is common in Netherland, Germany, and France. This butter has a low pH between 4.6 to 5.2, and the cultured salted variety is mainly produced in Scandinavia. Salt gives distinct taste to butter and acts as a preservative. In cultured varieties, low pH also provides better spreadability, microbial stability and distinct flavor (Mortensen & Denmark, 2011).

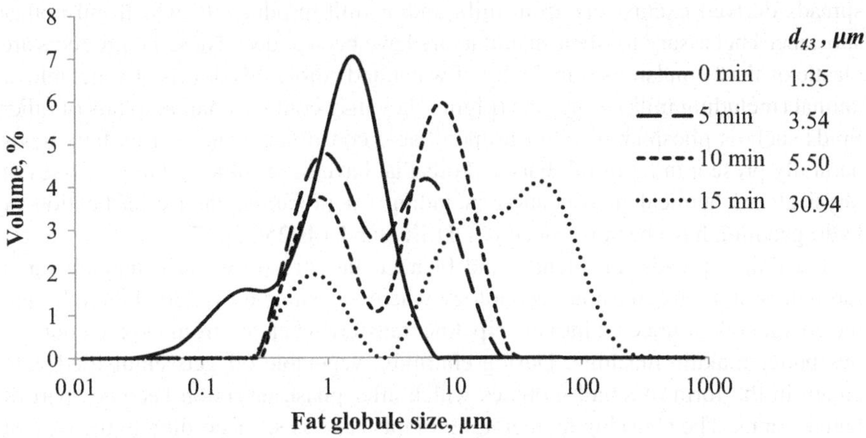

Fig. 21.2 Fat globule size (d_{43}—volume weighted mean diameter) of cream (38% fat) at different time interval during churning (10 °C) before phase inversion in batch type laboratory scale set up

Table 21.1 Essential fat composition of butter (CODEX STAN 279-1971) fat spreads and blended spreads (CODEX STAN 256-2007)

Product category	% total fat (m/m)	% milk fat of total fat (m/m)
Butter	Not less than 80	Not less than 80
Margarine	Not less than 80	Not more than 3
Fat spread	Less than 80	Not more than 3
Fat blend	Not less than 80	Not less than 3
Fat blend spread	Less than 80	Not less than 3

In fat blend and fat blend spread, higher minimum content of milk fat may be specified in accordance with the requirement of the country of the retail sale

2.2 Other Fat Blends and Spreads

According to Codex standards (253-2006), 'Dairy fat spread' is milk product relatively rich in fat in the form of a spreadable emulsion principally of the water in oil type that remains in solid state at a temperature of 20 °C. The milk fat content must be minimum 10 g 100 g^{-1} and maximum 80 g 100 g^{-1} and must represent at least two-thirds of the dry matter (Table 21.1). The raw materials should be milk and/or products obtained from milk including milk fat, may have been subjected to any appropriate processing (e.g., physical modification including fractionation) prior to their use. The permitted food additives that may be used includes colors, emulsifiers, preservatives, stabilizers or thickeners, acidity regulators, antioxidants, anti-foaming agents and flavor enhancers.

'Fat spreads' and 'blended spreads' have been defined as fat products containing not less than 10% and not more than 90% fat, intended primarily for use as spreads which includes margarine and products used for similar purposes and excludes products with a fat content of less than 2/3 of the dry matter. This also excludes fat spreads derived exclusively from milk and/or milk products to which only other substances necessary for their manufacture have been added. These fat products are plastic or fluid emulsions principally of water and edible fats or oils of vegetable or animal (including milk) or marine origin. They may contain small amounts of other lipids such as phosphatides, of unsaponifiable constituents and of free fatty acids naturally present in fat or oil. Fats and oils that have been subjected to processes of physical or chemical modification including fractionation, inter-esterification or hydrogenation have been included (CODEX standards 256-2007).

The dairy spreads, fat blends, and blended fat spreads are also manufactured through continuous churning technology where vegetable oil is added directly into the cream tank or injected into the pipeline transferring cream from tank to continuous butter making machine. During churning, vegetable oil gets emulsified with cream in the form of small globules which after phase inversion becomes part of butter grains. The churning temperature, in this case, is set according to the type of vegetable oil used, to avoid crystallization of oil. After separation of buttermilk and washing of grains, working is carried out at low shear rate and temperature to prevent overworking and greasiness. The emulsion thus formed imparts more softness. The buttermilk produced in this method contains residues of oil which makes it unapplicable in other dairy products (Mortensen, 2011).

2.3 Characteristics of Milk Fat-Based Products

Fat rich products of water in oil type are plastic emulsions that remain solid at room temperature. In stable emulsions, continuous phase of liquid fat/oil encases fat crystals, water droplets, air, salt, and other additives (Fox & McSweeney, 1998; Keogh, 2006). Proportion of both liquid and solid fat phase is of outmost important to attain the state of plasticity and desired consistency (Keogh, 2006; Narine & Marangoni, 1999; Varnam & Sutherland, 1994). In churned product like butter, there are usually two fat phases, one continuous fat phase squeezed out from denuded globules during churning and working, and other globular fat phase consisting of intact fat globules (Heertje, Leunis, Van Zeyl, & Berends, 1987; Mortensen & Denmark, 2011) as could be seen in microscopic image (Fig. 21.3). Approximate proportion and dimension of such microstructural components of conventional butter are presented in Table 21.2.

Solid fat exists as three-dimensional network of aggregated crystals which influences firmness, spreadability and mouthfeel of the butter (Juriaanse & Heertje, 1988; Rønholt, Kirkensgaard, Pedersen, Mortensen, & Knudsen, 2012). The number and size of these fat crystals depends on the temperature and temperature history. The fat crystals inside the globules do not participate in this network and hardly contribute to firmness. Because of this, butter contains more solid fat than margarine at equal firmness and so as feels cooler in the mouth. The presence of intact fat globules in butter is believed to influence mouthfeel and spreadability by interrupting crystal network and acting like microscopic ball bearings (Mulder & Walstra,

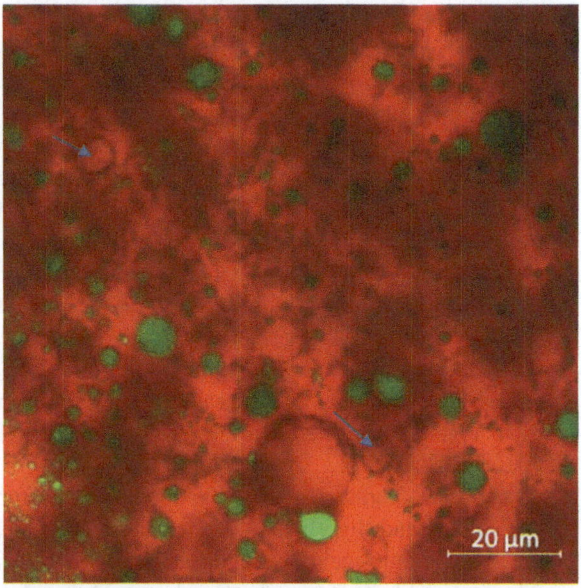

Fig. 21.3 Microstructure of butter analysed through confocal laser scanning microscopy (CLSM). The continuous liquid fat phase is in red and dispersed water droplets are in green color. Few of more or less intact fat globules are indicated by blue arrows. The black-grey shadows represent solid fat crystals network. (Produced in our laboratory-unpublished work)

Table 21.2 Structural elements of conventional butter (Walstra et al., 2005)

Structural element	Approx. number concentration (mL^{-1})	Proportion in butter (% v/v)	Dimension (μm)
Fat globule[a]	10^{10}	5–30[b]	1–5
Fat crystal[c]	10^{13}	10–40[d]	0.01–2
Moisture droplet	10^{10}	15	1–25[b]
Air cell	10^{6}	~2	>20

[a]With complete membrane
[b]Closely depends on intensity (both shear rate and time) of working
[c]At higher temperatures mainly inside the fat globules, at low temperatures forming solid networks
[d]Closely depends on the temperature

1974; Varnam & Sutherland, 1994; Walstra et al., 2005). Because of this, the absence of fat globules makes butter significantly more brittle compared to that containing fat globules (Rønholt et al., 2012).

In fat blends and spreads, addition of low melting oil or milk fat fraction increases the proportion of continuous fat phase and diminishes the globular fat proportionally (Mortensen, 2011; Varnam & Sutherland, 1994). Due to changes in washing, water addition and working in of salt, composition of moisture droplets in butter is not always uniform. Differences in osmotic pressure in proximity of salt particles leads to transport of water droplets forming larger droplet (Walstra et al., 2005). In some case, this causes physical defect called ‘leaky texture’. Size of moisture droplets is important with respect to release of flavor in mouth during melting and stability against microbial risks during storage (Heertje, 1993; Mortensen, 2011). The diameter of droplets in the range of 1–5 μm is desirable for good microbiological quality, and few larger droplets of 10–20 μm size are also beneficial in releasing typical butter flavor in mouth (Mortensen & Denmark, 2011).

In butter, margarine and full-fat blends, continuous network of solid and liquid fat (80% m/m) stabilizes dispersed aqueous phase (20% m/m), but in low fat spreads and blends, due to high water content tendency of droplet coalescence becomes high which sometimes results in emulsion destabilization (Rousseau, Ghosh, & Park, 2009). Therefore in spreads containing very low-fat content, addition of suitable stabilizers and emulsifiers is critical for even distribution and stabilization of the aqueous phase (Mortensen, 2011). In butter and margarine, these aqueous phase droplets are observed to be stabilized by surface active fat crystals present at water-oil interface. These crystals also referred as 'Pickering crystals' which form steric barrier between adjacent water droplets to hinder film drainage, droplet collision, and coalescence (Ghosh & Rousseau, 2011), whereas in low fat spreads interfacial layer of crystalline fat and/or surfactant is in common (Buchheim & Dejmek, 1997; Rousseau et al., 2009).

As per ideal characteristics, butter should have natural golden or creamy yellow colour, smooth and slightly matt surface (absence of greasiness, oiling off and leakiness), firm enough against sagging under its own weight, sufficiently soft to be readily spreadable (not too short and crumbly) and easily deformable in mouth without greasiness (Walstra et al., 2005). The characteristic flavor of sweet cream butter is imparted through short-chain fatty acids of milk fat, lactones, ketones, and phospholipids. In cultured cream butter, diacetyl, lactic acid, acetic acid, and acetic aldehyde also contribute significantly (Mortensen & Denmark, 2011). Margarines appear shinier and greasy than butter because of differences in the form of crystalline fat and size of moisture droplets (Lane, 1998).

2.4 Factors Affecting Microstructure and Rheology

Structural behavior is not determined by one single parameter, but by the interactions between many. Parameters such as chemical composition of milk fat, thermal treatment of cream prior to churning, and water content influence not only solid-liquid fat proportion but also crystal polymorphism, and number and size of fat crystals which consequently affect crystal-crystal interactions and rheology of product. Thus, to achieve desired product output, it is essential to evaluate how the parameters individually and in combination, contribute to the textural properties of the final product. Methods to modify the rheological behaviour of milk fat-based products by modifying chemical composition through either cattle-diet alteration, chemical interesterification, blending with specific milk fat fractions, or vegetable oils have also been an area of interest for researchers during the last 50 years (Wright et al., 2001; Wright & Marangoni, 2006). Table 21.3 lists most common methods of manipulating butter's textural and rheological properties specifically improving cold spreadability while maintaining stability at room temperatures.

Table 21.3 Different approaches to modify textural and rheological properties of butter (Wright et al., 2001)

Alteration of butter composition	Treatments applied during manufacture of butter	Treatments applied after manufacture of butter
Alteration to cow feed • Oil seed supplementation • Calcium salts of unsaturated fatty acids • Encapsulated unsaturated oils	Method of manufacture • Conventional batch method • Continuous process	Storage conditions Time and temperature
Blending with unsaturates	Cream agitation during cooling	Mechanical working
Milk fat fractionation	Rate of cooling	
Chemical and/or enzymatic milk fat interesterification	Recycling Addition of crystal seeds	
Changes in other components Water content, air content, addition of surfactants	Ripening of cream	

2.4.1 Composition

Dietary Alteration One factor is the chemistry of milk fat as defined by nature. Differences in fat composition, depending on the stage of lactation, can affect milk fat's consistency (Wright et al., 2001). Field studies have shown that seasonal variation in the feeding of cows influence the fat-water ratio in butter, ranging from 86:14 (fat:water) in winter time to 76:24 during summer (Rønholt et al., 2013), and also crystallization behaviour due to different triglyceride compositions (Shi, Smith, & Hartel, 2001). Also, winter butter is notoriously hard and brittle because of seasonal variation in cow diet leading fats with higher melting points and lower iodine numbers (Wright et al., 2001). Moreover, it is possible to relate different feeding regime to the fatty acid composition and melting behavior of milk fat but may not be possible to extract clear correlation due to individual cow variation (Buldo, Larsen, & Wiking, 2013). Dolby (1949) reported that 80% of the variations in softness of New Zealand factory butter was due to changes in butterfat composition. However, seasonal variation in milk quality can be avoided by using a standardized feeding regime year-round (Buldo et al., 2013; Couvreur, Hurtaud, Lopez, Delaby, & Peyraud, 2006; Shi et al., 2001).

Manipulation of the fatty acid profile has a potential to improve the nutritional properties and physical functionality of milk fat, and its acceptability in the market (Henning et al., 2006; Hillbrick & Augustin, 2002). From nutritional point of view, the aims of modifications are (a) reducing the ratio of saturated to unsaturated fatty acids; and (b) increasing the content of n-3 series fatty acids, including α-linolenic acid (ALA), which is recognized as minimizing the risk of cardiovascular disease, and found to improve functional development of the central nervous system (Hurtaud, Faucon, Couvreur, & Peyraud, 2010), and conjugated linoleic acid (CLA), which has been reported to have anticancer, antiatherogenic, antidiabetic, and antiobesity effects for human health (Williams, 2000). From a perspective of physical functionality, the outcome targeted has been an improvement in the spreadability of butter by reducing the hardness of milk fat. Both on-farm strategies and the application of appropriate post-farm processing technologies may be used to alter the milk fat composition to enhance its nutritional image and its physical functionality for a range of product application.

It is now well established that supplementation of cow diet with unsaturated fatty acids affects milk fatty acid profile (Hurtaud et al., 2010). Banks and Christie (1990) have reviewed feeding strategies to alter fatty acid composition to achieve butter with desirable cold-spreadability. Achieving more spreadable butter requires an increase in the low-melting triglycerides in milk fat that may be achieved by increasing the relative amount of short-chain fatty acids or by increasing the level of unsaturation, which in turn reduces the solid fat content of the butter at low temperature.

Many sources of unprotected fats such as cottonseeds, soybeans, and sunflower seeds have been used in dairy cattle rations. When cows were fed oil-seeds higher in long-chain unsaturated fatty acids, the milk fat, and therefore butter manufactured from this milk, contained increased amounts of stearic (C18:0), oleic (C18:1) and linoleic (C18:2) acid, whereas the amounts of short- and medium-chain fatty acids (C8:0 to C16:0) and solid fat content decreased (Henning et al., 2006; Murphy, Connolly, & McNeill, 1995; Murphy, McNeill, Connolly, & Gleeson, 1990); this would make butter made from it more spreadable at low temperature. Lipids from canola seeds have potentially interesting characteristics to achieve these modifications as they contain much higher cis-C18:1 to C16:0 ratio than other common oilseeds (Bayourthe, Enjalbert, & Moncoulon, 2000). Butter produced from cows fed with extruded blend of canola meal and canola seeds (2% of dietary dry matter) contained lower proportion of solid fat content (decreased by 4.7–7.5%) between 0 and 12 °C with no major difference at temperature 18 °C and above compared to control diet (based on corn silage) butter; which would induce improved spreadability at refrigeration temperature and sustain structural integrity at room temperature without oiling off (Bayourthe et al., 2000). Inclusion of fish oil (2% on a dry matter basis) in cow diet increased concentration of conjugated linoleic acid (0.70–2.58%), trans vaccenic acid (1.52–6.44%), total unsaturated fatty acids (31.79–42.95%), and decreased amount of total saturated fatty acids (68.22–56.66%) in butter compared to control diet (50:50 ratio of forage to concentrate) butter. These changes in fatty acid composition resulted in softer butter at 4 and 20 °C (as per penetrometer readings) in comparison to control diet butters without any significant differences in flavor characteristics (Baer et al., 2001). The butter produced from cows fed high oleic and regular sunflower seeds was softer, more unsaturated, and had acceptable flavor (Henning et al., 2006). Similarly, when calcium salts of unsaturated fatty acids were added to dairy cow diets, the butter produced from the milk had improved thermal properties and was more spreadable from the refrigerator (Hillbrick & Augustin, 2002; Lin, Sims, Staples, & O'Keefe, 1996). Several types of encapsulated (protected) fat have also been investigated including tallow, coconut oil, sunflower oil, and canola oil which allows fatty acids to be adsorbed directly by the small intestine escaping hydrogenation in the cow's rumen (Scott et al., 1970). Butter produced from cows fed a canola-based protected lipid had no off-flavors and was found to have increased spreadability compared with conventional butter (Cadden, Urquhart, & Jelen, 1984). Farmers in Ireland are producing milk through cow diet modification for manufacture of naturally spreadable butter (Fearon, 2001).

Chemical Modification Apart from altering milk fat composition by dietary means, fractionation and interesterification of fat has also broadened scope to tailor ingredients to meet specific applications. Fractionation aid to obtain fractions (groups of triglycerides) of lipids with characteristic chemical and physical properties and altered rheological behaviour (Wright & Marangoni, 2006). The spreadability of butter can be improved by fractionating milk fat and then recombining the fractions in various proportions (Wright et al., 2001). Very high melting fractions provide key structural integrity while low melting fractions serve to induce softness in recombined spreadable butters (Kaylegian & Lindsay, 1992; Wright & Marangoni, 2006). Enrichment of high melting triglycerides in butter provide higher solid fat content and viscosity, improved structural stability without oiling off, and reduced tendency of moisture migration at higher temperature (34 °C) (Shukla, Bhaskar, Rizvi, & Mulvaney, 1994). Incorporation of low-melting-point fraction to cream during the normal churning process improved the cold-spreadability of butter (Henning et al., 2006; Hillbrick & Augustin, 2002; Schäffer, Szakály, Lorinczy, & Belágyi, 2000). The addition of low melting milk fat fraction (LMP-20, melting point 20 °C) to cream before churning, decreased firmness of butter (i.e. improved spreadability) at a low (below 10 °C) temperature, proportionally to the increase of its amount. However, at room temperature (about 20 °C), the values of consistency firmness are almost the same (Schäffer et al., 2000).

Similarly, butter samples made from low melting liquid fractions or a combination of primarily low melting liquid fractions and a small amount of high melting solid fractions exhibited good spreadability at refrigerator temperature (4 °C). However, stability at room temperature was compromised, samples were almost melted at 21 °C (Kaylegian & Lindsay, 1992). Although this post-farm strategy appears to be an effective alternative to modify butter functionality (Shukla et al., 1994), the wholesome image and natural delicate flavor of butter is compromised (Hillbrick & Augustin, 2002). Additionally, as a result of the high cost associated with fractionation process, butter made from milk fat fractions are comparatively rare in the market (Wright et al., 2001). Using enzymatic or chemical interesterification, the solid fat content, polymorphism, and elastic behavior of spreadable fat products can be modified depending on the degree of interesterification (Zárubová, Filip, Kšandová, Šmidrkal, & Piska, 2010; Zhang, Smith, & Adler-Nissen, 2004). Use of 1,3-specific lipases for interesterification could improve the spreadability of modified butterfat compared to untreated butterfat by decreasing the softening temperature (by 2–5 °C) (Marangoni & Rousseau, 1998). However, the loss of buttery flavor due to the necessary refining procedures following interesterification is a serious drawback associated with this modification of butter-based products (Kontkanen et al., 2011; Marangoni & Rousseau, 1998).

Effect of Minor Components Minor components can be indigenous to fats or present as an extraneously added additives. The indigenous minor components of milk fat are free fatty acid (FFA), monoacylglycerol (MAG), diacylglycerol (DAG), and phospholipids, but these are also used as an additives for the influence they have, or are presumed to have on crystallization, surface gloss, temperature stability, rheology, polymorphic stability, etc. (Smith, Bhaggan, Talbot, & van Malssen, 2011). The crystallization may change in the presence of minor components by effect on either nucleation or crystal growth or both. With respect to nucleation, it can be alteration in nucleation time, shift of nucleation temperature, or change in number and nature of nuclei formed. In the palm oil system, presence of saturated MAG derived from palm oil itself found to promote nucleation whereas unsaturated MAG from sunflower oil did not affect on nucleation (Fredrick, Foubert, De Sype, & Dewettinck, 2008). Cerdeira, Martini, Hartel, and Herrera (2003) examined the effect of palmitic and stearic sucrose ester on crystallization of milk fat- sunflower oil blends. They found that both palmitic and stearic sucrose ester with HLB = 1 (mainly di-, tri-, and polyester) delay nucleation with decreased crystal size and narrow size distribution whereas palmitic sucrose ester with HLB=16 (greater proportion of monoester) does not influence the crystal size or distribution.

In milk fat crystallization, the influence of minor components has been studied and found significant. Vanhoutte, Dewettinck, Foubert, Vanlerberghe, and Huyghebaert (2002) reported delayed onset crystallization in milk fat upon addition of phospholipids up to 0.07%. Below 2% level, phospholipids act as a nucleation site and thereby increase the spherulite size of the crystals. This result in the microstructure fabricated with uniformly dispersed individual crystals making butter harder, less grainy at low temperatures and less prone to oiling-off above room temperature. But, at higher concentrations (2 wt.%), it inhibits spherulite formation which has been suggested possibly either due to the inhibition of secondary nucleation at crystal surface or insufficient supercooling as result of polymorphic transition from 3L α form to a 2L β' species (Fedotova & Lencki, 2010). Smith (2000) also noticed an increase in the size of palm oil spherulites in the presence of 0.2% phosphatidyl ethanolamine making them denser. These changes are due to the interference of phospholipid molecules with the crystallizing triglyceride molecules causing changes in the shape and size of the resultant fat crystal network (Smith, 2000).

Thus, changes in fat globule membrane composition can lead to different nucleation properties and thereby can affect the crystallization behavior and fat crystal network in end product (Lopez & Ollivon, 2009; Rønholt et al., 2012).

Incorporation of Carbon Dioxide Influence of carbon dioxide (CO_2) on crystallization and bulk properties of milkfat has been studied recently (Truong, Palmer, Bansal, & Bhandari, 2017; Truong, Palmer, Bansal, Bhandari, & Hub, 2018). Mechanisms and kinetics of nucleation and crystal growth can be influenced by dissolved or suspended entities present in the fat. Carbon dioxide is a hydrophobic gas that was noticed to be highly soluble in anhydrous milk fat (AMF) with a solubility coefficient of 1648 ppm kg fat^{-1} atm^{-1}. The solubility behaviour is dependent on both intrinsic (partial pressure and temperature) and extrinsic variables (fatty acid composition and physical state of the AMF). The presence of dissolved CO_2 at concentration of 1379 ppm was found to induce initiation of crystallization at relatively higher temperature (25.3 ± 1.6 °C) compared to that of non-carbonated AMF (19.2 ± 0.4 °C) when cooled from 35 to 5 °C at 0.5 °C min^{-1}. Due to rapid crystallization, unstable species of α-form crystals were formed together with β′ and β polymorphs whereas non-treated AMF exhibited only β′ and β fat polymorphs. Increasing dissolved CO_2 concentration further to 2000 ppm was associated with relatively smaller sized crystals with two- to threefold increase in crystal numbers and solid fat content during isothermal crystallization at 28 °C for 20 min (Truong et al., 2017). Carbonation in cream and butter has also been investigated to study its effect on microbial stability, flavour and keeping quality (Hunziker, 1924; Prucha, Brannon, & Ruehe, 1925). A recent study (Truong et al., 2018) showed that the dissolution of CO_2 in fat phase of cream could shorten the churning time and can be applied to reduce the aging time during production of butter. It was related to an improved whipping of cream in the presence of CO_2 bubbles. The resultant butter produced was relatively softer (lower elastic modulus) with higher melting point without change in the microstructure, color and sensory properties.

2.4.2 Manufacturing Conditions

Heat Treatment of Cream During pasteurisation of cream, all of the fat becomes liquid which commence to crystallise upon cooling. Most of the latent heat of crystallisation of the milk fat releases before the cream leaves the cooler depending on the time-temperature characteristics of the process and fat content of cream (Wilbey, 1994). The effect of cooling rate on fat crystallisation (Herrera & Hartel, 2000b; Rønholt et al., 2012; Rønholt, Madsen, Kirkensgaard, Mortensen, & Knudsen, 2014) consequently influences rheological properties (Heertje, Van Eendenburg, Cornelissen, & Juriaanse, 1988; Rønholt et al., 2012; Rønholt, Kirkensgaard, Mortensen, & Knudsen, 2014) and microstructure (Heertje et al., 1988; Herrera & Hartel, 2000a; Rønholt et al., 2012; Rønholt, Kirkensgaard, et al., 2014) of milk-fat based products. Upon fast cooling (strong super-cooling), fat crystallisation equilibrium does not reach, and nucleation predominates over crystal growth. As a consequence, many small homogenous crystals form, primarily of α-form that subsequently transform to *β′* (Campos, Narine, & Marangoni, 2002; Wiking, De Graef, Rasmussen, & Dewettinck, 2009). Formation of many small crystals upon fast cooling provide larger surface area facilitating more liquid fat to be adsorbed to the crystal surfaces. As a result, less liquid fat remains free to form the continuous oil phase during churning and working, resulting in firmer butter (Wright & Marangoni, 2006). In contrast to fast cooling, crystallisation upon slow cooling is near equilibrium. The crystals of *β′*-polymorph form in more stable lamellar structure (Rønholt et al., 2013). However, Rønholt et al. (2012) reported the presence of α- and *β′*-polymorph to be independent of cream cooling rate (fast 7.5 °C/min, slow 0.4 °C/min). This contrariety could be ascribed to the differences in the chemical composition of the sample studied. The discrepancy between the studies on the effect of cooling rate on SFC has also been reported. Campos et al. (2002) found higher SFC (40.9%) in AMF immediately after fast cooling (5 °C/min) than that of (37.5%) after slow cooling (0.1 °C/min). In another study, similar SFC (16%) was observed in milk fat at both fast (10 °C/min) and slow (0.1 °C/min) cooling rate when cooled to 20 °C and stored isothermal for 80 min. Although fast cooled milk fat was significantly firmer (higher complex modulus) than the slow cooled (Wiking et al., 2009). This can be related to differences in fat crystal size and morphology that influences the fat crystal network. Smaller crystals provide more contact sites between crystals which stabilize solid crystal network compared to larger crystals (Heertje et al., 1988; Wiking et al., 2009). Rønholt, Kirkensgaard, et al. (2014) reported increased hardness (higher elastic modulus) in butter produced from fast (7.5 °C/min) cooled cream compared to slow (0.4 °C/min) cooled up to seven days of isothermal storage at 5 °C. The variation of elastic modulus within fast and slow-cooled samples was also increased up to 7 days of storage. But, after 21 days of storage, the effect of cooling rate was eliminated with respect to SFC, elastic modulus, and microstructure possibly as a result of fully developed crystal size and network (Rønholt, Kirkensgaard, et al., 2014).

In addition to cooling rate, maturation of cream also governs the milk fat crystallisation in globules, either by separating or mixing high- and low-melting fractions of the milk fat, consequently affecting crystal microstructure and rheological behaviour (Rønholt et al., 2013). It has been used for nearly a century as an economical approach to modify butter consistency (Wright & Marangoni, 2006). Butter held before churning has more free liquid fat and a softer texture than butter that is churned immediately after cooling (Dolby, 1954). Ageing the cream overnight between 10 °C and 12.8 °C (Wright et al., 2001) and another type of ripening called Alnarp or cold-warm-cold (CWC) method (Schäffer et al., 2000) can improve the texture of winter butter. It is also well demonstrated that all butter samples (with/without added milk fat fraction) of traditionally ripened butters are firmer at a low temperature than the samples of heat-step ripened butters, and also that the slope of the straight lines is greater. Various explanations for the improved softness through this cold-warm-cold method, generally related to the effects of liquid fat on crystal formation (Wright et al., 2001). During cooling numerous crystal nuclei form, and upon warming crystals of high melting TAGs probably melt, and further recrystallise upon cooling. It appears that through this method butter produced will have a higher amount of liquid fat and decreased hardness by up to 25% as compared to butter produced from cream cooled directly to lower temperature (Rønholt et al., 2013; Wright & Marangoni, 2006). Tondhoosh, Nayebzadeh, Mohamadifar, Homayouni-Rad, and Hosseinoghli (2016) proposed that by increasing the holding time (from 3 h to 5 h at 18 °C) of Alnarp process and fat content (40–45%) of cream at lower churning temperature (from 12 °C to 10 °C) could achieve smoother texture and improved spreadability in butter. In particular, butter made according to the CWC method is sensitive to temperature fluctuations during storage (Walstra, 1999). Rønholt et al. (2012) found that maturing the slow cooled (0.4 °C/min) cream at 5 °C for 48 h, increased elastic modulus significantly (0.30–0.38 MPa) in butter as a result of enhanced crystal growth compared to that made from non-matured slow cooled cream. Interestingly, the effect of fast and slow cooling rate was diminished after maturation of cream (5 °C for 48 h), as no difference was observed in microstructure and rheology of butter. Likely, all crystal may have been reached the critical size for aggregation forming dense crystalline network and similar elastic modulus in all samples independent of cooling rate (Rønholt et al., 2012). Thus, as a result of post-crystallization during storage of the product, the favourable effect of a certain temperature treatment of cream on the spreadability of the butter may actually be disappointing (Walstra, 1999).

Churning There are differences in butter made by the conventional (batch churning in metal churns) or continuous method specifically in the degree of crystallinity of the fat and fat crystal morphologies (Wright et al., 2001). Additionally, in part because of differences in the structural integrity of the fat globules, continuously churned butter is typically harder than that made by the conventional process (Wright & Marangoni, 2006). In batch churning, a high proportion of fat (up to 46%) can remain in the globules (Wright et al., 2001). The globules formed upon CWC ripening are more stable as thick layer of high melting TAGs form on surface while crystal aggregates and liquid oil in the core. Hence, the large number of these globules survive during churning in contrast to globules formed upon 'cold-ripening' (Juriaanse & Heertje, 1988). In our recent study (Panchal et al., 2017), we observed that the stability of fat globules during the churning process in terms of churning time also depends on its size and membrane material. An increase in churning time of cream with reduction in average fat globule size (0.6 μm) was dramatic when sodium caseinate (32 min) was used as an emulsifier in comparison to Tween80 (3.5 min) at similar concentration (0.5%, w/w) and churning temperature (10 °C).

However, such relative stability of globules in high shearing continuous churning process could be of minor difference. In continuous Gold' n flow process, the globule structure is completely destroyed, and the major proportion of the fat exist as free fat (Wright et al., 2001; Wright & Marangoni, 2006). Today, the continuous butter-making technologies predominate, including accelerated phase inversion (Fritz-type), and emulsification method.

Usually, churning temperature employed is in the range from 10 °C to 15 °C. Upon increasing churning temperature from 10 °C to 22 °C, Rønholt, Madsen, et al. (2014) found that despite of differences in crystal stability, solid fat content, and water droplet size, no significant change was observed in microstructure and hardness or brittleness of butter during 28 days of isothermal storage. Moreover, crystal polymorphism was also similar as all butters contained primarily β′-crystals with traces of α- and β-form. However, it was concluded that low solid fat content at high churning temperature could limit the possibility of incorporating high amount of make-up water in the butter-matrix. Additionally, high churning temperature induce excessive liquid fat that all could not entrap in butter grains, hence loss of fat into buttermilk. Cooling also occur more rapidly in the scraped surface heat exchangers of a continuous method than traditional batch process (Wright & Marangoni, 2006).

Rapid cooling leads to higher solid fat content and a firmer product. Nevertheless, there are a number of interrelated parameters simultaneously influencing butter texture rather than just single factor cooling rate. Solid fat content, crystal size, crystal-crystal interaction, and crystal network formation are all affected by temperature treatment (Wright et al., 2001).

Mechanical Working Before packing, working or mixing serves to uniformly disperse the salt and the both internal and external water in the continuous oil phase of milkfat products (Wright et al., 2001; Wright & Marangoni, 2006). The internal water is that incorporated into the butter grains during phase inversion. The content of solid fat affects the water retention. Crystallised fat appears to be harder allowing more water to be present in the cavities, whereas soft fat globules are likely to get squeezed forming smaller cavities and therefore lower water content (Rønholt, Madsen, et al., 2014). After addition of external water (to make up desired moisture level), working step disperse the water over surface of butter grains and in the network of fat crystals where larger surface area of small butter grains allows to retain more water (Rønholt et al., 2013). In addition to that, working continues the release of fat crystal and liquid fat from the fat globules, consequently improve the texture (Wright et al., 2001; Wright & Marangoni, 2006). The shear applied during the working break down the crystal bridges within the crystal network, decreasing the hardness to about a quarter of the original value of fat spreads (Heertje et al., 1988). The intense working, however, destroys a large number of fat globules resulting in a more crystalline inter-globular phase and consequently a harder consistency (Juriaanse & Heertje, 1988). Reworking can be used to further improve softness of butter however original firmness is regained during storage because of thixotropic setting (post-crystallisation) (Wright & Marangoni, 2006) and effect of wide differences in the amount of working is eliminated (Dolby, 1941). Beyond a certain point, working has no effect on butter hardness; in fact, upon excessive working butter can become sticky (Wright et al., 2001).

Storage Conditions Post manufacturing storage and handling have a considerable effect on butter firmness (Walstra et al., 2005). From industrial point of view, the effect of storage conditions on spreadable fat products is important as the stability of rheological behaviour during storage is essential for the shelf-life of the products (Martini & Herrera, 2008; Pothiraj, Zuntilde, Simonin, Chevallier, & Le-Bail, 2012), the effect of fluctuating temperature on rheological behaviour during storage and transportation is crucial in relation to consumer acceptance (Martini & Herrera, 2008; Pothiraj et al., 2012), and the storage conditions can be optimised to alter the functionality of products (Segura, Herrera, & Añón, 1990).

The time and temperature of storage affect the setting (post-crystallization) of butter and as a result, texture. Setting occurs faster at higher temperature as the rearrangement of compound crystals and polymorphic changes proceed rapidly in the presence of liquid fat (Walstra et al., 2005). Setting is likely to be more extensive in conventionally made butter than that is churned by continuous method. Composition of fat and original firmness also determine the extent of setting (Tanaka, Isogai,

Miura, & Murakami, 2010; Wright et al., 2001). Solid fat content is strongly dependent on the time of storage. Upon isothermal storage (5 °C) of milk fat-based products, solid fat content increases significantly up to 14 days, as a result of post crystallisation (Rønholt, Kirkensgaard, et al., 2014). However, a similar increase was not observed in elastic modulus (Rønholt, Kirkensgaard, et al., 2014) which illustrates the complexity of crystal-crystal interaction and network formation. Solid fat content is not necessarily predictive for the rheological behaviour of milk fat-based products, as after a critical solid fat content is reached, the hardness of the milk fat-based products does not necessarily increase, even though solid fat content keeps increasing due to post-crystallization (Herrera & Hartel, 2000b). The initial differences in rheological behaviour between samples as a consequence of either different cream cooling rate (different size and number of crystals), water content, or fraction of milk fat globules is found to be diminished upon storage (Rønholt et al., 2012; Rønholt, Kirkensgaard, et al., 2014) as result of continuous fat crystal network formation till critical solid fat content (Rønholt et al., 2013).

Fluctuation of temperature during storage also cause differences in rheological behavior (Rønholt et al., 2012). For both butter and margarine-like emulsions, solid fat content is higher after storage at fluctuating temperature compared to isothermal storage. In contrast, elastic modulus decreases after storage at fluctuating temperature (Rønholt et al., 2012), as large crystal are formed, limiting the number of crystal-crystal interactions. However, hardness of the butter was shown to increase by 25% upon temporary increase in storage temperature from 8 to 20 °C, independent of cream cooling rate and fat composition (Rønholt et al., 2013). If the temperature is temporary raised, low-melting triglycerides melts which again adsorb on the surface of remaining crystals upon cooling, forming larger crystals. Further growth of crystals induces strong primary crystal-crystal interaction and more solid structure corresponding increase in hardness (Walstra et al., 2005). Crystals once grown to a certain size (>100 μm) can also impair the homogeneity and softness of water/oil type spreadable fats, and texture degradation (Tanaka et al., 2010). With the time of storage, increased changes in the rheological behavior of butter can be observed including increased brittleness as a consequence of evaporation of water and coalescence of moisture droplets (Rønholt, Kirkensgaard, et al., 2014).

Water, Air, Surfactants, and Stabilizers Consistency of milk fat-based products can also be adjusted by varying its moisture content as the strength of a fat crystal network depends crucially on the water droplets size as well along with the amount of crystallized fat (Juriaanse & Heertje, 1988). Changing the water content can cause both an increase and a decrease of water droplet sizes and/or number of water droplets in continuous fat phase (Van Dalen, 2002), which can subsequently change the contact points between crystals and so possibly the rheological behaviour of product (Rønholt, Kirkensgaard, et al., 2014). Increasing moisture content from 12 to 15% imparted softer texture to butter at both 5 and 15 °C. Further increase in moisture content up 35% completely changed the rheological properties of butter (Wright et al., 2001). It is not only solid fat content itself but the ratio between solid and liquid (both liquid oil and water) that strongly affects the hardness and spreadability of butter-like products (Pothiraj et al., 2012; Rønholt, Kirkensgaard, et al., 2014). When increasing water content, the ratio between solid and liquid phases (oil and water) is shifted toward more liquid (Rønholt, Kirkensgaard, et al., 2014), so less fat contributes to the fat crystal network consequently affecting product firmness and water droplet stability (Mulder & Walstra, 1974).

However, the disadvantage of adding moisture to soften the texture of butter include structural and microbial stability, hydrolytic rancidity, and violating standards of identity (Wright & Marangoni, 2006). Incorporation of air into butter has also been reported to reduce the hardness significantly. Effects are likely to be more at higher temperature and equivalent to the effects of working (Wright et al., 2001). Incorporation of air into butterfat system exhibited a greater decrease in firmness than incorporation of a similar amount of water (Foley & Cooney, 1982). Increase in overrun (between 10 and 125%) in whipped butter results in a significant increase in spreadability (Wright et al., 2001). Multiple factors like the manufacturing method, amount, type and duration of gas incorporation, and fat composition govern the effectiveness of whipping on achievement desired spreadability in whipped butter (Wright et al., 2001).

The addition of surfactants to cream or directly to butter may also improve the texture of butter (Wright & Marangoni, 2006). Addition of 1% monoglycerides to cream before churning reported to improve spreadability of butter by 30% (Wright et al., 2001). Kapsalis, Kristoffersen, Gould, and Betscher (1963) reported improvement in spreadability by 10–44% upon addition of surfactants (Tweens and Spans, lecithin, buttermilk solids, and skim milk solids) individually or in combination. The effect was attributed to surface active properties of additives. Drawbacks associated with application of additives were difficulty of incorporation and uniform distribution of additives into butter, off-flavors at concentration greater than 1% (Kapsalis et al., 1963), and loss of most of the surfactants in buttermilk (when added to cream before churning) (Wright et al., 2001). On account of these reasons, the application of surfactants for the practical significance of butter rheology is limited (Wright & Marangoni, 2006).

The low fat spreads contain a substantial amount of water compared to butter and margarine. Above a certain level of water, the rheological properties of a water-in-oil spread significantly changes (Mounsey et al., 2008). This introduces specific problems in low-fat spreads such as the coalescence and flow of disproportionate aqueous phase, and the occurrence of loose moisture on spreading (Alexa, Mounsey, O'Kennedy, & Jacquier, 2010). To achieve relatively firm consistency and a plastic rheology, a water phase is required to be structured somehow (Evageliou, Alevisopoulos, & Kasapis, 1997).

Mainly four types of structuring agents have been identified to obtain desired structure and plasticity. These are viscosifying (milk proteins or high-molecular-weight polysaccharides), gelling (hydrocolloids used to gel the aqueous phase), phase-separating (with thermodynamically incompatible hydrocolloids) and synergistic (exploiting known synergistic interactions between hydrocolloids) agents (Cheng, Lim, Chow, Chong, & Chang, 2008; Mounsey et al., 2008). Application of proteins (casein, whey, egg, soy or gelatin) in combination with polysaccharides (starch, carrageenan, alginate, pectin) is common for viscosity enhancement, water binding and gel formation, prevention of emulsion breakdown and syneresis, and better release of flavor in mouth (Cheng et al., 2008; Moran, 1994; Mounsey et al., 2008; Syrbe, Bauer, & Klostermeyer, 1998). Addition of k-carrageenan up to 0.75% significantly increased the viscosity of aqueous phase and modified the size and distribution of water droplets in 60% fat water-in-oil spread, but at concentration higher than that resulted in a less homogeneous distribution of aqueous phase and disrupted microstructure. However, the effect of concentration of k-carrageenan on the firmness of spreads was inconsistent (Alexa et al., 2010). The texture of low-fat W/O spreads is suggested to be less affected by the viscosity of the aqueous phase (Keogh, 2006; Mounsey et al., 2008). Cation mediated gelation (ionotropic gelation) of negatively charged polysaccharides at appropriate concentration of specific ions can also be employed to enhance thickening properties of hydrocolloids (Burey, Bhandari, Howes, & Gidley, 2008).

3 New Approaches to Improve the Functionality of Butter

In addition to continuous liquid fat phase and crystal fat network, structuring of globular phase containing more or less intact fat globules and dispersed water phase can also be used as a novel tool to design functionality of butter-like churned emulsions. Our research group studied the decisive role of fat globule size and its interfacial material on functionality of butter (not published). It appeared that sodium caseinate on the globule surface enhanced reinforcement of fat crystal network whereas in the presence of Tween80 interactions with surrounding crystals were weaker which enhanced flexibility of the butter microstructure as observed as improved softness (penetration test). Emulsifying components on water droplet surface have also been found determining solidity of water-in-oil emulsion (Rafanan & Rousseau, 2017). Such demonstrative effects of interfacial material, similar to the mechanism of active and inactive filler particles in composite gels (Dickinson & Chen, 1999), can be further studied to engineer the butter and continuous fat emulsions. Dissolution of CO_2 into cream has been found effective in reducing churning time, enhancing crystallization rate, and reducing the hardness (penetration test) of laboratory scale produced butter (Truong et al., 2018). This new approaches provide a scope of further inverstigation and application in novel food product development with desired end-use functionality.

4 Conclusion and Further Research

To summarise, this chapter underscores the importance of understanding of the processing and compositional factors in order to control and modify the structural properties of milk fat-based products. However, microstructural changes due to post-crystallization and crystal network formation are the biggest challenge to maintain the initially induced rheological differences in products. The strong temperature influence on milk-fat rheology, seasonal and regional variation in milk fat composition, the strict legal standards of identity for butter, and the fact that multiple parameters in milk fat-rich products are affected simultaneously makes characterization and optimization of milk fat-based products consistency complicated. For further study, it would be relevant to relate the processing and compositional parameters, and fat globule size variation to rheological and sensory attributes of butter and spreads. More work is needed to understand the role of surfactants and stabilizers to achieve desired multifunctional characteristics and stability in normal and low-fat butter and spreads at commercial scale. Overall, a better understanding of the factors controlling functionality and stability in a product environment, more refine analysis of factory unit operation, and accurate correlation of identified consumer preferences with product properties and the impact of process, all require further long-term studies.

Acknowledgements This research was supported under Australian Research Council's Industrial Transformation Research Hub (ITRH) funding scheme (IH120100005). The ARC Dairy Innovation Hub is a collaboration between the University of Melbourne, the University of Queensland and Dairy Innovation Australia Ltd.

References

Ahmed, S. H., & Luksas, A. J. (1988). *Spreadable butter-like composition and method for production thereof.* In Google Patents.

Alexa, R. I., Mounsey, J. S., O'Kennedy, B. T., & Jacquier, J. C. (2010). Effect of κ-carrageenan on rheological properties, microstructure, texture and oxidative stability of water-in-oil spreads. *LWT - Food Science and Technology, 43*(6), 843–848.

Baer, R., Ryali, J., Schingoethe, D., Kasperson, K., Donovan, D., Hippen, A., et al. (2001). Composition and properties of milk and butter from cows fed fish oil. *Journal of Dairy Science, 84*(2), 345–353.

Banks, W., & Christie, W. (1990). Feeding cows for the production of butter with good spreadability at refrigeration temperatures. *Outlook on Agriculture, 19*(1), 43–47.

Bayourthe, C., Enjalbert, F., & Moncoulon, R. (2000). Effects of different forms of canola oil fatty acids plus canola meal on milk composition and physical properties of butter. *Journal of Dairy Science, 83*(4), 690–696.

Buchheim, W., & Dejmek, P. (1997). *Milk and dairy-type emulsions. Food emulsions*. New York, NY: Marcel Dekker.

Buldo, P., Larsen, M. K., & Wiking, L. (2013). Multivariate data analysis for finding the relevant fatty acids contributing to the melting fractions of cream. *Journal of the Science of Food and Agriculture, 93*(7), 1620–1625.

Bumbalough, J. (2000). *Margarine types and preparation technology. Fats and oils technology* (2nd ed.pp. 452–462). Champaign, IL: AOCS.

Burey, P., Bhandari, B., Howes, T., & Gidley, M. (2008). Hydrocolloid gel particles: Formation, characterization, and application. *Critical Reviews in Food Science and Nutrition, 48*(5), 361–377.

Cadden, A.-M., Urquhart, A., & Jelen, P. (1984). Evaluation of milk and butter from commercial dairy herds fed canola-based protected lipid feed supplement. *Journal of dairy science, 67*(9), 2041–2044.

Campos, R., Narine, S., & Marangoni, A. (2002). Effect of cooling rate on the structure and mechanical properties of milk fat and lard. *Food Research International, 35*(10), 971–981.

Cerdeira, M., Martini, S., Hartel, R. W., & Herrera, M. L. (2003). Effect of sucrose ester addition on nucleation and growth behavior of milk fat-sunflower oil blends. *Journal of Agricultural and Food Chemistry, 51*(22), 6550–6557.

Cheng, L., Lim, B., Chow, K., Chong, S., & Chang, Y. (2008). Using fish gelatin and pectin to make a low-fat spread. *Food Hydrocolloids, 22*(8), 1637–1640.

Couvreur, S., Hurtaud, C., Lopez, C., Delaby, L., & Peyraud, J.-L. (2006). The linear relationship between the proportion of fresh grass in the cow diet, milk fatty acid composition, and butter properties. *Journal of Dairy Science, 89*(6), 1956–1969.

Dairy Australia. (2018). *Consumption summary*. Retreived from https://www.dairyaustralia.com.au/industry/production-and-sales/consumption-summary

Dickinson, E., & Chen, J. (1999). Heat-set whey protein emulsion gels: Role of active and inactive filler particles. *Journal of Dispersion Science and Technology, 20*(1-2), 197–213.

Dolby, R. (1941). The rheology of butter. III. The effect of variation in butter-making conditions on the hardness of the butter. *Journal of Dairy Research, 12*(3), 344–349.

Dolby, R. (1949). The properties of New Zealand butters and butterfats: II. The relation of hardness of New Zealand commercial butter to composition of the butterfat. *Journal of Dairy Research, 16*(3), 336–347.

Dolby, R. (1954). The effect of temperature treatment of cream before churning on the consistency of butter. *Journal of Dairy Research, 21*(1), 67–77.

Early, R. (1998). Technology of dairy products: Springer Science & Business Media.

Evageliou, V., Alevisopoulos, S., & Kasapis, S. (1997). Application of stress-controlled analysis to the development of low fat spreads. *Journal of Texture Studies, 28*(3), 319–335.

Fearon, A. (2001). Optimising milkfat composition and processing properties. *Australian Journal of Dairy Technology, 56*(2), 104.

Fedotova, Y., & Lencki, R. W. (2010). The effect of phospholipids on butter physical and sensory properties. *Journal of the American Oil Chemists' Society, 87*(1), 75–82.

Foley, J., & Cooney, C. (1982). *Quantification of factors influencing the firmness of butter*. In Paper presented at the proceedings of the 21th Int Dairy Congress.

Fox, P. F., & McSweeney, P. L. (1998). *Dairy chemistry and biochemistry*. New York, NY: Academic.

Fredrick, E., Foubert, I., De Sype, J. V., & Dewettinck, K. (2008). Influence of monoglycerides on the crystallization behavior of palm oil. *Crystal Growth and Design, 8*(6), 1833–1839.

Freeman, I. P., & Melnikov, S. M. (2005). Margarines. In *Ullmann's encyclopedia of industrial chemistry*. Weinheim, Germany: Wiley.

Ghosh, S., & Rousseau, D. (2011). Fat crystals and water-in-oil emulsion stability. *Current Opinion in Colloid & Interface Science, 16*(5), 421–431.

Ghotra, B. S., Dyal, S. D., & Narine, S. S. (2002). Lipid shortenings: A review. *Food Research International, 35*(10), 1015–1048.

Heertje, I. (1993). Structure and function of food products: A review. *Food Structure, 12*(3), 7.

Heertje, I., Leunis, M., Van Zeyl, W., & Berends, E. (1987). Product morphology of fatty products. *Food Structure, 6*(1), 2.

Heertje, I., Van Eendenburg, J., Cornelissen, J., & Juriaanse, A. (1988). The effect of processing on some microstructural characteristics of fat spreads. *Food Structure, 7*(2), 9.

Henning, D., Baer, R., Hassan, A., & Dave, R. (2006). Major advances in concentrated and dry milk products, cheese, and milk fat-based spreads. *Journal of Dairy Science, 89*(4), 1179–1188.

Herrera, M., & Hartel, R. (2000a). Effect of processing conditions on physical properties of a milk fat model system: Microstructure. *Journal of the American Oil Chemists' Society, 77*(11), 1197–1205.

Herrera, M., & Hartel, R. (2000b). Effect of processing conditions on physical properties of a milk fat model system: Rheology. *Journal of the American Oil Chemists' Society, 77*(11), 1189–1196.

Hillbrick, G., & Augustin, M. (2002). Milkfat characteristics and functionality: Opportunities for improvement. *Australian Journal of Dairy Technology, 57*(1), 45.

Hunziker, O. (1924). Facts about carbonated butter. *Journal of Dairy Science, 7*(5), 484–496.

Hurtaud, C., Faucon, F., Couvreur, S., & Peyraud, J.-L. (2010). Linear relationship between increasing amounts of extruded linseed in dairy cow diet and milk fatty acid composition and butter properties. *Journal of Dairy Science, 93*(4), 1429–1443.

Juriaanse, A., & Heertje, I. (1988). Microstructure of shortenings, margarine and butter-a review. *Food Structure, 7*(2), 8.

Kapsalis, J., Kristoffersen, T., Gould, I., & Betscher, J. (1963). Effect of chemical additives on the spreading quality of butter. II. Laboratory and plant churnings. *Journal of Dairy Science, 46*(2), 107–113.

Kaylegian, K. E., & Lindsay, R. C. (1992). Performance of selected milk fat fractions in cold-spreadable butter. *Journal of Dairy Science, 75*(12), 3307–3317.

Keogh, M. (2006). Chemistry and technology of butter and milk fat spreads. In *Advanced dairy chemistry volume 2: Lipids* (pp. 333–363). New York, NY: Springer.

Kontkanen, H., Rokka, S., Kemppinen, A., Miettinen, H., Hellström, J., Kruus, K., et al. (2011). Enzymatic and physical modification of milk fat: A review. *International Dairy Journal, 21*(1), 3–13.

Krause, A., Lopetcharat, K., & Drake, M. (2007). Identification of the characteristics that drive consumer liking of butter. *Journal of Dairy Science, 90*(5), 2091–2102.

Lane, R. (1998). Butter and mixed fat spreads. In *Technology of dairy products* (p. 158). London, UK: Blackie Academic & Professional.

Lin, M., Sims, C., Staples, C., & O'Keefe, S. (1996). Flavor quality and texture of modified fatty acid high monoene, low saturate butter. *Food Research International, 29*(3), 367–371.

Lopez, C., & Ollivon, M. (2009). Crystallisation of triacylglycerols in nanoparticles. *Journal of Thermal Analysis and Calorimetry, 98*(1), 29–37.

Marangoni, A., & Rousseau, D. (1998). Chemical and enzymatic modification of butterfat and butterfat-canola oil blends. *Food Research International, 31*(8), 595–599.

Martini, S., & Herrera, M. L. (2008). Physical properties of shortenings with low-trans fatty acids as affected by emulsifiers and storage conditions. *European Journal of Lipid Science and Technology, 110*(2), 172–182.

McClements, D. J. (2004). Protein-stabilized emulsions. *Current Opinion in Colloid & Interface Science, 9*(5), 305–313.

Moran, D. P. (1994). Fats in spreadable products. In *Fats in food products* (pp. 155–211). Cham, Switzerland: Springer.

Mortensen, B. (2011). Milk fat-based spreads. In *Encyclopedia of dairy sciences* (pp. 522–527). Oxford, UK: Academic.

Mortensen, B., & Denmark, T. (2011). Butter and other milk fat products, the product and its manufacture. In *Encyclopedia of dairy sciences* (pp. 492–499). Oxford, UK: Academic.

Mounsey, J. S., Stathopoulos, C. E., Chockchaisawasdee, S., O'Kennedy, B. T., Gee, V. L., & Doyle, J. (2008). Effect of zinc fortifications on rheological properties and micro-structure of water-in-oil spreads containing κ-carrageenan. *European Food Research and Technology, 227*(3), 675–681.

Mulder, H., & Walstra, P. (1974). The milk fat globule. In *Emulsion science as applied to milk products and comparable foods*. Wageningen, Netherlands: Centre for Agricultural Publishing and Documentation.

Murphy, J., Connolly, J., & McNeill, G. (1995). Effects on cow performance and milk fat composition of feeding full fat soyabeans and rapeseeds to dairy cows at pasture. *Livestock Production Science, 44*(1), 13–25.

Murphy, J. J., McNeill, G. P., Connolly, J. F., & Gleeson, P. A. (1990). Effect on cow performance and milk fat composition of including full fat soyabeans and rapeseeds in the concentrate mixture for lactating dairy cows. *Journal of Dairy Research, 57*(03), 295–306.

Narine, S. S., & Marangoni, A. G. (1999). Fractal nature of fat crystal networks. *Physical Review E, 59*(2), 1908.

Panchal, B., Truong, T., Prakash, S., Bansal, N., Bhandari, B., & Hub, A. D. I. (2017). Effect of fat globule size on the churnability of dairy cream. *Food Research International, 99*(1), 229–238.

Platt, B. L., & Gupta, B. B. (1989). Low fat spread and process for preparing same. U.S. Patent No. 4,865,867. Washington, DC: U.S. Patent and Trademark Office.

Pothiraj, C., Zuntilde, R., Simonin, H., Chevallier, S., & Le-Bail, A. (2012). Methodology assessment on melting and texture properties of spread during ageing and impact of sample size on the representativeness of the results. *Journal of Stored Products and Postharvest Research, 3*(10), 137–144.

Prucha, M., Brannon, J., & Ruehe, H. (1925). Carbonation of butter. *Journal of Dairy Science, 8*(4), 318–329.

Rafanan, R., & Rousseau, D. (2017). Dispersed droplets as active fillers in fat-crystal network-stabilized water-in-oil emulsions. *Food Research International, 99*, 355–362.

Rønholt, S., Kirkensgaard, J. J. K., Mortensen, K., & Knudsen, J. C. (2014). Effect of cream cooling rate and water content on butter microstructure during four weeks of storage. *Food Hydrocolloids, 34*, 169–176.

Rønholt, S., Kirkensgaard, J. J. K., Pedersen, T. B., Mortensen, K., & Knudsen, J. C. (2012). Polymorphism, microstructure and rheology of butter. Effects of cream heat treatment. *Food Chemistry, 135*(3), 1730–1739.

Rønholt, S., Madsen, A. S., Kirkensgaard, J. J., Mortensen, K., & Knudsen, J. C. (2014). Effect of churning temperature on water content, rheology, microstructure and stability of butter during four weeks of storage. *Food Structure, 2*(1), 14–26.

Rønholt, S., Mortensen, K., & Knudsen, J. C. (2013). The effective factors on the structure of butter and other milk fat-based products. *Comprehensive Reviews in Food Science and Food Safety, 12*(5), 468–482.

Rousseau, D., Ghosh, S., & Park, H. (2009). Comparison of the dispersed phase coalescence mechanisms in different tablespreads. *Journal of Food Science, 74*(1), E1–E7.

Sato, K., & Ueno, S. (2014). Physical properties of fats in food. In *Fats in food technology*. Chichester, UK: Wiley.

Schäffer, B., Szakály, S., Lorinczy, D., & Belágyi, J. (2000). Structure of butter. IV. Effect of modification of cream ripening and fatty acid composition on the consistency of butter. *Milchwissenschaft, 55*(3), 132–135.

Scott, T., Cook, L., Ferguson, K., McDonald, I., Buchanan, R., & Hills, G. L. (1970). Production of poly-unsaturated milk fat in domestic ruminants. *Australian Journal of Science, 32*, 291–293.

Segura, J., Herrera, M., & Añón, M. (1990). Storage of commercial margarine at different temperatures. *Journal of the American Oil Chemists' Society, 67*(12), 989–992.

Shi, Y., Smith, C., & Hartel, R. (2001). Compositional effects on milk fat crystallization. *Journal of Dairy Science, 84*(11), 2392–2401.

Shukla, A., Bhaskar, A., Rizvi, S., & Mulvaney, S. (1994). Physicochemical and rheological properties of butter made from supercritically fractionated milk fat. *Journal of Dairy Science, 77*(1), 45–54.

Smith, K. W., Bhaggan, K., Talbot, G., & van Malssen, K. F. (2011). Crystallization of fats: Influence of minor components and additives. *Journal of the American Oil Chemists' Society, 88*(8), 1085–1101.

Smith, P. R. (2000). The effects of phospholipids on crystallisation and crystal habit in triglycerides. *European Journal of Lipid Science and Technology, 102*(2), 122–127.

Syrbe, A., Bauer, W., & Klostermeyer, H. (1998). Polymer science concepts in dairy systems—an overview of milk protein and food hydrocolloid interaction. *International Dairy Journal, 8*(3), 179–193.

Tanaka, L., Isogai, T., Miura, S., & Murakami, M. (2010). Effect of triacylglycerol species on the crystallizing behavior of a model water/oil emulsion. *European Journal of Lipid Science and Technology, 112*(3), 304–309.
Tondhoosh, A., Nayebzadeh, K. A., Mohamadifar, M., Homayouni-Rad, A., & Hosseinoghli, H. (2016). Industrial application of different heat treatments and cream fat contents for improving the spreadability of butter. *Recent Patents on Food, Nutrition & Agriculture, 8*(2), 107–115.
Truong, T., Palmer, M., Bansal, N., & Bhandari, B. (2017). Effect of solubilised carbon dioxide at low partial pressure on crystallisation behaviour, microstructure and texture of anhydrous milk fat. *Food Research International, 95*, 82–90.
Truong, T., Palmer, M., Bansal, N., Bhandari, B., & Hub, A. D. I. (2018). Effects of dissolved carbon dioxide in fat phase of cream on manufacturing and physical properties of butter. *Journal of Food Engineering, 226*, 9–21.
Van Dalen, G. (2002). Determination of the water droplet size distribution of fat spreads using confocal scanning laser microscopy. *Journal of Microscopy, 208*(2), 116–133.
Vanhoutte, B., Dewettinck, K., Foubert, I., Vanlerberghe, B., & Huyghebaert, A. (2002). The effect of phospholipids and water on the isothermal crystallisation of milk fat. *European Journal of Lipid Science and Technology, 104*(8), 490–495.
Varnam, A. H., & Sutherland, J. P. (1994). Butter, margarine and spreads. In *Milk and milk products* (pp. 224–274). New York, NY: Springer.
Walstra, P. (1999). *Dairy technology: Principles of milk properties and processes*. New York: CRC Press.
Walstra, P., Walstra, P., Wouters, J. T., & Geurts, T. J. (2005). *Dairy science and technology*. New York: CRC Press.
Wiking, L., De Graef, V., Rasmussen, M., & Dewettinck, K. (2009). Relations between crystallisation mechanisms and microstructure of milk fat. *International Dairy Journal, 19*(8), 424–430.
Wilbey, R. (1994). Production of butter and dairy based spreads. In *Robinson: Modern dairy technology* (pp. 107–158). New York, NY: Springer.
Williams, C. M. (2000). *Dietary fatty acids and human health*. In Paper presented at the Annales de Zootechnie.
Wright, A., & Marangoni, A. (2006). Crystallization and rheological properties of milk fat. In *Advanced dairy chemistry volume 2: Lipids* (pp. 245–291). New York, NY: Springer.
Wright, A., Scanlon, M., Hartel, R., & Marangoni, A. (2001). Rheological properties of milkfat and butter. *Journal of Food Science, 66*(8), 1056–1071.
Zárubová, M., Filip, V., Kšandová, L., Šmidrkal, J., & Piska, I. (2010). Rheological and crystalline properties of trans-free model fat blends as affected by the length of fatty acid chains. *Journal of Food Engineering, 99*(4), 459–464.
Zhang, H., Smith, P., & Adler-Nissen, J. (2004). Effects of degree of enzymatic interesterification on the physical properties of margarine fats: Solid fat content, crystallization behavior, crystal morphology, and crystal network. *Journal of Agricultural and Food Chemistry, 52*(14), 4423–4431.

Chapter 22
Fat-Reduced Cream Cheeses

Dian Widya Ningtyas and Sangeeta Prakash

1 Introduction

Cream cheese is a fresh acid coagulated cheese product with soft and spreadable texture, which is acidified by mesophilic lactic acid starter culture. Variants of some of the soft, fresh cheese (e.g., Quark, Cottage cheese, Fromage frais, Bakers cheese, Queso Blanco, and Neufchatel) are also produced from acidification of milk to pH 4.6 which causes the casein to coagulate at their isoelectric point (Fox, Guinee, Cogan, & Mcsweeney, 2017). Regular cream cheese contains a higher percentage of fat, minimum of 33% in the US and 30% in Canada compared to other types of cheese (Phadungath, 2005). Due to high-fat content in cream cheese and the increased consumer awareness of the health risks associated with high dietary fat, the demand for low-fat foods, including cheese, has grown substantially. Although fat reduction may provide consumers with healthier products, the changes in sensory and textural characteristics of low-fat cream cheese, compared to its full-fat counterpart, may influence the consumer's response.

Cottage cheese and Quark are examples of low-fat soft cheeses which are accepted by consumers and have been manufactured for decades. In contrary, the difficulties arise when attempts are made to develop the low-fat variants of cream cheese which are already well-known as full-fat products. Several cream cheese products with varying fat content are also available in the market. Reducing the fat content modifies the composition (fat, protein, carbohydrate, and moisture) and structure of cream cheese which often improve the changes in color, flavor, and texture (Banks, 2004; Mistry, 2001).

Reduced-fat cream cheese is characterised by high moisture content (55–70%) and usually consumed fresh after processing without ripening. The total fat content

D. W. Ningtyas · S. Prakash (✉)
School of Agriculture and Food Sciences, The University of Queensland,
St Lucia, QLD, Australia
e-mail: s.prakash@uq.edu.au

T. Truong et al. (eds.), *Dairy Fat Products and Functionality*,
https://doi.org/10.1007/978-3-030-41661-4_22

of low-fat cream cheese is 16.5–20% and light/or very low-fat cream cheese is less than 16.5% (USDA, 1994). Many researchers have reported on cream cheese, but unfortunately, there is quite limited information published on the manufacturing of reduced-fat cream cheese. Crane (1992) described the process of making fat-free cream cheese based on the full-fat and Neufchatel cheese methods with some modification to the process (increased the moisture content and homogenised the milk). Cream cheese was prepared from concentrated skim milk and fat replacer (hydrocolloid or WPC), fermented to a pH of 4.8–5. To improve the texture of fat-free cream cheese similar to regular cream cheese, bulking agents and stabilisers were added during the mixing steps.

In most research, production of reduced-fat cream cheese is adapted from the original cream cheese making process, but with a reduction in the fat content of milk. Generally, the method include pre-treatment of cheese milk, incubation of the inoculated milk at 20–35 °C, breaking and heating the cheese curd, whey separation, and further treatments of the curd (addition of salt, thickener, fat replacers, flavor) with mixing at high shear rate for uniform dispersion of the added ingredients. The step-by-step process of cream cheese making is described in Fig. 22.1. The treated curd may be hot-packed (70–85 °C) or cold-packed that results in a difference in their shelf life. Examples of cold-pack cheese are Quarg, cottage cheese, and cream cheese, while Neufchatel is a hot-pack cheese (Fox et al., 2017; Phadungath, 2005).

In order to produce low-fat cream cheese with similar texture and sensory properties as their full-fat counterpart, some processing parameters must be altered substantially. The various physicochemical, microstructural and sensory changes that occur as a result of reducing the fat content have been reported extensively (Ningtyas, Bhandari, Bansal, & Prakash, 2017; Wendin, Langton, Caous, & Hall, 2000; Zalazar et al., 2002). Consistency, creaminess, smoothness are the parameters closely related to fat content (Brighenti, Govindasamy-Lucey, Lim, Nelson, & Lucey, 2008; Muir, Williams, Tamime, & Shenana, 1997; Wendin et al., 2000). Brighenti et al. (2008) reported that total solids and homogenisation in full-fat cream cheese influence the texture and create a more elastic, harder and more cohesive texture than Neufchatel or fat-free cheeses. In addition, the presence of gums/stabilizers and the difference in manufacturing process greatly influence the rheological profiles of fat-free cream cheese (Lashkari, Madadlou, & Alizadeh, 2014; Madadlou, Khosroshahi, & Mousavi, 2005; Ningtyas et al., 2017; Ningtyas, Bhandari, Bansal, & Prakash, 2018b). Cheese spreads with different fat contents also exhibit different texture parameters, as observed by Muir et al. (1997).

Successful attempts to increase the fullness of reduced-fat cheese were obtained by using homogenisation, that increases the number of small-sized (0.5–1.3 μm) fat droplets (Coutouly, Riaublanc, Axelos, & Gaucher, 2014; Karaman & AkalıN, 2013; Madadlou, Mousavi, Khosrowshahi Asl, Emam-Djome, & Zargaran, 2007; Rudan, Barbano, Guo, & Kindstedt, 1998). With reduced-fat content cream cheese, the interaction between salt and proteins is important for the aggregation of milk protein (Wendin et al., 2000). Other attempts at obtaining low-fat cream cheese with sensory attributes similar to the full-fat cream cheese are through modification of

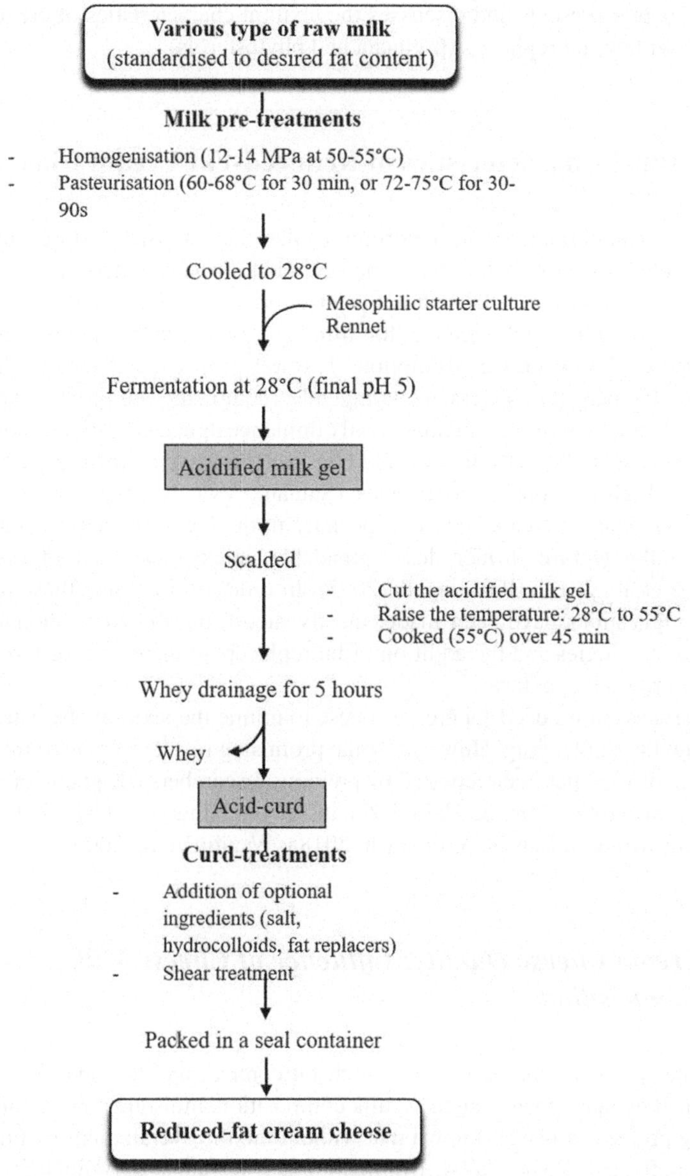

Fig. 22.1 Manufacture of reduced-fat cream cheese. Adapted from Ningtyas et al. (2017), Ningtyas et al. (2018b), Phadungath (2005)

the production parameters, like pH at breaking the milk gel and through the addition of thickeners or stabilisers such as hydrocolloid (Koca & Metin, 2004; Lashkari et al., 2014; Ningtyas et al., 2018b; Sahan, Yasar, Hayaloglu, Karaca, & Kaya, 2008; Zalazar et al., 2002).

This chapter presents and discusses the textural characteristics of cream cheese with and without fat replacers (β-glucan and phytosterols).

2 Textural Characteristics of Reduced-Fat Cream Cheese

Fat plays an important role in determining the characteristic texture, flavor, and aroma of cheese. Despite disturbing the acid gels (Frost & Janhoj, Frost, Prinz, & Ipsen, 2009; Janhoj et al., 2009), it has been theorized that fat globules have a "ball-bearing" effect which will create a thin film layer on the palate and give a lubrication property during oral consumption. Textural properties of cream cheese are influenced by many parameters, including gel structure, the condition of whey separation, and treatment of the curd that greatly influence its acceptability by consumers.

It is expected to be rich, smooth, and free from lumps and grittiness, attributes to which the high-fat content contributes (Sainani, Vyas, & Tong, 2004). Several defects can occur in cream cheese, especially in the low-fat types, such as grainy, sandy, chalky texture, firmer, less spreadable, sticky, and lack of creaminess (Ningtyas et al., 2018b; Phadungath, 2005). In order to overcome these problems, various suggestions have been made, mostly modifying the conventional cheese-making technologies and the addition of fat replacers to increase moisture content, which improves the texture.

The process of reduced-fat cream cheese to mimic the textural characteristics of the original is challenging. However, some promising results from textural and sensory point of view has been reported by previous researchers (Brighenti et al., 2008; Janhoj et al., 2009; Miri & Habibi Najafi, 2011; Ningtyas et al., 2017, 2018b; Ningtyas, Bhandari, Bansal, & Prakash, 2018a; Wendin et al., 2000).

2.1 Cream Cheese Quality: Influence of Cheese Milk Composition

Traditionally, cream cheese has very high-fat content, and the final fat content is determined by many factors such as milk composition, moisture content, and manufacturing process. It is well-known that without fat, the overall quality of the cream cheese is affected (Banks, 2004; Brighenti, 2009; Karaman & AkalıN, 2013; Koca & Metin, 2004; Romeih, Michaelidou, Biliaderis, & Zerfiridis, 2002). However, studies suggest it is possible to produce reduced-fat cream cheese with quality similar to high-fat cream cheese by implementing approaches as below:

a. Increase the moisture level or moisture to protein ratio: As an oil-in-water emulsion product, cream cheese texture is supported by a gel network of hydrated and emulsified casein proteins. The hydrated caseins are able to keep a large amount of water in cheese matrix due to hydrophilic interactions, and also allow them to interact with the fat phase via hydrophobic interactions (Zalazar et al., 2002). An interactive effect between the protein content and homogenisation pressure on cream [6.9 MPa (first stage) and 3.5 MPa (second stage)] also increases the moisture content of Cheddar cheese as reported by Oommen, Mistry, and Nair (2000).

b. Standardize the ratio of casein to fat (C/F) and % fat-in-dry matter (FDM): To increase the C/F ratio of the milk, the most common way is by removing fat from the milk or by adding reconstituted non-fat dry milk, hence the FDM of the cheese decreases (Johnson, Kapoor, Mcmahon, Mccoy, & Narasimmon, 2009).
c. Salt: The amount of salt added into the curd is another compositional parameters influencing the texture of cream cheese as reported by Andersen, Frøst, and Viereck (2010). Higher salt content yields larger fat droplets due to interactions between salt, protein, and fat (Wendin et al., 2000), contrary to the results obtained by Janhoj et al. (2009).

2.2 *Particle Size and Microstructure*

Previous research has reported mixed results about the particle size and microstructure of cream cheese as affected by reduced fat content. The particle size greatly depends on the method of making reduced-fat cream cheese. Results from Ningtyas et al. (2017) showed that an increased fat content resulted in a decrease in droplet size and particle size of aggregated droplets of cream cheese (Fig. 22.2). Similarly, Wendin et al. (2000), observed varied fat globule size in high-fat cream cheese that

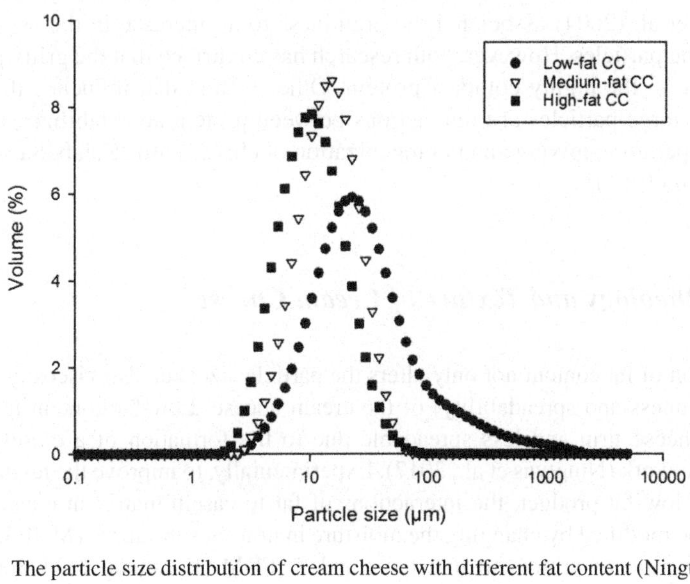

Fig. 22.2 The particle size distribution of cream cheese with different fat content (Ningtyas et al., 2017)

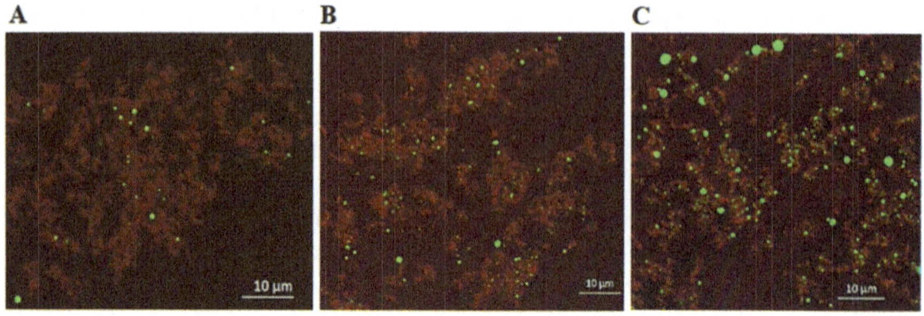

Fig. 22.3 Confocal micrographs of cream cheese samples: (**a**) low-fat, (**b**) medium-fat, (**c**) high-fat, with Nile red and Rhodamine B for staining fat (green) and protein (red), respectively (Ningtyas et al., 2017)

were dispersed in between the casein matrix (Fig. 22.3c), producing soft and spreadable cream cheese.

Conversely, as the fat content in cream cheese reduced, the fat globules are uniformly dispersed with less clumping, as observed in Fig. 22.3a. However, there are insufficient fat globules to keep the casein strands apart with a high degree of casein aggregation and a dense matrix as the curd is forming, resulting in increased particle size distribution (2.8–1995 μm), as can be seen from Fig. 22.2 (Ningtyas et al., 2017).

Several studies have reported the microstructural changes during cheese processing. A gritty or grainy mouthfeel perceived in low-fat cream cheese was attributed to the formation of small hard particles during the heating of cheese from 44 to 63 °C for 30 min (Modler, Kaláb, Yiu, & Bollinger, 1990). In contrary, Sainani et al. (2004) associated the graininess to an increase in the amount and size of the particles. However, both research has confirmed that the gritty particles were due to the highly compact protein. Other factors that influence the occurrence of large particles are interactions between protein and stabilizer, coagulation temperature, mixing and homogenization of cheese curd (Kalab, Sargant, and Froehlich (1981).

2.3 *Rheology and Texture of Cream Cheese*

Reduction of fat content not only alters the particle size, but also viscosity and texture (firmness and spreadability) of the cream cheese. Low-fat content makes the cream cheese firm and less spreadable due to the formation of a closely bound casein network (Ningtyas et al., 2017). Experimentally, to improve the textural attributes of low-fat product, the interactions of fat to casein matrix in cream cheese should be modified by changing the moisture in non-fat substances (MNFS, specifically the ratio of protein to moisture) and % FDM (Karaman & AkalıN, 2013; Lashkari et al., 2014). Karaman and Akalın (2013) reported that homogenisation of the cream during cheese making had a significant effect on increasing the dry matter and led to a larger number of fat particles being dispersed in the cheese matrix. In addition, at high concentrations of the fat replacer (guar gum), the elastic modulus (G′ values) decreases due to an increase in the moisture content in low-fat Iranian White cheese (Lashkari et al., 2014). However, an excessive amount of added stabilizers should be avoided as they can create a firmer texture due to an increase in the total solids content (Lucey, 2004).

Spreadability is the most important textural property for cream cheese and correlates well with the yield stress (σ_0). From market cream cheese with various fat levels and different brands, Breidinger and Steffe (2001) conclude that softer cream cheese was generally more spreadable with higher yield stress. Similar results were obtained by Monteiro, Tavares, Kindstedt, and Gigante (2009) and Ningtyas et al. (2017) who associate the low protein content to a less rigid network and thus softer cream cheese. From Fig. 22.4, the low-fat cream cheese is firmer and less spreadable (showed by higher spreadability value) than high-fat cream cheese.

Rheological behavior of cream cheese is not only determined by the total amount of fat in the product but also more by the amount of fat contained in crystal and liquid form. Also, Breidinger and Steffe (2001) found that the influence of stabilisers and texture modifiers could be a more important factor than emulsion characteristics and phase properties. A slight change in type and level of these ingredients can strongly influence the spreadability of cream cheese, which can also be confirmed with rheological properties (yield stress).

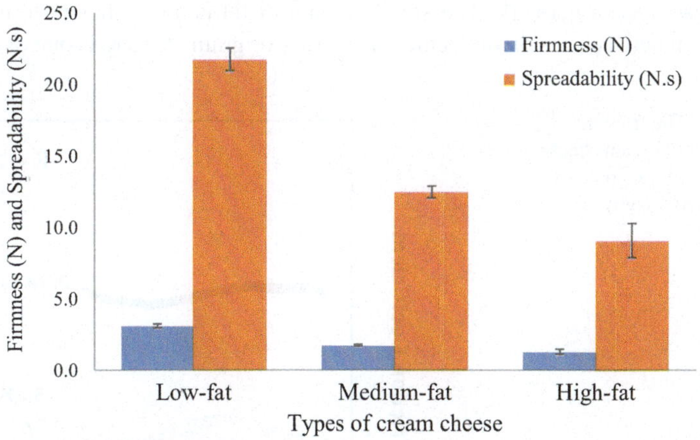

Fig. 22.4 Textural parameters (firmness and spreadability) of cream cheese samples with different fat content (low-fat, medium-fat, high-fat)

2.4 Tribology

Tribology refers to the measure of smoothness sensation related to the presence of fat/oil as lubricant and creaminess perception (Prakash, Tan, & Chen, 2013). The Stribeck curve of cream cheese samples with different fat content show the typical boundary, mixed and hydrodynamic regimes (Nguyen, Bhandari, & Prakash, 2016). As the fat content decreases, the friction coefficient of cream cheese increases due to the inability of fat to create a thin-film layer between two surfaces (Fig. 22.5). According to Lashkari et al. (2014), in the casein matrix of cream cheese, fat and moisture act as a filler and provide the lubricity and softness.

Ongoing research to improve the lubrication properties of cream cheese and relate to the dynamic oral perception is being undertaken. The use of fat replacers (phytosterols and β-glucan), single or in combination, significantly reduced the friction coefficient of cream cheese, similar to as high-fat counterpart, hence improve the spreadability of reduced-fat cream cheese (Ningtyas et al., 2018b). The mechanism on how fat replacers able to improve cream cheese texture will be discussed more detailed in Sect. 3.

2.5 Textural Defect in Cream Cheese as a Result of Reduced-Fat Content

A major concern in manufacturing low-fat cream cheese is the firm texture because of compact casein network. The small amount of fat is not sufficient to break the interaction between casein molecules. The gritty or grainy texture is one of the most

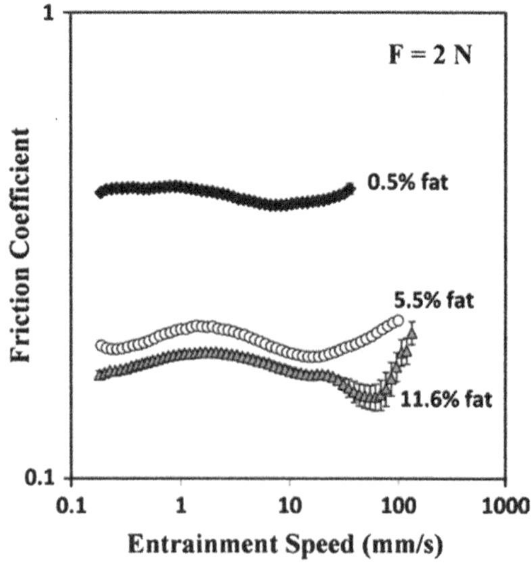

Fig. 22.5 The friction coefficient of cream cheese with different fat contents. Adapted from Nguyen et al. (2016)

detectable attributes when dealing with low-fat products. This could be due to the shape of casein particles formed during gelation (Sainani et al., 2004) and the interactions between protein and fat replacers which act as stabilisers or thickeners (Kalab et al., 1981). In addition, Kalab et al. (1981) also mention that the structure of the cream cheese particle (corpuscular) determines the spreadability and smoothness of the final product. Therefore, mixing and homogenisation of the cheese curd during manufacture is important to reduce the particle size and create a smooth and spreadable texture.

Another possible way to disrupt the association between casein molecules and reduce its density is by increasing the moisture content and adding the fat replacers such as starch, hydrocolloids, denatured whey protein or any other materials, which can interact with casein (Lashkari et al., 2014). These have been proposed as the main strategy to improve functional and textural attributes of low-fat cheese in different studies. The soft and smooth texture of cheese made with fat replacers, such as xanthan gum and β-glucan, was expected as fat act as a lubricant and impeded the formation of a solid protein matrix (Romeih et al., 2002; Volikakis, Biliaderis, Vamvakas, & Zerfiridis, 2004). Interestingly, the results from Zalazar et al. (2002) on low-fat soft cheese (Cremoso Argentino) showed that increasing moisture content up to 60% produced an acceptable texture and quality of cream cheese without using fat replacer.

3 Reduced-Fat Cream Cheese

In recent times, many researchers have been working on developing reduced-fat cream cheese with acceptable texture and sensory characteristics. This is achieved by either modifying the cream cheese making process or through the addition of fat replacers.

3.1 Modifying the Cream Cheese Making Process

Since the last decade, many research has been conducted in order to produce a high-quality cream cheese texture but with lower fat content. Fat has an important role in the development of textural properties of cream cheese. Modifying the processing technology on making reduced-fat cream cheese has been proposed to optimise the function of fat in textural attributes. Homogenisation of cream has significantly reduced the hardness of cream cheese by increasing the fat dry matter, thus producing softer, smoother and creamier cream cheese (Karaman & AkalıN, 2013). Through homogenisation, the fat globule size becomes smaller and hence distribute more evenly in the casein matrix. Moreover, Almena-Aliste and Kindstedt (2005) also reported that at pH value between 4.8 and 5.2 the texture of cream cheese is softer than at pH 4.6–4.7. Incorporation of stabiliser and fat replacer into cream cheese curd found to be the most effective method in manufacturing reduced-fat cream cheese. These compounds caused an increase in moisture content and cheese yield lead to a great improvement in cheese texture (Koca & Metin, 2004; Romeih et al., 2002). The fat replacers act as an active filler that interacts with the casein micelle in cream cheese matrix. By breaking the casein-casein bond, the firmness of cream cheese is reduced and the spreadability is increased.

3.2 *Fat Replacers to Improve Reduced-Fat Cream Cheese Texture*

An increasing number of degenerative health problems such as obesity, coronary heart disease, and elevated blood pressure is associated with higher consumption of dietary fat. Therefore, the fat replacers have a growing interest in developing low-fat products due to its ability to mimic the functional properties of fat in foods without having any negative effect on health. A reduction in fat can be achieved by replacing it with carbohydrate-, protein-, and/or lipid-based fat replacers, used individually or in combination). The health benefits from fat replacers such as β-glucan and phytosterols are well reported. As per regulations (FSANZ and FDA), daily consumption of 1.3 g plant sterols and/or 3 g β-glucan is favorable in reducing cholesterol levels and glycemic index.

According to Ognean, Darie, and Ognean (2006) and Shahidi and Senanayake (2000), most of the carbohydrate-based fat replacers are hydrocolloids with a greater ability to bind water, thus can act as thickening and gelling agent with the ability to mimic the mouthfeel and improve the lubrication properties similar to fat. Figure 22.6 represents the gelation mechanisms of hydrocolloids as fat replacers.

Protein-based mimetic, such as gelatine, whey protein concentrate, or microparticulated protein, have limited applications since these ingredients are sensitive to high temperature. Generally used in dairy products, salad dressings, frozen desserts, and margarine, they provide good emulsification properties. The fat substitutes are compounds that physically and chemically resemble triglycerides and widely used to replace fat on a one-to-one basis (Akoh, 1998).

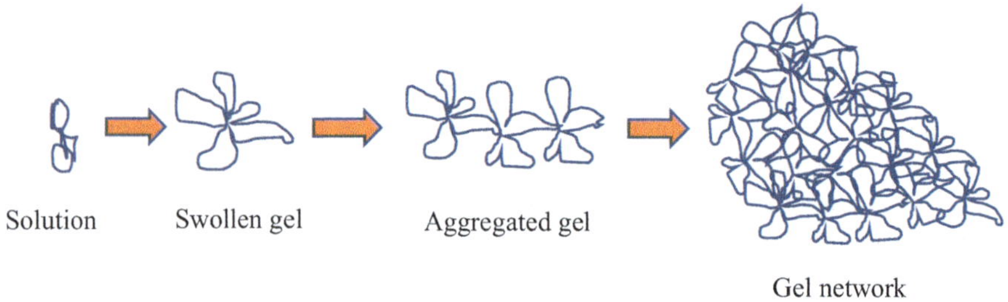

Fig. 22.6 The gelation mechanisms of hydrocolloid as carbohydrate fat-replacer

The role of fat replacers in improving textural properties of cream cheese depends on their interactions with the casein network. In the matrix where fat, water and other ingredients are entrapped or bind within the casein networks, fat globules act as a filler to break the casein-casein bond (Guinee & Mcsweeney, 2006; Malin & Tunick, 1995) as can be seen in Fig. 22.7. Lobato-Calleros et al. (2007) stated that the structural changes induced by WPC in the protein network of reduced-fat Manchego cheese were due to whey protein-casein and whey protein-water interactions.

Among fat replacers, water-soluble and polar fat substitutes have been widely recommended because of their ability to entrap water mechanically and their strong hydrophilicity. According to Fox et al. (2017), the addition of fat replacers may allow: (1) binding of water that increase the viscosity and minimise casein aggregation, (2) product structuring to improve the body of cream cheese. These can be achieved by forming a polymer gel network that reinforces and immobilizes the existing network, or binding to the protein and creating a particle network which limiting the mobility and re-arrangement of the gel structure. Neutral hydrocolloids such as guar gum, xanthan gum, and locust bean gum are commonly added as a stabiliser to the curd after whey separation with controlled shear mixing. In addition, xanthan gum provides lubrication property similar to fat yielding a soft and smooth texture of cheese (Romeih et al., 2002). Similarly, Malin and Tunick (1995) also found that microparticulated whey protein particles were distributed evenly in the cheese matrix, similarly like fat, and not bound to the matrix, resulting in a softer cheese.

Very limited research has been published on the production of reduced-fat cream cheese with the incorporation of fat replacers to improve the texture. Salari et al. (2017) developed cream cheese with reduced-fat content by adding different concentrations of xanthan gum and CMC. They observed that low-level (0.05%) of xanthan gum had a more desirable effect while 0.02–0.2% of CMC produces cream cheese with acceptable texture, flavour, odour, colour, and overall quality.

Ningtyas et al. (2018b) have utilized two functional ingredients, β-glucan and phytosterols, which act as fat replacer to produce low-fat cream cheese. Analyses

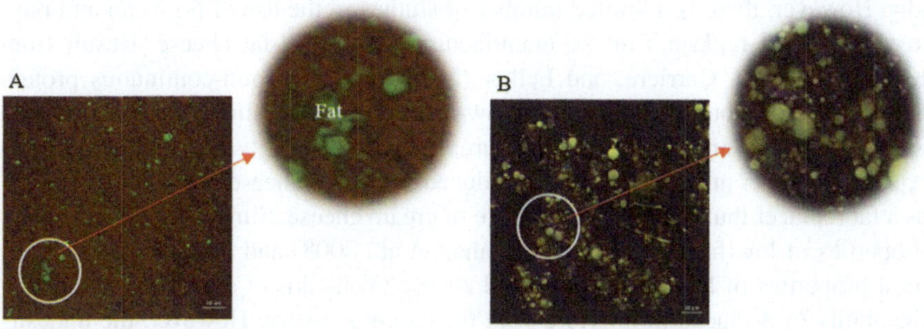

Fig. 22.7 Microstructure images of medium-fat cream cheese: (**A**) Without fat replacers and (**B**) with fat replacers (β-glucan & phytosterols). The CLSM images (magnification 60×) are representative of the overlay of signals obtained from Calcofluor (β-glucan/purple), Rhodamine B (protein/red) and Nile Red (fat/green). Modified from Ningtyas et al. (2018b)

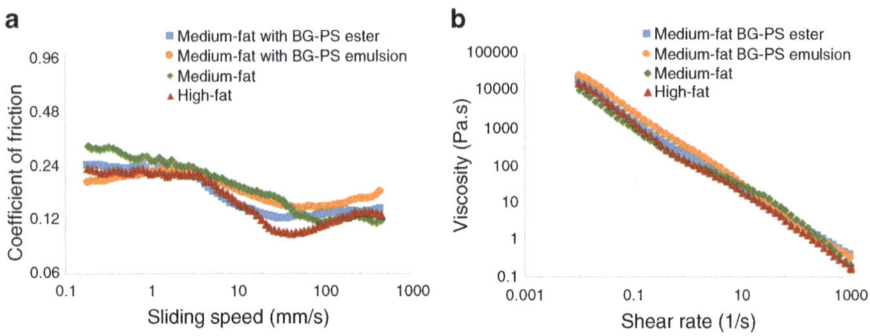

Fig. 22.8 Lubrication properties (**a**) and viscosity (**b**) of functional cream cheese with added fat replacer (β-glucan and phytosterols) compared to medium-fat and high-fat cream cheese

performed on the samples demonstrate that the combined use of both fat replacers exhibit a synergistic effect improving the cream cheese texture. β-glucan will bind the water and act as a filler to break the casein matrix, while phytosterols will provide the lubrication properties to create a more spreadable cream cheese (Fig. 22.8). A similar increase in spreadability of low-fat fresh Kashar cheese has been reported by Koca and Metin (2004) using a combination of protein- and carbohydrate-based fat replacers (Simplesse®D-100, Dairy-Lo™, Raftiline®HP). The used of these fat replacers have shown an increase in moisture content and moisture of non-fat substances thus soften the texture of low-fat fresh Kashar cheese.

3.3 Textural Characteristics of Cream Cheese with Fat Replacers

Research on the application of fat replacers in various low-fat food has grown rapidly. However, there is a limited number of studies in the use of β-glucan and phytosterols as fat replacers in the manufacture of reduced-fat cheese. Result from Konuklar, Inglett, Carriere, and Felker (2004) showed a non-continuous protein network with globular clusters from low-fat cheese containing Nutrim (β-glucan) resulting a softer and less rubbery texture, meanwhile Ningtyas et al. (2018b) also reported a higher moisture content in reduced-fat cream cheese containing β-glucan as a fat replacer thus altered the structure of cream cheese. Similarly, better-melting properties of low-fat Kashar cheese (Sahan et al., 2008) and improving in all textural properties of low-fat white brined cheese (Volikakis et al., 2004) suggesting the ability of β-glucan to interfere with the casein network. However, the undesirable flavor and mouthfeel in cheese because of β-glucan addition might occur and affect consumer preferences. Therefore, more research is needed to investigate the interactions of β-glucan with other ingredients, especially on the textural and sensory attributes.

Combination used of fat replacers to improve textural characteristics of cream cheese with reduced fat content has shown good results. A study by Ningtyas et al. (2018b) has shown that incorporating phytosterols together with β-glucan as fat replacers in reduced-fat cream cheese produce cream cheese with similar textural and lubrication properties as its full-fat counterpart. Phytosterols were plays an important role in lower the friction coefficient thus create a softer and more spreadable cream cheese. Another research by Koca and Metin (2004) also found that Simplesse®D-100 and Raftiline®HP successfully corrected all the defect appear from low-fat fresh Kashar Cheese. The results indicated that combination used of both fat replacers were effectively improved the textural and sensory properties of the cheese for over 60 days of storage. Similarly, Sahan et al. (2008) reported that low-fat Kashar cheese with Avicel Plus® CM 2159 (a microcrystalline mixture of cellulose, sodium carboxymethylcellulose, sucrose, carrageenan) as fat replacer had sensory attributes closest to full-fat Kashar cheese.

4 Conclusion

Cream cheese is an acid-fresh cheese with soft, smooth and spreadable texture. It is normally consumed as a spread or as an ingredient for baking and cooking. They normally contain a high amount of fat but in recent times, the consumers are increasingly interested in reduced-fat cream cheese. The processing steps in making reduced-fat cream cheese are modified from the original process mainly from the fat and protein ratio in milk, moisture content, and also physical treatments such as mixing, homogenisation, and heating. However, omitting the fat will significantly change the textural characteristics such as firmness, viscosity, smoothness, spreadability, and lubrication behavior. Fat plays an important role in casein matrix as an active filler to break the bond between casein-casein. Therefore, to improve the defect caused by reducing fat, more studies to add fat replacers has been going on. The type of fat replacers (carbohydrate-, protein-, or lipid-based) and the amount added influence the final texture of cream cheese. It has been demonstrated that these ingredients are able to entrap water and/or interact with casein and resulting in softer, smoother, and more spreadable cream cheese. A potential exists for a combination of β-glucan and plant sterols to be added for making reduced-fat cream cheese as well as provide the health benefits of reducing cholesterol levels. Further research on the bioavailability, digestibility, and its correlation with the dynamic change during oral processing in the mouth will require to be pursued.

References

Akoh, C. C. (1998). Fat replacers. *Technologists, 52*(3), 47–53.

Almena-Aliste, M., & Kindstedt, P. S. (2005). Effect of increasing pH on texture of full and reduced-fat cream cheese. *Australian Journal of Dairy Technology, 60*, 225–230.

Andersen, C. M., Frøst, M. B., & Viereck, N. (2010). Spectroscopic characterization of low- and non-fat cream cheeses. *International Dairy Journal, 20*, 32–39.
Banks, J. M. (2004). The technology of low-fat cheese manufacture. *International Journal of Dairy Technology, 57*, 199–207.
Breidinger, S. L., & Steffe, J. F. (2001). Texture map of cream cheese. *Journal of Food Science: Sensory and Nutritive Qualities of Food, 66*, 453–456.
Brighenti, M. (2009). *Texture and functionality of cream cheese*. Doctor of Philosophy, University of Wisconsin-Madison.
Brighenti, M., Govindasamy-Lucey, S., Lim, K., Nelson, K., & Lucey, J. A. (2008). Characterization of the rheological, textural, and sensory properties of samples of commercial US cream cheese with different fat contents. *Journal of Dairy Science, 91*, 4501–4517.
Coutouly, A., Riaublanc, A., Axelos, M., & Gaucher, I. (2014). Effect of heat treatment, final pH of acidification, and homogenization pressure on the texture properties of cream cheese. *Dairy Science and Technology, 94*, 125–144.
Crane, L. A. (1992). Method of manufacture of a non-fat cream cheese product. United States patent application 5,079,024.
Fox, P. F., Guinee, T. P., Cogan, T. M., & Mcsweeney, P. L. H. (2017). *Fundamentals of cheese science*. Boston, MA: Springer.
Guinee, T. P., & Mcsweeney, P. L. H. (2006). Significance of milk fat in cheese. In P. F. Fox & P. L. H. Mcsweeney (Eds.), *Advanced dairy chemistry volume 2: Lipids*. Boston, MA: Springer.
Janhoj, T., Frost, M. B., Prinz, J., & Ipsen, R. (2009). Sensory and instrumental characterization of low-fat and non-fat cream cheese. *International Journal of Food Properties, 12*, 211–227.
Johnson, M. E., Kapoor, R., Mcmahon, D. J., Mccoy, D. R., & Narasimmon, R. G. (2009). Reduction of sodium and fat levels in natural and processed cheeses: Scientific and technological aspects. *Comprehensive Reviews in Food Science and Food Safety, 8*, 252–268.
Kalab, M., Sargant, A. G., & Froehlich, D. A. (1981). Electron microscopy and sensory evaluation of commercial cream cheese. *Scanning Electron Microscopy, 3*, 473–482.
Karaman, A. D., & AkalıN, A. S. (2013). Improving quality characteristics of reduced and low fat Turkish white cheeses using homogenized cream. *Food Science and Technology, 50*, 503–510.
Koca, N., & Metin, M. (2004). Textural, melting and sensory properties of low-fat fresh kashar cheeses produced by using fat replacers. *International Dairy Journal, 14*, 365–373.
Konuklar, G., Inglett, G. E., Carriere, C. J., & Felker, F. C. (2004). Use of a β-glucan hydrocolloidal suspension in the manufacture of low-fat cheddar cheese: Manufacture, composition, yield and microstructure. *International Journal of Food Science & Technology, 39*, 109–119.
Lashkari, H., Madadlou, A., & Alizadeh, M. (2014). Chemical composition and rheology of low-fat Iranian white cheese incorporated with guar gum and gum arabic as fat replacers. *Journal of Food Science and Technology, 51*, 2584–2591.
Lobato-Calleros, C., Reyes-Hernández, J., Beristain, C. I., Hornelas-Uribe, Y., Sánchez-García, J. E., & Vernon-Carter, E. J. (2007). Microstructure and texture of white fresh cheese made with canola oil and whey protein concentrate in partial or total replacement of milk fat. *Food Research International, 40*, 529–537.
Lucey, J. A. (2004). Formation, structural properties and rheology of acid-coagulated milk gels. In P. F. Fox, P. L. H. Mcsweeney, T. M. Cogan, & T. P. Guinee (Eds.), *Cheese - Chemistry, physics and microbiology* (3rd ed.). London, UK: Elsevier Science and Technology.
Madadlou, A., Khosroshahi, A., & Mousavi, M. E. (2005). Rheology, microstructure, and functionality of low-fat iranian white cheese made with different concentrations of rennet. *Journal of Dairy Science, 88*, 3052–3062.
Madadlou, A., Mousavi, M. E., Khosrowshahi Asl, A., Emam-Djome, Z., & Zargaran, M. (2007). Effect of cream homogenization on textural characteristics of low-fat Iranian White cheese. *International Dairy Journal, 17*, 547–554.
Malin, E. L., & Tunick, M. H. (1995). *Chemistry of structure-function relationships in cheese*. New York, NY: Springer Science+Business Media, Llc..
Miri, M. A., & Habibi Najafi, M. B. (2011). The effect of adding enzyme-modified cheese on sensory and texture properties of low- and high-fat cream cheeses. *International Journal of Dairy Technology, 64*, 92–98.

Mistry, V. V. (2001). Low fat cheese technology. *International Dairy Journal, 11*, 413–422.

Modler, H. W., Kaláb, M., Yiu, S. H., & Bollinger, U. K. (1990). *Detection and prevention of grittiness in a cream cheese spread*. Brussels, Belgium: International Dairy Federation.

Monteiro, R. R., Tavares, D. Q., Kindstedt, P. S., & Gigante, M. L. (2009). Effect of pH on microstructure and characteristics of cream cheese. *Journal of Food Science, 74*, C112–C117.

Muir, D. D., Williams, S. A. R., Tamime, A. Y., & Shenana, M. E. (1997). Comparison of the sensory profiles of regular and reduced-fat commercial processed cheese spreads. *International Journal of Food Science & Technology, 32*, 279–287.

Nguyen, P. T. M., Bhandari, B., & Prakash, S. (2016). Tribological method to measure lubricating properties of dairy products. *Journal of Food Engineering, 168*, 27–34.

Ningtyas, D. W., Bhandari, B., Bansal, N., & Prakash, S. (2017). A tribological analysis of cream cheeses manufactured with different fat content. *International Dairy Journal, 73*, 155–165.

Ningtyas, D. W., Bhandari, B., Bansal, N., & Prakash, S. (2018a). Effect of homogenisation of cheese milk and high-shear mixing of the curd during cream cheese manufacture. *International Journal of Dairy Technology, 71*, 417–431.

Ningtyas, D. W., Bhandari, B., Bansal, N., & Prakash, S. (2018b). Texture and lubrication properties of functional cream cheese: Effect of beta-glucan and phytosterol. *Journal of Texture Studies, 49*, 11–22.

Ognean, C. F., Darie, N., & Ognean, M. (2006). Fat replacers - Review. *Journal of Agroalimentary Processes and Technologies, 7*, 433–442.

Oommen, B. S., Mistry, V. V., & Nair, M. G. (2000). Effect of homogenization of cream on composition, yield, and functionality of Cheddar cheese made from milk supplemented with ultrafiltered milk. *Le Lait, 80*, 77–91.

Phadungath, C. (2005). Cream cheese products: A review. *Songklanakarin Journal of Science and Technology, 27*(1), 191–199.

Prakash, S., Tan, D. D. Y., & Chen, J. (2013). Applications of tribology in studying food oral processing and texture perception. *Food Research International, 54*, 1627–1635.

Romeih, E. A., Michaelidou, A., Biliaderis, C. G., & Zerfiridis, G. K. (2002). Low-fat white-brined cheese made from bovine milk and two commercial fat mimetics: Chemical, physical and sensory attributes. *International Dairy Journal, 12*, 525–540.

Rudan, M. A., Barbano, D. M., Guo, M. R., & Kindstedt, P. S. (1998). Effect of the modification of fat particle size by homogenization on composition, proteolysis, functionality, and appearance of reduced fat mozzarella cheese. *Journal of Dairy Science, 81*, 2065–2076.

Sahan, N., Yasar, K., Hayaloglu, A. A., Karaca, O. B., & Kaya, A. (2008). Influence of fat replacers on chemical composition, proteolysis, texture profiles, meltability and sensory properties of low-fat Kashar cheese. *Journal of Dairy Research, 75*, 1–7.

Sainani, M. R., Vyas, H. K., & Tong, P. S. (2004). Characterization of particles in cream cheese. *Journal of Dairy Science, 87*, 2854–2863.

Salari, S., Zanganeh, M., Fadavi, A., & Ahmadi, Z. (2017). Effect of xanthan gum and carboxymethyl cellulose on chemical and sensory properties of cream cheese. *International Journal of Advancements in Technology, 8*, 175–180.

Shahidi, F., & Senanayake, S. P. J. N. (2000). Fat replacers. In *Kirk-Othmer encyclopedia of chemical technology*. Indianapolis, IN: John Wiley & Sons, Inc.

USDA. (1994). *USDA specifications for cream cheese*. Retrieved from http://www.ams.usda.gov/Amsv1.0/getfile?dDocName=Steldev3004555

Volikakis, P., Biliaderis, C. G., Vamvakas, C., & Zerfiridis, G. K. (2004). Effects of a commercial oat-β-glucan concentrate on the chemical, physico-chemical and sensory attributes of a low-fat white-brined cheese product. *Food Research International, 37*, 83–94.

Wendin, K., Langton, M., Caous, L., & Hall, G. (2000). Dynamic analyses of sensory and microstructural properties of cream cheese. *Food Chemistry, 71*, 363–378.

Zalazar, C. A., Zalazar, C. S., Bernal, S., Bertola, N., Bevilacqua, A., & Zaritzky, N. (2002). Effect of moisture level and fat replacer on physicochemical, rheological and sensory properties of low fat soft cheeses. *International Dairy Journal, 12*, 45–50.

Chapter 23
Dairy Fat Replacement in Low-Fat Cheese (LFC): A Review of Successful Technological Interventions

Bal Kumari Sharma Khanal and Nidhi Bansal

1 Introduction

Cheese is a fermented dairy product which features a variety of flavours and is produced all over the world. It is prepared by coagulation of milk proteins with the help of coagulant enzyme, acid or both. Lactic acid bacteria are commonly used to produce acid in situ during cheesemaking (McElhatton & El Idrissi, 2016). The earliest cheese was made in Central Asia by Nomadic tribes (Dalby, 2009; Scott, Robinson, & Wilbey, 1998). The cheese was believed to have evolved from Tigris and Euphrates rivers (current Iraq) 8000 years ago when people started to domesticate plants and animals as a source of food (Fox, 2011). More than 1400 kinds of cheeses are listed in a record of the University of Wisconsin Centre for Dairy Research (CDR, 2014). Nevertheless, there are 18 common types of cheese: Brick, Trappist, Camembert, Neufchatel, Roquefort, Edam, Gouda, Sapsago, Hand, Cheddar, Limburger, Provolone, Cottage, Cream, Parmesan, Romano, Swiss and Whey cheese (McSweeney, Ottogalli, & Fox, 2017). There are different classes of cheese. The majority of cheeses are classified according to their texture and moisture content, namely very hard, hard, semi-hard, semisoft and soft. Cheese can be prepared from whole milk, partially skimmed milk or skimmed milk, all of which can be obtained from cow, buffalo, camel, goat, yak, sheep, reindeer, horse, and donkey milk, in addition to milk powder (Fox & McSweeney, 2017).

Milk and dairy products including cheese provide vital nutrients for our body such as proteins, fat and other minerals like calcium and phosphorous (Johnson, 2016). Cheese contains 3–40% protein and 4–48% fat depending upon the type of cheese and method of manufacturing. Furthermore, cheese is a good source of vitamin A, B_{12}, riboflavin and folates. All the essential amino acids excluding cysteine

B. K. S. Khanal · N. Bansal (✉)
School of Agriculture and Food Sciences, The University of Queensland, St Lucia, QLD, Australia
e-mail: n.bansal@uq.edu.au

T. Truong et al. (eds.), *Dairy Fat Products and Functionality*,
https://doi.org/10.1007/978-3-030-41661-4_23

and methionine are present in cheese in an abundant amount (McCarthy, O'Connor, & O'Brien, 2014). Besides other nutrients present in cheese, fat plays a crucial role in sensory perception of cheese. Fat acts as filler particles in three dimensional protein network of cheese (Khanal, Bhandari, Prakash, & Bansal, 2017) and provide lubricity, creaminess and mouthfeel in cheese along with other sensory and functional properties. Besides, it also helps to develop the flavor and affects the storage stability of food products (Lashkari, Khosrowshahi Asl, Madadlou, & Alizadeh, 2014). Removal of fat from cheese leads to a hard and rubbery texture, off-flavours, poor colour and poor meltability as compared to their full-fat counterparts (Ibáñez, Waldron, & McSweeney, 2016; Zahra, Mohammad, Sahel, & Mohammad Ali, 2013). Hence, production of low-fat cheese that has comparable characteristics and consumer acceptance as its full-fat counterpart with more than 50% fat reduction has always being a challenging task for the researchers and manufacturers (Johnson, 2016; Jooyandeh, Goudarzi, Rostamabadi, & Hojjati, 2017).

Of late, preference for consumption of low-fat cheese (LFC) over the regular full-fat cheese has been increasing due to low risk of obesity and cardiovascular diseases. Owing to health benefits of LFC, its sale in the market has increased due to high consumer demand (Di Cagno et al., 2014; Ibáñez et al., 2016). Due to increase in obesity percentage (BMI > 30) in men and women from 1979 to 2012 along with increase in percentage of extremely obese (BMI > 40) population, the United States has recommended to limit calories from added sugars and saturated fats and consume less than 10 percent calories per day from saturated fats (Reedy, 2016). The US Department of Health and Human Services has given greater attention to consumption of fat-free or low-fat dairy products including milk, yoghurt, cheese and/or fortified soy beverages as a key recommendation in its five overarching guidelines (USDA, 2016). WHO has also given advice on maintaining a healthy diet by limiting the consumption of foods containing high amounts of saturated fats, such as cheese, ice-cream and fatty meat (WHO, 2015). One study conducted in the Netherlands found that cheese was the major contributor (17.4%) to the dietary saturated fat among other dairy products. Hence, dairy products like cheese, due to high levels of saturated fat and cholesterol need to be cut down from the diets of dyslipidaemia patents (Reedy, 2016).

1.1 Definition of LFC

According to Food and Drug Administration (FDA, 2017), cheese is described by the level of fat to other solid matters in/or permissible amount of moisture and fat in dry matter. CODEX has not given any legitimacy for the zero-fat Cheddar cheese for the international trade. Nevertheless, each country can fix their standards for internal trade from their corresponding referenced variety of cheese (CODEX, 2013). In the United States of America, the cheese with 6% fat is considered as a LFC while the cheese in which fat level is reduced to at least 25% from the original one is termed as reduced fat cheese (Johnson, 2011). Composition of cheese with different fat levels is shown in Table 23.1.

Table 23.1 Composition of cheese

Nutrients	Full-fat	Reduced-fat/100g	Low-fat
Moisture	36–39	42–48	50–57
Ash	3.2	3.5–3.9	4.9
Protein (g)	24–26	28–32	30–35
Fat (g)	32–37	15–18	5–6

Source: Kumar (2012), Kumar et al. (2011)

1.2 Issues with LFC

Although industrial-scale production of LFC has been started more than two decades ago, the massive growth in market for LFC has slowed down due to reduced consumer acceptance of the first generation LFC owing to their inadequate flavor and texture as compared to full fat cheese (Banks, 2004; Di Cagno et al., 2014). LFC exhibits hard or firm and grainy texture, dry, off- flavor, bitterness and dull colour as well as possess inferior mechanical and melting properties as compared to full-fat cheese (Costa et al., 2010; Drake, Miracle, & McMahon, 2010; Zahra et al., 2013; Zisu, 2005). The hardness and springiness of LFC is more than full-fat cheese, while adhesiveness and cohesiveness are less than that of full-fat cheese (Bryant, Ustunol, & Steffe, 1995; Rodriguez, 1998; Zahra et al., 2013). This is due to the larger structural protein matrix per unit cross-sectional area in low and reduced fat cheese (Katsiari & Voutsinas, 1994; Rodriguez, 1998). In LFC, there is a high amount of protein/casein with respect to the fat and as a result, high amount of calcium is retained in cheese as colloidal calcium. This higher amount of calcium is also responsible for the poor melt and stretch properties of some cheeses (Koca & Metin, 2004; Nauth & Hayashi, 1995).

Lactose crystallization is another factor responsible for the rubbery and firm body of LFC (Emmons, Emmons, Kalab, Larmond, & Lowrie, 1980). LFC cheese milk has less fat, which consequently leads to higher protein concentration. Colloidal calcium (bound form) is increased by a higher amount of protein. Hence, the concentration of soluble calcium will be increased by the increase in colloidal calcium if cheese has low pH. The soluble calcium can combine with lactose and form calcium lactate. Several factors such as milk composition and quality, cheese making techniques, starter culture used and cheese ripening conditions affect the formation of calcium lactate crystals, which thus leads to difference in textural and visual attributes (Agarwal, Powers, Swanson, Chen, & Clark, 2008). The degree of proteolysis is also a major factor in determining texture, i.e. the cheese firmness. During proteolysis, peptides and amino acids are produced by hydrolyzing casein network by the action of the protease enzymes. Amino acids and peptides produced by proteolysis are responsible for the development of improved characteristics, desirable and undesirable flavor in cheese (Johnson, 2016; Khosrowshahi, Madadlou, Ebrahim zadeh Mousavi, & Emam-Djomeh, 2006). Fat reduction in cheese leads to off flavours and bitterness, and is also due to high moisture content and low amount of available aromatic compounds formed by degradation of fat (Di Cagno et al., 2014; Rogers et al., 2009). Rheology and microstructure of cheese are also affected by the reduction of fat level. Open-intricate microstructure of cheese vanishes by reducing the fat content. Removal of fat in cheese decreases the yield of cheese as well (Farkye & Guinee, 2017; Katsiari & Voutsinas, 1994; Rodriguez, 1998).

2 LFC Technology: Processing Modifications During Cheese Making

Modern technical approaches have been established currently to overcome flavor and texture defects in the LFC through the application of sound knowledge to physicochemical and biochemical properties of LFC. Considering the important function of fat in relation to textural characteristics and acceptability of LFC, several modification approaches and different types of fat replacers or hydrocolloids have been recommended (Johnson, 2016; Jooyandeh et al., 2017). Fat replacers are used either to partially or fully replace milk fat. The common LFC prepared by different modification techniques are both soft and hard ripened varieties such as Cottage, Minash fresh, Gouda, Kasar, Cream, Cheddar, Keshire, Mozzarella and Feta cheese (Banks, Brechany, & Christie, 1989; Chatli, Gandhi, & Singh, 2017; Diamantino, Beraldo, Sunakozawa, & Penna, 2014; Felfoul, Bornaz, Baccouche, Sahli, & Attia, 2015; Sanli, Gursel, Sanli, Acar, & Benli, 2013).

Improvement in LFC can be made by increasing the moisture content of the cheese and by modifying cheese manufacturing procedure (Johnson, 2016). Some of the processing interventions are discussed below.

2.1 *Starter Microorganisms, Adjunct Cultures, Coagulants and Enzymes*

Starter bacteria used in cheese making play a crucial role during proteolysis and hence the development of texture and flavor in cheese during ripening. There is a high possibility of change in growth and microbial metabolism due to different nutritional environment in LFC (Farkye & Guinee, 2017). High moisture in LFC develops off flavor and bitterness due to the activity of bacteria and formation of bitter peptides. Use of heat treated lactobacilli enhances flavor production and decreases bitterness in 10% fat semi-hard, round-eye cheese (Ardö, Mansson, Hedenberg, & Larsson, 1989). It is recommended to use the starter with low proteolytic and high peptidolytic activity (Mistry, 2001). Reduction of fat in Cheddar cheese is associated with the changes in the number of bacterial population, whereas it does not change the predominant bacteria in cheese. Hence, this indicates the difference in ripening biochemistry in LFC and FFC, thus adjunct culture is needed to enhance the flavour in LFC (Broadbent et al., 2013). Studies carried out using *Lactobacillus casei* TINE36 and *L. plantaram* TINE18 (Skeie et al., 2013) in LFC resulted in improved firmness; and *L. plantarum* LP and *L. rhamnosus* LRA in low-fat Caciotta-type cheese (Di Cagno et al., 2014) showed higher levels of acetic acid, diacetyl and other volatile compounds, thus increasing overall acceptability of cheese.

Protease enzymes from the plant origin, such as *Cynara cardunculus*, are used instead of rennet to produce some cheese varieties such as Quwso Serra Cheese of Portugal and this enzyme improves the flavor of low fat (13%) Cheddar cheese. Nonetheless, it develops bitterness during ripening period (Banks, Roa, & Muir, 1998).

Proteolysis in cheese is triggered by increasing the amount of coagulant enzyme used, using coagulant with higher rate of proteolytic to milk clotting activity than calf rennet, and such proteases are extracted from *Rhizomucor miehei* and *R. pusillus* (McCarthy, Wilkinson, & Guinee, 2017b). Similarly, proteolysis in LFC can be altered by replacing the bovine rennet with camel rennet that has a higher ratio between clotting and general proteolytic (C/P) activity and has C/P ratio seven times higher than bovine rennet. Furthermore, camel rennet hydrolyses κ-CN Phe105-Met106 bond faster than bovine rennet in milk, and is less sensitive to calcium content, pH and heat (from 40 to 55 °C) (Børsting et al., 2012; Soodam, Ong, Powell, Kentish, & Gras, 2015a). However, other studies reported that due to slower rate of proteolysis, low-fat Cheddar cheese and ultra-filtered Iranian white cheese prepared by recombinant camel rennet as a coagulant had a thicker protein network as compared to the one made from microbial rennet. In addition, these cheeses had less protein breakdown (possessed a lower ratio of pH 4.6 soluble nitrogen/total nitrogen) than the cheese made from microbial rennet (Soltani, Boran, & Hayaloglu, 2016; Soodam, Ong, Powell, Kentish, & Gras, 2015b). Moreover, study conducted by Akkerman et al. (2017) in NaCl reduced semi-hard Danish cheese and (McCarthy et al., 2017b) in half-fat, half-salt Cheddar cheese, use of camel rennet resulted in a higher firmness than using bovine chymosin.

2.2 *Cooking Temperature, Calcium Reduction, Cutting pH, Pre-acidification and Ripening Temperature*

Calcium in cheese is reduced by lowering pH at coagulation and during whey drainage. Calcium-induced crosslinking would help to alleviate adverse effect by dense casein network in a cheese texture (McCarthy, Wilkinson, & Guinee, 2017c). Calcium reduction in half-fat, half-salt Cheddar style cheese is characterized by higher primary proteolysis and lower secondary proteolysis and higher water holding capacity. In addition, it counterbalances the detrimental effect of fat reduction in textural and cooking properties, provides more extensive flow on heating, lowers fracture stress and strain (Farkye & Guinee, 2017; McCarthy et al., 2017c). Both pH and temperature play an important role for dissolution of calcium and dissociation of casein from the casein micelles. Pre-acidification of milk to pH 5.4 at 4 °C solubilizes more calcium, which thus results in a maximum dissociation of casein (30%)

(Farkye & Guinee, 2017). Pre-acidification of milk or lowering pH of milk is achieved by different approaches such as using food grade acids (e.g. lactic), using higher level of starter culture, CO_2 injection and prolonging the curd holding time in a cheese vat. Moisture content of cheese is increased by the use of pre-acidified milk, which is accomplished by extending holding time prior to whey drainage (McCarthy et al., 2017c). Furthermore, to alter textural properties of cheese, pH during manufacturing can also be controlled by adding salt, altering the buffering capacity of cheese and decreasing the level of residual lactose (Chevanan, Muthukumarappan, Upreti, & Metzger, 2006).

In LFC manufacturing, low cooking temperature and high pH at draining have been practised to overcome possible defects. To manufacture 25 and 50% reduced fat Cheddar cheese, 37 °C and 35 °C cooking temperature have been recommended, respectively (Farkye & Guinee, 2017). Holding time during cooking, pH at milling and rate of salting are also considered to manipulate manufacturing steps to produce LFC (Mistry, 2001). Rise in pH (from 5.35 to 5.75) also results in increased moisture in reduced fat Cheddar cheese (Guinee et al., 1998). Furthermore, different ripening temperature has been used to manufacture reduced fat Cheddar cheese prepared from condensed milk. The cheese ripened at 12 °C was significantly softer than cheese ripened at 7 °C (Fenelon et al., 1999). However, elevated ripening temperature (11 °C) increases rate of proteolysis and off-flavor develops after 12 weeks of storage (Brandsma, Mistry, Anderson, & Baldwin, 1994). Another study by Nauth and Hayashi (1995) used cultured skim milk to lower the pH of cheese milk very slowly. They believed that ionic calcium would increase at the expense of colloidal calcium, but most of the ionic calcium was lost in the whey. Hence, there was a decrease in total calcium retained in the cheese. As a result, the cheese showed improved melt and stretch characteristics (Nauth & Hayashi, 1995).

2.3 *Moisture Control by Different Approaches*

Low-fat Cheddar cheese manufactured from the low-fat milk has a very firm and rubbery texture. The physical properties of cheese are determined by the casein gel network structure formed by renneting and moisture removal during syneresis. Firm structure of LFC is governed by the formation of compact para-casein network (Amelia, Drake, Nelson, & Barbano, 2013; Anderson, Mistry, Brandsma, & Baldwin, 1993; Drake, Herfett, Boylston, & Swanson, 1995). Hence, curd moisture needs to be increased to overcome textural defects and is achieved by manipulating stirring time and scalding temperature. Furthermore, the moisture content of LFC can be improved by increasing milk pasteurization temperature, short pressing time, decreasing the whey draining temperature (Johnson, 2011; Rodriguez, 1998).

To manufacture reduced fat Cheddar cheese, increase in pasteurization temperature interferes with rennet curd formation, increases setting time of the curd and thus increases the moisture in a curd (Guinee et al., 1998). Apart from pasteurization, high heat-treatment (HHT) of milk is also used to improve texture and rheology of unheated reduced fat cheese. Whey proteins are denatured in situ by HHT in milk resulting in the formation of filamentous appendages, which protrude from the micelle surface, thus preventing comprehensive fusion of casein micelles during gelation (Farkye & Guinee, 2017).

Curd washing is also another approach used to increase the moisture content in cheese. Many manufacturers in the USA use cold water (22 °C) to rinse or soak the cheese curd after whey draining to increase moisture and to decrease the rate of syneresis. The low rate of syneresis helps to allow absorption of more water by the curd (Drake et al., 1995; Johnson, 2016; Johnson & Chen, 1995). However, curd washing causes lack of flavor development and poor keeping quality leading to decrease in shelf life in a reduced fat cheese (Johnson, 2016; Rodriguez, 1998). Though young Cheddar cheese possess softer and smoother body with no flavor defects, it deteriorates into a weak pasty cheese (Johnson & Chen, 1995), and possess objectionable flavors such as meaty and brothy after a week if manufactured by curd washing process (Farkye & Guinee, 2017). So, researchers have recommended curd washing to limited kinds of starter cultures used in cheese making. There is also a possibility of acquiring high pH in a washed curd cheese due to less buffering capacity than the cheese made without curd washing (Johnson & Chen, 1995). In addition, there is an inverse relationship between decrease in yield of Gouda cheese (measured after 12 days of manufacture) and the amount of added water in a curd (El-Gawad & Ahmed, 2011).

Johnson and Chen (1995) used high pH (5.8–5.9) during cutting and milling with the curd washing at 22 and 35 °C. Their results showed that the cheese washed at 22 °C showed highest moisture content and lower pH than the curd washed at 35 °C. Nevertheless, curd washing prevents the formation of calcium lactate crystals by removing soluble calcium and lactic acid from the curd. Additionally, it restricts undesirable fermentation in cheese by removing the residual lactose from the curd (Farkye & Guinee, 2017). Hou, Hannon, McSweeney, Beresford, and Guinee (2014) used curd washing to achieve different target levels of lactose plus lactic acid (5.3, 4.5, 4.3 and 3.9%) in cheese and found curd washing had less effect on composition and proteolysis (measured as pH 4.6 soluble nitrogen) and resulted in less flavour development, firmer and less brittle cheese. Comparison of the composition of reduced fat Cheddar cheese with wash and non-wash curd procedures are given in Table 23.2. Other methods to increase moisture content in LFC are: increasing size of the curd, delaying the cutting time, matting the curd after whey separation, using less pressure during pressing, reducing the stir-out time before and after whey drainage, using low concentration brine or brining for the short time etc. (Johnson, 2016).

2.4 *Casein/Fat Ratio*

Since, fat content in cheese is associated with texture, lubricity and flavor development (Farkye & Guinee, 2017; Ohren & Tuckey, 1969), casein to fat ratio is also a very important parameter for LFC manufacturing. Kosikowski and Mistry (1990) suggested 1.58 casein to fat ratio to manufacture 33% reduced fat Cheddar cheese, whereas Merrill, Oberg and McMahon (1994) suggested 2.4 ratio to manufacture

Table 23.2 Composition of reduced fat cheddar cheese with respect to different curd wash processes

Process of manufacturing	% fat	% moisture	% salt
No-wash curd	17.7	43.60	1.67
22 °C wash	17.3	46.84	1.61
35 °C wash	17.8	43.87	1.60

Source: Modified from Johnson and Chen (1995)

50% reduced fat Mozzarella cheese. Ohren and Tuckey (1969) found typical flavor in Cheddar cheese when the fat in dry matter was 50% or above and flavour was not developed in 16% fat cheese.

2.5 Homogenization and Micro-Fluidization of Cheese Milk

Emulsification or homogenization can reduce the size of milk fat globules (2 μm or less) and can modify the milk fat globule membrane. During homogenization, caseins, the primary proteins of the milk, are adsorbed on the surface of newly formed fat globule (Kelly, Huppertz, & Sheehan, 2008; Vélez, Hynes, Meinardi, Wolf, & Perotti, 2017). This has the following effects on the characteristics of milk:

- Reduction in the amount of micellar casein but casein surface area in the milk is increased.
- Formation of two types of casein micelle surface layer, native casein micelles and casein cover for the fat globules.
- Increase in effective surface area and a decrease in surface density of κ-casein due to spreading of casein micelles over the surface of fat globules (Kelly et al., 2008).

Emulsification or homogenization reduces fat loss in whey, increases yield of cheese and higher moisture retention in cheese due to slower whey expulsion (Madadlou, Mousavi, Khosrowshahiasl, Emam-Djome, & Zargaran, 2007; Rudan, Barbano, Guo, & Kindstedt, 1998). A higher degree of fat emulsification reduces leakage of fat during melting, decreases the level of free oil in cheese and decreases the rate of hydrolysis during lipid oxidation. Homogenization decreases curd firmness and syneresis (Lemay, Paquin, & Lacroix, 1994). Homogenization of milk can not only increase the yield of cheese but also reduce the amount of fat loss into whey (Kelly et al., 2008; Zamora, Ferragut, Jaramillo, Guamis, & Trujillo, 2007). Homogenization has also been shown to increase the moisture content in cheese (Karaman & Akalın, 2013; Madadlou et al., 2007), reduce coagulation time and improve acid production rate, curd tension, curd fusion and elasticity of curd (Jana & Upadhyay, 1992; O'Mahony, Auty, & McSweeney, 2005).

Homogenization is carried out either for the whole milk or only for the cream part and/or addition of the homogenized cream to skim milk (Karaman & Akalın, 2013). Usually, homogenization of whole cheese milk is not suggested because of its deleterious effects on flavour and texture of cheese (Deegan, Holopainen, McSweeney, Alatossava, & Tuorila, 2014). Selective homogenization of cheese milk, such as homogenization of cream and its incorporation into the skim milk (Deegan et al., 2013; Karaman & Akalın, 2013), is utilized to preserve optimal flavour and texture. Selective or complete homogenization of cheese milk has been reported to impart high moisture content, improved flavour, texture, body, microstructure, sensory qualities and functional properties in baby Gouda, Cheddar, Roquefort, Blue, Edam, Nyamunas, Kariesh, low fat Iranian white and pickled cheeses (Emmons et al., 1980; Jana & Upadhyay, 1992; Madadlou et al., 2007). Selective homogenization has also been reported to reduce free oil in reduced-fat Cheddar and Mozzarella cheese (Metzger & Mistry, 1995).

Though, homogenization is considered as a good processing tool in manufacturing LFC, Mozzarella cheese (26% fat) prepared from homogenized milk reportedly showed less meltability than the cheese prepared by un-homogenized milk in one study (Rudan, Barbano, Guo, & Kindstedt, 1998) while textural properties of cheese in unmelted condition (at 10 °C) were not influenced by the homogenization of cream part only. Hence, homogenization of cream only is recommended by some authors (El-Gawad, Ahmed, El-Abd, & El-Rafee, 2012; Rudan, Barbano, Guo, & Kindstedt, 1998).

One study compared the rheological and proteolytic proprieties of full and low-fat Mozzarella cheese prepared from un-homogenized milk and milk homogenized at different pressures (10.3 and 17.2 Mpa) (Tunick et al., 1993). Cooking temperature of samples were also varied in this study. The cheese prepared at a lower temperature (cooked at 32.4 °C) produced high moisture in a non-fat product as compared to the cheese prepared at a higher temperature (cooked at 45.9 °C). The casein (α_{s1}) partly degraded to (α_{s1}-$_1$) casein in the cheese cooked at a lower temperature during 6 weeks storage. It was concluded that the LFC with equivalent textural and melting characteristics to full-fat cheese could be prepared using homogenized milk, lower cooking temperature and storage at refrigeration temperature (Tunick et al., 1993).

Although several studies describe the effect of milk homogenisation on cheese quality, much fewer studies have reported the effect of fat globule size. MFGs help to disrupt the dense protein network in cheese and act as inert fillers (Michalski, Cariou, Michel, & Garnier, 2002). During renneting, MFGs affect the rate of gelation, curd firmness and the elasticity of the curd, which in turn affect the texture and flavour characteristics of cheese (Logan et al., 2017; Lucey, Johnson, & Horne, 2003; Michalski et al., 2007). MFGs' size and the interactions of MFGs and/or free fat within the casein matrix may have a leading role in cheese meltability (Rowney, Hickey, Roupas, & Everett, 2003).

Most studies so far (Deegan et al., 2013; Karaman & Akalın, 2013; Logan et al., 2015; O'Mahony et al., 2005) have reported on the effect of size of MFGs on full and reduced fat cheeses only. Small MFGs (D[4,3] ~ 3 μm) have been reported to improve the textural, flavour and sensory properties of full-fat Cheddar cheese (Logan et al., 2017) and Emmental and Camembert cheese (Michalski et al., 2003, 2004, 2007). Additionally, Logan et al. (2014) reported firmer texture of the cheese curd prepared from large MFGs (D[4,3] = 3.88–5.78 μm) in combination of small casein micelles (D[4,3] = 153–159 nm) because the large MFGs acted as local filler particles and did not break the protein network. However, O'Mahony et al. (2005) reported lowest firmness (storage modulus) in rennet gel containing large size MFGs (D[4,3] = 4.68 μm) compared to smaller MFGs (D[4,3] = 3.45 μm) during miniature Cheddar-type cheese manufacture. Similarly, bigger MFGs (D[3,2] = 4.6 μm decreased G′ compared to the smaller (D [3,2] = 1.89 μm) and medium sized (1.46 μm) MFGs in rennet gels (Michalski et al., 2002). The porous area within a protein gel network is too large to fit the small and medium sized MFGs as filler particles to reinforce the overall gel firmness; while, MFGs larger than the size casein micelles are deleterious to the gel strength (Michalski et al., 2002). Hence, some conflicting results have been reported in literature on the effect of MFG size on textural properties of cheese. In our latest study, we analysed the physico-chemical and biochemical properties of low fat Cheddar cheese made from micron to nano sized milk fat emulsions (Khanal et al., 2019). We found that the emulsion size affected the textural, microstructure, compositional, proteolysis and color properties of LFCs, but the textural properties were not significantly improved and were not found to be comparable to control full fat cheese and textural properties did not change significantly during cheese ripening. At size smaller than 1 μm, emulsions might not have been effective in disrupting the protein network. Submicron emulsions probably acted as inert particles and did not improve moisture retention in LFCs. This might be the reason for differences in results compared to studies showing textural improvement using homogenised milk. Homogenisation of milk not only reduces the fat particle size but also leads to high moisture retention leading to significant improvement in texture of LFC.

Microfluidization, due to its high emulsification and energy input, gives more stable emulsion and smaller droplet size than the conventional high-pressure homogenization (Dissanayake, Kelly, & Vasiljevic, 2010; Iordache & Jelen, 2003). One study has reported the effect of microfluidization of cheese milk on microstructure of full-fat and low-fat Mozzarella cheese. The micro-fluidization caused poor meltability and did not improve rheological properties of LFC (Tunick, Van Hekken, Cooke, Smith, & Malin, 2000). Later, Van Hekken, Tunick, Malin, and Holsinger (2007) demonstrated changes in melting and rheological properties of full- fat and low-fat Mozzarella cheese by micro-fluidization of cheese milk at different pressure and temperature. The LFC made from milk micro-fluidized at 10 °C (103 or 172 Mpa) and the full-fat control cheese could be stretched during hot water stretching. However, other cheese did not fuse together due to short curds (Van Hekken et al., 2007) leading to poor stretchability.

2.6 Membrane Filtration

Membrane filtration is commonly used in the dairy industry to remove bacteria from the milk without application of heat (using microfiltration, MF) (Amornkul & Henning, 2007) and to standarise the protein content in milk (using ultrafiltration,

UF) (Salvatore, Pirisi, & Corredig, 2011). It has also been used to modify α_s-/βcasein ratio or enrichment of caseins (Huppertz, Düsterhöft, & Engels, 2017; Mistry, Mistry, & Maubois, 1993). The existing cheese manufacturing technologies can be strengthened by membrane filtration of the cheese milk (Thomann, Schenkel, & Hinrichs, 2008). Membrane filtration can be used in cheese making to manipulate the texture of cheese by fractionating globular milk fat and to manufacture high-solids dairy products such as hard cheese, soft cheese and cheese base for further food processing (Henning, Baer, Hassan, & Dave, 2006).

The separation processes, such as MF and UF, may impart several advantages during cheese making such as increased cheese yield, enhanced coagulation properties and improved quality and safety of cheese (Heino, Uusi-Rauva, & Outinen, 2010; Kosikowski & Mistry, 1990). Researchers have claimed that cheese prepared from MF milk are hygienically safer compared to cheese made from pasteurized milk due to removal of a large amount of spore-forming bacteria, *Clostridium tyrobutyricum* (Saboya & Maubois, 2000).

MF of cheese milk (skim milk) through a 0.1 μm ceramic membrane can selectively concentrate micellar casein and calcium phosphate in the MF concentrate while removing water, α-lactalbumin, non-protein nitrogen, β-lactoglobulin, lactose and soluble minerals in the permeate (Neocleous, Barbano, & Rudan, 2002a). Hence, MF concentrate was successfully used to prepare ripened Cheddar cheese without any negative impact on cheese yield, texture, proteolysis and flavor.

While the retained, undenatured serum proteins (Lawrence, Creamer, & Gilles, 1987) and other minor milk serum proteins (Lelievre et al., 1990) during UF may have a negative impact on the normal proteolytic process during the aging, resulting in cheese texture and flavor development defects in some hard and semi-hard ripened cheeses, UF technique has been widely used for preparation of several varieties of fresh cheeses, namely Camembert, Feta, Cream, Quarg and Cottage cheese (Heino et al., 2010; Salvatore et al., 2011; Waungana, Singh, & Bennett, 1998). Cheese made from UF milk is firmer than those made from un-concentrated milk as minerals associated with caseins are concentrated in the retentate during UF (Caron, St-Gelais, & Pouliot, 1997).

Currently, UF has gained popularity in research because of the possibility of enriching cheese milk and to increase yield of a final product (Boivin-Piché, Vuillemard, & St-Gelais, 2016). Better rennet coagulation properties can be achieved by increasing the concentration of the protein with UF and or MF process (Heino et al., 2010). There is also a higher moisture retention and recovery of whey proteins in UF cheese depending on the heat treatment of milk (Lo & Bastian, 1998). Due to different coagulation properties of UF milk and un-concentrated milk, clotting time is less in UF milk and gel is firmer (Caron et al., 1997; Kelly et al., 2008).

In one study, low-fat Cheddar cheese was prepared from low mineral MF retentate (LMR) and high mineral retentate (HMR) powders with different sized milk fat globules. In this study, the cheese prepared from the large fat globule (LFG) milk (surface weighted diameter ~2.4 μm) resulted in higher scores in terms of flavor, texture and color than the cheese prepared from small milk fat globules (surface weighted diameter ~1.6 µm). The overall quality (texture, flavor and color) of low-fat Cheddar cheese was best for the LFG fraction combined with LMR (St-gelais, Piette, & Belanger, 1995; St-Gelais, Roy, & Audet, 1998). Flavor and aroma of LFC prepared from UF milk are acceptable though the texture is regarded as too soft due to retention of whey proteins (Rodriguez, Requena, Fontecha, Goudedranche, & Juarez, 1999). However, some authors reported flavor defects such as bitterness are common for the long-ripened cheese prepared from UF milk. Flavor intensity also decreases and the hardness of the Cheddar cheese also increases using MF cheese milk (Thomann et al., 2008). While another study reported acceptable texture and sensory score for the cheese made from UF milk (Miočinović et al., 2011).

In another study, low-moisture part-skim Mozzarella cheese was prepared from highly concentrated skim milk MF retentate and reducing the rennet by 93%. MF cheese in this study achieved 66–71% depletion of whey proteins(Ardisson-Korat & Rizvi, 2004). The composition of cheese prepared from MF retentate was similar to the commercial samples. However, proteolysis (measured as pH 4.6 and 12% TCA soluble nitrogen) was significantly slower than the commercial samples (Ardisson-Korat & Rizvi, 2004). This was due to low moisture in cheese prepared from MF milk, low residual chymosin and inhibition of chymosin activity by high molecular weight whey protein retentates (Neocleous et al., 2002a; Neocleous, Barbano, & Rudan, 2002b). Mistry, Metzger, and Maubois (1996) used UF sweet buttermilk to produce reduced fat Cheddar cheese and after 4 weeks of storage, the cheese made with the buttermilk concentrate was softer than the cheese prepared from the control milk. Furthermore, body of the cheese was also improved as compared to the control cheese. This was due to the presence of large amounts of milk fat globule membrane material in the buttermilk which acts as a surface-active agent and improves texture and flavour of the cheese (Mistry et al., 1996).

3 Use of Fat Replacers

Fat replacer is a substance responsible for delivering full or partial functions of the fat so as to provide lower calories than the regular fat, or the substance which provides one or more sensory and physical functions of the fat in food (Awad, Hassan, & Halaweish, 2005; Chavan, Khedkar, & Bhatt, 2016; Jooyandeh et al., 2017; Koca & Metin, 2004). Fat replacers can be divided into two categories: fat mimetic and fat substitute. Fat substitute comprises similar functional and physical characteristics of conventional fat molecules, such as triglycerides. Fat mimetics can mimic some of the organoleptic and physical characteristics of conventional fat molecules, but they cannot replace fat molecules in foods on a weight-for-weight basis, for example, protein or carbohydrate-based molecules used to replace fat (Chavan et al., 2016; Jooyandeh et al., 2017; Koca & Metin, 2004; O'Connor & O'Brien, 2011). Fat replacers used in LFC production can also be categorized into three groups based on their source: materials based on protein, carbohydrate and lipids source. Microparticulated whey protein is an example of protein-based fat replacers. Similarly, carbohydrate-based fat replacers include starch, fiber, modified starch, pectin, dextrins, maltodextrins, polydextrose, cellulose and bacterial gums such as guar gums and xanthan gums (Khanal et al., 2017). FDA has categorized the fat replacers into two types: one is as GRAS (generally recognized as a safe), and other is as food additive. Most of the fat replacers used in food at present fall under GRAS category (Artz, Lai, & Hansen, 2007).

Fat replacers have now been effectively used for the production of different food items including bakery items, salad dressings, processed cheese and confectionary products. Their use has been restricted for frying but can be used for baking or other purposes (McMahon, Alleyne, Fife, & Oberg, 1996; Position of the American Dietetic Association: fat replacers, 2005). Fat replacers have been used to increase moisture content in cheese by holding water but type and characteristics of fat replacers determine the level of increased functionality in cheese (McMahon et al., 1996; Plug & Haring, 1993; Setser & Racette, 1992; Shamil, Wyeth, & Kilcast, 1991; Zisu, 2005). The cheese prepared with fat content less than 10% and without adding fat replacers is tough in texture, dull in colour and poor in flavor, hence, these properties can be improved by using whey proteins or other fat replacers that have high water retention capacity and resemble fat. Fat replacers can modify functional, structural and sensory characteristics of LFC with improved sensory and mechanical properties besides increasing the yield (Drake, Boylston, & Swanson, 1996; Lobato-Calleros, Lobato-Calleros, Robles-Martinez, Caballero-Perez, & Vernon-Carter, 2000). On the other hand, they may produce undesirable flavor, reduce shredability and escalate stickiness in cheese, if used in enormous level (Johnson, 2016).

3.1 Carbohydrate-Based Fat Replacers

Some examples of carbohydrate-based fat replacers are pectin, alginate, potato starch, lecithin, starch guar gum, gum arabic, gum tragacanth, κ- and λ-carrageenans, β-glucan, maize starch and pre-gelatinize starch (Drake, Drake, Truong, & Daubert, 1999; Khanal et al., 2017: Khanal et al., 2018; Kavas, Oysun, Kinik, & Uysal, 2004; Konuklar, Inglett, Carriere, & Felker, 2004; Lashkari et al., 2014; Lobato-Calleros et al., 2000; Mounsey & O'Riordan, 2007; Oliveira et al., 2010; Rahimi, Khosrowshahi, Madadlou, & Aziznia, 2007; Sipahioglu, Alvarez, & Solano-Lopez, 1999; Totosaus & Guemes-Vera, 2008; Volikakis, Biliaderis, Vamvakas, & Zerfiridis, 2004). Use of all these aforementioned fat replacers provide softer texture in LFC. Other commonly used synthetic carbohydrate fat replacers are inulin, oatrim, Z-trim, polydextrose, maltodextrin, NovagelTM NC-200, Avicel® RCN 30, Raftuline® and Slendid™ (Akoh, 1994; Drake et al., 1996; Lashkari et al., 2014; McMahon et al., 1996; Plug & Haring, 1993; Position of the American Dietetic Association: fat replacers, 2005; Romeih, Michaelidou, Biliaderis, & Zerfiridis, 2002; Zisu, 2005). Good textural properties, bulkiness, thickness and creaminess can be achieved if two or more fat replacers' mixtures combined with other modification of procedure are used to prepare LFC (Position of the American Dietetic Association: fat replacers, 2005). Commercially available novel fat replacers are also a mixture of two or more types, and they balance functions of each other. For example, Rice Trin 3 (Zumbro Inc., Hayfield, MN) is a mixture of rice protein and maltodextrin. The combined effect of these two substances provides Rice Trin 3 good gelling properties similar to microparticulated proteins (Lucca & Tepper, 1994; Plug & Haring, 1993; Position of the American Dietetic Association: Fat Replacers, 2005; Zisu, 2005).

Several studies have been conducted using different types of carbohydrate-based fat replacers. McMahon et al. (1996) used water soluble microparticulated carbohydrate (Stellar™ and Novagel™) and microparticulated protein (Simplesse® and Dairy-Lo®) as fat replacers to manufacture low-fat (4–5% fat) Mozzarella cheese and determined its composition and functionality. They did not find any difference in the functionality of experimental Mozzarella cheese stored until 28 days at 4 °C. However, they found a significant effect ($p < 0.05$) on meltability. The heating behaviour of carbohydrate and protein based materials are different to that of fat. This is a major problem in Mozzarella cheese as it is consumed in a melted form in pizza. Additionally, moisture content of LFC containing fat replacers was increased significantly as compared to the control cheese without fat replacers (McMahon et al., 1996). Use of tragacanth gum as a fat replacer improved the texture of low-fat Iranian white brined cheese. On the other hand, the cheese containing a greater level of tragacanth gum exhibited objectionable characteristics after 6 weeks of ageing (Rahimi et al., 2007). Another study has found greater moisture retention and softer texture in reduced fat Cheddar cheese prepared from pectin (amidated pectin) and overall increase in functionality (Ibáñez et al., 2016).

In a recent study, we (Khanal et al., 2018) prepared LFC with up to 91% fat reduction using sodium alginate as a fat replacer. Cheese yield was directly proportional to the fat and alginate levels in milk, whereas the moisture and total protein in cheese were inversely proportional to the fat in milk. Texture profile analysis illustrated a significant improvement in the texture of alginate added LFC as compared to control low fat cheese (CLFC). The textural attributes of alginate added LFC ripened for 30 days were comparable to control full fat cheese (CFFC) ripened for 60 days and beyond. A close resemblance in textural attributes between alginate added LFC and CFFC, not previously reported when using other fat replacers, was observed. Scanning electron micrograph (SEM) images revealed that alginate added LFCs had smoother surfaces as compared to CFFC and CLFC, and the dense and compact protein matrix characteristic of CLFC was not observed. Confocal laser scanning microscopy (CLSM) suggested that the fat particle size, area and volume were affected in all LFCs due to their lower fat level and these parameters increased during ripening in CFFC. Nuclear magnetic resonance (NMR) results revealed an increase in higher mobility water fraction in alginate added cheese compared to CFFC and CLFC. Magnetic Resonance Imaging (MRI) results verified the presence of more water in alginate added cheese, and signal intensity and moisture were directly proportional with each other. Hunter L, a and b values for alginate added LFCs indicated that they were whiter than CLFC and less yellowish than CFFC at the beginning of ripening; the color of some of the alginate added LFCs was comparable to CFFC after 120 days of ripening. Overall, the addition of alginate significantly improved the textural, microstructural properties and color of LFCs, affirming its potential as a promising texture modifier.

Furthermore, coefficient of friction (CoF_{min}) of CFFC and LFCA were not different from each other and were lower than CLFC. Rate of protein release and cheese matrix disintegration were directly influenced by cheese composition (moisture and fat content and presence of alginate) and textural properties of cheese. Presence of fat and sodium alginate in cheese exhibited more lubrication and enhanced digestion compared to CLFC, whereas sensory properties of all cheeses were similar to each other (Khanal et al., 2020).

3.2 *Protein-Based Fat Replacers*

Protein-based fat replacers are usually prepared from whey, milk, vegetables or eggs. They provide 1–4 kcal/g. They are small globular particles and provide a creamy mouthfeel comparable to fat particles. One gram of protein-based fat mimetics can be used instead of 3 g of fat in the cream (Hahn, 1997; Position of the American Dietetic Association: Fat Replacers, 2005). Some protein-based fat replacers and their functions in food have been compiled in Table 23.3.

Microparticulated whey protein (MWP) is a commonly used protein based fat replacer in cheese (Stankey et al., 2017). MWP provides only 1–2 kcal/g and is approved as GRAS (Hahn, 1997). Microparticulation is a process of exposing the whey proteins to high-pressure shearing with or without heating to form uniform, microscopic spheres which are analogous to fat particles (Dissanayake et al., 2010; Hahn, 1997). MWP gives several textural benefits in various food preparations

Table 23.3 Examples of protein-based fat replacers

Replacers	Trade name	Source
Milk-derived solids (micro-particulated protein)	Dairylight	Mainly proteins
Milk-derived solids	Simplesse®100, DairyLo®	23% whey proteins, 17% carbohydrates, 2% fat and 2% ash
Egg-derived solids	Simplesse®300	12% proteins, 10% carbohydrates, 0.1% fat
Zein	Lita®	Protein from corn

Source: Modified from Plug and Haring (1993), Position of the American Dietetic Association: Fat Replacers (2005), Setser and Racette (1992), Zisu (2005)

(Dissanayake et al., 2010). Microparticulated protein or carbohydrate materials provide lubricity and creaminess by entrapping the water mechanically. However, they cannot substitute non-polar and flavor carrying properties of the fat (Amelia et al., 2013; Romeih et al., 2002). Commercial MWP used in cheese and other foods are Simplesse®100 and DairyLo®, they contain whey particles/aggregates of different size (0.5–5 μm) (Farkye & Guinee, 2017). Texture of low-fat Kashar cheese, White brined cheese and Cheddar cheese is improved by adding Simplesse®100, but less effect on flavor is observed. Improvement in texture is due to increase in moisture content in cheese as a result of combined effect of water binding and occlusion of the para-casein by whey protein particles (Fenelon & Guinee, 1997; Koca & Metin, 2004; Romeih et al., 2002; Sahan, Yasar, Hayaloglu, Karaca, & Kaya, 2008; Schenkel, Samudrala, & Hinrichs, 2013). MWP used up to 0.5% in low fat Cheddar cheese resulted in higher moisture, yield and lower G′ (Stankey et al., 2017). Other types of synthetic protein based replacers used in dairy foods and cheese are ALACO PALS™ and Lita® (Plug & Haring, 1993; Position of the American Dietetic Association: Fat Replacers, 2005; Setser & Racette, 1992; Zisu, 2005).

A study on the effect of incorporation of a mixture of low methyl pectin (LMP) and whey protein concentrate (WPC) on microstructure of LFC proved that LMP and WPC help to form looser and open protein network. The textural characteristics of LFC (prepared from LMP and WPC) such as hardness, springiness, cohesiveness and chewiness were not significantly different from FFC through their structural characteristics were different from FFC. Similarly, sensory characteristics of LFC containing WPC were also similar to the full-fat Manchego cheese (Lobato-Calleros, Lobato Calleros, Vernon Carter, Sanchez Garcia, & Garcia Galindo, 1999). Texture and rheological properties of cheese are inter-related and rely on the network composition, structure and molecular interactions (Lucey et al., 2003). Similar to microparticulated protein, soy-protein isolate has been used as a fat replacer in paneer and 0.2% soy-protein isolate containing paneer was found to have superior quality (Kumar et al., 2011).

3.3 Fat-Based Fat Replacers

Fat-based substitutes are derived from either chemical modifications of fatty acids or from synthetic fats. In modified fats, their configuration is changed by altering the fatty-acids on glycerol molecules (O'Connor & O'Brien, 2011). Caprenin and Salatrim (short- and long-chain acyl triglyceride molecule) are common examples of fat-based substitutes. Salatrim is produced by interesterification of tripropionin, triacetin, tributyrin and hydrogenated vegetable oils. Caprenin contains caprylic (8:0), capric (10:0) and behinic (22:0) fatty acids esterified to glycerol and is formed after esterification of glycerol with these fatty acids (Artz et al., 2007; O'Connor & O'Brien, 2011).

Table 23.4 Examples of lipid-based fat replacers (synthetic fat substitute)

Replacer	Trade name	Source
Altered triglycerides (caprocaprylobehenin)	Caprenin®, benefat	Cocoa butter replacer
Dialkyidihexadecyl	DDM	For high-temperature applications
Esterified propoxylated glycerol	EPG	For high-temperature and cold applications
Synthetic sucrose polyesters/ polymers	Olestra, prolestra, colestra	For high-temperature and cold applications
Trialkoxy citrate	TAC	For margarine and mayonnaise
Trialkoxytricarballylate	TATCA	For margarine and mayonnaise

Source: Modified from Position of the American Dietetic Association: Fat Replacers (2005), Setser and Racette (1992), Zisu (2005)

Olestra (sucrose fatty acid polyester) is the only synthetic fat-based substitute recommended for food use (Hahn, 1997; Position of the American Dietetic association: fat replacers, 2005). Olestra contains mixture of hexa, hepta and octa fatty acid esters of saturated and unsaturated fatty acids having a chain length of ≥12 carbons with sucrose (Artz et al., 2007). Olestra has similar mouthfeel, appearance, flash point, heat and oxidative stability as conventional fats. It is applicable to deep frying or baking. Though Olestra provides no calories to the diet, a few side effects like abdominal cramping, reduced absorption of fat soluble vitamins and loose stools have been reported (O'Connor & O'Brien, 2011). Other fat-based fat replacers used in food are colestra, trialkoxy citrate (TAC) esterified propoxylated glycerol (EPG), raffinose and sorbestin (fatreplacers from net.pdf; Position of the American Dietetic Association: fat replacers, 2005; Setser & Racette, 1992; Zisu, 2005). Table 23.4 shows examples of fat-based fat replacers.

Salatrim (short and long acyl triglyceride molecule) belongs to the group of structured triglycerides and has at least one short chain fatty acid (2:0, 3:0, 4:0) and at least one long chain fatty acid (generally 18:0) haphazardly esterified to glycerol. Hardness of salatrim is governed by the types and levels of short and long chain fatty acids present on it (O'Connor & O'Brien, 2011). Researchers used Salatrim to manufacture reduced fat Mozzarella cheese. They found a considerably lower degree of hardness, higher moisture content and more proteolysis than the control low-fat cheese. However, there was less effect of salatrim on meltability of Mozzarella cheese (Rudan, Barbano, & Kindstedt, 1998). The degree of lipolysis and proteolysis were increased significantly in low-fat-white-brined cheese prepared with fat replacers than in FFC. The maximum stress (σ max) (at 80% deformation on uniaxial compression) and the mean value of texture profile parameters of the LFC containing fat replacers were considerably less than full-fat cheese. Hence, they suggested a significant improvement in a cheese texture made with fat replacers (Romeih et al., 2002).

3.4 Use of Exopolysaccharide (EPS)-Producing Bacteria

Some strains of lactic acid bacteria (LAB) used in dairy products processing can produce EPS. EPS can protect the microbial cell from harsh conditions such as drying of the cells, osmotic stress, toxic compounds and phagocytosis (De Vuyst & Degeest, 1999; Zisu, 2005). EPS generally denotes both types of extracellular polysaccharides; cell bound EPS (it conglutinate to the surface of bacterial cell) and EPS that is released into the surrounding environment (Tang et al., 2017). This polysaccharide can sometimes be secreted in the cell wall of the capsule or in the growth medium as a slime. Hence, capsular EPS is extracellular material of bacterial metabolites covering the cell surface and remains attached to the cell surface, while ropy EPS is a slime produced by the bacteria and these polysaccharides are released from the microbial cells (Duguid & Duguid, 1951; Lynch, Coffey, & Arendt, 2017).

EPS produced by different microorganisms vary in their size, types of linkage, structure and composition (Zafar, Siddiqui, Shahid, Qader, & Aman, 2017). Composition of EPS produced by EPS-producing microorganism depends on the environmental conditions such as pH, temperature and available carbon sources (De Vuyst et al., 2003). A high carbon to nitrogen ratio at neutral pH is the best condition for EPS production (Gancel & Novel, 1994). Though the chemical composition of EPS produced by different microorganisms differs from one another, the basic composition is the same i.e. D-galactose and D-glucose are predominant in different proportions (De Vuyst & Degeest, 1999).

Table 23.5 shows some EPS-producing LAB. Some strains of EPS-producing LAB have already been used to increase the moisture retention, yield and functional properties of Mozzarella cheese. Use of EPS in LFC helps to increase moisture (up to 5%) retention, and affects gel strength, rheology and curd syneresis during coagulation of milk during cheesemaking, which then helps to improve texture and organoleptic properties of cheese (Di Cagno et al., 2014; Lynch et al.,

Table 23.5 Microorganisms used for milk fermentations and EPS production

Mesophiles	Thermophiles
Lactococcus lactis ssp. *lactis*	*Lactobacillus delbrueckii* ssp. *bulgaricus*
Lactococcus lactis ssp. *cremoris*	*Streptococcus thermophilus*
Leuconostoc sp.	*Lactobacillus helveticus*
Lactobacillus kefir	*Lactobacillus acidophilus*
Lactobacillus casei	*Bifidobacterium sp.*
Saccharomyces cerevisiae	
Candida kefir	

Source: Marshall and Tamime (1997), Zisu (2005)

2017). Use of EPS producing microorganisms (*Weissella cibaria* MG1 and *Lactobacillus reuteri* cc2) in low fat Cheddar cheese resulted in a higher moisture content and its microstructure was comparable to control full fat cheese (Lynch et al., 2014). Manufacture of half-fat Cheddar cheese by EPS producing *Lactococcus lactis* starter and it's isogenic Non-EPS producing varient resulted in a significant increase in moisture in cheese produced by EPS prodcing *L. lactis*. Increase in moisture was attributed due to retention of water by the EPS through hydrogen bonding (Costa et al., 2010).

Incorporation of EPS-producing capsular type microorganisms in low-fat Mozzarella cheese enhances the textural and functional properties and cheese yield. EPS-producing *S. thermophilus* strains 285 and 1275 in low-fat Mozzarella cheese decrease processing time (Zisu, 2005). The moisture retention, hardness, springiness, chewiness, stretch and melt characteristics are significantly improved by the use of capsular EPS-producing *S. thermophilus* 285 and ropy EPS- producing *S. thermophile* (Zisu, 2005). Use of EPS-producing starter and the adjunct cultures in low-fat Mozzarella cheese help to increase moisture content by 4% compared to the control cheese sample (Perry, McMahon, & Oberg, 1997). Yield (0.29–1.19%) and moisture retention (3.6–4.8%) in Cheddar cheese were also increased using capsular and ropy EPS-producing *Lactobacillus lactis* subsp. *lactis* (RBL259) and *Lactobacillus lactis* subsp. *cremoris* (RBL132) (Dabour, Kheadr, Fliss, & LaPointe, 2005). Higher rate of moisture retention and primary proteolysis were also observed by Awad et al. (2005) using ropy strain *Lactobacillus lactis* subsp. *cremoris* (JFR1) and capsule forming nonropy (and moderately ropy) strains of *Streptococcus thermophilus* in reduced-fat Cheddar cheese.

Furthermore, Hassan, Awad, and Mistry (2007) manufactured reduced fat Cheddar cheese from young reduced fat Cheddar cheese base containing EPS. They reported softer texture, less chewiness and gumminess and better flow in cheese containing EPS bacteria. On the other hand, they did not find any correlation between the physical and melting properties of cheese with EPS base and control cheese samples. Other previous studies indicated improvements in texture and quality of LFC by incorporating EPS-producing strains in various kinds of cheese varieties (Ayyash, Abu-Jdayil, Hamed, & Shaker, 2018; Jimenez-Guzman, Flores-Najera, Cruz-Guerrero, & Garcia-Garibay, 2009; Lynch et al., 2014; Nepomuceno, Costa Junior, & Costa, 2016; Oluk, Guven, & Hayaloglu, 2014; Sanli et al., 2013). It is recommended to use EPS-producing organism along with cheese starter in Mozzarella cheese to increase moisture retention and yield. However, its use sometimes is limited as the EPS cultures increases viscosity of whey by accumulating EPS and is not desirable during whey concentration process because it decreases the efficiency of membrane used for concentration (Cerning, 1995; Petersen, Dave, McMahon, Oberg, & Broadbent, 2000).

Previously used different approaches for the manufacture of low fat Cheddar cheese are compiled in Table 23.6 and approaches used for production of other types of LFC are given in Table 23.7.

Table 23.6 Different approaches used for manufacturing low fat cheddar cheese

Types of cheese	Approaches used	Aim	Sources
Half-fat and low-fat cheddar cheese	1. EPS producing bacteria (*L. lactis* ssp. C*remories* (DPC6532 and DPC6533, 1.5% inoculum) 2. High pasteurization temperautre (72 °C, 82 °C, 87 °C for 26 s) 3. High pressure homogenization (200 Mpa/5 min cycles) 4. Liposome-encapsulated nisin Z producing *L. lactis* ssp. *lactis* biovar. *diacetylactis* UL719, or *L. casei-casei* L2A adjunct culture	To increase the yield, to improve microbiological quality, structural characteristics, physicochemical and sensory attributes during ripening, textural and cooking qualities; to change rheological property and flavour	Costa et al. (2010), Rynne et al. (2004), Kheadr et al. (2002), Benech et al. (2003)
Reduced-fat and low-fat cheddar cheese	1. Homogenization of cream (17.3 and 3.4 MPa at 57 °C, 13.8 MPa and 3.45 MPa at 70 °C) 2. Addition of sweet or ultrafiltered butter milk 3. Combining reduced fat Cheddar cheese (33%) with micellar casein concentrate (20%) 4. Addition of condensed skim milk (10.3, 15.4 18.3, 22.2% total solids) 5. Combining condensed skim milk (15.4, 18.33 and 22.23% total solids) and ripening treatments (6–7, 11 °C and elevated temperature) 6. β-glucan (Nutrim) (5.2 and 2.5%) 7. Fat mimetics (Novagel™ NC200, Dairy Lo®, Alaco Pals™) 8. Different ageing time 9. Skim mik enriched with 5% total protein with a low or a high mineral content UF milk retentate powder and different fractions of fat globules (size: D = 2.4 and 1.6 μm) 10. Caseins enriched milk with different milk protein concentrate powders (diafiltered microfiltered retentate, calcium caseinate and ultrafilterd retentate powder)	To improve texture, microstructure and flavour, cheesemaking peformance, appearance, moisture retention, salt retention,yield, sensory and rheological characteristicss, fat retention and composition in cheese, to reduce fat loss in whey	Metzger and Mistry (1995), Amelia et al. (2013, 1995), Mistry et al., 1996, Rudan, Barbano, and Kindstedt (1998), Anderson et al. (1993), Brandsma et al. (1994), Konuklar et al. (2004), Drake et al. (1996), Rogers et al. (2009), St-Gelais et al. (1997), St-Gelais et al. (1998)

Table 23.7 Different approaches used for manufacturing LFC other than low fat cheddar cheese

Types of cheese	Approaches used	Aim	Sources
LFC	1. Mixture of microparticulated whey protein (MWP) (0, 3 and 6%), butter milk (0 and 15%) and *L. plantarun* TINE18 (1%), *L. casei* TINE36 (1%) 2. Mixed culture of L. acidophilus and bifidobacteria	To make a cheese with bio-active properties; to improve flavour and texture	Ryhänen, Pihlanto-Leppälä and Pahkala (2001), Skeie et al. (2013)
Reduced fat processed cheese	Granular hydrogenated lecithin (0.025, 0.05,0.1 or 0.2% (w/w)	To improve textural and sensory properties	Drake et al. (1999)
Low fat white pickled cheese	Fat replacers 1. Carbohydrate based: perfectsmyl gel MB (0.5%) (modified potato starch) and Satiagel ME4 (0.4%) 2. Protein based: Simplesse® (0.5%) D100, Dairy-Lo™ (0.5%)	To improve chemical, physical and sensory attributes	Kavas et al. (2004)
Low fat feta cheese	1. Tapioca and lecithin as a fat replacers 2. Ewe's milk (1.5% fat) 3. Tapioca starch and lecithin (0.1and 0.2%) as a fat replacer	To determine optimum amount of fat for the acceptable quality; to improve flavour, texture and overall acceptability	Katsiari and Voutsinas (1994), Sipahioglu et al. (1999)
Semi-hard LFC	Ultra filtered and microfiltered milk	To increase whey retention, to improve texture and sensory characteristics	Rodriguez et al. (1999)
Low fat green edam cheese	Guar gum (0.0025–0.01% w/v)	To enhance physicochemical, thermal, rheological and textural properties	Oliveira et al. (2010)
Low fat white-brined cheese	Different types of fat replacers Oat-β-glucan (0.7 and 1.4% w/w), Simplesse® D-100 (1% w/w), Novagel™ NC-200(0.125% w/w)	To improve textural qualities, rheology, yield, composition, proteolysis and sensory attributes	Volikakis et al. (2004), Romeih et al. (2002)
Low moisture partly-skimmed mozzarella cheese	Chamel chymosin (0.05 and 0.037 IMCU/ml)	To compare bovine and camel chymosin on texture, functionality and sensory properties	Moynihan et al. (2013)

(continued)

Table 23.7 (continued)

Types of cheese	Approaches used	Aim	Sources
Low fat Iranian white cheese, low and reduced fat Turkish white cheese	1. Cream homogenisation (for Iranian white cheese: two stage homogenizer) 2. Gum tragacanth as a fat replacer (0.25, 0.5, 0.75 or 1 g/L) 3. Guar gum (0, 150 and 300 ppm) and gum arabic (0, 75, 150 and 300 ppm) as a fat replacer 4. Turkish white: one stage homogenizer 5. Different rennet concentration (one-, two- and threefold of normal usage)	To improve textural, functional, rheology, meltability, microstructure properties, fat globule distribution and sensory qualities	Madadlou et al. (2007), Karaman and Akalın (2013), Rahimi et al. (2007), Karaman, Benli, and Akalin (2012), Lashkari et al. (2014)
Low fat soft cheese	Fat replacer (Dairy-Lo® 2% w/w)	To improve sensory, rheological and physiochemical properties	Zalazar et al. (2002)
Low fat oaxaca cheese	κ- and λ-carrageenans mixture as fat replacers (0.17–1 g/L)	To enhance melting properties, moisture and yield	Totosaus and Guemes-Vera (2008)
Reduced-fat Havarti-type cheese	Native and denatured whey proteins	To improve composition, texture, flavour and acceptability	Lo and Bastian (1998)
Low fat mozzarella cheese	1. Pre-acidification by citric acid (pH 6 and 5.8) and acetic acid 2. Ultrafiltered sweet buttermilk and homogenized cream (3 or 5%) 3. Denatured whey protein (0.1 and 1%) 4. Microfludized milk (34, 103 or 172 MPa/10, 43 and 54 °C) 5. Fat replacers: carbohydrate based (Stellar 100X™ and Novagel™ RCN-15) and protein based (Simplesse® D100 and Dairy-Lo®)	To improve composition (calcium, fat, protein), body, texture and microstructure, sensory properties, accepatability, meltability and rheological properties, to increase moisture content and functional properties (melting, hot water stretching)	Metzgeret et al. (2000), Poduval and Mistry (1999), Ismail, Ammar and El-Metwally (2011), Van Hekken et al. (2007), McMahon et al. (1996)
Low fat paneer	Soy protein isolate (0, 0.1, 0.2, 0.3%)	To increase chemical composition and yield	Kumar et al. (2011)
Low fat Mexican manchego cheese	Fat replacers [low methooxy pectin (LMP), whey protein concentrate (WPC), Dairy-Lo®, MWP, Simplesse®	To improve textural,microstructural and sensory properties	Lobato-Callers et al. (2000)

Reduced fat monterey jack type cheese	Decreased cooking temperature (35 °C), decreased ripening time (decreased upto 30 min), decresed starter (0.22% lactic starter), homogenization (6900 KPa), added milk-solid-non fat (1.2% prior to pasteurzation), added flavour cultures (0.1% w/w CR210 and 0.045% w/w 15× starter distillate) and washed curd (draining half whey and addition of water at 35 °C for 5 min)	To improve sensory quality	Drake et al. (1995)
Low fat Italian caciotta-type cheese	MWP (0.5%), EPS producing *S. thermophilus* St446 (7.0 ± 0.2 $\log_{10}$ cfu/mL) and adjunt cutures *L. plantarum* and *L. rhamnosus* as adjunct cultures (6.0 ± 0.1 $\log_{10}$ cfu/mL)	To improve textural, compositional, sensory, microbiological, biochemical,rheological and cooking properties	Di Cagno et al. (2014), Mistry et al. (1996)
Low fat fresh kashar cheese	Fat replacers Protein based: Dairy-Lo® (1% w/w), Simplesse® (1% w/w) Carbohydrate based: 5% Raftiline® HP	To improve textural, melting and sensory properties	Koca and Metin (2004)
Low fat soft cheese (Cremoso Argentino)	Protein based fat replacer: Dairy-Lo® (2% w/w)	To improve chemical, rheological, sensory and melting properties	Zalazar et al. (2002)
Reduced fat mozzarella cheese	Stirred curd procedure and fat replacer (Salatrim®) type 1 and 3 (20% w/w)	To improve chemical composition, yield, functionality and appearance	Rudan, Barbano, and Kindstedt (1998)
Iranian white brined cheese	Sodium alginate (2% in distilled water) and resistant starch microencapsulation to detect survival rate of *L. acidophilus* (0.1%)	To increase survival rate of *L. acidophilus* as a probiotic bacteria	Mirzaei, Pourjafar, and Homayouni (2012)
Minas fresh cheese	Octenyl succinylated waxy starch as a fat replacer	To enhance textural, microstructure and physicochemical properties	Diamantino et al. (2014)
Rennet-casein-based model processed cheese	Rennet casein (ALAREN™), normal maize (25% amylose) starch, waxy maize starch (99% amylopectin) and high amylose (70% amylose) maize starch, soya oil (25%) as a fat replacers	To improve rheological, meltabilities and microstructure	Ye, Hewitt, and Taylor (2009)
Imitation cheese	Pre-gelatinised strach (3% w/w maize, waxy maize, wheat, potato or rice starches) as a fat replacers	To improve texture, rheology and melting properties	Mounsey and O'Riordan (2007)

4 Conclusion

With the increasing demand of LFC, several factors need to be considered during its production to make it comparable to the full-fat counterpart. Properties of LFC are entirely different to the FFC in terms of texture, body, appearance and flavor. Improvement in only one parameter may not be sufficient to meet consumer's expectations of the quality of LFC. Hence, sound knowledge of technology, chemistry, structure, functionality and microbiology is necessary to prepare an ideal LFC. In future, genomics and proteomics will be helpful in understanding the key role of microbiology and biochemical reactions in cheese. Texture and functionality of the cheese will be understood by molecular interactions. Use of some of the fat replacers has provided satisfactory results for the texture in up to 30% reduced-fat cheese. Further research is required to produce the LFC with more than 90% fat reduction. Choice of appropriate starter adjunct could improve and optimize the aroma and flavor of LFC. In addition, most of the studies have been conducted on the hard and semi-hard varieties of commercial cheese. Hence, work is necessary on other traditional varieties too.

References

Agarwal, S., Powers, J., Swanson, B., Chen, S., & Clark, S. (2008). Influence of salt-to-moisture ratio on starter culture and calcium lactate crystal formation. *Journal of Dairy Science, 91*(8), 2967–2980.

Akkerman, M., Kristensen, L. S., Jespersen, L., Ryssel, M. B., Mackie, A., Larsen, N. N., et al. (2017). Interaction between sodium chloride and texture in semi-hard Danish cheese as affected by brining time, dl-starter culture, chymosin type and cheese ripening. *International Dairy Journal, 70*(Suppl C), 34–45.

Akoh, C. C. (1994). *Carbohydrate polyesters as fat substitutes* (Vol. 62). Boca Raton, FL: CRC Press.

Amelia, I., Drake, M., Nelson, B., & Barbano, D. M. (2013). A new method for the production of low-fat Cheddar cheese. *Journal of Dairy Science, 96*(8), 4870–4884.

Amornkul, Y., & Henning, D. R. (2007). Utilization of microfiltration or lactoperoxidase system or both for manufacture of cheddar cheese from raw milk. *Journal of Dairy Science, 90*(11), 4988–5000.

Anderson, D. L., Mistry, V. V., Brandsma, R. L., & Baldwin, K. A. (1993). Reduced fat cheddar cheese from condensed milk. 1. Manufacture, composition, and ripening. *Journal of Dairy Science, 76*(10), 2832–2844.

Ardisson-Korat, A. V., & Rizvi, S. S. H. (2004). Vatless manufacturing of low-moisture part-skim mozzarella cheese from highly concentrated skim milk microfiltration retentates. *Journal of Dairy Science, 87*(11), 3601–3613.

Ardö, Y., Mansson, H. L., Hedenberg, A., & Larsson, P. O. (1989). Studies of peptidolysis during early maturation and its influence on low-fat cheese quality. *Milchwissenschaft-Milk Science International, 44*(8), 485–495.

Artz, W. E., Lai, L. L., & Hansen, S. L. (2007). Chapter 11 - The chemistry and nutrition of non-nutritive fats A2. In M. D. Erickson (Ed.), *Deep frying* (2nd ed., pp. 229–249). Urbana, IL: AOCS Press.

Awad, S., Hassan, A. N., & Halaweish, F. (2005). Application of exopolysaccharide-producing cultures in reduced-fat cheddar cheese: Composition and proteolysis. *Journal of Dairy Science, 88*(12), 4195–4203.

Ayyash, M., Abu-Jdayil, B., Hamed, F., & Shaker, R. (2018). Rheological, textural, microstructural and sensory impact of exopolysaccharide-producing Lactobacillus plantarum isolated from camel milk on low-fat akawi cheese. *LWT - Food Science and Technology, 87*(Suppl C), 423–431.

Banks, J. M. (2004). The technology of low-fat cheese manufacture. *International Journal of Dairy Technology, 57*(4), 199–207.

Banks, J. M., Brechany, E. Y., & Christie, W. W. (1989). The production of low fat cheddar type cheese. *International Journal of Dairy Technology, 42*(1), 6–9.

Banks, J. M., Roa, I., & Muir, D. D. (1998). Manipulation of the texture of low-fat chedder using a plant protease extracted from cynara cardunculus. *Australian Journal of Dairy Technology, 53*(2), 105.

Benech, R. O., Kheadr, E. E., Lacroix, C., & Fliss, I. (2003). Impact of Nisin producing culture and liposome-encapsulated Nisin on ripening of Lactobacillus added-Cheddar cheese. *Journal of Dairy Science, 86*(6), 1895–1909.

Boivin-Piché, J., Vuillemard, J.-C., & St-Gelais, D. (2016). Technical note: Vitamin D-fortified cheddar type cheese produced from concentrated milk. *Journal of Dairy Science, 99*(6), 4140–4145.

Børsting, M. W., Qvist, K. B., Rasmussen, M., Vindeløv, J., Vogensen, F. K., & Ardö, Y. (2012). Impact of selected coagulants and starters on primary proteolysis and amino acid release related to bitterness and structure of reduced-fat cheddar cheese. *Dairy Science & Technology, 92*(5), 593–612.

Brandsma, R. L., Mistry, V. V., Anderson, D. L., & Baldwin, K. A. (1994). Reduced fat cheddar cheese from condensed milk. 3. Accelerated ripening. *Journal of Dairy Science, 77*(4), 897–906.

Broadbent, J. R., Brighton, C., McMahon, D. J., Farkye, N. Y., Johnson, M. E., & Steele, J. L. (2013). Microbiology of cheddar cheese made with different fat contents using a Lactococcus lactis single-strain starter. *Journal of Dairy Science, 96*(7), 4212–4222.

Bryant, A., Ustunol, Z., & Steffe, J. (1995). Texture of cheddar cheese as influenced by fat reduction. *Journal of Food Science, 60*(6), 1216–1219.

Caron, A., St-Gelais, D., & Pouliot, Y. (1997). Coagulation of milk enriched with ultrafiltered or diafiltered microfiltered milk retentate powders. *International Dairy Journal, 7*(6-7), 445–451.

CDR. (2014). College of agriculture and life sciences, University of Wisconsin-Madison.

Cerning, J. (1995). Production of exopolysaccharides by lactic acid bacteria and dairy propionibacteria. *Le Lait, 75*(4-5), 463–472.

Chatli, M. K., Gandhi, N., & Singh, P. (2017). Efficacy of sodium alginate as fat replacer on the processing and storage quality of buffalo mozzarella cheese. *Nutrition & Food Science, 47*, 3.

Chavan, R. S., Khedkar, C. D., & Bhatt, S. (2016). Fat replacer. In B. Caballero, P. M. Finglas, & F. Toldrá (Eds.), *Encyclopedia of food and health* (pp. 589–595). Oxford: Academic Press.

Chevanan, N., Muthukumarappan, K., Upreti, P., & Metzger, L. (2006). Effect of calcium and phosphorus, residual lactose and salt to moisture ratio on textural properties of cheddar cheese during ripening. *Journal of Texture Studies, 37*(6), 711–730.

CODEX. (2013). *General standard for cheese*. Retrieved from http://www.google.com.au/url?sa=t&rct=j&q=&esrc=s&source=web&cd=2&ved=0CCUQFjAB&url=http%3A%2F%2Fwww.codexalimentarius.org%2Finput%2Fdownload%2Fstandards%2F186%2FCXS_263e.pdf&ei=YYKLV1XzPIqk8QXP7oLYCQ&usg=AFQjCNH3B5pBiab-pEBV1UdI9c9fnllYJw&bvm=bv.81828268

Costa, N. E., Hannon, J. A., Guinee, T. P., Auty, M. A., McSweeney, P. L., & Beresford, T. P. (2010). Effect of exopolysaccharide produced by isogenic strains of Lactococcus lactis on half-fat cheddar cheese. *Journal of Dairy Science, 93*(8), 3469–3486.

Dabour, N., Kheadr, E. E., Fliss, I., & LaPointe, G. (2005). Impact of ropy and capsular exopolysaccharide-producing strains of Lactococcus lactis subsp. cremoris on reduced-fat cheddar cheese production and whey composition. *International Dairy Journal, 15*(5), 459–471.

Dalby, A. (2009). *Cheese a global history*. London, UK: Reaktion Books.

De Vuyst, L., & Degeest, B. (1999). Heteropolysaccharides from lactic acid bacteria. *FEMS Microbiology Reviews, 23*(2), 153–177.

De Vuyst, L., Zamfir, M., Mozzi, F., Adriany, T., Marshall, V., Degeest, B., et al. (2003). Exopolysaccharide-producing Streptococcus thermophilus strains as functional starter cultures in the production of fermented milks. *International Dairy Journal, 13*(8), 707–717.

Deegan, K. C., Heikintalo, N., Ritvanen, T., Putkonen, T., Rekonen, J., McSweeney, P. L. H., et al. (2013). Effects of low-pressure homogenisation on the sensory and chemical properties of Emmental cheese. *Innovative Food Science & Emerging Technologies, 19*, 104–114.

Deegan, K. C., Holopainen, U., McSweeney, P. L. H., Alatossava, T., & Tuorila, H. (2014). Characterisation of the sensory properties and market positioning of novel reduced-fat cheese. *Innovative Food Science & Emerging Technologies, 21*, 169–178.

Di Cagno, R., De Pasquale, I., De Angelis, M., Buchin, S., Rizzello, C. G., & Gobbetti, M. (2014). Use of microparticulated whey protein concentrate, exopolysaccharide-producing Streptococcus thermophilus, and adjunct cultures for making low-fat Italian Caciotta-type cheese. *Journal of Dairy Science, 97*(1), 72–84.

Diamantino, V. R., Beraldo, F. A., Sunakozawa, T. N., & Penna, A. L. B. (2014). Effect of octenyl succinylated waxy starch as a fat mimetic on texture, microstructure and physicochemical properties of Minas fresh cheese. *LWT - Food Science and Technology, 56*(2), 356–362.

Dissanayake, M., Kelly, A. L., & Vasiljevic, T. (2010). Gelling properties of microparticulated whey proteins. *Journal of Agricultural and Food Chemistry, 58*(11), 6825–6832.

Drake, M. A., Boylston, T. D., & Swanson, B. G. (1996). Fat mimetics in low-fat cheddar cheese. *Journal of Food Science, 61*(6), 1267–1270.

Drake, M. A., Drake, M. A., Truong, V. D., & Daubert, C. R. (1999). Rheological and sensory properties of reduced-fat processed cheeses containing lecithin. *Journal of Food Science, 64*(4), 744–747.

Drake, M. A., Herfett, W., Boylston, T. D., & Swanson, B. G. (1995). Sensory evaluation of reduced fat cheeses. *Journal of Food Science, 60*(5), 898–901.

Drake, M. A., Miracle, R. E., & McMahon, D. J. (2010). Impact of fat reduction on flavor and flavor chemistry of Cheddar cheeses. *Journal of Dairy Science, 93*(11), 5069–5081.

Duguid, J. P., & Duguid, J. P. (1951). The demonstration of bacterial capsules and slime. *Journal of Pathology and Bacteriology, 63*(4), 673–685.

El-Gawad, A., Ahmed, N. S., El-Abd, M., & El-Rafee, S. A. (2012). Effect of homogenization on the properties and microstructure of Mozzarella cheese from buffalo milk. Acta scientiarum polonorum. *Technologia Alimentaria, 11*(2), 121–135.

El-Gawad, M. A. A., & Ahmed, N. S. (2011). Cheese yield as affected by some parameters review. *Acta Scientiarum Polonorum. Technologia Alimentaria, 10*(2), 131–153.

Emmons, D. B., Emmons, D. B., Kalab, M., Larmond, E., & Lowrie, R. J. (1980). Milk gel structure. X. Texture and Microstructure in Cheddar cheese made from whole milk and from homogenized low-fat milk. *Journal of Texture Studies, 11*(1), 15–34.

Farkye, N. Y., & Guinee, T. P. (2017). *Chapter 28 - Low-fat and low-sodium cheeses cheese* (4th ed.pp. 699–714). San Diego, CA: Academic Press.

FDA. (2017). Code of federal regulations, chapter 21.

Felfoul, I., Bornaz, S., Baccouche, A., Sahli, A., & Attia, H. (2015). Low-fat Gouda cheese made from bovine milk-olive oil emulsion: Physicochemical and sensory attributes. *Journal of Food Science and Technology, 52*(10), 6749–6755. https://doi.org/10.1007/s13197-015-1736-0

Fenelon, M., & Guinee, T. (1997). The compositional, textural and maturation characteristics of reduced-fat cheddar made from milk containing added dairy-Lo. *Milchwissenschaft, 52*(7), 385–389.

Fenelon, M. A., Ryan, M. P., Rea, M. C., Guinee, T. P., Ross, R. P., Hill, C., et al. (1999). Elevated temperature ripening of reduced fat cheddar made with or without lacticin 3147-producing starter culture. *Journal of Dairy Science, 82*(1), 10–22.

Fox, P. F. (2011). Cheese: overview. In J. W. Fuquay (Ed.), *Encyclopedia of dairy sciences* (2nd ed., pp. 533–543). San Diego, CA: Academic Press.

Fox, P. F., & McSweeney, P. L. H. (2017). Chapter 1 - Cheese: An overview. In P. L. H. McSweeney, P. F. Fox, P. D. Cotter, & D. W. Everett (Eds.), *Cheese* (4th ed., pp. 5–21). San Diego, CA: Academic Press.

Gancel, F., & Novel, G. (1994). Exopolysaccharide production by Streptococcus salivarius ssp. thermophilu cultures. 1. Conditions of production. *Journal of Dairy Science, 77*(3), 685–688.

Guinee, T. P., Fenelon, M. A., Mulholland, E. O., O'Kennedy, B. T., O'Brien, N., & Reville, W. J. (1998). The influence of milk pasteurization temperature and pH at curd milling on the composition, texture and maturation of reduced fat cheddar cheese. *International Journal of Dairy Technology, 51*(1), 1–10.

Hahn, N. I. (1997). Replacing fat with food technology. *Journal of the American Dietetic Association, 97*(1), 15–16.

Hassan, A. N., Awad, S., & Mistry, V. V. (2007). Reduced fat process cheese made from young reduced fat cheddar cheese manufactured with exopolysaccharide-producing cultures. *Journal of Dairy Science, 90*(8), 3604–3612.

Heino, A., Uusi-Rauva, J., & Outinen, M. (2010). Pre-treatment methods of Edam cheese milk. Effect on cheese yield and quality. *LWT - Food Science and Technology, 43*(4), 640–646.

Henning, D. R., Baer, R. J., Hassan, A. N., & Dave, R. (2006). Major advances in concentrated and dry milk products, cheese, and milk fat-based spreads. *Journal of Dairy Science, 89*(4), 1179–1188.

Hou, J., Hannon, J. A., McSweeney, P. L., Beresford, T. P., & Guinee, T. P. (2014). Effect of curd washing on cheese proteolysis, texture, volatile compounds, and sensory grading in full fat cheddar cheese. *International Dairy Journal, 34*(2), 190–198.

Huppertz, T., Düsterhöft, E.-M., & Engels, W. (2017). *Pre-treatment of cheese-milk reference module in food science*. London, UK: Elsevier.

Ibáñez, R. A., Waldron, D. S., & McSweeney, P. L. (2016). Effect of pectin on the composition, microbiology, texture, and functionality of reduced-fat cheddar cheese. *Dairy Science & Technology, 96*(3), 297–316.

Iordache, M., & Jelen, P. (2003). High pressure microfluidization treatment of heat denatured whey proteins for improved functionality. *Innovative Food Science & Emerging Technologies, 4*(4), 367–376.

Ismail, M., Ammar, E.-T., & El-Metwally, R. (2011). Improvement of low fat mozzarella cheese properties using denatured whey protein. *International Journal of Dairy Technology, 64*(2), 207–217.

Jana, A. H., & Upadhyay, K. G. (1992). Homogenization of milk for cheese making – A review. *Australian Journal of Dairy Technology, 47*(1), 72–79.

Jimenez-Guzman, J., Flores-Najera, A., Cruz-Guerrero, A. E., & Garcia-Garibay, M. (2009). Use of an exopolysaccharide-producing strain of Streptococcus thermophilus in the manufacture of Mexican Panela cheese. *LWT- Food Science and Technology, 42*(9), 1508–1512.

Johnson, M. E. (2011). Cheese: Low-fat and reduced-fat cheese. In J. W. Fuquay (Ed.), *Encyclopedia of dairy sciences* (2nd ed., pp. 833–842). San Diego, CA: Academic Press.

Johnson, M. E. (2016). *Cheese: Low-fat and reduced-fat cheese reference module in food science*. London, UK: Elsevier.

Johnson, M. E., & Chen, C. M. (1995). Technology of manufacturing reduced-fat cheddar cheese. *Advances in Experimental Medicine and Biology, 367*, 331–337.

Jooyandeh, H., Goudarzi, M., Rostamabadi, H., & Hojjati, M. (2017). Effect of Persian and almond gums as fat replacers on the physicochemical, rheological, and microstructural attributes of low-fat Iranian White cheese. *Food Science & Nutrition, 5*(3), 669–677.

Karaman, A. D., & Akalın, A. S. (2013). Improving quality characteristics of reduced and low fat Turkish white cheeses using homogenized cream. *LWT - Food Science and Technology, 50*(2), 503–510.

Karaman, A. D., Benli, M., & Akalin, A. S. (2012). Microstructure of industrially produced reduced and low fat Turkish white cheese as influenced by the homogenization of cream. *Grasas y Aceites, 63*(3), 267–273.

Katsiari, M. C., & Voutsinas, L. P. (1994). Manufacture of low-fat feta cheese. *Food Chemistry, 49*(1), 53–60.

Kavas, G., Oysun, G., Kinik, O., & Uysal, H. (2004). Effect of some fat replacers on chemical, physical and sensory attributes of low-fat white pickled cheese. *Food Chemistry, 88*(3), 381–388.

Kelly, A. L., Huppertz, T., & Sheehan, J. J. (2008). Pre-treatment of cheese milk: Principles and developments. *Dairy Science and Technology, 88*(4-5), 549–572.

Khanal, B. K. S., Bhandari, B., Prakash, S., & Bansal, N. (2017). Effect of sodium alginate addition on physical properties of rennet milk gels. *Food Biophysics, 12*(2), 141–150.

Khanal, B. K. S., Bhandari, B., Prakash, S., Liu, D., Zhou, P., & Bansal, N. (2018). Modifying textural and microstructural properties of low fat Cheddar cheese using sodium alginate. *Food Hydrocolloids, 83*, 97–108.

Khanal, B. K. S., Budiman, C., Hodson, M. P., Plan, M. R. R., Prakash, S., Bhandari, B., et al. (2019). Physico-chemical and biochemical properties of low fat cheddar cheese made from micron to nano sized milk fat emulsions. *Journal of Food Engineering, 242*, 94–105.

Khanal, B. K. S., Bhandari, B., Prakash, S., & Bansal, N. (2020). Simulated oral processing, in vitro digestibility and sensory perception of low fat Cheddar cheese containing sodium alginate. *Journal of Food Engineering, 270*, 109749.

Kheadr, E. E., Vachon, J. F., Paquin, P., & Fliss, I. (2002). Effect of dynamic high pressure on microbiological, rheological and microstructural quality of Cheddar cheese. *International Dairy Journal, 12*(5), 435–446.

Khosrowshahi, A., Madadlou, A., Ebrahim zadeh Mousavi, M., & Emam-Djomeh, Z. (2006). Monitoring the Chemical and textural changes during ripening of Iranian white cheese made with different concentrations of starter. *Journal of Dairy Science, 89*(9), 3318–3325.

Koca, N., & Metin, M. (2004). Textural, melting and sensory properties of low-fat fresh kashar cheeses produced by using fat replacers. *International Dairy Journal, 14*(4), 365–373.

Konuklar, G., Inglett, G. E., Carriere, C. J., & Felker, F. C. (2004). Use of a beta-glucan hydrocolloidal suspension in the manufacture of low-fat cheddar cheese: manufacture, composition, yield and microstructure. *International Journal of Food Science and Technology, 39*(1), 109–119.

Kosikowski, F. V., & Mistry, V. V. (1990). Microfiltration, ultrafiltration, and centrifugation separation and sterilization processes for improving milk and cheese quality. *Journal of Dairy Science, 73*(6), 1411–1419.

Kumar, R. (2012). *An investigation into improvement of low fat cheddar cheese by the addition of hydrocolloids*. (Masters Thesis), University of Minnesota, USA.

Kumar, S. S., Balasubramanian, S., Biswas, A. K., Chatli, M. K., Devatkal, S. K., & Sahoo, J. (2011). Efficacy of soy protein isolate as a fat replacer on physico-chemical and sensory characteristics of low-fat paneer. *Journal of Food Science and Technology, 48*(4), 498–501.

Lashkari, H., Khosrowshahi Asl, A., Madadlou, A., & Alizadeh, M. (2014). Chemical composition and rheology of low-fat Iranian white cheese incorporated with guar gum and gum arabic as fat replacers. *Journal of Food Science and Technology, 51*(10), 2584–2591.

Lawrence, R. C., Creamer, L. K., & Gilles, J. (1987). Texture development during cheese ripening. *Journal of Dairy Science, 70*(8), 1748–1760.

Lelievre, J., Creamer, L. K., & Tate, K. L. (1990). Inhibition of calf vell and microbial rennet action by whey-protein concentrate. *Milchwissenschaft-Milk Science International, 45*(2), 71–75.

Lemay, A., Paquin, P., & Lacroix, C. (1994). Influence of microfluidization of milk on Cheddar cheese composition, color, texture, and yield. *Journal of Dairy Science, 77*(10), 2870–2879.

Lo, C. G., & Bastian, E. D. (1998). Incorporation of native and denatured whey proteins into cheese curd for manufacture of reduced fat, havarti-type cheese. *Journal of Dairy Science, 81*(1), 16–24.

Lobato-Calleros, C., Lobato Calleros, C., Vernon Carter, E. J., Sanchez Garcia, J., & Garcia Galindo, H. S. (1999). Textural characteristics of cheese analogs incorporating fat replacers. *Journal of Texture Studies, 30*(5), 533–548.

Lobato-Calleros, C., Lobato-Calleros, C., Robles-Martinez, J. C., Caballero-Perez, J. F., & Vernon-Carter, E. J. (2000). Fat replacers in low-fat Mexican manchego cheese. *Journal of Texture Studies, 32*(1), 1–14.

Logan, A., Day, L., Pin, A., Auldist, M., Leis, A., Puvanenthiran, A., et al. (2014). Interactive effects of milk fat globule and casein micelle size on the renneting properties of milk. *Food and Bioprocess Technology, 7*(11), 3175–3185.

Logan, A., Leis, A., Day, L., Øiseth, S. K., Puvanenthiran, A., & Augustin, M. A. (2015). Rennet gelation properties of milk: Influence of natural variation in milk fat globule size and casein micelle size. *International Dairy Journal, 46*, 71–77.

Logan, A., Xu, M., Day, L., Singh, T., Moore, S. C., Mazzonetto, M., et al. (2017). Milk fat globule size affects cheddar cheese properties. *International Dairy Journal, 70*(Suppl C), 46–54.

Lucca, P. A., & Tepper, B. J. (1994). Fat replacers and the functionality of fat in foods. *Trends in Food Science & Technology, 5*(1), 12–19.

Lucey, J., Johnson, M., & Horne, D. (2003). Invited review: Perspectives on the basis of the rheology and texture properties of cheese. *Journal of Dairy Science, 86*(9), 2725–2743.

Lynch, K. M., Coffey, A., & Arendt, E. K. (2017). Exopolysaccharide producing lactic acid bacteria: Their techno-functional role and potential application in gluten-free bread products. *Food Research International, 110*, 52–61.

Lynch, K. M., McSweeney, P. L. H., Arendt, E. K., Uniacke-Lowe, T., Galle, S., & Coffey, A. (2014). Isolation and characterisation of exopolysaccharide-producing Weissella and Lactobacillus and their application as adjunct cultures in Cheddar cheese. *International Dairy Journal, 34*(1), 125–134.

Madadlou, A., Mousavi, M. E., Khosrowshahiasl, A., Emam-Djome, Z., & Zargaran, M. (2007). Effect of cream homogenization on textural characteristics of low-fat Iranian white cheese. *International Dairy Journal, 17*(5), 547–554.

Marshall, V. M., & Tamime, A. Y. (1997). Starter cultures employed in the manufacture of biofermented milks. *International Journal of Dairy Technology, 50*(1), 35–41.

McCarthy, A. L., O'Connor, T. P., & O'Brien, N. M. (2014). Cheese in the context of diet and nutrition. In V. R. Preddy, R. R. Watson, & V. B. Patel (Eds.), *Handbook of cheese in health: Production, nutrition and medical sciences* (Vol. 1, pp. 15–26). Wageningen, Netherlands: Wageningen Academic Publishers.

McCarthy, C. M., Wilkinson, M. G., & Guinee, T. P. (2017b). Effect of coagulant type and level on the properties of half-salt, half-fat Cheddar cheese made with or without adjunct starter: Improving texture and functionality. *International Dairy Journal, 75*(Suppl C), 30–40.

McCarthy, C. M., Wilkinson, M. G., & Guinee, T. P. (2017c). Effect of calcium reduction on the properties of half-fat Cheddar-style cheeses with full-salt or half-salt. *International Dairy Journal, 73*(Suppl C), 38–49.

McElhatton, A., & El Idrissi, M. M. (2016). Traditional polish curd cheeses. In *Modernization of traditional food processes and products* (pp. 4–12). Boston, MA: Springer.

McMahon, D. J., Alleyne, M. C., Fife, R. L., & Oberg, C. J. (1996). Use of fat replacers in low fat mozzarella cheese. *Journal of Dairy Science, 79*(11), 1911–1921.

McSweeney, P. L. H., Ottogalli, G., & Fox, P. F. (2017). Chapter 31. In *Diversity and classification of cheese varieties: An overview cheese* (4th ed., pp. 781–808). San Diego, CA: Academic Press.

Merrill, R. K., Oberg, C. J., & McMahon, D. J. (1994). A method for manufacturing reduced fat mozzarella Cheese1. *Journal of Dairy Science, 77*(7), 1783–1789.

Metzger, L. E., & Mistry, V. V. (1995). A new approach using homogenization of cream in the manufacture of reduced fat cheddar cheese. 2. Microstructure, fat globule distribution, and free oil. *Journal of Dairy Science, 78*(9), 1883–1895.

Metzger, L. E., Barbano, D. M., Rudan, M. A., & Kindstedt, P. S. (2000). Effect of Milk Preacidification on low fat mozzarella cheese I. Composition and Yield. *Journal of Dairy Science, 83*(4), 648–658.

Michalski, M.-C., Camier, B., Briard, V., Leconte, N., Gassi, J.-Y., Goudédranche, H., et al. (2004). The size of native milk fat globules affects physico-chemical and functional properties of Emmental cheese. *Le Lait, 84*(4), 343–358.

Michalski, M.-C., Camier, B., Gassi, J.-Y., Briard-Bion, V., Leconte, N., Famelart, M.-H., et al. (2007). Functionality of smaller vs control native milk fat globules in Emmental cheeses manufactured with adapted technologies. *Food Research International, 40*(1), 191–202.

Michalski, M. C., Cariou, R., Michel, F., & Garnier, C. (2002). Native vs. damaged milk fat globules: Membrane properties affect the viscoelasticity of milk gels. *Journal of Dairy Science, 85*(10), 2451–2461.

Michalski, M.-C., Gassi, J.-Y., Famelart, M.-H., Leconte, N., Camier, B., Michel, F., et al. (2003). The size of native milk fat globules affects physico-chemical and sensory properties of Camembert cheese. *Le Lait, 83*(2), 131–143.

Miočinović, J., Puđa, P., Radulović, Z., Pavlović, V., Miloradović, Z., Radovanović, M., et al. (2011). Development of low fat UF cheese technology. *Mljekarstvo, 61*(1), 33.

Mirzaei, H., Pourjafar, H., & Homayouni, A. (2012). Effect of calcium alginate and resistant starch microencapsulation on the survival rate of Lactobacillus acidophilus La5 and sensory properties in Iranian white brined cheese. *Food Chemistry, 132*(4), 1966–1970.

Mistry, V. V. (2001). Low fat cheese technology. *International Dairy Journal, 11*(4–7), 413–422.

Mistry, V. V., Metzger, L. E., & Maubois, J. L. (1996). Use of ultrafiltered sweet buttermilk in the manufacture of reduced fat cheddar cheese. *Journal of Dairy Science, 79*(7), 1137–1145.

Mistry, V. V., Mistry, V. V., & Maubois, J. L. (1993). Application of membrane separation technology to cheese production. In *Cheese: Chemistry, physics and microbiology* (pp. 261–267). London, UK: Elsevier Science & Technology Books.

Mounsey, J. S., & O'Riordan, E. D. (2007). Modification of imitation cheese structure and rheology using pre-gelatinised starches. *European Food Research and Technology, 226*(5), 1039–1046.

Moynihan, A. C., Govindasamy-Lucey, S., Jaeggi, J. J., Johnson, M. E., Lucey, J. A., & McSweeney, P. L. (2014). Effect of camel chymosin on the texture, functionality, and sensory properties of low-moisture, part-skim mozzarella cheese. *Journal of Dairy Science, 97*(1), 85–96.

Nauth, K. R., & Hayashi, D. K. (1995). Method for manufacture of low fat pasta filata cheese: United States Patent, 5431931.

Neocleous, M., Barbano, D. M., & Rudan, M. A. (2002a). Impact of low concentration factor microfiltration on the composition and aging of cheddar cheese. *Journal of Dairy Science, 85*(10), 2425–2437.

Neocleous, M., Barbano, D. M., & Rudan, M. A. (2002b). Impact of low concentration factor microfiltration on milk component recovery and cheddar cheese yield. *Journal of Dairy Science, 85*(10), 2415–2424.

Nepomuceno, R. S. A. C., Costa Junior, L. C. G., & Costa, R. G. B. (2016). Exopolysaccharide-producing culture in the manufacture of Prato cheese. *LWT - Food Science and Technology, 72*, 383–389.

O'Connor, T. P., & O'Brien, N. M. (2011). Butter and other milk fat products: Fat replacers. In J. W. Fuquay (Ed.), *Encyclopedia of dairy sciences* (2nd ed., pp. 528–532). San Diego, CA: Academic Press.

O'Mahony, J. A., Auty, M. A., & McSweeney, P. L. (2005). The manufacture of miniature cheddar-type cheeses from milks with different fat globule size distributions. *Journal of Dairy Research, 72*(3), 338–348.

Ohren, J. A., & Tuckey, S. (1969). Relation of flavor development in cheddar cheese to chemical changes in the fat of the cheese. *Journal of Dairy Science, 52*(5), 598–607.

Oliveira, N. M., Dourado, F. Q., Peres, A. M., Silva, M. V., Maia, J. M., & Teixeira, J. A. (2010). Effect of guar gum on the physicochemical, thermal, rheological and textural properties of green edam cheese. *Food and Bioprocess Technology, 4*(8), 1414–1421.

Oluk, A. C., Guven, M., & Hayaloglu, A. A. (2014). Proteolysis texture and microstructure of low-fat Tulum cheese affected by exopolysaccharide-producing cultures during ripening. *International Journal of Food Science and Technology, 49*(2), 435–443.

Perry, D. B., McMahon, D. J., & Oberg, C. J. (1997). Effect of exopolysaccharide-producing cultures on moisture retention in low fat mozzarella cheese. *Journal of Dairy Science, 80*(5), 799–805.

Petersen, B. L., Dave, R. I., McMahon, D. J., Oberg, C. J., & Broadbent, J. R. (2000). Influence of capsular and ropy exopolysaccharide-producing Streptococcus thermophilus on mozzarella cheese and cheese whey. *Journal of Dairy Science, 83*(9), 1952–1956.

Plug, H., & Haring, P. (1993). The role of ingredient-flavour interactions in the development of fat-free foods. *Trends in Food Science & Technology, 4*(5), 150–152.

Poduval, V. S., & Mistry, V. V. (1999). Manufacture of reduced fat mozzarella cheese using Ultrafiltered sweet buttermilk and homogenized cream. *Journal of Dairy Science, 82*(1), 1–9.

Position of the American Dietetic Association: Fat Replacers. (2005). *Journal of the American Dietetic Association, 105*(2), 266–275.

Rahimi, J., Khosrowshahi, A., Madadlou, A., & Aziznia, S. (2007). Texture of low-fat Iranian White cheese as influenced by gum tragacanth as a fat replacer. *Journal of Dairy Science, 90*(9), 4058–4070.

Reedy, J. (2016). *How the US low-fat diet recommendations of 1977 contributed to the declining health of Americans*. (Honors Scholar Thesis), University of Connecticut, Storrs, CT.

Rodriguez, J. (1998). Recent advances in the development of low-fat cheeses. *Trends in Food Science & Technology, 9*(6), 249–254.

Rodriguez, J., Requena, T., Fontecha, J., Goudedranche, H., & Juarez, M. (1999). Effect of different membrane separation technologies (ultrafiltration and microfiltration) on the texture and microstructure of semihard low-fat cheeses. *Journal of Agriculture and Food Chemistry, 47*(2), 558–565.

Rogers, N. R., Drake, M. A., Daubert, C. R., McMahon, D. J., Bletsch, T. K., & Foegeding, E. A. (2009). The effect of aging on low-fat, reduced-fat, and full-fat cheddar cheese texture. *Journal of Dairy Science, 92*(10), 4756–4772.

Romeih, E. A., Michaelidou, A., Biliaderis, C. G., & Zerfiridis, G. K. (2002). Low-fat white-brined cheese made from bovine milk and two commercial fat mimetics: Chemical, physical and sensory attributes. *International Dairy Journal, 12*(6), 525–540.

Rowney, M. K., Hickey, M. W., Roupas, P., & Everett, D. W. (2003). The effect of homogenization and milk fat fractions on the functionality of mozzarella cheese. *Journal of Dairy Science, 86*(3), 712–718.

Rudan, M. A., Barbano, D. M., Guo, M. R., & Kindstedt, P. S. (1998). Effect of the modification of fat particle size by homogenization on composition, proteolysis, functionality, and appearance of reduced fat mozzarella cheese. *Journal of Dairy Science, 81*(8), 2065–2076.

Rudan, M. A., Barbano, D. M., & Kindstedt, P. S. (1998). Effect of fat replacer (Salatrim®) on chemical composition, proteolysis, functionality, appearance, and yield of reduced fat mozzarella cheese. *Journal of Dairy Science, 81*(8), 2077–2088.

Ryhänen, E.-L., Pihlanto-Leppälä, A., & Pahkala, E. (2001). A new type of ripened, low-fat cheese with bioactive properties. *International Dairy Journal, 11*(4–7), 441–447.

Rynne, N. M., Beresford, T. P., Kelly, A. L., & Guinee, T. P. (2004). Effect of milk pasteurization temperature and in situ whey protein denaturation on the composition, texture and heat-induced functionality of half-fat Cheddar cheese. *International Dairy Journal, 14*(11), 989–1001.

Saboya, L. V., & Maubois, J.-L. (2000). Current developments of microfiltration technology, in the dairy industry. *Le Lait, 80*(6), 541–553.

Sahan, N., Yasar, K., Hayaloglu, A. A., Karaca, O. B., & Kaya, A. (2008). Influence of fat replacers on chemical composition, proteolysis, texture profiles, meltability and sensory properties of low-fat Kashar cheese. *Journal of Dairy Research, 75*(01), 1–7.

Salvatore, E., Pirisi, A., & Corredig, M. (2011). Gelation properties of casein micelles during combined renneting and bacterial fermentation: Effect of concentration by ultrafiltration. *International Dairy Journal, 21*(11), 848–856.

Sanli, T., Gursel, A., Sanli, E., Acar, E., & Benli, M. (2013). The effect of using an exopolysaccharide-producing culture on the physicochemical properties of low-fat and reduced-fat Kasar cheeses. *International Journal of Dairy Technology, 66*(4), 535–542.

Schenkel, P., Samudrala, R., & Hinrichs, J. (2013). The effect of adding whey protein particles as inert filler on thermophysical properties of fat-reduced semihard cheese type Gouda. *International Journal of Dairy Technology, 66*(2), 220–230.

Scott, R., Robinson, R. K., & Wilbey, R. A. (1998). Introduction to cheese making. In R. Scott (Ed.), *Cheesemaking practice* (2nd ed., pp. 37–43). Gaithersburg, MD: Aspen Publication.

Setser, C. S., & Racette, W. L. (1992). Macromolecule replacers in food products. *Critical Reviews in Food Science and Nutrition, 32*(3), 275–297.

Shamil, S., Wyeth, L. J., & Kilcast, D. (1991). Flavour release and perception in reduced-fat foods. *Food Quality and Preference, 3*(1), 51–60.

Sipahioglu, O., Alvarez, V., & Solano-Lopez, C. (1999). Structure, physico-chemical and sensory properties of feta cheese made with tapioca starch and lecithin as fat mimetics. *International Dairy Journal, 9*(11), 783–789.

Skeie, S., Alseth, G. M., Østlie, H., Abrahamsen, R. K., Johansen, A. G., & Øyaas, J. (2013). Improvement of the quality of low-fat cheese using a two-step strategy. *International Dairy Journal, 33*(2), 153–162.

Soltani, M., Boran, O. S., & Hayaloglu, A. A. (2016). Effect of various blends of camel chymosin and microbial rennet (Rhizomucor miehei) on microstructure and rheological properties of Iranian UF White cheese. *LWT - Food Science and Technology, 68*(Supplement C), 724–728.

Soodam, K., Ong, L., Powell, I. B., Kentish, S. E., & Gras, S. L. (2015a). Effect of calcium chloride addition and draining pH on the microstructure and texture of full fat cheddar cheese during ripening. *Food Chemistry, 181*, 111–118.

Soodam, K., Ong, L., Powell, I. B., Kentish, S. E., & Gras, S. L. (2015b). Effect of rennet on the composition, proteolysis and microstructure of reduced-fat cheddar cheese during ripening. *Dairy Science & Technology, 95*(5), 665–686.

Stankey, J. A., Lu, Y. J., Abdalla, A., Govindasamy-Lucey, S., Jaeggi, J. J., Mikkelsen, B. O., et al. (2017). Low-fat Cheddar cheese made using microparticulated whey proteins: Effect on yield and cheese quality. *International Journal of Dairy Technology, 70*(4), 481–491.

St-gelais, D., Piette, M., & Belanger, G. (1995). Production of cheddar cheese using milk enriched with microfluidized milk retentate - A preliminary study. *Milchwissenschaft-Milk Science International, 50*(11), 614–619.

St-Gelais, D., Roy, D., & Audet, P. (1998). Manufacture and composition of low fat cheddar cheese from milk enriched with different protein concentrate powders. *Food Research International, 31*(2), 137–145.

St-Gelais, D., Passey, C. A., Haché, S., & Roy, P. (1997). Production of low-fat Cheddar cheese from low and high mineral retentate powders and different fractions of milkfat globules. *International Dairy Journal, 7*(11), 733–741.

Tang, W., Dong, M., Wang, W., Han, S., Rui, X., Chen, X., et al. (2017). Structural characterization and antioxidant property of released exopolysaccharides from Lactobacillus delbrueckii ssp. bulgaricus SRFM-1. *Carbohydrate Polymers, 173*, 654–664.

Thomann, S., Schenkel, P., & Hinrichs, J. (2008). The impact of homogenization and microfiltration on rennet-induced gel formation. *Journal of Texture Studies, 39*(4), 326–344.

Totosaus, A., & Guemes-Vera, N. (2008). Effect of κ- and λ-Carrageenans as fat-replacers in low-fat oaxaca cheese. *International Journal of Food Properties, 11*(3), 656–668.

Tunick, M. H., Mackey, K. L., Shieh, J. J., Smith, P. W., Cooke, P., & Malin, E. L. (1993). Rheology and microstructure of low-fat mozzarella cheese. *International Dairy Journal, 3*(7), 649–662.

Tunick, M. H., Van Hekken, D. L., Cooke, P. H., Smith, P. W., & Malin, E. L. (2000). Effect of high pressure microfluidization on microstructure of mozzarella cheese. *LWT - Food Science and Technology, 33*(8), 538–544.

USDA. (2016). *New Dietary guidelies to encourage healthy eating patterns to prevent chronic diseases*. Washington, DC: HHS Press.

Van Hekken, D. L., Tunick, M. H., Malin, E. L., & Holsinger, V. H. (2007). Rheology and melt characterization of low-fat and full fat Mozzarella cheese made from microfluidized milk. *LWT - Food Science and Technology, 40*(1), 89–98.

Vélez, M. A., Hynes, E. R., Meinardi, C. A., Wolf, V. I., & Perotti, M. C. (2017). Cheese milk low homogenization enhanced early lipolysis and volatiles compounds production in hard cooked cheeses. *Food Research International, 96*, 215–225.

Volikakis, P., Biliaderis, C. G., Vamvakas, C., & Zerfiridis, G. K. (2004). Effects of a commercial oat-β-glucan concentrate on the chemical, physico-chemical and sensory attributes of a low-fat white-brined cheese product. *Food Research International, 37*(1), 83–94.

Waungana, A., Singh, H., & Bennett, R. J. (1998). Rennet coagulation properties of skim milk concentrated by ultrafiltration: effects of heat treatment and pH adjustment. *Food Research International, 31*(9), 645–651.

WHO. (2015). *Healthy diet*. Fact sheet No. 394. Retrieved from http://www.who.int/mediacentre/factsheets/fs394/en/

Ye, A., Hewitt, S., & Taylor, S. (2009). Characteristics of rennet–casein-based model processed cheese containing maize starch: Rheological properties, meltabilities and microstructures. *Food Hydrocolloids, 23*(4), 1220–1227.

Zafar, S. B., Siddiqui, N. N., Shahid, F., Qader, S. A. U., & Aman, A. (2017). Bioprospecting of indigenous resources for the exploration of exopolysaccharide producing lactic acid bacteria. *Journal, Genetic Engineering & Biotechnology, 16*(1), 17–22.

Zahra, S., Mohammad, H., Sahel, S., & Mohammad Ali, S. (2013). The influence of malt extraction adding to UF fresh low fat cheese on its textural properties. *International Journal of Science and Engineering, 6*(1), 52–55.

Zalazar, C. A., Zalazar, C. S., Bernal, S., Bertola, N., Bevilacqua, A., & Zaritzky, N. (2002). Effect of moisture level and fat replacer on physicochemical, rheological and sensory properties of low fat soft cheeses. *International Dairy Journal, 12*(1), 45–50.

Zamora, A., Ferragut, V., Jaramillo, P. D., Guamis, B., & Trujillo, A. J. (2007). Effects of ultra-high pressure homogenization on the cheese-making properties of milk. *Journal of Dairy Science, 90*(1), 13–23.

Zisu, B. (2005). *Impact of pre-acidification, fat replacers and exopolysaccharide producing starter cultures on functionality of low fat mozzarella cheese*. (PhD Thesis), Victoria University, Werribee Campus, VIC, Australia.

Chapter 24
Influence of Milk Fat on Foam Formation, Foam Stability and Functionality of Aerated Dairy Products

Thao M. Ho, Bhesh Bhandari, and Nidhi Bansal

1 Introduction

A foam is typically defined as a two-phase system in which the pockets of gas are dispersed in a continuous emulsified phase. The dispersed gases (internal phase) could be any type of inert gases such as air, carbon dioxide (CO_2), nitrogen (N_2), nitrous oxide (N_2O) or steam depending on application while the continuous phase (external phase) can be in liquid state or change into a gel or (semi-)solid phase once the dispersion of gas is completed (Walstra, 1989). In this chapter, hereafter, the terms of "liquid foam" and "solid foam" are used to denote dairy foam systems in which the continuous phase is in liquid state (e.g. cappuccino-style beverages) or solid state (e.g. whipped cream and ice cream), respectively. In dairy production, the foam plays an important role in providing desirable characteristics to many products such as cappuccino-style beverages, milkshakes, whipped cream, ice cream, dairy desserts or mousses. In such products which are also known as aerated dairy products, the foam properties such as mouthfeel, texture, appearance and stability are the decisive factors to the product quality and attraction to the consumers. However, in many cases, for example, the transportation, pumping and container filling of milk, and reconstitution of milk powders (e.g. instant formula), the formation of foam is undesirable (Campbell & Mougeot, 1999; Gamboa & Barraquio, 2013). The foaming of dairy products, although has been studied over the past several decades, still is of interest to many researchers in dairy production sectors nowadays when the consumer's requirement for good and consistent quality foam-based products is rapidly increasing, aiming to control the foaming on demand.

The foam is produced as a result of an incorporation of air into the emulsified solution, which can be accomplished by supersaturating the solution with gas (e.g.

T. M. Ho · B. Bhandari · N. Bansal (✉)
ARC Dairy Innovation Hub, School of Agriculture and Food Sciences, The University of Queensland, St Lucia, QLD, Australia
e-mail: n.bansal@uq.edu.au

T. Truong et al. (eds.), *Dairy Fat Products and Functionality*,
https://doi.org/10.1007/978-3-030-41661-4_24

dissolving gas under pressure) or by mechanical means (e.g. injection, agitation, beating, whipping or shaking) (Walstra, 1989). The rate of foam formation and the foam stabilization are greatly dependent on the surface tension of liquid in the emulsion. The higher surface tension of liquid renders it more difficult to foam and leads to less stable foam. As soon as the air bubbles are generated in the liquid, the film at air-water interface should be formed and stabilized sufficiently enough to prevent them from collapsing (Huppertz, 2010). During foaming, active surfactants adsorb rapidly on the air-water interface to reduce the surface tension, resulting in a significant enhancement of foamability and foam stability (Walstra, 1989). In milk, there are two types of such surfactants including low and high molecular weight substances, of which only high molecular weight substances such as caseins and whey proteins have ability to form a highly viscoelastic film on the interfacial regions. In contrast, low molecular weight surfactants such as milk fat can either compete with proteins to adsorb at the air-water interface, resulting in either unfavourable effects on the foam properties of liquid foams, or positively contribute to the stabilization of air bubbles in solid foams, depending on the properties, especially components and physical state of milk fat (Pei & Schmidt, 2010). Moreover, the formation and stabilization of foam in the dairy products are also determined by many other factors ranging from the properties of milk (e.g. origin, age, composition, protein concentration and presence of surfactants) to the foaming conditions (e.g. method, temperature, pH, heat treatment and homogenization) (Gamboa & Barraquio, 2013; Goh, Kravchuk, & Deeth, 2009; Huppertz, 2010; Zayas, 1997). Although there are many studies performed on the effects of these factors on the foaming properties of dairy products over last few years (Huppertz, 2010; Walstra, 1989), there has been no comprehensive report on the role of milk fat in the development and stabilization of foam in liquid and solid foams. Thus, the main focus of this chapter is the discussion on the effects of milk fat on the foaming process of liquid and solid foams. In addition, the importance and mechanism of foaming process in the production of such dairy products are also described.

2 The Significance of the Foaming Process in Dairy Products

As mentioned, the foaming process in dairy products can be wanted or unwanted. In the production of many dairy products, the foaming of milk is an inherent property of many processes, which leads to the product loss or serious engineering problems. This is due to the excellent foaming ability of milk proteins such as caseins and whey proteins. These proteins have a high water solubility, which allows them to diffuse quickly to the air-water interface where they can unfold, rearrange the hydrophilic and hydrophobic groups and interact with each other to form a stable film around the air bubbles (Zayas, 1997). Following are several examples of unwanted and unavoidable foaming processes which are typically found in dairy production. During transportation of milk in the pipe systems for heat treatment processes, especially in pasteurization or sterilization of milk, the formation of air bubbles prevents the fluid from contacting the heating surface, resulting in a sharp decrease in the efficiency of heat transfer, subsequently the effectiveness of destruction of harmful microorganisms and the possibility of localised burning of the products due to uneven heating. In heat exchanger, the air bubbles can act as nuclei for fouling of proteins, reducing the performance of the equipment (Bogdan & Chandrapala, 2015; Chandan, Kilara, & Shah, 2009). Moreover, during the filling of milk into bottles or containers, the foaming possibly leads to uneven liquid level in the bottle or improper sealing (Gamboa & Barraquio, 2013). The pumping of milk from the bulk tanks in the farm to the trucks for transportation to factory as well as a flushing of milk in the tanks during transportation also creates the foam, resulting in the loss of the product. Other undesirable foaming processes are during the reconstitution of milk powder (especially infant formula) and blending of other ingredients to milk. During these processes, the agitation inevitably incorporates a large amount of air into the bulk of the products, by which the foam is created. In order to avoid foaming during the processing of dairy products, several approaches such as proper design of equipment, use of anti-foaming and defoaming agents (oils or silicon-based compounds) as processing aids, or use of mechanical foam breakers, can be applied (Berk, 2008).

However, the formation of the stable and good quality foam is highly expected in many dairy products including liquid and solid foams. In liquid foams, such as cappuccino-style coffees in which steam is directly injected into the milk to create the foam and to heat the milk to 65 °C prior to mixing with extracted coffee, the foam layer on the surface is a critical element to the product due to its light and soft texture as well as its pleasant feeling in the mouth. Moreover, such a foam layer also helps to slow down the release rate of coffee aroma (Khezri, Shahriari, & Shahsavani, 2017). Therefore, the good quality foam should be fine and have uniform size of air bubbles, and stable for at least a half-hour during which half a cup of coffee is typically consumed (Ranjith & Wijewardene, 2006). Similarly, in solid foams such as milkshake, ice cream and whipped cream, the foam has a critical functional property to determine the consumer appeal to the products. In such products, an overrun (a term referring to the percentage of incorporated air) should be within a range of 80–150%, depending on the product formula. Typically, a maximum overrun of 100% is easily obtained for most common formula (Chee & Chow, 2006). The amount of air incorporated has a great effect on the product texture, especially sensory properties. A sufficient amount of added air will provide the products with a smooth and light texture. A high overrun causes the products to be too crumbly and fluffy while a low overrun leads to dense and heavy texture of the products. Moreover, the overrun affects the melting behavior of the products. A low thermal conductivity of air results in overall low thermal conductivity of the products, as a result of which the melting of the products is significantly delayed (Pei & Schmidt, 2010). The slow melting is a desirable characteristic of these products as they are supposed to be consumed in the hot weather. It was reported that thermal conductivity of air and ice is 0.024 and 2.18 W/(m.K), respectively, and thus thermal conductivity of ice cream with 100% overrun is only 0.30 W/(m.K) (Clarke, 2006).

3 Foaming Mechanisms and the Role of Milk Proteins

The formation of foam is a very complicated process with an involvement of gas, liquid and solid, and includes two main stages which occur almost simultaneously, namely air incorporation and stabilization of air bubbles. There are several approaches employed to incorporate air into the products. Common ones consist of mechanical agitation, injection and supersaturation. Their main characteristics are illustrated in Table 24.1 (Huppertz, 2010; Walstra, 1989).

Unlike the mechanical agitation and injection methods which are typically used in the preparation of both liquid and solid foams, the supersaturation method is only used to produce the foam structure of aerosol cream. In this product, cream is packed in a can with gas under high pressure. As the pressure is released through a small valve, the aeration of gas causes the cream to foam out, resulting in the appearance and texture of whipped cream (NIIR Board, 2013).

Immediately after the gas is introduced, the gas-liquid interfacial film is formed and its stability is dependent on the surface tension between gas and liquid. A decrease in surface tension facilitates the formation of the interfacial film, subsequently promoting the formation of foam. In foaming of milk, proteins (caseins or whey proteins) play an important role in the initial formation and subsequent stabilization of foam. As mentioned, the milk proteins dissolve well in water and possess both hydrophilic and hydrophobic groups in the molecule, and the ability to reorient these groups at the air-water interface. These properties allow them to diffuse and adsorb rapidly on the interfacial regions during foaming, leading to a reduction in

Table 24.1 The common foaming methods in the dairy industry

Foaming methods	Characteristics
Mechanical agitation	– The formation of foam is a result of mixing gas and the product together with an input of mechanical energy – It is accomplished by stirring, pouring, whipping or beating using milk-whisking devices, high-shear mixers or blenders – The foam properties are mainly influenced by intensity and duration of agitation and product temperature
Injection	– It includes cold aeration and steam injection – Gas, air or steam is directly injected into the products through very small openings – Cold aeration has no effect on the product temperature whereas steam injection leads to an increase in the product temperature and the condensation of steam in the product – Foam properties are primarily determined by duration of aeration, gas flow rate, dimensions of openings and product temperature
Supersaturation	The foam formation is a result of dissolving a gas in a liquid under pressure and then releasing of pressure

the surface tension. At the interfacial regions, the proteins unfold and rearrange the polar and nonpolar groups towards the aqueous and non-aqueous phases, respectively. This is followed by the interactions among unfolded proteins by primarily electrostatic and hydrophobic interactions and hydrogen bonds, resulting in the formation of a strong, high viscous and high elastic interfacial film which helps to stabilize the air bubbles (Huppertz, 2010; Zayas, 1997). The mechanism of this process is demonstrated in Fig. 24.1. It should be noted that this section only illustrates the role of proteins in the formation of a high viscoelastic interfacial film without competing with other surfactants such as milk fat. The effect of milk fat on the formation of interfacial film is described in the following section. In foam, the air bubbles are separated by liquid films, also known as lamellae. The point at which the lamellae meet each other is called as a plateau border. A sketch illustrating the structure of foam is shown in Fig. 24.2a. At the plateau border, pressure (P_P) is small due to the curvature of air-liquid interface while pressure along the thin liquid film

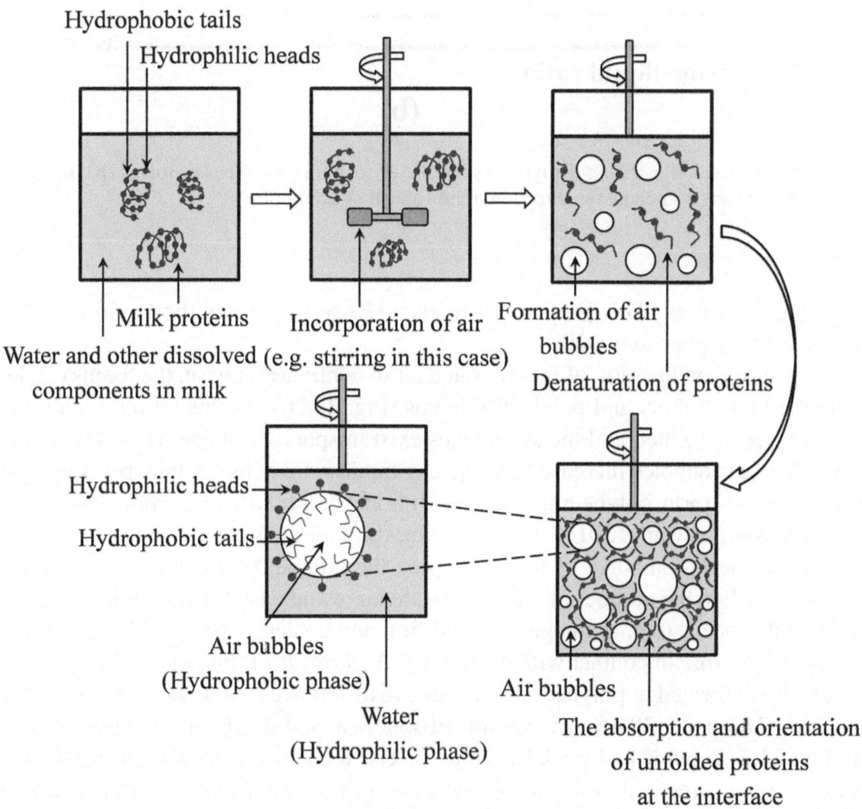

Fig. 24.1 A sketch to illustrate the role of proteins in the formation of milk foams

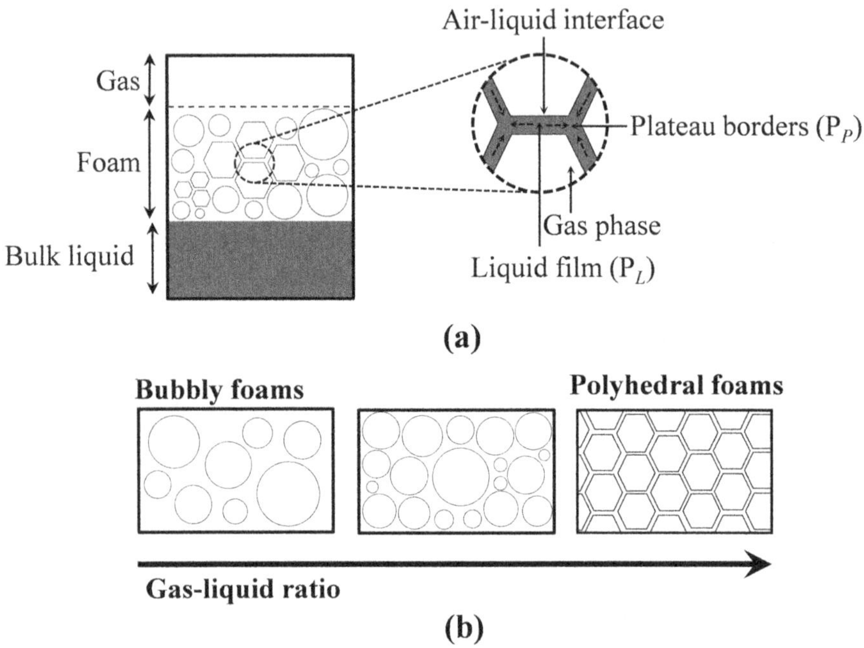

Fig. 24.2 A sketch illustrating the typical structure of foam (**a**) and types of foams (**b**). In figure (**a**), the dotted arrows denote the direction of movement of liquid film

region (P_L) is high. This pressure difference ($P_P > P_L$) causes the liquid to move towards to the plateau border.

Depending on the ratio of incorporated air to continuous phase, the foams can be classified into bubbly and polyhedral foams (Fig. 24.2b). As this ratio is small, the former type is formed and the air bubbles exist in spherical shape. However, as the number of air bubbles increase (e.g. the gas-liquid ratio is high), they press against each other to form polyhedral foams with honeycomb structure (Venkatachalam, John, & Kuppuswamy, 2015). In milk foam, the spherical foams are usually found shortly in the initial stage of foaming when the number of air bubbles are small. However, when the number of air bubbles are large and destabilization processes of air bubbles occur (e.g. drainage of liquid film and coalescence of air bubbles), the air bubbles come in contact with each other. As a result of this, the air bubbles are gradually deformed to polyhedral shape due to differences in buoyancy forces of air bubbles (Huppertz, 2010). An example of spherical and polyhedral foams is shown in Fig. 24.3 (unpublished work by authors). These images were taken from the surface of foam prepared from the whole cow milk using steam injection foaming approach. These images illustrate that when air bubbles were small and well separated, most of them have spherical shape, however as the liquid films between air

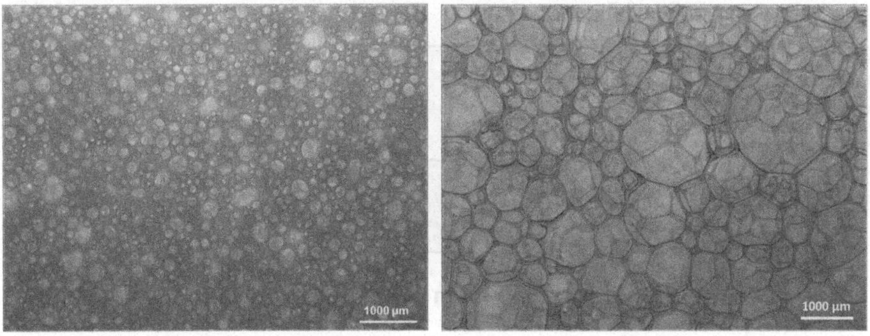

Fig. 24.3 Light microscopic images of foam surface prepared from full cream milk by steam approach at 65 °C immediately after foaming (on the left) and at half-life (on the right)

bubbles no longer existed, the air bubbles aggregate with each other, leading to a distortion of the foam shape.

As soon as the air bubbles are formed, there are several changes which greatly affect the stability and the properties of foam. These changes include the drainage of liquid film from the foam, the coalescence and disproportionation of air bubbles. These phenomena have been well described by Huppertz (2010) and Walstra (1989) and are illustrated in Fig. 24.4.

Drainage of Liquid Films from the Foam It is also known as creaming of air bubbles (Fig. 24.4a). Due to the differences in density between air bubbles and surrounding liquid, and the differences in pressure at the plateau border and along the thin liquid film regions, the air bubbles move upwards whereas the liquid flows towards to the plateau borders from which it drains through these channels to the bulk of liquid. A result of which is the thinning of liquid film between the air bubbles and movement of air bubbles. Eventually, the liquid films are ruptured, and the air bubbles are collapsed or coalesced.

Coalescence of Air Bubbles It is the fusion of air bubbles as a result of the collapsing of liquid films between air bubbles, followed by the formation of air bubbles with larger size (Fig. 24.4b). Thus, this phenomenon declines the number of air bubbles and increases the size of air bubbles in foam.

Disproportionation of Gas Bubbles (Ostwald Ripening) It is the diffusion of gas from the small air bubbles to the large ones through the continuous phase (Fig. 24.4c). The gas diffusion between air bubbles is caused by the difference in the internal pressure between the small and large air bubbles. The smaller air bubbles have the higher internal pressure due to the higher Laplace pressure difference over the curved surface of an air bubble. Once the air bubbles in different size are in a close contact, the air from the small air bubbles diffuses to the large ones, causing the small air bubbles to become smaller and eventually disappear, whereas the large ones become bigger.

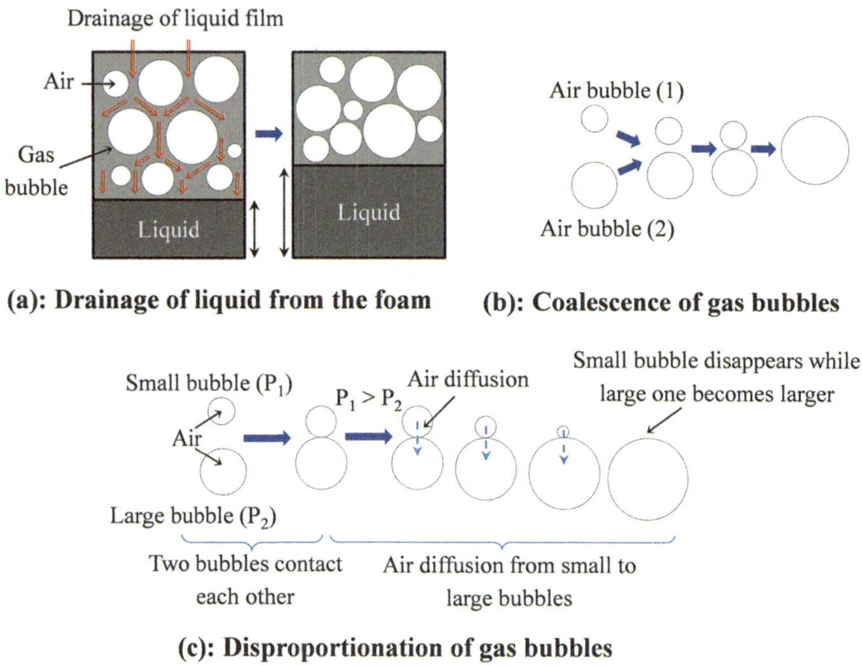

Fig. 24.4 The destabilization processes of foam; (**a**) drainage of liquid from the foam, (**b**) coalescence of gas bubbles, and (**c**) disproportionation of gas bubbles

4 Effect of Milk Fat on Foaming Properties in Dairy Products

4.1 Milk Fat and Its Role in the Dairy Products

Lipid in milk exists in the form of emulsified globules (2–4 μm in diameter) coated with a membrane mainly made of phospholipids, cholesterol, lipoproteins, glycoproteins and proteins (Kontkanen et al., 2011). This membrane prevents lipase enzymes from accessing the fat globules to initiate lipolysis (Deeth, 2006) and acts as an emulsion stabilizer through preventing the fat globules from coalescence (Jensen, 2002). In milk, the fat accounts for 3.0–5.0% (w/w) and consists of a majority of triacylglycerols (esters of glycerol and fatty acids); a small quantity of diacylglycerols, monoacylglycerols, free fatty acids, phospholipids, and cholesterols, which are mostly located in the globule membrane; and a trace of fat-soluble vitamins, β-carotene and fat-soluble/favouring compounds (MacGibbon & Taylor, 2006). The content and composition of lipid vary markedly according to the species, breed and diet of animals; the stage of lactation; and the health of the animals (Gordon, 2013). Because milk fat constitutes more than 98% (w/w) triacylglycerols, its physical properties such as density, crystallization and melting behavior are determined by the properties of triacylglycerols. However, triacylglycerols are a complex mixture, composed of a large number and wide variety of fatty acids which have a significant difference in molecular weight and degree of unsaturation. This heterogeneity and complexity provides milk fat with a very wide range of melting temperature, ranging from −40 to +40 °C. Between this range, milk fat exists as a mixture of solid and liquid fats, resulting a wide range of plastic and rheological properties (Wright & Marangoni, 2006).

Milk fat plays an important role in providing distinctive nutritional, textural and organoleptic properties, and consequently the eating pleasure for many dairy products (MacGibbon & Taylor, 2006). Milk fat greatly contributes to consumer's perception about the dairy products in terms of smoothness, thickness, viscosity, moistness, flavour, taste and cooling feeling (Drewnowski, 1997). Many flavour compounds (e.g. lactones, fatty acids, aldehydes and methyl ketones) have been identified in milk fat. These compounds, together with the ones produced during manufacture, storage and utilization of dairy products, contribute to the overall flavour profile of dairy products (Day, 1966). In addition, the lipid content has a great effect on the sensory properties in terms of appearance, texture and flavour of milk. A comparison of sensory properties of milk with different fat levels revealed that the whole milk with the highest fat level got the highest score in taste, creaminess and mouthfeel perceptions, whitest colour, most viscous and highest intensity in flavour (Golding & Pelan, 2008).

Other components in milk fat which have an important contribution to the stabilization of milk itself are phospholipids. In milk, the concentration of phospholipids is about 0.5–1.0% (w/w) of total fat, in which about 60–65% (w/w) of phospholipids are located on the fat globule membrane, the remaining amount is found in the aqueous phase in forms of complex with proteins or fragment materials of the membrane. Owing to both lipophilic and hydrophilic properties, they are naturally excellent emulsifying agents in milk to prevent the fat globules from aggregating and coalescing in the aqueous environment of the milk, resulting in avoiding of the milk creaming for a certain time after milking (MacGibbon & Taylor, 2006). In addition, phospholipids are an integral part of the fat globule membrane which protects triacylglycerols from the hydrolysis by lipolytic enzymes (Gantner, Mijić, Baban, Škrtić, & Turalija, 2015). There are two types of lipolysis in milk, namely spontaneous and induced lipolysis. The former refers to the lipolysis of milk during cooling of milk shortly after secretion regardless of the integrity of the membrane due to a balance between lipolysis activating and inhibiting factors while the latter is the result of physical damage to the membrane due to the processing conditions such as cooling, blending, pumping or homogenizing (Deeth, 2006). The products of milk lipolysis are mainly free fatty acids and partially glycerides (monoglycerides and diglycerides) which lead to the formation of off-flavour and markedly change the functionality of milk fat.

In foam-based products, the formation and stabilization of the foam depends highly on content, composition, size and physical state (e.g. ratio of solid-liquid fat) of milk fat (Pilhofer, Lee, McCarthy, Tong, & Bruce German, 1994). Effects of these factors of milk fat on the creation and stabilization of foam in aerated dairy products (liquid and solid foams) are discussed in detail in the following sections.

4.2 Effect of Milk Fat on the Foaming Properties of Aerated Dairy Products

4.2.1 Liquid Foam-Based Products

Typical products of liquid foam-based products are cappuccino-style beverages prepared from extracted coffee (espresso), hot milk and milk foam on the top (Hidden, Boomsma, Schins, & Van den Berg, 2012). The quality of such cappuccino cups is largely decided by the properties of the top foam layer. The air bubbles in the foam layer must be small and uniform in size, and last long until about a half of cup is consumed (~30 min). During foaming, the presence of milk fat or lipids, even a very small amount, leads to detrimental effects on the creation and stabilization of the foam. The differences in the adsorption mechanism at the interfacial region between molecules of milk fat and proteins, which counteract each other, resulting in the rupture of interfacial film of foam (Nylander, Arnebrant, Bos, & Wilde, 2008). As shown in Fig. 24.5a, as there is only the lipids involved in foaming process, lipid molecules with high mobility (due to low molecular weight) congregate quickly at the interface to form a thin film (lipid-stabilized thin film) but are unable to form a viscous film around the air bubble. The stretching of this film causes local thinning of the film, followed by the generation of surface tension gradient across the local thin film. As a result of this, there is a migration of lipid molecules from the interlamellar to the thinning regions of the film to restore the surface tension and film thickness. On the contrary, the proteins possess an excellent ability to form a strong, highly viscous and highly elastic interfacial film although they have low mobility (due to high molecular weight) as compared to lipids molecules (Fig. 24.5b). The role of proteins in the formation and stabilization of such an interfacial film has been illustrated in Fig. 24.1. The integrity of the protein-stabilized films is maintained during the film stretching because of the cohesiveness of adsorbed protein layers and deformability of the adsorbed protein molecules. When the foaming systems contain a mixture of lipids and proteins, which is commonly found in a very wide range of food products, lipids compete for adsorption, co-adsorption or complex formation with proteins on the interfacial regions because they are more surface-active than proteins. This interferes with the formation of intermolecular interactions of proteins, resulting in weakening or destroying the integrity, cohesiveness and viscoelastic properties of the interfacial film, and collapse of the foam (Fig. 24.5c).

The incompatibility in the adsorption mechanism of lipids and proteins at the interfacial region explains the unfavourable effects of milk fat on the creation and stabilization of liquid milk foam. These negative effects of milk fat can be due to

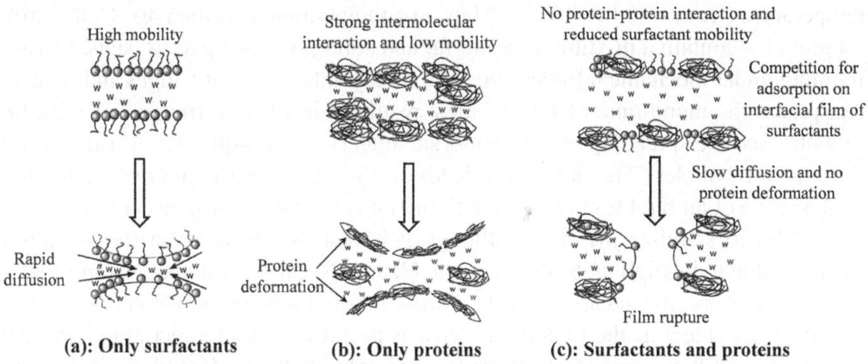

Fig. 24.5 Possible mechanism for the stabilization-destabilization of foams with polar lipids (**a**), proteins (**b**) and mixtures of the two components (**c**). Cross-sections of the thin films are shown where the aqueous interlamellar spacing is marked with (w). Recreated with permission from Nylander et al. (2008)

either (1) the content and physical state of milk fat, or (2) the products of milk fat lipolysis occurring during transportation, storage and processing of milk.

Effects of Milk Fat Content and Its Physical Properties on Foaming Properties of Liquid Milks Many studies have reported that milk with a high fat content exhibits a very poor foaming ability and stability as compared to ones with a low fat level, and that skim milk, even reconstituted skim milk powders, created stronger and more stable foams than whole or full-cream milk (Anderson & Brooker, 1988; Gamboa & Barraquio, 2013; Goh et al., 2009; Huppertz, 2010; Walstra, 1989). Apart from the extremely low fat content, skim milk, especially skim milk powders also contain a high amount of lactose which has a great water holding capacity. This property of lactose contributes to an increase in the viscosity of milk by which the stability of adsorbed protein films is markedly reinforced (Gamboa & Barraquio, 2013). The destructive effects of milk fat on the foaming properties of milks have been reported. The foamability of milk containing less than 0.04% (w/w) fat were nearly double to that of milk containing 1.5 and 3.0–3.9% fat (w/w) (Gamboa & Barraquio, 2013). Similarly, Anderson and Brooker (1988) reported that the foamability of milk significantly decreased with an increase in fat concentration from 0 to 1.5%. Goh et al. (2009) stated that regardless of the foaming method (e.g. mechanical agitation, steam injection and air bubbling) and heat treatment approaches (pasteurization and ultra-high temperature treatment), the foams obtained from skim milk were significantly stronger and lasted longer those preparing from full-cream milk.

The influence of fat level on the foaming properties of milk are dependent on the temperature at which the foam is created because the temperature affects the physical state of milk fat. As mentioned, the milk fat has a wide range of the melting temperature (from −40 to +40 °C). Thus, at a temperature less than 40 °C, the milk fat globules contain a mixture of liquid fat and solid fat existing as crystals. During foaming under mechanical forces, the solid fat crystals with sharp edges and corners can pierce the membrane of fat globules, resulting in the deformation of milk fat globules and the spreading of the membrane materials and liquid fat over the liquid films of air bubbles. This spreading leads to the disruption of foam lamella and coalescence of air bubbles because the liquid fat is unable to form viscoelastic interfacial film to stabilize the foam although it is readily adsorbed on the air-liquid interface due to its high mobility (Walstra, Geurts, Noomen, Jellema, & vanBoekel, 1999). Therefore, the negative effects of milk fat on the foaming become more predominantly evident as the foaming process is performed at a temperature less than 40 °C, especially in a range of 10–40 °C. Kamath, Huppertz, Houlihan, and Deeth (2008) reported that at a temperature range of 5–45 °C, there was a large difference in foamability and foam stability between skim and whole milk in which the former showed a significantly higher foaming capacity and foam stability than the latter. Nevertheless, as the foaming temperatures were higher than 45 °C, a similarity in the foamability between them was observed, and other factors such as heat treatment and homogenization, rather than fat content had a major effect on the foam stability. This is because, at temperatures higher than 40 °C, milk fat globules contain only liquid fat, which is far less influential on the deformation of the foam lamella. At temperatures less than 10 °C, the milk fat globules are predominantly occupied by the solid fat. This provides the milk fat globules with a markedly high resistance to the deformation, thus the foamability of milk is less effected at those temperatures (Huppertz, 2010). However, at the whole range of foaming temperatures (5–85 °C), the foam produced from skim milk still was much more stable than that prepared from the whole ones. The reasons are possibly that regardless of physical state of milk fat (liquid or solid fats), it is unable to contribute to the formation of stable interfacial films as proteins do. Similar findings were also reported by Pilhofer et al. (1994) in which the creation and stabilization of milk foam were highly dependent on the fractions of milk fat. In conclusion, for both whole and skim milk, it is better to perform the foaming process at high temperature (50–60 °C) at which milk fat, containing only liquid fat, has less influence on the development and stability of foam (Borcherding, Hoffmann, Lorenzen, & Schrader, 2008; Kamath, Huppertz, et al., 2008). Moreover, the milk fat also affects the appearance and size distribution of air bubbles. It was reported by Kamath, Huppertz, et al. (2008) that the foam prepared from whole milk was smaller in size and narrower in size distribution than those of the foam prepared from skim milk. A higher viscosity of whole milk enhances the formation of smaller bubble size and prevents air bubbles from coalescence. However, the air bubbles in the foam prepared from whole milk ruptured with a markedly high rate.

In addition, the particle size of milk fat globules greatly affects the formation and stabilization of foam, especially for those prepared from whole milk at high temperatures. It was reported that at temperature under 40 °C, both unhomogenized and homogenized whole milk were unable to form the foam which can last for several minutes. At higher than 40 °C temperatures, a size reduction of milk fat globules accomplished by the homogenization improved the foam stability of whole milk three to four times, but almost did not significantly affect the foamability (Kamath, Huppertz, et al., 2008). This increases in foam stability could be due to several reasons. Firstly, after homogenized, the solid fat crystals become too small to disrupt the foam lamella. In addition, the homogenization substantially increases the surface area and the number of fat globules. This results in a deficiency of the milk fat globule membrane material and liquid fat available to stabilize the newly generated surface of homogenized fat globules. Thus, the amount of liquid fat released from the homogenized fat globules can be insufficient to cause the thinning of liquid films. Moreover, as there is a shortage of membrane material of fat globules, the newly formed surface of the homogenized ones is reinforced by casein micelles and whey proteins, making it much more resistant to disruption as compared to the natural fat globule membrane (Huppertz, 2010; Truong, Palmer, Bansal, & Bhandari, 2016). Similarly, Borcherding et al. (2008) reported that whole milk with 3.0% fat (w/w) homogenised by a two-stage homogenization at 200 bar (first stage) and 50 bar (second stage) exhibited a very poor foamability and foam stability when the foaming process was performed at low temperatures, especially at 20 °C at which milk fat globules contained about 20% (w/w) solid fat crystals which were apparently detrimental to the foaming process. However, the whole milk which was homogenized under similar conditions had only marginal effect on the formation and stability of foam when the foaming process was performed at 50 °C, at which the fat globules contained only liquid fat. These research results demonstrate that the physical state of milk fat in terms of the ratio of solid and liquid fat affected by the foaming temperature is more important than the size reduction of milk fat globules accomplished by homogenization in determination of the creation and stabilization of foam produced from whole milk.

Similar negative effects of residual lipids on the foaming properties (capacity and stability) of whey protein concentrates have been reported (Karleskind, Laye, Mei, & Morr, 1995; Kim, Morr, Seo, & Surak, 1989; Patel & Kilara, 1990; Peltonen-Shalaby & Mangino, 1986; Rinn, Morr, Seo, & Surak, 1990; Vaghela & Kilara, 1996). In those studies, it was reported that the whey protein concentrates with a higher lipid content display a poorer foaming capacity and a lower foam stability than those with a lower lipid content. Moreover, it was reported that the hydrolysis of phospholipids by lipase from *Mucor miehei* or *Fusarium venenatum* phospholipase A1 showed a significant improvement in the formation and stability of the foam generated from both milk and whey protein concentrates (Blecker et al., 1997; Lilbaek, Fatum, Ipsen, & Sorensen, 2007).

Effects of Products of Lipolysis It has been reported that the poor foaming ability of milk is associated with products of milk lipolysis (Buchanan, 1965; Deeth & Smith, 1983; Huppertz, 2010; Kamath, Wulandewi, & Deeth, 2008; Kitchen & Cranston, 1969). These products include mostly free fatty acid (FFA), monoglycerides and diglycerides, which are the result of the breaking down of triglycerides (triacylglycerol), the main component in milk fat, under the action of lipase enzymes

(Deeth, 2006). Actually, these products are present in milk in a small quantity as a result of incomplete synthesis of triglycerides (Huppertz, 2010). However, many stages of dairy processing can cause physical damage to the fat globule membrane (cooling, agitation, mixing, homogenization and freezing/thawing), from which a large proportion of lipolyzed products are produced. Naturally, milk fat is protected by the membrane against the approach of lipase enzymes, therefore the rupture of the membrane allows lipase enzymes to access the milk fat by which the lipolysis is initiated. The effects of lipolyzed products on frothing ability of milk was firstly reported by Buchanan (1965). It was stated that monoglycerides and diglycerides, not FFAs, had a strongly detrimental effect on the foamability of milk. Moreover, an addition of a mixture of commercial monoglycerides and diglycerides into milk also showed a similar depressing effect on the foamability to that caused by those produced from lipolyzed milk.

In subsequent studies (Deeth & Smith, 1983; Kamath, Wulandewi, & Deeth, 2008; Kitchen & Cranston, 1969), it was found that the formation and stabilization of foam substantially reduced with an increase in FFA content. Deeth and Smith (1983) reported that an increase in FFA from about 0.8 to 1.2 μ.equiv/mL led to a decrease of about 60% frothing value. At concentrations higher than 2.0 μ.equiv/mL, the frothing ability of milk was negligible. Moreover, mixing of a lipolyzed milk (3.5 μ.equiv/mL of FFA) with an unlipolyzed one (<0.75 μ.equiv/mL of FFA) significantly declined the foamability of the latter. Thus, on a milk farm, mixing of several types of milk, in which at least one has been lipolyzed, into a tanker can cause poor foaming problems for the whole bulk of the milk. Similarly, Kamath, Wulandewi, and Deeth (2008) illustrated the depressing effects of FFA on the foamability and stability of foam (Fig. 24.6). However, the starting point of FFA level inducing the negative effects on foaming ability was different from those reported by Deeth and Smith (1983). High volume of stable foam was still obtained at the FFA levels higher than 2.0 μ.equiv/mL although the same foaming method (steam injection at 65–70 °C) was employed (Fig. 24.6a, b). The differences in the results of two studies are possibly explained by the dissimilarities in heat treatment and homogenization conditions of milk prior to foaming, both of which affect the particle size of fat globules and the integrity of proteins, and in methods to induce the lipolysis of milk fat. In our recent study (unpublished work), it was confirmed that the different ways to induce lipolysis in milk had different effects on foaming properties. Unlike ultra-turrax-induced lipolysis which caused the foaming ability of milk to drop rapidly as FFA content was higher than 2 μ.equiv/mL, microfluidised-induced lipolysis led to a steady decrease in the foaming ability of milk with a correlation coefficient of 0.9 (Fig. 24.6d).

As displayed in Fig. 24.6c, FFA content also affected the appearance of foam in which a foam with smooth and creamy texture was only obtained at low FFA content and an increase in FFA concentration resulted in an increase in foam coarseness. In addition, it was also reported that in air aeration foaming method, the depressing effects of FFA on foaming stability were dependent on temperature at which foam was produced. At <3.0 μ.equiv/mL, the negative effects of FFA on the

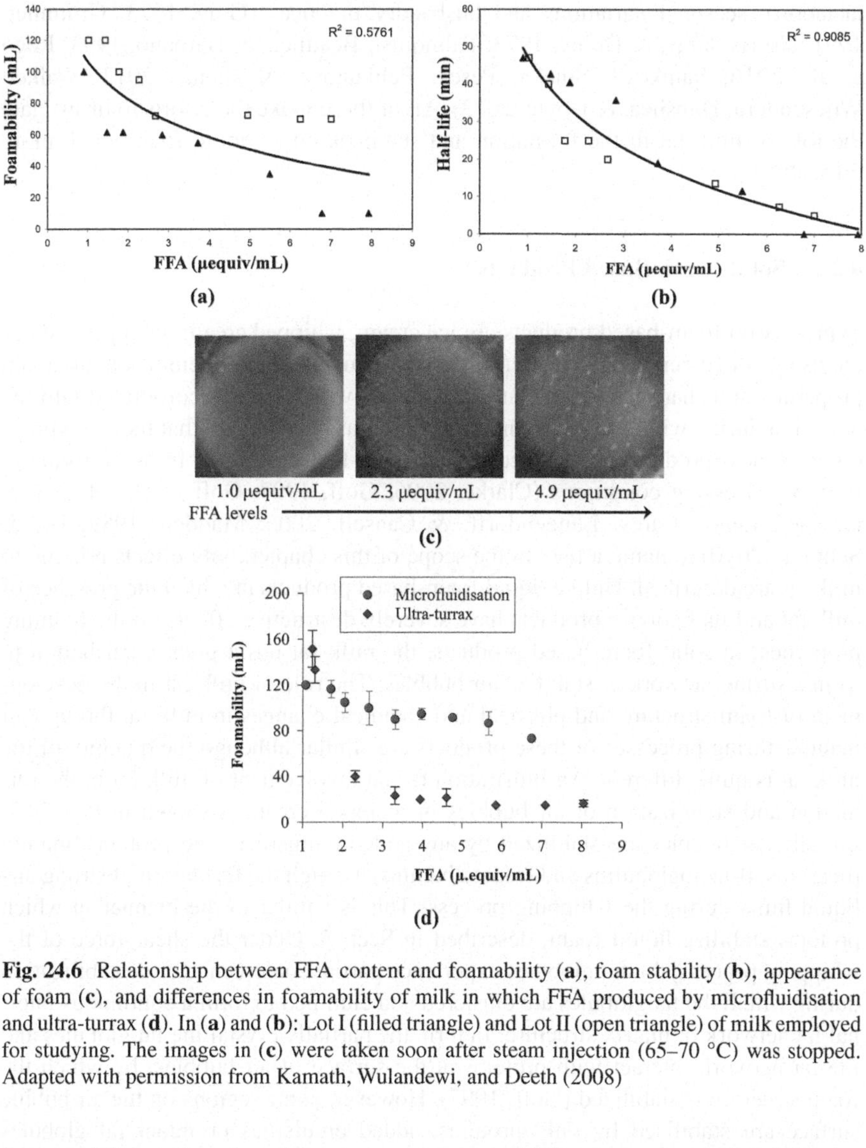

Fig. 24.6 Relationship between FFA content and foamability (**a**), foam stability (**b**), appearance of foam (**c**), and differences in foamability of milk in which FFA produced by microfluidisation and ultra-turrax (**d**). In (**a**) and (**b**): Lot I (filled triangle) and Lot II (open triangle) of milk employed for studying. The images in (**c**) were taken soon after steam injection (65–70 °C) was stopped. Adapted with permission from Kamath, Wulandewi, and Deeth (2008)

foam stability at low foaming temperature (e.g. 5 °C) was more profound than at high foaming temperature (e.g. 65 °C) while at >3.0 μ.equiv/mL, the adverse effects of FFA were irrespective of the foaming temperature. A rapid absorption ability of proteins at high temperature, making them more effective in competing with FFA to adsorb on air-liquid interface, resulted in a higher foam stability at high foaming temperature than that at low foaming temperature (Kamath, Wulandewi, & Deeth, 2008).

It is important to note that the content, composition and properties of milk fat are highly influenced by animal factors (breed, cow individuality, parity and stage of lactation), seasonal variations and husbandry practices (Gray, 1973; Grummer, 1991; Norris, Gray, & Dolby, 1973; Palmquist, Beaulieu, & Barbano, 1993; Rego et al., 2016; Samkova, Spicka, Pesek, Pelikanova, & Hanus, 2012; Walker, Wijesundera, Dunshea, & Doyle, 2013). All of these make the efforts to investigate the role of milk fat in the formation and stabilization of liquid foam much more difficult.

4.2.2 Solid Foam-Based Products

Typical solid foam-based products are ice cream, whipped cream, whippable dairy emulsions or frozen aerated desserts. These products have both emulsion and foam properties and share a similar foam structure in which air is incorporated into the emulsion during whipping/churning process. It has been known that the foam properties of these products are decided by numerous factors ranging from the formulation to processing conditions (Clarke, 2006; Goff, 1997; Goff & Hartel, 2013a; Granger, Leger, Barey, Langendorff, & Cansell, 2005; Madden, 1989; Pei & Schmidt, 2010) to name a few. In the scope of this chapter, only effects relating to milk fat are described. Unlike liquid foam-based products in which the presence of milk fat and its lipolysis products have severely destructive effects on the foaming properties, in solid foam-based products, the milk fat has a great contribution to form a strong network to stabilize air bubbles. The role of milk fat in the development of foam structure and physical and chemical changes in milk fat throughout manufacturing processes of these products are similar although the quantity of the milk fat is quite different. An illustration for an involvement of milk fat in the formation and stabilization of air bubbles in whipped cream is shown in Fig. 24.7. Initially, air bubbles are stabilized by any added emulsifiers and proteins, mainly β-caseins, β-lactoglobulins and α-lactalbumins, through the formation of strong air-liquid films during the whipping process. This is similar to the manner in which proteins stabilize liquid foam, described in Sect. 3. Under the shear force of the whipping process, the fat globules experience partial coalescence or destabilization during which the fat globules are clustered and clumped to form a continuous internal fat network or matrix structure. In turn, the partially crystalline fat globules and the fat network interact with proteins at the surface of air bubbles, by which the foam structure is stabilized (Goff, 1997). However, some regions on the air bubble surface are stabilized by only proteins, added emulsifies or intact fat globules (Koxholt, Eisenmann, & Hinrichs, 2001). The coalescence of fat globules is primarily caused by puncturing of interfacial membrane of the fat globules by fat crystals under high shear conditions during whipping process, leading to the spreading of liquid fat promoting the interactions between fat globules. It was suggested that the mechanism described in Fig. 24.7 is also applicable to explain the structure development of ice cream or other whippable dairy emulsions (Madden, 1989).

The stability of foam structure and the product properties depend greatly on the amount of fat present in the emulsion during the whipping step. The higher degree of adsorbed fat globules on the air bubble surface, consequently the higher rigidity

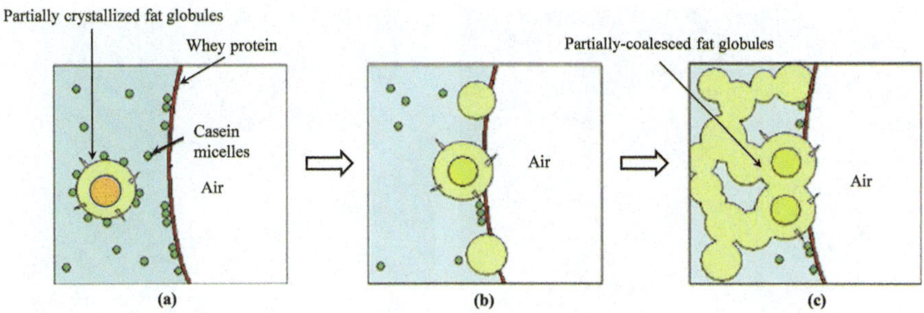

Fig. 24.7 An illustration of the structure development of whipped cream during the whipping process. (**a**) Initial stabilization of air bubble film by proteins, (**b**) Further stabilization of air bubbles by adsorption of fat globules, and (**c**) Formation of partially coalesced fat network in the external phase. Adapted with permission from Golding and Pelan (2008)

of foam structure, is found in the products prepared from an emulsion containing the higher fat content. This is easily observed in the structure of whipped cream and ice cream. The main difference between whipped cream and ice cream is the fat content in the former (approximately 30–40% (w/w)) is markedly higher than that in the latter (around 1–20% (w/w)), depending on desirable characteristics, regulations or price (Euston, 2008; Madden, 1989). As a consequence of this, the amount of the fat alone in whipped cream is much more sufficient to form a fat globule network to stabilize all air bubbles in the product for at least several weeks of storage at ambient temperature. As illustrated in Fig. 24.8a, b, whipped cream had a markedly higher degree of the adsorption of fat globules and formation of partially coalesced fat globule network on the air bubble surface than that observed for ice cream. In fact, the formation of a such fat globule network is unnecessary for ice cream because the foam structure of ice cream is not only stabilized by fat globule network but also by a significant amount of ice crystals formed in the freezing and hardening stages and high viscosity of freeze-concentrated unfrozen phase (Fig. 24.8c). Many studies on effects of different fat content (0.1–10%, w/w) on texture and sensory properties of ice cream showed that ice cream prepared from a higher fat content formulation was much better in flavour retention, softer and smoother in texture, slower in melting rate, less visible in air bubbles, and even whiter in colour than that produced from a lower fat level one (Aime, Arntfield, Malcolmson, & Ryland, 2001; Koeferli, Piccinali, & Sigrist, 1996; Li, Marshall, Heymann, & Fernando, 1997; Prindiville, Marshall, & Heymann, 1999; Roland, Phillips, & Boor, 1999). This relationship can be possibly explained by the ability of the milk fat to interact with flavour molecules through Van der Waals and hydrophobic interactions, to dissolve flavour molecules, and to contribute to foam structure development. All these provide the products with physical barriers to flavour transport and melting rate, and with desirable texture and sensory properties in terms of coldness, smoothness, viscosity and mouth coating. Therefore, the product with a high fat level is typically preferred by consumers to the one with a low fat content.

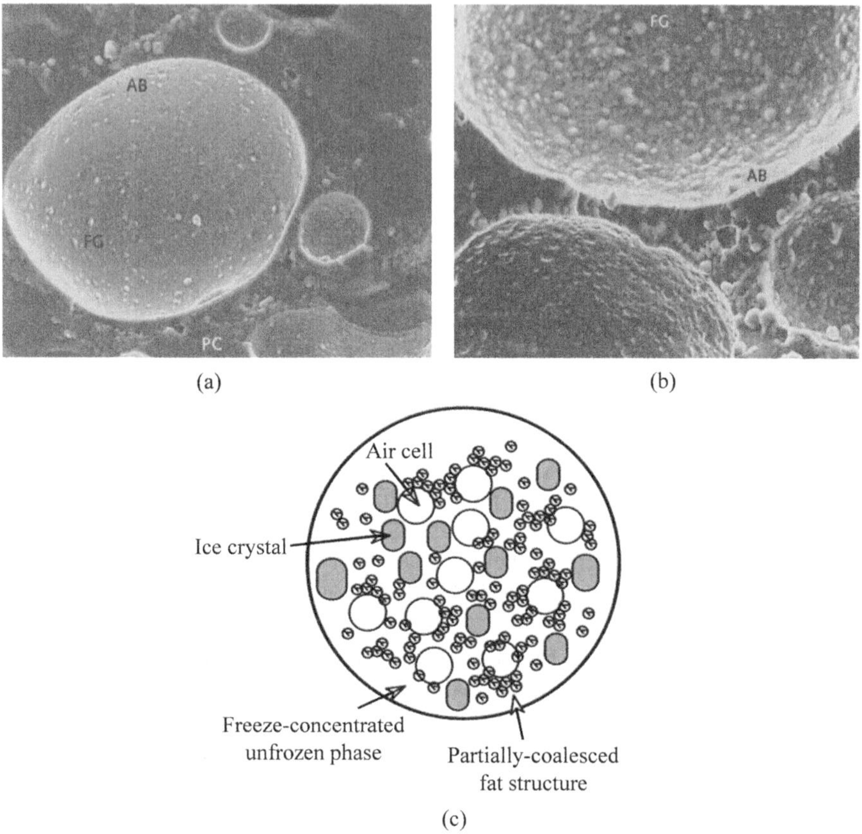

Fig. 24.8 Cryo-scanning electron micrograph of ice cream (**a**) and whipped cream (**b**), and structure of ice cream (**c**). In (**a**) and (**b**), *AB* air bubble, *FG* fat globule, *PC* partially coalesced fat globules. (**a**) and (**b**) adapted with permission from Euston (1997), (**c**) adapted with permission from Goff and Hartel (2013a)

One of the important properties of the milk fat which has strong effects on its role in the structure development of solid foam-based products is the particle size of fat globules which is typically controlled through homogenization conditions, especially homogenization pressure, of the emulsion. The purpose of the homogenization is to form a stable and finely dispersed emulsion by which the size of fat globules is decreased to less than 2 μm. When fat globules are reduced in size, the number and surface area of fat globules are markedly increased while there is a limitation of the membrane material to cover the newly formed surface of homogenized fat globules. As a result, the newly formed surface is adsorbed by added emulsifiers and proteins, leading to a significant increase in hydration properties of surface area of homogenized fat globules, providing the smoothness to the products (Goff & Hartel, 2013b). An optimal temperature range during homogenization of most emulsions is of 60–80 °C because this temperature helps to melt completely all the fat, inactivate lipase enzymes (preventing lipolysis) and maintain viscosity of the emulsion (limiting of clumping of fat globules in the emulsion) (Goff & Hartel, 2013b; Madden, 1989). In the meantime, the homogenization pressure has a great effect on the properties of the emulsion and the whipped products prepared thereof because it brings about marked changes in the fat globule size in an emulsion and the size of fat globule agglomerates in the whipping stage. Conventional homogenization is typically performed in a range of pressure less than 50 MPa (Marshall & Arbuckle, 1996). It was reported that fat globules with a size range of 0.44–3.33 μm could be achieved at homogenization pressure of 30 MPa and that homogenization of ice cream mixes containing less than 10% (w/w) of fat at pressures of 10 MPa with only one stage was sufficient to reduce the fat globule size suitable for production of stable ice cream. An increase in homogenization pressure resulted in a decrease in the size of fat globules and agglomerates, which subsequently had a great effect on the meltdown of the foam structure. As the size of fat globules and agglomerates was within a range of the width of foam lamellas (known as critical diameter of fat globule, assumed about 0.85 μm) or larger, they obstructed foam lamellas or attached to the air bubble surface through calcium bridges of caseins in the membranes of fat globules, respectively by which the drainage of foam was inhibited. However, the ones with smaller than the critical diameter of fat globule drained out of the foam matrix easily, making ice cream to be unstable and melt rapidly (Koxholt et al., 2001).

Similarly, it was found that an increase in homogenization pressure from 20 to 50 MPa decreased the droplet size distribution of whipping cream emulsion from 0.16–17.38 to 0.12–11.48 μm. This resulted in more profound gel-like behaviour and reduction in the surface protein concentration of the emulsion, especially after double homogenization, as the droplet size distribution was 0.16–3.80 μm and 0.08–3.31 μm, respectively (Long, Zhao, Zhao, Yang, & Liu, 2012). For milk-based emulsions, Truong, Bansal, and Bhandari (2014) reported that a reduction in the emulsion droplet size from microscale (1.2 μm) to nanoscale (0.2 μm) resulted in a lower percentage of overrun and shorter half-life of the foam due to an increase in apparent viscosity and a decrease in solid fat content of the emulsion. Both these changes led to the restriction of air incorporation and the oil wetting of foam surface due to the spreading of liquid fat during the whipping process. Recently, high pressure homogenization (80–400 MPa) and microfluidization (50–200 MPa) are utilized for ice cream mixes to not only reduce fat globule size but also enhance rheological properties of colloidal phase, and induce changes in the milk proteins in serum phase, aiming to improve sensory properties and slow down melting rate of ice cream (Biasutti, Venir, Marino, Maifreni, & Innocente, 2013; Hayes, Lefrancois, Waldron, Goff, & Kelly, 2003; Huppertz, Smiddy, Goff, & Kelly, 2011; Innocente, Biasutti, Venir, Spaziani, & Marchesini, 2009; Olson, White, & Watson, 2003). However, the effects of homogenization conditions on the properties of the emulsions as well as the final products are highly dependent on the formulation, especially the presence of emulsifiers.

Following homogenization, the emulsions are cooled and aged at 4 °C to initiate complete crystallization of fat, together with re-distribution of proteins on of the fat-serum interfacial membrane and further hydration of the polysaccharides and hydrocolloid stabilizers (Goff & Hartel, 2013b). Partial crystallization of fat is very important to enhance fat destabilization, to incorporate stable air bubbles into the ice cream mix, and to churn out of the ice cream mix under high shear during whipping. It is noted that crystallization of fat globules in an emulsion is much slower than that in the bulk fat due to the boundaries of fat globule interfacial membrane, and that crystallization rate is highly dependent on the type of fat and emulsifiers used in the emulsion formulation. However, Barford, Krog, Larsen, and Buchheim (1991) reported that the crystallization of fat globules in ice cream mix almost completed at the first four hour of aging irrespective to types of emulsifiers. Crystallization mechanisms of fat globules during aging of ice cream mix is a complicated process and could be described in several ways: (1) low molecular weight fat molecules orient themselves on the interface and act as nuclei for crystallization of higher molecular weight fat molecules, (2) small crystals are formed inside the fat globules and then orient themselves (grow in size) tangentially at the globule boundary, and (3) at supercooled state, the fat globules with small size (less than 2 μm) nucleates so quickly that the crystal layers, rather than the individual crystals, are formed (Madden, 1989). Insufficient aging of ice cream mix, resulting in incomplete crystallization of fat, causes ice cream product to be less stable in the foam structure and rapid meltdown (Goff & Hartel, 2013b).

In summary, regardless of type of solid foam-based products, the milk fat globules in the emulsion experience several structural alterations such as reduction in size, complete melting into liquid fat and partial crystallization (all of which take place during homogenization and aging at cold temperature steps) prior to contributing to development of stable foam structure during whipping process.

5 Conclusion

Although the foaming of milk can cause technical troubles for several processing stages of dairy products, the high foamability and high foam stability are greatly expected in many aerated dairy products such as cappuccino-style drinks, ice cream, whipped cream, milkshake, etc. In the foaming process, milk proteins such as caseins and whey proteins determine the creation and stabilization of foam due to their ability to form a strong and highly viscoelastic film at the interfacial regions. Among foam quality-driving factors (milk properties, processing conditions, storage temperature and duration, etc.), the effects of milk fat on foaming properties of aerated dairy products are probably the most complicated because of heterogeneity and complexity of milk fat in the structure and properties. In liquid foam-based products the presence of milk fat, even at a very small amount, and the products of its lipolysis leads to the depressing effects on the foaming properties because they highly compete with proteins to adsorb on the interfacial films but are unable to form a strong and high viscoelastic film like proteins. Therefore, the removal of milk fat accomplished by centrifugation or enzyme hydrolysis helps to improve the foaming ability and foam stability of milk. In this regard, skim milk can produce a high volume of stable foam easily as compared to whole or full cream milk. The degree of influence of milk fat is mainly dependent on the particle size of fat globules and especially temperature at which the foam is created because the temperature affects the physical state of milk fat (e.g. ratio of solid and liquid fats). On the contrary, milk fat plays an essential role in the formation of the structure of solid foam-based products. The foam structure of those products is mostly created at low temperature at which milk fat exists as partially crystalline fat globules. Under mechanical force, these fat globules adsorb to the surface of protein which initially stabilizes air bubbles, followed by the aggregation of fat droplets and subsequently the formation of a fat globule network, by which the drainage of the stabilized foam is prevented, and the long-lasting foam structure of the products is achieved.

Acknowledgements This work was funded through Australian Research Council's Industrial Transformation Research Hub (ARC-ITRH) grant with Lion Dairy and Drinks as an industry partner. The ARC Dairy Innovation Hub is a collaboration between The University of Melbourne, The University of Queensland and Dairy Innovation Australia Ltd. The authors acknowledge the facilities, and the scientific and technical assistance, of the School of Agriculture and Food Sciences at The University of Queensland.

References

Aime, D., Arntfield, S., Malcolmson, L., & Ryland, D. (2001). Textural analysis of fat reduced vanilla ice cream products. *Food Research International, 34*, 237–246.

Anderson, M., & Brooker, B. E. (1988). Dairy foams. In E. Dickinson & G. Stainsby (Eds.), *Advances in food emulsions and foams* (pp. 221–225). London, UK: Elsevier Applied Science Publisher.

Barford, N. M., Krog, N., Larsen, G., & Buchheim, W. (1991). Effects of emulsifiers on protein-fat interaction in ice cream mix during ageing. I: Quantitative analyses. *Lipids, 93*, 24–29.

Berk, Z. (2008). *Food process engineering and technology*. Oxford, UK: Academic Press.

Biasutti, M., Venir, E., Marino, M., Maifreni, M., & Innocente, N. (2013). Effects of high pressure homogenisation of ice cream mix on the physical and structural properties of ice cream. *International Dairy Journal, 32*, 40–45.

Blecker, C., Paquot, M., Lamberti, I., Sensidoni, A., Lognay, G., & Deroanne, C. (1997). Improved emulsifying and foaming of whey proteins after enzymic fat hydrolysis. *Journal of Food Science, 62*, 48.

NIIR Board. (2013). *Modern technology of milk processing & dairy products* (4th ed.). Delhi, India: NIIR Project Consultancy Services.

Bogdan, Z., & Chandrapala, J. (2015). High power ultrasound processing in milk and dairy products. In N. Datta & P. M. Tomasula (Eds.), *Emerging dairy processing technologies: Opportunities for the dairy industry* (pp. 149–180). Chichester, UK: John Wiley & Sons.

Borcherding, K., Hoffmann, W., Lorenzen, P. C., & Schrader, K. (2008). Effect of milk homogenisation and foaming temperature on properties and microstructure of foams from pasteurised whole milk. *LWT - Food Science and Technology, 41*, 2036–2043.

Buchanan, R. (1965). Lipolysis and the frothing of milk. *Australian Journal of Dairy Technology, 20*, 62–66.

Campbell, G. M., & Mougeot, E. (1999). Creation and characterisation of aerated food products. *Trends in Food Science & Technology, 10*, 283–296.

Chandan, R. C., Kilara, A., & Shah, N. P. (2009). *Dairy processing and quality assurance* (2nd ed.). Chichester, UK: John Wiley & Sons.

Chee, C. H., & Chow, M. C. (2006). Edible emulsions. In P. Somasundaran (Ed.), *Encyclopedia of surface and colloid science* (Vol. 3, pp. 1846–1856). Boca Raton, FL: Taylor & Francis.

Clarke, C. (2006). *The science of ice cream*. Cambridge, UK: RSC Publishing, The Royal Chemistry Society.

Day, E. A. (1966). *Role of milk lipids in flavors of dairy products* (Advances in chemistry series) (p. 94). Washington, DC: American Chemical Society.

Deeth, H., & Smith, R. (1983). Lipolysis and other factors affecting the steam frothing capacity of milk. *Australian Journal of Dairy Technology, 38*, 14.

Deeth, H. C. (2006). Lipoprotein lipase and lipolysis in milk. *International Dairy Journal, 16*, 555–562.

Drewnowski, A. (1997). Why do we like fat? *Journal of the American Dietetic Association, 97*, S58–S62.

Euston, S. R. (1997). Emulsifiers in dairy products and dairy substitutes. In G. L. Hasenhuettl & R. W. Hartel (Eds.), *Food emulsifiers and their applications* (pp. 173–210). Boston, MA: Springer US.

Euston, S. R. (2008). Emulsifiers in dairy products and dairy substitutes. In G. L. Hasenhuettl & R. W. Hartel (Eds.), *Food emulsifiers and their applications* (2nd ed., pp. 195–232). New York, NY: Springer.

Gamboa, G. V., & Barraquio, V. L. (2013). Foaming properties at different fat levels and age of milk. *The Philippine Agricultural Scientist, 95*, 416–421.

Gantner, V., Mijić, P., Baban, M., Škrtić, Z., & Turalija, A. (2015). The overall and fat composition of milk of various species. *Mljekarstvo/Dairy, 65*, 223–231.

Goff, H. D. (1997). Colloidal aspects of ice cream - A review. *International Dairy Journal, 7*, 363–373.

Goff, H. D., & Hartel, R. W. (2013a). Ice cream structure. In *Ice cream* (pp. 313–352). New York, NY: Springer.

Goff, H. D., & Hartel, R. W. (2013b). Mix processing and properties. In *Ice cream* (pp. 121–154). New York, NY: Springer.

Goh, J., Kravchuk, O., & Deeth, H. (2009). Comparison of mechanical agitation, steam injection and air bubbling for foaming milk of different types. *Milchwissenschaft, 64*, 121–124.

Golding, M., & Pelan, E. (2008). Application of emulsifiers to reduce fat and enhance nutritional quality. In G. L. Hasenhuettl & R. W. Hartel (Eds.), *Food emulsifiers and their applications* (2nd ed., pp. 327–348). New York, NY: Springer.

Gordon, M. H. (2013). Milk lipids. In Y. W. Park & G. F. W. Haenlein (Eds.), *Milk and dairy products in human nutrition: Production, composition and health* (pp. 65–79). Oxford, UK: John Wiley & Sons.

Granger, C., Leger, A., Barey, P., Langendorff, V., & Cansell, M. (2005). Influence of formulation on the structural networks in ice cream. *International Dairy Journal, 15*, 255–262.

Gray, I. (1973). Seasonal variations in the composition and thermal properties of New Zealand milk fat: I. Fatty-acid composition. *Journal of Dairy Research, 40*, 207–214.

Grummer, R. R. (1991). Effect of feed on the composition of milk fat. *Journal of Dairy Science, 74*, 3244–3257.

Hayes, M. G., Lefrancois, A. C., Waldron, D. S., Goff, H. D., & Kelly, A. L. (2003). Influence of high pressure homogenisation on some characteristics of ice cream. *Milchwissenschaft-Milk Science International, 58*, 519–523.

Hidden, F., Boomsma, J., Schins, A., & Van den Berg, E. (2012). Cappuccino and specific heat versus heat of vaporization. *The Physics Teacher, 50*, 103–104.

Huppertz, T. (2010). Foaming properties of milk: A review of the influence of composition and processing. *International Journal of Dairy Technology, 63*, 477–488.

Huppertz, T., Smiddy, M. A., Goff, H. D., & Kelly, A. L. (2011). Effects of high pressure treatment of mix on ice cream manufacture. *International Dairy Journal, 21*, 718–726.

Innocente, N., Biasutti, M., Venir, E., Spaziani, M., & Marchesini, G. (2009). Effect of high-pressure homogenization on droplet size distribution and rheological properties of ice cream mixes. *Journal of Dairy Science, 92*, 1864–1875.

Jensen, R. G. (2002). The composition of bovine milk lipids: January 1995 to December 2000. *Journal of Dairy Science, 85*, 295–350.

Kamath, S., Huppertz, T., Houlihan, A. V., & Deeth, H. C. (2008). The influence of temperature on the foaming of milk. *International Dairy Journal, 18*, 994–1002.

Kamath, S., Wulandewi, A., & Deeth, H. (2008). Relationship between surface tension, free fatty acid concentration and foaming properties of milk. *Food Research International, 41*, 623–629.

Karleskind, D., Laye, I., Mei, F. I., & Morr, C. V. (1995). Foaming properties of lipid-reduced and calcium-reduced whey-protein concentrates. *Journal of Food Science, 60*, 738–741.

Khezri, M., Shahriari, S., & Shahsavani, L. (2017). The effect of xanthan gum and temperature on foam stability of milk-based espresso coffees. *Journal of Food Biosciences and Technology, 7*, 15–22.

Kim, S. H., Morr, C. V., Seo, A., & Surak, J. G. (1989). Effect of whey pretreatment on composition and functional-properties of whey-protein concentrate. *Journal of Food Science, 54*, 25–29.

Kitchen, B., & Cranston, K. (1969). Lipase activation in farm milk supplies. *Australian Journal of Dairy Technology, 24*, 107–112.

Koeferli, C. R. S., Piccinali, P., & Sigrist, S. (1996). The influence of fat, sugar and non-fat milk solids on selected taste, flavor and texture parameters of a vanilla ice-cream. *Food Quality and Preference, 7*, 69–79.

Kontkanen, H., Rokka, S., Kemppinen, A., Miettinen, H., Hellstrom, J., Kruus, K., et al. (2011). Enzymatic and physical modification of milk fat: A review. *International Dairy Journal, 21*, 3–13.

Koxholt, M. M. R., Eisenmann, B., & Hinrichs, J. (2001). Effect of the fat globule sizes on the meltdown of ice cream. *Journal of Dairy Science, 84*, 31–37.

Li, Z., Marshall, R., Heymann, H., & Fernando, L. (1997). Effect of milk fat content on flavor perception of vanilla ice cream. *Journal of Dairy Science, 80*, 3133–3141.

Lilbaek, H. M., Fatum, T. M., Ipsen, R., & Sorensen, N. K. (2007). Modification of milk and whey surface properties by enzymatic hydrolysis of milk phospholipids. *Journal of Agricultural and Food Chemistry, 55*, 2970–2978.

Long, Z., Zhao, M. M., Zhao, Q. Z., Yang, B., & Liu, L. Y. (2012). Effect of homogenisation and storage time on surface and rheology properties of whipping cream. *Food Chemistry, 131*, 748–753.

MacGibbon, A. K. H., & Taylor, M. W. (2006). Composition and structure of bovine milk lipids. In P. F. Fox & P. L. H. McSweeney (Eds.), *Advanced dairy chemistry volume 2: Lipids* (pp. 1–42). Boston, MA: Springer.

Madden, J. (1989). Ice cream. In A. Wilson (Ed.), *Foams: Physics, chemistry and structure* (pp. 185–196). London, UK: Springer.

Marshall, R. T., & Arbuckle, W. S. (1996). *Ice cream*. New York, NY: International Thomson Publishing.

Norris, G. E., Gray, I., & Dolby, R. (1973). Seasonal variations in the composition and thermal properties of New Zealand milk fat: II. Thermal properties of milk fat and their relation to composition. *Journal of Dairy Research, 40*, 311–321.

Nylander, T., Arnebrant, T., Bos, M., & Wilde, P. (2008). Protein/emulsifier interactions. In G. L. Hasenhuettl & R. W. Hartel (Eds.), *Food emulsifiers and their applications* (2nd ed., pp. 89–171). New York, NY: Springer.

Olson, D. W., White, C. H., & Watson, C. E. (2003). Properties of frozen dairy desserts processed by microfluidization of their mixes. *Journal of Dairy Science, 86*, 1157–1162.

Palmquist, D., Beaulieu, A. D., & Barbano, D. (1993). Feed and animal factors influencing milk fat composition. *Journal of Dairy Science, 76*, 1753–1771.

Patel, M. T., & Kilara, A. (1990). Studies on whey protein concentrates. 2. Foaming and emulsifying properties and their relationships with physicochemical properties. *Journal of Dairy Science, 73*, 2731–2740.

Pei, Z., & Schmidt, K. (2010). Ice cream: Foam formation and stabilization - a review. *Food Reviews International, 26*, 122–137.

Peltonen-Shalaby, R., & Mangino, M. (1986). Compositional factors that affect the emulsifying and foaming properties of whey protein concentrates. *Journal of Food Science, 51*, 91–95.

Pilhofer, G. M., Lee, H.-C., McCarthy, M. J., Tong, P. S., & Bruce German, J. (1994). Functionality of milk fat in foam formation and stability. *Journal of Dairy Science, 77*, 55–63.

Prindiville, E., Marshall, R., & Heymann, H. (1999). Effect of milk fat on the sensory properties of chocolate ice cream. *Journal of Dairy Science, 82*, 1425–1432.

Ranjith, P. H. M., & Wijewardene, U. (2006). Lipid emulsifiers and surfactants in dairy and bakery products. In F. D. Gunstone (Ed.), *Modifying lipids for use in food* (pp. 393–428). Cambridge, UK: Woodhead Publishing.

Rego, O. A., Cabrita, A. R., Rosa, H. J., Alves, S. P., Duarte, V., Fonseca, A. J., et al. (2016). Changes in milk production and milk fatty acid composition of cows switched from pasture to a total mixed ration diet and back to pasture. *Italian Journal of Animal Science, 15*, 76–86.

Rinn, J.-C., Morr, C., Seo, A., & Surak, J. (1990). Evaluation of nine semi-pilot scale whey pretreatment modifications for producing whey protein concentrate. *Journal of Food Science, 55*, 510–515.

Roland, A. M., Phillips, L. G., & Boor, K. J. (1999). Effects of fat content on the sensory properties, melting, color, and hardness of ice cream. *Journal of Dairy Science, 82*, 32–38.

Samkova, E., Spicka, J., Pesek, M., Pelikanova, T., & Hanus, O. (2012). Animal factors affecting fatty acid composition of cow milk fat: A review. *South African Journal of Animal Science, 42*, 83–100.

Truong, T., Bansal, N., & Bhandari, B. (2014). Effect of emulsion droplet size on foaming properties of milk fat emulsions. *Food and Bioprocess Technology, 7*, 3416–3428.

Truong, T., Palmer, M., Bansal, N., & Bhandari, B. (2016). Methodologies to vary milk fat globule size. In *Effect of milk fat globule size on the physical functionality of dairy products* (pp. 15–30). New York, NY: Springer International Publishing.

Vaghela, M. N., & Kilara, A. (1996). Foaming and emulsifying properties of whey protein concentrates as affected by lipid composition. *Journal of Food Science, 61*, 275–280.

Venkatachalam, S., John, S. G., & Kuppuswamy, K. (2015). Foam mat drying of food materials: A review. *Journal of Food Processing and Preservation, 39*, 3165–3174.

Walker, G. P., Wijesundera, C., Dunshea, F. R., & Doyle, P. T. (2013). Seasonal and stage of lactation effects on milk fat composition in northern Victoria. *Animal Production Science, 53*, 560–572.

Walstra, P. (1989). Principles of foam formation and stability. In A. Wilson (Ed.), *Foams: Physics, chemistry and structure* (pp. 1–15). London, UK: Springer.

Walstra, P., Geurts, T. J., Noomen, A., Jellema, A., & vanBoekel, M. A. J. S. (1999). *Dairy technology: Principles of milk properties and processes*. New York, NY: Marcel Dekker, Inc..

Wright, A. J., & Marangoni, A. G. (2006). Crystallization and rheological properties of milk fat. In P. F. Fox & P. L. H. McSweeney (Eds.), *Advanced dairy chemistry volume 2: Lipids* (pp. 245–291). Boston, MA: Springer.

Zayas, J. F. (1997). Foaming properties of proteins. In *Functionality of proteins in food* (pp. 260–309). New York, NY: Springer.

The manufacturer's authorised representative in the EU is Springer Nature Customer Service Centre GmbH, Europaplatz 3, 69115 Heidelberg, Germany. If you have any concerns regarding our products, please contact ProductSafety@springernature.com

Printed and bound by CPI Group (UK) Ltd, Croydon, CR0 4YY
28/09/2026
02224065-0017